Laurent Gizon · Paul Cally · John Leibacher
Editors

Helioseismology, Asteroseismology, and MHD Connections

Previously published in *Solar Physics* Volume 251, Issues 1–2, 2008

 Springer

Laurent Gizon
Max Planck Institute for Solar System Research
Katlenburg-Lindau, Germany

John Leibacher
National Solar Observatory
Tucson AZ, USA

Paul Cally
Monash University
Clayton VIC, Australia

Cover illustration: Map of horizontal mass flows (arrows) just beneath the solar surface, inferred using local helioseismology and SOHO/MDI observations. The sunspot at the center of the image is surrounded by an outflow with an amplitude of about 500 m/s.
Courtesy of Laurent Gizon, Max Planck Institute for Solar System Research.

All rights reserved.

Library of Congress Control Number: 2008940334

ISBN-978-0-387-89481-2   e-ISBN-978-0-387-89482-9

Printed on acid-free paper.

© 2008 Springer Science+Business Media, BV

No part of this work may be reproduced, stored in a retrieval system, or transmitted in any form or by any means, electronic, mechanical, photocopying, microfilming, recording or otherwise, without the written permission from the Publisher, with the exception of any material supplied specifically for the purpose of being entered and executed on a computer system, for the exclusive use by the purchaser of the work.

springer.com

Contents

Preface
L. Gizon · P. Cally · J. Leibacher **1**

1. ASTEROSEISMOLOGY

The Current Status of Asteroseismology
C. Aerts · J. Christensen-Dalsgaard · M. Cunha · D.W. Kurtz **3**

The Solar-Stellar Connection: Magneto-Acoustic Pulsations in $1.5 M_\odot - 2 M_\odot$ Peculiar A Stars
D.W. Kurtz **21**

Fourier Analysis of Gapped Time Series: Improved Estimates of Solar and Stellar Oscillation Parameters
T. Stahn · L. Gizon **31**

2. GLOBAL HELIOSEISMOLOGY

Perspectives in Global Helioseismology and the Road Ahead
W.J. Chaplin · S. Basu **53**

Meridional Circulation and Global Solar Oscillations
M. Roth · M. Stix **77**

Global Effects of Local Sound-Speed Perturbations in the Sun: A Theoretical Study
S.M. Hanasoge · T.P. Larson **91**

Angular Momentum Transport in the Sun's Radiative Zone by Gravito-Inertial Waves
S. Mathis · S. Talon · F.-P. Pantillon · J.-P. Zahn **101**

Influence of Low-Degree High-Order p-Mode Splittings on the Solar Rotation Profile
R.A. García · S. Mathur · J. Ballot · A. Eff-Darwich · S.J. Jiménez-Reyes · S.G. Korzennik **119**

Can We Constrain Solar Interior Physics by Studying the Gravity-Mode Asymptotic Signature?
R.A. García · S. Mathur · J. Ballot **135**

Zonal Velocity Bands and the Solar Activity Cycle
K.R. Sivaraman · H.M. Antia · S.M. Chitre · V.V. Makarova **149**

Acoustic Radius Measurements from MDI and GONG
S. Kholikov · F. Hill **157**

Spatio-Temporal Analysis of Photospheric Turbulent Velocity Fields Using the Proper Orthogonal Decomposition
A. Vecchio · V. Carbone · F. Lepreti · L. Primavera · L. Sorriso-Valvo · T. Straus · P. Veltri **163**

Calculation of Spectral Darkening and Visibility Functions for Solar Oscillations
C. Nutto · M. Roth · Y. Zhugzhda · J. Bruls · O. von der Lühe **179**

Uncovering the Bias in Low-Degree p-Mode Linewidth Fitting
W.J. Chaplin · Y. Elsworth · B.A. Miller · R. New · G.A. Verner **189**

Analysis of MDI High-Degree Mode Frequencies and their Rotational Splittings
M.C. Rabello-Soares · S.G. Korzennik · J. Schou **197**

3. LOCAL HELIOSEISMOLOGY

Recent Developments in Local Helioseismology
M.J. Thompson · S. Zharkov **225**

Observation and Modeling of the Solar-Cycle Variation of the Meridional Flow
L. Gizon · M. Rempel **241**

Three-Dimensional MHD Wave Propagation and Conversion to Alfvén Waves near the Solar Surface. I. Direct Numerical Solution
P.S. Cally · M. Goossens **251**

Surface-Focused Seismic Holography of Sunspots: I. Observations
D.C. Braun · A.C. Birch **267**

Helioseismology of Sunspots: Confronting Observations with Three-Dimensional MHD Simulations of Wave Propagation
R. Cameron · L. Gizon · T.L. Duvall Jr. **291**

Time–Distance Modelling in a Simulated Sunspot Atmosphere
H. Moradi · P.S. Cally **309**

Modelling the Coupling Role of Magnetic Fields in Helioseismology
B. Pintér **329**

Physical Properties of Wave Motion in Inclined Magnetic Fields within Sunspot Penumbrae
H. Schunker · D.C. Braun · C. Lindsey · P.S. Cally **341**

On Absorption and Scattering of P Modes by Small-Scale Magnetic Elements
R. Jain · M. Gordovskyy **361**

Time–Distance Analysis of the Emerging Active Region NOAA 10790
S. Zharkov · M.J. Thompson **369**

High-Resolution Mapping of Flows in the Solar Interior: Fully Consistent OLA Inversion of Helioseismic Travel Times
J. Jackiewicz · L. Gizon · A.C. Birch **381**

Structure and Evolution of Supergranulation from Local Helioseismology
J. Hirzberger · L. Gizon · S.K. Solanki · T.L. Duvall Jr. **415**

Probing the Subsurface Structures of Active Regions with Ring-Diagram Analysis
R.S. Bogart · S. Basu · M.C. Rabello-Soares · H.M. Antia **437**

Effects of Random Flows on the Solar f Mode: I. Horizontal Flow
N. Mole · A. Kerekes · R. Erdélyi **451**

Effects of Random Flows on the Solar f Mode: II. Horizontal and Vertical Flow
A. Kerekes · R. Erdélyi · N. Mole **467**

Instrumental Response Function for Filtergraph Instruments
R. Wachter **489**

4. ATMOSPHERIC AND FLARE-EXCITED OSCILLATIONS

Seismology of a Sunspot Atmosphere
Y.D. Zhugzhda **499**

Enhanced p-Mode Absorption Seen Near the Sunspot Umbral – Penumbral Boundary
S.K. Mathew **513**

Global Acoustic Resonance in a Stratified Solar Atmosphere
Y. Taroyan · R. Erdélyi **521**

DOT Tomography of the Solar Atmosphere VII. Chromospheric Response to Acoustic Events
R.J. Rutten · B. van Veelen · P. Sütterlin **531**

Velocity and Intensity Power and Cross Spectra in Numerical Simulations of Solar Convection
G. Severino · T. Straus · M. Steffen **547**

3D MHD Coronal Oscillations about a Magnetic Null Point: Application of WKB Theory
J.A. McLaughlin · J.S.L. Ferguson · A.W. Hood **561**

Nonlinear Numerical Simulations of Magneto-Acoustic Wave Propagation in Small-Scale Flux Tubes
E. Khomenko · M. Collados · T. Felipe **587**

Seismic Emissions from a Highly Impulsive M6.7 Solar Flare
J.C. Martínez-Oliveros · H. Moradi · A.-C. Donea **611**

Mechanics of Seismic Emission from Solar Flares
C. Lindsey · A.-C. Donea **625**

Preface

Laurent Gizon · Paul Cally · John Leibacher

Originally published in the journal Solar Physics, Volume 251, Nos 1–2, 1–2.
DOI: 10.1007/s11207-008-9248-y © The Author(s) 2008

The seismology of the Sun and stars has come a long way in a short time. The "original" *Global Helioseismology* has reached a level of maturity that allows many internal properties of the Sun to be probed with exquisite precision, although it currently faces a severe challenge to reconcile interior models with helioseismic inversions near the base of the convection zone in the age of the new solar chemical abundances. *Asteroseismology* suffers in comparison by being restricted to very low spherical harmonic degree (ℓ), but it makes up for this by providing many more subjects for study (including solar-like stars) and many cases of well-identified g modes. Where once we were restricted to stellar spectra in studying individual stars, asteroseismology now provides a crucial tool with which we may explore their deep structure. Its natural synergy with planet-search programs also invigorates it. *Local Helioseismology* has seen the development of an exciting array of techniques and insights over the two decades since observations of surface oscillations in and around active regions gave the first clues that something different was happening there, and it has been particularly important in mapping flows of various types in shallow subsurface layers. Its current challenge is to take better account of surface magnetism and to illuminate the coupling between interior and atmospheric oscillations where magnetic fields clearly play an important

---

Helioseismology, Asteroseismology, and MHD Connections
Guest Editors: Laurent Gizon and Paul Cally

L. Gizon (✉)
Max-Planck-Institut für Sonnensystemforschung, 37191 Katlenburg-Lindau, Germany
e-mail: gizon@mps.mpg.de

P. Cally
Centre for Stellar and Planetary Astrophysics, School of Mathematical Sciences, Monash University, Clayton, Victoria 3800, Australia
e-mail: paul.cally@sci.monash.edu.au

J. Leibacher
National Solar Observatory, Tucson, AZ, USA
e-mail: jleibacher@nso.edu

role. The crucial search for deep stored magnetic field associated with the solar dynamo also provides a challenge for the coming years to both global and local helioseismologists. *Coronal Helioseismology*, still in its infancy, is making rapid progress driven by the astounding high-resolution data and images from spacecraft such as TRACE and *Hinode*. Helio- and asteroseismology benefit from and inform modern numerical simulations of magnetoconvection and wave propagation through complex media. Testing of inversion techniques in the "numerical laboratory" provided by these models is becoming increasingly valuable.

This volume presents a timely snapshot of the state of helio- and asteroseismology in the era when SOHO/MDI is about to be replaced by SDO/HMI and CoRoT is yielding its first long-duration light curves of thousands of stars. It is inspired by two seminal conferences: HELAS II "Helioseismology, Asteroseismology and MHD Connections" in August 2007 in Göttingen, Germany, and SOHO 19/GONG 2007 "Seismology of Magnetic Activity" held at Monash University in Australia in July 2007.

Many of the papers included here represent work presented at one or other of the meetings, but this Topical Issue was thrown open for general submission on their core topics. All papers were refereed to the usual high standards of *Solar Physics*. Three papers describing the current status of asteroseismology, global helioseismology, and local helioseismology were specially commissioned for the volume, and these set the context for the other contributions.

HELAS II was supported in part by the European Helio- and Asteroseismology Network (HELAS, a major collaboration funded by the European Union's sixth framework programme and coordinated by Oskar von der Lühe and Markus Roth) and in part by the Max Planck Institute for Solar System Research through Ulrich Christensen and Sami Solanki. SOHO 19/GONG 2007 was generously supported by the SOHO Project Science Team (through Bernhard Fleck), GONG (Global Oscillation Network Group, through Frank Hill), and Monash University.

The Current Status of Asteroseismology

C. Aerts · J. Christensen-Dalsgaard · M. Cunha ·
D.W. Kurtz

Originally published in the journal Solar Physics, Volume 251, Nos 1–2, 3–20.
DOI: 10.1007/s11207-008-9182-z © Springer Science+Business Media B.V. 2008

Abstract Stellar evolution, a fundamental bedrock of modern astrophysics, is driven by the physical processes in stellar interiors. While we understand these processes in general terms, we lack some important ingredients. Seemingly small uncertainties in the input physics of the models (*e.g.*, the opacities or the amount of mixing and of interior rotation) have large consequences for the evolution of stars. The goal of asteroseismology is to improve the description of the interior physics of stars by means of their oscillations, just as global helioseismology led to a huge step forward in our knowledge about the internal structure of the Sun. In this paper we present the current status of asteroseismology by considering case studies of stars with a variety of masses and evolutionary stages. In particular, we outline how the confrontation between the observed oscillation frequencies and those predicted by

Invited review

Helioseismology, Asteroseismology, and MHD Connections
Guest Editors: Laurent Gizon and Paul Cally

C. Aerts (✉)
Instituut voor Sterrenkunde, Katholieke Universiteit Leuven, Celestijnenlaan 200 D, 3001 Leuven, Belgium
e-mail: conny@ster.kuleuven.be

C. Aerts
Afdeling Sterrenkunde, Radboud University Nijmegen, Post Office Box 9010, 6500 GL Nijmegen, The Netherlands

J. Christensen-Dalsgaard
Institut for Fysik og Astronomi, Aarhus Universitet, Aarhus, Denmark
e-mail: jcd@phys.au.dk

M. Cunha
Centro de Astrofísica da Universidade do Porto, Rua das Estrelas, 4150-762 Porto, Portugal
e-mail: mcunha@astro.up.pt

D.W. Kurtz
Centre for Astrophysics, University of Central Lancashire, Preston PR1 2HE, UK
e-mail: dwkurtz@uclan.ac.uk

the models allows us to pinpoint limitations of the input physics of current models and improve them to a level that cannot be reached with any other current method.

Keywords Oscillations, stellar · Interior, convective zone · Interior, core · Instrumentation and data management

1. Introduction

Despite extensive research in recent decades, we lack detailed knowledge of some important physical processes relevant for the description of stellar interiors. The reason is that, in general, the existing observations do not yet allow a detailed confrontation with the description of the physical properties of either the stellar material in the deepest internal layers or of the dynamics of the outer stellar envelope. At first sight, seemingly small uncertainties in the input physics of the models have large consequences for the whole duration and end of the stellar life cycle. The lack of a good understanding of interior transport processes, caused by different phenomena such as rotation, gravitational settling, and radiative levitation, magnetic diffusion, is particularly acute when it comes to precise predictions of stellar evolution and the galactic chemical enrichment accompanying it.

Given that global helioseismology led to a huge step forward in the accuracy of the internal structure model and of the transport processes inside the Sun, asteroseismology aims to obtain similar improvements for different types of stars by means of their oscillations. Stellar oscillations indeed offer a unique opportunity to probe the internal properties and processes, because these affect the observable frequencies. Confronting the frequencies measured with high accuracy with those predicted by models gives insight into the limitations of the input physics of models and improves them. In fact, stellar oscillation frequencies are the best diagnostic known that can reach the required precision in the derivation of interior stellar properties.

At present, the unknown aspects of the physics and dynamics are dealt with by using parameterized descriptions, where the parameters are tuned from observational constraints. These concern, for example, the treatments of convection, the equation of state, diffusion, and settling of elements. When a lack of observational constraints occurs, solar values are often assigned (*e.g.*, to the mixing length parameter in the description of convection) or phenomena are ignored (*e.g.*, convective overshooting and the diffusion of heavy elements). It is hard to imagine, however, that one single set of parameters is appropriate for very different types of stars. Similarly, rotation is either not included or is included only with a simplified treatment of the evolution of the rotation law in stellar models (*e.g.*, Maeder and Meynet, 2000). Fortunately, rotation also modifies the frequencies of the star's modes of oscillation (*e.g.*, Gough, 1981; Saio, 1981). An adequate seismic modeling of rotation inside stars therefore is within reach with high-precision measurements and mode identifications of stellar oscillations.

The origin and physical nature of stellar oscillations, as well as their mathematical properties, were thoroughly discussed by Cunha *et al.* (2007), to which we refer the reader for the theoretical considerations of asteroseismology. Extensive recent overviews of the occurrence of stellar oscillations across the entire HR diagram, and asteroseismic applications thereof, are already available in Kurtz (2004, 2006), Cunha *et al.* (2007), and Aerts, Christensen-Dalsgaard, and Kurtz (in preparation). Rather than repeating such an observational overview here in a far more concise format, we have opted to outline the current status of asteroseismology by focusing on a few carefully chosen examples that show the merits this research field has brought to the improvement of stellar modeling, that is, we confine ourselves to cases where quantitative measures of internal structure parameters have been achieved. We start by considering examples of stars that oscillate similarly to the Sun, and then move on to pulsators excited by a heat mechanism for a discussion of convective overshooting and rotation inside stars. The rapidly oscillating Ap stars, being pulsators with very strong magnetic fields, are of particular interest to solar astronomers, so are discussed in a separate paper in this volume (Kurtz, 2008).

2. From the Sun to Stars: The Properties of Solar-Like Pulsators

As the oscillations of the Sun are caused by turbulent convective motions near its surface, we expect such oscillations to be excited in all stars with significant outer convection zones. Solar-like oscillations are indeed predicted for the lowest mass main-sequence stars up to objects near the cool edge of the classical instability strip with masses near $1.6 M_\odot$ (*e.g.*, Christensen-Dalsgaard, 1982; Christensen-Dalsgaard and Frandsen, 1983; Houdek *et al.*, 1999) as well as in red giants (Dziembowski *et al.*, 2001). Such stochastically excited oscillations have very tiny amplitudes, which makes them hard to detect, particularly for the low-mass stars. Velocity amplitudes were predicted to scale roughly as L/M, where L and M are the luminosity and mass of the star, before the first firm discoveries of such oscillations in stars other than the Sun (Kjeldsen and Bedding, 1995). This scaling law was later modified to $(L/M)^{0.8}$ from excitation predictions based on 3D computations of the outer convection zones of the stars (Samadi *et al.*, 2005), resulting in lower amplitudes compared with those found for 1D models.

The modes observed in the Sun at low spherical harmonic degree ℓ, and hence solar-like oscillations observable in distant stars, are high-order acoustic modes. They satisfy an approximate asymptotic relation (*e.g.*, Vandakurov, 1967; Tassoul, 1980; Gough, 1993), according to which, to leading order,

$$v_{n\ell} \sim \Delta v \left(n + \frac{\ell}{2} + \epsilon \right), \tag{1}$$

where $v_{n\ell}$ is the cyclic frequency of a mode of radial order n and degree ℓ and ϵ is a function of frequency determined mainly by the conditions near the stellar surface. Also,

$$\Delta v = \left(2 \int_0^R \frac{dr}{c} \right)^{-1} \tag{2}$$

is a measure of the inverse sound travel time over a stellar diameter, with r being the distance to the stellar center, R the surface radius of the star, and c the adiabatic sound speed. From simple considerations, it follows that *the large frequency separation* satisfies $\Delta v \propto (M/R^3)^{1/2}$ and hence is a measure of the mean density of the star. Departures from Equation (1) can be characterized by *the small frequency separation*: $\delta v_{n\ell} = v_{n\ell} - v_{n-1\,\ell+2}$. This quantity is mainly sensitive to the sound speed in the core of the star and hence provides a measure of the evolutionary state. The sensitivity of Δv and δv on stellar properties allows a calibration of stellar models in terms of $(\Delta v, \delta v)$ to estimate the mass and evolutionary stage of the star (*e.g.*, Christensen-Dalsgaard, 1984, 1988; Ulrich, 1986).

The search for solar-like oscillations in stars in the solar neighborhood has been ongoing since the early 1980s. The first indication of stellar power with a frequency dependence similar to that of the Sun was obtained by Brown *et al.* (1991) in α CMi (Procyon, F5 IV). The first detection of individual frequencies of solar-like oscillations was achieved from high-precision time-resolved spectroscopic measurements only in 1995 for the G5 IV star η Boo (Kjeldsen *et al.*, 1995; Brown *et al.*, 1997); however, a confirmation of this detection from

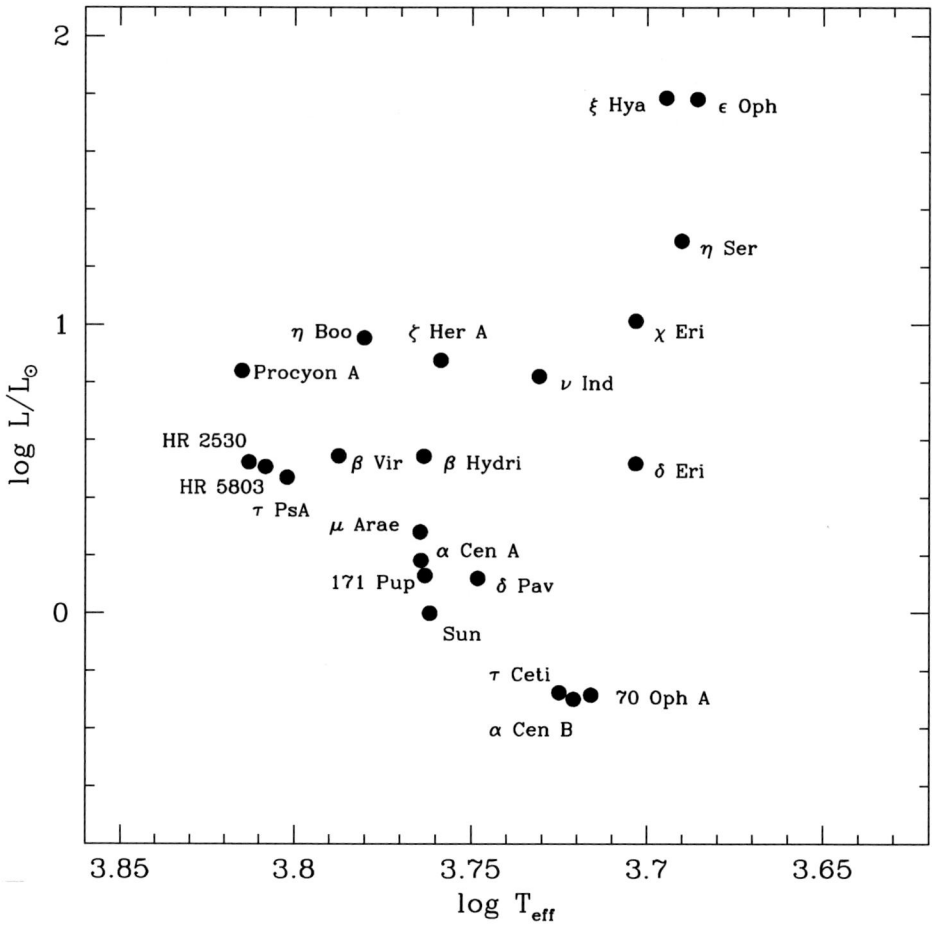

Figure 1 HR diagram showing the stars in which solar-like oscillations have been detected (with HR 2530 = HD 49933). The discoveries for 171 Pup, HR 5803 (HD 139211), and τ PsA are unpublished (Carrier *et al.*, in preparation). (Figure courtesy of Fabien Carrier.)

independent measurements could not be established, but it was subsequently confirmed by Carrier, Bouchy, and Eggenberger (2003) and Kjeldsen *et al.* (2003). It took another four years before solar-like oscillations were definitely established in Procyon (Martić *et al.*, 1999). Although there was a recent controversy about this detection (Matthews *et al.*, 2004; Bedding *et al.*, 2005), which we do not discuss in detail here, the results of Martić *et al.* (1999) have been confirmed (Mosser *et al.*, 2008) and an in-depth asteroseismic investigation based on a large multisite campaign is presently being conducted. Subsequent to Martić *et al.*'s results, solar-like oscillations were found in two more stars: the G2 IV star β Hyi (Bedding *et al.*, 2001) and the solar twin α Cen A (Bouchy and Carrier, 2001). These important discoveries led to several more subsequent detections, a summary of which was provided by Bedding and Kjeldsen (2007). The positions of confirmed solar-like pulsators in the HR diagram are displayed in Figure 1. The detected frequencies and frequency separations for all stars behave as expected from theoretical predictions and from scaling relations based on extrapolations from helioseismology.

The oscillation frequencies and frequency separations detected in solar-like pulsators provide additional constraints with which to test models of stellar structure and evolution in conditions slightly different from those provided by the Sun. Such studies generally involve a fit of theoretical models, characterized by a number of model parameters, to the set of seismic and nonseismic data available for a given pulsator. Theoretical modeling of solar-like pulsators using this direct fitting approach has been carried out for several stars, including η Boo (Carrier, Eggenberger, and Bouchy, 2005; Guenther *et al.*, 2005), Procyon (Eggenberger *et al.*, 2004a; Eggenberger, Carrier, and Bouchy, 2005; Provost *et al.*, 2006), and α Cen A and B (*e.g.*, Eggenberger *et al.*, 2004b; Miglio and Montalbán, 2005; Yıldız, 2007). So far, the main results of these fits are estimates of the stellar masses, ages, and initial metallicities, even though in some cases the results are still controversial and call for better sets of data.

Among the solar-like pulsators, the binary star α Cen A and B provides a particularly interesting test bed for studies of stellar structure and evolution, owing to the numerous and precise seismic and nonseismic data that are available for both components of the binary. Studies of α Cen A and B, including seismic and nonseismic data for both components, indicate that the age of the system is likely to be between 5.6 and 7.0 Gyr, the value derived being dependent, in particular, on the seismic observables that are included in the fits. Moreover, the same studies point to a significant difference in the values of the mixing-length parameter (α_{MLT}; see Section 3 for a definition) for the two stars, although the sign is uncertain and possibly dependent on the detailed treatment of the effects of the near-surface layers in the analysis of the observed frequencies. Eggenberger *et al.* (2004b) and Miglio and Montalbán (2005) found that the value for α Cen B is larger than that for α Cen A, whereas Teixeira *et al.* (in preparation) found that α_{MLT} was slightly *smaller* for α Cen B than for α Cen A. The latter study also found that the best-fitting model for α Cen A was on the border of having a convective core (see Christensen-Dalsgaard, 2005): even a slight increase in the mass of the model led to a significant convective core and hence a model that was quite far from matching the observed properties.

Despite the successful case studies just outlined, the detailed seismic studies of stars with stochastically excited modes are currently still in their infancy compared with global helioseismology. However, given the recent detections and the continuing efforts to improve them, we expect very substantial progress in the seismic interpretation of such targets in the coming years. In particular, the CoRoT (*e.g.*, Michel *et al.*, 2006) and *Kepler* (*e.g.*, Christensen-Dalsgaard *et al.*, 2007) missions will give data of very high quality on solar-like oscillations. As seen in the example of α Cen A, it is noteworthy that the class of main-sequence solar-like oscillators encompasses transition objects regarding the development of a convective core on the main sequence ($1 M_\odot < M < 1.5 M_\odot$). Asteroseismology will surely refine the details of the yet poorly understood physics that occurs near the core of the objects in this transition region. Also, data of the expected quality will provide information about the depth and helium content of the convective envelope (*e.g.*, Houdek and Gough, 2007), as well as more reliable determinations of stellar ages (Houdek and Gough, 2008).

Additional information from solar-like oscillations is available in the cases of relatively evolved stars, beyond the stage of central hydrogen burning. Here the frequency range of stochastically excited modes may encompass *mixed modes* behaving as standing internal gravity waves, or *g* modes, in the deep chemically inhomogeneous regions, thus providing much higher sensitivity to the properties of this region. In fact, there is some evidence that such modes have been found in the subgiant η Boo (Christensen-Dalsgaard, Bedding, and Kjeldsen, 1995).

Evidence for rotational splitting (see Section 4 for a definition) has been found in α Cen A (Fletcher *et al.*, 2006; Bazot *et al.*, 2007). However, it has not yet been possible to map the interior rotation of a solar-like pulsator, since the present frequency multiplet detections are insufficient. We note that Lochard, Samadi, and Goupil (2004) found, with simulated data, that the presence of mixed modes in a star such as η Boo may allow some information to be derived about the variation of the internal rotation with position.

For a few pulsators excited by the heat mechanism, data are already available that provide such information, albeit only very roughly. We discuss this further in Section 4, but first we highlight in the next section some case studies through which the properties of core convection have been tuned by asteroseismology.

3. Seismic Derivation of Convective Overshooting inside Stars

The standard description of convection used in stellar modeling is the mixing length theory (MLT) of Böhm-Vitense (1958). In this theory, the convective motions are treated as being time-independent. In the absence of a rigorous theory of convective motions based on first principles, the convective cells are assumed to have a mean-free-path length of $\alpha_{\mathrm{MLT}} H_\mathrm{p}$, where H_p is the local pressure scale height. The mixing-length parameter depends on the physics considered in the model and on the specific formulation of the MLT used. Its value for Model S for the Sun of Christensen-Dalsgaard et al. (1996) is $\alpha_{\mathrm{MLT}} \approx 1.99$, in the Böhm-Vitense (1958) MLT formulation.

In the context of stellar evolution, it is of crucial importance to quantify the amount of matter in the fully mixed central region of the star. This amount is usually derived from the Schwarzschild criterion, which states that convection occurs in regions where the adiabatic temperature gradient is smaller than the radiative gradient. However, from a physical point of view, it is highly unlikely that convective elements stop abruptly at the boundary set by the Schwarzschild criterion. Rather, their inertia causes them to overshoot into the adjacent stable area where radiative energy transport takes place. The amount of such overshooting is, however, largely unknown. For this reason, it is customary to express it as $\alpha_{\mathrm{ov}} H_\mathrm{p}$ where α_{ov} is expected to be a small fraction of α_{MLT}.

The inability to derive a value for α_{MLT} and α_{ov} from a rigorous theoretical description is highly unsatisfactory, particularly for stars with a convective core, because the total mass of the well-mixed central region of the star determines its stellar lifetime. This is the reason why great effort has been, and is being, made to quantify α_{ov}, while keeping in mind that we have already a fairly good estimate of α_{MLT} from the Sun. We describe here the power of asteroseismology to determine α_{ov}.

In the solar case, helioseismic analyses have provided constraints on the overshoot from the solar convective envelope, under the assumption that this results in a nearly adiabatic extension of the convection zone followed by an abrupt transition to the radiative temperature gradient (Zahn, 1991). If we assume also, as usual, a spherically symmetric model, such a behavior introduces a characteristic pattern in the frequencies in the form of an oscillatory variation of the frequencies as functions of the mode order. From the observed amplitude of this signal, an overshoot region of the nature considered must have an extent less than around $0.1 H_\mathrm{p}$ (Basu, Antia, and Narasimha, 1994; Monteiro, Christensen-Dalsgaard, and Thompson, 1994; Christensen-Dalsgaard, Monteiro, and Thompson, 1995). Monteiro, Christensen-Dalsgaard, and Thompson (2000) found that a similar analysis can be carried out on the basis of just low-degree modes, such as will be observed in distant stars.

Owing to their sensitivity to the core structure, the low-degree solar-like oscillations should in principle be sensitive to overshoot from convective cores. Models of η Boo without and with overshoot were considered by Di Mauro et al. (2003, 2004). Although the present observed frequencies are not sufficiently accurate to provide direct information about the properties of the core, it was found that for $\alpha_{\rm ov} \geq 0.2$ models could be found in the central hydrogen-burning stage that matched the observed location on the HR diagram. In such models, mixed modes are not expected; thus the definite identification of mixed modes would constrain the extent of overshoot in the star. Straka, Demarque, and Guenther (2005) considered models of Procyon with various types of core overshoot to determine the extent to which overshoot could be asteroseismically constrained.

For the p-mode diagnostics considered, little sensitivity to overshoot was found, whereas the, perhaps unlikely, detection of g modes in Procyon, such as have been claimed in the Sun, would provide much stronger constraints on the overshoot distance. As in the case of η Boo, the definite identification of the star as being on the subgiant branch (*e.g.*, from the properties of the oscillation frequencies) would provide strict constraints on the extent of overshoot during the central hydrogen-burning phase. Mazumdar *et al.* (2006a) made a detailed analysis of the sensitivity of suitable frequency combinations to the properties of stellar cores and found that the mass of the convective core, possibly including overshoot, could be determined with substantial precision, given frequencies with errors that should soon be reached. Cunha and Metcalfe (2007) developed diagnostics of small convective cores that may in principle also provide information about the properties of overshoot; the detailed sensitivity still needs investigation, however.

Quantitative measures of the core convective overshooting parameter have been achieved by fitting the frequencies of some of the β Cep stars. This group of young, Population I, near-main-sequence pulsating B stars has been known for more than a century. They have masses from $8M_\odot$ to $18M_\odot$, and they oscillate in low-order p and g modes with periods in the range two – eight hours. These oscillations are excited by a heat mechanism acting through opacity features associated with elements of the iron group (*e.g.*, Dziembowski and Pamiatnykh, 1993; Pamyatnykh, 1999; Miglio, Montalbán, and Dupret, 2007). A recent overview of the observational properties of the class was provided by Stankov and Handler (2005). Most of the β Cep stars show multiperiodic light and line-profile variations and most rotate at only a small fraction of the critical velocity.

Significant progress in the detailed seismic modeling of the β Cep stars has occurred over the past few years and has led to quantitative estimates of the core overshooting parameter $\alpha_{\rm ov}$ for several class members with slow rotation (see, *e.g.*, Aerts, 2006, for a summary). We illustrate this here for the star θ Oph, whose frequency spectrum was determined from a multisite photometric campaign and is represented in Figure 2 (Handler, Shobbrook, and Mokgwetsi, 2005). An additional long-term, high-resolution spectroscopic campaign revealed that this star is a member of a spectroscopic binary with an orbital period of 56.71 days and an eccentricity of 0.17 (Briquet *et al.*, 2005) and allowed the identification of the spherical wavenumbers (ℓ, m) of the seven detected frequencies from the line-profile variations induced by the oscillations (see Table 1, reproduced from Briquet *et al.*, 2007).

Because the frequency spectra of β Cep stars are so sparse for low-order p and g modes compared with those of solar-like pulsators (see Figure 2), one does not have many degrees of freedom to fit the securely identified modes. This led to the identification of the radial order of the modes of θ Oph as g_1 for the frequency quintuplet containing ν_1, ν_5, ν_2, the radial fundamental for ν_3, and p_1 for the triplet ν_4, ν_6, ν_7. Fitting the three independent $m = 0$ frequencies results in a relation between the metallicity and the core-overshooting parameter, because the stellar models for main-sequence B stars typically depend on the five parameters (X, $\alpha_{\rm ov}$, Z, M, and age) if we ignore effects of diffusion. Note that $\alpha_{\rm MLT}$ is usually fixed to the solar value; for B stars, with their extremely thin and inefficient outer

Table 1 The identification of the pulsation modes of the β Cep star θ Oph derived from multicolor photometric and high-resolution spectroscopic data. Positive m values represent prograde modes. The amplitudes of the modes are given for the Strömgren u filter and for the radial velocities. (Table reproduced from Briquet et al., 2007).

ID	Frequency (d^{-1})	(ℓ, m)	u amplitude (mmag)	RV amplitude (km s^{-1})
ν_1	7.1160	(2, −1)	12.7	2.54
ν_5	7.2881	(2, +1)	2.1	–
ν_2	7.3697	(2, +2)	3.6	–
ν_3	7.4677	(0, 0)	4.7	2.08
ν_4	7.7659	(1, −1)	3.4	–
ν_6	7.8742	(1, 0)	2.3	–
ν_7	7.9734	(1, +1)	2.4	–

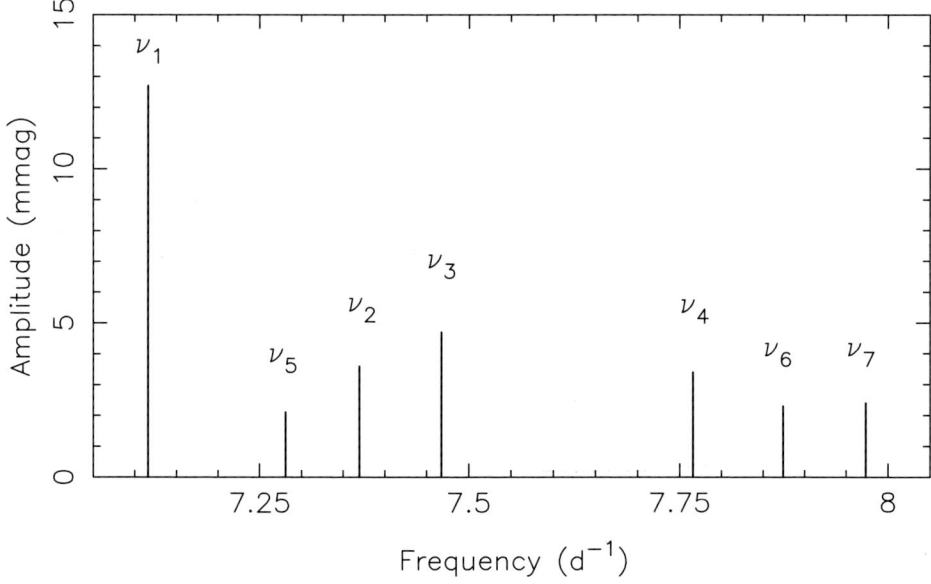

Figure 2 The schematic frequency spectrum of the β Cep star θ Oph for the Strömgren u filter as derived from a multisite photometric campaign. The measured photometric amplitude ratios led to an identification of the frequencies ν_1, ν_2, ν_3, and ν_4 as, respectively, $\ell = 2, 2, 0, 1$. (Figure reproduced from Handler, Shobbrook, and Mokgwetsi, 2005.)

convection zones, changing α_{MLT} within reasonable limits does not change the characteristics of the models. In this way, one finds $\alpha_{\mathrm{ov}} = 0.44 \pm 0.07$ from a detailed high-precision abundance determination for θ Oph (Briquet et al., 2007).

The reason why we can derive the core overshooting and the rotation (see Section 4), and provide a quantitative measure of these parameters for this star, is the different probing ability of the nonradial modes. This can be illustrated by plotting probing kernels of the modes. Different types of such kernels are used, depending on the kind of behavior under investigation. This is illustrated in Figure 3, where we show the rotational splitting kernels $K(x)$ [which will be defined in Equation (4) later] of θ Oph for the two nonradial modes. It can be seen that the g_1 mode's kernel behaves differently near the boundary of the core region, and thus it probes that region in a different way than the p_1 mode, allowing the derivation of details of the rotational properties as explained in the following. A similar figure holds for

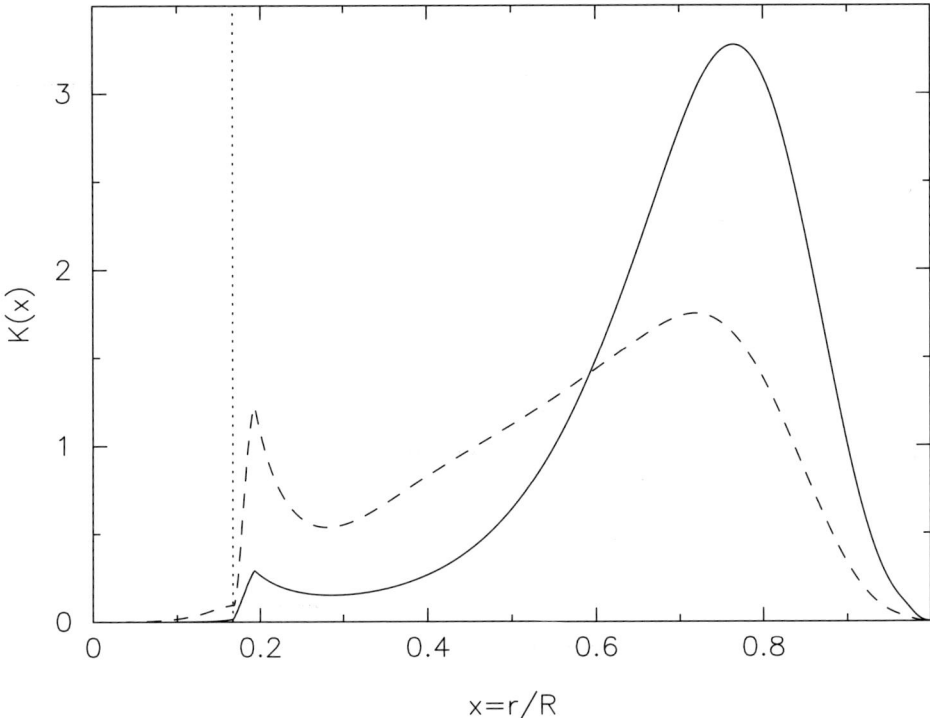

Figure 3 The rotational kernels defined in Equation (4) as a function of radial distance inside the star ($x = r/R$), for the identified $\ell = 1$, p_1 mode (solid line) and $\ell = 2$, g_1 mode (dashed line) of the β Cep star θ Oph. The vertical dotted line marks the position of the boundary of the convective core, including the overshoot region. Note that the kernels also approximately represent the relative sensitivity of the mode frequencies to other aspects of the stellar interior. (Figure reproduced from Briquet et al., 2007.)

the energy distribution, which allows probing the extent of the core region. A comparable result was obtained for V836 Cen, whose frequency spectrum is almost a copy of that of θ Oph (see Figure 5), and also for ν Eri (Pamyatnykh, Handler, and Dziembowski, 2004).

The combination of low-order p and g modes thus turns out to be a very powerful tool to derive the internal structure parameters of massive stars. Additional measures of the core overshooting have been obtained for the β Cep stars β CMa (Mazumdar et al., 2006b) and δ Ceti (Aerts et al., 2006). For all these β Cep stars, α_{ov} ranges from 0.1 to 0.5, although these values depend somewhat on the adopted metal mixture (Thoul et al., 2004). It is remarkable that the frequencies of just two well-identified oscillation modes that have sufficiently different kernels allow one to derive the overshooting parameter with a precision of typically 0.05 expressed in H_p. Adding just a few more well-identified modes should drastically reduce this error for specific input physics of the models.

The seismically derived estimates of core overshooting in β Cep stars are compatible with the quantitative results for eight detached double-lined eclipsing binaries obtained by Ribas, Jordi, and Giménez (2000), who found α_{ov} to range from 0.1 to 0.6 for primary masses ranging from $1.5 M_\odot$ to $9 M_\odot$. Another way of determining the amount of overshooting from data is by fitting stellar evolutionary tracks to the dereddened color-magnitude diagrams of clusters; for example, $\alpha_{ov} = 0.20 \pm 0.05$ for the intermediate age open cluster NGC3680 (Kozhurina-Platais et al., 1997) and $\alpha_{ov} \approx 0.07$ for the old open cluster M67 (VandenBerg and Stetson, 2004). In these two methods, essentially the same five unknown structure parameters occur as for the seismic modeling, since stellar evolution models are used to fit the position of the binary components and of the cluster main-sequence turnoff point in the HR diagram, respectively. The uncertainty on the overshoot distance derived from the light curve analysis of an accurately modeled eclipsing binary or from fitting of a cluster turnoff point is typically between 0.05 and 0.1 H_p provided that the metallicities are known. It is interesting, although perhaps fortuitous, that all of these quantitative measures of the amount of overshooting are in agreement with the theoretical predictions by Deupree (2000) from 2D hydrodynamic simulations of zero-age main-sequence stars with a convective core.

4. Seismic Derivation of the Internal Rotation Profile of Stars

The rotation of a star implies a splitting of the oscillation frequencies compared with the case without rotation. Hence, rotation becomes apparent in frequency spectra as multiplets of $2\ell + 1$ components for each mode of degree ℓ. By ignoring rotational effects higher than order one in the rotational frequency as well as the influence of a magnetic field, the frequency splitting becomes

$$\nu_m = \nu_0 + m \int_0^R K(r) \frac{\Omega(r)}{2\pi} \frac{dr}{R}, \qquad (3)$$

where ν_m is the cyclic frequency of a mode of azimuthal-order m, and Ω is the angular velocity, which we here assume to depend only on the distance (r) to the center. The rotational kernels are defined as

$$K(r) = \frac{\{\xi_r^2 - 2\xi_r\xi_h + [\ell(\ell+1) - 1]\xi_h^2\}r^2\rho}{\int_0^R \frac{dr}{R}[\xi_r^2 + \ell(\ell+1)\xi_h^2]r^2\rho}, \qquad (4)$$

with ξ_r and ξ_h the radial and tangential components of the displacement vector

$$\boldsymbol{\xi} = (\xi_r \mathbf{e}_r + \xi_h \nabla_h) Y_\ell^m. \qquad (5)$$

If such multiplets are observed, their structures are a great help in mode identification, as nonradial modes with a given value of ℓ have $2\ell + 1$ multiplet peaks corresponding to the different values of m, although not all peaks may be visible owing to the geometry of the modes or excitation of the components, whereas radial modes show no multiplet structure. The recognition of the multiplet structure is far easier for very slow rotators, where "slow" here means that the rotational frequency is far lower than the frequency spacing for $m = 0$ components of modes of adjacent radial order n.

Next, we describe two types of pulsators for which a quantitative measure of differential interior rotation has been established.

4.1. White Dwarfs

The first detection of differential (*i.e.*, nonrigid) rotation inside a star besides the Sun was achieved for the DBV white dwarf GD 358 from a multisite campaign by the Whole Earth Telescope organization (Winget *et al.*, 1994). The multiperiodic variations of DBV white dwarfs are due to low-degree, high-order *g* modes, excited by the heat mechanism active in the second partial ionization zone of helium. Their oscillation periods range from 4 to

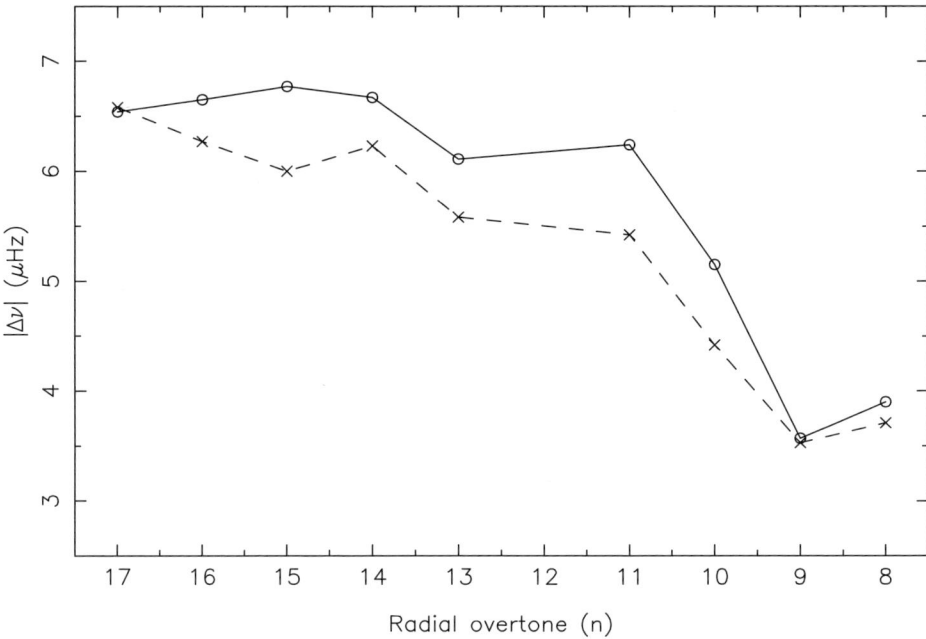

Figure 4 Frequency splittings $|\Delta\nu| = |\nu_m - \nu_0|$ of the triplets detected for GD 358 as a function of radial overtone (n). The full line connects the values for the $m = +1$ components corresponding to prograde modes and the dotted line those for $m = -1$ representing retrograde modes. (Figure reproduced from Winget et al., 1994.)

12 minutes and their photometric amplitudes are relatively large, from a few mmag to 0.2 mag (*e.g.*, Bradley, 1995). Among the more than 180 significant frequency peaks detected in the white-light photometric lightcurve of GD 358 covering 154 hours of data, 27 are the components of well-identified triplets. These frequency splittings are shown as a function of radial order n in Figure 4. It can be seen that larger splittings occur for higher radial order, whereas one would expect these splittings to be constant for rigid rotation inside the white dwarf. Since the modes of higher radial order probe predominantly the outer layers and those of lower radial order the inner parts, the rotation of GD 358 must be radially differential. The mean splitting for the modes of $n = 16$ and 17 leads to a rotation period of 0.89 days through Equations (3) and (4), and for $n = 8, 9$ the rotation period is 1.6 days. Winget et al. (1994) therefore concluded that the inner parts of GD 358 rotate 0.6 times slower than its outer layers, where "inner" and "outer" refer to those regions probed by the detected triplets. However, Kawaler, Sekii, and Gough (1999) found that the data were not yet of sufficient quality to allow a more detailed inversion for the variation of the internal rotation with depth.

The detailed seismic modeling of GD 358 followed that achieved previously for the prototypical DOV white dwarf PG1159-035 (GW Vir) described in the seminal work by Winget et al. (1991), which was again based on data collected by the Whole Earth Telescope consortium (Nather et al., 1990). This led to 125 significant frequencies for GW Vir, of which 101 were identified as components of rotationally split triplets and quintuplets. Unlike the case for GD 358, all multiplets for a given ℓ showed the same frequency splitting within the measurement errors, allowing Winget et al. (1991) to deduce a constant rotation period of approximately 1.38 ± 0.01 day throughout the white dwarf. Classical spectroscopy can in no way reveal the rotation periods of single compact stellar remnants with such high precision, not even in the case where the inclination angle can be estimated from independent information.

The deviation of GD 358's splittings for $m = +1$ with respect to those for $m = -1$ in Figure 4 was interpreted by Winget *et al.* (1994) in terms of a weak magnetic field of 1300 ± 300 G, which causes splittings $\sim |m^2|$ in addition to the rotational splitting given in Equation (3) (*e.g.*, Dziembowski and Goode, 1984; Jones *et al.*, 1989). The effect of a magnetic field could not be established for the frequency multiplets of PG 1159-035, which led to an upper limit of 6000 G for that object's magnetic field (Winget *et al.*, 1991). It is noteworthy that the magnetic-field strength that can be probed by classical spectroscopy of white dwarfs through the Zeeman effect requires fields roughly a factor of 1000 stronger than what can be found from asteroseismology.

The case studies of GD 358 and PG 1159-035 by the Whole Earth Telescope consortium implied not only a first test case for the technique of asteroseismology but at the same time a real breakthrough in the derivation of white dwarf structure models. It led to estimates of internal rotation and magnetic field strength and also allowed a high-precision mass estimate [$(0.586 \pm 0.003) M_\odot$ for PG1159-035 and $(0.61 \pm 0.03) M_\odot$ for GD 358]. It also proved that the outer layers of white dwarfs are compositionally stratified. This was derived from deviations of the frequency spacings from mode trapping compared with spacings for unstratified models. Mass estimates with such high precision cannot be achieved from other means, except for relativistic effects in binary pulsars. These two seismic studies of white dwarfs paved the road for many others of their kind, but none of the more recent ones have led to more accurate internal rotation rates than those for PG1159-035 and GD 358. We refer to Kepler (2007) and Fontaine and Brassard (in preparation) for recent review papers on white-dwarf seismology.

4.2. Main-Sequence Stars

There are presently only three main-sequence stars, besides the Sun, for which an observational constraint on the internal-rotation profile has been derived. In all three cases, it was achieved through asteroseismology of β Cep stars. Several of these are suitable targets to attempt mapping of their interior rotation because their rotational frequencies are well below the frequency spacing between multiplets. These stars are particularly interesting targets for this purpose, because the largest uncertainty in stellar evolution models for massive stars is precisely concerned with rotational mixing effects.

The first seismic proof of differential rotation in a massive star was obtained for the B3V star V836 Cen (HD 129929; Aerts *et al.*, 2003). This result was derived from the well-identified parts of one rotationally split frequency triplet and one quintuplet, as shown in Figure 5. This star's detected frequency spectrum is obviously very similar to the one of θ Oph (compare Figures 2 and 5), except that V836 Cen is a slower rotator than θ Oph. Given that only two multiplets were available for V836 Cen, Dupret *et al.* (2004) assumed a linear rotation law and concluded that the rotational frequency near the stellar core is 3.6 times higher than at the surface. It was possible to derive this because the kernels of the g_1 and p_1 modes probe the rotational behavior near the stellar core differently, just as for θ Oph (see Figure 3). A very similar result, the rotation of the deep interior exceeding the surface rotation by a factor of between three and five, was obtained by Pamyatnykh, Handler, and Dziembowski (2004) from the g_1 and p_1 $\ell = 1$ modes of the B2III β Cep star ν Eri (HD 29248). These results are compatible with the assumption of local angular momentum conservation. Both V836 Cen and ν Eri are – for upper main-sequence stars – very slow rotators, with surface rotation velocities of 2 km s^{-1} (V836 Cen) and 6 km s^{-1} (ν Eri). This

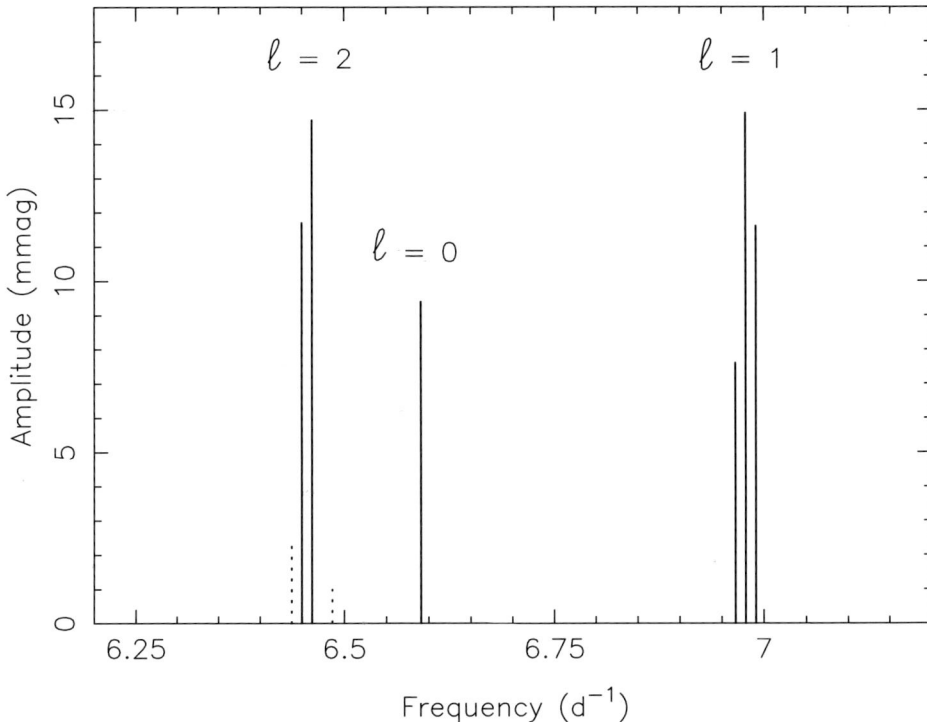

Figure 5 The schematic frequency spectrum of the β Cep star V836 Cen derived from single-site Geneva U data spanning 21 years. The dotted lines are frequencies that are not yet firmly established; these were not used in the seismic modeling. (Figure reproduced from Aerts *et al.*, 2004.)

made the seismic derivation of the interior rotation possible, because the splitting of the multiplets does not interfere with the frequency separation between different multiplets. The rotation profile itself could not be tuned further, given that only parts of very few multiplets were available. Classical spectroscopy, even at extremely high resolution, could never have led to the proof of differential interior rotation, as it can only measure the surface rotation. Moreover, the intrinsic line broadening of such stars is typically of order ≈ 10 km s^{-1}, which is larger than the surface rotation velocity of these two stars, preventing a derivation of the projected equatorial rotation velocity to better than 1 km s^{-1}.

For the star θ Oph, which is a twin of V836 Cen as far as the detected frequency spectrum is concerned, rigid interior rotation could not be excluded from comparison of the frequency spacing in its triplet and its quintuplet (Briquet *et al.*, 2007). Its frequency precision is two orders of magnitude lower that for V836 Cen and one order of magnitude lower than for ν Eri. In any case, strong differential rotation is excluded for that star as well.

5. Expected Future Improvements

5.1. Compact Pulsators and the Tuning of Atomic Diffusion

A field within asteroseismology, which we did not discuss extensively here but which is undergoing rapid growth and may tune our knowledge of microscopic diffusion for stellar structure and of binary star evolution, is the application to pulsating subdwarf B stars (sdBVs). Although sdBVs with p modes were discovered a decade ago (Kilkenny et al., 1997), those with g modes were discovered more recently (Green et al., 2003). The existence of sdBVs was predicted independently and simultaneously with their observational discovery (Charpinet et al., 1996). An opacity bump associated primarily with iron-group elements turns out to be an efficient driving mechanism. The atomic diffusion processes that are at work in sdB stars – radiative levitation and gravitational settling – cause iron (and also zinc) to become overabundant in the driving zone, thus exciting low-order p and g modes (Charpinet et al., 1997; Jeffery and Saio, 2006). The details of the diffusion processes are, however, still uncertain. These may also be relevant for the slowly pulsating B stars (SPBs) and β Cep stars (Bourge et al., 2006), for which diffusion processes have been ignored so far in the seismology. Such processes are dominant in the atmospheres of the roAp stars, as is discussed by Kurtz (2008).

From an evolutionary point of view, the sdB stars are poorly understood. Their effective temperatures are in the range 23 000 – 32 000 K, and their $\log g$ values in the range 5 – 6. They all have masses below $0.5 M_\odot$, which implies that they have lost almost their entire hydrogen envelope at the tip of the red-giant branch. Their thin hydrogen layer does not contain enough mass to burn hydrogen, making them evolve immediately from the giant branch toward the extreme horizontal branch. Although it is clear they will end their lives as low-mass white dwarfs, it is as yet unclear how they expelled their envelopes. All scenarios that have been proposed involve close binary interaction (Han et al., 2003; Hu et al., 2007).

The currently known sdB pulsators have multiple periods in the range 80 – 600 seconds and amplitudes up to 0.3 mag. Their amplitude variability and faintness have prevented unambiguous mode identifications so far, limiting the power of seismic inference to tune the diffusive and rotational processes. Rapidly rotating cores have been claimed for some of the sdBVs to explain their dense frequency spectra in terms of low-degree modes (Kawaler and Hostler, 2005). Firm observational proof of that is not yet available, but, given the impressive efforts undertaken to understand the internal and atmospheric structure of these stars as well as their evolutionary status, we expect rapid progress in the near future. For a recent overview of the status of sdB seismology, we refer to Charpinet et al. (2007).

Recently, a seven-year study of the sdBV star V391 Peg (Silvotti et al., 2007) used the extreme frequency stability of two independent pulsation frequencies to show the presence of a $\approx 3.2 M_{Jupiter}$ planet that had moved from about 1 AU out to 1.7 AU during the red giant phase of the sdB star precursor, allowing the planet to survive, much as the Earth may survive the Sun's red-giant phase in about 7 Gyr. This novel application of asteroseismology highlights the close relation and mutual interests of helioseismology, asteroseismology, planet-finding, and solar system studies.

5.2. Heat-Driven Pulsators along the Main Sequence

Overshooting parameters and internal rotation profiles have not yet been determined for the other heat-driven pulsators known along the main sequence (besides the β Cep stars), such as the SPBs and the A- and F-type δ Sct and γ Dor stars. The main obstacles to overcome are the limited number of detected oscillation frequencies of the g modes for the SPB stars and γ Dor stars and reliable mode identification for those as well as for the δ Sct stars. Although the pioneering space missions WIRE (Buzasi, 2002; Bruntt and Southworth, 2008) and MOST (Matthews et al., 2004; Walker, 2008) led to an impressive and unprecedented number of oscillation modes for several such stars, the time base of the data was limited to a few weeks and unique mode identifications are not available for these mission's target stars. It is to be expected that the uninterrupted photometry obtained by the CoRoT (launched on 26 December 2006 and having a five-month time base) and *Kepler* (which will have a 3.5-year time base and will be launched in 2009) space missions, along with their ground-based spectroscopy programs, will result in the necessary frequency precision and mode identification. This should imply big steps forward for the seismic modeling of these type of stars.

Thus, even with several space missions in operation, ground-based efforts to increase the number of heat-driven pulsators with (preferably simultaneous) long-term multicolor photometric and high-resolution spectroscopic data for mode identification should definitely be intensified. It was this type of extensive data that yielded sudden and immense progress in the β Cep star seismology discussed in this paper and that also advanced significantly the interpretation of the oscillation spectrum of the prototypical δ Sct star FG Vir (Zima *et al.*, 2006). Only systematic and dedicated observing programs can bring us to the stage of mapping and calibrating the internal-mixing processes inside stars across the HR diagram.

5.3. Solar-Like Pulsators

A new dimension in the progress for stochastically excited pulsators is expected from the combination of asteroseismic and interferometric data, as explained by Cunha *et al.* (2007). The CoRoT and *Kepler* missions will also provide a large improvement in the data for solar-like oscillators. In particular, *Kepler* will yield data over several years for more than a hundred stars, together with three-month surveys of many more stars. The very extended observations may reveal possible frequency variations associated with stellar magnetic cycles, as has been observed in the Sun, and hence improve our understanding of such cycles. Also, the identification and interpretation of mixed modes, which have both a p-mode and a g-mode character owing to a highly condensed stellar core, would be a great help to tune stellar evolution models toward the end of, and after, the central hydrogen-burning phase.

Data of even higher quality on solar-like oscillators can be obtained with dedicated meter-per-second precision radial-velocity campaigns, since the intrinsic stellar noise background is much lower, relative to the oscillations, in velocity than in photometry (*e.g.*, Harvey, 1988). This is the goal of the SIAMOIS (Mosser *et al.*, 2007) and SONG (Grundahl *et al.*, 2007) projects. SIAMOIS will operate from the South Pole, and SONG aims at establishing a global network of moderate-sized telescopes. Both are dedicated to obtain high-precision radial-velocity observations. This will increase substantially the number of (rotationally split) detected modes, particularly at relatively low frequency where the mode lifetime is longer and the potential frequency accuracy is higher. With the high-quality data expected from CoRoT, *Kepler*, SIAMOIS, and SONG, we may hope to carry out inverse analyses (*e.g.*, Basu, Christensen-Dalsgaard, and Thompson, 2002; Roxburgh and Vorontsov, 2002) to infer the detailed properties of stellar cores.

These projects, and others further into the future, covering stars across the HR diagram, will provide an extensive observational basis for investigating stellar interiors. Together with the parallel development of stellar modeling techniques we may finally approach the point, in the words of Eddington (1926), of being "competent to understand so simple a thing as a star."

Acknowledgements C.A. is supported by the Research Council of the Catholic University of Leuven under Grant No. GOA/2003/04. M.C. is supported by the EC's FP6, FCT, and FEDER (POCI2010) and through the project POCI/CTE-AST/57610/2004. The authors acknowledge support from the FP6 Coordination Action HELAS.

References

Aerts, C.: 2006, In: Fletcher, K., Thompson, M. (eds.) *Proceedings of SOHO 18/GONG 2006/HELAS I, Beyond the Spherical Sun*, SP-**624**, ESA, Noordwijk, 131.
Aerts, C., Thoul, A., Daszyńska, J., Scuflaire, R., Waelkens, C., Dupret, M.A., Niemczura, E., Noels, A.: 2003, *Science* **300**, 1926.
Aerts, C., Waelkens, C., Daszyńska-Daszkiewicz, J., Dupret, M.-A., Thoul, A., Scuflaire, R., Uytterhoeven, K., Niemczura, E., Noels, A.: 2004, *Astron. Astrophys.* **415**, 241.
Aerts, C., Marchenko, S.V., Matthews, J.M., Kuschnig, R., Guenther, D.B., Moffat, A.F.J., Rucinski, S.M., Sasselov, D., Walker, G.A.H., Weiss, W.W.: 2006, *Astrophys. J.* **642**, 470.
Basu, S., Antia, H.M., Narasimha, D.: 1994, *Mon. Not. Roy. Astron. Soc.* **267**, 209.
Basu, S., Christensen-Dalsgaard, J., Thompson, M.J.: 2002, In: Favata, F., Roxburgh, I.W., Galadí-Enríquez, D. (eds.) *Stellar Structure and Habitable Planet Finding*, SP-**485**, ESA, Noordwijk, 249.
Bazot, M., Bouchy, F., Kjeldsen, H., Charpinet, S., Laymand, M., Vauclair, S.: 2007, *Astron. Astrophys.* **470**, 295.
Bedding, T.R., Kjeldsen, H.: 2007, *Commun. Asteroseismol.* **150**, 106.
Bedding, T.R., Butler, R.P., Kjeldsen, H., Baldry, I.K., O'Toole, S.J., Tinney, C.G., Marcy, G.W., Kienzle, F., Carrier, F.: 2001, *Astrophys. J.* **549**, L105.
Bedding, T.R., Kjeldsen, H., Bouchy, F., Bruntt, H., Butler, R.P., Buzasi, D.L., Christensen-Dalsgaard, J., Frandsen, S., Lebrun, J.-C., Martić, M., Schou, J.: 2005, *Astron. Astrophys.* **432**, L43.
Böhm-Vitense, E.: 1958, *Z. Astrophys.* **46**, 108.
Bouchy, F., Carrier, F.: 2001, *Astron. Astrophys.* **374**, L5.
Bourge, P.-O., Ałecian, G., Thoul, A., Scuflaire, R., Theado, S.: 2006, *Commun. Asteroseismol.* **147**, 105.
Bradley, P.A.: 1995, *Baltic Astron.* **4**, 311.
Briquet, M., Lefever, K., Uytterhoeven, K., Aerts, C.: 2005, *Mon. Not. Roy. Astron. Soc.* **362**, 619.
Briquet, M., Morel, T., Thoul, A., Scuflaire, R., Miglio, A., Montalbán, J., Dupret, M.-A., Aerts, C.: 2007, *Mon. Not. Roy. Astron. Soc.* **381**, 1482.
Brown, T.M., Gilliland, R.L., Noyes, R.W., Ramsey, L.W.: 1991, *Astrophys. J.* **368**, 599.
Brown, T.M., Kennelly, E.J., Korzennik, S.G., Nisenson, P., Noyes, R.W., Horner, S.D.: 1997, *Astrophys. J.* **475**, 322.
Bruntt, H., Southworth, J.: 2008, In: Gizon, L., Roth, M. (eds.) *Proc. HELAS II International Conference: Helioseismology, Asteroseismology, and MHD Connections, J. Phys. Conf. Ser.*, in press.
Buzasi, D.: 2002, In: Aerts, C., Bedding, T.R., Christensen-Dalsgaard, J. (eds.) *IAU Colloq. 185: Radial and Nonradial Pulsations as Probes of Stellar Physics*, CS-**259**, Astron. Soc. Pacific, San Francisco, 616.
Carrier, F., Bouchy, F., Eggenberger, P.: 2003, In: Thompson, M.J., Cunha, M.S., Monteiro, M.J.P.F.G. (eds.) *Asteroseismology Across the HR Diagram*, Kluwer, Dordrecht, 315.
Carrier, F., Eggenberger, P., Bouchy, F.: 2005, *Astron. Astrophys.* **434**, 1085.
Charpinet, S., Fontaine, G., Brassard, P., Dorman, B.: 1996, *Astrophys. J.* **471**, L103.
Charpinet, S., Fontaine, G., Brassard, P., Chayer, P., Rogers, F.J., Iglesias, C.A., Dorman, B.: 1997, *Astrophys. J.* **483**, L123.
Charpinet, S., Fontaine, G., Brassard, P., Chayer, P., Green, E.M., Randall, S.K.: 2007, *Commun. Asteroseismol.* **150**, 241.
Christensen-Dalsgaard, J.: 1982, *Adv. Space Res.* **2**, 11.
Christensen-Dalsgaard, J.: 1984, In: Mangeney, A., Praderie, F. (eds.) *Space Research Prospects in Stellar Activity and Variability*, Paris Observatory Press, Paris, 11.
Christensen-Dalsgaard, J.: 1988, In: Christensen-Dalsgaard, J., Frandsen, S. (eds.) *Advances in Helio- and Asteroseismology*, Proc. IAU Symp. **123**, Reidel, Dordrecht, 295.
Christensen-Dalsgaard, J.: 2005, In: Favata, F., Hussain, G., Battrick, B. (eds.) *13th Cambridge Workshop on Cool Stars, Stellar Systems and the Sun*, SP-**560**, ESA, Noordwijk, 81.
Christensen-Dalsgaard, J., Frandsen, S.: 1983, *Solar Phys.* **82**, 469.
Christensen-Dalsgaard, J., Bedding, T.R., Kjeldsen, H.: 1995, *Astrophys. J.* **443**, L29.
Christensen-Dalsgaard, J., Monteiro, M.J.P.F.G., Thompson, M.J.: 1995, *Mon. Not. Roy. Astron. Soc.* **276**, 283.
Christensen-Dalsgaard, J., Dappen, W., Ajukov, S.V., Anderson, E.R., Antia, H.M., Basu, S., Baturin, V.A., Berthomieu, G., Chaboyer, B., Chitre, S.M., Cox, A.N., Demarque, P., Donatowicz, J., Dziembowski, W.A., Gabriel, M., Gough, D.O., Guenther, D.B., Guzik, J.A., Harvey, J.W., Hill, F., Houdek, G., Iglesias, C.A., Kosovichev, A.G., Leibacher, J.W., Morel, P., Proffitt, C.R., Provost, J., Reiter, J., Rhodes, E.J. Jr., Rogers, F.J., Roxburgh, I.W., Thompson, M.J., Ulrich, R.K.: 1996, *Science* **272**, 1286.
Christensen-Dalsgaard, J., Arentoft, T., Brown, T.M., Gilliland, R.L., Kjeldsen, H., Borucki, W.J., Koch, D.: 2007, *Commun. Asteroseismol.* **150**, 350.
Cunha, M.S., Metcalfe, T.S.: 2007, *Astrophys. J.* **666**, 413.

Cunha, M.S., Aerts, C., Christensen-Dalsgaard, J., Baglin, A., Bigot, L., Brown, T.M., Catala, C., Creevey, O.L., Domiciano de Souza, A., Eggenberger, P., Garcia, P.J.V., Grundahl, F., Kervella, P., Kurtz, D.W., Mathias, P., Miglio, A., Monteiro, M.J.P.F.G., Perrin, G., Pijpers, F.P., Pourbaix, D., Quirrenbach, A., Rousselet-Perraut, K., Teixeira, T.C., Thevenin, F., Thompson, M.J.: 2007, *Astron. Astrophys. Rev.* **14**, 217.
Deupree, R.G.: 2000, *Astrophys. J.* **543**, 395.
Di Mauro, M.P., Christensen-Dalsgaard, J., Kjeldsen, H., Bedding, T.R., Paternò, L.: 2003, *Astron. Astrophys.* **404**, 341.
Di Mauro, M.P., Christensen-Dalsgaard, J., Paternò, L., D'Antona, F.: 2004, *Solar Phys.* **220**, 185.
Dupret, M.-A., Thoul, A., Scuflaire, R., Daszyńska-Daszkiewicz, J., Aerts, C., Bourge, P.-O., Waelkens, C., Noels, A.: 2004, *Astron. Astrophys.* **415**, 251.
Dziembowski, W., Goode, P.R.: 1984, *Mem. Soc. Astron. Ital.* **55**, 185.
Dziembowski, W.A., Pamiatnykh, A.A.: 1993, *Mon. Not. Roy. Astron. Soc.* **262**, 204.
Dziembowski, W.A., Gough, D.O., Houdek, G., Sienkiewicz, R.: 2001, *Mon. Not. Roy. Astron. Soc.* **328**, 601.
Eddington, A.S.: 1926, *The Internal Constitution of the Stars*, Cambridge University Press, Cambridge.
Eggenberger, P., Carrier, F., Bouchy, F.: 2005, *New Astron.* **10**, 195.
Eggenberger, P., Carrier, F., Bouchy, F., Blecha, A.: 2004a, *Astron. Astrophys.* **422**, 247.
Eggenberger, P., Charbonnel, C., Talon, S., Meynet, G., Maeder, A., Carrier, F., Bourban, G.: 2004b, *Astron. Astrophys.* **417**, 235.
Fletcher, S.T., Chaplin, W.J., Elsworth, Y., Schou, J., Buzasi, D.: 2006, *Mon. Not. Roy. Astron. Soc.* **371**, 935.
Green, E.M., Fontaine, G., Reed, M.D., Callerame, K., Seitenzahl, I.R., White, B.A., Hyde, E.A., Østensen, R., Cordes, O., Brassard, P., Falter, S., Jeffery, E.J., Dreizler, S., Schuh, S.L., Giovanni, M., Edelmann, H., Rigby, J., Bronowska, A.: 2003, *Astrophys. J.* **583**, L31.
Guenther, D.B., Kallinger, T., Reegen, P., Weiss, W.W., Matthews, J.M., Kuschnig, R., Marchenko, S., Moffat, A.F.J., Rucinski, S.M., Sasselov, D., Walker, G.A.H.: 2005, *Astrophys. J.* **635**, 547.
Gough, D.O.: 1981, *Mon. Not. Roy. Astron. Soc.* **196**, 731.
Gough, D.O.: 1993, In: Zahn, J.-P., Zinn-Justin, J. (eds.) *Astrophysical Fluid Dynamics – Les Houches Session XLVII, 1987*, Elsevier, Amsterdam, 399.
Grundahl, F., Kjeldsen, H., Christensen-Dalsgaard, J., Arentoft, T., Frandsen, S.: 2007, *Commun. Asteroseismol.* **150**, 300.
Han, Z., Podsiadlowski, P., Maxted, P.F.L., Marsh, T.R.: 2003, *Mon. Not. Roy. Astron. Soc.* **341**, 669.
Handler, G., Shobbrook, R.R., Mokgwetsi, T.: 2005, *Mon. Not. Roy. Astron. Soc.* **362**, 612.
Harvey, J.W.: 1988, In: Christensen-Dalsgaard, J., Frandsen, S. (eds.) *Advances in Helio- and Asteroseismology*, *Proc. IAU Symp.* **123**, Reidel, Dordrecht, 497.
Houdek, G., Gough, D.O.: 2007, *Mon. Not. Roy. Astron. Soc.* **375**, 861.
Houdek, G., Gough, D.O.: 2008, In: Stancliffe, R.J., Dewi, J., Houdek, G., Martin, R.G., Tout, C.A. (eds.) *Unsolved Problems in Stellar Physics*, American Institute of Physics, Melville, in press [arXiv:0710.0762].
Houdek, G., Balmforth, N.J., Christensen-Dalsgaard, J., Gough, D.O.: 1999, *Astron. Astrophys.* **351**, 582.
Hu, H., Nelemans, G., Østensen, R., Aerts, C., Vučković, M., Groot, P.J.: 2007, *Astron. Astrophys.* **473**, 569.
Jeffery, C.S., Saio, H.: 2006, *Mon. Not. Roy. Astron. Soc.* **372**, L48.
Jones, P.W., Hansen, C.J., Pesnell, W.D., Kawaler, S.D.: 1989, *Astrophys. J.* **336**, 403.
Kawaler, S.D., Hostler, S.R.: 2005, *Astrophys. J.* **621**, 432.
Kawaler, S.D., Sekii, T., Gough, D.: 1999, *Astrophys. J.* **516**, 349.
Kepler, S.O.: 2007, *Commun. Asteroseismol.* **150**, 221.
Kilkenny, D., Koen, C., O'Donoghue, D., Stobie, R.S.: 1997, *Mon. Not. Roy. Astron. Soc.* **285**, 640.
Kjeldsen, H., Bedding, T.R.: 1995, *Astron. Astrophys.* **293**, 87.
Kjeldsen, H., Bedding, T.R., Viskum, M., Frandsen, S.: 1995, *Astron. J.* **109**, 1313.
Kjeldsen, H., Bedding, T.R., Baldry, I.K., Bruntt, H., Butler, R.P., Fischer, D.A., Frandsen, S., Gates, E.L., Grundahl, F., Lang, K., Marcy, G.W., Misch, A., Vogt, S.S.: 2003, *Astron. J.* **126**, 1483.
Kozhurina-Platais, V., Demarque, P., Platais, I., Orosz, J.A., Barnes, J.: 1997, *Astron. J.* **113**, 1045.
Kurtz, D.W.: 2004, *Solar Phys.* **220**, 123.
Kurtz, D.W.: 2006, In: Sterken, C., Aerts, C. (eds.) *Astrophysics of Variable Stars*, *CS-349*, Astron. Soc. Pacific, San Francisco, 101.
Kurtz, D.W.: 2008, *Solar Phys.*, in press. doi:10.1007/s11207-008-9139-2.
Lochard, J., Samadi, R., Goupil, M.-J.: 2004, *Solar Phys.* **220**, 199.
Maeder, A., Meynet, G.: 2000, *Ann. Rev. Astron. Astrophys.* **38**, 143.
Martić, M., Schmitt, J., Lebrun, J.-C., Barban, C., Connes, P., Bouchy, F., Michel, E., Baglin, A., Appourchaux, T., Bertaux, J.-L.: 1999, *Astron. Astrophys.* **351**, 993.
Matthews, J.M., Kuschnig, R., Guenther, D.B., Walker, G.A.H., Moffat, A.F.J., Rucinski, S.M., Sasselov, D., Weiss, W.W.: 2004, *Nature* **430**, 51.

Mazumdar, A., Basu, S., Collier, B.L., Demarque, P.: 2006a, *Mon. Not. Roy. Astron. Soc.* **372**, 949.
Mazumdar, A., Briquet, M., Desmet, M., Aerts, C.: 2006b, *Astron. Astrophys.* **459**, 589.
Michel, E., Baglin, A., Auvergne, M., Catala, C., Aerts, C., Alecian, G., Amado, P., Appourchaux, T., Ausseloos, M., Ballot, J., Barban, C., Baudin, F., Berthomieu, G., Boumier, P., Bohm, T., Briquet, M., Charpinet, S., Cunha, M.S., De Cat, P., Dupret, M.A., Fabregat, J., Floquet, M., Fremat, Y., Garrido, R., Garcia, R.A., Goupil, M.-J., Handler, G., Hubert, A.-M., Janot-Pacheco, E., Lambert, P., Lebreton, Y., Lignieres, F., Lochard, J., Martin-Ruiz, S., Mathias, P., Mazumdar, A., Mittermayer, P., Montalban, J., Monteiro, M.J.P.F.G., Morel, P., Mosser, B., Moya, A., Neiner, C., Nghiem, P., Noels, A., Oehlinger, J., Poretti, E., Provost, J., Renan de Medeiros, J., de Ridder, J., Rieutord, M., Roca-Cortes, T., Roxburgh, I., Samadi, R., Scuflaire, R., Suarez, J.C., Theado, S., Thoul, A., Toutain, T., Turck-Chieze, S., Uytterhoeven, K., Vauclair, G., Vauclair, S., Weiss, W.W., Zwintz, K.: 2006, In: Fridlund, M., Baglin, A., Lochard, J., Conroy, L. (eds.) *The CoRoT Mission, SP*-**1306**, ESA, Noordwijk, 39.
Miglio, A., Montalbán, J.: 2005, *Astron. Astrophys.* **441**, 615.
Miglio, A., Montalbán, J., Dupret, M.-A.: 2007, *Mon. Not. Roy. Astron. Soc.* **375**, L21.
Monteiro, M.J.P.F.G., Christensen-Dalsgaard, J., Thompson, M.J.: 1994, *Astron. Astrophys.* **283**, 247.
Monteiro, M.J.P.F.G., Christensen-Dalsgaard, J., Thompson, M.J.: 2000, *Mon. Not. Roy. Astron. Soc.* **316**, 165.
Mosser, B., The Siamois Team: 2007, *Commun. Asteroseismol.* **150**, 309.
Mosser, B., Bouchy, F., Martić, M., Appourchaux, T., Barban, C., Berthomieu, G., Garcia, R.A., Lebrun, J.C., Michel, E., Provost, J., Thévenin, F., Turck-Chièze, S.: 2008, *Astron. Astrophys.* **478**, 197.
Nather, R.E., Winget, D.E., Clemens, J.C., Hansen, C.J., Hine, B.P.: 1990, *Astrophys. J.* **361**, 309.
Pamyatnykh, A.A.: 1999, *Acta Astron.* **49**, 119.
Pamyatnykh, A.A., Handler, G., Dziembowski, W.A.: 2004, *Mon. Not. Roy. Astron. Soc.* **350**, 1022.
Provost, J., Berthomieu, G., Martić, M., Morel, P.: 2006, *Astron. Astrophys.* **460**, 759.
Ribas, I., Jordi, C., Giménez, Á.: 2000, *Mon. Not. Roy. Astron. Soc.* **318**, L55.
Roxburgh, I.W., Vorontsov, S.V.: 2002, In: Favata, F., Roxburgh, I.W., Galadi, D. (eds.) *Stellar Structure and Habitable Planet Finding, SP*-**485**, ESA, Noordwijk, 341.
Saio, H.: 1981, *Astrophys. J.* **244**, 299.
Samadi, R., Goupil, M.-J., Alecian, E., Baudin, F., Georgobiani, D., Trampedach, R., Stein, R., Nordlund, Å.: 2005, *J. Astrophys. Astron.* **26**, 171.
Silvotti, R., Schuh, S., Janulis, R., Solheim, J.-E., Bemabei, S., Ostensen, R., Oswalt, T.D., Bruni, I., Gualandi, R., Bonanno, A., Vauclair, G., Reed, M., Chen, C.-W., Leibowitz, E., Paparo, M., Baran, A., Charpinet, S., Dolez, N., Kawaler, S., Kurtz, D., Moskalik, P., Riddle, R., Zola, S.: 2007, *Nature* **448**, 189.
Stankov, A., Handler, G.: 2005, *Astrophys. J. Suppl. Ser.* **158**, 193.
Straka, C.W., Demarque, P., Guenther, D.B.: 2005, *Astrophys. J.* **629**, 1075.
Tassoul, M.: 1980, *Astrophys. J. Suppl. Ser.* **43**, 469.
Thoul, A., Scuflaire, R., Ausseloos, M., Aerts, C., Noels, A.: 2004, *Commun. Asteroseismol.* **144**, 35.
Ulrich, R.K.: 1986, *Astrophys. J.* **306**, L37.
Vandakurov, Y.V.: 1967, *Astron. Zh.* **44**, 786 (English translation: *Sov. Astron. AJ* **11**, 630).
VandenBerg, D.A., Stetson, P.B.: 2004, *Publ. Astron. Soc. Pac.* **116**, 997.
Walker, G.A.H.: 2008, In: Gizon, L., Roth, M. (eds.) *Proc. HELAS II International Conference: Helioseismology, Asteroseismology, and MHD Connections, J. Phys. Conf. Ser.*, in press.
Winget, D.E., Nather, R.E., Clemens, J.C., Provencal, J., Kleinman, S.J., Bradley, P.A., Wood, M.A., Claver, C.F., Frueh, M.L., Grauer, A.D., Hine, B.P., Hansen, C.J., Fontaine, G., Achilleos, N., Wickramasinghe, D.T., Marar, T.M.K., Seetha, S., Ashoka, B.N., O'Donoghue, D., Warner, B., Kurtz, D.W., Buckley, D.A., Brickhill, J., Vauclair, G., Dolez, N., Chevreton, M., Barstow, M.A., Solheim, J.E., Kanaan, A., Kepler, S.O., Henry, G.W., Kawaler, S.D.: 1991, *Astrophys. J.* **378**, 326.
Winget, D.E., Nather, R.E., Clemens, J.C., Provencal, J.L., Kleinman, S.J., Bradley, P.A., Claver, C.F., Dixson, J.S., Montgomery, M.H., Hansen, C.J., Hine, B.P., Birch, P., Candy, M., Marar, T.M.K., Seetha, S., Ashoka, B.N., Leibowitz, E.M., O'Donoghue, D., Warner, B., Buckley, D.A.H., Tripe, P., Vauclair, G., Dolez, N., Chevreton, M., Serre, T., Garrido, R., Kepler, S.O., Kanaan, A., Augusteijn, T., Wood, M.A., Bergeron, P., Grauer, A.D.: 1994, *Astrophys. J.* **430**, 839.
Yıldız, M.: 2007, *Mon. Not. Roy. Astron. Soc.* **374**, 1264.
Zahn, J.-P.: 1991, *Astron. Astrophys.* **252**, 179.
Zima, W., Wright, D., Bentley, J., Cottrell, P.L., Heiter, U., Mathias, P., Poretti, E., Lehmann, H., Montemayor, T.J., Breger, M.: 2006, *Astron. Astrophys.* **455**, 235.

The Solar-Stellar Connection: Magneto-Acoustic Pulsations in $1.5 M_\odot - 2 M_\odot$ Peculiar A Stars

D.W. Kurtz

Originally published in the journal Solar Physics, Volume 251, Nos 1–2, 21–30.
DOI: 10.1007/s11207-008-9139-2 © Springer Science+Business Media B.V. 2008

Abstract Stellar astronomers look on in envy at the wealth of data, the incredible spatial resolution, and the maturity of the theoretical understanding of the Sun. Yet the Sun is but one star, so stellar astronomy is of great interest to solar astronomers for its range of different conditions under which to test theoretical understanding gained from the study of the Sun. The rapidly oscillating peculiar A stars are of particular interest to solar astronomers. They have strong, global, dipolar magnetic fields with strengths in the range 1 – 25 kG, and they pulsate in high-overtone p modes similar to those in the Sun; thus they offer a unique opportunity to study the interaction of pulsation, convection, and strong magnetic fields, as is now done in the local helioseismology of sunspots. Some of them even pulsate in modes with frequencies above the acoustic cutoff frequency, in analogy with the highest frequency solar modes, but with mode lifetimes up to decades in the roAp stars, very unlike the short mode lifetimes of the Sun. They offer the most extreme cases of atomic diffusion, a small, but important ingredient of the standard solar model with wide application in stellar astrophysics. They are compositionally stratified and are observed and modelled as a function of atmospheric depth and thus can inform plans to expand helioseismic observations to have atmospheric depth resolution. Study of this unique class of pulsating stars follows the advanced state of studies of the Sun and offers more extreme conditions for the understanding of physics shared with the Sun.

Keywords Helioseismology, theory · Helioseismology, observations · Magnetic fields, photosphere · Oscillations, solar · Oscillations, stellar · Sunspots, magnetic fields · Waves, acoustic · Waves, Alfvén

1. Introduction

Global and local helioseismology have had resounding successes in providing unprecedented – and not-so-long-ago even unimagined – detailed knowledge of the physics of the

---

Guest Editors: Laurent Gizon and Paul Cally

D.W. Kurtz (✉)
Centre for Astrophysics, University of Central Lancashire, Preston, PR1 2HE, UK
e-mail: dwkurtz@uclan.ac.uk

solar interior and atmosphere. Even geoseismologists look admiringly at helioseismology for its surface resolution of the Sun, which is far better than for Earth, given the dearth of seismic stations over the global oceans. With the stunning detail of helioseismic studies (*e.g.*, from the internal-rotation profiles to far-side views of active regions), what in asteroseismology could be of direct interest to the helioseismologist or other solar astronomer? After all, there is no stellar equivalent (yet) of GONG,[1] BiSON,[2] or the other solar observational networks, so there are no long, continuous, high-duty-cycle stellar asteroseismic observations. From a solar astronomer's point of view, stars are faint, point sources with no spatial resolution, low spectral resolution, and low signal-to-noise ratio. Therefore, what is of interest in asteroseismology for the solar astronomer?

The answer is a very great deal. As is shown in the invited review of asteroseismology in this volume (Aerts *et al.*, 2008), asteroseismology of solarlike pulsators is now an active field with first data sets and theoretical interpretations. The reason for this recent activity is that observational radial-velocity precision for the brightest solarlike oscillators is now as good as radial-velocity precision was for the Sun two decades ago. Compare the power spectra of α Cen (Bazot *et al.*, 2007; Bedding *et al.*, 2004) and the Sun from the South Pole (Fossat, Grec, and Pomerantz, 1981). The stellar observations have a precision of 2–4 cm s^{-1}; the solar observations from 25 years earlier are similarly precise to better than 10 cm s^{-1}. Although a solar astronomer's initial reaction might be, "asteroseismology is 25 years behind the times", remember that α Cen A has an apparent magnitude of $M_V = 0.0$ compared with the Sun's $M_V = -26.8$, for a flux ratio of $F_\odot/F_{\alpha \text{Cen}} = 5.2 \times 10^{10}$! In this light (or lack of it, I should say), observational asteroseismology is stunningly advanced.

The observational radial-velocity precision that can now be obtained with highly stabilised spectrographs on large telescopes was pioneered by the planet hunters with their need for ultraprecise Doppler-shift measurement in the quest to find – at first – any planets, and now to detect Earth-mass planets. Thus asteroseismology and planet finding go hand in hand observationally, both with ground-based high-precision radial velocity measurements and space-based photometry with the MOST[3] (*e.g.*, Croll *et al.*, 2007; Walker *et al.*, 2003), CoRoT[4] (*e.g.*, Michel *et al.*, 2007), and *Kepler*[5] (*e.g.*, Borucki *et al.*, 2003; Christensen-Dalsgaard *et al.*, 2007) missions.

The radial velocity studies that have been done and the photometric data about to come from satellite missions give a variety of new stars similar to the Sun, but not identical to it, to test the well-developed solar modelling tools and theory. This provides more than an opportunity to apply solar tools to new problems; it proves those tools by extending the physical conditions under which they are tested. While Aerts *et al.* (2008) review asteroseismology in general, it is my purpose here to discuss just one class of pulsating stars – the rapidly oscillating peculiar A stars (roAp stars) – that I believe are particularly fascinating for many helioseismologists. These stars have masses and radii about twice that of the Sun; they have core convection zones and very thin surface convection zones. So at first look the physics seems to be significantly different from the solar case. But there are some striking similarities that allow tests of physics important in the Sun in ways that can be done for no other type of stars:

---

[1] Global Oscillation Network Group; see http://gong.nso.edu/.

[2] Birmingham Solar Oscillation Network; see http://bison.ph.bham.ac.uk/.

[3] See http://www.astro.ubc.ca/MOST/.

[4] See http://smsc.cnes.fr/COROT/GP_satellite.htm.

[5] See http://kepler.nasa.gov/.

1. The roAp stars were the first main-sequence stars observed to have high radial-overtone p-mode pulsations showing asymptotic frequency spacings from which the large and small separations can be measured and modelled. Although this is now possible for the solarlike oscillators, the roAp stars are still the only other known main sequence p-mode pulsators with frequencies in the asymptotic p-mode regime.
2. The roAp stars have strong, global (basically dipolar) magnetic fields with surface field strengths up to 25 kG – far stronger that that seen in sunspots. These strong fields perturb the pulsation frequencies significantly – by more than the large separation in some cases – thus providing observational data to test theories of the interaction of p modes with strong magnetic fields. Solar observations and theoretical models of p-mode interaction with sunspots is highly relevant to the understanding of the pulsations in roAp stars, and conversely these stars provide an alternative environment and many more objects on which to test theory. The pulsation modes are magneto-acoustic in the roAp stars – throughout most of the observable atmosphere the magnetic pressure exceeds the gas pressure, and the Alfvén speed exceeds the sound speed – thus emphasising the similarity to p modes within sunspots in the Sun.
3. The roAp stars are spectroscopically exceedingly peculiar; arguably, they are the most peculiar stars in the sky. Their extreme overabundances of many rare-earth elements – up to a factor of 10^5 compared to the Sun – and milder deficiencies of a few lighter elements are primarily caused by atomic diffusion (with possible contributions from other mechanisms, *i.e.*, the competition between radiative levitation and gravitational settling of ions). This is a minor, but significant, process in the standard solar model. It has to be considered in the current discrepancy between the helioseismic sound speed with the standard solar model and the new solar abundances, although it is not the solution to this problem. The theory of atomic diffusion was originally developed to explain the peculiar atmospheres of the Ap stars, and those stars remain the most significant (and perhaps most difficult) test bed for that theory.
4. Atomic diffusion has stratified the abundance distributions in the atmospheres of roAp stars, particularly of some rare-earth elements. High-time-resolution, high-spectral-resolution, high-signal-to-noise ratio (by stellar standards) observations with the VLT and other large telescopes are giving significant depth resolution of the pulsation modes in these stars, which may inform discussion in the helioseismology community of future network observations using many spectral lines to give depth information for the pulsations in the Sun. Here is a case where stellar astronomy is in the lead. The roAp are also "spotted" – not as the Sun is with short-lived, small sunspots, but with large, permanent horizontal abundance variations that give rise to intensity variations, so are also intensity spots as are sunspots. Those horizontal abundance variations can be mapped with Doppler imaging for stars with moderate rotational velocities, so there is now a real prospect of 3D maps of abundances and pulsation field geometries in the roAp stars. These are the only stars other than the Sun where this can be done.

For these reasons the magnetic roAp stars hold considerable interest for solar astronomers, and, of course, solar astronomy (and helioseismology in particular) guides the study of these stars.

2. Mode Identification in roAp Stars: The Oblique Pulsator Model

Fundamental to the application of asteroseismic techniques is mode identification. The ray path of an acoustic mode is a function of the geometry of that mode: Each mode penetrates to a particular depth in the star and the pulsation frequency is a measure of the sound travel time along the ray path. The frequency thus depends on a integral including the sound speed through the "acoustic cavity" of the mode. With enough modes, the sound speed profile can be recovered and much physical information inferred from it, such as temperature and density and, with an equation of state, pressure and mean molecular weight and hence chemical composition.

Mode identification is therefore a requirement for progress in asteroseismology. This is relatively straightforward for the Sun with the high angular resolution that is possible and large number of modes observed. It is notoriously difficult for spatially unresolved stars. The roAp stars are a special case where mode geometry can be determined better than for other types of pulsating stars. This is because the roAp stars are "oblique pulsators", in which the pulsation axis of each mode is aligned, or nearly aligned, with the strong global magnetic field, which is itself inclined to the rotation axis (*e.g.*, Kurtz, 1982; Bigot and Dziembowski, 2002; Saio, 2005). Thus with rotation we observe the pulsation mode from varying aspect. For all nonradial modes in roAp stars we therefore see modulation of the pulsation amplitude and/or phase with rotation and have strong constraints on the mode geometry and hence can identify the mode.

Thus in one sense for the roAp stars we can see the pulsation modes even better than in the Sun. Although the surface of the Sun is well resolved, all Earth-based observations are with a line of sight near the solar equator, and thus all nonradial modes are observed from that viewpoint. For the roAp stars the pulsation modes are inclined to the rotation axis, and the rotation axes of the many stars are variously inclined to the line of sight from Earth. With many stars to study, with many different rotation periods, and with many different inclination angles for the magnetic-pulsation axes to the rotation axes, we can observe the nonradial modes from all aspects with important advantages over our equator-only view of helioseismology.

3. Asymptotic *p*-Mode Frequency Spectra

For high-radial-overtone *p*-mode pulsations a fundamental observational goal of asteroseismology is to determine the large and small separations, and from modelling these obtain the stellar mass and age. The large separation ($\Delta\nu_0$, when unperturbed by a magnetic field), is a measure of the sound crossing time of the star, which in turn is determined by the star's mean density and radius. With a typical mass of roAp stars of about $2M_\odot$, $\Delta\nu_0$ to first order reflects the radius of the star. In the asymptotic limit, the number of nodes in the radial direction (n) is much larger than the spherical degree (ℓ). By assuming adiabatic pulsations in spherically symmetric stars, the pulsation frequencies are

$$\nu_{n,\ell} \sim \Delta\nu_0(n + \ell/2 + \epsilon) + \delta\nu, \tag{1}$$

where ϵ is a slowly varying function of frequency of order unity (Tassoul, 1980, 1990) and $\delta\nu$, the small separation (again, when unperturbed by a magnetic field), is a measure of the age of the star, since it is sensitive to the central condensation and hence the core H mass fraction. As can be seen from Equation (1), the basic spacing expected in the amplitude spectra is half the large separation ($\Delta\nu_0/2$) with peaks for alternating even and odd ℓ modes. This is easily seen in amplitude spectra of observations of the Sun as a star with BiSON and for low-degree modes with GONG (Howe *et al.*, 2003), and now for solarlike oscillators such as α Cen (Bazot *et al.*, 2007; Bedding *et al.*, 2004).

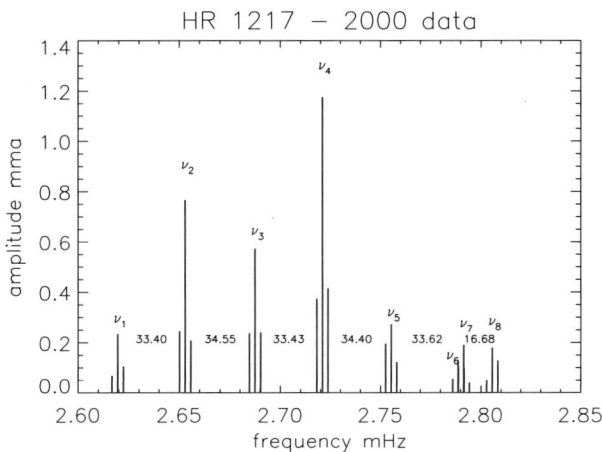

Figure 1 A schematic amplitude spectrum for the roAp star HR 1217 showing a set of frequencies basically spaced by $\Delta\nu_0/2$ with alternating even and odd ℓ modes, but with an unusual spacing of about $\Delta\nu_0/4$ for the highest frequencies never seen in the Sun or solarlike oscillators. This is a result of the perturbations to the pulsation frequencies by the strong magnetic field (From Kurtz et al., 2005).

The Sun and the solarlike oscillators are relatively easy to model for mass and age in terms of the large and small spacings. The roAp stars pose a far greater challenge because of the interaction of the pulsation modes with the strong magnetic field, as is discussed in Section 4. HR 1217 demonstrates this, as can be seen in Figure 1 from data obtained with by WET[6] (Kurtz et al., 2005). The basic spacing seen is half the large separation ($\Delta\nu_0/2$) stretched by the magnetic frequency perturbations; hence we are seeing the frequencies of a series of alternating even- and odd-ℓ modes, as is seen in the frequency spectrum of the Sun. However, the two highest frequencies have a spacing that is never seen in helioseismic observations or for solarlike oscillators. That spacing of about $\Delta\nu_0/4$ is not possible for the asymptotic spacings given in Equation (1) and hence is a signal of the importance of the effect of the magnetic field on the pulsation frequencies. It indicates that the roAp star frequency spectra provide unique information to probe the interaction of pulsation and strong magnetic fields.

4. Interaction of the Pulsation with the Magnetic Field

In the Sun, the interaction of p modes with the strongly magnetic environment of sunspots has been observed and discussed for 20 years (e.g., Braun, Duvall, and LaBonte, 1987; Spruit, 1991). It is now known from both observation and theory that solar p modes undergo significant phase shifts, energy losses, and mode conversion to magneto-acoustic modes when encountering sunspots (e.g., Cally, 2005; Gizon, Hanasoge, and Birch, 2006; Schunker and Cally, 2006; McDougall and Hood, 2007). As the modes rise though a magnetic sunspot region, they are converted from acoustic p modes to a combination of standing wave p modes, outward running wave acoustic modes, and magnetic slow and fast waves. The acoustic running waves and inwardly travelling magnetic slow waves dissipate energy from the pulsation modes. Mode conversion is most efficient at the atmospheric level where the sound speed and Alfvén speed are equal, significantly altering the acoustic modes coming from below. Higher in the atmosphere where the Alfvén speed exceeds the sound speed and where observational helioseismology is yet to be studied in great detail, the modes have

---

[6]Whole Earth Telescope; see http://www.physics.udel.edu/darc/wet/.

been converted mostly to magnetic-acoustic waves. A full understanding of these interactions gives a small improvement for the use of p modes for global helioseismic inference; more importantly, it allows local helioseismic study of the magnetic regions. The roAp stars provide the only other known stellar environment where similar physics can be studied in detail.

In the roAp stars the atmospheric level where the sound speed and Alfvén speed are equal is of the order of continuum optical depth unity (*e.g.*, Saio, 2005). Thus the observable atmospheric layers of roAp stars are regions where significant mode conversion has taken place and the modes are expected to be magneto-acoustic. With *global* magnetic field strengths of the order of 1 – 25 kG, the magnetic field significantly alters the pulsation modes – both directly and indirectly (see Cunha, 2006; Gough, 2005). Many studies show that the mode frequencies are significantly perturbed by the magnetic field; the shifts in frequency can even be as large as the asymptotic large frequency separation, although the perturbations to the large spacing (Δv_0) are never this large. These frequency shifts depend on the interaction of the mode geometry and magnetic geometry; in particular, they are a function of radial overtone with a smoothly changing magnetic perturbation over a certain frequency range, then an abrupt jump in frequency (Dziembowski and Goode, 1996; Bigot *et al.*, 2000; Cunha and Gough, 2000; Saio and Gautschy, 2004; Cunha, 2006). Understanding this led Cunha (2001) to a prediction regarding the frequency spectrum of HR 1217 that was confirmed by the WET amplitude spectrum shown in Figure 1: The magnetic frequency perturbations provide understanding of, and an explanation for, the previously inexplicable $\approx \Delta v_0/4$ spacing seen for the highest frequencies in Figure 1.

In roAp stars the odd-degree modes appear to be primarily dipole modes, and the even-degree modes are probably either radial or quadrupole modes. However, the magnetic field distorts each mode so that it can no longer be described by a single spherical harmonic. The best observed example of this is HR 3831. Kurtz, Kanaan, and Martinez (1993) showed that the pulsation mode in this singly periodic roAp star is predominantly a dipole ($\ell = 1$) mode, but with axisymmetric radial ($\ell = 0$), quadrupole ($\ell = 2$), and octupole ($\ell = 3$) components needed to account for the rotational amplitude and phase variability. Kochukhov (2004, 2006) similarly found both dipole and octupole components in his detailed high-resolution spectroscopic studies of HR 3831. His map of the pulsation velocity field in this star is in good agreement with the oblique-pulsator model.

The magnetic field also directly alters the pulsation eigenmodes so that higher-degree modes ($\ell \geq 3$) that would normally not be observable in stars because of cancellation over the unresolved surface may have components that appear as lower degree, observable modes, or distorted modes (Cunha, 2006; Saio, 2005). This, too, is consistent with the observations of HR 3831 described here and of HD 6532 (Kurtz *et al.*, 1996). It also may be a clue to the recent discovery that there are many pulsation modes, observed high in the atmospheres of roAp stars where the Alfvén speed exceeds the sound speed, that have not been detected in decades of photometric observations (Kurtz, Elkin, and Mathys, 2006a; Kurtz *et al.*, 2007).

A major indirect effect of strong magnetic fields in roAp stars is the suppression of convection in the envelope of the star, probably particularly near the magnetic poles. This is an important ingredient of ideas for mode excitation by driving in the H I ionisation zone (Balmforth *et al.*, 2001; Cunha, 2002; Saio, 2005). It is easy to see the common interest between studies of the Sun and of roAp stars, given the importance of convection to the excitation of solar p modes and the importance of magnetic fields in sunspots to the suppression of energy transport there and the dissipation of acoustic modes. The roAp stars give a global environment with much stronger magnetic fields in which to study the interaction of convection, pulsation, and magnetic field. Interestingly, in the context of roAp stars, calculations for sunspots (*e.g.*, Schunker and Cally, 2006) and for roAp stars (Sousa and Cunha, 2008) show that the mode conversion is strongly dependent on the angle between the acoustic wave and the magnetic field; there is efficient transmission and little conversion when the angle is small, and strong conversion when the angle is large. Sousa and Cunha (2008) find, in a simple model, maximum pulsational energy losses to slow running Alfvén waves in an annulus about $30°$ from the magnetic poles. It will be interesting to see the developments of these models and their impact on our understanding of mode selection, distortion, and orientation with respect to the magnetic field.

5. The Critical Cutoff Frequency

Many of the roAp stars pulsate with frequencies higher than the acoustic cutoff frequency of nonmagnetic standard A star models. The highest frequencies (shortest periods) are around 2.6–3.0 mHz ($5.65 \leq P \leq 6.2$ minutes) in the stars HD 134214, HD 203932, HD 24712 (HR 1217 – Figure 1), and HD 86181. There has been considerable discussion of this phenomenon (*e.g.*, Shibahashi and Saio, 1985; Audard *et al.*, 1998; Cunha, 1998; Gautschy, Saio, and Harzenmoser, 1998; Sousa and Cunha, 2008). Gautschy, Saio, and Harzenmoser (1998) put forward the idea of a chromospheric temperature inversion to provide a sharper boundary to trap higher-frequency acoustic modes, but there is no evidence of chromospheres in roAp stars (or A stars in general). A temperature inversion large enough to account for the observed high frequencies would produce observable chromospheric signatures. It is easy to see the commonality of this problem and solar physics.

With pulsation frequencies near or above the cutoff frequency, it would be expected that energy losses would be great for the roAp stars and that for some of them (HD 134214 in particular with its shortest pulsation period of 5.65 minutes) modes might be expected to have short lifetimes. After all, the solar pulsation frequencies have lifetimes that are frequency dependent, with mode lifetimes being longer for the lowest-frequency modes and very short for the highest-frequency modes above the solar cutoff frequency of 5.3 mHz. Yet HD 134214 has been studied for more than two decades and seems to have a basically stable pulsation over that time span; that is, its mode lifetime is greater than decades (but it does show very small frequency and amplitude variability). Obviously, here is a problem – pulsation above the acoustic cutoff frequency – that is common to the Sun and roAp stars. This commonality is emphasised by studies such as those of McIntosh and Jefferies (2006), who observed running magneto-acoustic gravity waves in the solar chromosphere that appear to have a cutoff frequency that depends on magnetic field inclination, as predicted theoretically. Unlike the Sun, mode driving in roAp stars must overcome the expected energy losses for such high-frequency modes, giving very different physical conditions to those in the Sun to study a similar problem.

6. Resolution of Pulsation Modes as a Function of Atmospheric Depth in the Sun and roAp Stars

Although there are short-duration helioseismic studies at higher atmospheric levels, helioseismic network observations are at present confined to a single atmospheric level. For example, GONG observes with a narrow band centred on the single line of Ni I at 6768 Å; SOHO/MDI[7] scans a narrow range centred on the same Ni I line; BiSON observes with a

---

[7] See http://soi.stanford.edu/.

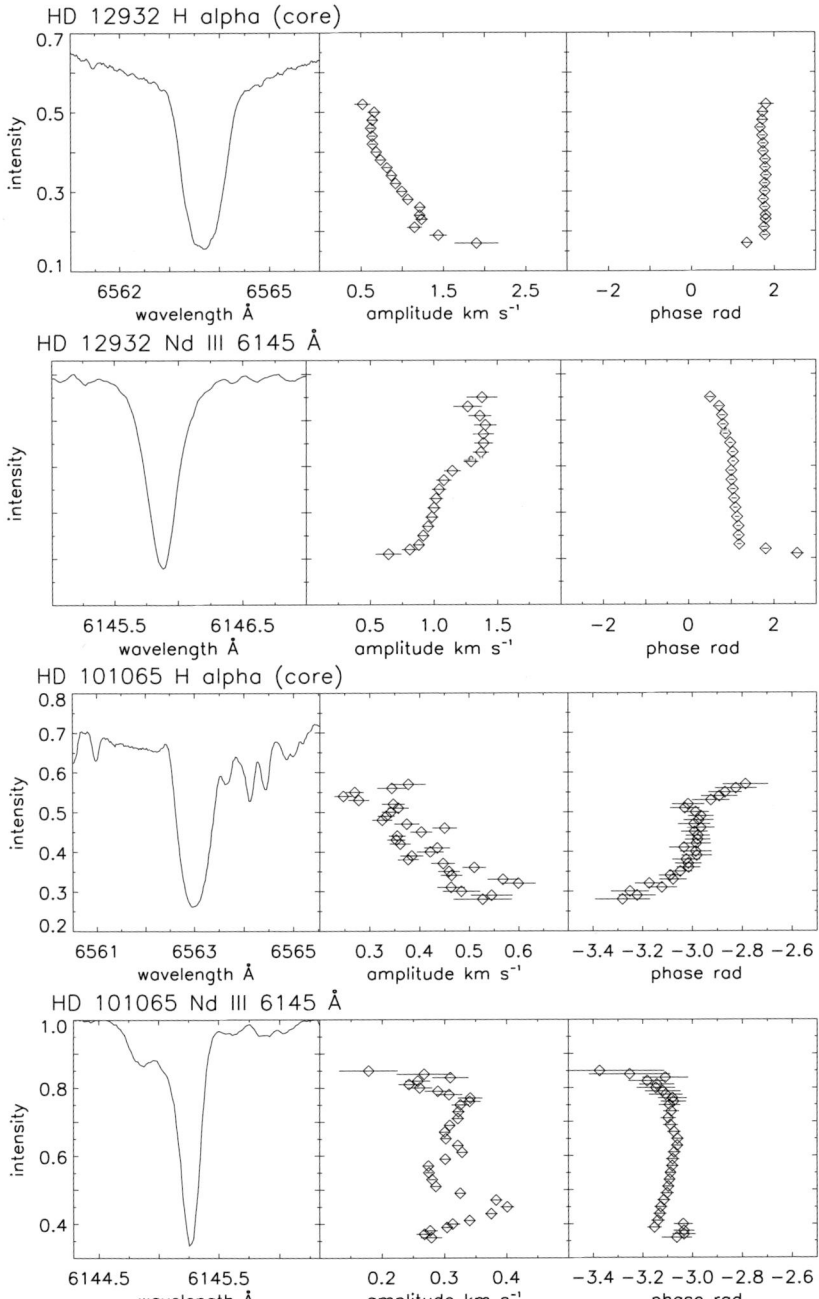

Figure 2 Line profiles for Hα and Nd III 6145 Å are shown on the left of each panel, the amplitudes of the radial-velocity variations at different line depths in the middle, and the phases for those on the right. Deeper in the line is higher in the atmosphere, and larger phases mean earlier times of maximum. Both stars have magnetic fields of the order of 1 kG, and pulsation periods near 12 minutes (From Kurtz, Elkin, and Mathys, 2006b).

potassium vapour cell to provide stabilised radial velocities for the K I line at 7699 Å. Looking to the future of helioseismology, Hill (private communication, 2007) has pointed out the desirability of helioseismic observations with as many as eight spectral lines that sample different depths in the solar atmosphere. The roAp stars are already being studied in this way, and they offer unique conditions found in no other stars, or the Sun, to view pulsation modes "in depth".

Atomic diffusion in roAp stars has stratified their atmospheres so that observations of Fe peaks and certain other elements typically sample depths around $\log \tau_{5000} \approx 0.5$, whereas ions of Nd II, Nd III, Pr II, Pr III, Tb III, and other rare-earth elements are radiatively levitated to around $\log \tau_{5000} \approx -5$ – levels that are high in the chromosphere of the Sun. The Hα line core forms in the range $-2 \geq \log \tau_{5000} \geq -4$. The pulsation modes have amplitudes and phases that are strikingly a function of atmospheric depth (see, *e.g.*, Kurtz, Elkin, and Mathys, 2005a; Ryabchikova *et al.*, 2007). They show a large range of behaviour with apparent standing waves and running waves at different levels; with amplitudes increasing and decreasing with height in the atmosphere; and even with radial nodes clearly resolved by ions with line-forming regions on opposite sides of the radial node, as in 33 Lib (Mkrtichian, Hatzes, and Kanaan, 2003; Kurtz, Elkin, and Mathys, 2005b). Additionally, line bisector studies of roAp stars provide stunning depth resolution in individual lines. Baldry and Bedding (2000) pioneered this technique in a study of the Hα line in α Cir. Kurtz, Elkin, and Mathys (2006b) show the amplitude and phase behaviour for the core of the Hα line and the 6145-Å line of Nd III for ten roAp stars. We show similar diagrams here for two stars in Figure 2 to illustrate how easy it is to observe the range of amplitude and phase behaviour as a function of atmospheric height in these stars. We expect that solar observations using many lines at different atmospheric depth will similarly lead to important new discoveries.

7. Conclusion

There are many similarities between the physics studied by helioseismology and that studied by asteroseismology in roAp stars. Solar astronomers with interests in atomic diffusion, the local helioseismic study of the interaction of p modes, and convection with magnetic fields, pulsation above the critical cutoff frequency, and the future resolution of solar p modes with atmospheric depth may find observational and theoretical studies of the roAp stars intriguing and enlightening.

Acknowledgements I thank Margarida Cunha for enlightening comments and suggestions.

References

Aerts, C., Christensen-Dalsgaard, J., Cunha, M., Kurtz, D.: 2008, *Solar Phys.*, submitted.
Audard, N., Kupka, F., Morel, P., Provost, J., Weiss, W.W.: 1998, *Astron. Astrophys.* **335**, 954.
Baldry, I.K., Bedding, T.R.: 2000, *Mon. Not. Roy. Astron. Soc.* **318**, 341.
Balmforth, N.J., Cunha, M.S., Dolez, N., Gough, D.O., Vauclair, S.: 2001, *Mon. Not. Roy. Astron. Soc.* **323**, 362.
Bazot, M., Bouchy, F., Kjeldsen, H., Charpinet, S., Laymand, M., Vauclair, S.: 2007, *Astron. Astrophys.* **470**, 295.
Bedding, T.R., Kjeldsen, H., Butler, R.P., McCarthy, C., Marcy, G.W., O'Toole, S.J., Tinney, C.G., Wright, J.T.: 2004, *Astrophys. J.* **614**, 380.
Bigot, L., Dziembowski, W.A.: 2002, *Astron. Astrophys.* **391**, 235.
Bigot, L., Provost, J., Berthomieu, G., Dziembowski, W.A., Goode, P.R.: 2000, *Astron. Astrophys.* **356**, 218.

Borucki, W.J., Koch, D.G., Basri, G., *et al.*: 2003, In: Fridlund, M., Henning, T. (eds.) *Proceedings of the Conference on Towards Other Earths: DARWIN/TPF and the Search for Extrasolar Terrestrial Planet*, *ESA SP* **539**, ESA Publications Division, Noordwijk, 69.
Braun, D.C., Duvall, T.L. Jr., LaBonte, B.J.: 1987, *Astrophys. J.* **319**, L27.
Cally, P.S.: 2005, *Mon. Not. Roy. Astron. Soc.* **358**, 353.
Christensen-Dalsgaard, J., Arentoft, T., Brown, T.M., Gilliland, R.L., Kjeldsen, H., Borucki, W.J., Koch, D.: 2007, *Comm. Asteroseismol.* **150**, 350.
Croll, B., Matthews, J., Rowe, J., *et al.*: 2007, *Astrophys. J.* **658**, 1328.
Cunha, M.S.: 1998, *Contrib. Astron. Obs. Skaln. Pleso* **27**, 272.
Cunha, M.S.: 2001, *Mon. Not. Roy. Astron. Soc.* **325**, 373.
Cunha, M.S.: 2002, *Mon. Not. Roy. Astron. Soc.* **333**, 47.
Cunha, M.S.: 2006, *Mon. Not. Roy. Astron. Soc.* **365**, 153.
Cunha, M.S., Gough, D.: 2000, *Mon. Not. Roy. Astron. Soc.* **319**, 1020.
Dziembowski, W.A., Goode, P.R.: 1996, *Astrophys. J.* **458**, 338.
Fossat, E., Grec, G., Pomerantz, M.: 1981, *Solar Phys.* **74**, 59.
Gautschy, A., Saio, H., Harzenmoser, H.: 1998, *Mon. Not. Roy. Astron. Soc.* **301**, 31.
Gizon, L., Hanasoge, S.M., Birch, A.C.: 2006, *Astrophys. J.* **643**, 549.
Gough, D.: 2005, In: *The Roger Taylor Memorial Lectures*, *Astron. Geophys.*, special issue, 16.
Howe, R., Chaplin, W.J., Elsworth, Y.P., Hill, F., Komm, R., Isaak, G.R., New, R.: 2003, *Astrophys. J.* **588**, 1204.
Kochukhov, O.: 2004, *Astrophys. J.* **615**, L149.
Kochukhov, O.: 2006, *Astron. Astrophys.* **446**, 1051.
Kurtz, D.W.: 1982, *Mon. Not. Roy. Astron. Soc.* **200**, 807.
Kurtz, D.W., Elkin, V.G., Mathys, G.: 2005a, *Eur. Astron. Soc.* **17**, 91.
Kurtz, D.W., Elkin, V.G., Mathys, G.: 2005b, *Mon. Not. Roy. Astron. Soc.* **358**, L6.
Kurtz, D.W., Elkin, V.G., Mathys, G.: 2006a, *Mon. Not. Roy. Astron. Soc.* **370**, 1274.
Kurtz, D.W., Elkin, V.G., Mathys, G.: 2006b, *ESA SP* **624**, 33.
Kurtz, D.W., Kanaan, A., Martinez, P.: 1993, *Mon. Not. Roy. Astron. Soc.* **260**, 343.
Kurtz, D.W., Cameron, C., Cunha, M., *et al.*: 2005, *Mon. Not. Roy. Astron. Soc.* **358**, 651.
Kurtz, D.W., Elkin, V.G., Mathys, G., van Wyk, F.: 2007, *Mon. Not. Roy. Astron. Soc.* **381**, 1301.
Kurtz, D.W., Martinez, P., Koen, C., Sullivan, D.J.: 1996, *Mon. Not. Roy. Astron. Soc.* **281**, 883.
McDougall, A.M.D., Hood, A.W.: 2007, *Solar Phys.* **246**, 259.
McIntosh, S.W., Jefferies, S.M.: 2006, *Astrophys. J.* **647**, L77.
Michel, E., Baglin, A., Samadi, R., Baudin, F., Auvergne, M.: 2007, *Comm. Asteroseismol.* **150**, 341.
Mkrtichian, D.E., Hatzes, A.P., Kanaan, A.: 2003, *Mon. Not. Roy. Astron. Soc.* **345**, 781.
Ryabchikova, T., Sachkov, M., Kochukhov, O., Lyashko, D.: 2007, *Astron. Astrophys.* **473**, 907.
Saio, H.: 2005, *Mon. Not. Roy. Astron. Soc.* **360**, 1022.
Saio, H., Gautschy, A.: 2004, *Mon. Not. Roy. Astron. Soc.* **350**, 485.
Schunker, H., Cally, P.S.: 2006, *Mon. Not. Roy. Astron. Soc.* **372**, 551.
Shibahashi, H., Saio, H.: 1985, *Pub. Astron. Soc. Japan* **37**, 2.
Sousa, S.G., Cunha, M.: 2008, *Mon. Not. Roy. Astron. Soc.*, arXiv:0802.0783.
Spruit, H.C.: 1991, In: Gough, D., Toomre, J. (eds.) *Challenges to Theories of the Structure of Moderate–Mass Stars*, *Lec. Notes Phys.* **388**, Springer, Berlin, 121.
Tassoul, M.: 1980, *Astrophys. J. Suppl.* **43**, 469.
Tassoul, M.: 1990, *Astrophys. J.* **358**, 313.
Walker, G., Matthews, J., Kuschnig, R., *et al.*: 2003, *Pub. Astron. Soc. Pac.* **115**, 1023.

Fourier Analysis of Gapped Time Series: Improved Estimates of Solar and Stellar Oscillation Parameters

Thorsten Stahn · Laurent Gizon

Originally published in the journal Solar Physics, Volume 251, Nos 1–2, 31–52.
DOI: 10.1007/s11207-008-9181-0 © The Author(s) 2008

Abstract Quantitative helioseismology and asteroseismology require very precise measurements of the frequencies, amplitudes, and lifetimes of the global modes of stellar oscillation. The precision of these measurements depends on the total length (T), quality, and completeness of the observations. Except in a few simple cases, the effect of gaps in the data on measurement precision is poorly understood, in particular in Fourier space where the convolution of the observable with the observation window introduces correlations between different frequencies. Here we describe and implement a rather general method to retrieve maximum likelihood estimates of the oscillation parameters, taking into account the proper statistics of the observations. Our fitting method applies in complex Fourier space and exploits the phase information. We consider both solar-like stochastic oscillations and long-lived harmonic oscillations, plus random noise. Using numerical simulations, we demonstrate the existence of cases for which our improved fitting method is less biased and has a greater precision than when the frequency correlations are ignored. This is especially true of low signal-to-noise solar-like oscillations. For example, we discuss a case where the precision of the mode frequency estimate is increased by a factor of five, for a duty cycle of 15%. In the case of long-lived sinusoidal oscillations, a proper treatment of the frequency correlations does not provide any significant improvement; nevertheless, we confirm that the mode frequency can be measured from gapped data with a much better precision than the $1/T$ Rayleigh resolution.

Keywords Helioseismology: observations · Oscillations: solar · Oscillations: stellar

1. Introduction

Solar and stellar oscillations are a powerful tool to probe the interior of stars. In this paper we classify stellar oscillations as solar-like or deterministic. Solar-like oscillations are

---

Helioseismology, Asteroseismology, and MHD Connections
Guest Editors: Laurent Gizon and Paul Cally

T. Stahn · L. Gizon (✉)
Max-Planck-Institut für Sonnensystemforschung, 37191 Katlenburg-Lindau, Germany
e-mail: gizon@mps.mpg.de

stochastically excited by turbulent convection and are present in the Sun and other main-sequence, subgiant, and giant stars (see, *e.g.*, Bedding and Kjeldsen, 2007, and references therein). Deterministic oscillations are seen in classical pulsators and have mode lifetimes much longer than any typical observational run; one of the best studied objects in this class is the pre-white dwarf PG1159-035, also known as GW Vir (Winget *et al.*, 1991). In practice, observations of solar-like or deterministic pulsations always have an additional stochastic component owing to instrumental, atmospheric, stellar, or photon noise.

An important aspect of helioseismology and asteroseismology is the determination of the parameters of the global modes of oscillation, especially the mode frequencies. In the case of the Sun, it is known (Woodard, 1984) that the measurement precision is limited by the stochastic nature of the oscillations (realization noise). Libbrecht (1992) and Toutain and Appourchaux (1994) have shown that realization noise is expected to scale as $1/\sqrt{T}$, where T is the total duration of the observation. A common practice is to extract the solar mode parameters from the power spectrum by using maximum likelihood estimation (MLE; see, *e.g.*, Anderson Duvall, and Jefferies, 1990; Schou, 1992; Toutain and Appourchaux, 1994; Appourchaux, Gizon, and Rabello-Soares, 1998; Appourchaux *et al.*, 2000). In its current form, however, this method of analysis is only valid for uninterrupted time series. This is a significant limitation because gaps in the data are not uncommon (because of the day–night cycle, bad weather, or technical problems). The gaps complicate the analysis in Fourier space: The convolution of the data with the observation window leads to correlations between the different Fourier components. The goal of this paper is to extend the Fourier analysis of solar and stellar oscillations to time series with gaps, by using appropriate maximum likelihood estimators based on the correct statistics of the data.

Section 2 poses the problem of the analysis of gapped time series in Fourier space. In Section 3 we derive an expression for the joint probability density function (PDF) of the observations, taking into account the frequency correlations. Our answer is consistent with an earlier (independent) derivation by Gabriel (1994). Based on this PDF, we derive maximum likelihood estimators in Section 4. In Section 5 we recall the "old method" of maximum likelihood estimation based on the unjustified assumption that frequency bins are statistically independent. Section 6 explains the setup of the Monte Carlo simulations used to test the fitting methods on artificial data sets. In Section 7 we present the results of the Monte Carlo simulations and compare the new and old fitting methods. For the sake of simplicity, we consider only one mode of oscillation at a time (solar-like or deterministic). We present several cases for which our new fitting method leads to a significant improvement in the determination of oscillation parameters, and in particular the mode frequency.

2. Statement of the Problem

2.1. The Observed Signal in Fourier Space

Let us denote by $\tilde{y} = \{\tilde{y}_i\}$ the time series that we wish to analyze. It is sampled at times $t_i = i\Delta t$, where i is an integer in the range $0 \leq i \leq N - 1$ and $\Delta t =$ one minute is the sampling time. All quantities with a tilde are defined in the time domain. The total duration of the time series is $T = (N-1)\Delta t$. By choice, all of the missing data points were assigned the value zero: This enables us to work on a regularly sampled time grid. Formally, we write

$$\tilde{y}_i = \tilde{w}_i \tilde{x}_i, \quad i = 0, 1, \ldots, N-1, \tag{1}$$

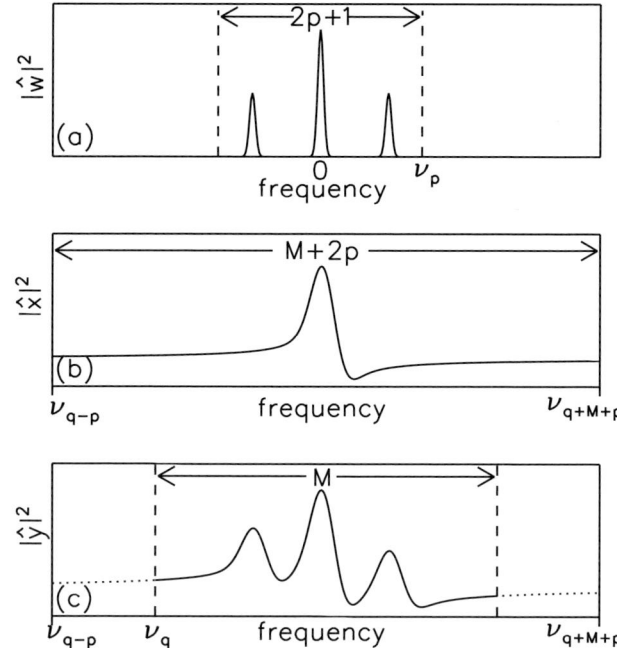

Figure 1 Schematic representation in Fourier space of the convolution of the signal [$\hat{x}$] with the window function [$\hat{w}$]. For the sake of simplicity, only the power spectra of the different quantities are shown here. Panel (a) shows the window function [$\hat{w}$] and its cutoff frequency [ν_p], panel (b) shows the unconvolved signal [$\hat{x}$], and panel (c) shows the observed signal [$\hat{y}$]. Note that the selected section of the observed signal, starting at frequency ν_p, is of length M, whereas the unconvolved signal is of length $M + 2p$.

where $\tilde{x}$ is the uninterrupted time series that we *would* have observed if there had been no gaps, and $\tilde{w}$ is the window function defined by $\tilde{w}_i = 1$ if an observation is recorded at time t_i and $\tilde{w}_i = 0$ otherwise. The time series $\tilde{x}$ is drawn from a random process, whose statistical properties will be discussed later.

We define the discrete Fourier transform $\hat{y}$ of $\tilde{y}$ by

$$\hat{y}_j = \frac{1}{N} \sum_{i=0}^{N-1} \tilde{y}_i e^{-i2\pi \nu_j t_i} \quad \text{for } j \in \mathbb{N}, \qquad (2)$$

where $\nu_j = j \Delta\nu$ is the frequency and $\Delta\nu = 1/N\Delta t$. Note that $\hat{y}_j = \hat{y}^*_{N-j}$ and $\hat{y}_j = \hat{y}^*_{-j}$, where the star denotes the complex conjugate. The Fourier transform has periodicity $1/\Delta t$ or twice the Nyquist frequency.

Our intention is to fit not the complete Fourier spectrum but a rather small interval that contains one (or a few) modes of stellar oscillation. Thus, we extract a section of the data of length M starting from a particular frequency ν_q, as shown in Figure 1(c). This subset of the data is represented by the vector $y = [y_0, y_1, \ldots, y_{M-1}]^T$ with components

$$y_i = \hat{y}_{q+i}, \quad i = 0, 1, \ldots, M-1. \qquad (3)$$

By using the definition of the Fourier transform [Equation (2)], the vector y is given by the convolution of $\hat{x}$ with the window $\hat{w}$:

$$y_i = \sum_{j=-p}^{M+p-1} \hat{w}_{i-j} \hat{x}_{q+j}. \qquad (4)$$

The integer p in Equation (4) refers to the cutoff frequency ν_p beyond which the observation window has no significant power. Truncating the window function at frequency ν_p is a simplification of the general problem. Our main goal, however, is, given a known window function, to study its effects on the determination of the parameters of stellar oscillations. Figure 1 is a schematic representation in Fourier space of the convolution of a single mode of oscillation by the window function. The observed signal is spread over some frequency range and, as we shall see later, its statistical properties are affected.

We note that, in practice, one can never completely isolate one single mode of oscillation in the power spectrum. In particular, other modes with frequencies outside of the fitting range can leak into it after convolution by the temporal window function. Hence, fitting one mode of oscillation is a simplification. But our first objective is to try to study the effects of gaps, independently from the complications associated with a badly specified model.

Equation (4) can be rewritten in matrix form as

$$y = Wx, \qquad (5)$$

where the vector $x = [x_0, x_1, \ldots, x_{M+2p-1}]^T$ of length $M + 2p$ is defined by

$$x_i = \hat{x}_{q-p+i}, \quad i = 0, 1, \ldots, M + 2p - 1, \qquad (6)$$

and $W = [W_{ij}]$ is the $M \times (M + 2p)$ rectangular window matrix with elements $W_{ij} = \hat{w}_{i-j+p}$, where $i = 0, 1, \ldots, M - 1$ and $j = 0, 1, \ldots, M + 2p - 1$:

$$W = \begin{bmatrix} \hat{w}_p & \cdots & \hat{w}_0 & \cdots & \hat{w}_{-p} & & & & \\ & \ddots & & \ddots & & \ddots & & 0 & \\ & & \hat{w}_p & \cdots & \hat{w}_0 & \cdots & \hat{w}_{-p} & & \\ & 0 & & \ddots & & \ddots & & \ddots & \\ & & & & \hat{w}_p & \cdots & \hat{w}_0 & \cdots & \hat{w}_{-p} \end{bmatrix}. \qquad (7)$$

Note that $\hat{w}_i = \hat{w}_{-i}^*$ and that W is of rank M.

Equation (5) is the master equation. Our goal is to extract the stellar oscillation parameters (contained in x), given the uncomplete information y.

2.2. Statistics of the Unconvolved Signal

Here we describe the basic assumptions that we make about the statistics of the data in the Fourier domain. The unconvolved signal $[x]$ consists of a deterministic component $[d]$ and a zero-mean stochastic component $[e]$ such that

$$x = d + e. \qquad (8)$$

The deterministic component $[d]$ may include deterministic stellar oscillations that are long-lived compared to the total length of the observation. The stochastic component $[e]$ may include various sources of noise (*e.g.*, stellar convection, photon noise, atmospheric noise, *etc.*) and stochastically excited pulsations as observed on the Sun.

We assume that the e_i are $M + 2p$ independent random variables in the Fourier domain. This is equivalent to saying that the stochastic component of the signal in the time domain is stationary. We further assume that e is a Gaussian random vector with independent real and imaginary parts and covariance matrix

$$E[e_i^* e_j] = \sigma_i^2 \delta_{ij}, \quad i, j = 0, 1, \ldots, M + 2p - 1, \qquad (9)$$

where E denotes the expectation value and σ_i is the standard deviation of e_i at frequency ν_i. One may invoke the central limit theorem to justify the choice of Gaussian distributions. The quantity σ_i^2 is the expected power spectrum at frequency ν_i, which may include background noise and peaks corresponding to the modes of oscillations (Duvall and Harvey, 1986; Appourchaux, Gizon, and Rabello-Soares, 1998). In terms of a complex Gaussian random vector g with unit covariance matrix, $E[g^*g^T] = I_{M+2p}$, we can rewrite e as

$$e = Sg, \tag{10}$$

where S is the $(M+2p) \times (M+2p)$ diagonal matrix

$$S = \text{diag}(\sigma_0, \sigma_1, \ldots, \sigma_{M+2p-1}). \tag{11}$$

We emphasize that, although the e_i are uncorrelated random variables, the y_i are correlated because of the multiplication of x by the window matrix (Equation (5)).

3. Joint PDF of the Complex Fourier Spectrum

In this section we derive an expression for the joint probability density function of the observed signal y. This problem had already been solved by Gabriel (1994). We reach the same conclusion, independently and with more compact notation. We start by rewriting the master equation, Equation (5), as

$$y = Wd + Cg, \tag{12}$$

where

$$C = WS \tag{13}$$

is an $M \times (M+2p)$ matrix with rank M and singular value decomposition (Horn and Johnson, 1985, Chapter 7.3) given by

$$C = U\Sigma V^H. \tag{14}$$

Here the superscript H denotes the Hermitian conjugate, and U and V are unitary matrices of dimensions $M \times M$ and $(M+2p) \times (M+2p)$, respectively (*i.e.*, $U^H U = I_M$ and $V^H V = I_{M+2p}$). The $M \times (M+2p)$ matrix Σ can be written as

$$\Sigma = [\Lambda \,|\, 0], \quad \Lambda = \text{diag}(\lambda_0, \lambda_1, \ldots, \lambda_{M-1}), \tag{15}$$

where $\lambda_0, \lambda_1, \ldots, \lambda_{M-1}$ are the M (positive) singular values of the matrix C. Thus, there exists a vector $\xi = V^H g$ such that

$$y = Wd + U[\Lambda \,|\, 0]\xi. \tag{16}$$

Since g has unit covariance matrix and V is unitary, the vector ξ is a complex Gaussian random vector of size $M + 2p$ with unit covariance matrix. It is obvious from Equation (16) that there exists a lower rank complex Gaussian random vector of length M, $\eta = [\xi_0, \xi_1, \ldots, \xi_{M-1}]^T$, such that

$$y = Wd + U\Lambda\eta. \tag{17}$$

The variables $\xi_M, \xi_{M+1}, \ldots, \xi_{M+2p-1}$ are dummy variables, which do not enter in the description of y. Equation (17) is an important step, as the vector y of length M is now expressed in terms of M independent complex Gaussian variables. This enables us to write the PDF of y as

$$p_y(y) = \frac{1}{J} p_\eta\big((U\Lambda)^{-1}(y - Wd)\big), \tag{18}$$

where $p_\eta(\eta)$ denotes the PDF of η and J is the Jacobian of the linear transformation $\eta \to y$. Since η is a complex Gaussian random vector with unit covariance (*i.e.*, $E[\eta^* \eta^T] = I_M$), we have

$$p_\eta(\eta) = \frac{\exp(-\|\eta\|^2)}{\pi^M}, \tag{19}$$

where we used the notation $\|\eta\|^2 = \eta^H \eta$. Since U is unitary and Λ is diagonal and real, the Jacobian of the transformation is given by

$$J = \big|\det(U\Lambda)\big|^2 = (\det \Lambda)^2 = \prod_{i=0}^{M-1} \lambda_i^2. \tag{20}$$

Combining Equations (18), (19), and (20), we get the joint PDF of the observed vector y:

$$p_y(y) = \frac{\exp(-\|\Lambda^{-1} U^H (y - Wd)\|^2)}{\pi^M (\det \Lambda)^2}. \tag{21}$$

This expression is, perhaps, more elegantly written as

$$p_y(y) = \frac{\exp(-\|C^\dagger (y - Wd)\|^2)}{\pi^M (\det \Lambda)^2} \tag{22}$$

in terms of $C^\dagger$, the $(M + 2p) \times M$ Moore–Penrose generalized inverse of C (Horn and Johnson, 1985, Chapter 7.3), given by

$$C^\dagger = V \Sigma^\dagger U^H = C^H \big(C C^H\big)^{-1}, \tag{23}$$

where $\Sigma^\dagger$ is the transpose of Σ in which the singular values are replaced by their inverse. One may ask, after the fact, if the quantity $(U\Lambda)^{-1}$ in Equation (18) is always defined. The answer would appear to be yes since the Moore–Penrose generalized inverse of C is perfectly well defined. It is not excluded, however, that some singular values λ_i could be infinitesimally small. We have not encountered any such difficulty with the test cases given in Section 7. Should C be ill-conditioned in other cases, a simple truncated SVD would help in avoiding a numerical problem.

Before discussing the implementation of the method in Section 4, we should like to draw attention to a parallel between fitting data with temporal gaps and fitting data with spatial gaps. To understand this analogy, we refer the reader to the work of Appourchaux, Gizon, and Rabello-Soares (1998, Section 3.3.4), who discuss how to interpret the spatial leaks of nonradial oscillations that arise from the fact that only half of the solar disk can be observed from Earth. Their approach is similar to the one developed in this paper.

4. Maximum Likelihood Estimation of Stellar Oscillation Parameters

Let us assume that the stellar oscillation model that we are trying to fit to the data depends on a set of k parameters $\mu = (\mu_0, \mu_1, \ldots, \mu_{k-1})$. These parameters may be the amplitude, the phase, the frequency, the line asymmetry, the noise level, *etc*. The basic idea of maximum likelihood estimation is to pick the estimate $[\mu_\star]$ that maximizes the likelihood function. The likelihood function is another name for the joint PDF (Equation (22)) evaluated for the sample data. In practice, one minimizes

$$\mathcal{L}(\mu) = -\ln p_y = \left\| C^\dagger (y - Wd) \right\|^2 + 2 \sum_{i=0}^{M-1} \ln \lambda_i + \text{constant}, \tag{24}$$

rather than maximizing the likelihood function itself. In this expression, the quantities $C^\dagger$ and λ_i all depend implicitly on the model parameters μ through the covariance matrix S. The vector d also depends on the model parameters in the case of deterministic oscillations. The probability of observing the sample data is greatest if the unknown parameters are equal to their maximum likelihood estimates $\mu_\star$:

$$\mu_\star = \arg\min_\mu \mathcal{L}(\mu). \tag{25}$$

The method of maximum likelihood has many good properties (Brandt, 1970). In particular, in the limit of a large sample size (M large), the maximum likelihood estimator is unbiased and has minimum variance.

What is particularly new about our work is the minimization of the likelihood function given by Equation (24). We use the direction set method, or Powell's algorithm, to solve the minimization problem with a computer. In practice, the result of the fit depends on the initial guess and the fractional tolerance of the minimization procedure (the relative decrease of $\mathcal{L}$ in one iteration). The fitted parameters depend on the initial guess because the function $\mathcal{L}$ may have local minima in addition to the global minimum. We will address this issue in more detail in Section 7.

4.1. Special Case: Solar-Like Oscillations

In the case of solar-like oscillations, there is no deterministic component and the log-likelihood becomes

$$\mathcal{L}(\mu) = \left\| C^\dagger y \right\|^2 + 2 \sum_{i=0}^{M-1} \ln \lambda_i + \text{constant}. \tag{26}$$

4.2. Special Case: Deterministic Oscillations plus White Noise

If background white noise is the only stochastic component then

$$\sigma_i = \sigma_0 = \text{constant}, \quad i = 0, 1, \ldots, M + 2p - 1. \tag{27}$$

The log-likelihood function becomes

$$\mathcal{L}(\mu) = \frac{1}{\sigma_0^2} \left\| W^\dagger (y - Wd) \right\|^2 + M \ln \sigma_0^2 + \text{constant}. \tag{28}$$

Splitting the unknowns $\mu = (\check{\mu}, \sigma_0)$ into the parameters describing the oscillations, $\check{\mu} = (\mu_0, \mu_1, \ldots, \mu_{k-2})$, and the noise level σ_0, reduces the minimization problem to finding the most likely estimates

$$\check{\mu}_\star = \arg\min_{\check{\mu}} \|W^\dagger(y - Wd)\|^2, \tag{29}$$

where $d = d(\check{\mu})$. The noise level is explicitly given by

$$\sigma_{0\star} = M^{-1/2} \|W^\dagger[y - Wd(\check{\mu}_\star)]\|. \tag{30}$$

5. The Old Way: Fitting the Power Spectrum and Ignoring the Correlations

Maximum likelihood estimation has been used in the past to infer solar and stellar oscillation parameters, even in the case of gapped time series. The joint PDF of the observations was assumed to be the product of the PDFs of the individual y_i, as if the frequency bins were uncorrelated. For comparison purposes, we briefly review this (unjustified) approach.

According to Equation (12), the PDF of y_i is a normal distribution

$$p_{y_i}(y_i) = \frac{\exp(-|y_i - \overline{y}_i|^2/v_i)}{\pi v_i} \tag{31}$$

with mean

$$\overline{y}_i = \sum_{j=0}^{M+2p-1} W_{ij} d_j \tag{32}$$

and variance

$$v_i = \sum_{j=0}^{M+2p-1} |W_{ij}|^2 \sigma_j^2. \tag{33}$$

Under the (incorrect) assumption that the y_i are independent random variables, the joint PDF of y becomes

$$p_y^{\text{nc}}(y) = \prod_{i=0}^{M-1} p_{y_i}(y_i), \tag{34}$$

where the superscript "nc" stands for "no correlation." This joint PDF uses the correct mean $[\overline{y}_i]$ and variance $[v_i]$ of the data, but it ignores all the nonvanishing cross-terms $E[y_i^* y_j]$. In other words, the spread of power implied by the convolution with the window is taken care of, but not the proper statistics.

Under the same simplifying "no-correlation" assumption, the log-likelihood function is

$$\mathcal{L}^{\text{nc}}(\mu) = \sum_{i=0}^{M-1} \frac{|y_i - \overline{y}_i|^2}{v_i} + \sum_{i=0}^{M-1} \ln v_i + \text{constant}, \tag{35}$$

where the $\overline{y}_i$ and v_i are implicit functions of the model parameters μ.

5.1. Special Case: Solar-Like Oscillations

If the signal has no deterministic component ($d = 0$), then the power spectrum $[P_i(\mu) = |y_i|^2]$ has the expectation value $\overline{P}_i = E[P_i] = v_i$. Thus, in the case of purely solar-like oscillations, we recover the standard expression (Toutain and Appourchaux, 1994)

$$\mathcal{L}^{\mathrm{nc}}(\mu) = \sum_{i=0}^{M-1}\left(\frac{P_i}{\overline{P}_i} + \ln \overline{P}_i\right) + \text{constant} \quad \text{when } d = 0. \tag{36}$$

Although this expression is perfectly valid for uninterrupted data, it is not justified when gaps are present. The parameters $\mu_\star^{\mathrm{nc}}$ that minimize $\mathcal{L}^{\mathrm{nc}}(\mu)$ are not optimal, as will be shown later by using Monte Carlo simulations.

5.2. Special Case: Deterministic Oscillations plus White Noise

When $\sigma_i = \sigma_0 = $ constant, the "no-correlation" log-likelihood function simplifies to

$$\mathcal{L}^{\mathrm{nc}}(\mu) = \frac{1}{\sigma_0^2}\frac{\|y - Wd\|^2}{\sum_{j=-p}^{p}|\hat{w}_j|^2} + M\ln\sigma_0^2 + \text{constant}. \tag{37}$$

The minimization problem is to find the $(k-1)$ parameters $\check{\mu} = (\mu_0, \mu_1, \ldots, \mu_{k-2})$ such that

$$\check{\mu}_\star^{\mathrm{nc}} = \arg\min_{\check{\mu}}\|y - Wd(\check{\mu})\|^2. \tag{38}$$

The noise level is explicitly given by

$$\sigma_{0\star}^{\mathrm{nc}} = \left(M\sum_{j=-p}^{p}|\hat{w}_j|^2\right)^{-1/2}\|y - Wd(\check{\mu}_\star^{\mathrm{nc}})\|. \tag{39}$$

6. Simulation of Artificial Time Series

So far we have considered a general signal, which includes a deterministic component and a stochastic component. The parametrization of each component depends on prior knowledge about the physics of the stellar oscillations. Solar-like pulsations are stochastic in nature and no deterministic component is needed in this case. However, long-lived stellar pulsations are treated as deterministic. Some stars may support both deterministic and stochastic oscillations. In this section, we model the two cases separately.

We want to test the fitting method (Equations (24) and (25)) by applying it to simulated time series with gaps. For comparison, we also want to apply the old fitting method (Section 5) to the same time series. We need to generate many realizations of the same random process to test the estimators for bias and precision: This is called Monte Carlo simulation. In Section 6.1 we discuss the generation of the synthetic window functions. We then discuss the parametrization of the solar-like oscillations (Section 6.2) and the deterministic oscillations (Section 6.3) used to simulate the unconvolved signal.

Figure 2 Square root of the power spectra of the synthetic window functions [$\hat{w}$] used in this paper for duty cycles of (a) 100%, (b) 66%, (c) 30%, and (d) 15%. The main periodicity of the window is 24 hours for cases (b) and (c) and 48 hours for window (d). All windows are truncated at frequency $\nu_p = 34.3$ µHz.

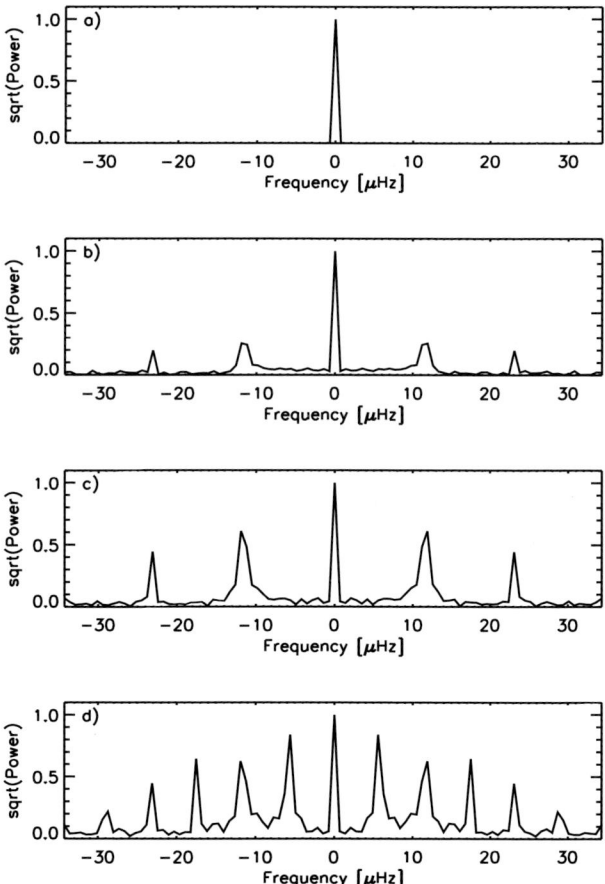

6.1. Synthetic Window Functions

We generate three different observation windows, corresponding to different duty cycles. The observation windows are first constructed in the time domain. By definition, $\tilde{w}_i$ is set to one if an observation is available and zero otherwise. The total length of all time series is fixed at $T = 16.5$ days (frequency resolution $\Delta \nu = 0.7$ µHz). A window function is characterized by two main properties: the duty cycle (fraction of ones) and the average periodicity. A typical window function for a single ground-based site has a 24-hour periodicity. To deviate slightly from purely periodic window functions we introduce some randomness for the end time of each observation block.

Figure 2(b)–(d) shows the power spectra of the three window functions. The first and second window functions have a main periodicity of 24 hours and duty cycles of 66% and 30%, respectively. Two side lobes occur at frequencies of 11.6 µHz and 23.1 µHz. The nonvanishing power between the side lobes is due to the deviation from a purely periodic window. The third window function has a main periodicity of 48 hours and a duty cycle of only 15%. All of these window functions are not unrealistic.

We apply a sharp low-pass filter at frequency $\nu_p = 34.3$ µHz ($p = 49$) to all window functions. The power at higher frequencies corresponds to about 5% of the total power in the windows. This truncation is needed to apply the fitting algorithm, which assumes that there exists a frequency ν_p beyond which the power in the window vanishes (*i.e.,* that the window function is band limited).

6.2. Modeling Solar-Like Oscillations

We generate the realizations of the unconvolved solar-like oscillation signal directly in the Fourier domain. We consider a purely stochastic signal ($d = 0$) and a single mode of oscillation. Since we assumed stationarity in the time domain, the Fourier spectrum of the unconvolved signal for one single mode can be written as

$$x_i = e_i = \left[\mathcal{S} L(\nu_i) + \mathcal{N} \right]^{1/2} \eta_i, \quad i = 0, 1, \ldots, M + 2p - 1, \qquad (40)$$

where L describes the line profile of the mode in the power spectrum, $\mathcal{S}$ is the mode's maximum power, $\mathcal{N}$ is the variance of the background noise, and η_i is a centered complex Gaussian random variable with unit variance and independent real and imaginary parts. Solar-like oscillations are stochastically excited and intrinsically damped by turbulent convection (Goldreich and Keeley, 1977; Stein *et al.*, 2004). The expectation value of the power spectrum is nearly Lorentzian, except for some line asymmetry (*e.g.,* Duvall *et al.*, 1993). Here we use a simple asymmetric line profile:

$$L(\nu) = \frac{(1 + bX)^2 + b^2}{1 + X^2} \quad \text{with} \quad X = \frac{\nu - \nu_0}{\Gamma/2}, \qquad (41)$$

where ν_0 is the resonant frequency, b is the asymmetry parameter of the line profile ($|b| \ll 1$), and Γ is a measure of the width of the line profile. We refer to $\mathcal{S}/\mathcal{N}$ as the signal-to-noise ratio in the power spectrum. As b tends to zero, Γ becomes the full width at half maximum (FWHM) of the power spectrum and $1/(\pi \Gamma)$ the mode lifetime. There are five model parameters, $\mu = (\nu_0, \Gamma, b, \mathcal{S}, \mathcal{N})$.

Once the unconvolved signal x has been generated in the Fourier domain, the observed signal y is obtained by multiplication with the window matrix W, as previously explained.

6.3. Modeling Deterministic Sinusoidal Oscillations plus White Noise

In the time domain, we consider a purely sinusoidal function on top of white background noise:

$$\tilde{x}_i = A \sin(2\pi \nu_0 t_i + \varphi) + \sigma_t \eta_i, \quad i = 0, 1, \ldots, N - 1. \qquad (42)$$

The first term describes the deterministic component of the signal, where A is the amplitude, ν_0 the mode frequency, and φ the phase of the mode. The second term is stochastic noise with standard deviation σ_t. The η_i are N normally distributed, independent, real random variables with zero mean and unit variance. The observed signal is obtained by multiplying $\tilde{x}_i$ by the window $\tilde{w}_i$ in the time domain. The model parameters are $\mu = (\nu_0, \varphi, A, \sigma_t)$.

We have defined the signal and the noise in the time domain, but a definition of signal-to-noise ratio in the Fourier domain is desirable. On the one hand, the variance of the noise in the Fourier domain is

$$\sigma_n^2 = \sigma_0^2 \sum_{i=-p}^{p} |\hat{w}_i|^2 = \frac{\sigma_t^2}{N} \sum_{i=-p}^{p} |\hat{w}_i|^2, \qquad (43)$$

where $\sum_i |\hat{w}_i|^2$ is the total power in the window. On the other hand, the maximum power of the signal in Fourier space is $P_{\max} = A^2 |\hat{w}_0|^2 / 4$, where $|\hat{w}_0|^2$ is the power of the window at zero frequency. Thus, by analogy with the solar-like case, it makes sense to define the signal-to-noise ratio in the Fourier domain as

$$\mathcal{S}/\mathcal{N} = \left(\frac{N |\hat{w}_0|^2}{4 \sum_i |\hat{w}_i|^2} \right) \frac{A^2}{\sigma_t^2}. \tag{44}$$

In practice we fix A and $\mathcal{S}/\mathcal{N}$ and deduce the corresponding noise level σ_t.

7. Testing the Fitting Methods

Several hundreds of realizations are needed to assess the quality of a fitting method. We do not intend to test all possible combinations of mode parameters but we want to show a few cases for which the new fitting method provides a significant improvement compared to the old fitting method.

7.1. Solar-Like Oscillations: Window Function with a 30% Duty Cycle

Figure 3 shows one realization of a simulated mode of solar-like oscillation with input parameters $\nu_0 = 3000$ µHz, $\Gamma = 3.2$ µHz, $\mathcal{S} = 0.9$, $\mathcal{N} = 0.15$, and $b = 0.1$. The signal-to-noise ratio is $\mathcal{S}/\mathcal{N} = 6$ and the window function is 30% full (see Figure 2(c)). The mode lifetime is $1/(\pi \Gamma) = 27.6$ hours. Figure 3(a) displays the real and imaginary parts of the Fourier transform y, together with the standard deviation of the data (nc fit in blue, new fit in red, and expectation value in green). Figure 3(b) shows the power spectrum and the fits. Notice the side lobes introduced by the convolution of the signal with the window functions. The "no-correlation" fit is done to the power spectrum (Equation (36)), whereas the new fit is performed in complex Fourier space (Equation (24)).

Each fit shown in Figure 3 corresponds in fact to the best fit out of five fits with different initial guesses. For each realization, we use the frequency guesses $3000 + (0, \pm 5.5, \pm 11.9)$ µHz for ν_0. The last two frequency guesses correspond to the frequencies of the two main side lobes of the window function (Figure 2(c)). For the other parameters, we choose random guesses within $\pm 20\%$ of the input values. The reason for using several guesses is to ensure that the fit converges to the global maximum of the likelihood, not to a nearby local maximum (*i.e.*, that the estimates returned by the code are the MLE estimates defined by Equation (25)). In some cases, the global maximum coincides with a side lobe at ± 11.57 µHz from the main peak. We note that the new fitting method requires a much longer computing time than the old "nc" method: typically, three hours on a single CPU core for a single realization (five guesses and five fits).

For the particular realization of Figure 3, the new fit is closer to the expectation value (*i.e.*, is closer to the answer) than the old "nc" fit. No conclusions should be drawn, however, from looking at a single realization.

To test the reliability of each fitting method, we computed a total of 750 realizations with the same input parameters as in Figure 3 and the same window function (30% full). The quality (bias and precision) of the estimators can be studied from the distributions of the inferred parameters. As shown by the distributions of Figure 4 the new fitting method is superior to the old "nc" method. This is true for all of the parameters, in particular the mode frequency ν_0. The distributions for the mode frequency (Figure 4(a)) are quite symmetric

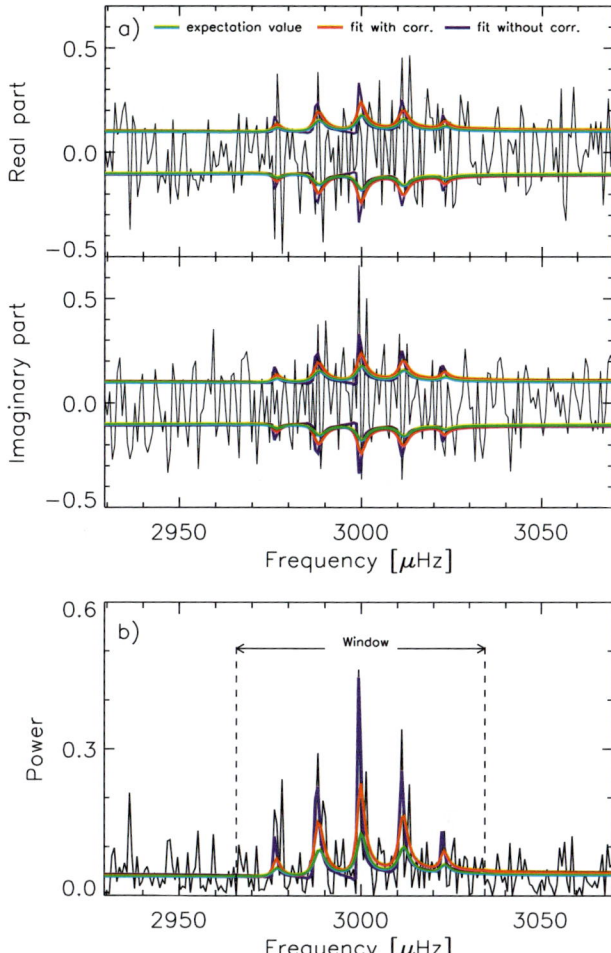

Figure 3 Example of a realization of one mode of a solar-like oscillation (black line) with input frequency $\nu_0 = 3000$ µHz, line width $\Gamma = 3.2$ µHz, and $S/N = 6$. The window function is 30% full. Panel (a) shows the real and imaginary parts of the Fourier spectrum. Panel (b) shows the power spectrum. The vertical dashed lines represent the width of the window function. Also shown are the new fit (red), the old fit (blue), and the expectation value (green).

and Gaussian-like, although the old fitting method leads to a significant excess of values beyond the two-σ mark. We note that, in general, the old fitting method is more sensitive to the initial frequency guess. Also the estimates of the line width [Γ] and the mode power [S] are significantly more biased with the old fitting method than with the new one (Figures 4(b) and (c)). It is worth noting that the fits return a number of small Γ/large S estimates away from the main peaks of the distributions, but less so for the new fits. These values correspond to instances when the signal barely comes out of the noise background. The new fit returns the noise level [N] with a higher precision and a lower number of underestimated outliers than the old method (where the outliers are represented by the vertical bars in Figure 4(d)). Although the estimation of the asymmetry parameter is unbiased with the new fitting method (Figure 4(e)), the uncertainty on b is so large that it probably could have been ignored in the model.

Quantitative estimates of the mean and the dispersion of the estimators are provided in Table 1. Because the distributions of the estimated parameters are not always Gaussian and may contain several outliers, we compute the median (instead of the mean) and the lower and upper bounds corresponding to $\pm 34\%$ of the points on each side of the median

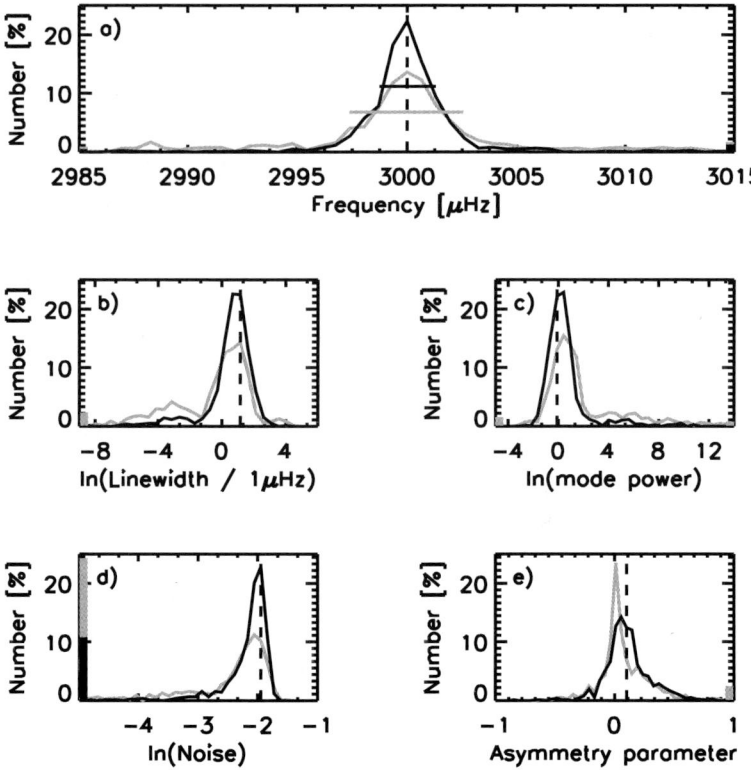

Figure 4 Distributions of the inferred oscillation parameters from fits to 750 realizations of a single mode of solar-like oscillations. The input parameters are given in Table 1 and the window function is 30% full. The five panels show the distributions of the inferred (a) mode frequency ν_0, (b) line width Γ, (c) mode power $\mathcal{S}$, (d) noise level $\mathcal{N}$, and (e) asymmetry parameter b. The black lines show the results obtained with the new fitting method and the gray lines show the old "no-correlation" fits. The vertical dashed line in each plot indicates the input value. The horizontal lines in panel (a) are intervals containing 68% of the fits for the new (black line) and the old (gray line) fitting methods. The thick black and gray vertical lines in panel (d) give the numbers of outliers with $\ln N < -5$.

(instead of the one-σ dispersion). This definition has the advantage of being robust with respect to the outliers. The notation 3000.0^{+b}_{-a} µHz in the first row of Table 1 means that the median mode frequency is 3000.0 µHz and that 68% of the fits belong to the interval $[3000.0 - a, 3000.0 + b]$ µHz. We emphasize that the subscript $-a$ and the superscript $+b$ do not refer to an uncertainty in the determination of the median: The median is known to a much higher precision because of the large number of realizations. Later we relax the language and refer to the "one-σ uncertainty" to mean the average $\sigma = (a + b)/2$.

The numbers from the last two columns in Table 1 confirm the analysis of Figure 4. The mode frequency can be measured with a precision of 1.4 µHz, and so the precision of the new fitting method is exactly twice that of the old one. This gain in precision is very significant and potentially important. Since measurement uncertainty scales as $T^{-1/2}$ (Libbrecht, 1992), one may equate the gain in using the proper fitting procedure to an effective increase in the total length of the time series by a factor of four. As seen in Table 1, the line width, the mode power, the background noise, and the line asymmetry parameter are all less biased and

Table 1 Medians and scatters of the distributions of the estimated parameters of solar-like oscillations (see Figure 4). The window function is 30% full, the input line width is 3.2 µHz, and the input signal-to-noise ratio is $S/\mathcal{N} = 6$. The new and old MLE estimates are given in the last two columns. By definition, 68% of the fits fall within the bounds set by the subscripts and superscripts (with the notation explained in detail in the text).

Mode parameter	Input value	New fitting	Old fitting
ν_0 (µHz)	3000.0	$3000.0^{+1.4}_{-1.4}$	$3000.0^{+2.8}_{-2.8}$
$\ln[\Gamma$ (µHz)]	1.2	$0.8^{+0.8}_{-1.0}$	$0.2^{+1.1}_{-3.7}$
$\ln S$	-0.1	$0.2^{+0.9}_{-0.9}$	$0.9^{+4.3}_{-1.2}$
$\ln \mathcal{N}$	-1.9	$-2.1^{+0.2}_{-0.9}$	$-2.4^{+0.4}_{-6.8}$
b	0.1	$0.1^{+0.2}_{-0.1}$	$0.0^{+0.2}_{-0.1}$

more precise with the new fitting method than the old one. Notice that the larger dispersions in the old-fit case are due in part to non-Gaussian distributions with extended tails.

7.2. Solar-Like Oscillations: Different Window Functions

Here we study how bias and precision change as the window function changes, in particular as the duty cycle changes. We consider the four window functions defined in Section 6.1 with duty cycles [α] equal to 15%, 30%, 66%, and 100%. First we consider input parameters of solar-like oscillations that are exactly the same as in the previous section: $\nu_0 = 3000$ µHz, $\Gamma = 3.2$ µHz, $S = 0.9$, $S/\mathcal{N} = 6$, and $b = 0.1$. Figure 5 shows the distributions of the inferred mode frequencies and line widths, using the old (Figures 5(a) and (b)) and the new (Figures 5(c) and (d)) fitting methods. Each fit is the best fit from five different ν_0 guesses (see Section 7.1). The distributions for the 100% window are identical for the two fitting methods; this is expected since the old and new fitting methods are equivalent in the absence of gaps.

The precision on ν_0 using the old "no-correlation" MLE drops fast as the duty cycle decreases (Figure 5(a)). This drop is much faster than in the case of the fits that take the frequency correlations into account (Figure 5(c)). When the duty cycle is 15%, the frequency estimate is five times better with the new method than with the old one. The difference is perhaps even more obvious for the line width. For the 15% window, it is almost impossible to retrieve Γ with the old fitting method (Figure 5(b)), whereas the new method gives estimates that are almost as precise as in the no-gap case (Figure 5(d)). The estimates of Γ are significantly less biased with the new method. Figure 5 confirms the importance of using the correct expression for the likelihood function.

Table 2 gives the medians and half-widths of the ν_0 distributions. The one-σ dispersions are plotted as a function of the duty cycle α in the left panel of Figure 6. The improvement in the fits is quite spectacular. For example, when $\alpha = 15\%$ the dispersion on ν_0 is reduced by nearly 80% with the new fitting method (1.5 µHz vs. 7.4 µHz).

With the old method, the uncertainty on ν_0 increases much faster than $\alpha^{-1/2}$ as the duty cycle α drops ($\sim \alpha^{-1}$ between the 30% and 15% windows). This steep dependence on α is worse than "predicted" by Libbrecht (1992). In his paper, Libbrecht suggested using the uncertainty $\sigma_{\nu_0} = \sqrt{f\Gamma/(4\pi T)}$, where $f(\beta) = (1 + \beta)^{1/2}[(1 + \beta)^{1/2} + \beta^{1/2}]^3$ and β is an "effective" noise-to-signal ratio. He suggested that the main effect of the gaps is to increase the noise-to-signal ratio $\mathcal{N}/S$, presumably by a factor $\sum_i |\hat{w}_i|^2/|\hat{w}_0|^2$, itself proportional to

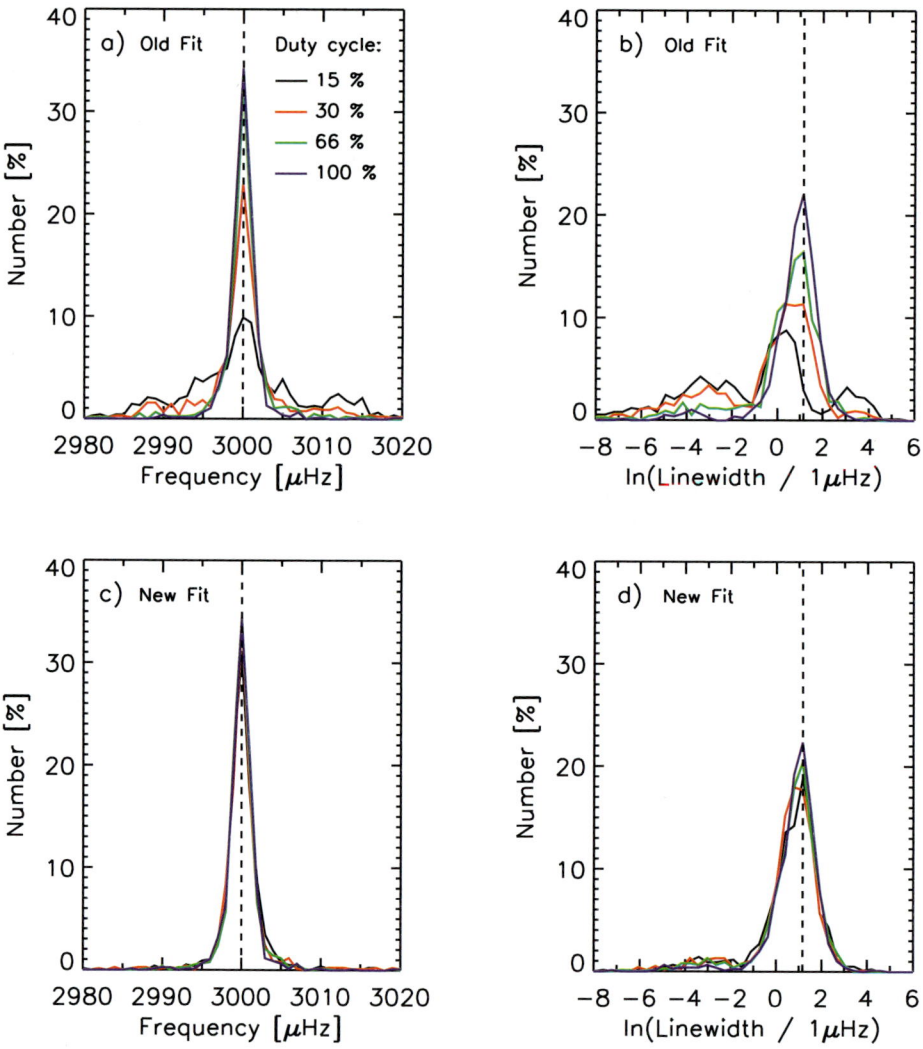

Figure 5 Distributions of the mode frequency and the line width for 750 realizations of solar-like oscillations, using the old fitting method (panels (a) and (b)) and the new fitting method (panels (c) and (d)). The observation windows have duty cycles of 15%, 30%, 66%, and 100%. The vertical dashed lines represent the input values. The input line width is $\Gamma = 3.2$ µHz.

$1/\alpha$. This leads however to a dependence of σ_{ν_0} on α, which, in our particular case, is closer to $\alpha^{-1/2}$ than α^{-1}. We suspect that the Libbrecht formula underestimates the dispersion because it ignores the frequency correlations.

The new fitting method returns a ν_0 uncertainty that is much less sensitive to the duty cycle, with a variation as $\sim \alpha^{-0.15}$ (red curve, left panel of Figure 6). This is quite remarkable. That the frequency uncertainty could remain nearly constant for $\alpha > 30\%$ is not really surprising since the average gap (see numbers in Table 2) is less than the mode lifetime $\tau = 1/(\pi \Gamma) = 27.6$ hours. This regime was studied by Fossat et al. (1999) using a gap-filling method: As long as the signal-to-noise ratio is large enough, the signal can be recon-

Table 2 Medians and scatters of the mode frequency estimates (solar-like oscillations) for the window functions defined in Section 6.1. The input mode frequency is $\nu_0 = 3000$ μHz, the input line width is $\Gamma = 3.2$ μHz, and the signal-to-noise ratio is fixed at $\mathcal{S}/\mathcal{N} = 6$. The mode lifetime is 27.6 hours.

Duty cycle	Window function		Frequency estimate (μHz)	
	Main period	Average gap	New fitting	Old fitting
100%	–	–	$3000.0^{+1.1}_{-1.2}$	$3000.0^{+1.1}_{-1.2}$
66%	24 hours	7.4 hours	$3000.0^{+1.1}_{-1.3}$	$3000.1^{+1.5}_{-1.4}$
30%	24 hours	16.4 hours	$3000.0^{+1.4}_{-1.4}$	$3000.0^{+2.8}_{-2.8}$
15%	48 hours	40.7 hours	$3000.0^{+1.7}_{-1.3}$	$3000.0^{+8.3}_{-6.5}$

structed. Why the new fit is doing such a good job for duty cycles $\alpha \leq 30\%$ is, however, puzzling (at first sight), since the average gap (40.7 hours) is larger than the mode lifetime. This can be understood as follows. For small duty cycles, the time series is effectively a collection of nearly independent blocks of data, which, for the 30% window function, are eight-hours long on average. Since MLE simulations tell us that the uncertainty on the mode frequency for an uninterrupted series of eight hours is about 5.5 μHz, we would expect for the gapped time series ($T = 16.5$ days, with 24-hour periodicity) to be able to reach the uncertainty $5.5/\sqrt{16} = 1.375$ μHz. This value, represented by the box with a cross in Figure 6, is found to be very close to the MLE estimate from the new fits. Hence, what matters at very low duty cycle is the number of independent blocks of continuous data. The new fitting method captures this very well, which is satisfying. By comparison, the old no-correlation fitting method does poorly (black line).

To further investigate this last point, we ran another set of simulations using a mode line width ($\Gamma = 10$ μHz) corresponding to a mode lifetime $\tau = 8.8$ hours, which is significantly smaller than the average gap lengths of the 30% and the 15% windows. The other input parameters remained the same as before. We computed and fitted 1350 realizations. The results are shown in the right panel of Figure 6. For the new fitting method, the dependence of the frequency uncertainty on the duty cycle is about $\alpha^{-0.12}$, which is comparable to the previous simulations with $\Gamma = 3.2$ μHz. We conclude that it is really worth solving for the correct minimization problem and that fitting for the phase information in complex Fourier space is important to get a good match between the model and the data. Of course, this can only be done properly when we have perfect knowledge of the model, which is the case with these numerical simulations but is rarely the case with real observations.

7.3. Solar-Like Oscillations: Cramér–Rao Lower Bounds

Monte Carlo simulations are very useful for assessing the variance and the bias of a particular estimator. When fitting real observations, however, the variance of the estimator cannot be computed directly by Monte Carlo simulation since the input parameters are, by definition, not known. Hopefully, the fit can return a formal error from the shape of the likelihood function in the neighborhood of the global maximum.

The Cramér–Rao lower bound (Kendall and Stuart, 1967) achieves minimum variance among unbiased estimators. It is obtained by expanding $\mathcal{L}$ about its minimum. The formal error σ_{μ_i} of the parameter μ_i is given by

$$\sigma_{\mu_i} = \sqrt{K_{ii}}, \qquad (45)$$

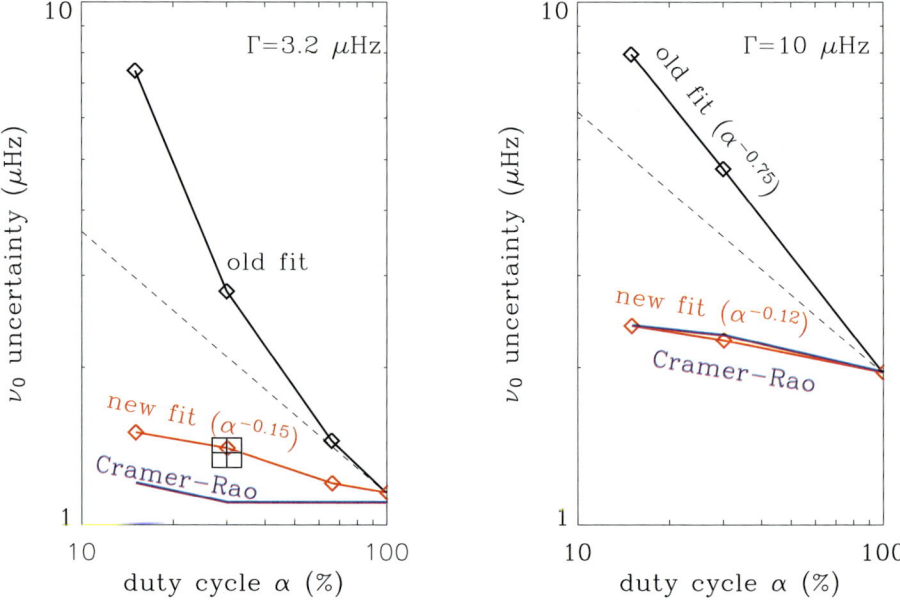

Figure 6 Uncertainty of estimates of the mode frequency [ν_0] as a function of the window duty cycle [α]. The window functions are as defined in Section 6.1. The red curve shows the 1-σ Monte Carlo MLE uncertainties for the new fitting method. The black curve shows the 1-σ Monte Carlo MLE uncertainties for the old no-correlation fitting method. The blue curves show the mean Cramér–Rao lower bounds (formal error bars). The square symbol with a cross at $\alpha = 30\%$ in the left panel is a rough estimate (see text). In the left panel, the input line width is $\Gamma = 3.2$ µHz (see also numbers in Table 1). In the right panel, the input line width is $\Gamma = 10$ µHz, with all of the other parameters being the same as in the left panel. In both panels the signal-to-noise ratio is $\mathcal{S}/\mathcal{N} = 6$. For reference, the dashed lines have slope $\alpha^{-1/2}$.

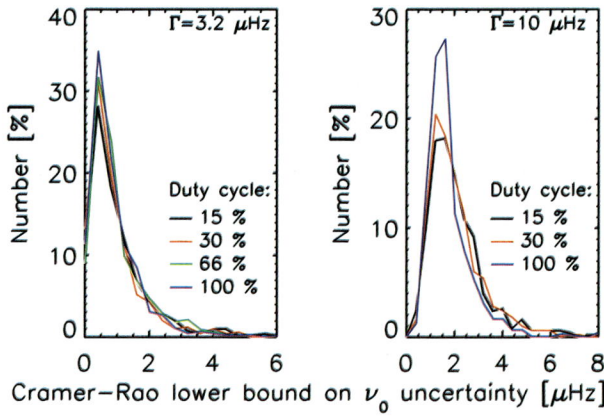

Figure 7 Distributions of formal errors of the mode frequency obtained by inverting the Hessian. The left panel is for the simulation with $\Gamma = 3.2$ µHz (see Figure 5(c)) and the right panel is for $\Gamma = 10$ µHz. The different curves correspond to different window functions, as indicated in the legend. The means of these distributions (Cramér–Rao lower bounds) give the blue curves plotted in Figure 6.

where K_{ii} is the ith element on the diagonal of the inverse ($K = H^{-1}$) of the Hessian matrix with elements

$$H_{ij} = \frac{\partial^2 \mathcal{L}}{\partial \mu_i \partial \mu_j} \quad \text{for } i, j = 0, 1, \ldots, k-1. \tag{46}$$

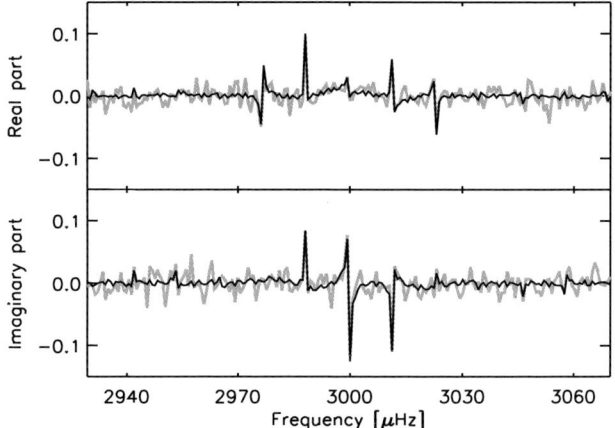

Figure 8 Real and imaginary parts of the Fourier transform of a simulated gapped time series containing a sinusoid on top of white background noise. The signal-to-noise ratio is $\mathcal{S}/\mathcal{N} = 100$. The observation window has a duty cycle of 30%. The simulated data are shown by the thick gray line. The thin black line shows the fit to the data using the new fitting method. The fit with the old method is not shown as it is almost identical.

The Cramér–Rao formal errors have been used in helioseismology by, for example, Toutain and Appourchaux (1994), Appourchaux, Gizon, and Rabello-Soares (1998), and Gizon and Solanki (2003).

We have computed the formal error on the mode frequency for many realizations and for all window functions. The resulting distributions are shown in Figure 7. The mean formal error from each distribution is plotted in Figure 6. Overall the Cramér–Rao lower bound is remarkably close to the Monte Carlo MLE uncertainty using the new fitting method; they are even undistinguishable when $\Gamma = 10$ µHz.

This is useful information as it means that, on average, the Hessian method provides reasonable error estimates. It should be clear, however, that the distributions shown in Figure 7 show a significant amount of scatter: The formal error from the Hessian may be misleading for particular realizations.

7.4. Sinusoidal Deterministic Oscillation plus White Noise

Figure 8 shows the Fourier spectrum of a simulated time series containing a sinusoidal mode of oscillation on top of a white noise background as described in Section 6.3. In this particular case the observation window with a duty cycle of 30% is used (see Figure 2(c)). The input parameters of the sinusoidal function are the mode frequency $\nu_0 = 3000$ µHz, the amplitude $A = 1.1$, and the phase $\varphi = 60°$. The signal-to-noise ratio is $\mathcal{S}/\mathcal{N} = 100$. The fit shown in Figure 8 was obtained with the new fitting method. Since we found no significant difference between the old and the new fitting methods in this case, the old fitting method is not shown. Differences between the data and the fit are essentially due to the noise.

We computed 500 realizations of sinusoidal oscillations with the same mode parameters (frequency, amplitude, and phase) as before, the same observation window (30% full), but with a signal-to-noise ratio $\mathcal{S}/\mathcal{N} = 46$. The resulting distributions of the inferred parameters obtained with the two fitting methods are shown in Figure 9. For this simulation, the known input values were used as an initial guess to speed up the minimization; we checked that it is acceptable to do so on several realizations when the signal-to-noise ratio is large. The distributions of the inferred parameters (Figure 9) show that, for sinusoidal oscillations, the new fitting method does not provide any significant improvement compared to the old fitting method.

We emphasize that the fitting parameters can be determined with a very high precision when the noise level is small. In particular, we confirm that the uncertainty of the frequency

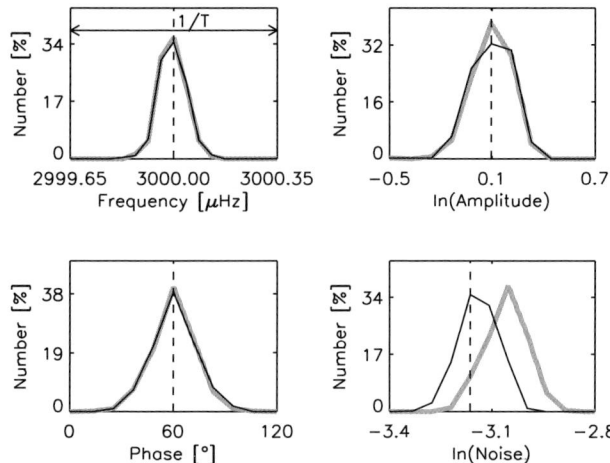

Figure 9 Distributions of the inferred oscillation parameters for a set of 500 realizations of long-lived sinusoidal oscillations with $S/N = 46$. The window function with a duty cycle of 30% is used. The black and the gray lines are for the new and old fitting methods, respectively. The vertical dashed line in each plot indicates the input value. The parameters shown are (a) the mode frequency $[\nu_0]$, (b) the logarithm of the mode amplitude $[\ln A]$, (c) the phase of the oscillation $[\phi]$, and (d) the logarithm of the noise level $[\ln \sigma_0]$ (see Section 6.3). Notice that the estimate of the noise is biased when frequency correlations are ignored (old "nc" fit), although by a very small amount.

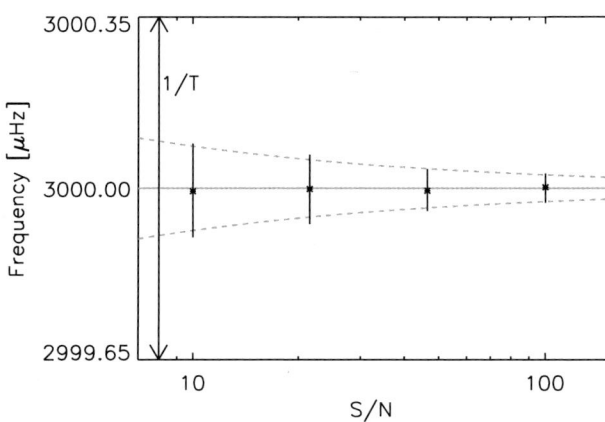

Figure 10 Median (cross) and standard deviation (vertical bar) of the inferred frequency of sinusoidal oscillations $[\nu_0]$ as a function of signal-to-noise ratio S/N. The duty cycle is 30%. Only the results obtained with the new fitting method are shown. The horizontal gray line shows the input mode frequency. The dashed gray lines show the theoretical value of frequency uncertainty, σ_{ν_0}, given by Equation (47). The vertical axis of the plot spans the interval $\Delta \nu = 1/T = 0.7$ μHz.

estimator can be much smaller than $1/T$ (see Figure 9(a)). Figure 10 shows the median and the standard deviation of the mode frequency for different signal-to-noise ratios. Each symbol and its error bar in Figure 10 is based on the computation of 500 realizations of sinusoidal oscillations with the same mode parameters as before, the same observation window (30% full), but various signal-to-noise ratios. Since we did not find any significant difference between the two fitting methods, only the results obtained with the new fitting method are shown. Figure 10 illustrates that, even for a relatively low signal-to-noise ratio of $S/N = 10$, the standard deviation of the inferred mode frequency is smaller than $1/T$ by a factor of four. For higher signal-to-noise ratios the precision is even more impressive: When $S/N = 100$, the standard deviation of the mode frequency is about $1/20$ that of $1/T$.

The theoretical value of the standard deviation of the mode frequency obtained by Cuypers (1987) can be extended to the case of gapped data (Cuypers, 2008, private communication) as follows:

$$\sigma_{\nu_0} = \frac{\sqrt{6}\sigma_t}{\pi A T \sqrt{n}}, \qquad (47)$$

where A is the amplitude of the sinusoid in the time domain, σ_t is the rms value of the noise, $n = \alpha N$ is the number of recorded data points, and T is the total observation length. This theoretical uncertainty is overplotted in Figure 10. The match with our Monte Carlo measurements is excellent. This confirms that, in this case, it is equivalent to perform the fits in the temporal or in the Fourier domains. Note that Equation (47) is only valid under the assumption that the noise is uncorrelated in the time domain, a condition fulfilled by our simulations. The main reason why the measurement precision is only limited by the noise-to-signal ratio is because perfect knowledge of the model is assumed.

8. Conclusion

In this paper we derived an expression for the joint PDF of solar or stellar oscillations in complex Fourier space, in agreement with the work of Gabriel (1994). This joint PDF explicitly takes into account frequency correlations introduced by the convolution with the window function. We implemented a maximum likelihood estimation method to retrieve the parameters of stellar oscillations. Both stochastic solar-like oscillations and deterministic sinusoidal oscillations were considered.

In the case of solar-like oscillations, we performed Monte Carlo simulations to show that the improvement provided by our fitting method can be very significant in comparison with a fitting method that ignores the frequency correlations. The results are summarized in Figure 6. In one particular example, by using an observation window with a duty cycle $\alpha = 30\%$ and a signal-to-noise ratio $\mathcal{S}/\mathcal{N} = 6$, the new fitting method increased the precision of the mode frequency by a factor of two and the estimates of the line width and mode power were less biased and more precise. For a window with a duty cycle $\alpha = 15\%$, the precision of the mode frequency estimate was increased by a factor of five. We also found that the Cramér–Rao lower bounds (formal errors) can provide reasonable estimates of the uncertainty on the MLE estimates of the oscillation parameters.

In the case of long-lived, purely sinusoidal oscillations, we did not find any significant improvement in using this new fitting method. Yet, we confirm that the standard deviation of the mode frequency can be measured in Fourier space with a precision much better than $1/T$ for large signal-to-noise ratios, in accordance with a previous time-domain calculation (Cuypers, 1987; Cuypers, 2008, private communication).

The analysis of time series containing many gaps can benefit from our work. Applications may include, for example, the reanalysis of solar oscillations from the early days of the BiSON network (Miller et al., 2004) or the solar-like oscillations of α Centauri observed from the ground with two telescopes (Butler et al., 2004).

The MLE source code is available from the Internet platform of the European Helio- and Asteroseismology Network (HELAS, funded by the European Union) at http://www.mps.mpg.de/projects/seismo/MLE_SoftwarePackage/.

Acknowledgements We thank T. Appourchaux for useful discussions, in particular for the suggestion to compute the Cramér–Rao lower bounds. T. Stahn is a member of the International Max Planck Research School on Physical Processes in the Solar System and Beyond at the Universities of Göttingen and Braunschweig.

References

Anderson, E.R., Duvall, T.L., Jefferies, S.M.: 1990, *Astrophys. J.* **364**, 699.
Appourchaux, T., Gizon, L., Rabello-Soares, M.-C.: 1998, *Astron. Astrophys. Suppl. Ser.* **132**, 107.
Appourchaux, T., Chang, H.-Y., Gough, D.O., Sekii, T.: 2000, *Mon. Not. Roy. Astron. Soc.* **319**, 365.
Bedding, T.R., Kjeldsen, H.: 2007, *Commun. Asteroseismol.* **150**, 106.
Brandt, S.: 1970, *Statistical and Computational Methods in Data Analysis*, North-Holland, Amsterdam.
Butler, R.P., Bedding, T.R., Kjeldsen, H., McCarthy, C., O'Toole, S.J., Tinney, C.G., Marcy, G.W., Wright, J.T.: 2004, *Astrophys. J.* **600**, L75.
Cuypers, J.: 1987, *Bull. Acad. Roy. Sci. Belg. (Cl. Sci.)* **49**, 21.
Duvall, T.L., Harvey, J.W.: 1986, In: Gough, D.O. (ed.) *NATO Advanced Research Workshop*, Reidel, Dordrecht, 105.
Duvall, T.L., Jefferies, S.M., Harvey, J.W., Osaki, Y., Pomerantz, M.A.: 1993, *Astrophys. J.* **410**, 829.
Fossat, E., Kholikov, S., Gelly, B., Schmider, F.X., Fierry-Fraillon, D., Grec, G., Palle, P., Cacciani, A., Ehgamberdiev, S., Hoeksema, J.T., Lazrek, M.: 1999, *Astron. Astrophys.* **343**, 608.
Gabriel, M.: 1994, *Astron. Astrophys.* **287**, 685.
Gizon, L., Solanki, S.K.: 2003, *Astrophys. J.* **589**, 1009.
Goldreich, P., Keeley, D.A.: 1977, *Astrophys. J.* **212**, 243.
Horn, R.A., Johnson, C.R.: 1985, *Matrix Analysis*, Cambridge University Press, Cambridge.
Kendall, M.G., Stuart, A.: 1967, *The Advanced Theory of Statistics. Inference and Relationship*, vol. **2**, 2nd edn., Butler and Tanner, London.
Libbrecht, K.G.: 1992, *Astrophys. J.* **387**, 712.
Miller, B.A., Hale, S.J., Elsworth, Y., Chaplin, W.J., Isaak, G.R., New, R.: 2004, In: Danesy, D. (ed.) *Proc. SOHO 14/GONG 2004 Workshop: Helio- and Asteroseismology: Towards a Golden Future*, **SP-559**, ESA, Noordwijk, 571.
Schou, J.: 1992, Ph.D. Dissertation, University of Aarhus.
Stein, R., Georgobiani, D., Trampedach, R., Ludwig, H.-G., Nordlund, Å, 2004, *Solar Phys.* **220**, 229.
Toutain, T., Appourchaux, T.: 1994, *Astron. Astrophys.* **289**, 649.
Winget, D.E., Nather, R.E., Clemens, J.C., Provencal, J., Kleinman, S.J., Bradley, P.A., Wood, M.A., Claver, C.F., Frueh, M.L., Grauer, A.D., Hine, B.P., Hansen, C.J., Fontaine, G., Achilleos, N., Wickramasinghe, D.T., Marar, T.M.K., Seetha, S., Ashoka, B.N., O'Donoghue, D., Warner, B., Kurtz, D.W., Buckley, D.A., Brickhill, J., Vauclair, G., Dolez, N., Chevreton, M., Barstow, M.A., Solheim, J.E., Kanaan, A., Kepler, S.O., Henry, G.W., Kawaler, S.D.: 1991, *Astrophys. J.* **378**, 326.
Woodard, M.F.: 1984, Ph.D. Dissertation, University of California, San Diego.

Perspectives in Global Helioseismology and the Road Ahead

William J. Chaplin · Sarbani Basu

Originally published in the journal Solar Physics, Volume 251, Nos 1–2, 53–75.
DOI: 10.1007/s11207-008-9136-5 © Springer Science+Business Media B.V. 2008

Abstract We review the impact of global helioseismology on key questions concerning the internal structure and dynamics of the Sun and consider the exciting challenges the field faces as it enters a fourth decade of science exploitation. We do so with an eye on the past, looking at the perspectives global helioseismology offered in its earlier phases, in particular the mid-to-late 1970s and the 1980s. We look at how modern, higher quality, longer datasets coupled with new developments in analysis have altered, refined, and changed some of those perspectives and opened others that were not previously available for study. We finish by discussing outstanding challenges and questions for the field.

Keywords Sun: helioseismology · Abundances · Activity · Magnetic fields

1. Introduction

The field of global helioseismology – the use of accurate and precise observations of the globally coherent modes of oscillation of the Sun to make inference on the internal structure and dynamics of our star – is about to enter its fourth decade. The observational starting point for global helioseismology was marked in the mid-to-late 1970s by several key papers. First, the observational confirmation by Deubner (1975), and independently by Rhodes, Ulrich, and Simon (1977), of the standing-wave nature of the five-minute oscillations observed on the surface of the Sun, which was proposed by Ulrich (1970) and Leibacher and Stein (1971), and then the discovery that the oscillations displayed by the Sun were truly global

---

Invited Review.

Helioseismology, Asteroseismology, and MHD Connections
Guest Editors: Laurent Gizon and Paul Cally.

W.J. Chaplin (✉)
School of Physics and Astronomy, University of Birmingham, Edgbaston, Birmingham B15 2TT, UK
e-mail: w.j.chaplin@bham.ac.uk

S. Basu
Department of Astronomy, Yale University, P.O. Box 208101, New Haven, CT 06520-8101, USA
e-mail: sarbani.basu@yale.edu

whole-Sun, core-penetrating, radial-mode pulsations (Claverie et al., 1979). Previous observations of pulsating stars had revealed many objects that were oscillating in one, or at most a few, modes. The rich spectrum of oscillations displayed by the Sun was a different matter entirely. Christensen-Dalsgaard and Gough (1976) pointed out the great potential that a multimodal spectrum could offer: The information content of the observations could potentially be so great as to allow a reconstruction of the internal structure of the star. More than 30 years on, exquisite observational reconstructions of the internal structure and dynamics of the Sun are in everyday use, thanks to helioseismology.

Several thousand modes of oscillation of the Sun have to date been observed, identified, and studied. The oscillations are standing acoustic waves, for which the gradient of pressure (p) is the principal restoring force. The modes are excited stochastically, and damped intrinsically, by the turbulence in the outermost layers of the subsurface convection zone. The stochastic excitation mechanism limits the amplitudes of the p modes to intrinsically weak values. However, it gives rise to an extremely rich spectrum of modes, the most prominent generally being high-order overtones. Detection of individual gravity modes – for which buoyancy acts as the principal restoring force – remains an important goal for the field. The g modes are confined in cavities in the radiative interior, and if observed would provide a much more sensitive probe of conditions in the core than the p modes.

The small-amplitude solar oscillations may be described in terms of spherical harmonics $Y_\ell^m(\theta, \phi)$, where θ and ϕ are the co-latitude and longitude, respectively:

$$Y_\ell^m(\theta, \phi) = (-1)^m c_{\ell m} P_\ell^m(\cos\theta) \exp(im\phi). \qquad (1)$$

In Equation (1), the P_ℓ^m are Legendre functions and the $c_{\ell m}$ are normalization constants. The p modes probe different interior volumes, with the radial and other low-degree (low-ℓ) modes probing as deeply as the core. This differential penetration of the modes allows the internal structure and dynamics to be inferred, as a function of position, to high levels of precision not usually encountered in astrophysics. The Sun has not surprisingly been the exemplar for the development of seismic methods for probing stellar interiors. Extension of the observations to other Sun-like stars (asteroseismology) has demonstrated that the Sun-like oscillations are a ubiquitous feature of stars with subsurface convection zones.

In this review our aim is to look at some of the recent advances that global helioseismology has made for studies of various aspects of the internal structure of the Sun. We discuss the observational challenge posed by the detection and identification of individual g modes. We also seek to provide in each section some historical context for discussion of the contemporary results and challenges. We round out the review by looking to the future, in particular the need for continued multi-instrument, multinetwork observations of the global modes; and we finish by listing some important questions and challenges for global helioseismology.

2. The Standard Solar Model and the Abundance Problem

The first important inference to be made by helioseismology on the internal structure concerned the depth of the solar convection zone. Gough (1977) and (independently) Ulrich and Rhodes (1977) realized that a mismatch between computed p-mode frequencies (Ando and Osaki, 1975) and the observed frequencies (actually the locations in frequency of ridges in the $k-\omega$ diagram) could be reconciled by increasing the depth of the convection zone by about 50% compared to typical values used at the time.

Investigations soon followed into the compatibility of solar models, having different heavy-element abundances, with the observed p-mode frequencies. An important aim was to see whether models having low initial heavy-element abundances, but significant accretion rates, could reconcile the "solar neutrino problem" and at the same time be consistent with the seismic data (Christensen-Dalsgaard, Gough, and Morgan, 1979). The only seismic frequencies that were available when this work was done were those for modes that penetrated the near-surface layers. As already noted, these seismic data favored a deep convection zone. This meant that the seismic data were also at odds with a low surface abundance of helium: A deeper zone goes hand in hand with a higher helium abundance. Given that the amounts of helium and heavier elements are inexorably tied together, this seemed to suggest that the heavy-element abundance could not be low as well.

More robust conclusions were possible once frequencies of the core-penetrating, low-ℓ modes became available, from the early 1980s onward (e.g., see Christensen-Dalsgaard and Gough, 1980, 1981). Once the Kitt Peak data (Duvall and Harvey, 1983) had "filled the gap" between the high-ℓ and low-ℓ data that were already available, inversions for the internal sound speed were possible. These data, and more modern data, have resulted in detailed inference on the solar structure. For instance, the position of the convection zone (e.g., Basu and Antia, 1997) and the convection-zone helium abundance (e.g., Däppen et al., 1991; Basu and Antia, 1995) are both known to high precision. These inferences on the solar structure, and the ability to determine sound-speed and density differences between solar models and the Sun, provide a means to test how well solar models fare against the Sun and thereby allow tests of the input physics to be made. For example, the inversions of Christensen-Dalsgaard et al. (1985) suggested there were problems with computation of the astrophysical opacities. Significant improvements were made in the OPAL opacity tables (e.g., Iglesias and Rogers, 1991, 1996). Further improvements to the standard models followed with the routine inclusion of diffusion and settling (e.g., Demarque and Guenther, 1988; Cox, Guzik, and Kidman, 1989; Christensen-Dalsgaard, Proffitt, and Thompson, 1993). Tests are also possible for the equation of state (Christensen-Dalsgaard and Däppen, 1992; Basu and Christensen-Dalsgaard, 1997; Elliott and Kosovichev, 1998), and as a result of the improvements that followed most modern solar models are now constructed with the OPAL2001 equation of state (Rogers and Nayfonov, 2002).

Early inversions, such as those by Christensen-Dalsgaard et al. (1985), appeared to rule out the possibility of a low helium and low heavy-element abundance. Although the issue of the solar helium abundance has been settled because it can be inferred from the very precise p-mode frequencies, the heavy-element abundance question is alive again, and it may be said that the field is confronted with a "solar abundance problem." Whereas in the late 1970s and early 1980s, the question of which abundances fitted the helioseismic data best was open – the helioseismic data were new, and the community was still feeling its way on fully exploiting the data – today we know the answer to that question. The issue is instead very much open because estimates of the photospheric abundances, provided by spectroscopy, have been revised downward.

One of the important inputs into these models is the heavy-element abundance (Z) or, alternatively, the ratio of the heavy element to the hydrogen abundance (Z/X). Solar models that have shown a good agreement with results from helioseismology were constructed with the solar abundance as given by Grevesse and Noels (1993), or more recently by Grevesse and Sauval (1998; henceforth GS98). The GS98 table shows that $Z/X = 0.0229$ (i.e., $Z = 0.0181$) for the Sun. The situation has changed recently. Asplund et al. (2000, 2004), Allende Prieto, Lambert, and Asplund (2001, 2002), and Asplund, Grevesse, and Sauval (2005) find that the solar heavy-element abundances need to be reduced drastically,

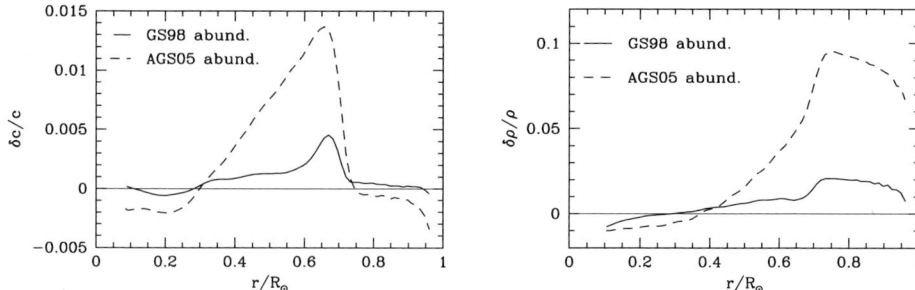

Figure 1 The relative sound-speed difference (left) and density difference (right) between the Sun and a standard solar model constructed with the GS98 metallicity [model BP04(Garching)], and also a standard solar model constructed with the AGS05 metallicity [model BS05(AGS, OPAL)] of Bahcall, Basu, and Serenelli (2005). The model with GS98 Z/X has a CZ He abundance of $Y_{CZ} = 0.243$ and a CZ base at $R_{CZ} = 0.715\,r/R_\odot$. The AGS05 model has $Y_{CZ} = 0.230$ and $R_{CZ} = 0.729\,r/R_\odot$.

based on what they claim are better calculations with improved models of the solar photosphere. This led Asplund, Grevesse, and Sauval (2005; henceforth AGS05) to compile a table of solar abundances, with $Z/X = 0.0166$ (i.e., $Z = 0.0122$). This has resulted in considerable discussion in the community since the sound-speed and density profiles of models constructed with AGS05 do not agree well with the Sun. This disagreement can be seen in Figure 1, where we show the density and sound-speed differences between the Sun and two solar models, one constructed with the GS98 abundances and the other with the AGS05 abundances.

The mismatch between the models with low Z and the Sun is most striking in the outer parts of the radiative interior, a result of the fact that the low-Z models have a much shallower convection zone than the Sun. There are, however, differences in other regions too: All standard models with AGS05 abundances have low helium abundance in the convection zone (e.g., Montalbán et al., 2004; Guzik, Watson, and Cox, 2005; Bahcall, Basu, and Serenelli, 2005); the seismic signatures of the ionization zones do not match observations (Lin, Antia, and Basu, 2007); and the helioseismic signatures from the core do not match observations either (Basu et al., 2007).

There have been several attempts to reconcile low-Z solar models with helioseismic data. Given that the largest discrepancy is at the base of the convection zone, the first attempts involved modifying the input opacities. It was found that large changes in opacity, in the range 11% to 21%, would be needed at temperatures relevant to the base of the convection zone to resolve the problem (Montalbán et al., 2004; Basu and Antia, 2004; Bahcall et al., 2005). However, later recalculation of the opacities by the OP group (Badnell et al., 2005) showed an opacity increase of only 2%. Other attempts included increasing the diffusion coefficient (e.g., Montalbán et al., 2004; Basu and Antia, 2004; Guzik, Watson, and Cox, 2005). A large change was needed to get the correct position of the convection-zone base, which resulted in an extremely low convection-zone helium abundance. Attempts were also made to increase the metallicity of the models by increasing the abundance of uncertain elements such as neon, which does not have any photospheric lines (Antia and Basu, 2005; Bahcall et al., 2005). However it is not clear whether such an increase is justified in the case of the Sun (Schmelz et al., 2005; Young, 2005). Other attempts involve ad hoc prescriptions of mixing at the tachocline (Turck-Chieze et al., 2004b; Montalbán et al., 2006) or mixing by gravity waves (Young and Arnett, 2005). Late accretion of low-Z material by the zone has been tried as well (Guzik, Watson, and Cox, 2005; Castro, Vauclair, and Richard, 2007).

None of the models match the Sun unless two or more modifications are used (Montalbán *et al.*, 2004) and even in those cases the changes in physics have to be fine-tuned carefully.

Given that attempts to adjust physical inputs in an *ad hoc* manner have not resulted in low-Z solar models that agree with the Sun, Antia and Basu (2006) tried to derive Z for the Sun using signatures of the heavy-element ionization zones. The method was similar to that used by Basu and Antia (1995) to determine the He abundance of the Sun. They found a solar Z of 0.0172 ± 0.002, a value close to the GS98 value and much larger than the AGS05 value. The errors in the result are not affected by errors in opacity. Although the Antia and Basu (2006) result was obtained from helioseismic signatures in the upper convection zone, using data from the solar core Chaplin *et al.* (2007a) also concluded that Z in the Sun has to be high. They found that the mean molecular weight averaged over the inner 20% by radius of the Sun is in the range 0.7209 to 0.7231 and that the corresponding surface Z is in the range 0.0187 to 0.0239.

The obvious discrepancy between the low-Z models and the Sun creates a problem that has not yet been resolved. If the new, lower abundances are correct then the obvious culprit is missing or incorrect physics in the solar models. If, however, the lower abundances are incorrect, then we can be confident that the input physics in our models is within errors. This supposition is supported by the seismically determined value of Z. The seismic Z determinations have been achieved by using techniques that depend on different inputs, and despite the differences in techniques and dependencies on different inputs, all seismic estimates of Z/X are consistent with the higher GS98 abundances, and they agree with each other as well.

If the convection-zone abundances of the Sun are indeed consistent with the low abundances compiled by AGS05, then almost all of the input physics that goes into construction of stellar models must be much more uncertain than has been assumed to be the case. It is also possible that some fundamental process is missing in the theory of stellar structure and evolution, but it is difficult to speculate what that could be. However, if the GS98 abundances are correct then the currently known input physics is consistent with seismic data, and the AGS05 abundances need to be revised upward. It is easier to list reasons why the new abundances could be incorrect. These include the fact that the three-dimensional convection simulation used in the atmospheric models may not have the correct thermal structure (Ayres, Plymate, and Keller, 2006). There may be some effects from having a grid of finite resolution: Scott *et al.* (2006) found significant changes in the line bisectors high in the atmosphere as they changed their resolution. In addition there could be problems with the non-LTE effects used in the line-formation calculations and atomic physics.

In conclusion, disagreements between helioseismic estimates and recent spectroscopic estimates of the solar heavy-element abundance call for continuation of the careful examination of solar atmospheric models and also of models of the solar interior.

3. Internal Rotation and Dynamics

The time-averaged internal-rotation profile revealed by global helioseismology has in many respects been something of a surprise (see Thompson *et al.*, 2003, for an excellent review on the internal rotation). The known pattern of differential rotation at the surface was observed to penetrate the interior, down to the base of the convection zone, but not in the manner expected. The seismic data then revealed the solar tachocline, a narrow region in the stably stratified layer just beneath the base of the convective envelope, which mediates the transition from differential rotation above to a solid-body-like profile in the radiative zone below.

This solid-body-like profile, with its "slow" (*i.e.*, surface-like) rotation rate, was of course the other surprise.

The paradigm for the spin-down of Sun-like stars involves the action of a dynamo in the outer envelope. Magnetic braking slows the rate of rotation in the envelope. The question then arises as to the degree of coupling between the radiative interior and the envelope. Coupling allows the envelope to draw on the large reservoir of angular momentum residing in the core. This has two consequences. First, it will delay the rate at which the envelope is spun down. Second, it will bleed momentum from the core, thereby altering the rotation in the deep interior. The extent to which the core and envelope are coupled therefore plays an important role in the dynamic evolution of a star. In the pre-helioseismic era, the conventional wisdom was that the core and deep radiative interior of the Sun would be expected to rotate much more rapidly than the layers above.

Further to answering questions posed by stellar evolution theory, models and conjectures on the rotation were also of considerable interest for attempts to constrain one of the well-known tests of Einstein's general theory of relativity, the advance in the perihelion of Mercury test. Observations of the surface oblateness of the Sun, by Dicke and Goldenberg (1967), had suggested that the solar gravitational quadrupole moment was large enough to give a significant contribution to the perihelion advance. This created something of a problem, for it reduced the fraction of the contribution that could be set aside as being relativistic in origin to the point where what was left over conflicted with the prediction for the general theory of relativity. Interest in competing theories of relativity was reinvigorated (*e.g.*, the theory of Brans and Dicke (1961) which could be "tuned" into agreement with the oblateness observations).

The result of Dicke and Goldenberg had important implications for the Sun's interior structure, for it suggested a rapidly spinning core might be needed to account for the apparently large surface oblateness. The concept of rapid internal rotation also happened to be in vogue at the time for another reason: It offered a possible way to solve the solar neutrino problem. In the presence of rapid internal rotation, thermal pressure would no longer be required to carry the full burden of support to maintain the star in equilibrium, meaning temperatures in the core could be lower than previously thought. A nice snapshot is provided by the papers of Ulrich (1969), Demarque, Mengel, and Sweigart (1973), and Roxburgh (1974).

What of the outer layers? Pre-helioseismic models of rotation in the convection zone gave a pattern in which the rotation was constant on cylinders wrapped around the rotation axis (*e.g.*, Glatzmaier, 1985; Gilman and Miller, 1986). Small-diameter cylinders intersect the solar surface at high latitudes, whereas larger cylinders do so at low latitudes, in the vicinity of the solar Equator. To match to the surface differential rotation, material lying on the surface of a small cylinder must rotate less rapidly than plasma on a larger cylinder. An important consequence of the rotation models was that they therefore predicted an increase of rotation rate with increasing radius.

3.1. Observed Rotation: The Deep Interior

Helioseismic inference of the rotation of the deep interior demanded estimates of the rotational frequency splittings of the core-penetrating low-ℓ p modes. The first estimates of the rotational frequency splittings (Claverie *et al.*, 1981) suggested the core might indeed be spinning more rapidly than the outer layers, but by nowhere near enough to give the surface oblateness claimed by Dicke and Goldenberg. The first inversion for the internal rotation

(Duvall et al., 1984) showed that the outer parts of the radiative interior were actually rotating at a rate not dissimilar to the surface, a result that all but ruled out the possibility of a significant gravitational quadrupole moment.

The intervening years have seen a steady, downward revision of the magnitudes of the quoted estimates of the low-ℓ rotational frequency splittings (*e.g.*, see discussion in Chaplin, 2004). By the mid 1990s this downward trend had flattened out. Alas, the trend was not solar in origin. It was the result of having longer, higher quality datasets available, coupled to a better understanding of the subtleties and pitfalls involved in extracting the frequency splittings (*e.g.*, Appourchaux et al., 2000a; Chaplin et al., 2001b). In short, estimates from short datasets tend to overestimate the true splittings because there is insufficient resolution in frequency to properly resolve individual components (which is why the initial estimates of Claverie et al. were high). This is surely one of the best examples helioseismology can offer on how accumulation of data from long-term observations (coupled with a better feel for the analysis) can give significant improvement on accuracy of inference on the internal structure.

Inversions made with the modern, high-quality data (*e.g.*, Eff-Darwich, Korzennik, and Jiménez-Reyes, 2002; Couvidat et al., 2003; García et al., 2004) give well-constrained estimates on the rotation down to $r/R_\odot \simeq 0.25$, where the rotation rate is observed to be similar to that in the mid-latitude near-surface layers. We comment later in Section 3.3 on how the solid-body-like rotation might be enforced in the radiative interior.

The conclusion that the quadrupole moment is of insufficient size to give a significant contribution to the perihelion advance of Mercury has been upheld by the modern observations of slow rotation in the interior (*e.g.*, Pijpers, 1998; Roxburgh, 2001). Contemporary measurements of the solar shape – made, for example, with MDI data (Emilio et al., 2007) – show a minute oblateness. The possibility of rapid internal rotation providing a solution to the solar neutrino problem was therefore moot, not only because of the slow rotation, but also because agreement between the sound-speed profiles of solar models and the Sun showed the problem was not one in solar physics (*e.g.*, Bahcall et al., 1997), a result confirmed by observations made by the Sudbury Neutrino Observatory (*e.g.*, Ahmad et al., 2001, 2002; Ahmed et al., 2004).

But what of the rotation rate in the core itself? This remains very uncertain. Use of the p modes presents several difficulties because only a small number of the modes penetrate the core, and those that do have only a modest sensitivity to the rotation. It will be through the measurement of the rotational frequency splittings of g modes that a clear picture of the rotation in the core will properly emerge. Mathur et al. (2007) have demonstrated that by augmenting the p-mode splittings with splittings of a small number of g modes, it should be possible to obtain precise, and reasonably accurate, estimates of the rotation profile throughout a substantial fraction of the core. This is surely sufficient reason alone to redouble our effects to detect individual g modes. We discuss the current status of the observational claims in Section 5.

3.2. Observed Rotation: The Convection Zone and Tachocline

Initial glimpses of the rotation in the near-surface layers were provided by Rhodes, Ulrich, and Deubner (1979). However, it was Brown (1985) who presented the first evidence demonstrating that the surface differential rotation penetrated the convection zone. Further studies, using inversion of the rotational frequency splittings, followed (*e.g.*, Brown and Morrow, 1987; Kosovichev, 1988; Brown et al., 1989; Dziembowski, Goode, and Libbrecht, 1989; Rhodes et al., 1990; Thompson, 1990). By the end of the 1980s, the rotation inversions

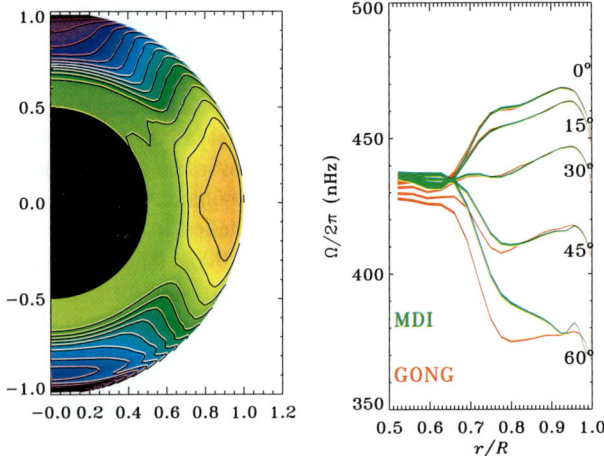

Figure 2 The mean, time-averaged rotation profile in the convection zone and outer parts of the radiative zone. Left: Two-dimensional cutaway, showing the mean profile obtained from GONG (upper half) and MDI (lower half) observations. Right: Mean rotation profile at different latitudes. (Courtesy of R. Howe.)

were able to show that the differential rotation underwent a marked transition at the base of the convection zone to something resembling a solid-body-like profile below (Christensen-Dalsgaard and Schou, 1988). The tachocline (Spiegel and Zahn, 1992) had been discovered.

Analysis with the more extensive modern data (*e.g.*, Basu and Antia, 2003) indicates that the characteristic thickness of the tachocline is only a few percent of the solar-radius. The tachocline is oblate and is slightly thicker at the solar Equator. The steep gradient in rotation present across the tachocline – much stronger than anything present elsewhere in the outer layers – means it is of considerable interest to the dynamo modelers and is an attractive site in which to locate stretching, and winding-up, of magnetic field (poloidal to toroidal) by the Ω effect (*e.g.*, see Tobias, 2002).

What of the rotation in the convection zone itself? The pattern revealed by analysis of the modern data (Figure 2) does not match the rotation-on-cylinders prediction. Rather, the rotation is approximately constant on lines inclined some $27°$ to the rotation axis (*e.g.*, Gilman and Howe, 2003). Furthermore, the rotation rate decreases with radius in the low- to mid-latitude layers very close to the surface (*e.g.*, Corbard and Thompson, 2002). Although there is a general consensus that the differential rotation in the convection zone is driven by thermal perturbations, the challenge remains to understand in detail the mean observed profile (*e.g.*, see Rempel, 2005; Miesch, Brun, and Toomre, 2006).

Some of the most striking results of helioseismology have come from the detection of small, but significant, temporal variations of the rotation rate in the convection zone, which carry signatures of the solar activity cycle. We shall discuss these variations in Section 4.

3.3. How Is the Tachocline Confined and Solid-Body Rotation Enforced?

The existence of the tachocline has raised several fundamental questions regarding the dynamic evolution of the Sun. This thin layer matches the transition in rotational behavior above and below, and it must mediate or act as the intermediary for the transfer of angular momentum from the immense reservoir in the core to the outer envelope and beyond, as the star evolves. For the rotation to change its character, something must be acting in the radiative interior to mix angular momentum in latitude, so that the differential rotation from the convective zone above is smoothed out, or removed, below. The mechanism must be anisotropic, in the sense that it must be much less efficient at mixing angular momentum in the radial – as opposed to the latitudinal – direction to explain the narrow width of the tachocline. Mixing by anisotropic turbulence is one possibility. Another possibility is the effect of a fossil magnetic field threading the radiative interior. Gough and McIntyre (1998) considered circulations that penetrate from the convection zone into the tachocline, which are then diverted by a weak magnetic field further down. The magnetic field acts to prevent the tachocline spreading out in radius, in effect forming a firm lower boundary. The pattern of rotation at low and high latitudes acts to keep field "bottled up" in the radiative interior. This magnetic field need only have a strength that is a tiny fraction of that at the surface to give the required effect. Moreover, the magnetic field is then also a prime candidate for enforcing the solid-body-like rotation that is present in the radiative zone (see also Eggenberger, Maeder, and Meynet, 2005). Brun and Zahn (2006) have also recently looked at the problem of confinement of the tachocline by a magnetic field. However, their results imply that a fossil field in the radiative interior cannot prevent the radial spread of the tachocline and that, furthermore, it also cannot prevent penetration of the differential rotation from the convection zone into the radiative interior.

Another mechanism that has received attention as a possible means to enforce solid-body rotation in the deep interior is angular momentum transport by internal gravity (buoyancy) waves, which are excited at the base of the convection zone. It has recently been demonstrated (Charbonnel and Talon, 2005) that such models can in principle redistribute angular momentum efficiently over time from the core to the outer envelope. In the presence of shear turbulence, the gravity waves also give rise to shear-layer oscillations that resemble the "quasi-biennial" oscillations observed in the Earth's atmosphere (Talon, 2006).

4. The Changing Sun

A rich and diverse body of observational data is now available on temporal variations of the properties of the global p modes. The signatures of these variations are correlated strongly with the well-known 11-year cycle of surface activity, and as such the accepted paradigm is that the "seismic" solar cycle is associated with changes taking place in the outer layers, not the deep radiative interior, of the Sun.

Evolutionary changes to the equilibrium structure of the Sun will of course also leave their imprint on the p modes, by virtue of a very slow adjustment of the interior structure as the star ages sedately on the main sequence. The frequencies of the low-ℓ p modes are predicted by the standard solar models to decrease by ≈ 1 µHz every 6×10^6 years owing to the evolutionary effects. If we alter the time scale to something more practical from an observer's point of view – say ten years – the evolutionary change is reduced to only $\approx 10^{-6}$ µHz. Measurement of such tiny frequency changes, against the backdrop of variations resulting from the solar cycle and instrumental noise properties, is beyond the current scope of the data. The observed variations of the p modes, owing to changes in the outer layers, are some five orders of magnitude larger than the predicted evolutionary variations.

The search for temporal variations of the properties of the p modes began in the early 1980s, following accumulation of several years of global seismic data. The first positive result was uncovered by Woodard and Noyes (1985), who found evidence in observations by ACRIM for a systematic decrease of the frequencies of low-ℓ p modes between 1980 and 1984. The first year coincided with high levels of global surface activity, whereas during the latter period activity levels were much lower. The modes appeared to be responding to the Sun's 11-year cycle of magnetic activity. The uncovered shifts were, on average, about 0.4 µHz. This meant that the frequencies of the most prominent modes had decreased by roughly 1 part in 10 000 between the activity maximum and minimum of the cycle. By the

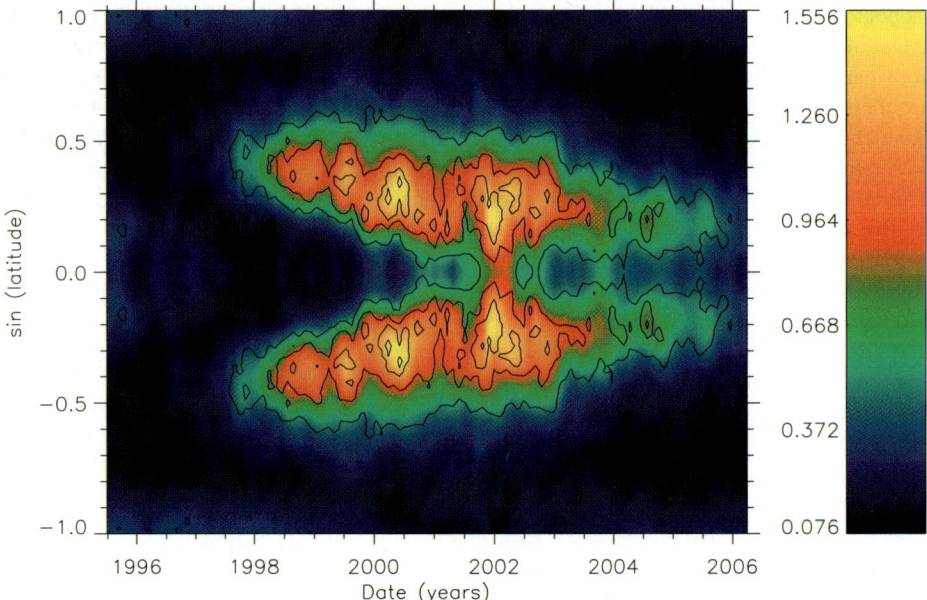

Figure 3 Mode frequency shifts (in µHz) as a function of time and latitude. The values come from analysis of GONG data. The contour lines indicate the surface magnetic activity. (Courtesy of R. Howe.)

late 1980s, an in-depth study of frequency variations of global p modes, observed in the Big Bear data, had demonstrated that the agent of change was confined to the outer layers of the interior (Libbrecht and Woodward, 1990).

The passage of time and accumulation of data from the new networks and instruments has allowed us to study the frequency variations in unprecedented detail and to uncover signatures of subtle, structural change in the subsurface layers. It has led to the discovery of solar-cycle variations in the mode parameters associated with the excitation and damping (*e.g.*, power, damping rate, and peak asymmetry). Patterns of flow that penetrate a substantial fraction of the convection zone have been uncovered as well as possibly (but controversially) signatures of changes in the rotation rate of the layers that straddle the tachocline. Let us say a little more about these observations and what they might mean for our understanding of the solar variability.

4.1. Structural Changes

The modern seismic data give unprecedented precision on measurements of the p-mode frequency shifts. From observations of the medium-ℓ frequency shifts – for example in GONG and MDI data – it is possible to produce surface maps (Figure 3) showing the strength of the solar-cycle shifts as a function of latitude and time (Howe, Komm, and Hill, 2002). These maps bear a striking resemblance to the butterfly diagrams that show variations in the strength of the surface magnetic field over time. The implication is that the frequency shift of a given mode depends on the strength of that component of the surface magnetic field that has the same spherical harmonic projection on the surface. This dependence is also observed in studies of the frequency shifts of the less numerous low-ℓ modes (see Chaplin, 2004, and references therein). The precision in the medium-ℓ data is such that significant frequency changes can now be tracked on time scales as short as nine days (see Tripathy *et al.*, 2007). Meanwhile, current results on frequency shifts of high-ℓ modes ($\ell > 100$) – extracted by using global helioseismology techniques (Rabello-Soares, Korzennik, and Schou, 2008) – show trends that match those of the medium-ℓ and low-ℓ shifts (*e.g.*, Chaplin *et al.*, 2001a). In particular, when the frequency shifts are multiplied by the mode inertia, and then normalized by the inertia of a radial mode of the same frequency, the modified shifts are found to be a function of frequency alone. The high-ℓ modes provide important information since they are confined in the layers close to the surface where the physical changes responsible for the frequency shifts are also located.

Detailed comparison of the low-ℓ frequency shifts with changes in various disk-averaged proxies of global surface activity provides further tangible input to the solar-cycle studies. This is because different proxies show differing sensitivity to various components of the surface activity. Chaplin *et al.* (2007b) recently compared frequency changes in 30 years of BiSON data with variations in six well-known activity proxies. Interestingly, they found that only activity proxies having good sensitivity to the effects of weak-component magnetic flux – which is more widely distributed in latitude than the strong flux in the active regions – were able to follow the frequency shifts consistently over the three cycles.

What is the physical mechanism behind the frequency shifts? Broadly speaking, the magnetic fields can affect the modes in two ways. They can do so directly, by the action of the Lorentz force on the plasma. This provides an additional restoring force, the result being an increase of frequency, and the appearance of new modes. Magnetic fields can also influence matters indirectly, by affecting the physical properties in the mode cavities and, as a result, the propagation of the acoustic waves within them. This indirect effect can act both ways, to either increase or decrease the frequencies. The exact nature of the physical changes is still somewhat controversial, although Dziembowski and Goode (2005) have recently made important headway on the problem. Their analysis of MDI p-mode frequency shifts suggests that indirect effects dominate, in particular changes to the near-surface stratification resulting from the suppression of convection by the magnetic field. They suggest that the magnetic fields are too weak in the near-surface layers where the p-mode shifts originate for the direct effect to contribute significantly.

It is also interesting to note that Dziembowski and Goode found small, but significant, departures for the lower-frequency p modes of a simple scaling of the frequency shifts with the inverse mode inertia. The nature of these small departures suggests that there is a contribution to the low-frequency shifts from deeper layers, owing to the direct effect of the magnetic fields. Similar departures in behavior had also been seen and noted by Chaplin *et al.* (2001a).

Variations of global f-mode frequencies reveal information on changes in a thin layer that extends some 15 Mm below the base of the photosphere. The most recent observations (*e.g.*, Lefebvre and Kosovichev, 2005) suggest that as activity rises there is an expansion between $r/R_\odot \sim 0.97$ and 0.99, and possibly a contraction above $r/R_\odot \approx 0.99$. It is, however, not yet possible to reconcile the observations with theoretical predictions of the variations (see Lefebvre, Kosovichev, and Rozelot, 2007, and also Sofia *et al.*, 2005).

Variations very close to the surface, in the He II ionization zone at a depth $\approx 0.98\,r/R_\odot$, have also been revealed by analysis of medium-ℓ p modes. From appropriate combinations of mode frequencies, Basu and Mandel (2004) uncovered apparent solar-cycle variations in the amplitude of the depression in the adiabatic index (Γ_1) in the He II zone. These variations presumably reflect the impact of the changing activity on the equation of state of the gas in the layer. These results have since been confirmed, by using a different method to extract the acoustic signatures of the He II zone, and with only low-ℓ frequencies (Verner, Chaplin, and Elsworth, 2006).

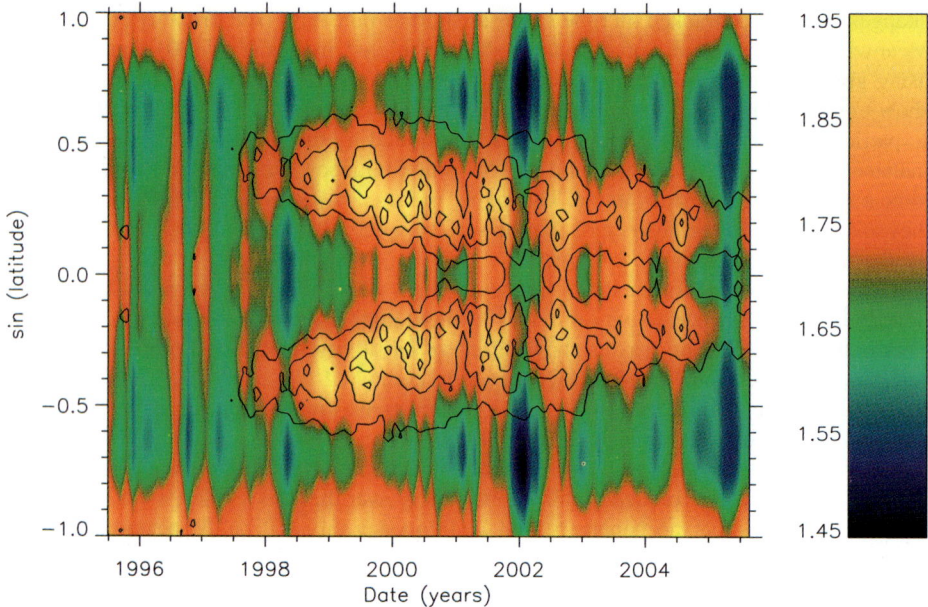

Figure 4 Mode line width (in μHz) as a function of time and latitude. The values come from analysis of GONG data. The contour lines indicate the surface magnetic activity. (Courtesy of R. Howe.)

The results discussed so far pertain to changes taking place very close to the surface. What of possible changes deeper down? Chou and Serebryanskiy (2005) and Serebryanskiy and Chou (2005) have found intriguing evidence of signatures in the p-mode frequency shifts that may reflect changes taking place near the base of the convection zone. The authors suggest that the signatures they uncover are consistent with a fractional perturbation to the sound speed, at depth 0.65 to $0.67\,r/R_\odot$, of size a few parts in 10^5 (assuming the perturbation may be described as a Gaussian with a FWHM of $0.05\,r/R_\odot$ in radius).

Surface maps, such as the frequency-shift map shown in Figure 3, may also be made for variations observed in the mode powers and damping rates (Komm, Howe, and Hill, 2002), which, like the frequency maps, show a close spatial and temporal correspondence with the evolution of active-region field (Figure 4). Meanwhile, peak asymmetry is the most recent addition to the list of parameters that show solar-cycle variations (Jiménez-Reyes *et al.*, 2007). Careful measurement of variations in the powers, damping rates, and peak asymmetries – all parameters associated with the excitation and damping – allows studies to be made of the impact of the solar cycle on the convection properties in the near-surface layers.

4.2. Torsional Oscillations

One of the most striking results from helioseismology has been the discovery that the so-called torsional oscillations – which modulate the observed pattern of surface differential rotation – penetrate a substantial fraction of the convection zone. The surface torsional oscillations were first observed by Howard and La Bonte (1980). The observations showed bands of plasma at particular latitudes rotating either slightly faster or slower (by a few percent) than the level expected from the smooth, underlying differential rotation. Moreover,

Figure 5 Variations in the solar internal rotation, relative to the rotation at the epoch of solar minimum, as determined by analysis of MDI observations. (Courtesy of S. Vorontsov.)

the bands shifted position as the solar cycle progressed, tracking toward the Equator on a time scale that suggested they carried the signature of the effects of the cycle.

Modern, high-quality seismic observations (Figure 5) reveal that these bands of flow are present within the convection zone (Howe *et al.*, 2000a; Antia and Basu, 2000, 2001; Vorontsov *et al.*, 2002). Furthermore, an additional strong, poleward branch has been revealed; this branch appears to penetrate the entire convection zone. The amplitudes and phases of the signals show a systematic variation with position in the convection zone (Howe *et al.*, 2005). The observed behavior of the flows is strongly suggestive of being a signature of magnetic effects from the solar cycle.

An obvious candidate is the back-reaction of the magnetic field on the solar plasma, via the Lorentz force. Lorentz force feedback was proposed originally by Schüssler (1981) and Yoshimura (1981) as a means to explain the surface torsional oscillations. Later models looked at the effect of the Lorentz force from small-scale magnetic field on the turbulent Reynolds stresses (*e.g.*, Küker, Rüdiger, and Pipin, 1996). A thermal mechanism was also proposed by Spruit (2003) to explain the low-latitude branch, having its origin in small gradients of temperature caused by the magnetic field. Incorporation of the Lorentz force feedback into dynamo models (which are then termed "dynamic") can reproduce observed features of the torsional oscillations (*e.g.*, Covas, Moss, and Tavakol, 2005). Rempel (2007) has recently forced torsional oscillations in a mean-field differential rotation model, which includes the effect of the Lorentz force feedback in meridional planes. He found that although the poleward-propagating high-latitude branch could be explained by Lorentz-force feedback, or thermal driving, the low-latitude branch is most likely not due to the Lorentz feedback and probably has a thermal origin.

Howe *et al.* (2006) conducted experiments using artificial data containing migrating flows like those seen in the real observations. Analysis of these artificial data suggests inferences made on the depth of penetration and the amplitude and phase of the solar torsional oscillations are likely to be real, and not artifacts of the analysis. With the collection of more data, coupled to a better understanding of the subsurface torsional oscillations, it should be possible to constrain the perturbations driving the flows (*e.g.*, Lanza, 2007). This might lead to the possibility of obtaining indirect measures of the strength of the magnetic field with depth in the convection zone (since direct measurement of the field is not yet possible).

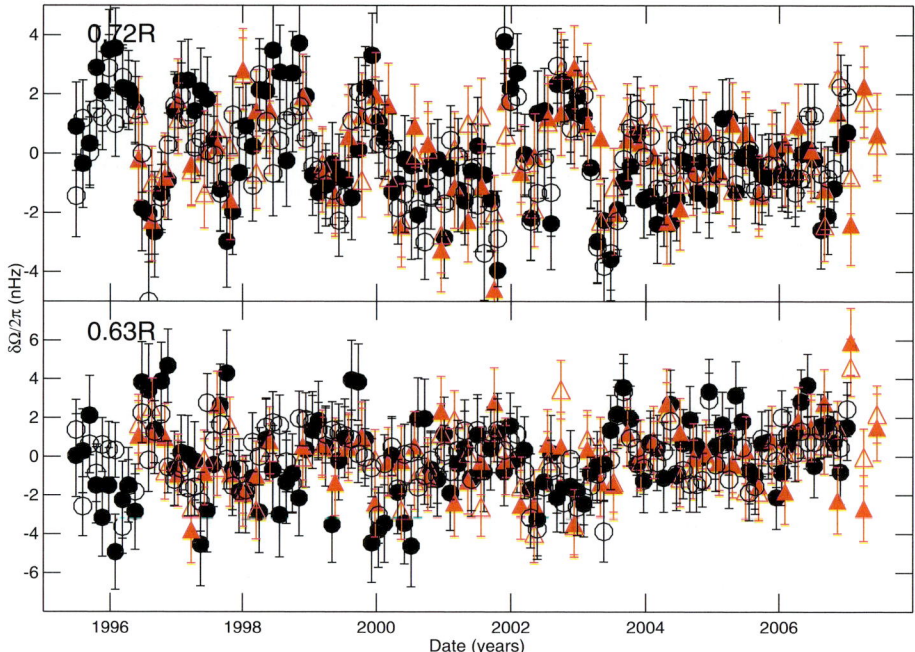

Figure 6 The temporal variation of the rotation rate in the equatorial regions near the base of the convection zone at $0.72 r/R_\odot$ (top) and at $0.63 r/R_\odot$ (bottom) from GONG (circles) and MDI data (triangles). The average rotation rate has been subtracted. (Courtesy of R. Howe.)

4.3. The 1.3-Year Periodicities Near the Tachocline

Claims that the rotation rate in the layers just above and below the tachocline varies on a time scale of ≈ 1.3 years (Howe *et al.*, 2000b; see Figure 6) remain controversial. When they were first uncovered by the analysis, the changes appeared to be most prominent in the low-latitude regions just above the base of the convection zone. At the same time there were suggestions of variations in antiphase some $\approx 60\,000$ km deeper down, in the outer parts of the radiative zone. The variations uncovered by the analysis then all but disappeared when mid-latitude regions were tested, but a periodic-looking signal of period closer to one year was found when attention was focused at a latitude of $60°$.

The result is controversial principally for two reasons. First, independent analyses have failed to reveal the same quasi-periodic variation of the rotation rate (Antia and Basu, 2001, 2004; Corbard *et al.*, 2001). Second, the quasi-periodic signal appears to have all but disappeared in more recent data, collected from 2004 onward, having also been absent over the period from ≈ 2000 to ≈ 2002 (Howe, 2006). The intermittency need not necessarily imply the phenomenon is an artifact, and the claims continue to draw considerable interest. The fact that the signals uncovered above and below the tachocline are in antiphase – meaning as one region speeds up the other slows down – suggests that, if real, they may be signatures of some form of angular momentum exchange between the interior and envelope, mediated by the tachocline. Intriguingly, there are also reports in the literature of quasi-1.3-year periodicities in observations of sunspots and geomagnetic indices (see Howe, 2006, and references therein).

5. The Observer's Holy Grail: *g* Modes and Very Low Frequency *p* Modes

As the title of this section suggests, the drive to detect the gravity (*g*) modes (and also the very low frequency *p* modes) has assumed an added significance as time has passed. Detection of the *g* modes presents a major observational challenge, because the amplitudes of the modes are predicted to be extremely weak at the photospheric level. Early claims of detections of low-ℓ *g* modes (*e.g.*, Delache and Scherrer, 1983; Fröhlich and Delache, 1984) were to prove unfounded, and it has become increasingly apparent that unambiguous detection of the modes will demand very long datasets, excellent instrumental noise performance at low frequencies, and ingenuity in both the observations (*e.g.*, possibly from new approaches) and the analysis.

Upper limits on the amplitudes of individual *g* modes – as set from analysis of the long, high-quality modern datasets (*e.g.*, Appourchaux *et al.*, 2000b; Gabriel *et al.*, 2002; Wachter *et al.*, 2003; Turck-Chieze *et al.*, 2004a) – are far superior (*i.e.*, much lower) than those set in the early analyses (where the datasets were much shorter, and usually the quality of the data was inferior to contemporary levels). For some methods of analysis the limits are approaching the level of 1 mm s^{-1} per mode. Whereas predictions of the *g*-mode amplitudes, based on the assumption of stochastic excitation in the convection zone (*e.g.*, Gough, 1985; Andersen, 1996; Kumar, Quataert, and Bahcall, 1996), may be rather uncertain – predictions for the same range in frequency can differ by more than an order of magnitude – it is worth noting that some of the predictions are not too dissimilar from the current best observational upper limits (*e.g.*, see Elsworth *et al.*, 2006).

Early attempts to detect low-ℓ *g* modes concentrated largely on the very low frequency asymptotic regime (*e.g.*, see Provost and Berthomieu, 1986), where the near-constant spacing in period offered potential advantages for detection algorithms. However, amplitudes are expected to be appreciably larger in the higher frequency part of the *g*-mode spectrum. This is why the lack of any convincing detections lower down shifted the focus of attempts in the late 1990s toward the higher frequency (nonasymptotic) range (*e.g.*, Appourchaux, 2003), where modes with mixed *g* and *p* characteristics are also expected. Searches were made, unsuccessfully, for signatures of individual *g* modes (although a few potential candidates were claimed by Turck-Chieze *et al.*, 2004a).

Now, things have gone full circle. Searches in the very low frequency asymptotic regime are back in fashion. As previously noted, analysis methods can take advantage of the near-regular spacing in period and increase the effective signal-to-noise ratio by looking for the cumulative effect of several *g*-mode overtones. Searches by García *et al.* (2007) for the cumulative signature of $\ell = 1$ *g* modes in almost ten years of GOLF data have yielded what may be regarded as the first serious claim of a detection. The analysis technique has the advantage that it is both elegant and simple. The low-frequency part of the frequency power spectrum is first prewhitened. The prewhitened part is then presented in the form of power spectral density against period, and its periodogram is computed. The cumulative signature of the $\ell = 1$ overtones is manifested as a peak in the periodogram, at a period corresponding to the period spacing between the overtones.

García *et al.* claim a statistically significant peak in the GOLF periodogram, that lies at roughly the period predicted by the standard solar models. Is this really the signature of $\ell = 1$ *g* modes? Experiments performed by the authors with artificial data suggest that the individual mode peaks must have widths in the frequency power spectrum that are commensurate with damping times of several months. Excitation of gravity waves at the base of the convection zone might lead to the comparatively heavy damping implied by these values (Dintrans *et al.*, 2005), although the theoretical work is not yet sufficiently developed to enable accurate damping-rate predictions to be made for the global low-ℓ *g* modes.

The validity of conclusions drawn on all searches for low-frequency modes rests on a robust and proper use of statistics. Some methods assess the likelihood that prominent features are part of a broadband noise source, the so-called H0 hypothesis; others test against the likelihood that signal (*e.g.*, a sine wave or a damped wave) is buried in broadband noise, the so-called H1 hypothesis (see Appourchaux, 2003; Chaplin, 2004, for brief discussions on low-frequency detection methods). Some of the tests bring more prior information to bear than others: For example, tests are often predicated on the assumption that the g modes are very lightly damped, like the low-frequency p modes, meaning individual components will appear as spikes in the frequency power spectrum. It is also important to recognize that quoted likelihoods depend on the question being asked of the data. For example, does one flag a prominent spike as a candidate mode if it has less than a 1% likelihood of appearing by chance in a range of 10 µHz of the spectrum? Or does one perhaps demand that it has less than a 1% chance of appearing in a range of 100 µHz of the spectrum? Quoting a 1% likelihood in these two cases means two different things (the second criterion being a more demanding limit). And should one fold in the fact that the number of bins in a fixed range of frequency increases as more data are collected on the Sun?

The use of prior information, and the need to fix *a priori* choices for hypothesis testing (which has an element of subjectivity), suggests that a Bayesian (and not a frequentist) approach is the best route to assessing the likelihoods associated with searches for the g modes. This is the approach currently advocated by the Phoebus Collaboration, which is leading the way on development of analysis techniques in this area (Appourchaux, 2008).

The application of low-frequency detection algorithms is now yielding multiple detections on p modes at frequencies below 1000 µHz, though at $\ell \geq 4$ (*e.g.*, see Salabert, Leibacher, and Appourchaux, 2008). The lowest frequency detection at $\ell \leq 3$ – which has received independent confirmation by different analyses of more than one dataset – is the $\ell = 0$, $n = 6$ overtone at ≈ 973 µHz (*e.g.*, see García *et al.*, 2001; Broomhall *et al.*, 2007).

6. Driving and Damping the Modes

The first calculations of the excitation rates of the p modes suggested the modes were unstable (*e.g.*, see the discussion in Christensen-Dalsgaard, 2004). We now know they are in fact stable, being stochastically excited and intrinsically damped by the convection (see Houdek, 2006, for a recent review). Theoretical modeling of global excitation and damping based on an analytical (or semianalytical) approach (*e.g.*, Houdek *et al.*, 1999; Dupret *et al.*, 2005; Samadi *et al.*, 2005) can give predictions of two independent quantities: the damping rates (η) and acoustic powers (P) (the latter corresponding to the rates at which energy is pumped into, and then dissipated by, the modes). It is therefore incumbent on the observers to provide as accurate and precise measures of these parameters as possible.

The parameters that are usually extracted directly by the observers are the widths (Δ) and heights (maximum power spectral densities) (H) of the peaks in the frequency power spectrum. The linear damping constants (η) are related to the peak widths via

$$\Delta = \eta/\pi. \qquad (2)$$

If observations are of sufficient length to resolve mode peaks in the frequency power spectrum, the observed heights of the mode peaks are given by

$$H = \frac{2V^2}{\pi \Delta} = \frac{P}{\eta^2 I}, \qquad (3)$$

where the V are mode amplitudes (written here for Doppler velocity) and the I are the mode inertias. There has recently been a shift toward making comparisons of observational and theoretical mode amplitudes by using H (e.g., Chaplin et al., 2005; Belkacem et al., 2006a), as opposed to V (or V^2) as had previously been the practice. The height H is after all what one "sees" for the vast majority of modes in the frequency power spectrum (i.e., provided they are well resolved).

Measurement of the acoustic powers (P) from the peak-bagging estimates of H and Δ is fraught with potential pitfalls. Equations (2) and (3) imply that

$$P = \pi^2 I H \Delta^2. \tag{4}$$

There is a strong anticorrelation of the fitted H and Δ in fits made to peaks in the frequency power spectrum. The effect cancels when estimates of V^2 are sought, since $V^2 \propto H \Delta$; but estimates of P contain another factor of Δ. Some other means of estimating the damping, that is, one much less strongly correlated with H, would offer a way around the problem.

The appearance of the mode inertia (I) in Equation (4) gives rise to further complications. Because different instruments show different Doppler velocity responses with height in the photosphere, the I are instrument (i.e., observation) dependent. Baudin et al. (2005) have demonstrated the importance of attempting to correct for this effect. Without proper normalization between results of different instruments, differences in estimates of P arise.

The frequency dependence of the acoustic powers P is a particularly important diagnostic. The most recent comparisons of theory and observation have shown good agreement over the main part of the low-ℓ mode spectrum (e.g., Chaplin et al., 2005; Belkacem et al., 2006a, 2006b). Samadi et al. (2007) have also found that absolute predictions of P by three-dimensional numerical simulations tend to be lower than the P given by the semi-analytical models. With regard to the damping, the comparison by Chaplin et al. (2005) of the observed and theoretical H of the radial modes has demonstrated much more clearly than before the shortcomings in the theoretical computations of the low-frequency damping rates.

An interesting question for the analytical models concerns the description of the temporal behavior of the dynamics of the small-scale turbulence. A Gaussian or Lorentzian function is usually adopted. Chaplin et al. (2005) found that when they adopted a Lorentzian description the predicted H of the low-frequency modes severely overestimated observed values. They concluded that a Gaussian description gave a much better match to the observations. The results of Samadi et al. (2003, 2007) in contrast tend to favor the Lorentzian description. Results from numerical simulations indicate that neither description is strictly correct (Georgobiani, Stein, and Nordlund, 2006): Variation in the behavior of the small-scale turbulence is observed with both temporal frequency and depth in the convection zone.

The two sources of excitation are the fluctuating turbulent pressure (Reynolds stresses) and gas pressure. In the numerical simulations (e.g., Stein et al., 2004) there is some cancellation between the sources. This cancellation is not shown by the analytical models, which may explain why the analytical models overestimate the p-mode amplitudes of stars hotter than the Sun (see discussion in Houdek, 2006).

7. The Road Ahead

Global helioseismic studies continue to make great progress, but many outstanding questions and challenges remain. If there is one thing that 30 years of global helioseismology has taught us, it is that there are two highly desirable requirements on the observational data:

that they should provide long-term, high-duty-cycle monitoring of modes from low to high ℓ and that they should offer observational redundancy and complementary data.

The first requirement enables us to use the global p modes to "sound" the solar activity cycle (*i.e.*, to monitor in detail the temporal behavior of the seismic Sun) on time scales commensurate with the cycle period, and preferably longer to facilitate comparisons of one cycle with another. It also enables us to obtain extremely precise and accurate estimates of fundamental mode parameters, since long datasets are vital for detection of the very low frequency modes, and to measure other parameters that would not otherwise be determined robustly (*e.g.*, multiplet frequency asymmetries in low-ℓ modes). We are therefore in a position to be able to make precise and accurate inference on the internal structure and dynamics (both the time-averaged properties and the properties as a function of time).

The second requirement enables us to confirm the solar origins of subtle, but potentially important, phenomena in the data. For example, detection of weak very low frequency modes in two or more contemporaneous datasets significantly lowers the probability of a false detection having been made. Complementary Doppler velocity and intensity observations, and observations by Doppler velocity instruments in different atmospheric absorption lines, create opportunities for studies of the physics of the photosphere, studies that can in turn be used to obtain more accurate estimates of mode frequencies from better understanding and modeling of the peak asymmetry.

To fully exploit the potential science benefits that global helioseismology has to offer, we need continuation of operations of the two main ground-based networks, GONG and BiSON. As new science results drive the need for different data products, the ground-based networks are in a position to implement "responsive" changes to their instrumentation. (We shall return to the issue of new observational requirements later.) In the post-SOHO era, BiSON will continue to provide, and will then be the only dedicated source of, high-quality low-ℓ data from its Sun-as-a-star observations. Continuation of GONG, in its current multi-site configuration, would provide high-quality, high-duty-cycle resolved-Sun products, particularly on higher-ℓ modes, to go alongside the HMI resolved-Sun data (due for launch on SDO in early 2009).

It is important to remember that to make optimal use of the low-ℓ modes for probing the solar core we need contemporaneous medium- and high-ℓ data It is worth stressing the important role the high-ℓ modes can play in this regard, in that they can be used to constrain the hard-to-model near-surface layers, thereby cleaning things up for more accurate inference on the structure deeper down. However, reliable measurement of the high-ℓ frequencies presents something of a challenge, because of the sensitivity of the frequencies to instrumental effects (*e.g.*, see Korzennik, Rabello-Soares, and Schou, 2004; Rabello-Soares, Korzennik, and Schou, 2008).

To further improve the accuracy of the inversions we must continue studies into optimizing combinations of frequencies from different instruments (*e.g.*, the low-ℓ Sun-as-a-star BiSON and GOLF data with the resolved-Sun MDI and GONG data and the future HMI data). As datasets get longer, and quality improves, so new subtle effects come to light that must be properly allowed for when the datasets are analyzed. In the past few years we have developed a much better understanding of the underlying frequency bias between resolved-Sun and Sun-as-a-star frequencies. But more work is needed. New instrument combinations inevitably present their own unique problems.

Bias comes not only from instrumental effects but also from the analysis pipelines (*e.g.*, see Schou *et al.*, 2002; Basu *et al.*, 2003). Hare-and-hounds exercises on realistic artificial data are a valuable tool for uncovering and understanding such effects. The solarFLAG group is currently concluding a second round of hare-and-hounds exercises testing peak-bagging on low-ℓ modes in Sun-as-a-star data (see Chaplin *et al.*, 2006, for results on the

Scott, P.C., Asplund, M., Grevesse, N., Sauval, A.J.: 2006, *Astron. Astrophys.* **456**, 675.
Serebryanskiy, A., Chou, D.-Yi.: 2005, *Astrophys. J.* **633**, 1187.
Sofia, S., Basu, S., Demarque, P., Li, L., Thuillier, G.: 2005, *Astrophys. J.* **632**, L147.
Spiegel, E.A., Zahn, J.-P.: 1992, *Astron. Astrophys.* **265**, 106.
Spruit, H.C.: 2003, *Solar Phys.* **213**, 1.
Stein, R.F., Georgobiani, D., Trampedach, R., Ludwig, H.-G., Nordlund, Å.: 2004, *Solar Phys.* **220**, 229.
Talon, S.: 2006, In: Dansey, D., Thompson, M.J. (eds.) *Beyond the Spherical Sun, SOHO18/GONG 2006/HELAS I*, **SP-624**, ESA, Noordwijk, 73.1.
Thompson, M.J.: 1990, *Solar Phys.* **125**, 1.
Thompson, M.J., Christensen-Dalsgaard, J., Miesch, M.S., Toomre, J.: 2003, *Ann. Rev. Astron. Astrophys.* **41**, 599.
Tobias, S.M.: 2002, *Roy. Soc. Lond. Phil. Trans. A.* **360**, 2741.
Tripathy, S.C., Hill, F., Jain, K., Leibacher, J.W.: 2007, *Solar Phys.* **243**, 105.
Turck-Chieze, S., *et al.*: 2004a, *Astrophys. J.* **604**, 455.
Turck-Chieze, S., Couvidat, S., Piau, L., Ferguson, J., Lambert, P., Ballot, J., García, R.A., Nghiem, P.: 2004b, *Phys. Rev. Lett.* **93**, 211102.
Ulrich, R.K.: 1969, *Astrophys. J.* **158**, 427.
Ulrich, R.K.: 1970, *Astrophys. J.* **162**, 993.
Ulrich, R.K., Rhodes, E.J. Jr.: 1977, *Astrophys. J.* **218**, 581.
Verner, G.A., Chaplin, W.J., Elsworth, Y.: 2006, *Mon. Not. Roy. Astron. Soc.* **640**, L95.
Vorontsov, S.V., Christensen-Dalsgaard, J., Schou, J., Strakhov, V.N., Thompson, M.J.: 2002, *Science* **296**, 101.
Wachter, R., Schou, J., Kosovichev, A.G., Scherrer, P.H.: 2003, *Astrophys. J.* **588**, 1199.
Woodard, M.F., Noyes, R.W.: 1985, *Nature* **318**, 449.
Yoshimura, H.: 1981, *Astrophys. J.* **247**, 1102.
Young, P.R.: 2005, *Astron. Astrophys.* **444**, L45.
Young, P.A., Arnett, D.: 2005, *Astrophys. J.* **618**, 908.

Meridional Circulation and Global Solar Oscillations

M. Roth · M. Stix

Originally published in the journal Solar Physics, Volume 251, Nos 1–2, 77–89.
DOI: 10.1007/s11207-008-9232-6 © The Author(s) 2008

Abstract We investigate the influence of large-scale meridional circulation on solar p modes by quasi-degenerate perturbation theory, as proposed by Lavely and Ritzwoller (*Roy. Soc. Lond. Phil. Trans. Ser. A* **339**, 431, 1992). As an input flow we use various models of stationary meridional circulation obeying the continuity equation. This flow perturbs the eigenmodes of an equilibrium model of the Sun. We derive the signatures of the meridional circulation in the frequency multiplets of solar p modes. In most cases the meridional circulation leads to negative average frequency shifts of the multiplets. Further possibly observable effects are briefly discussed.

Keywords Helioseismology, direct modeling · Interior, convection zone · Oscillations, solar · Velocity fields, interior · Waves, acoustic, modes

1. Introduction

Meridional circulation is a large-scale flow observed on both hemispheres of the solar surface (Duvall, 1979; Hathaway, 1996; Komm, Howard, and Harvey, 1993). Its predominant direction is from the Equator to the Poles, and its amplitude is of the order of $15 \, \mathrm{m \, s^{-1}}$. As mass does not accumulate in the polar regions, a return flow from the Poles to the Equator is suspected deeper within the solar interior. The top half of the convection zone contains approximately 0.25% of the solar mass, the mass of the bottom half is approximately five times larger. Consequently, a poleward flow of $10 \, \mathrm{m \, s^{-1}}$ in the top half of the convection zone

---

Helioseismology, Asteroseismology, and MHD Connections
Guest Editors: Laurent Gizon and Paul Cally.

M. Roth (✉)
Max-Planck-Institut für Sonnensystemforschung, Katlenburg-Lindau, Germany
e-mail: roth@mps.mpg.de

M. Stix
Kiepenheuer-Institut für Sonnenphysik, Freiburg, Germany
e-mail: stix@kis.uni-freiburg.de

could be compensated by an equatorward flow of $2\,\mathrm{m\,s^{-1}}$ in the lower half. The transport of magnetic flux from mid to low latitudes by such a flow at the bottom of the convection zone would last approximately ten years, which is close to the period of the solar magnetic cycle.

In addition to magnetic flux, the meridional flow also transports angular momentum. Indeed, the circulation played a key role in early theories of the solar non-uniform rotation as well as of the magnetic cycle (Bjerknes, 1926; Kippenhahn, 1963). More recently, differential rotation has been explained in mean-field models as a consequence of the Reynolds stresses (Rüdiger, 1980; Küker and Stix, 2001; Küker and Rüdiger, 2005), and in three-dimensional numerical models by the influence of the Coriolis force on global convection (Miesch et al., 2000; Miesch, Brun, and Toomre, 2006). Nevertheless, meridional circulation occurs as well in these models, and in solar-cycle models it has regained popularity, since the mean-field latitude migration along the surfaces of isorotation that occurs in the traditional $\alpha\Omega$ dynamo (Parker, 1955) does not seem to suffice. The effect of the circulation on the butterfly diagram had been demonstrated by Roberts and Stix (1972); it is considered to be essential in more recent versions of the $\alpha\Omega$ dynamo, which therefore have been termed "flux-transport dynamos" (Choudhuri, Schüssler, and Dikpati, 1995; Dikpati and Charbonneau, 1999; Nandy and Choudhuri, 2002; Rempel, 2006a, 2006b).

Local helioseismology has investigated the strength of the meridional flow in the solar interior. By means of ring-diagram analysis (Hill, 1988), Haber et al. (2002) inverted data for the circulation in a 15 Mm deep region below the solar surface. In data from 1998 they found a flow emerging at high northern latitudes with equatorward orientation. This was interpreted as an evolving second cell of circulation. Further studies on the evolution of the flow either by ring-diagrams and by time-distance helioseismology (e.g., Zhao and Kosovichev, 2004; Zaatri et al., 2006) show predominantly a poleward flow with a strong variability in the outer 15 Mm of the Sun. The velocity reaches $40\,\mathrm{m\,s^{-1}}$. These findings were interpreted as the upper parts of meridional circulation cells in the two hemispheres.

Theoretically, the influence of global-scale stationary flows on solar p modes was studied in detail by quasi-degenerate perturbation theory (Lavely and Ritzwoller, 1992). These studies were successfully used to solve the forward problem for the influence of differential rotation on the p modes. The results lead to an improved inversion method for determining the radial dependence of the differential rotation (Ritzwoller and Lavely, 1991). Following Lavely and Ritzwoller (1992), Roth and Stix (1999) studied the influence of large-scale sectoral poloidal flow components that could be related to giant convection cells. They found that such flows yield additional frequency shifts that can only be described with quasi-degenerate perturbation theory, as these frequency shifts are effects of higher order. In a subsequent study Roth, Howe, and Komm (2002) were able to show that sectoral poloidal flows could only be found with the current inversion methods of global helioseismology as long as they exceed an amplitude of $10\,\mathrm{m\,s^{-1}}$. The meridional flow was found to be not detectable by the current inversion methods as the frequency splittings are fitted by an incomplete set of basis functions, which are tailored to measure only zonal toroidal flows. However, no detailed study on the effect of the meridional circulation on the oscillation frequencies was given, and few other attempts to derive observable signatures of the meridional flow in global helioseismology data exist (Woodard, 2000).

In this contribution we concentrate on a theoretical study of the influence of the meridional circulation on the solar p-mode frequencies and describe a possibly observable effect. This effect is significantly smaller than the frequency splitting caused by solar differential rotation. But as time series of global oscillation data exist that cover more than ten years, the necessary frequency resolution might be available to detect it. The advantage of studying the meridional circulation by global helioseismic techniques is a possible inference of information from greater depths.

2. Frequency Shifts Caused by Meridional Circulation

The effect of the meridional circulation on solar oscillations shall be investigated by solving the forward problem. We use quasi-degenerate perturbation theory as proposed by Lavely and Ritzwoller (1992) to calculate shifts of the oscillation frequencies. A short outline of the mathematics is given in the following.

As described by Lavely and Ritzwoller (1992) and Roth and Stix (1999), we consider a spherically symmetric equilibrium model of the Sun. For this purpose "Model S" from Christensen-Dalsgaard *et al.* (1996) is used. The solar p modes are adiabatic eigenoscillations ($\boldsymbol{\xi}_k$) of small amplitude, where k stands for the three indices: harmonic degree (ℓ), azimuthal order (m), and radial order (n). We calculate the eigenfrequencies and eigenmodes numerically with the http://www.phys.au.dk/~jcd/adipack.n/ ADIPACK code (Christensen-Dalsgaard and Berthomieu, 1991).

The equation governing the eigenmodes is

$$\mathcal{L}_0 \boldsymbol{\xi}_k = -\rho_0 \omega_k^2 \boldsymbol{\xi}_k, \tag{1}$$

where ρ_0 is the density and ω_k the oscillation frequency. The operator $\mathcal{L}_0$ acts on the eigenoscillation by

$$\mathcal{L}_0 \boldsymbol{\xi} = -\nabla P' + \rho_0 \boldsymbol{g}' + \rho' \boldsymbol{g}_0, \tag{2}$$

where the primed quantities are the Eulerian variations of pressure, gravitational acceleration, and density caused by an oscillation mode (*e.g.*, Stix, 2004; Unno *et al.*, 1989).

Describing the disturbing effect of the meridional circulation, we replace in Equations (1) the operator $\mathcal{L}_0$, the squared eigenfrequency ω_k^2, and the eigenfunction $\boldsymbol{\xi}_k$ with

$$\begin{aligned} \mathcal{L}_0 &\rightarrow \mathcal{L}_0 + \mathcal{L}_1, \\ \omega_k^2 &\rightarrow \tilde{\omega}_j^2, \\ \boldsymbol{\xi}_k &\rightarrow \tilde{\boldsymbol{\xi}}_j, \end{aligned} \tag{3}$$

where the perturbation operator $\mathcal{L}_1$ is defined in terms of the meridional circulation $\boldsymbol{u}_0$ and acts on the eigenmodes as

$$\mathcal{L}_1(\boldsymbol{\xi}_k) = -2i\omega_{\text{ref}} \rho_0 (\boldsymbol{u}_0 \cdot \nabla) \boldsymbol{\xi}_k. \tag{4}$$

The perturbed eigenfunctions are expressed in terms of the unperturbed normal modes of the equilibrium model

$$\tilde{\boldsymbol{\xi}}_j = \sum_{k \in \mathcal{K}} a_k^j \boldsymbol{\xi}_k, \tag{5}$$

where the index k and the subscript j are elements of the subspace $\mathcal{K}$ that is spanned by the eigenfunctions which are quasi-degenerate. We identify j with that k for which the expansion coefficients a_k^j are maximal in magnitude. Following Lavely and Ritzwoller (1992), we use

$$\tilde{\omega}^2 = \omega_{\text{ref}}^2 + \lambda \tag{6}$$

for the perturbed frequency. The reference frequency (ω_{ref}) should be chosen in the vicinity of the eigenfrequencies of the modes in the subspace $\mathcal{K}$.

According to Lavely and Ritzwoller (1992), the problem can be turned into an algebraic eigenvalue problem by substituting Equations (5) and (6) into the perturbed eigenvalue problem and making use of the orthogonality of the eigenfunctions

$$\sum_{k \in \mathcal{K}} a_k^j Z_{k'k} = \sum_{k \in \mathcal{K}} a_k^j \lambda_j \delta_{k'k} \quad \text{for } k' \in \mathcal{K}. \tag{7}$$

The elements of the matrix **Z** are given by

$$Z_{k'k} = \frac{1}{N_{k'}} \{ H_{n'n,\ell'\ell}^{m'm} - (\omega_{\text{ref}}^2 - \omega_k^2) N_k \delta_{k'k} \}, \tag{8}$$

with

$$H_{n'n,\ell'\ell}^{m'm} = -\int \boldsymbol{\xi}_{k'}^* \cdot \mathcal{L}_1 \boldsymbol{\xi}_k \, \mathrm{d}^3 r. \tag{9}$$

The normalization of the eigenfunctions is given by

$$N_k = \int_0^{R_\odot} \rho_0 [\xi_r^2 + \ell(\ell+1)\xi_h^2] r^2 \, \mathrm{d}r. \tag{10}$$

2.1. The Model of the Meridional Circulation

We use a linear superposition of Legendre polynomials $[P_s(\cos\theta)]$ with degree s for our representation for the meridional velocity field

$$\boldsymbol{u}_0 = \sum_s u_s(r) c_s P_s(\cos\theta) \boldsymbol{e}_r + v_s(r) \nabla_\mathrm{h} c_s P_s(\cos\theta), \tag{11}$$

where ∇_h is a horizontal gradient and θ is the colatitude. The factor c_s is used for normalization of the Legendre polynomials and is determined by $c_s^2 = (2s+1)/4\pi$. The expansion coefficients $u_s(r)$ and $v_s(r)$ are functions of the radial coordinate (r), and represent the radial and horizontal strength of the flow component with degree s. In this model the meridional velocity field is a zonal poloidal flow which is independent of longitude.

The meridional circulation shall be free of divergence, i.e., $\nabla \cdot (\rho_0 \boldsymbol{u}_0) = 0$, which enables us to express v_s in terms of u_s,

$$\rho_0 r s(s+1) v_s = \partial_r (r^2 \rho_0 u_s). \tag{12}$$

We adopt a simple model for the depth dependence of $u_s(r)$, which is given by

$$u_s(r) = A \sin\left(n_c \pi \frac{r - r_\mathrm{b}}{R_\odot - r_\mathrm{b}}\right) \quad \text{for } r_\mathrm{b} \leq r \leq R_\odot,$$
$$u_s(r) = 0 \quad \text{otherwise.} \tag{13}$$

The return flow of the meridional flow is closed at the radius given by the parameter r_b, which we set equal to the bottom of the convection zone $r_\mathrm{b} = 0.713 R_\odot$ (Basu and Antia, 1997). The number of circulation cells in depth is defined by the parameter n_c. We select the amplitude (A) such that the horizontal flow component (v_s) at the solar surface has a maximum amplitude of $15 \, \mathrm{m \, s^{-1}}$.

Below we present results of the effects on the solar p-mode frequencies of 15 various models of the meridional circulation. We vary s between two and six and n_c between one and three. The higher degrees s represent more than one cell in each hemisphere; such multi-cell circulation occurs in numerical simulations (Brun and Toomre, 2002; Miesch, Brun, and Toomre, 2006) and has been observed by means of sunspot tracing on the solar surface (Tuominen, Tuominen, and Kyröläinen, 1983; Wöhl and Brajša, 2001). The models with odd s have an equator-crossing flow; we include these as such asymmetric flows have been found on the Sun as well (Haber *et al.*, 2000; Zhao and Kosovichev, 2004). Figure 1 illustrates the 15 models.

Due to Equation (12), models with higher degree s show stronger radial flows in the down- and upflow channels whereas the horizontal flow strength is the same in all models.

2.2. Theoretical Frequency Shifts

We evaluate the integration of Equation (9) by using the model of the meridional circulation (11) and (12). The integral over longitude does not vanish only if $m' = m$. The result is

$$H_{n'n,\ell'\ell}(m) = 8i\omega_{\text{ref}}\pi(-1)^m \gamma_\ell \gamma_{\ell'}$$
$$\times \sum_s \gamma_s \int_0^{R_\odot} u_s \left[R_s - \partial_r \left(\frac{H_s}{s(s+1)} \right) \right] \rho_0 r^2 \, dr \begin{pmatrix} s & \ell & \ell' \\ 0 & m & -m \end{pmatrix}. \quad (14)$$

The integral kernels $R_s(r)$ and $H_s(r)$ are the equivalents of the more general poloidal-flow kernels given in Lavely and Ritzwoller (1992)

$$R_s(r) = \frac{1}{4}\left(\xi'_r \frac{\partial \xi_r}{\partial r} - \frac{\partial \xi'_r}{\partial r}\xi_r\right)\left(1+(-1)^{s+\ell+\ell'}\right)\begin{pmatrix} s & \ell & \ell' \\ 0 & 0 & 0 \end{pmatrix}$$
$$+ \frac{1}{2}\left(\xi'_h \frac{\partial \xi_h}{\partial r} - \frac{\partial \xi'_h}{\partial r}\xi_h\right)\left(1+(-1)^{s+\ell+\ell'}\right)\Omega_\ell \Omega_{\ell'} \begin{pmatrix} s & \ell & \ell' \\ 0 & 1 & -1 \end{pmatrix},$$

$$H_s(r) = \frac{1}{2}\left[\ell(\ell+1) - \ell'(\ell'+1)\right]\left(1+(-1)^{s+\ell+\ell'}\right)$$
$$\times \left[\xi'_r \xi_r \begin{pmatrix} s & \ell & \ell' \\ 0 & 0 & 0 \end{pmatrix} - \xi'_h \xi_h \Omega_\ell \Omega_{\ell'} \begin{pmatrix} s & \ell & \ell' \\ 0 & 1 & -1 \end{pmatrix}\right] \quad (15)$$
$$- \xi'_h \xi_r \left(1+(-1)^{s+\ell+\ell'}\right)\Omega_{\ell'} \Omega_s \begin{pmatrix} s & \ell & \ell' \\ 1 & 0 & -1 \end{pmatrix}$$
$$+ \xi'_r \xi_h \left(1+(-1)^{s+\ell+\ell'}\right)\Omega_\ell \Omega_s \begin{pmatrix} s & \ell & \ell' \\ 1 & -1 & 0 \end{pmatrix},$$

where $\gamma_x = \sqrt{(2x+1)/4\pi}$ and $\Omega_x = \sqrt{x(x+1)/2}$. The 3×2 arrays are Wigner-$3j$ symbols that result from the coupling of the angular momenta of the oscillations and the flow involved (Edmonds, 1974). Due to this coupling of angular momenta, the matrix element $H_{n'n,\ell'\ell}(m)$ is non-vanishing if certain selection rules are fulfilled. The first rule arises from properties of the Wigner-$3j$ symbols which vanish except when the harmonic degrees ℓ, ℓ', and s satisfy a triangular condition. The second rule follows from Equation (15); the sum of the degrees must be even otherwise the factor $(1+(-1)^{s+\ell+\ell'})$ in R_s and H_s vanishes,

$$s + \ell + \ell' \equiv 0 \mod 2. \quad (16)$$

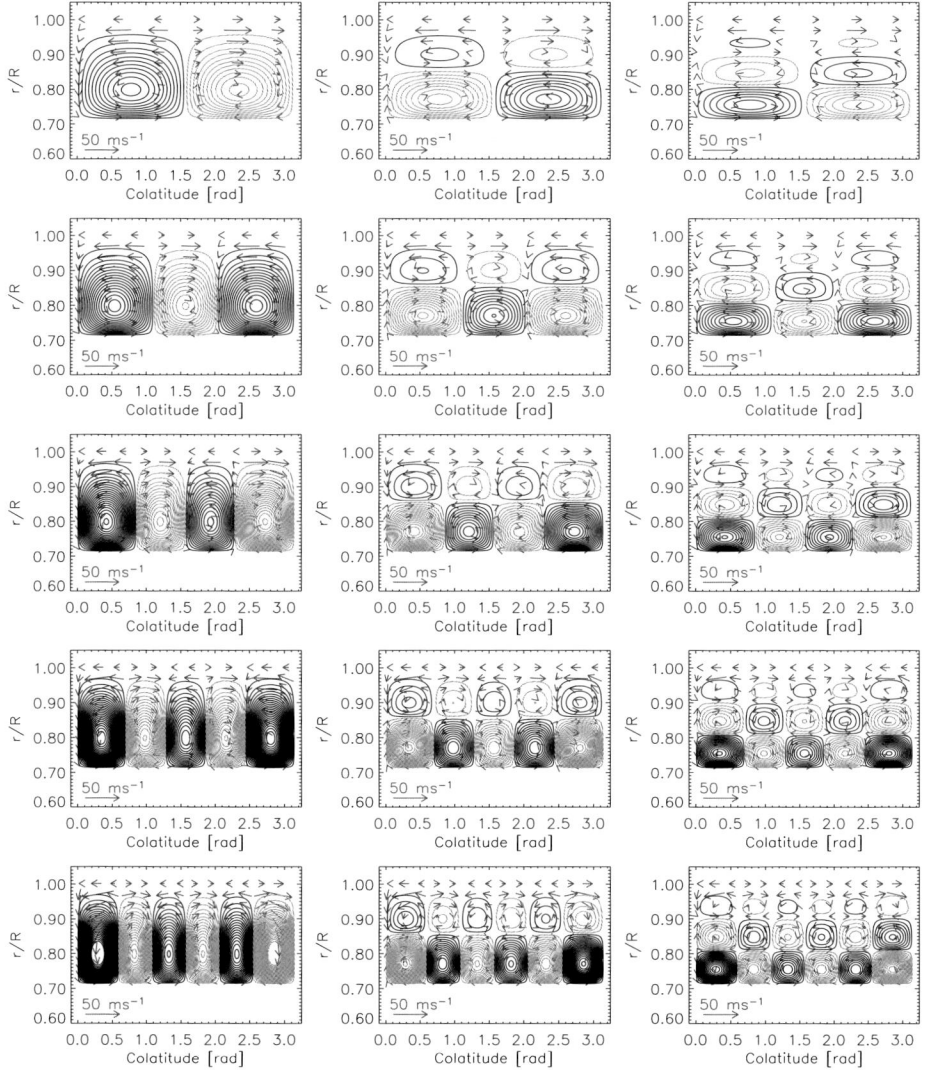

Figure 1 Streamlines for the flows $\rho\boldsymbol{u}_0$ of the 15 meridional-circulation models as functions of solar colatitude and radius. Black indicates counterclockwise flow, gray clockwise. The arrows indicate the flow velocity at some positions. The number of cells in horizontal direction changes from top to bottom between two and six. The number of cells in depth changes from left to right between one and three.

We can simplify Equation (14) by defining a poloidal flow kernel $K_s(r)$

$$H_{n'n,\ell'\ell}(m) \equiv 8i\omega_{\text{ref}}\pi(-1)^m \gamma_\ell \gamma_{\ell'} \sum_s \gamma_s \int_0^{R_\odot} u_s K_s(r) \rho_0 r^2 \, dr \begin{pmatrix} s & \ell & \ell' \\ 0 & m & -m \end{pmatrix}. \quad (17)$$

According to this result, the meridional circulation leads to an effective shift of the eigenvalues, *i.e.*, the squared mode frequencies, if the selection rules are fulfilled. These shifts are given by the eigenvalues λ of the matrix $\boldsymbol{Z}$, or in more detail, the corrected mode frequency

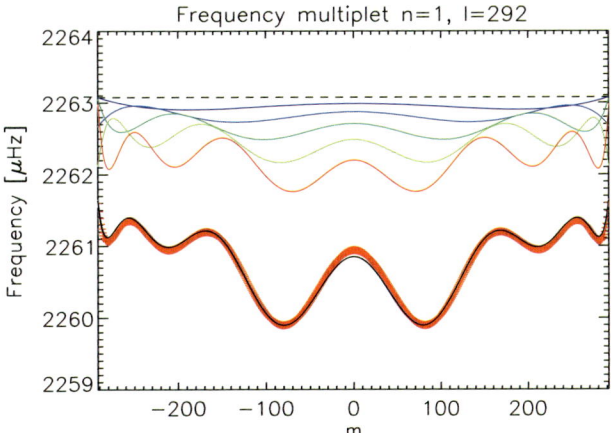

Figure 2 Splittings of the p-mode multiplet $n = 1$, $\ell = 292$, $\nu = 2263.08\,\mu\text{Hz}$ as a function of azimuthal order m. The splittings are caused by different meridional flow models: $s = 2$, $n_c = 1$ (purple), $s = 3$, $n_c = 1$ (blue), $s = 4$, $n_c = 1$ (green), $s = 5$, $n_c = 1$ (yellow), $s = 6$, $n_c = 1$ (red). The total effect of a sum of these five flow models is displayed by the thick red line, the black line on top of this gives the result for a sum of the five flow models but with $n_c = 3$. The unperturbed mode frequency is given by the dashed line.

$\tilde{\omega}_{nlm}$ is given by

$$\tilde{\omega}^2_{nlm} = \omega^2_{\text{ref}} + \lambda_{nlm}. \tag{18}$$

Following the concept of quasi-degenerate perturbation theory, the coupling of only two modes in the presence of only one flow component was investigated by Roth and Stix (1999). They found that due to coupling of two modes by any poloidal velocity field the absolute magnitudes of the frequency shifts are equal for both couplers, but the sign differs: the shift is negative for the mode with the lower frequency. In the case of two multiplets where pairs of modes are coupling, the frequency shifts are a function of the azimuthal order m. The functional dependence on m is given by the Wigner-$3j$ symbol as a function of m, and by the factor $(-1)^m$. It then follows that the shifts within one multiplet are either always negative or always positive. From this it follows for the case of the meridional circulation that, due to the dependence of the Wigner-$3j$ symbol on m, the frequency shifts are always symmetric about the mode with $m = 0$ which in general is also shifted.

We would also like to point out that due to the second selection rule (16), the flow kernel K_s vanishes if $\ell' = \ell$, i.e. there are no contributions to the diagonal entries on the matrix $\mathbf{Z}$ from the matrix elements $H_{n'n,\ell'\ell}(m)$. As the matrix elements $H_{n'n,\ell'\ell}(m)$ are complex and $\mathbf{Z}$ is Hermitian, a change of the orientation of $\boldsymbol{u}_0$ does not change the frequency shifts.

3. Results

The results of our investigation are frequency shifts for the mode multiplets with $0 \leq \ell \leq 300$ and $1 \leq n \leq 30$ calculated numerically by setting up the full coupling matrix (8), i.e., taking all possible couplings of modes into account and evaluating the eigenvalues.

We first focus on the frequency shifts in the multiplets. According to the properties of the Wigner-$3j$ symbols, the shifts within a multiplet are symmetric about the mode with $m = 0$. Figure 2 displays an example. The frequency splitting is stronger for meridional circulation

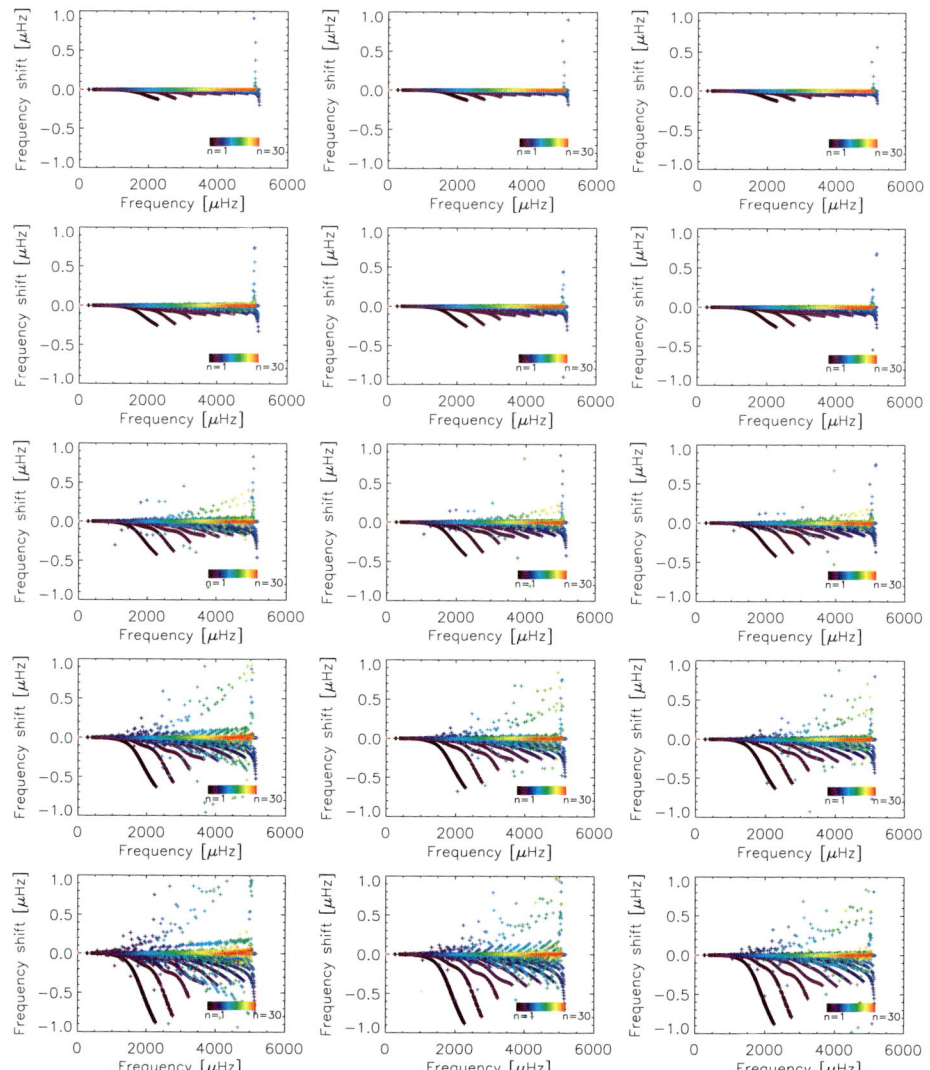

Figure 3 The mean frequency shifts of the multiplets caused by the 15 different configurations of the meridional circulation (cf. Figure 1) as functions of the unperturbed mode frequencies. The color code gives the radial order (n) of the modes. The degree of the meridional circulation affecting the modes changes from $s = 2$ (top) to $s = 6$ (bottom), the order n_c of the meridional circulation changes from one (left) to three (right).

components with higher degree s. In the example of Figure 2 the shifts are on the order of 1 μHz for $s = 6$. The overall effect of the superposition of the meridional circulation components results in shifts on the order of 3 μHz away from the unperturbed frequency. The difference between the summed effect of models with $n_c = 1$ cell in depth and models with $n_c = 3$ cells in depth is on the order of 0.1 μHz.

To give a better overview of the effect on all modes, Figure 3 displays the mean frequency shifts of the multiplets as a function of the unperturbed mode frequencies. The frequency

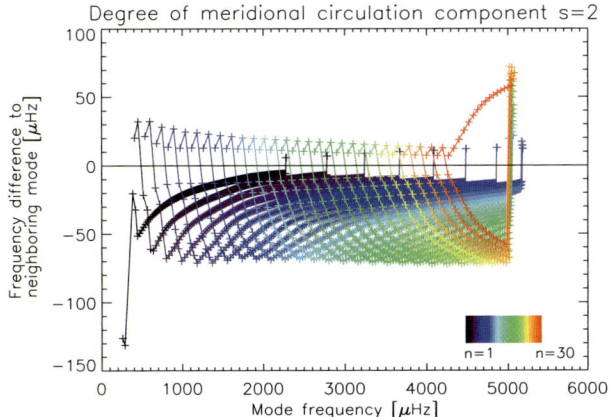

Figure 4 The frequency difference to the next nearest mode fulfilling the selection rule for the meridional circulation component with $s = 2$. The color code gives the radial order of the modes.

shifts were obtained for the single meridional flow components. One common result is a negative frequency shift of most modes. We find positive frequency shifts only in a few multiplets. The origin of this effect is given by the shape of the $\ell - \nu$ diagram. According to quasi-degenerate perturbation theory, the multiplet nearest in frequency fulfilling the selection rules causes the strongest shift (Roth and Stix, 1999). As shown for the case $s = 2$ in Figure 4, in most cases these nearest neighboring modes have a higher frequency. This is explained by the first selection rule and the curvature of the ridges in the $\ell - \nu$ diagram. Because of that rule, the difference $|\ell - \ell'|$ of the harmonic degrees of two coupling modes can not be larger than s, which is a small number as we consider low degrees s of the flow components. Possible coupling partners for a particular mode can then only come from a very narrow region in ℓ. As the slopes of the ridges are steeper towards lower harmonic degrees, the frequency spacing between the modes decreases, with increasing ℓ, along a ridge. In addition, the frequency difference between two ridges increases with ℓ. Therefore the nearest neighboring mode fulfilling the selection rules usually lies on the same ridge and has a higher frequency and a higher harmonic degree. Figure 5 shows that in most of the investigated cases the coupling partners lie on the same ridge. Interestingly there is some structure visible in these plots. There are areas where all coupling partners are coming from the same ridge or from one ridge above or below. These areas are defined by the curvature of the ridges. With higher degree s of the meridional flow, the coupling of modes from different ridges becomes more frequent. These plots indicate only the position of the nearest coupling partner, because this partner causes the strongest effect. The effects of coupling partners with greater distances in frequency are weaker.

Another common result is that the frequency shifts are strongest for the modes with low radial order (n) and with high harmonic degree (ℓ). This is again explained by the location of the nearest neighbor in frequency (Figure 4). The lower the radial order n the smaller the frequency difference to the next mode. Along a ridge, the frequency difference to the next mode decreases causing a stronger splitting of the multiplets. On the ridges with high radial order, the frequency spacing between two neighboring modes is larger than on the ridges with low radial order.

These two common results apply in principle to all poloidal flows, not only to the meridional circulation. Therefore we can conclude that the overall effect of large-scale poloidal flows in the convection zone is a lowering of the mode frequencies. This effect could contribute to reduce the discrepancy between the theoretically calculated mode frequencies and

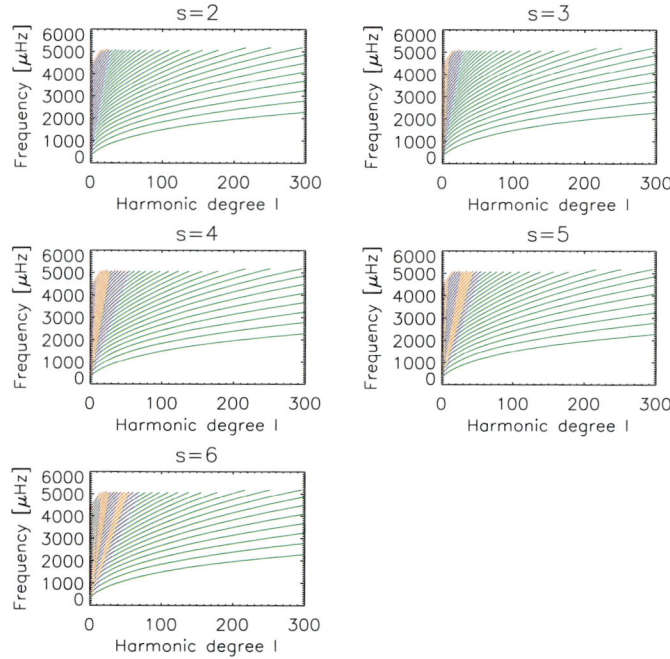

Figure 5 Theoretical $\ell - \nu$ diagrams with colors indicating the origin of the coupling partner with nearest frequency. Green: coupling partner from the same ridge; blue: coupling partner from one ridge above; red: coupling partner from one ridge below, black: coupling partner is more than one ridge away. The degree (s) of the meridional circulation components changes from two to six.

the observed frequencies. A small frequency reduction has also been found, for a different cause, by Zhugzhda and Stix (1994) and Stix and Zhugzhda (1998).

Investigating the results shown in Figure 3 in detail for the meridional circulation models used, we find that the frequency shifts are increasing from $s = 2$ to $s = 6$. In the case of $s = 2$, $n_c = 1$ the total average of these shifts *over* all modes affected by the meridional circulation is 0.01 µHz, whereas in the case $s = 6$, $n_c = 1$ the magnitude of the average shift is 0.1 µHz.

There is also a small difference noticeable in the frequency shift if the number (n_c) of cells changes. The difference of the average shifts calculated for the models $s = 2$, $n_c = 1$ and $n_c = 3$ is -0.002 µHz. Whereas the difference of the average shifts determined for the models $s = 6$, $n_c = 1$ and $n_c = 3$ is -0.03 µHz. However, the scatter in the frequency shifts is high. In the case of $s = 2$, $n_c = 1$ the largest shift observed is 22.5 µHz, and in the case of $s = 6$, $n_c = 1$ the largest shift observed is 79.4 µHz (outside the range shown in Figure 3). Strong shifts above 10 µHz occur for the modes with high harmonic degree ℓ.

4. Discussion

In this paper we used simple models of the meridional circulation to investigate their influence on the solar p-mode frequencies. The simplest model consisted of one cell per hemisphere and depth with a maximum horizontal flow velocity of 15 m s^{-1} on the solar surface. The most complicated model we used had three cells per hemisphere and three cells in depth with a horizontal flow velocity of 15 m s^{-1} at the surface, too. We performed numerical calculations based on quasi-degenerate perturbation theory to obtain the frequency splittings of the solar p modes due to these meridional circulation models. We find that the meridional circulation lifts the degeneracy of the multiplets. For the simplest model the shifts are on average only 0.01 µHz with a few shifts up to several µHz. Models with more cells per hemisphere affect the p modes more. For the model with three cells per hemisphere we find an average shift of 0.1 µHz, with many shifts of the order of 1 µHz. In most cases the shifts are negative due to the fact that the next neighboring mode in frequency dominates the shifting; due to structure of the $\ell - \nu$ diagram this nearest neighbor has usually a higher frequency, causing therefore negative shifts.

Comparing models with the same number of cells in latitude but different number of cells in depth, we find only tiny differences in the resulting frequency splittings. In the case of $s = 6$, the difference in the frequency splitting is on average 0.03 µHz. However a number of differences of the order of 1 µHz can occur.

On the Sun, the meridional circulation probably consists of a superposition of flow components with different numbers of cells in depth and latitude. In contrast to the models used in our work, the amplitudes of these flow components are in reality likely to be very different and highly variable in time. Nevertheless, based on our results we are optimistic that the meridional circulation on the Sun could leave an observable signature in the p-mode frequencies. In order to detect the effect, a frequency resolution of at least 0.1 µHz must be achieved. This could be done by a several-month-long time series. In order to be able to distinguish not only the various components of the meridional circulation in the orders s but also in the radial orders n_c, several years of data need to be averaged to obtain the required precision in the frequencies. As such long data sets exist it will be worth while to search for this effect in the p-mode frequencies.

Compared to the splitting caused by the differential rotation, the effect of meridional circulation is small. But as the effect of the rotational splitting is odd in m it can be separated from the even meridional splitting. However, e.g., asphericities and the magnetic field might cause symmetrical frequency shifts, as well. Therefore, a lot of forward modeling will be necessary before the effect of the meridional circulation can be disentangled from these other effects. In this sense, one possible future extension of our work is the determination of the frequency shifts due to more sophisticated models of the meridional circulation, e.g., from three-dimensional numerical models. This might allow tailoring inversion routines for estimating the meridional circulation in the Sun from global solar oscillation frequencies.

Acknowledgements M.R. acknowledges support from the European Helio- and Asteroseismology Network (HELAS) which is funded by the European Commission's Sixth Framework Programme.

References

Basu, S., Antia, H.M.: 1997, Seismic measurement of the depth of the solar convection zone. *Mon. Not. Roy. Astron. Soc.* **287**, 189–198.
Bjerknes, V.: 1926, Solar hydrodynamics. *Astrophys. J.* **64**, 93–121.
Brun, A.S., Toomre, J.: 2002, Turbulent convection under the influence of rotation: Sustaining a strong differential rotation. *Astrophys. J.* **570**, 865–885.
Choudhuri, A.R., Schüssler, M., Dikpati, M.: 1995, The solar dynamo with meridional circulation. *Astron. Astrophys.* **303**, L29–L32.
Christensen-Dalsgaard, J., Berthomieu, G.: 1991, Theory of solar oscillations. In: Cox, A.N., Livingston, W.C., Matthews, M.S. (eds.) *Solar Interior and Atmosphere*, University of Arizona Press, Tucson, 401–478.

Christensen-Dalsgaard, J., Däppen, W., Ajukov, S.V., Anderson, E.R., Antia, H.M., Basu, S., Baturin, V.A., Berthomieu, G., Chaboyer, B., Chitre, S.M., Cox, A.N., Demarque, P., Donatowicz, J., Dziembowski, W.A., Gabriel, M., Gough, D.O., Guenther, D.B., Guzik, J.A., Harvey, J.W., Hill, F., Houdek, G., Iglesias, C.A., Kosovichev, A.G., Leibacher, J.W., Morel, P., Proffitt, C.R., Provost, J., Reiter, J., Rhodes, E.J., Rogers, F.J., Roxburgh, I.W., Thompson, M.J., Ulrich, R.K.: 1996, The current state of solar modeling. *Science* **272**, 1286–1292.

Dikpati, M., Charbonneau, P.: 1999, A Babcock-Leighton flux transport dynamo with solar-like differential rotation. *Astrophys. J.* **518**, 508–520.

Duvall, T.L. Jr.: 1979, Large-scale solar velocity fields. *Solar Phys.* **63**, 3–15.

Edmonds, A.R.: 1974, *Angular Momentum in Quantum Mechanics*, Princeton University Press, Princeton.

Haber, D.A., Hindman, B.W., Toomre, J., Bogart, R.S., Thompson, M.J., Hill, F.: 2000, Solar shear flows deduced from helioseismic dense-pack samplings of ring diagrams. *Solar Phys.* **192**, 335–350.

Haber, D.A., Hindman, B.W., Toomre, J., Bogart, R.S., Larsen, R.M., Hill, F.: 2002, Evolving submerged meridional circulation cells within the upper convection zone revealed by ring-diagram analysis. *Astrophys. J.* **570**, 855–864.

Hathaway, D.H.: 1996, Doppler measurements of the Sun's meridional flow. *Astrophys. J.* **460**, 1027–1033.

Hill, F.: 1988, Rings and trumpets – Three-dimensional power spectra of solar oscillations. *Astrophys. J.* **333**, 996–1013.

Kippenhahn, R.: 1963, Differential rotation in stars with convective envelopes. *Astrophys. J.* **137**, 664–678.

Komm, R.W., Howard, R.F., Harvey, J.W.: 1993, Meridional flow of small photospheric magnetic features. *Solar Phys.* **147**, 207–223.

Küker, M., Rüdiger, G.: 2005, Differential rotation on the lower main sequence. *Astron. Nachr.* **326**, 265–268.

Küker, M., Stix, M.: 2001, Differential rotation of the present and the pre-main-sequence Sun. *Astron. Astrophys.* **366**, 668–675.

Lavely, E.M., Ritzwoller, M.H.: 1992, The effect of global-scale, steady-state convection and elastic-gravitational asphericities on helioseismic oscillations. *Roy. Soc. Lond. Phil. Trans. Ser. A* **339**, 431–496.

Miesch, M.S., Brun, A.S., Toomre, J.: 2006, Solar differential rotation influenced by latitudinal entropy variations in the tachocline. *Astrophys. J.* **641**, 618–625.

Miesch, M.S., Elliott, J.R., Toomre, J., Clune, T.L., Glatzmaier, G.A., Gilman, P.A.: 2000, Three-dimensional spherical simulations of solar convection. I. Differential rotation and pattern evolution achieved with laminar and turbulent states. *Astrophys. J.* **532**, 593–615.

Nandy, D., Choudhuri, A.R.: 2002, Explaining the latitudinal distribution of sunspots with deep meridional flow. *Science* **296**, 1671–1673.

Parker, E.N.: 1955, Hydromagnetic dynamo models. *Astrophys. J.* **122**, 293–314.

Rempel, M.: 2006a, Flux-transport dynamos with Lorentz force feedback on differential rotation and meridional flow: Saturation mechanism and torsional oscillations. *Astrophys. J.* **647**, 662–675.

Rempel, M.: 2006b, Transport of toroidal magnetic field by the meridional flow at the base of the solar convection zone. *Astrophys. J.* **637**, 1135–1142.

Ritzwoller, M.H., Lavely, E.M.: 1991, A unified approach to the helioseismic forward and inverse problems of differential rotation. *Astrophys. J.* **369**, 557–566.

Roberts, P.H., Stix, M.: 1972, α-Effect dynamos, by the Bullard–Gellman formalism. *Astron. Astrophys.* **18**, 453–466.

Roth, M., Stix, M.: 1999, Coupling of solar p modes: Quasi-degenerate perturbation theory. *Astron. Astrophys.* **351**, 1133–1138.

Roth, M., Howe, R., Komm, R.: 2002, Detectability of large-scale flows in global helioseismic data – A numerical experiment. *Astron. Astrophys.* **396**, 243–253.

Rüdiger, G.: 1980, Reynolds stresses and differential rotation. I – On recent calculations of zonal fluxes in slowly rotating stars. *Geophys. Astrophys. Fluid Dyn.* **16**, 239–261.

Stix, M.: 2004, *The Sun: An Introduction*, 2nd edn. Springer, Berlin.

Stix, M., Zhugzhda, Y.D.: 1998, Waves in structured media: Non-radial solar p modes. *Astron. Astrophys.* **335**, 685–690.

Tuominen, J., Tuominen, I., Kyröläinen, J.: 1983, Eleven-year cycle in solar rotation and meridional motions as derived from the positions of sunspot groups. *Mon. Not. Roy. Astron. Soc.* **205**, 691–704.

Unno, W., Osaki, Y., Ando, H., Saio, H., Shibahashi, H.: 1989, *Nonradial Oscillations of Stars*, 2nd edn. University of Tokyo Press, Tokyo.

Wöhl, H., Brajša, R.: 2001, Meridional motions of stable recurrent sunspot groups. *Solar Phys.* **198**, 57–77.

Woodard, M.F.: 2000, Theoretical signature of solar meridional flow in global seismic data. *Solar Phys.* **197**, 11–20.

Zaatri, A., Komm, R., González Hernández, I., Howe, R., Corbard, T.: 2006, North south asymmetry of zonal and meridional flows determined from ring diagram analysis of GONG++ data. *Solar Phys.* **236**, 227–244.

Zhao, J., Kosovichev, A.G.: 2004, Torsional oscillation, meridional flows, and vorticity inferred in the upper convection zone of the sun by time-distance helioseismology. *Astrophys. J.* **603**, 776–784.

Zhugzhda, Y.D., Stix, M.: 1994, Acoustic waves in structured media and helioseismology. *Astron. Astrophys.* **291**, 310–319.

Global Effects of Local Sound-Speed Perturbations in the Sun: A Theoretical Study

S.M. Hanasoge · T.P. Larson

Originally published in the journal Solar Physics, Volume 251, Nos 1–2, 91–100.
DOI: 10.1007/s11207-008-9208-6 © Springer Science+Business Media B.V. 2008

Abstract We study the effect of localized sound-speed perturbations on global mode frequencies by applying techniques of global helioseismology to numerical simulations of the solar acoustic wave field. Extending the method of realization-noise subtraction (*e.g.*, Hanasoge, Duvall, and Couvidat, *Astrophys. J.* **664**, 1234, 2007) to global modes and exploiting the luxury of full spherical coverage, we are able to achieve very highly resolved frequency differences that are then used to study sensitivities and the signatures of the thermal asphericities. We find that *i*) global modes are almost twice as sensitive to sound-speed perturbations at the bottom of the convection zone in comparison to anomalies well inside the radiative interior ($r \lesssim 0.55 R_\odot$), *ii*) the m degeneracy is lifted ever so slightly, as seen in the a coefficients, and *iii*) modes that propagate in the vicinity of the perturbations show small amplitude shifts. Through comparisons with error estimates obtained from Michelson Doppler Imager (MDI; Scherrer *et al.*, *Solar Phys.* **162**, 129, 1995) observations, we find that the frequency differences are detectable with a sufficiently long time series (70–642 days).

Keywords Helioseismology: Direct modeling · Interior: Tachocline · Interior: Convective zone · Waves: Acoustic

1. Introduction

Global helioseismology has proven very successful at allowing us to infer large-scale properties of the Sun (for a review, see Christensen-Dalsgaard, 2002, 2003). Because they are very robust, the extension of these methods to studies of localized variations in the structure and dynamics of the solar interior has been of some interest (*e.g.*, Swisdak and Zweibel,

---

Helioseismology, Asteroseismology, and MHD Connections
Guest Editors: Laurent Gizon and Paul Cally.

S.M. Hanasoge (✉) · T.P. Larson
W.W. Hansen Experimental Physics Laboratory, Stanford University, Stanford, CA 94305, USA
e-mail: shravan@stanford.edu

T.P. Larson
e-mail: tplarson@sun.stanford.edu

1999). However, the precise sensitivities of global modes to local perturbations are difficult to estimate through analytical means, especially in cases where the flows or thermal asphericities of interest possess complex spatial dependencies. To address questions relating to sensitivities and with the hope of perhaps discovering hitherto unknown phenomena associated with global modes, we introduce here a technique to study the effects of arbitrary perturbations on global mode parameters in the linear limit of small wave amplitudes. In addition, this method can be employed to study the interactions of waves with arbitrary magnetic structures. Since this is a first step along these lines, we choose a fairly simple set of problems to study, ones that are somewhat amenable to analytical methods as well.

Global modes attain resonant frequencies as a consequence of differentially sampling the entire region of propagation, making it somewhat more difficult (in comparison to local helioseismology) to pinpoint local thermal asphericities at depth. Exactly how difficult is one of the questions we have attempted to answer in this article. Jets in the tachocline (*e.g.*, Christensen-Dalsgaard *et al.*, 2005) are a subject of considerable interest since their existence (or lack thereof) is very important in understanding the angular momentum balance of the Sun. Studying the sensitivities and signatures of waves to flows at depth may open up possibilities for their detection.

Forward modeling as a means of studying wave interactions in a complex medium like the Sun has become quite favored (*e.g.*, Rosenthal *et al.*, 1999; Georgobiani *et al.*, 2004; Hanasoge *et al.*, 2006; Parchevsky and Kosovichev, 2007; Hanasoge, Duvall, and Couvidat, 2007; Cameron, Gizon, and Daifallah, 2007). The discovery of interesting phenomena, especially in the realm of local helioseismology (*e.g.*, Hanasoge *et al.*, 2007; Birch, Braun, and Hanasoge, 2008), adds motivation to the pursuit of direct calculations. With the application of noise subtraction (Werne, Birch, and Julien, 2004; Hanasoge, Duvall, and Couvidat, 2007), we can now study the signatures of a wide range of perturbations in a realistic multiple source picture. Here, we attempt to place bounds on the detectability of thermal asphericities at various depths in the Sun. We introduce and discuss the method of simulation with a description of the types of perturbations introduced in the model in Section 2. The estimation of mode parameters can prove somewhat difficult because of restrictions on the duration of the simulation (< 24 hours, owing to the expensive nature of the computation). The data analysis techniques used to characterize the modes are presented in Section 3. We then discuss the results from the analyses of the simulated data in Section 4 and summarize this work in Section 5.

2. Simulations and Perturbations

The linearized 3D Euler equations in spherical geometry are solved in the manner described in Hanasoge *et al.* (2006). The computational domain is a spherical shell extending from $0.24 R_\odot$ to $1.002 R_\odot$, with damping sponges placed adjacent to the upper and lower radial boundaries to allow the absorption of outgoing waves. These sponges are Newton-cooling-like terms in the Euler equations. They serve to damp the outgoing waves and prevent reflections at the boundaries of the computational domain. The background stratification is a convectively stabilized form of model S (Christensen-Dalsgaard *et al.*, 1996; Hanasoge *et al.*, 2006); the highly (convectively) unstable near-surface layers ($r > 0.98 R_\odot$) are altered while the interior is the same as model S. In contrast to the Sun where waves are excited by the vigorous near-surface convective activity, we utilize a simple phenomenological linear substitute in our calculations. The waves are stochastically excited over a 200-km-thick subphotospheric spherical envelope, through the application of a dipolar source function in the vertical (radial) momentum equation (Hanasoge *et al.*, 2006;

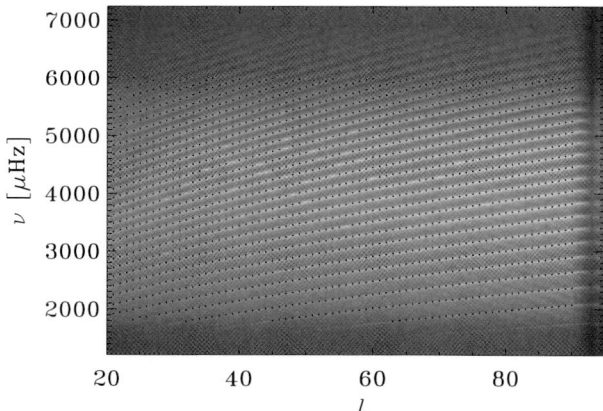

Figure 1 Example power spectrum from a simulation with $\ell_{\max} = 95$ and the corresponding frequency fits (symbols). Apart from damping sponges on the upper and lower boundaries, there is very little numerical damping – this results in very thin line widths, especially at lower frequencies. This is from a 24-hour simulation. The fitting algorithm is described in Section 3. Missing modes indicate that the fit for these did not converge. We do not understand why this occurs in the center of the power spectrum, but these modes can be made to converge by perturbing their initial guesses. Modes with $\ell < 20$ mostly disappear from the computational domain because the lower boundary is placed at $r = 0.24 R_\odot$. We also do not excite the highest l modes because we wish to avoid any issues related to numerical aliasing. A frequency cut through the spectrum will show frequency-dependent asymmetric mode profiles (Hanasoge and Duvall, 2007).

Hanasoge and Duvall, 2007). The forcing function is uniformly distributed in spherical harmonic space (ℓ, m); in frequency, a solar-like power variation is imposed. Any damping of the wave modes away from the boundaries is entirely of numerical origin. The radial velocities associated with the oscillations are extracted 200 km above the photosphere and used as inputs to the peak-bagging analyses. Data over the entire 360° extent of the sphere are utilized in the analyses, thus avoiding issues related to mode leakage. We show an example power spectrum in Figure 1 along with the fits.

The technique of realization-noise subtraction (*e.g.*, Hanasoge, Duvall, and Couvidat, 2007) is extensively applied in this work. Because of the relatively short durations of the simulations (the shortest time series we have yet worked with is 500 minutes long), the power spectrum is not highly resolved and it would seem that the resulting uncertainty in the mode parameter fits might constrain our ability to study small perturbations. To overcome this limit, we perform two simulations with identical realizations of the forcing function: a "quiet" run with no perturbations and a "perturbed" run that contains the anomaly of interest. Fits to the mode parameters in these two datasets are then subtracted, thus removing nearly all traces of the realization and retaining only effects arising from mode-perturbation interactions (see Section 3). As an example, we show in Figure 2 how a localized sound-speed perturbation placed at the bottom of the convection zone scatters waves that then proceed to refocus at the antipode (the principle of far-side holography; Lindsey and Braun, 2000). The presence of the sound-speed perturbation is not seen in Figure 2(a), whereas it is clearly seen in the noise-subtracted images of Figures 2(b) and (c).

In these calculations, we only consider time-stationary perturbations. The sound-speed perturbations are taken to be solely the result of changes in the first adiabatic index (Γ_1); we do not study sound-speed variations arising from changes in the background pressure or density since altering these variables can create hydrostatic instabilities. Lastly, the amplitudes of all perturbations are taken to be much smaller than the local sound speed ($\lesssim 5\%$).

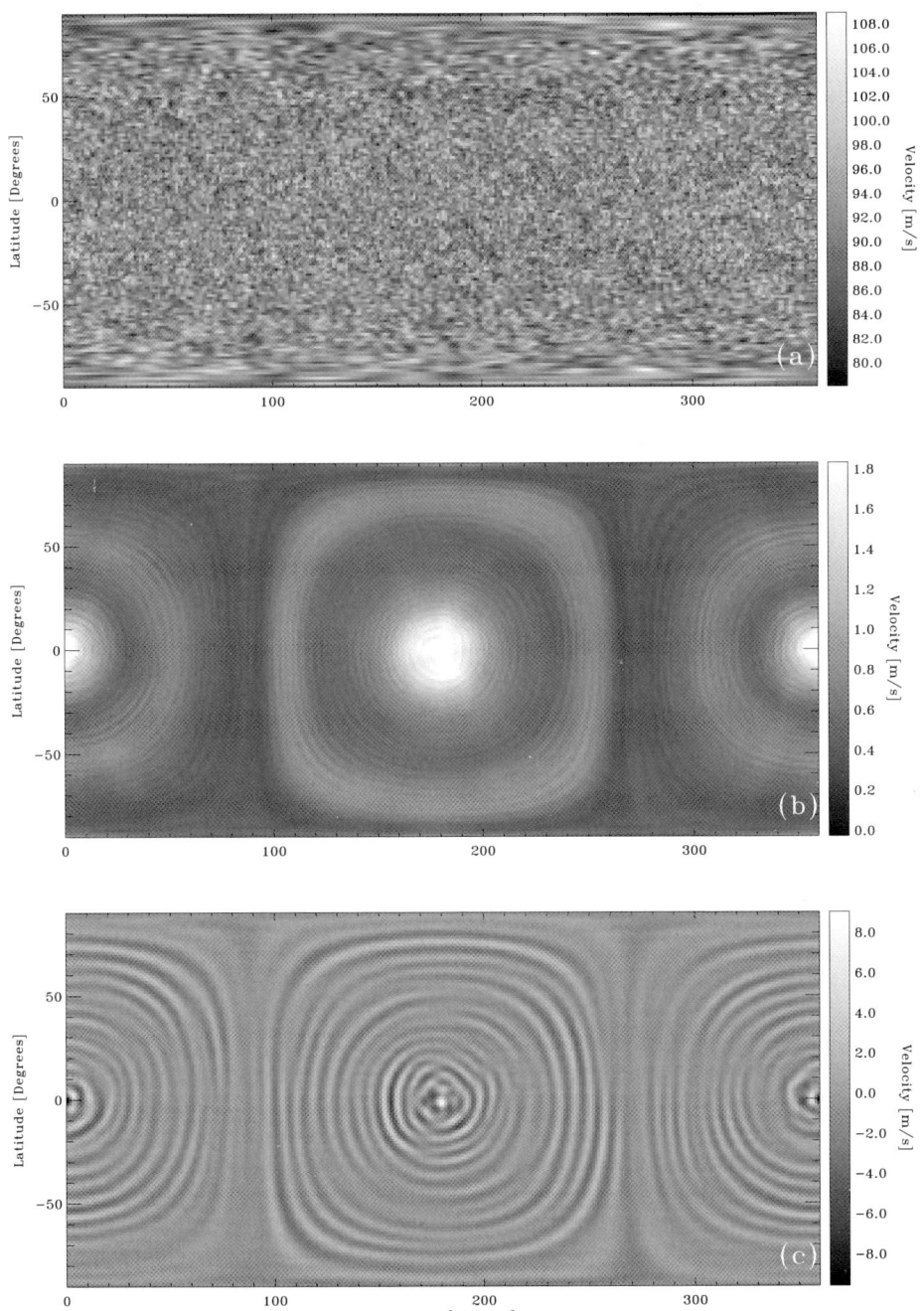

Figure 2 Noise subtraction. (a) Time-averaged RMS of the radial velocities of the perturbed simulation; the sound-speed perturbation (located along the Equator at $r = 0.7R_\odot$ and $180°$ longitude) is invisible. (b) Time-averaged RMS of the difference between the quiet and perturbed simulations. (c) The instantaneous difference. The scattering of waves and their refocusing at the antipode is clearly seen in (b).

3. Peak-Bagging Analysis

Our first round of peak-bagging is performed on the m-averaged power spectrum for the quiet simulation. For each ℓ that we attempt to fit, we search for minima in the second derivative of the power. Unlike the power itself, which has a background, the second derivative has the advantage of having an approximately zero baseline. The search is accomplished by finding the frequency at which the minimum value of the second derivative occurs, estimating the mode parameters using a frequency window of width 100 µHz centered on this peak frequency, zeroing the second derivative in this interval, and iterating. If the range of power in the frequency window is not above a certain threshold, we check the peak frequency found; if it is too close to a frequency found on a previous iteration, that maximum is rejected, the same interval is again zeroed, and iteration continues. Note that such a simple algorithm is feasible only because the simulation data contain no leaks. Once we have found as many peaks as possible with this procedure, we assign a value of n to each one based on a model computed using ADIPACK (Christensen-Dalsgaard and Berthomieu, 1991; Hanasoge, 2007).

The next step is to perform an actual fit to the power spectrum in the vicinity of each peak that we identified. For the line profile we use a Lorentzian of the form

$$P = \frac{A}{\pi} \frac{w}{(\nu - \nu_0)^2 + w^2} + B, \qquad (1)$$

where A is the total power, w is the half width at half maximum, ν_0 is the peak frequency, and B is the background power density. The initial guesses for these parameters are obtained as follows: B is set to the minimum value of the power density in the frequency window around the peak, A is set to the integral under the power density curve minus B times the width of the window, and w is set to $1/(\pi P_{\max})$, where $P_{\max}$ is the maximum value of the power density in the frequency window. The fitting interval extends halfway to the adjacent peaks, or 100 µHz beyond the peak frequency of the modes at the edge. The fitting itself is done by using the IDL routine curvefit. This function uses a gradient-expansion algorithm to compute a nonlinear least squares fit (for more details, see the IDL online help).

Once we have fit these mode parameters for the m-averaged spectrum, we use them as the initial guesses for fitting the individual m spectra. Then for each ℓ and n we can fit a set of a-coefficients to the frequencies as functions of m/ℓ. The a-coefficients are measures of the degree of departure from spherical symmetry and are described by

$$\omega_{nlm} = \omega_{nl} + \sum_{j=1}^{j_{\max}} a_j(n,l) \mathcal{P}_j^{(l)}(m), \qquad (2)$$

where $\mathcal{P}_j^{(l)}(m)$ are polynomials of degree j (Schou, Christensen-Dalsgaard, and Thompson, 1994). Although for the quiet Sun we would expect all of the a-coefficients to be zero, this calculation is still necessary to perform the noise subtraction.

We also use the mode parameters from the m-averaged spectrum of the quiet simulation as initial guesses for fitting the (unshifted) m-averaged spectrum of the perturbed simulation. Although the perturbations may lift the degeneracy in m, we expect the splitting to be very small, so that the peaks in the m-averaged spectrum can still be well represented by a Lorentzian. We also use those same initial guesses for fitting the individual m spectra of the perturbed simulation and recalculate the a-coefficients.

An empirical estimate of the error in frequency differences (not taking fit errors into account) for the sound-speed perturbation at $r = 0.7 R_\odot$ (see Section 4.1) is computed in the following manner. We look at the difference in mode parameters only for those modes that do not penetrate to the depth of the perturbation (all modes with $\nu/(\ell + 1/2) < 60$ µHz). We then make a histogram of these differences with a bin size of one and fit a Gaussian to the resulting distribution. With this method we find a standard deviation of 0.47 nHz. This result is confirmed by also computing the standard deviation of 95% of the closest points to the mean. Note that only the remnants of the noise-subtraction procedure contribute to this estimate. The error in estimating the frequency differences applies to the other cases (*i.e.*, the perturbations at $r = 0.55 R_\odot$, $1.0 R_\odot$) as well.

4. Results and Discussion

4.1. Sound-Speed Anomalies

We place three equatorially centered perturbations of horizontal size $8° \times 8°$ (in longitude and latitude) with a full width at half maximum in radius of 2% $R_\odot$ (13.9 Mm) at depths of $r = 0.55 R_\odot$, $0.7 R_\odot$, and $1.0 R_\odot$, each with an amplitude (α) of +5% of the local sound speed. Because of the fixed angular size, the perturbations grow progressively smaller in physical size with depth; our intention was to keep the perturbation as localized and non-spherically symmetric as possible. We were unable to perform a more exhaustive exploration of parameter space in terms of sizes and magnitudes of the perturbations because of the expensive nature of the calculation. Despite the fact that the perturbations at depth are subwavelength in size (with the wavelength at $r = 0.7 R_\odot$ being 76 Mm or 11% $R_\odot$), we notice that for these (relatively) small amplitude anomalies, the global-mode frequency shifts are predominantly a function of the spherically symmetric component of the spatial structure of the perturbation. In other words, what matters most is the contribution from the $\ell = 0$ coefficient in the spherical harmonic expansion of the horizontal spatial structure of the perturbation. We verify this by computing the frequency shifts associated with a spherically symmetric area-averaged version of the localized perturbation (with an amplitude of $0.05 A_{\text{local}}/(4\pi)$, where A_{local} is the solid angle subtended by the localized perturbation, with 0.05 referring to the 5% increase in sound speed). We were careful to ensure that the radial dependence of the magnitude of the perturbation was unchanged. The frequency shifts associated with the spherically symmetric perturbations were calculated independently through simulation and the oscillation frequency package ADIPACK (Christensen-Dalsgaard and Berthomieu, 1991) and were seen to match accurately, as shown in Figure 3.

Because of the nonspherically symmetric nature of the perturbation, we expect to see shifts in the a-coefficients. Similarly, it is likely that the reduction in the acoustic cavity size will reduce the mode mass, thereby resulting in a slight increase in the amplitudes of modes that propagate in regions close to and below the locations of the perturbation. We display these effects for the cases with the perturbations located at $r = 0.7 R_\odot$ and $1.0 R_\odot$ in Figure 4. The change in a_1 is extremely small and possibly well below detection thresholds. In addition, solar rotation creates far more significant changes in a_1, of the order of 400 µHz or so, making it all but impossible to study thermal asphericities using the a-coefficients.

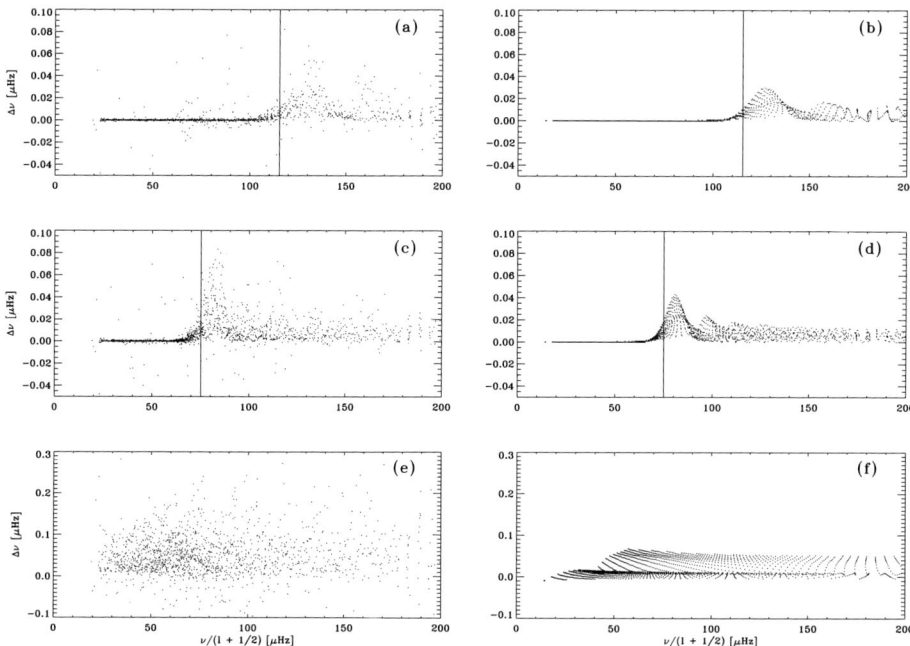

Figure 3 Frequency shifts ($\Delta \nu$) plotted against phase speed ($\nu/(\ell + 1/2)$) for the sound-speed perturbations of Section 4.1. The rows from upper to lower show frequency differences caused by anomalies at $r = 0.55 R_\odot$, $0.7 R_\odot$, and $1.0 R_\odot$, respectively. Solid lines indicate the phase speed of waves that have $r = 0.55 R_\odot$ and $0.7 R_\odot$ as inner turning points. Panels (a), (c), and e are results from the global mode analyses of the simulations; panels (b), (d), and (f) show frequency shifts obtained from ADIPACK for the spherically symmetric component of the corresponding perturbations. The perturbation reduces the size of the acoustic cavity – consequently modes whose inner turning points are just below the radial location of the anomaly are the ones most sensitive to it. In panels (e) and (f), it is apparent that all of the modes feel the presence of the relatively large near-surface hot spot. Noise subtraction does not remove the realization noise associated with the scattering process itself; therefore the spread in the frequency shifts of the simulated data is greater than ADIPACK ones. Of course, it may be that the scatter in the frequency difference is associated with the local nature of the anomaly itself.

4.2. Scattering Extent

We introduce a nondimensional measure (κ) to characterize the degree of scattering exhibited by the anomaly:

$$\kappa = 10^{-3} \frac{4\pi R_\odot^3}{3} \frac{1}{\alpha V} \sqrt{\frac{1}{N} \sum_{n,\ell} \left(\frac{\delta \nu}{\sigma}\right)^2}, \tag{3}$$

where $\alpha = \delta c/c$, the amplitude of the sound-speed perturbation expressed in fractions of the local sound speed, $\sigma = \sigma_{n,l}(\nu)$ is the formal observational frequency fit error, V is the volume of the perturbation, and N is the number of modes in the summation term. Essentially, this parameter tells us how clearly the perturbation registers in the modal frequency shifts, with larger κ implying a greater degree of detectability and *vice versa*. Because it is independent of perturbation size or magnitude, κ can be extended to study flow perturbations as well. This measure is meaningful only in the regime where the frequency shifts are presumably linear functions of the perturbation magnitude. It is reasonable to say that

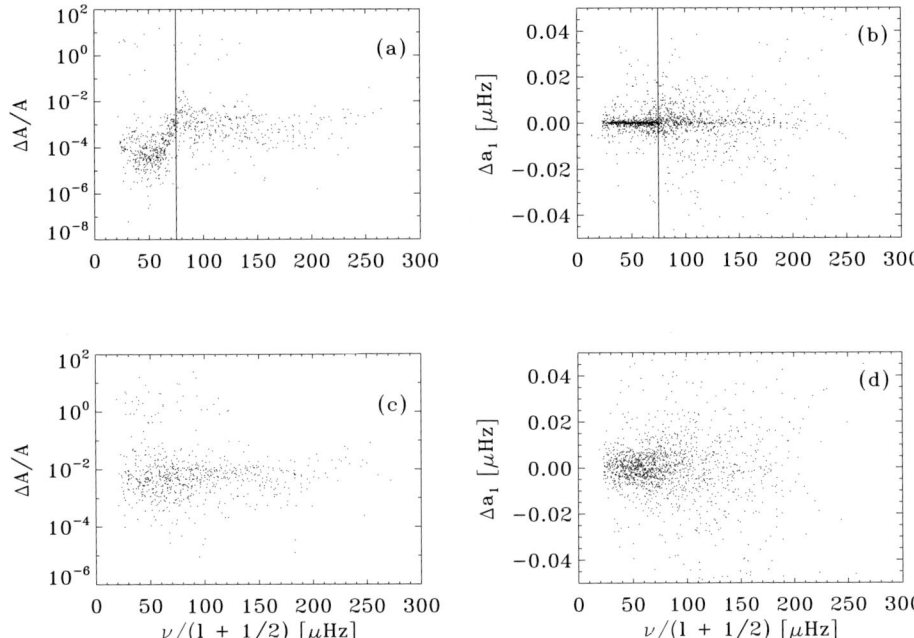

Figure 4 Changes in mode amplitudes ((a) and (c)) and the a_1-coefficients ((b) and (d)) from the sound-speed perturbation located at $r = 0.7R_\odot$ and $1.0R_\odot$, respectively, shown as functions of the phase speed of the waves. In panel (a) it seen that only modes that have turning points close to and below the location of the perturbation show changes in the amplitude (on the order of 0.5% or so). Owing to the spatially localized nature of the sound-speed perturbation, the m degeneracy is lifted, causing a dispersion in the a_1-coefficient for modes that propagate in the vicinity of the perturbation. Although not shown here, we observe that several a-coefficients, even and odd, show the presence of the perturbation. The near-surface perturbation affects all the modes, altering the amplitudes and a_1-coefficients.

κ is almost exclusively dependent on the radial location of the perturbation since different parts of the spectrum see different regions of the Sun (and hence it is an index of detectability). For example, placing an anomaly at the surface will likely affect the entire spectrum of global modes, as seen in Figure 3. By using values of σ derived from observations (72-day series, MDI medium-ℓ data, Figure 5; Schou, 1999) for modes common to the simulation, we are able to directly estimate the degree of detectability of comparable perturbations in the Sun. Results for κ shown in Table 1 contain no surprises; for a given size and magnitude of the perturbation, the effect on the global frequencies increases strongly with its location in radius. The signature of a perturbation at the bottom of the convection zone on the global modes is twice as strong as an anomaly in the radiative interior ($r = 0.55R_\odot$). The surface perturbation is a little more difficult to compare with the others because, in contrast to the two deeper perturbations, it is locally far larger than the wavelengths of the modes. The result however is in line with expectation; the near-surface scatterer is far more potent than the other two anomalies. Since the fit errors scale inversely with the square root of the length of the time series, we also place an estimate on the observational time length T_{req} required to detect these perturbations.

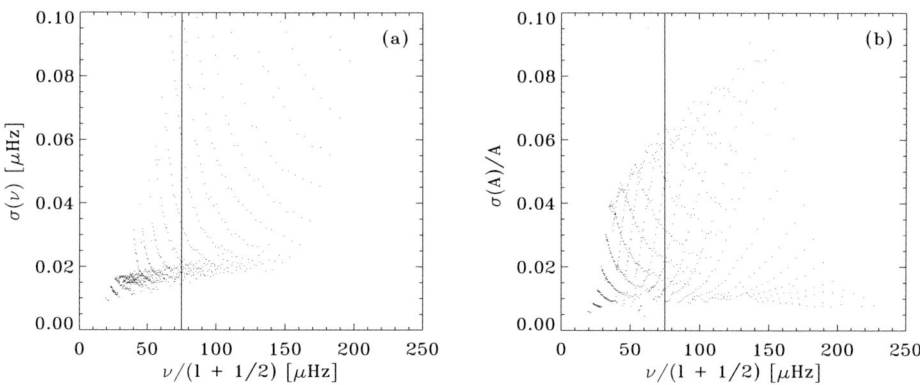

Figure 5 Formal one-σ errors in the fits to the frequencies (a) and amplitudes (b) of global modes from an analysis of an MDI 72-day medium-ℓ dataset. By sheer coincidence, the ranges on the two plots are identical. We only display modes that are approximately common to both the simulation and observations. From comparisons between the results of Figures 3(a), (c), and 4(a) and the formal errors in panel (a), it appears that these deep perturbations are marginally detectable. Of course, a longer time series would yield lower errors ($\propto T^{-1/2}$, where T is the duration) and the frequency differences resulting from the anomalies at $r = 0.55 R_\odot$ and $0.7 R_\odot$ may become visible.

Table 1 The scattering extents (κ) of various perturbations. The root mean square (RMS) variation in $\delta\nu/\sigma(\nu)$ is shown as well. T_{req} is the duration of observations required to cross the detectability threshold.

Depth ($r/R_\odot$)	RMS ($\delta\nu/\sigma(\nu)$)	κ	T_{req} (days)
0.55	0.7676	3.4061	642
0.70	2.6252	7.1916	144
1.00	4.0348	10.3645	70

5. Conclusion

We have introduced a method to systematically study the effects of various local perturbations on global mode frequencies. Techniques of mode finding and parameter fitting are applied to artificial data obtained from simulations of wave propagation in a solar-like stratified spherical shell. We are able to beat the issue of poor frequency resolution by extending the method of realization noise subtraction (Hanasoge, Duvall, and Couvidat, 2007) to global-mode analysis. These methods can prove very useful in the study of shifts caused by perturbations of magnitudes beyond the scope of first-order perturbation theory; moreover, extending this approach to investigate systematic frequency shifts in other stars may prove exciting. The study of wave interactions with complex flow fields and magnetic fields also becomes accessible. However, a downside of these computations is their expensive nature – the numerical experiments must be chosen somewhat carefully.

In relation to the perturbations studied here, we find that *i*) global modes are sensitive to the spherically symmetric component of localized thermal asphericities, *ii*) the duration of observations required to detect anomalies in the Sun comparable to those studied here is anywhere between 70 and 700 days, and *iii*) the asphericity registers, albeit well below the detectability threshold, in the a-coefficients. Note that the estimates of Table 1 could be scaled to evaluate the strength of the signatures of other such local thermal perturbations. We are currently studying the impact of complex flows such as convection and localized jets on the global frequencies. Preliminary results seem to indicate that flows are stronger scatterers (have larger κ) than sound-speed perturbations although more work needs to be done to confirm and characterize these effects.

Acknowledgements S.M.H. and T.P.L. were funded by Grant Nos. HMI NAS5-02139 and MDI NNG05GH14G. We would like to thank Jesper Schou, Tom Duvall Jr., Phil Scherrer, and an anonymous referee for useful discussions and suggestions. The simulations were performed on the Columbia supercomputer at NASA Ames.

References

Birch, A.C., Braun, D.C., Hanasoge, S.M.: 2008, *Solar Phys.* submitted.
Cameron, R., Gizon, L., Daifallah, K.: 2007, *Astron. Nach.* **328**, 313.
Christensen-Dalsgaard, J.: 2002, *Rev. Mod. Phys.* **74**, 1073.
Christensen-Dalsgaard, J.: 2003, *Lecture Notes on Stellar Oscillations*. http://astro.phys.au.dk/jcd/oscilnotes/.
Christensen-Dalsgaard, J., Berthomieu, G.: 1991, In: *Solar Interior and Atmosphere*, University of Arizona Press, Tucson, 401.
Christensen-Dalsgaard, J., Dappen, W., Ajukov, S.V., Anderson, E.R., Antia, H.M., Basu, S., Baturin, V.A., Berthomieu, G., Chaboyer, B., Chitre, S.M., Cox, A.N., Demarque, P., Donatowicz, J., Dziembowski, W.A., Gabriel, M., Gough, D.O., Guenther, D.B., Guzik, J.A., Harvey, J.W., Hill, F., Houdek, G., Iglesias, C.A., Kosovichev, A.G., Leibacher, J.W., Morel, P., Proffitt, C.R., Provost, J., Reiter, J., Rhodes, E.J., Jr., Rogers, F.J., Roxburgh, I.W., Thompson, M.J., Ulrich, R.K.: 1996, *Science* **272**, 1286.
Christensen-Dalsgaard, J., Corbard, T., Dikpati, M., Gilman, P.A., Thompson, M.J.: 2005, In: Sankarasubramanian, K.S., Penn, M.J., Pevtsov, A.A. (eds.) *Large-scale Structures and their Role in Solar Activity* **CS-346**, Astron. Soc. Pac., San Francisco, 115.
Georgobiani, D., Stein, R.F., Nordlund, Å., Kosovichev, A.G., Mansour, N.N.: 2004, In: Danesy, D. (ed.) *SOHO 14 Helio- and Asteroseismology: Towards a Golden Future* **SP-559**, ESA, Noordwijk, 267.
Hanasoge, S.M.: 2007, Theoretical studies of wave propagation in the Sun. Ph.D. thesis, Stanford University. http://soi.stanford.edu/papers/dissertations/hanasoge/.
Hanasoge, S.M., Duvall, T.L., Jr.: 2007, *Astron. Nachr.* **323**, 319.
Hanasoge, S.M., Duvall, T.L., Jr., Couvidat, S.: 2007, *Astrophys. J.* **664**, 1234.
Hanasoge, S.M., Larsen, R.M., Duvall, T.L., Jr., DeRosa, M.L., Hurlburt, N.E., Schou, J., Roth, M., Christensen-Dalsgaard, J., Lele, S.K.: 2006, *Astrophys. J.* **648**, 1268.
Hanasoge, S.M., Couvidat, S., Rajaguru, S.P., Birch, A.C.: 2007, ArXiv:0707.1369.
Lindsey, C., Braun, D.C.: 2000, *Science* **287**, 1799.
Parchevsky, K., Kosovichev, A.G.: 2007, *Astrophys. J.* **666**, 547.
Rosenthal, C.S., Christensen-Dalsgaard, J., Nordlund, Å., Stein, R.F., Trampedach, R.: 1999, *Astron. Astrophys.* **351**, 689.
Scherrer, P.H., Bogart, R.S., Bush, R.I., Hoeksema, J.T., Kosovichev, A.G., Schou, J., Rosenberg, W., Springer, L., Tarbell, T.D., Title, A., Wolfson, C.J., Zayer, I., MDI Engineering Team: 1995, *Solar Phys.* **162**, 129.
Schou, J.: 1999, *Astrophys. J.* **523**, L181.
Schou, J., Christensen-Dalsgaard, J., Thompson, M.J.: 1994, *Astrophys. J.* **433**, 389.
Swisdak, M., Zweibel, E.: 1999, *Astrophys. J.* **512**, 442.
Werne, J., Birch, A., Julien, K.: 2004, In: Danesy, D. (ed.) *SOHO 14 Helio- and Asteroseismology: Towards a Golden Future* **SP-559**, ESA, Noordwijk, 172.

Angular Momentum Transport in the Sun's Radiative Zone by Gravito-Inertial Waves

S. Mathis · S. Talon · F.-P. Pantillon · J.-P. Zahn

Originally published in the journal Solar Physics, Volume 251, Nos 1–2, 101–118.
DOI: 10.1007/s11207-008-9157-0 © Springer Science+Business Media B.V. 2008

Abstract Internal gravity waves constitute an efficient process for angular momentum transport over large distances. They are now seen as an important ingredient in understanding the evolution of stellar rotation and can explain the Sun's quasi-flat internal-rotation profile. Because the Sun's rotation frequency is of the same order as that of the waves, it is now necessary to refine our description of wave propagation and to take into account the action of the Coriolis acceleration in a coherent way. To achieve this goal, we adopt the traditional approximation, which can be applied to stellar radiation zones under conditions that are given. We present the modified transport equations and their numerical evaluation in a parameter range that is significant for the Sun. Consequences for the transport of angular momentum inside solar and stellar radiative regions are discussed.

Helioseismology, Asteroseismology, and MHD Connections
Guest Editors: Laurent Gizon and Paul Cally.

S. Mathis (✉)
Laboratoire AIM, CEA/DSM-CNRS, Université Paris Diderot, DAPNIA/SAp, 91191 Gif-sur-Yvette Cedex, France
e-mail: stephane.mathis@cea.fr

S. Mathis · J.-P. Zahn
LUTH, Observatoire de Paris, CNRS, Université Paris Diderot, 5 Place Jules Janssen, 92195 Meudon, France

J.-P. Zahn
e-mail: jean-paul.zahn@obspm.fr

S. Talon · F.-P. Pantillon
Département de Physique, Université de Montréal, Montréal, PQ H3C 3J7, Canada

S. Talon
e-mail: talon@astro.umontreal.ca

F.-P. Pantillon
e-mail: pantillon@astro.umontreal.ca

F.-P. Pantillon
Observatoire de Genève, Université de Genève, 51 ch. des Maillettes, 1290 Sauverny, Switzerland

Keywords Hydrodynamics · Turbulence · Waves · Methods: numerical · Stars: interiors · Stars: rotation

1. Introduction and Context

In standard models of stellar interiors, radiation zones that are convectively stable are postulated to be without motion other than rotation. But various observational results (*e.g.*, surface abundances of light elements and helioseismology) show that these regions are the seat of transport and of mild mixing. The most likely cause of such mixing is stellar rotation. First, it causes thermal imbalance that drives a large-scale meridional circulation. Second, since in general the star does not rotate as a solid body, shear instabilities may appear (for a review of these processes, see *e.g.*, Talon, 2007). A series of models have been built that include a self-consistent evolution of the internal-rotation profile, and for massive stars they agree rather well with the observations (see Maeder and Meynet, 2000).

The case of the Sun is somewhat different: Like all other stars that have a deep surface convection zone, it has been spun down during its infancy. When only the meridional circulation and the "classical" hydrodynamic instabilities are invoked, models predict a Sun with a core rotating much faster than the surface (Pinsonneault *et al.*, 1989; Chaboyer, Demarque, and Pinsonneault, 1995; Talon, 1997; Matias and Zahn, 1997) with a gradient that is incompatible with helioseismology (Brown *et al.*, 1989; Turck-Chièze *et al.*, 2004; García *et al.*, 2007). One must conclude therefore that another, more powerful process is operating, at least in slow rotators. The most plausible candidates are magnetic torquing (Gough and McIntyre, 1998; Garaud, 2002, and references therein; Brun and Zahn, 2006) and momentum transport by internal gravity waves (hereafter IGWs; Schatzman, 1993; Zahn, Talon, and Matias, 1997; Talon, Kumar, and Zahn, 2002; Talon and Charbonnel, 2005).

The treatment of these waves suffers from two major weaknesses. The first is the crude description of their generation by turbulent convection. The second is that the action of rotation on the waves is not taken into account. This is why we have undertaken to improve the modeling of the transport by internal waves by introducing the effects of the Coriolis acceleration. Indeed, the low-frequency internal waves that are responsible for the deposition or the extraction of angular momentum (*cf.* Talon, Kumar, and Zahn, 2002) are strongly influenced by the rotation because the frequencies of the waves are of the same order as the inertial frequency (2Ω). Thus, internal waves become gravito-inertial waves (*cf.* Berthomieu *et al.*, 1978; Lee and Saio, 1997; Dintrans, Rieutord, and Valdettaro, 1999), and one has to treat the action of the rotation on the waves and their feedback on its profile.

2. Structure of Low-Frequency Waves Influenced by the Coriolis Acceleration

In stellar radiative zones, the transport of angular momentum is dominated by low-frequency waves with $\sigma \ll N$, where σ and N are, respectively, the wave frequency and the Brunt–Väisälä frequency, which relates to buoyancy. In a (differentially) rotating stellar radiative zone, one must also consider the Coriolis acceleration, which is characterized by the inertial frequency (2Ω), where Ω is the angular velocity of the star. One then has to quantify the relative importance of each restoring force in the wave dynamics and determine whether the effects of the Coriolis acceleration can be treated in a perturbative way.

In the Sun, the answer is very clear for the acoustic waves, which have frequencies much greater than $\Omega_\odot$. The effects of the Coriolis acceleration can be treated as a perturbation

Figure 1 Rotation parameter $\nu = 2\Omega/\sigma$ for the Sun (in which we use $\Omega_\odot/2\pi = 430$ nHz) in the frequency range relevant for the calculation of angular momentum transport.

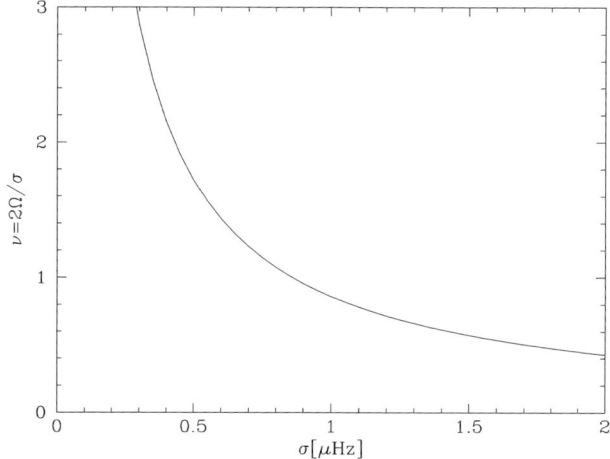

Figure 2 Wave types in a differentially rotating stellar radiative zone and associated frequencies (where f_s is the acoustic cutoff frequency).

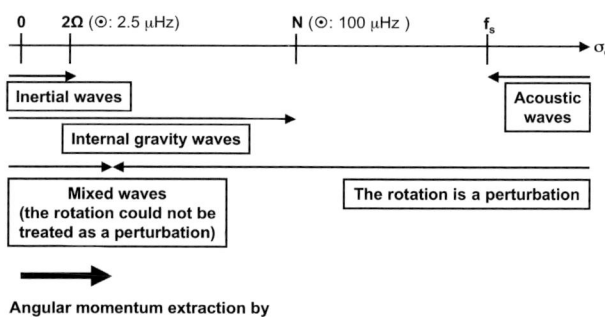

(*e.g.*, rotationally split frequencies). However, in the case of low-frequency IGWs, which have frequencies around 1 µHz, the spin parameter (ν), which measures the relative importance of rotation and stratification and is given by

$$\nu = \frac{2\Omega}{\sigma} = R_o^{-1}, \qquad (1)$$

where R, the Rossby number, is of the order of unity (see Figure 1). In this case, as illustrated in Figure 2, Coriolis effects cannot be treated as a perturbation.

The aim of this work is to examine how the improved description of wave-induced transport of angular momentum that takes into account the effect of the Coriolis acceleration modifies the spatial structure of IGWs and, consequently, the angular-momentum extraction from the solar interior by IGWs. We first recall the main assumptions of our model and give the corresponding dynamical equations. The derivation of these equations has been presented in Mathis (2005) and Pantillon, Talon, and Charbonnel (2007).

2.1. Dynamical Equations

We expand the macroscopic internal-velocity field in the radiative region as

$$\mathbf{V}(\mathbf{r},t) = r\sin\theta\, \Omega(r,\theta)\widehat{\mathbf{e}}_\varphi + \mathbf{u}(r,\theta,\varphi,t), \qquad (2)$$

where the first azimuthal term is the velocity associated with the differential rotation and **u** is the wave velocity field. The coordinates r, θ, and φ are the classical spherical coordinates with their associated unit vectors $\{\widehat{\mathbf{e}}_r, \widehat{\mathbf{e}}_\theta, \widehat{\mathbf{e}}_\varphi\}$ and t is the time. We ignore any large-scale velocity field that could be superposed.

The perturbations representing the waves obey the linearized momentum equation

$$D_t \mathbf{u} + \left[2\Omega \widehat{\mathbf{e}}_z \times \mathbf{u} + r\sin\theta \mathbf{u} \cdot \nabla \Omega \widehat{\mathbf{e}}_\varphi\right] = -\frac{\nabla P'}{\overline{\rho}} + \frac{\rho'}{\overline{\rho}}\mathbf{g}. \qquad (3)$$

The terms in brackets correspond to the Coriolis acceleration in the case of a general differential rotation with the extra acceleration $r\sin\theta \mathbf{u} \cdot \nabla \Omega \widehat{\mathbf{e}}_\varphi$ (Talon, 1997); $\overline{\rho}$ is the density (with $\overline{X}$ the mean value of X on the isobar, which is the generalization of the equipotential in the case of a differential rotation that disturbs the hydrostatic balance through the centrifugal force), **g** is the gravity, and ρ' and P' are, respectively, the Eulerian perturbations of density and pressure. $D_t = (\partial_t + \Omega \partial_\varphi)$ is the Lagrangian derivative that accounts for the Doppler shift owing to the differential rotation. The density fluctuations are linked to the velocity field by the continuity equation

$$D_t \rho' + \nabla \cdot (\overline{\rho}\mathbf{u}) = 0, \qquad (4)$$

and the conservation of energy equation is written here in the adiabatic limit:

$$D_t \left(\frac{\rho'}{\overline{\rho}} - \frac{1}{\Gamma_1}\frac{P'}{\overline{P}}\right) + \left[\frac{d\ln\overline{\rho}}{dr} - \frac{1}{\Gamma_1}\frac{d\ln\overline{P}}{dr}\right]u_r = 0, \qquad (5)$$

where $\Gamma_1 = (\partial \ln \overline{P}/\partial \ln \overline{\rho})_{\overline{S}}$ is the adiabatic exponent and $\overline{P}$ and $\overline{S}$ are, respectively, the mean pressure and macroscopic entropy. The nonadiabaticity of the waves will be treated in the next section by using the quasi-adiabatic approximation (*cf.* Press, 1981).

2.2. Main Assumptions

Let us begin by looking at the differential rotation law. First, we consider here a "shellular" rotation $\overline{\Omega}[r(P)]$ (in other words, the angular velocity is constant on an isobar) caused by anisotropic turbulence in a highly stratified star (Zahn, 1992); $r(P)$ is the radius of the isobar, which is the generalization of the equipotential in the case of differential rotation. Next, we split this "shellular" rotation law into a solid-body rotation $(\overline{\Omega})$ and a (small) differential rotation fluctuation $(\delta\overline{\Omega}(r))$. This hypothesis will allow us to separate the variables in the treatment of the dynamical equations. The formalism presented here remains valid only in the case of "reasonable" values of the fluctuations $\delta\overline{\Omega}$ of the angular velocity around its mean value $\overline{\Omega}_s$ and of the radial gradient of Ω. Thus, we write

$$\Omega(r,\theta) \approx \overline{\Omega}(r) = \overline{\Omega}_s + \delta\overline{\Omega}(r), \quad \text{where } \delta\overline{\Omega}(r) \ll \overline{\Omega}_s. \qquad (6)$$

$\overline{\Omega}_s$ will be taken into account for the calculation of the structure of the low-frequency adiabatic waves, and $\delta\overline{\Omega}$ will be accounted for only in the treatment of the damping from dissipative processes. This is the "weak differential-rotation case."

We now define the frequencies we shall use in this work:

$$\begin{cases} \sigma_s = \sigma_0 + m\overline{\Omega}_s, \\ \sigma(r) = \sigma_0 + m\overline{\Omega}(r) = \sigma_s + m\delta\overline{\Omega}(r), \end{cases} \qquad (7)$$

where σ_0 is the wave's frequency in an inertial frame and σ_s is given in the corotating frame (with the rotation angular velocity $\overline{\Omega}_s$). $\sigma(r)$ is the local frequency, Doppler shifted by the differential rotation. Note that we chose the sign of m such that the prograde waves have $m < 0$ whereas the retrograde waves have $m > 0$.

In the general case, the operator that governs the spatial structure of the waves, the Poincaré operator, is of mixed type (elliptic and hyperbolic) and not separable. (For a detailed discussion we refer the reader to Friedlander and Siegmann (1982), Dintrans (1999), Dintrans, Rieutord, and Valdettaro (1999), and Dintrans and Rieutord (2000).) This leads to the appearance of detached shear layers associated with the underlying singularities of the adiabatic problem, which could be crucial for transport and mixing processes in stellar radiation zones since they are the seat of strong dissipation (Stewartson and Richard, 1969; Stewartson and Walton, 1976; Dintrans, Rieutord, and Valdettaro, 1999; Dintrans and Rieutord, 2000).

However, in the largest part of stellar radiation zones, we are in a regime where $2\Omega \ll N$. Since we are interested here in low-frequency waves where $\sigma \ll N$, the traditional approximation. which consists in neglecting the horizontal component of the rotation vector $\Omega \hat{\mathbf{e}}_z$ in the momentum equation, can be adopted in the case where $2\Omega < \sigma (\nu < 1)$ (see, e.g., Eckart, 1960; for a modern description in a stellar context see Bildsten, Ushomirsky, and Cutler, 1996; Lee and Saio, 1997; Talon, 1997). In this frequency domain, it has been shown by Friedlander (1987) that variable separation in radial and horizontal components remains possible. This approximation has to be used carefully, as it changes the nature of the Poincaré operator, and it removes the singularities and associated shear layers that appear. It is valid only in the super-inertial regime $2\Omega < \sigma \ll N$ ($\nu < 1$) where the stratification dominates, which corresponds to the ergodic (regular) elliptic gravito-inertial mode family (the E_1 modes in Dintrans, Rieutord, and Valdettaro, 1999, and Dintrans and Rieutord, 2000). In the sub-inertial regime where $\sigma \leq 2\Omega$ ($\nu \geq 1$), which corresponds to the equatorially trapped hyperbolic modes (the H_2 modes in Dintrans, Rieutord, and Valdettaro, 1999, and Dintrans and Rieutord, 2000), the traditional approximation fails to reproduce the wave behavior and the complete dynamical equation has to be solved (detailed examples of which are given in Gerkema and Shrira, 2005, and Gerkema et al., 2008).

Therefore, we thus restrict ourselves here to the regular super-inertial modes ($\sigma > 2\Omega$, $\nu < 1$) for which the traditional approximation is applicable.

Finally, we neglect the fluctuations of the gravitational potential associated with the waves (thereby adopting the Cowling approximation; see, e.g., Cowling, 1941) as well as the effects of the centrifugal force since we are interested here in slow rotators.

2.3. Wave Velocity and Pressure Fields

Two additional approximations are made to simplify the problem. The first is the JWKB approximation, which is adopted for the radial dependency of pressure fluctuations and all three velocity components; it is justified because we are studying the low-frequency regime where $\sigma \ll N$. The second approximation is that dissipative processes such as thermal and turbulent diffusions are treated by using the quasi-adiabatic approximation (cf. Press, 1981; Zahn, Talon, and Matias, 1997). The velocity field of the waves is expanded as

$$\mathbf{u} = \begin{cases} \sum_{m,k} u_{r;k,m}(r,\theta,\varphi,t) \\ \sum_{m,k} u_{\theta;k,m}(r,\theta,\varphi,t), \\ \sum_{m,k} u_{\varphi;k,m}(r,\theta,\varphi,t) \end{cases} \qquad (8)$$

where the monochromatic radial, latitudinal, and azimuthal components are, respectively, given by

$$u_{r;k,m}(r,\theta,\varphi,t) = \mathcal{E}_{k,m}(r)\sin[\Phi_{k,m}(r,\varphi,t)]\Theta_{k,m}(\cos\theta;\nu_s)$$
$$\times \exp[-\tau_{k,m}(r,\delta\overline{\Omega}(r);\nu_s)/2], \tag{9}$$

$$u_{\theta;k,m}(r,\theta,\varphi,t) = -\frac{rk_{V;k,m}}{\Lambda_{k,m}(\nu_s)}\mathcal{E}_{k,m}(r)\cos[\Phi_{k,m}(r,\varphi,t)]\mathcal{H}^{\theta}_{k,m}(\cos\theta;\nu_s)$$
$$\times \exp[-\tau_{k,m}(r,\delta\overline{\Omega}(r);\nu_s)/2], \tag{10}$$

$$u_{\varphi;k,m}(r,\theta,\varphi,t) = \frac{rk_{V;k,m}}{\Lambda_{k,m}(\nu_s)}\mathcal{E}_{k,m}(r)\sin[\Phi_{k,m}(r,\varphi,t)]\mathcal{H}^{\varphi}_{k,m}(\cos\theta;\nu_s)$$
$$\times \exp[-\tau_{k,m}(r,\delta\overline{\Omega}(r);\nu_s)/2]. \tag{11}$$

The Hough functions ($\Theta_{k,m}$, $\mathcal{H}^{\theta}_{k,m}$, and $\mathcal{H}^{\varphi}_{k,m}$) will be discussed later in this section, and the attenuation ($\tau_{k,m}$) will be made explicit in Equation (22). For the problem of angular-momentum transport by IGWs, we are interested in propagating waves. In the JWKB regime, the phase function ($\Phi_{k,m}$) is given by

$$\Phi_{k,m}(r,\varphi,t) = \sigma_0 t + \int_r^{r_c} k_{V;k,m}\,dr' + m\varphi, \tag{12}$$

where the vertical wave vector

$$k^2_{V;k,m} = \left(\frac{N^2}{\sigma^2(r)} - 1\right)\frac{\Lambda_{k,m}(\nu_s)}{r^2} \tag{13}$$

has been drawn from the equation for the radial component of the Lagrangian displacement ξ, where $\mathbf{u} = d\xi/dt$:

$$\frac{d^2}{dr^2}(\overline{\rho}^{1/2}r^2\xi_{r;k,m}) + \left[\left(\frac{N^2}{\sigma^2(r)} - 1\right)\frac{\Lambda_{k,m}(\nu_s)}{r^2}\right](\overline{\rho}^{1/2}r^2\xi_{r;k,m}) = 0. \tag{14}$$

This equation, obtained after variable separation in r and θ in Equations (3), (4), and (5), has been derived by using the anelastic approximation where sonic waves are filtered. The JWKB amplitude function [$\mathcal{E}_{k,m}(r)$] is given by

$$\mathcal{E}_{k,m}(r) = \mathcal{A}_{k,m} r^{-\frac{3}{2}} \rho^{-\frac{1}{2}} \left(\frac{N^2}{\sigma^2} - 1\right)^{-\frac{1}{4}}, \tag{15}$$

where the amplitude of the wave ($\mathcal{A}_{k,m}$) must be determined from boundary conditions. The Brunt–Väisälä frequency takes into account the effects of both the thermal and the chemical composition gradients, with the classical notations $N^2 = N_T^2 + N_\mu^2$, where $N_T^2 = (\overline{g}\delta/H_P)(\nabla_{\text{ad}} - \nabla)$ and $N_\mu^2 = (\overline{g}\phi/H_P)\nabla_\mu$, with $H_P = |dr/d\ln\overline{P}|$, $\delta = -(\partial\ln\overline{\rho}/\partial\ln\overline{T})_{\overline{P},\overline{\mu}}$, $\phi = (\partial\ln\overline{\rho}/\partial\ln\overline{\mu})_{\overline{P},\overline{T}}$, $\nabla_{\text{ad}} = (\partial\ln\overline{T}/\partial\ln\overline{P})_{\text{ad}}$, $\nabla = \partial\ln\overline{T}/\partial\ln\overline{P}$, and $\nabla_\mu = \partial\ln\overline{\mu}/\partial\ln\overline{P}$, with $\overline{T}$ and $\overline{\mu}$ being, respectively, the mean temperature and mean molecular weight.

As already mentioned, this separation of variables is allowed by the traditional approximation and by the "weak differential-rotation approximation"; it leads to an equation that depends only on θ for the angular function of $u_{r;k,m}$, $\Theta_{k,m}$:

$$\mathcal{L}_{\nu_s;m}[\Theta_{k,m}(x;\nu_s)] = -\Lambda_{k,m}(\nu_s)\Theta_{k,m}(x;\nu_s) \tag{16}$$

with

$$\mathcal{L}_{\nu_s;m} = \frac{d}{dx}\left(\frac{1-x^2}{1-\nu_s^2 x^2}\frac{d}{dx}\right) - \frac{1}{1-\nu_s^2 x^2}\left(\frac{m^2}{1-x^2} + m\nu_s\frac{1+\nu_s^2 x^2}{1-\nu_s^2 x^2}\right), \tag{17}$$

where $x = \cos\theta$ and $\nu_s = 2\overline{\Omega}_s/\sigma_s$ is the spin parameter associated with $\overline{\Omega}_s$. The differential rotation fluctuation $\delta\Omega$ is only taken into account for the Doppler shift of the waves and their dissipation, but not in the derivation of their horizontal spatial structure. In fact, in the more general case where $\Omega(r,\theta) = \overline{\Omega}(r)$, $\nu = 2\overline{\Omega}(r)/\sigma$ depends on r as well as the associated eigenvalues ($\Lambda_{k,m}$) and eigenfunctions ($\Theta_{k,m}$) of $\mathcal{L}_{\nu;m}$ and the variables no longer separate. Equation (16) is the so-called Laplace equation (cf. Laplace, 1799) and the $\Theta_{k,m}$ are the Hough functions (cf. Hough, 1898; Longuet-Higgins, 1968). In the nonrotating case, the Laplace operator ($\mathcal{L}_{\nu_s;m}$) is equivalent to the classical horizontal spherical Laplacian, and the Hough functions reduce to the associated Legendre polynomials.

Let us briefly describe the main features of these Hough functions. First, since $\mathcal{L}_{\nu_s;m}$ depends explicitly on m, we have $\Lambda_{k,-m} \neq \Lambda_{k,m}$ and $\Theta_{k,-m} \neq \Theta_{k,m}$. In other words, for a given k, prograde and retrograde waves have a different horizontal spatial structure (see Figure 4). This is crucial for the transport of angular momentum by those waves, which depends on the subtle balance between prograde and retrograde waves (cf. Talon and Charbonnel, 2005, and references therein). Moreover, as will be shown later, this modifies the transmission of the kinetic energy flux of the turbulent motions, which are at the origin of the generation of the waves at the interface between the convective envelope and the radiative core.

Now, by using the latitudinal and the azimuthal components of the momentum equation (Equation (3)), and eliminating the pressure fluctuation, we obtain the respective angular functions for $u_{\theta;k,m}$ and $u_{\varphi;k,m}$, $\mathcal{H}^\theta_{k,m}(x;\nu_s)$ and $\mathcal{H}^\varphi_{k,m}(x;\nu_s)$:

$$\mathcal{H}^\theta_{k,m}(x;\nu_s) = \mathcal{L}^\theta_{\nu_s;m}[\Theta_{k,m}(x;\nu_s)], \tag{18}$$

where

$$\mathcal{L}^\theta_{\nu_s;m} = \frac{1}{(1-x^2\nu_s^2)\sqrt{1-x^2}} \times \left[-(1-x^2)\frac{d}{dx} + m\nu_s x\right] \tag{19}$$

and

$$\mathcal{H}^\varphi_{k,m}(x;\nu_s) = \mathcal{L}^\varphi_{\nu_s;m}[\Theta_{k,m}(x;\nu_s)], \tag{20}$$

where

$$\mathcal{L}^\varphi_{\nu_s;m} = \frac{1}{(1-x^2\nu_s^2)\sqrt{1-x^2}} \times \left[-\nu_s x(1-x^2)\frac{d}{dx} + m\right]. \tag{21}$$

As we had for the Hough functions, we get $\mathcal{H}^\theta_{k,-m} \neq \mathcal{H}^\theta_{k,m}$ and $\mathcal{H}^\varphi_{k,-m} \neq \mathcal{H}^\varphi_{k,m}$ since $\mathcal{L}^\theta_{\nu_s;-m} \neq \mathcal{L}^\theta_{\nu_s;m}$, $\mathcal{L}^\varphi_{\nu_s;-m} \neq \mathcal{L}^\varphi_{\nu_s;m}$, and $\Theta_{k,-m} \neq \Theta_{k,m}$.

Those properties are illustrated in Figures 3 and 4, where the eigenfunctions $\Theta_{k,m}$, $\mathcal{H}^\theta_{k,m}$, $\mathcal{H}^\varphi_{k,m}$ and their associated eigenvalues $\Lambda_{k,m}$ are given for prograde and retrograde waves in the case where $\nu_s = 0.86$ ($\sigma = 1$ μHz).

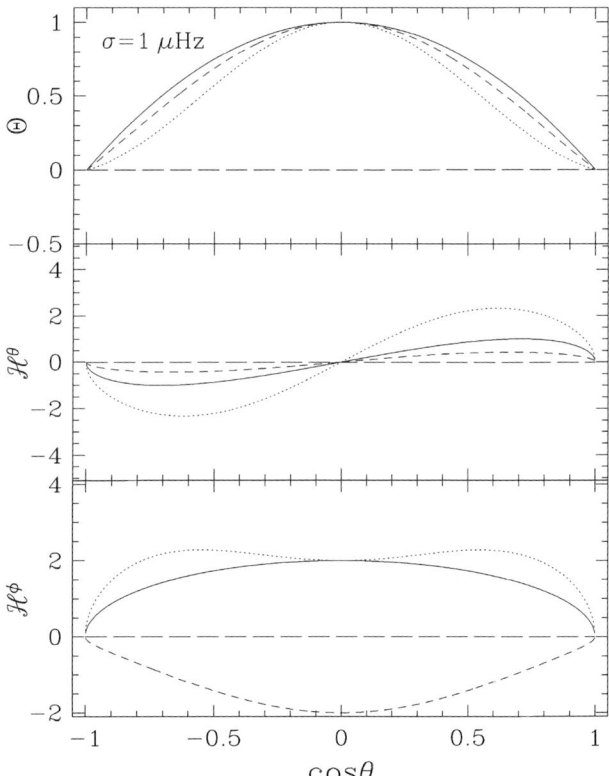

Figure 3 Hough functions: $\nu = 0$, $m = 2$ (continuous lines); $m = 2$, $\sigma = 1$ μHz, $\nu = 0.86$ (dotted lines); $m = -2$, $\sigma = 1$ μHz, $\nu = 0.86$ (dashed lines).

Finally, using the quasi-adiabatic approximation, we derive the radiative damping as

$$\tau_{k,m}\left(r; \delta\overline{\Omega}(r); \nu_s\right) = \Lambda_{k,m}^{3/2}(\nu_s) \int_r^{r_c} K \frac{N N_T^2}{\sigma^4} \sqrt{\frac{N^2}{N^2 - \sigma^2}} \frac{dr'}{r'^3}, \quad (22)$$

where K is the thermal diffusivity. As has been previously emphasized, $\Lambda_{k,-m} \neq \Lambda_{k,m}$; the respective radiative damping associated with the Doppler shift caused by the differential rotation is thus modified as well as the deposition and extraction of angular momentum.

Following Pantillon, Talon, and Charbonnel (2007), we define a horizontal wave number given by

$$k_{H;k,m} = \frac{\lambda_{k,m}^{1/2}(\nu_s)}{r}, \quad (23)$$

where

$$\lambda_{k,m}^2(\nu_s) = \frac{\langle |r^2 \nabla_H^2 \Theta_{k,m}(\cos\theta; \nu_s)|^2 \rangle_\theta}{\langle |\Theta_{k,m}(\cos\theta, \nu_s)|^2 \rangle_\theta}, \quad (24)$$

and where $\langle \cdots \rangle_\theta = \frac{1}{2} \int_0^\pi \cdots \sin\theta\, d\theta$ and ∇_H^2 is the horizontal spherical Laplacian. In the absence of rotation, we recover $\lambda_{k,m}(\nu_s) = \Lambda_{k,m}(\nu_s) = l(l+1)$.

Finally, keeping the radial component of the momentum equation (Equation (3)) and the energy equation (Equation (5)) and using once again the anelastic approximation, we can

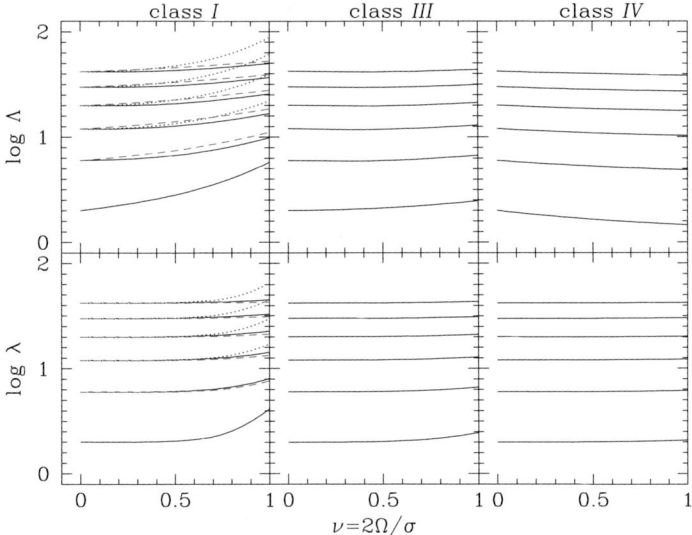

Figure 4 (Top) Eigenvalues (Λ) of Laplace's tidal equation in the presence of rotation, in the range relevant for a solar modal and the traditional approximation. In the absence of rotation, one has $\Lambda = \ell(\ell+1)$. (Bottom) Equivalent horizontal eigenvalue (λ) (see text for details). (Left) Class I waves with $1 \leq \ell \leq 6$, and $m = -\ell + 2$ (black continuous line), $m = 0$ (blue long dashed line), and $m = \ell$ (red dashed line). (Middle) Class III waves ($s = 0$) with negatives values of m ($m = 0, \ldots, -5$) are present in the relevant range of ν. (Right) Class IV waves ($s = -1$) have indices $m = -1, \ldots, -6$.

derive the monochromatic Eulerian pressure fluctuation ($P'_{k,m}$) as

$$P'_{k,m}(r, \theta, \varphi, t) = -\overline{\rho} \frac{\sigma}{k_{V;k,m}} \left(\frac{N^2}{\sigma^2} - 1 \right) \mathcal{E}_{k,m}(r) \sin\left[\Phi_{k,m}(r, \varphi, t)\right]$$

$$\times \Theta_{k,m}(\cos\theta; \nu_s) \exp\left[-\tau_{k,m}\left(r, \delta\overline{\Omega}(r); \nu_s\right)/2\right]. \quad (25)$$

When the Coriolis acceleration is taken into account in a rotating stably stratified radiative region, low-frequency internal gravity waves are thus modified and become gravito-inertial waves.

Under the traditional approximation, four types of gravito-inertial waves may be identified (Miles, 1977; Pedlosky, 1987; Townsend, 2003):

- *Class I waves* are internal gravity waves, which exist in the nonrotating case, that are modified by the Coriolis acceleration; rotation increases their eigenvalues ($\Lambda_{k,m}$) and hence their radial wavenumber and their damping (*cf.* Figure 4). These waves are thus damped closer to their excitation region, namely, at the bottom of the convective envelope, than when the Coriolis acceleration is ignored. They can be treated by using the traditional approximation as long as their frequencies are super-inertial ($\sigma > 2\overline{\Omega}_s$, $\nu_s < 1$).
- *Class II waves* are purely retrograde waves ($m > 0$), which exist only in the case of rapid rotation. Their dynamics is driven by the conservation of specific vorticity combined with the effects of curvature. However, because of their sub-inertial frequency range ($\sigma \leq 2\overline{\Omega}_s$, $\nu_s \geq 1$), they cannot be treated by using the traditional approximation. They are sometimes called "quasi-inertial" waves, which correspond to the geophysical Rossby waves (*cf.* Provost, Berthomieu, and Rocca 1981).

- *Class III waves* are mixed Class I and Class II waves. The $m \leq 0$ waves exist in the absence of rotation. The $m > 0$ waves appear when $\nu_s = m + 1$ with small eigenvalues and their horizontal eigenfunctions are $\Theta_{k,m}\,(\nu_s = m+1;x) = P_{m+1}^m(x)$. When they appear and have small eigenvalues, they behave mostly like Class II waves; $m \leq 0$ and $m > 0$ waves with large eigenvalues behave rather like Class I waves. Their eigenvalues are much smaller than those of Class I waves. Thus, they will be damped farther from the bottom of the convective zone and over a larger portion of the solar radiative region. As for Class I waves, they can be treated by using the traditional approximation as long as their frequencies are super-inertial. They may be identified with the geophysical Yanai waves (Yanai and Maruyama, 1966).
- *Class IV waves* are purely prograde waves ($m < 0$) whose characteristics change little with rotation, their displacement in the θ direction being very small. Like Class II waves, their dynamics is driven by the conservation of specific vorticity, but here it is combined with the stratification effects; their eigenvalues are smaller than those of either Class I or Class III waves. Hence, they are damped somewhat deeper in the solar radiation zone where they deposit their positive angular momentum. As for Class I and Class III waves, they can be treated by using the traditional approximation as long as their frequencies are super-inertial. They may be identified with the geophysical Kelvin waves.

If we define a new index s by

$$s = \ell - m + 1 \quad \text{for } m > 0 \quad \text{or} \quad s = \ell + m - 1 \quad \text{for } m \leq 0, \tag{26}$$

then Class IV waves have $s = -1$, Class III waves have $s = 0$, and Class I waves have $s = 1, 2, 3, \ldots$.

In the presence of Coriolis acceleration, the spatial structure and damping of the low-frequency waves are thus modified. We shall now introduce expressions for the mean energy and momentum fluxes associated with these waves.

3. Mean Vertical Fluxes Transported by a Monochromatic Wave

3.1. Mean Vertical Flux of Kinetic Energy

First we consider the mean flux of kinetic energy associated with a monochromatic wave. Following Press (1981), we can express this as the product of the mean volumetric density of kinetic energy of the wave on an isobar with its vertical group velocity:

$$\mathcal{F}_{V;k,m}^{K}(r) = \frac{1}{2}\overline{\rho}\langle u_{k,m}^2 \rangle V_{g;k,m}^V, \tag{27}$$

where the group velocity as derived from Equation (13) is

$$V_{g;k,m}^V = \frac{d\sigma}{dk_{V;k,m}} = -\frac{\sigma}{k_{V;k,m}}\frac{N^2 - \sigma^2}{N^2}, \tag{28}$$

$\langle u_{k,m}^2 \rangle$ is the mean value of the squared speed on an isobar,

$$\langle u_{k,m}^2 \rangle = \frac{1}{4\pi}\int_{\Omega=4\pi}\left(u_{r;k,m}^2 + u_{\theta;k,m}^2 + u_{\varphi;k,m}^2\right)d\Omega, \tag{29}$$

and $\int_{\Omega=4\pi} d\Omega = \int_0^{2\pi} \int_0^{\pi} \sin\theta \, d\theta \, d\varphi$ is the horizontal average over colatitudes (θ) and longitudes (φ). Using the final expression of the wave velocity field given in Equations (8)–(11), we finally find the following expressions as in Mathis (2005) and Pantillon, Talon, and Charbonnel (2007):

$$\mathcal{F}^{K}_{V;k,m}(r) = -\frac{1}{2}\bar{\rho}\frac{\mathcal{E}^2_{k,m}(r)}{2}\left(\langle\Theta^2_{k,m}(\cos\theta;\nu_s)\rangle_\theta + \frac{r^2 k^2_{V;k,m}}{\Lambda^2_{k,m}(\nu_s)}\mathcal{J}_{H;k,m}(\nu_s)\right)$$

$$\times \frac{\sigma^2}{N^2}\frac{(N^2-\sigma^2)^{\frac{1}{2}}}{k_{H;k,m}} \exp\left[-\tau_{k,m}(r,\delta\overline{\Omega}(r);\nu_s)\right], \tag{30}$$

where

$$\mathcal{J}_{H;k,m}(\nu_s) = \langle[\mathcal{H}^\theta_{k,m}(\cos\theta;\nu_s)]^2\rangle_\theta + \langle[\mathcal{H}^\varphi_{k,m}(\cos\theta;\nu_s)]^2\rangle_\theta. \tag{31}$$

3.2. Mean Vertical Flux of Angular Momentum

Let us now examine the mean flux of angular momentum carried by a monochromatic wave. In the rotating case, it is defined as the horizontal average of the Reynolds stresses across an Eulerian surface (term I) plus a Lagrangian contribution (term II) (see Bretherton, 1969):

$$\mathcal{F}^{AM}_{V;k,m}(r) = \frac{1}{4\pi}\int_{\Omega=4\pi}\Bigg\{\underbrace{\overline{\rho}r\sin\theta u_{\varphi;k,m}u_{r;k,m}}_{\text{I}}$$

$$+ \underbrace{\overline{\rho}r\sin\theta 2\overline{\Omega}_s\cos\theta\xi_{r;k,m}u_{\theta;k,m}}_{\text{II}}\Bigg\}d\Omega. \tag{32}$$

Using, once again, the final form of the wave velocity field given in Equations (8)–(11) and the definition of the Lagrangian displacement $\mathbf{u} = d\xi/dt$, we obtain for the angular momentum flux

$$\mathcal{F}^{AM}_{V;k,m}(r) = \frac{1}{2}\bar{\rho}r\frac{rk_{V;k,m}}{\Lambda_{k,m}(\nu_s)}\mathcal{E}^2_{k,m}(r)[\mathcal{J}_{I;k,m}(\nu_s) - \nu_s\mathcal{J}_{II;k,m}(\nu_s)]$$

$$\times \exp[-\tau_{k,m}(r,\delta\overline{\Omega}(r);\nu_s)], \tag{33}$$

where the angular integrals $\mathcal{J}_{I;k,m}(\nu_s)$, $\mathcal{J}_{II;k,m}(\nu_s)$ are given by

$$\begin{cases}\mathcal{J}_{I;k,m}(\nu_s) = \langle\Theta_{k,m}(\cos\theta;\nu_s)\mathcal{H}^\varphi_{k,m}(\cos\theta;\nu_s)\sin\theta\rangle_\theta,\\ \mathcal{J}_{II;k,m}(\nu_s) = \langle\Theta_{k,m}(\cos\theta;\nu_s)\mathcal{H}^\theta_{k,m}(\cos\theta;\nu_s)\cos\theta\sin\theta\rangle_\theta.\end{cases} \tag{34}$$

Next, following Pantillon, Talon, and Charbonnel (2007) and using the approximation $r^2 k_{V;k,m} \gg \Lambda_{k,m}$, we derive the relation between $\mathcal{F}^{AM}_{V;k,m}$ and $\mathcal{F}^{K}_{V;k,m}$:

$$\mathcal{F}^{AM}_{V;k,m}(r) = -\frac{2}{\sigma}m'(\nu_s)\mathcal{F}^{K}_{V;k,m}(r), \tag{35}$$

where m' is given by

$$m'(\nu_s) = \Lambda_{k,m}(\nu_s)\frac{\mathcal{J}_I(\nu_s) - \nu_s\mathcal{J}_{II}(\nu_s)}{\mathcal{J}_{H;k,m}(\nu_s)}. \tag{36}$$

Figure 5 Integrand for the calculation of angular momentum transport (see Equation (34)): (continuous lines) $\nu = 0$, $m = 2$; (dotted lines) $m = 2$, $\sigma = 1$ μHz, $\nu = 0.86$; (dashed lines) $m = -2$, $\sigma = 1$ μHz, $\nu = 0.86$.

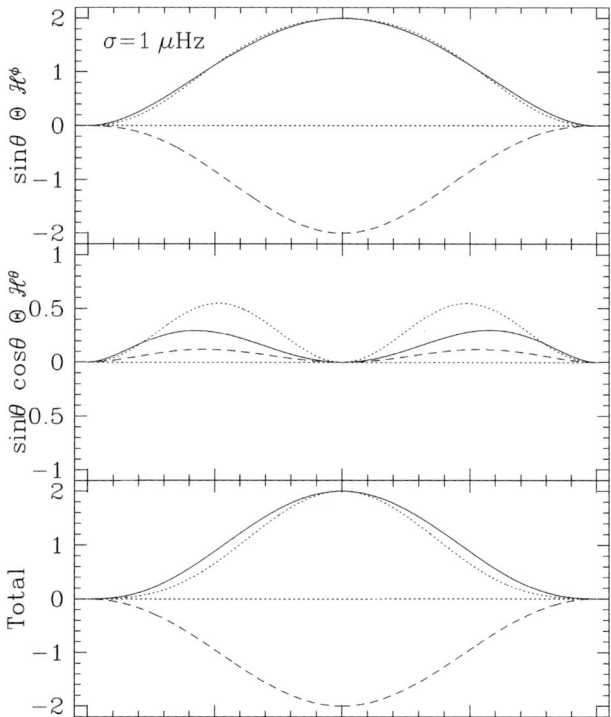

This relation links the mean flux of angular momentum carried by a monochromatic wave on an isobar to that of the kinetic energy. In the nonrotating case, we retrieve $m'(\nu_s = 0) = m$.

$m'(\nu_s)$ is given for each type of waves in Fig. 6. In the case of Class I gravito-inertial waves, $m'(\nu_s)$ is decreased compared to the nonrotating case. Heuristically, this is a consequence of the lesser horizontal extent of the eigenfunctions. Hence, the ability of those waves to transport angular momentum is decreased compared to the nonrotating case. However, we can see that if the prograde and retrograde waves are equally excited, we should expect that the damping of these waves could produce a shear-layer oscillation similar to the one obtained in the case where the Coriolis acceleration is not taken into account (*cf.* Talon and Charbonnel, 2005). In the case of Class III waves, we obtain the same behavior, but with a slower convergence rate. Finally, in the case of Class IV waves, $m'(\nu_s)$ varies only slightly with rotation and remains close to m, their angular momentum being always positive so that they could induce a deposition of angular momentum where they are damped.

3.3. Mean Vertical Action of Angular Momentum

In addition to the mean vertical flux of angular momentum, we now define the mean vertical action of angular momentum, also called the angular-momentum luminosity, a term coined by Goldreich and Nicholson (1989b):

$$\mathcal{L}^{AM}_{V;k,m}(r) = 4\pi r^2 \mathcal{F}^{AM}_{V;k,m}(r). \tag{37}$$

In the adiabatic limit, this momentum luminosity is conserved, as the monochromatic wave propagates (*cf.* Hayes, 1970; Goldreich and Nicholson, 1989a), but when radiative damping

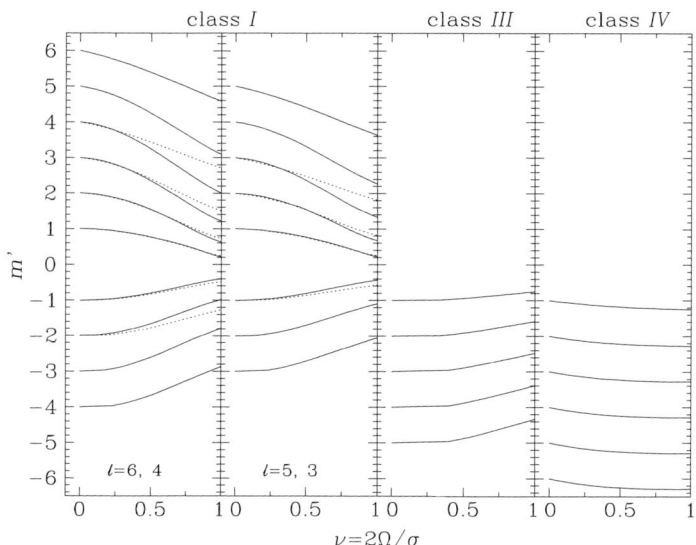

Figure 6 Ratio $m' = -\frac{\sigma}{2}\mathcal{F}_V^{AM}/\mathcal{F}_V^K$ for various modes. Class I waves corresponding to $\ell = 5, 6$ (continuous lines) and $\ell = 3, 4$ (dotted lines). Class III waves of order $m = -5, \ldots, -1$. Class IV waves of order $m \geq -6$.

is taken into account, it decreases according to

$$\mathcal{L}_{V;k,m}^{AM}(r) = \mathcal{L}_{V;k,m}^{AM}(r_c) \exp\left[-\tau_{k,m}\left(r, \delta\overline{\Omega}(r); \nu_s\right)\right], \tag{38}$$

where r_c is the radius of the layer where the waves are generated by the turbulent convective motions. Using Equation (35), we can explicitly write the initial luminosity $\mathcal{L}_{V;k,m}^{AM}(r_c)$ as

$$\mathcal{L}_{V;k,m}^{AM}(r_c) = -4\pi r_c^2 \left[\frac{2}{\sigma} m'(\nu_s) \mathcal{F}_{V;k,m}^K(r_c)\right]. \tag{39}$$

$\overline{\Omega}_s$ thus corresponds to the mean rotation rate at r_c: $\overline{\Omega}_s = \overline{\Omega}(r_c)$.

A prescription is needed to assess the kinetic energy flux of the waves that are emitted at $r = r_c$. As has been emphasized in the introduction, this is the main weakness of the theory of internal wave transport. In this work, we follow Pantillon, Talon, and Charbonnel (2007) and the procedure first given by García-López and Spruit (1991) to obtain

$$\mathcal{F}_{V;k,m}^K(r_c) = \frac{\overline{\rho}\lambda_{k,m}(\nu_s)}{2\sqrt{\Lambda_{k,m}(\nu_s)}} \frac{V_c^2 \sigma_c^2}{r_c k_c^2 N_c} \left(\frac{\sigma}{\sigma_c}\right)^{-2}, \tag{40}$$

where V_c, σ_c, and k_c are, respectively, the convective velocity, frequency, and wavenumber and N_c is the Brunt–Väisälä frequency at the border between the convection and radiation zones, which is nonzero owing to convective penetration. In this very crude treatment of the excitation, waves are generated at the boundary of the radiative zone by interface deformations caused by convective eddies. We plan to implement a more realistic description (Belkacem *et al.*, in preparation) that will convolve the wave eigenfunctions with Reynolds stresses and entropy perturbations to treat the volumetric excitation in the convective region, as was done by Goldreich, Murray, and Kumar (1994) and Samadi and Goupil (2001).

4. Transport of Angular Momentum

4.1. Transport of Angular Momentum by the Waves: Shear Layer Oscillation and Secular Effects

Waves deposit their angular momentum inside the star as they are damped. Locally, the total action of angular momentum is the sum of the contribution of all of the waves, each one being damped separately:

$$\mathcal{L}_V^{AM}(r) = \sum_{\sigma,m,k} \mathcal{L}_{V;k,m}^{AM}(r_c) \exp[-\tau_{k,m}(r, \delta\overline{\Omega}(r); \nu_s)]. \tag{41}$$

The deposition of angular momentum is then given by the radial derivative of this action of the angular momentum. The evolution of angular momentum by waves then follows:

$$\overline{\rho}\frac{d}{dt}[r^2\overline{\Omega}] = \pm\frac{3}{8\pi}\frac{1}{r^2}\partial_r[\mathcal{L}_V^{AM}(r)] \tag{42}$$

(Talon and Zahn, 1998). The "+" ("−") sign in front of the action of angular momentum corresponds to a wave traveling inward (outward).

Let us first take a look at the damping integral given in Equation (22) and assume that both prograde and retrograde waves are excited with the same amplitude and have the same eigenvalue $\Lambda_{k,m}$. In solid-body rotation, both waves are equally dissipated when traveling inward and there is no impact on the distribution of angular momentum. In the presence of differential rotation, the situation is different. If the interior is rotating faster than the convection zone, the local frequency of prograde waves decreases, which enhances their dissipation; the corresponding retrograde waves are then dissipated further inside. This produces an increase of the local differential rotation and creates a double-peaked shear layer because local shears are amplified by waves and even a small perturbation can trigger this (with the prograde waves transporting a positive flux of angular momentum and the retrograde waves transporting a negative one). In the presence of shear turbulence, this layer oscillates, producing a "shear layer oscillation" or SLO (cf. Ringot, 1998; Kumar, Talon, and Zahn, 1999). This is the first important feature of wave-mean flow interaction.

This SLO acts as a filter, through which most low-frequency waves cannot pass. However, if the core is rotating faster than the surface, this filter is not quite symmetric, and retrograde waves will be favored. As a result, a net *negative* flux of angular momentum will result, which produces a spin down of the core (Talon, Kumar, and Zahn, 2002). This is the filtered angular-momentum action flux that contributes to the secular evolution of angular momentum (for details, see Talon and Charbonnel, 2005). It plays a key role in flattening the rotation profile as observed in the present Sun (Charbonnel and Talon, 2005).

4.2. Toward a Complete Description of the Transport of Angular Momentum in the Radiative Region

We are now ready to derive the final equation for the vertical transport of angular momentum on an isobar in the radiation zone, namely those for the temporal evolution of $\delta\overline{\Omega}(r)$ under the combined action of the advection by the first mode of the meridional circulation [$U_2(r)$], the vertical turbulent transport associated with the different instabilities that are modeled through a vertical diffusion coefficient (ν_V), the torque of the Lorentz force associated with the magnetic field (*cf.* Mathis and Zahn, 2005) [$\overline{\Gamma}_{\mathcal{F}_\mathcal{L}}(r)$], and finally the filtered action of angular momentum [$\mathcal{L}_V^{\mathrm{AM,fil}}(r)$]. Following Mathis (2005), we thus obtain

$$\overline{\rho}\frac{\mathrm{d}}{\mathrm{d}t}\left(r^2\overline{\Omega}\right) - \frac{1}{5r^2}\partial_r\left(\overline{\rho}r^4\overline{\Omega}U_2\right)$$
$$= \frac{1}{r^2}\partial_r\left(\overline{\rho}v_V r^4\partial_r\overline{\Omega}\right)$$
$$+ \overline{\Gamma}_{\mathcal{F}_\mathcal{L}}(r) - \frac{3}{8\pi}\frac{1}{r^2}\partial_r\left[\mathcal{L}_V^{\mathrm{AM,fil}}(r)\right]. \tag{43}$$

We recall that the meridional circulation is expanded in spherical harmonics as

$$\mathcal{U}(r,\theta) = \sum_{l>0}\left\{U_l(r)P_l(\cos\theta)\widehat{\mathbf{e}}_r + V_l(r)\frac{\mathrm{d}P_l(\cos\theta)}{\mathrm{d}\theta}\widehat{\mathbf{e}}_\theta\right\}, \tag{44}$$

where $V_l = 1/[l(l+1)\overline{\rho}r]\,\mathrm{d}(\overline{\rho}r^2 U_l)/\mathrm{d}r$ is obtained with the anelastic approximation, and that U_2 is the only mode that leads to a net transport of angular momentum on an isobar as demonstrated in Mathis and Zahn (2004).

The associated boundary conditions at $r=r_\mathrm{b}$ and $r=r_\mathrm{t}$, where r_b and r_t are, respectively, the radius of the base and of the top of the considered radiative region, are given by

$$\frac{\mathrm{d}}{\mathrm{d}t}\left[\int_0^{r_\mathrm{b}} r^4\overline{\rho}\overline{\Omega}\,\mathrm{d}r\right] = \frac{1}{5}r^4\overline{\rho}\overline{\Omega}U_2 - \mathcal{F}_B(r_\mathrm{b}) - \mathcal{L}_V^{\mathrm{AM}}(r_\mathrm{b}) \tag{45}$$

and

$$\frac{\mathrm{d}}{\mathrm{d}t}\left[\int_{r_\mathrm{t}}^R r^4\overline{\rho}\overline{\Omega}\,\mathrm{d}r\right] = -\frac{1}{5}r^4\overline{\rho}\overline{\Omega}U_2 - \mathcal{F}_\Omega + \mathcal{F}_B(r_\mathrm{t}) + \mathcal{L}_V^{\mathrm{AM}}(r_\mathrm{t}), \tag{46}$$

where $\mathcal{F}_\Omega$ and $\mathcal{F}_B$ are, respectively, the flux of angular momentum loss at the surface and the magnetic angular-momentum flux through the interfaces. In the solar case, $r_\mathrm{b}=0$ and $r_\mathrm{t}=r_\mathrm{SLO}$.

We are thus now in a position where we are able to get a coherent picture of solar and stellar radiative-zone dynamics, taking into account the highly nonlinear interaction among the differential rotation, the associated meridional circulation, the vertical and horizontal shear-induced turbulence, a potential fossil magnetic field, and the low-frequency waves where the action of the rotation on the waves through the Coriolis acceleration and their feedback on the angular velocity distribution are treated in a coherent way. We now summarize those interactions with the action of each process (Figure 7):

- The *meridional circulation*, which is due to the thermal imbalance induced by perturbing forces, namely the centrifugal and the Lorentz forces, and by the extraction of angular momentum at the surface by the wind, advects angular momentum, chemical elements, and the magnetic field.
- The *shear-induced turbulence* acts to suppress its cause, namely the vertical and the horizontal gradients of angular velocity.
- The *fossil magnetic field* is advected by the meridional circulation, diffused by ohmic effects, and transports angular momentum through the large-scale Lorentz torque and the Maxwell stresses associated with MHD instabilities.
- The *low-frequency internal waves* generated by the convective movements transport angular momentum as well as the magnetic field, modifying the angular-velocity distribution and the associated mixing.

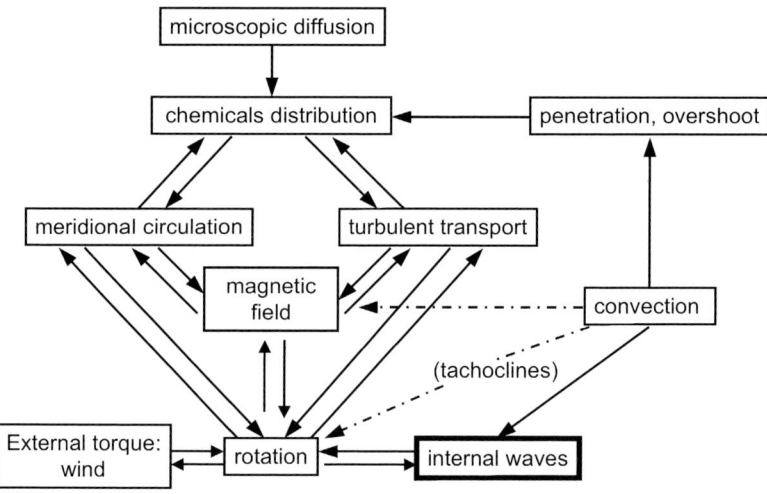

Figure 7 Dynamical transport processes in stellar radiation zones: these regions are the seat of highly non-linear interactions among differential rotation, meridional circulation, turbulence, magnetic field, and IGWs.

- The *braking caused by the wind* in the early phases forces an extraction of angular momentum at the surface, which drives the behavior of the meridional circulation and the potential wave fronts of angular momentum extraction.

5. Discussion

We have examined the transport of angular momentum by low-frequency waves excited at the bottom of the solar convection envelope and influenced by the Coriolis acceleration. The traditional approximation has been used to treat its impact on the waves in its domain of validity where $2\Omega < \sigma \ll N$ ($\nu < 1$). Two main effects are observed: First, the horizontal structure of the wave is modified, the dynamics being now driven simultaneously by the stratification, as in the nonrotating case, and by the Coriolis acceleration. Hence, the amount of angular momentum carried by a wave and its damping are modified. In the case of Class I gravito-inertial waves, the main effect of the Coriolis force is to modify the horizontal functions, the prograde and the retrograde waves being now different. Furthermore, a reduction of the effectiveness of those waves in transporting angular momentum is obtained. Moreover, rotation increases their damping, these waves being thus deposited closer to their excitation region than gravity waves in the nonrotating case. The Class III waves have the same behavior for m' but with a slower convergence rate. Their eigenvalues are however much smaller and hence they are damped farther from the convection zone basis. Finally, Class IV waves have a different behavior. For these waves the main restoring force is the conservation of vorticity combined with stratification. They have $m' = m$ and their eigenvalues are smaller than those of either Class I or Class III waves. Hence, they are damped deeper in the solar core where they deposit their positive angular momentum. Complete numerical simulations remain to be performed to verify and understand the net effects of all those gravito-inertial waves on the angular momentum transport in the solar interior and to obtain a coherent dynamical vision of it. Furthermore, the sub-inertial regime $\sigma \leq 2\Omega \ll N$ ($\nu \geq 1$), which has to be treated by taking into account the complete Coriolis acceleration, has to be examined to get a coherent picture of the gravito-inertial wave transport.

Acknowledgements The authors thank the referee M. Rieutord for his clarifying remarks on the applicability of the traditional approximation. This work was partially supported by the European Helio- and Asteroseismology Network (HELAS, http://www.helas-eu.org), a major international collaboration funded by the European Commission's Sixth Framework Program.

References

Bildsten, L., Ushomirsky, G., Cutler, C.: 1996, Ocean g-modes on rotating neutron stars. *Astrophys. J.* **460**, 827.
Berthomieu, G., Gonczi, G., Graff, P., Provost, J., Rocca, A.: 1978, Low-frequency gravity modes of a rotating star. *Astron. Astrophys.* **70**, 597.
Brown, T.M., Christensen-Dalsgaard, J., Dziembowski, W.A., Goode, P., Gough, D.O., Morrow, C.A.: 1989, Inferring the Sun's internal angular velocity from observed p-modes frequency splittings. *Astrophys. J.* **343**, 526.
Brun, A.-S., Zahn, J.-P.: 2006, Magnetic confinement of the solar tachocline. *Astron. Astrophys.* **457**, 665.
Bretherton, F.P.: 1969, Momentum transport by gravity waves. *Q. J. Roy. Meteorol. Soc.* **95**, 213.
Chaboyer, B., Demarque, P., Pinsonneault, M.H.: 1995, Stellar models with microscopic diffusion and rotational mixing. I. Application to the Sun. *Astrophys. J.* **441**, 865.
Charbonnel, C., Talon, S.: 2005, Influence of gravity waves on the internal rotation and Li abundance of solar-type stars. *Science* **309**, 2189.
Cowling, T.G.: 1941, The non-radial oscillations of polytropic stars. *Mon. Not. Roy. Astron. Soc.* **101**, 367.
Dintrans, B.: 1999, Ph.D. Thesis, Université Toulouse III.
Dintrans, B., Rieutord, M.: 2000, Oscillations of a rotating star: a non-perturbative theory. *Astron. Astrophys.* **354**, 86.
Dintrans, B., Rieutord, M., Valdettaro, L.: 1999, Gravito-inertial waves in a rotating stratified sphere or spherical shell. *J. Fluid Mech.* **398**, 271.
Eckart, C.: 1960, *Hydrodynamics of Oceans and Atmospheres*, Pergamon Press, Oxford.
Friedlander, S.: 1987, Internal waves in a rotating stratified spherical shell: asymptotic solutions. *Geophys. J. Roy. Astron. Soc.* **89**, 637.
Friedlander, S., Siegmann, W.L.: 1982, Internal waves in a rotating stratified fluid in an arbitrary gravitational field. *Geophys. Astrophys. Fluid Dyn.* **19**, 267.
García, R.A., Turck-Chièze, S., Jiménez-Reyes, S.J., Ballot, J., Pallé, P.L., Eff-Darwich, A., Mathur, S., Provost, J.: 2007, Tracking solar gravity modes: the dynamics of the solar core. *Science* **316**, 1591.
García-López, R.J., Spruit, H.C.: 1991, Li depletion in F stars by internal gravity waves. *Astrophys. J.* **377**, 268.
Garaud, P.: 2002, Dynamics of the solar tachocline. I. An incompressible study. *Mon. Not. Roy. Astron. Soc.* **329**, 1.
Gerkema, T., Shrira, V.I.: 2005, Near-inertial waves in the ocean: beyond the 'traditional approximation'. *J. Fluid Mech.* **529**, 195.
Gerkema, T., Zimmerman, J.T.F., Mass, L.R.M., Van Haren, H.: 2008, Geophysical and astrophysical fluid dynamics beyond the traditional approximation, *Rev. Geophys.* in press. doi:10.1029/2006RG000220.
Goldreich, P., Nicholson, P.D.: 1989a, Tides in rotating fluids. *Astrophys. J.* **342**, 1075.
Goldreich, P., Nicholson, P.D.: 1989b, Tidal friction in early-type stars. *Astrophys. J.* **342**, 1079.
Goldreich, P., Murray, N., Kumar, P.: 1994, Excitation of solar p-modes. *Astrophys. J.* **424**, 466.
Gough, D.O., McIntyre, M.E.: 1998, Inevitability of a magnetic field in the Sun's radiative interior. *Nature* **394**, 567.
Hayes, W.D.: 1970, Conservation of action and modal wave action. *Proc. Roy. Soc. Lond. A* **320**, 187.
Hough, S.S.: 1898, On the application of harmonic analysis to the dynamical theory of the tides. Part II: On the general integration of Laplace's dynamical equations. *Philos. Trans. Roy. Soc. A* **191**, 139.
Kumar, P., Talon, S., Zahn, J.-P.: 1999, Angular momentum redistribution by waves in the Sun. *Astrophys. J.* **520**, 859.
Laplace, P.-S.: 1799, *Mécanique Céleste*, Bureau des Longitudes, Paris.
Lee, U., Saio, H.: 1997, Low-frequency non-radial oscillations in rotating stars: I. Angular dependence. *Astrophys. J.* **491**, 839.

Longuet-Higgins, F.R.S.: 1968, The eigenfunctions of Laplace's tidal equations over a sphere. *Phil. Trans. Roy. Soc. A* **262**, 511.
Maeder, A., Meynet, G.: 2000, The evolution of rotating stars. *Ann. Rev. Astron. Astrophys.* **38**, 143.
Mathis, S.: 2005, Ph.D. Thesis, Université Paris XI.
Mathis, S., Zahn, J.-P.: 2004, Transport and mixing in the radiation zones of rotating stars: I. Hydrodynamical processes. *Astron. Astrophys.* **425**, 229.
Mathis, S., Zahn, J.-P.: 2005, Transport and mixing in the radiation zones of rotating stars: II. Axisymmetric magnetic field. *Astron. Astrophys.* **440**, 653.
Matias, J., Zahn, J.-P.: 1997, In: Provost J., Schmider F.-X. (eds.) *IAU Symposium 18*, Poster Volume. Observatoire de Nice, 103.
Miles, J.W.: 1977, Asymptotic eigensolutions of Laplace's tidal equation. *Proc. Roy. Soc. Lond. A* **353**, 377.
Pantillon, F.P., Talon, S., Charbonnel, C.: 2007, Angular momentum transport by internal gravity waves: IV. Wave excitation by core convection and the Coriolis effect. *Astron. Astrophys.* **474**, 155.
Pedlosky, J.: 1987, *Geophysical Fluid Dynamics*, 2nd edn., Springer, New York.
Pinsonneault, M.H., Kawaler, S.D., Sofia, S., Demarque, P.: 1989, Evolutionary models of the rotating sun. *Astrophys. J.* **338**, 424.
Press, W.H.: 1981, Radiative and other effects from internal waves in solar and stellar interiors. *Astrophys. J.* **245**, 286.
Provost, J., Berthomieu, G., Rocca, A.: 1981, Low frequency oscillations of a slowly rotating star: quasi-toroidal modes. *Astron. Astrophys.* **94**, 126.
Ringot, O.: 1998, About the role of gravity waves in the angular momentum transport inside the radiative zone of the Sun. *Astron. Astrophys.* **335**, L89.
Samadi, R., Goupil, M.-J.: 2001, Excitation of stellar p-modes by turbulent convection. I. Theoretical formulation. *Astron. Astrophys.* **370**, 136.
Schatzman, E.: 1993, Transport of angular momentum and diffusion by the action of internal waves. *Astron. Astrophys.* **279**, 431.
Stewartson, K., Richard, J.: 1969, Pathological oscillations of a rotating fluid. *J. Fluid Mech.* **35**, 759.
Stewartson, K., Walton, I.C.: 1976, On waves in a thin shell of stratified rotating fluid. *Proc. Roy. Soc. Lond. A* **349**, 141.
Talon, S.: 1997, Ph.D. Thesis, Université Paris VII.
Talon, S.: 2007, Transport processes in stars: diffusion, rotation, magnetic fields and internal waves. In: Charbonnel C., Zahn J.-P. (eds.) *Stellar Nucleosynthesis: 50 years after BBFH, Eur. Astron. Soc., Les Ulis.* in press (arXiv:0708.1499).
Talon, S., Charbonnel, C.: 2005, Hydrodynamical stellar models including rotation, internal gravity waves and atomic diffusion. I. Formalism and tests on Pop I dwarfs. *Astron. Astrophys.* **440**, 981.
Talon, S., Zahn, J.-P.: 1998, Towards a hydrodynamical model predicting the observed solar rotation profile. *Astron. Astrophys.* **329**, 315.
Talon, S., Kumar, P., Zahn, J.-P.: 2002, Angular momentum extraction by gravity waves in the Sun. *Astrophys. J.* **574**, L175.
Townsend, R.H.D.: 2003, Asymptotic expressions for the angular dependence of low-frequency pulsation modes in rotating stars. *Mon. Not. Roy. Astron. Soc.* **340**, 1020.
Turck-Chièze, S., Couvidat, S., Piau, L., Ferguson, J., Lambert, P., Ballot, J., García, R.A., Nghiem, P.: 2004, Surprising Sun: a new step towards a complete picture? *Phys. Rev. Lett.* **93**, id. 211102.
Yanai, M., Maruyama, T.: 1966, *J. Meteorol. Soc. Japan* **44**, 291.
Zahn, J.-P.: 1992, Circulation and turbulence in rotating stars. *Astron. Astrophys.* **265**, 115.
Zahn, J.-P., Talon, S., Matias, J.: 1997, Angular momentum transport by internal waves in the solar interior. *Astron. Astrophys.* **322**, 320.

Influence of Low-Degree High-Order p-Mode Splittings on the Solar Rotation Profile

R.A. García · S. Mathur · J. Ballot · A. Eff-Darwich ·
S.J. Jiménez-Reyes · S.G. Korzennik

Originally published in the journal Solar Physics, Volume 251, Nos 1–2, 119–133.
DOI: 10.1007/s11207-008-9144-5 © Springer Science+Business Media B.V. 2008

Abstract The solar rotation profile is well constrained down to about $0.25 R_\odot$ thanks to the study of acoustic modes. Since the radius of the inner turning point of a resonant acoustic mode is inversely proportional to the ratio of its frequency to its degree, only the low-degree p modes reach the core. The higher the order of these modes, the deeper they penetrate into the Sun and thus they carry more diagnostic information on the inner regions. Unfortunately, the estimates of frequency splittings at high frequency from Sun-as-a-star measurements have higher observational errors because of mode blending, resulting in weaker

---

Helioseismology, Asteroseismology, and MHD Connections
Guest Editors: Laurent Gizon and Paul Cally.

R.A. García (✉) · S. Mathur
Laboratoire AIM, CEA/DSM-CNRS, U. Paris Diderot, IRFU/SAp, 91191 Gif-sur-Yvette Cedex, France
e-mail: rgarcia@cea.fr

S. Mathur
e-mail: smathur@cea.fr

J. Ballot
Max-Planck-Institut für Astrophysik, Karl-Schwarzschild-Strasse 1, 85748 Garching, Germany
e-mail: jballot@mpa-garching.mpg.de

A. Eff-Darwich
Departamento de Edafología y Geología, Universidad de La Laguna, La Laguna, Tenerife, Spain
e-mail: adarwich@ull.es

A. Eff-Darwich · S.J. Jiménez-Reyes
Instituto de Astrofísica de Canarias, 38205 La Laguna, Tenerife, Spain

A. Eff-Darwich
e-mail: adarwich@iac.es

S.J. Jiménez-Reyes
e-mail: sjimenez@iac.es

S.G. Korzennik
Harvard-Smithsonian Center for Astrophysics, 60 Garden Street, Cambridge, MA 02138, USA
e-mail: skorzennik@cfa.harvard.edu

constraints on the rotation profile in the inner core. Therefore inversions for the solar internal rotation use only modes below 2.4 mHz for $\ell \leq 3$. In the work presented here, we used an 11.5-year-long time series to compute the rotational frequency splittings for modes $\ell \leq 3$ using velocities measured with the GOLF instrument. We carried out a theoretical study of the influence of the low-degree modes in the region from 2 to 3.5 mHz on the inferred rotation profile as a function of their error bars.

Keywords Helioseismology · Observations · Inverse modeling · Interior · Radiative zone · Core · Rotation

1. Introduction

Our knowledge of the solar rotation profile has been derived from the study of the resonant acoustic modes, which are trapped in the solar interior. Since the solar rotation lifts the azimuthal degeneracy of these resonant modes, their eigenfrequencies ($\nu_{n\ell m}$) are split into their m components, where ℓ is the angular degree, n the radial order, and m the azimuthal order. This separation $\Delta \nu_{n\ell m}$ – usually called rotational splitting (or just splitting) – depends on the rotation rate in the region sampled by the mode. Using inversion techniques, one can infer the rotation rate at different locations inside the Sun from a suitable linear combination of the measured rotational splittings

Today, the rotation rate inside the Sun is rather well known above $0.4R_\odot$ (Thompson et al., 1996; Schou et al., 1998; Howe et al., 2000; Antia and Basu, 2000). The convective zone is characterized by a differential rotation extending from the surface down to the tachocline, located around $0.7R_\odot$. Below the tachocline, inside the radiative region, the Sun appears to rotate as a rigid body with a nearly constant rate of ≈ 433 nHz down to the solar core (i.e., $\approx 0.25R_\odot$). The rotation rate inside the core derived from p modes is still uncertain (Jiménez et al., 1994; Elsworth et al., 1995; Chaplin et al., 2001; Chaplin et al., 2004). Recent measurements of the asymptotic properties of the dipole g modes and their comparison with solar models favors a faster rotation rate inside the solar core (García et al., 2007).

For p modes of a given degree ℓ, the radius at the inner turning point is a decreasing function of frequency, given by

$$r_\mathrm{t} = c_\mathrm{t} L/(2\pi \nu_{n\ell}),$$

where $L = \ell + 1/2$, $\nu_{n\ell}$ is the frequency of the mode, and $c_\mathrm{t} = c(r_\mathrm{t})$ the sound-speed at the radius r_t (see, for example, Lopes and Turck-Chièze, 1994). Thus the modes with increasing frequencies – higher radial order n – penetrate deeper inside the Sun. Unfortunately, when fitting Sun-as-a-star observations the uncertainties on the rotational splittings that are the most sensitive to low-degree p modes are very large. Indeed, as the mode lifetimes decrease with frequency their line widths increase. Therefore, for frequencies above ≈ 2.3 mHz, there is a substantial blending among the visible m components of the p modes. This blending makes it difficult to extract precisely the rotational splitting. At higher frequencies (≈ 3.9 mHz) even the successive pairs of $\ell = 0, 2$ and $\ell = 1, 3$ modes blend together. As a result, with today's fitting methods, it is still not possible to obtain values of the rotational splittings with an accuracy good enough to be useful in any rotation inversion.

By contrast, at low frequency – below $n = 16$, or about 2.4 mHz – the lifetime of the modes increases and thus their line width is very small. This allows us to measure their rotational splittings with very high precision. However, these modes have inner turning points at shallower depths than the high-frequency modes (above 0.08 and $0.12 R_\odot$ for the $\ell = 1$ and 2 modes, respectively). Therefore, even though these modes do not carry any information below $\approx 0.1 R_\odot$, they help improve our knowledge of the inner rotation rate because their inclusion contributes to an increase of the precision of the inversions since they have smaller error bars (Eff-Darwich, Korzennik, and Jiménez-Reyes, 2002; Eff-Darwich *et al.*, 2008).

For all of these reasons, rotation inversion methodologies usually limit the input data set of low-degree p modes to low-frequency modes. For example, Couvidat *et al.* (2003) used a limited number of splittings, corresponding to modes with $\ell \leq 3$ and frequencies below 2.4 mHz ($n = 15$) resulting from fitting GOLF[1] and MDI[2] 2243-day-long velocity time series, to infer the solar rotation profile. They concluded that the uncertainties in the rotation rate below $0.3 R_\odot$ were still quite large. Therefore, to obtain a better and more reliable rotation profile in the inner core, we need, on one hand, to include low-frequency acoustic, mixed, and gravity modes in the inversions (see discussions in Provost, Berthomieu, and Morel, 2000, and Mathur *et al.*, 2008). On the other hand, we also need to measure more accurately low-degree high-order p modes to further push the frequency limit of the modes used in the inversions.

We present, in Section 2, rotational splittings of $\ell \leq 3$ modes computed using 11.5 years of GOLF data, paying special attention that no bias is introduced by the solar activity or by the length of the fitting window. The length of these new time series improves the splitting error bars by $\sqrt{T}$ when compared to previous and shorter analyses with the same instrument. In Section 3, we study the sensitivity of the rotation rate below $0.25 R_\odot$ to the available p modes and their error bars. We then discuss, in Section 4, the influence of these modes on the inversions, based on the resulting resolution kernels, and we finish by discussing the inverted rotation profile using the newly computed GOLF splittings.

2. Data Analysis

We analyzed a 4182-day-long time series of GOLF observations that spans the 11 April 1996 to 22 September 2007 epoch ($\approx$ 11.5 years). These observations were calibrated by following the methods described in García *et al.* (2005). The duty cycle of that time series is 94.5%. The GOLF instrument is a resonant-scattering spectrophotometer onboard the SOHO[3] spacecraft. A standard fast Fourier transform algorithm was used to compute the power spectral density. This periodogram estimator was used because our primary interest is the study of modes in the medium-frequency range. We have therefore decided to privilege the simplest estimator, avoiding the use of multitapers or zero-padding estimators that are better suited to look for weak and narrow peaks at low frequency.

To obtain the mode parameters, each pair of modes ($\ell = 0, 2$ and $\ell = 1, 3$) was fitted to a set of asymmetrical Lorentzians, as defined in Nigam and Kosovichev (1998), by using a maximum-likelihood method since the power-spectrum estimator follows a χ^2 distribution with two degrees of freedom. For additional details on this method see, for example, Toutain and Appourchaux (1994). For each mode, we fitted the following set of parameters: the amplitude, the line width, the central frequency, the peak asymmetry, the rotational splitting (if $\ell > 0$), and the background noise.

---

[1] Global Oscillations at Low Frequency (Gabriel *et al.*, 1995).

[2] Michelson Doppler Imager (Scherrer *et al.*, 1995).

[3] *Solar and Heliospheric Observatory* (Domingo, Fleck, and Poland, 1995).

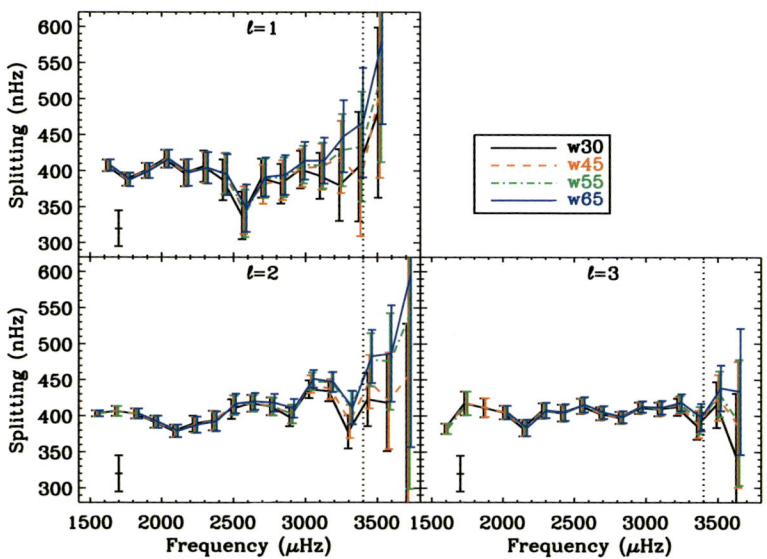

Figure 1 Synodic splittings for the $\ell = 1, 2$, and 3 modes fitted by using increasing fitting window widths from 30 to 65 µHz. The vertical dotted lines mark the upper limit of the modes whose splittings are used in this work. The error bar in the bottom-left corner of each plot corresponds to ± 20 nHz and is plotted as reference.

It has been shown that the amplitude ratios of the m components inside a multiplet could produce a bias in the splitting determination if they are not correctly set (Chaplin *et al.*, 2006). We have therefore fixed them to the averaged values obtained directly from the GOLF data set itself (for details see Henney, 1999). Finally, to reduce the parameter space, we used the same asymmetry and line width for both modes in any given fitting window. Indeed Chaplin *et al.* (2006) demonstrated that using the same width for the modes fitted simultaneously in a given fitting window reduces the bias in the splitting determination at high frequencies. Finally, the uncertainties on the fitted parameters were derived from the square root of the diagonal elements of the inverted Hessian matrix.

It has been shown, by using simulations, that the determination of the rotational splitting could be biased by the width of the fitting window (Chaplin *et al.*, 2006). This is caused by the leakage of the neighboring modes. For example, the splittings of the $\ell = 2$ modes could be noticeably modified (increased) at frequencies above 2.5 mHz owing to the leakage of the $\ell = 4$ modes. We have studied this effect using the GOLF data by selecting four different windows for the fit: 30, 45, 55, and 65 µHz. The results are plotted in Figure 1. As expected, the values fitted for the $\ell = 2$ splittings increase at high frequencies when wider fitting windows are used. This behavior is consistent with the results obtained by the solarFLAG group (see for comparison Figure 9 of Chaplin *et al.*, 2006). However, the smallest window (30 µHz), also biases the fittings at high frequency because fitting such narrow spectral range prevents the fitting code from properly constraining the background level. Therefore, we have adopted the 45-µHz window as the best compromise between these two effects. In the case of the $\ell = 1$ modes, the splittings computed using the smaller windows are roughly constant up to 3.4 mHz and then increase. For higher frequencies the splittings increase whatever window width is used, and it is not possible to distinguish between a bias induced by the blending of the two adjacent m components or a real increase. Therefore we have

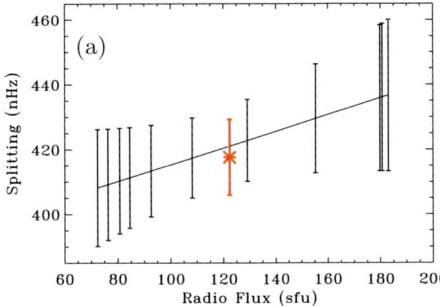 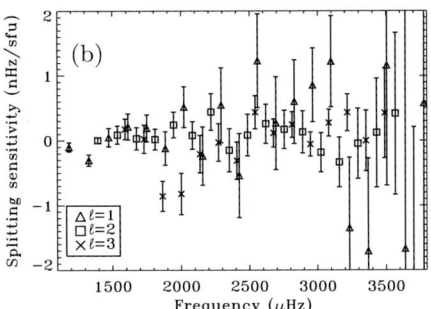

Figure 2 (a) Fitted splittings for the mode ($\ell = 2$, $n = 17$) as a function of the solar activity level measured by the solar radio flux at 10.7 cm (1 sfu = 10^{-22} Wm^{-2}Hz^{-1}). The asterisk with the thick error bar indicates the result of fitting the full time series. (b) Sensitivity of the splittings to the activity for all of the fitted modes ($\ell = 1$, 2, and 3).

limited our study to the splittings of modes below 3.4 mHz (a limit indicated with a vertical dotted line in Figure 1) where all of the splittings are roughly constant within the error bars. Another approach to limit the leakage of the neighbor peaks could be to fit the whole spectrum at once with a simple model of the background noise (Roca Cortés et al., 1998).

It is also important to notice that some splittings (for example, the $\ell = 1$, $n = 17$ mode) are slightly out of the general trend of the rest of the splittings (less than 2σ) as a consequence of the stochastic nature of the excitation. This deviation is in quantitative agreement with the results obtained from artificial simulations (see Chaplin et al., 2006).

We have also verified that there is no noticeable influence of solar activity on the extracted splittings. To do this, we followed an approach similar to the one described in Jiménez-Reyes et al. (2001): We divided the time series into 12 independent 350-day-long segments[4] and fitted them simultaneously. A linear dependence on a solar activity index was added to each parameter defining the spectral profiles, with two exceptions: First, we fixed the asymmetries to the values found by analyzing the full time series; second, we included a quadratic dependence on activity for the frequencies of the modes. We used the integrated 10.7-cm radio flux measurements from the National Geophysical Data Center,[5] averaged over the duration of each subseries as indicator of solar activity. The 10.7-cm radio flux has been shown to be the indicator that best correlates with the measured changes resulting from solar activity.

Figure 2(a) illustrates the variation of the splitting with solar activity for one mode ($\ell = 2$, $n = 17$), and Figure 2(b) shows the sensitivity of the splittings to the radio flux for all of the fitted modes (i.e., the slope in Figure 2(a)). This plot shows that there is no clear systematic dependence like the one seen for the frequency (Jiménez, Roca Cortés, and Jiménez-Reyes, 2002; Gelly et al., 2002; García et al., 2004b). Moreover variations during the solar cycle are generally marginally significant. The splittings obtained by fitting the whole time series are fully consistent with those obtained when using this activity-dependent method and correspond to the Sun at its mean activity level, as expected. The resulting rotational splittings for modes $\ell \leq 3$ are listed in Table 1.

---

[4]One 350-day-long segment was not fitted since its duty cycle is very small because of the loss of contact with the SOHO spacecraft during that epoch.

[5]http://www.ngdc.noaa.gov/ngdc.html.

Table 1 Central frequencies (ν_o), synodic sectoral splittings ($\Delta \nu_{n\ell m}$), and their respective 1σ error bars (σ_o and $\sigma_{n\ell m}$) computed by using 4182 days of GOLF velocity time series.

ℓ	n	ν_o (μHz)	σ_o (nHz)	$\Delta \nu_{n\ell m}$ (nHz)	$\sigma_{n\ell m}$ (nHz)
1	7	1185.5893	4.0	402.2	3.2
	8	1329.6368	3.3	405.2	2.9
	9	1472.8475	5.0	400.0	4.2
	10	1612.7268	9.9	407.6	8.0
	11	1749.2887	10.5	388.1	9.2
	12	1885.0853	12.2	400.4	10.6
	13	2020.8228	14.1	416.5	12.5
	14	2156.8158	18.5	397.3	18.4
	15	2292.0340	19.0	404.7	21.6
	16	2425.6381	20.0	396.0	26.3
	17	2559.2441	19.3	345.2	30.7
	18	2693.4385	18.5	381.5	27.0
	19	2828.2585	18.0	387.1	26.0
	20	2963.4225	17.7	403.8	24.9
	21	3098.2937	19.0	405.8	29.3
	22	3233.2855	22.2	419.2	41.6
	23	3368.6923	26.9	385.1	75.6
2	8	1394.6851	14.2	401.8	5.9
	9	1535.8642	7.0	403.1	4.1
	10	1674.5434	12.3	406.5	6.9
	11	1810.3293	13.3	403.0	6.9
	12	1945.8173	15.9	391.6	8.4
	13	2082.1175	18.5	379.2	8.6
	14	2217.6968	22.2	387.4	11.3
	15	2352.2626	23.7	393.4	13.5
	16	2485.9310	22.8	414.7	12.9
	17	2619.7165	20.6	417.6	11.7
	18	2754.5761	20.4	413.5	12.2
	19	2889.6857	20.5	402.7	12.7
	20	3024.8519	20.3	443.9	11.7
	21	3159.9752	22.3	437.2	13.6
	22	3295.2381	28.1	392.3	21.4
	23	3430.9280	38.6	447.5	33.4
3	9	1591.4891	19.5	382.3	7.1
	10	1729.1460	47.7	417.5	16.3
	11	1865.3041	36.9	411.6	12.9
	12	2001.2566	28.6	404.9	9.4
	13	2137.7634	38.2	383.3	11.4
	14	2273.4932	32.7	407.4	10.6
	15	2407.6669	36.2	403.8	11.5
	16	2541.7150	32.7	414.1	10.4
	17	2676.2540	27.1	403.7	8.9
	18	2811.4796	24.1	397.5	8.3
	19	2947.1109	23.1	411.6	7.8
	20	3082.4893	27.2	409.9	9.3
	21	3217.9177	31.3	415.0	11.4
	22	3353.6765	46.1	388.6	19.0
	23	3489.6852	66.4	425.2	30.0

Before inverting these rotational frequency splittings, we completed the set of splittings of low-degree low-order p modes down to 1 mHz with those extracted from the analysis of combined GOLF and MDI time series (García et al., 2004a). Then we added high-degree modes using splittings of modes between $\ell = 4$ and $\ell = 25$ from the analysis of the 2088-day-long time series of MDI observations fitted by Korzennik (2005).

3. Sensitivity of the Splittings to the Rotation Rate below $0.25 R_\odot$

We have shown in the introduction that the acoustic modes of higher radial order penetrate deeper in the solar interior and that they are potentially of great interest to better constrain the rotation inside the solar core. This is illustrated in Figure 3, where the modes listed in Table 1 are plotted as a function of the radius of their inner turning point.

The vertical error bars correspond to the 1σ error bars of the splittings. These uncertainties are smaller for the $\ell = 2$ and 3 than for the $\ell = 1$ modes because the visible m components of the latter are closer in frequency and therefore blend together at lower frequencies than for higher degree modes.

The amount of information on the rotation rate inside the core ($r \leq 0.25 R_\odot$) present in these modes depends on the precision of the measured splittings. The sensitivity of the splittings to the rotation rate below $0.25 R_\odot$ is plotted in Figure 4(a). This sensitivity is defined as the ratio of the contribution of the rotation rate below $0.25 R_\odot$ to the splitting ($\Delta \nu_{n\ell m}$) and the uncertainty ($\sigma_{n\ell m}$) in its determination [see Equation (1) for sectoral modes, $\ell = m$]:

$$\text{Sensitivity} = \left(\frac{\int_0^{0.25 R_\odot} K_{n,\ell}(r) \Omega(r) \, dr}{\int_0^{R_\odot} K_{n,\ell}(r) \Omega(r) \, dr} \Delta \nu_{n,\ell} \right) \Big/ \sigma_{n,\ell}. \quad (1)$$

For a given mode, if the sensitivity function is greater than one, this particular mode provides useful information on the core rotation. Up to ≈ 3.4 mHz, all of the modes $\ell \leq 3$ are potentially interesting for the inversions. The situation has evolved drastically since 2003 when the previous analysis of 2243-day-long GOLF and MDI time series showed that below $0.2 R_\odot$ only those modes below 2.2 mHz had a sensitivity greater than one (see the left panel of Figure 1 in Couvidat et al., 2003). In fact, when we compute the sensitivity function below the latter radius using the new computed GOLF splittings we obtain the same dependence

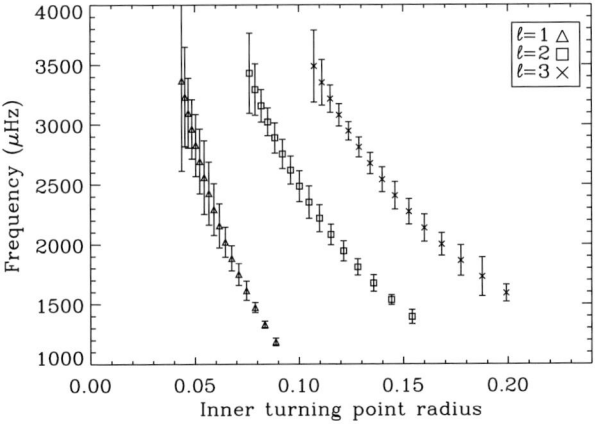

Figure 3 Fitted modes listed in Table 1 as a function of the inner turning point radii. The error bars are the splitting error bars in nanohertz magnified by a factor of 10^4.

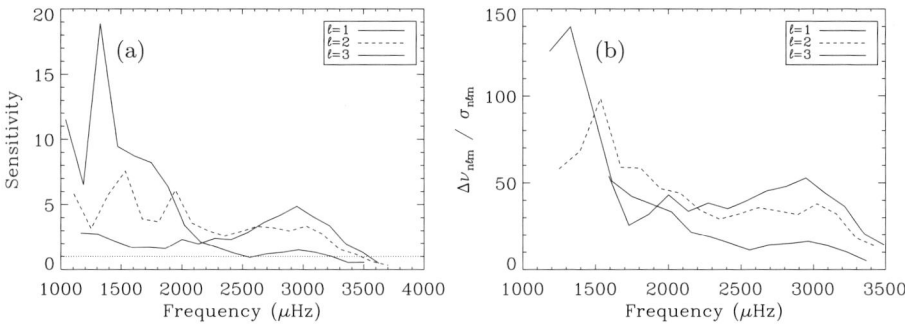

Figure 4 (a) Sensitivity of the splittings to the rotation rate below $0.25 R_\odot$ as defined in the text. The horizontal dotted line is the lower limit below which the splittings would not carry any useful information on the solar core rotation. (b) Splittings normalized by their error bars.

with the frequency below 0.25 $R_\odot$ but rescaled to lower sensitivity. For the $\ell = 2$ and 3 modes the sensitivity reduction is about a factor of two. In the case of the $\ell = 1$ modes, the factor is smaller and the sensitivity curve crosses unity at 3.1 instead of 3.3 mHz.

Figure 4(b) shows the resulting splittings normalized by their error bars. The precision clearly increases toward the low-frequency range for the $\ell = 1$ modes whereas for the $\ell = 2$ and 3 modes the curves are flatter, indicating that the error bars are nearly constant all along the analyzed frequency range.

4. Discussion

To study the effect of the $\ell \leq 3$ modes at medium-range frequencies on the inferred rotation rate we carry out some 2D inversions.

The rotational splittings ($\Delta \nu_{n\ell m}$) are the integral of the product of a sensitivity function – or a kernel – $K_{n\ell m}(r, \theta)$, a known function based on a solar model, with the rotation rate $[\Omega(r, \theta)]$ over the radius (r) and the co-latitude (θ) (Hansen, Cox, and van Horn, 1977):

$$\Delta \nu_{n\ell m} = \frac{1}{2\pi} \int_0^R \int_0^\pi K_{n\ell m}(r, \theta) \Omega(r, \theta) \, dr \, d\theta + \epsilon_{n\ell m}, \qquad (2)$$

where $\epsilon_{n\ell m}$ is the effective error on the measured value $\Delta \nu_{n\ell m}$. We assume that the errors ($\epsilon_{n\ell m}$) follow a normal distribution with a standard deviation $\sigma_{n\ell m}$, which is estimated by the fitting procedure (see Section 2), as listed in Table 1. This set of equations defines a classical inverse problem for the solar rotation. The inversion of this set of M integral equations – one for each measured $\Delta \nu_{n\ell m}$ – allows us to infer the rotation rate profile as a function of radius and latitude from a set of observed splittings.

The inversion method we use here is based on the regularized least-squares methodology (RLS) following the prescription described in Eff-Darwich and Pérez Hernández (1997). In this implementation, the regularization function is weighted differently for each model grid point (see also Eff-Darwich *et al.*, 2008). In summary, Equation (2) is transformed into a matrix relation,

$$\mathbf{D} = \mathbf{A}\mathbf{x} + \epsilon, \qquad (3)$$

where **D** is the data vector, with elements $\Delta \nu_{n\ell m}$ and dimension M, **x** is the solution vector to be determined at N model grid points, A is the matrix with the kernels, of dimension $M \times N$, and ϵ is the vector containing the error bars.

The RLS solution for the vector **x** is given by

$$\mathbf{x}_{\text{est}} = \left(A^{\text{T}} A + \gamma H\right)^{-1} A^{\text{T}} \mathbf{D}, \tag{4}$$

where γ is a scalar introduced to give a suitable weight to the constraint matrix H on the solution. Replacing **D** from Equation (3) we obtain

$$\mathbf{x}_{\text{est}} = \left(A^{\text{T}} A + \gamma H\right)^{-1} A^{\text{T}} A \mathbf{x} \equiv R\mathbf{x} \tag{5}$$

and hence

$$R = \left(A^{\text{T}} A + \gamma H\right)^{-1} A^{\text{T}} A. \tag{6}$$

The matrix R is referred to as the resolution or sensitivity matrix. Ideally, R would be the identity matrix, which corresponds to perfect resolution. However, if we try to find an inverse with a resolution matrix R close to the identity matrix, the solution is generally dominated by noise magnification. The individual columns of R display how anomalies in the corresponding model are imaged by the combined effect of measurement and inversion. In this sense, each element R_{ij} reveals how much of the anomaly in the jth inversion model grid point is transferred into the ith grid point. Consequently, the diagonal elements R_{ii} indicate how much of the information is saved in the model estimate and may be interpreted as the resolvability or sensitivity of x_i. We defined the sensitivity λ_i of the grid point x_i to the inversion process as follows:

$$\lambda_i = \frac{R_{ii}}{\sum_{j=1}^{N} R_{ij}}. \tag{7}$$

With this definition, a lower value of λ_i means a lower sensitivity of x_i to the inversion of the solar rotation. We define a smoothing vector **W** with elements $w_i = \lambda_i^{-1}$ that is introduced in Equation (4) to complement the smoothing parameter γ, namely

$$\mathbf{x}_{\text{est}} = \left(A^{\text{T}} A + \gamma W H\right)^{-1} A^{\text{T}} \mathbf{D}. \tag{8}$$

Such substitution allows us to apply different regularizations to different model grid points x_i, whose sensitivities depend on the data set that is used in the inversions. A set of results can be calculated for different values of γ, the optimal solution being the one with the best trade-off between error propagation and the quadratic difference $\chi^2 = |A\mathbf{x}_{\text{est}} - \mathbf{D}|^2$ as discussed in Eff-Darwich and Pérez Hernández (1997).

We have performed a theoretical study to determine the effect of adding the low-degree high-order p modes in the inversions. To do so, we have computed, using Equation (2), the splittings corresponding to an artificial rotation profile [$\Omega(r, \theta)$]. This artificial profile has a differential rotation in the convection zone (that mimics the real one), a rigid rotation from 0.7 down to $0.2 R_\odot$ equal to $\Omega_{\text{rz}} = 433$ nHz, and a steplike profile in the core having a rate of 350 nHz in the $0.1-0.2 R_\odot$ region and a rate three times larger than the rest of the radiative zone below $0.1 R_\odot$. Although this profile with steep changes and a small drop followed by an increase is unlikely to be realistic, it enables us to characterize the quality of the inversions as the inversion code has difficulty reproducing these steep gradients.

Table 2 Description of the artificial data sets used to study the sensitivity of low-degree high-order p modes.

Data set	$\ell = 1$ (mHz)	$\ell = 2, 3$ (mHz)	$\ell > 3$ (mHz)
1	$1 \leq \nu \leq 2.3$	$1 \leq \nu \leq 2.3$	$1 \leq \nu \leq 3.9$
2	$1 \leq \nu \leq 2.5$	$1 \leq \nu \leq 3.4$	$1 \leq \nu \leq 3.9$
3	$1 \leq \nu \leq 3.4$	$1 \leq \nu \leq 3.4$	$1 \leq \nu \leq 3.9$

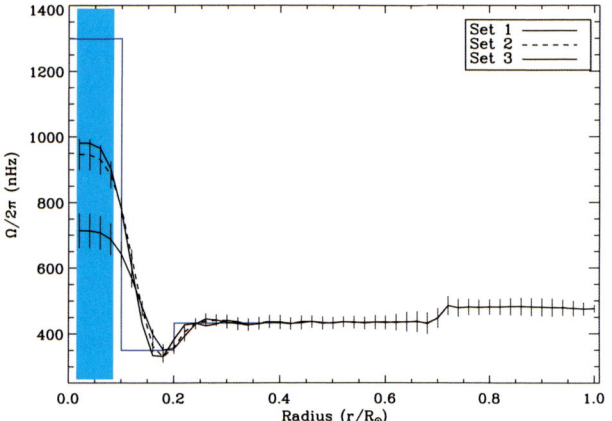

Figure 5 Equatorial rotation rate obtained from *ideal* inversions (all the modes with the same error bars) of an artificial step profile (continuous line) by using the three data sets described in Table 2. The inversion error bars are plotted only for the profiles resulting from inverting Sets 1 and 2. The shaded area corresponds to the region where the solution of the inversion is not reliable because the lack of sensitivity of the acoustic modes to this region of the Sun gives poor-quality resolution kernels.

Three artificial data sets have been built (see Table 2). The first, Set 1, is the reference and contains low-degree p modes below 2.3 mHz whereas p modes for $\ell > 3$ extend up to 3.9 mHz. The second set, Set 2, adds the $\ell = 2$ and 3 modes up to 3.4 mHz. As we have shown in the previous section, these modes could have some sensitivity to the rotation in the core (see Figure 4(a)). Finally, the third set, Set 3, also adds the $\ell = 1$ modes up to 3.4 mHz that seem to be at the sensitivity limit (see again Figure 4(a)).

Figure 5 shows the resulting inversions for these three artificial data sets. We shall qualify these inversions as *ideal* because all of the splittings were given the same error bars and are noiseless. The inferred rotation rate represents the best result that we can obtain for each set of modes.

As expected, the differences between the profiles appear below $0.3 R_\odot$ but they are only significant below $\approx 0.15 R_\odot$. All through the solar interior, the differences between Sets 2 and 3 are very small and are within the inversion uncertainties. Comparing these two data sets with Set 1, we obtain an improvement up to $\approx 30\%$ in the deepest region when the sets with more modes are used. Unfortunately, in all of the data sets, the recovered profiles are not accurate enough. This is easily understood by looking at the corresponding resolution kernels, which remain, indeed, very broad. Figure 6 shows some resolution kernels for the deepest model grid points and for Sets 1 and 3. There are no resolution kernels centered below $\approx 0.13 R_\odot$ for Set 1. Comparing the resolution kernels of Sets 1 and 3, we see an improvement in the second case. Indeed the resolution kernels are narrower (better resolution in the solution of the inversion) and they are slightly shifted to deeper layers (by about 3% of the solar radius).

We have shown the improvements of adding low-degree high-order p modes in the *ideal* inversions. However, the observed splittings have nonuniform error bars. Therefore, we inverted the same sets of artificial splittings but using the actual, measured error bars. To be more realistic, Gaussian noise was also added to the artificial splittings, commensurate with their corresponding uncertainty. The resulting rotation profiles are plotted in Figure 7. As

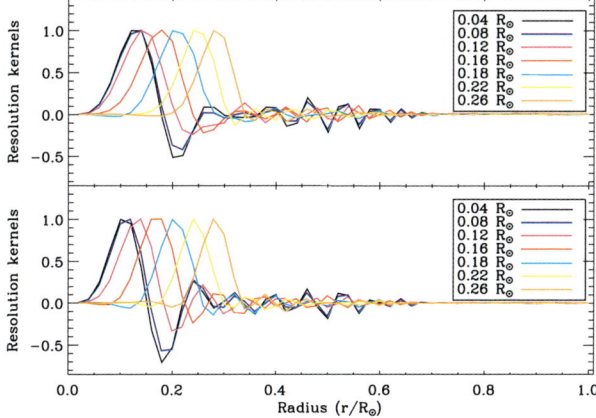

Figure 6 Resolution kernels of the inner model grid for Set 1 (top) and 3 (bottom) in the *ideal* case (where all the modes have the same error bars).

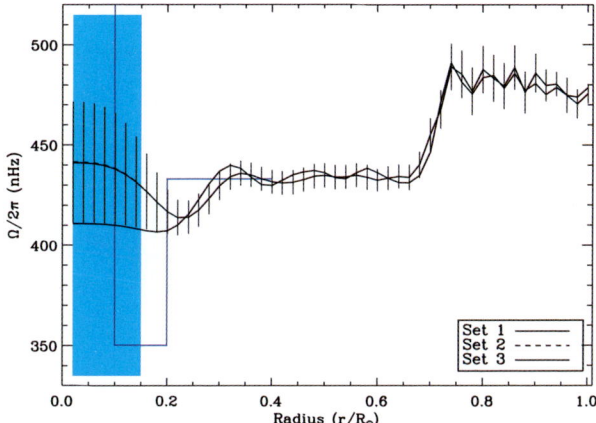

Figure 7 Equatorial rotation rate obtained from the inversions of the artificial step profile (continuous line) by using the three data sets described in Table 2 but with the real error bars as described in Section 2. For clarity, the inversion error bars are plotted only on the inversion of the Set 2. The shaded area corresponds to the region where the solution of the inversion is not reliable, as explained in Figure 5.

for the *ideal* case, Sets 2 and 3 give the same qualitative results and both are slightly better in the core than Set 1, even though the differences are within the inversion error bars. Nevertheless, the solution is improved in all of the radiative region where the ripples are reduced and the solution is smoother when more modes are added. Therefore, even with the realistic uncertainties, the recovered rotation rate is more accurate when Sets 2 or 3 are used. Unfortunately, the contribution of the $\ell = 1$ modes in the 2.5–3.4 mHz range remains negligible with the present magnitude of their uncertainties.

Including realistic uncertainties results in significant changes in the resolution kernels (see Figure 8). Indeed, the coefficients of the linear combination given to each splitting (inversely proportional to the error bars) changes the linear combination of the modes used to compute those resolution kernels. Therefore the inner resolution kernels shift outward: The maxima of the deepest one are now at around $0.16 R_\odot$ for all sets and, of course, are broader than the ones of the *ideal* case. The differences between the resolution kernels deduced from Sets 1 and 3 are smaller, explaining why the solutions in both cases are so similar.

After having studied the improvements in the rotation rate that we can expect by adding the low-degree high-order p modes, we computed the same set of inversions using the real data. Figure 9 shows the inferred rotation rate for the real data using the Sets 1, 2, and 3. The results are similar to what we found for the artificial cases with realistic uncertainties.

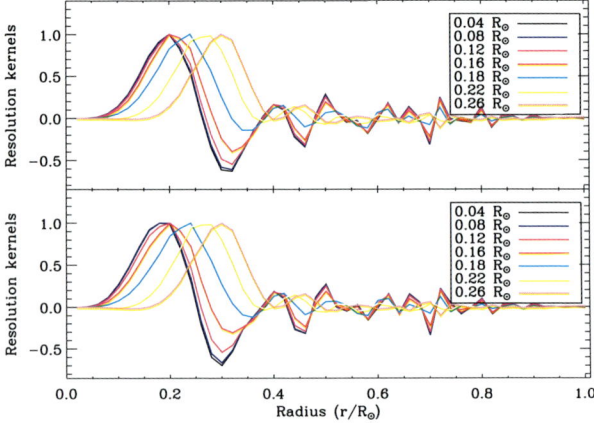

Figure 8 Resolution kernels of the inner model grid points for Set 1 (top) and 3 (bottom) but by using splittings with realistic uncertainties.

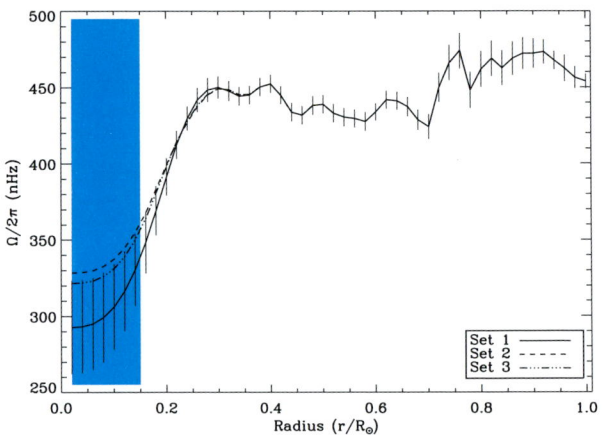

Figure 9 Equatorial rotation rate of real data by using only low-degree low-order p modes up to 2.3 and 3.4 mHz and the same set of high-degree modes. For clarity, the inversion error bars are shown only for the result of inverting Set 1. The shaded area corresponds to the region where the solution of the inversion is not reliable as explained in Figure 5.

Indeed, the rotation rate below $\approx 0.25 R_\odot$ changes slightly when one compares Sets 2 and 3 (which cannot be distinguished within the error bars) to Set 1. The rotation profiles show a small increase inside the core when the new modes are added (Sets 2 and 3). The difference is barely significant down to $0.16 R_\odot$, which is the reliability limit of the inversion. As we already knew, we need gravity modes to get more information about the dynamics of the deepest layers of the core (Mathur *et al.*, 2008). In the rest of the radiative zone (above $0.25 R_\odot$), the profiles inferred by using the three sets are the same.

We have also tested the effect of increasing the amplitude ratio of the visible m components of the $\ell = 2$ and 3 modes by 20% and 10% respectively. The resulting splittings are systematically increased by ≈ 10 nHz for the $\ell = 2$ modes and by 3 to 4 nHz for the $\ell = 3$ modes. The inversion computed with these new splittings gives the same qualitative results as those shown in Figure 9: The inferred profiles are improved when more modes are used (Sets 2 and 3). For these two profiles, the rotation rate obtained at $0.16 R_\odot$ has increased by some 7%, which is still within the inversion error bars.

5. Conclusion

We have analyzed a 4182-day-long time series of GOLF velocity observation and have derived the frequencies and the splittings of the acoustic modes up to 3.4 mHz. We used a 45-µHz window to fit the $\ell = 0, 2$ and $1, 3$ pairs to limit any bias in the fitted splittings. By fitting shorter time series simultaneously and including a linear dependence of the splitting with solar activity, we have checked that for most of the modes the variation of this parameter with the activity is within the error bars estimated by using the whole time series.

We then carried out a study of the sensitivity of the rotation of the core (below $0.25 R_\odot$) to the $\ell \leq 3$ modes up to 3.4 mHz. We show that the contribution of the rotation rate in the core to the splittings of the $\ell = 2$ and 3 modes is several times the size of their current error bars. By contrast the $\ell = 1$ modes in the 2.5 – 3.4 mHz range have a contribution from the core of the same order of magnitude as their error bars. Therefore, they cannot provide valuable information on the core's rotation.

By using artificial inversions of a steplike profile we have shown that, ideally, the new modes would improve the inferred rotation rate in the core but with a small accuracy on the profile itself owing to a lack of resolution in the inner regions of the Sun – as indicated by the corresponding resolution kernels. The effect of including the $\ell = 1$ modes between 2.5 and 3.4 mHz is negligible with the current size of their error bars. Using the real error bars degrades the inferred solution and the differences in the core profile are within the inversion uncertainties. However, the profile in the rest of the radiative region is smoother when more modes are used, demonstrating that the inclusion of those modes improves the inversion results.

Finally, the inversion of the real data, including the larger set of low-degree low-order p modes, shows no significant differences between Sets 2 and 3 as for the artificial data. There is a small increase of the rotation rate below $0.25 R_\odot$ compared to the reference Set 1 but it is within the inversion error bars. The resolution kernels are improved when the low-degree high-order p modes are included and the resolution kernels at the deepest locations are shifted by 3% of the solar radius toward the center. This result suggests that with the present error bars it is better to use such modes in the inversions (at least for the high-order $\ell = 2$ and 3 modes).

To make progress in the inference of the rotation rate in the solar core we would need to measure the low-degree modes – acoustic, mixed, and gravity modes – that will bring a direct signature of the dynamics of the still hidden core of the Sun.

Acknowledgements The GOLF experiment is based upon a consortium of institutes (IAS, CEA/Saclay, Nice, and Bordeaux Observatories from France and IAC from Spain) involving numerous scientists and engineers, as enumerated in Gabriel *et al.* (1995). SOHO is a mission of international cooperation between ESA and NASA. This work has been partially funded by Grant No. AYA2004-04462 of the Spanish Ministry of Education and Culture and partially supported by the European Helio- and Asteroseismology Network (HELAS[6]), a major international collaboration funded by the European Commission's Sixth Framework Programme.

References

Antia, H.M., Basu, S.: 2000, Temporal variations of the rotation rate in the solar interior. *Astrophys. J.* **541**, 442 – 448. doi:10.1086/309421.

---

[6]http://www.helas-eu.org/.

Chaplin, W.J., Elsworth, Y., Isaak, G.R., Marchenkov, K.I., Miller, B.A., New, R.: 2001, Rigid rotation of the solar core? On the reliable extraction of low-l rotational p-mode splittings from full-disc observations of the sun. *Mon. Not. Roy. Astron. Soc.* **327**, 1127–1136. doi:10.1046/j.1365-8711.2001.04805.x.

Chaplin, W.J., Sekii, T., Elsworth, Y., Gough, D.O.: 2004, On the detectability of a rotation-rate gradient in the solar core. *Mon. Not. Roy. Astron. Soc.* **355**, 535–542. doi:10.1111/j.1365-2966.2004.08338.x.

Chaplin, W.J., Appourchaux, T., Baudin, F., Boumier, P., Elsworth, Y., Fletcher, S.T., Fossat, E., García, R.A., Isaak, G.R., Jiménez, A., Jiménez-Reyes, S.J., Lazrek, M., Leibacher, J.W., Lochard, J., New, R., Pallé, P., Régulo, C., Salabert, D., Seghouani, N., Toutain, T., Wachter, R.: 2006, Solar FLAG hare and hounds: on the extraction of rotational p-mode splittings from seismic, sun-as-a-star data. *Mon. Not. Roy. Astron. Soc.* **369**, 985–996. doi:10.1111/j.1365-2966.2006.10358.x.

Couvidat, S., García, R.A., Turck-Chièze, S., Corbard, T., Henney, C.J., Jiménez-Reyes, S.: 2003, The rotation of the deep solar layers. *Astrophys. J.* **597**, L77–L79. doi:10.1086/379698.

Domingo, V., Fleck, B., Poland, A.I.: 1995, The SOHO mission: an overview. *Solar Phys.* **162**, 1–37.

Eff-Darwich, A., Pérez Hernández, F.: 1997, A new strategy for helioseismic inversions. *Astron. Astrophys. Suppl.* **125**, 391–398.

Eff-Darwich, A., Korzennik, S.G., Jiménez-Reyes, S.J.: 2002, Inversion of the internal solar rotation rate. *Astrophys. J.* **573**, 857–863. doi:10.1086/340747.

Eff-Darwich, A., Korzennik, S.G., Jiménez-Reyes, S.J., García, R.A.: 2008, The rotation of the solar radiative interior after 2088 days of helioseismic observations from GONG, GOLF and MDI. *Astrophys. J.* in press. http://arXiv.org/abs/0802.3604.

Elsworth, Y., Howe, R., Isaak, G.R., McLeod, C.P., Miller, B.A., New, R., Wheeler, S.J., Gough, D.O.: 1995, Slow rotation of the sun's interior. *Nature* **376**, 669–672. doi:10.1038/376669a0.

Gabriel, A.H., Grec, G., Charra, J., Robillot, J.M., Cortés, T.R., Turck-Chièze, S., Bocchia, R., Boumier, P., Cantin, M., Céspedes, E., Cougrand, B., Cretolle, J., Dame, L., Decaudin, M., Delache, P., Denis, N., Duc, R., Dzitko, H., Fossat, E., Fourmond, J.J., García, R.A., Gough, D., Grivel, C., Herreros, J.M., Lagardere, H., Moalic, J.P., Pallé, P.L., Petrou, N., Sanchez, M., Ulrich, R., van der Raay, H.B.: 1995, Global oscillations at low frequency from the SOHO mission (GOLF). *Solar Phys.* **162**, 61–99.

García, R.A., Corbard, T., Chaplin, W.J., Couvidat, S., Eff-Darwich, A., Jiménez-Reyes, S.J., Korzennik, S.G., Ballot, J., Boumier, P., Fossat, E., Henney, C.J., Howe, R., Lazrek, M., Lochard, J., Pallé, P.L., Turck-Chièze, S.: 2004a, About the rotation of the solar radiative interior. *Solar Phys.* **220**, 269–285. doi:10.1023/B:SOLA.0000031395.90891.ce.

García, R.A., Jiménez-Reyes, S.J., Turck-Chièze, S., Ballot, J., Henney, C.J.: 2004b, Solar low-degree p-mode parameters after 8 years of velocity measurements with SOHO. In: Danesy, D. (ed.) *ESA SP-559: SOHO 14 Helio- and Asteroseismology: Towards a Golden Future* **14**, ESA, Noordwijk, 436.

García, R.A., Turck-Chièze, S., Boumier, P., Robillot, J.M., Bertello, L., Charra, J., Dzitko, H., Gabriel, A.H., Jiménez-Reyes, S.J., Pallé, P.L., Renaud, C., Roca Cortés, T., Ulrich, R.K.: 2005, Global solar Doppler velocity determination with the GOLF/SoHO instrument. *Astron. Astrophys.* **442**, 385–395. doi:10.1051/0004-6361:20052779.

García, R.A., Turck-Chièze, S., Jiménez-Reyes, S.J., Ballot, J., Pallé, P.L., Eff-Darwich, A., Mathur, S., Provost, J.: 2007, Tracking solar gravity modes: The dynamics of the solar core. *Science* **316**, 1591–1593. doi:10.1126/science.1140598.

Gelly, B., Lazrek, M., Grec, G., Ayad, A., Schmider, F.X., Renaud, C., Salabert, D., Fossat, E.: 2002, Solar p-modes from 1979 days of the GOLF experiment. *Astron. Astrophys.* **394**, 285–297. doi:10.1051/0004-6361:20021106.

Hansen, C.J., Cox, J.P., van Horn, H.M.: 1977, The effects of differential rotation on the splitting of nonradial modes of stellar oscillation. *Astrophys. J.* **217**, 151–159.

Henney, C.J.: 1999, Comparison between simultaneous golf and mdi observations in search of low frequency solar oscillations. Ph.D. thesis, University of California, Los Angeles.

Howe, R., Christensen-Dalsgaard, J., Hill, F., Komm, R.W., Larsen, R.M., Schou, J., Thompson, M.J., Toomre, J.: 2000, Dynamic variations at the base of the solar convection zone. *Science* **287**, 2456–2460.

Jiménez, A., Roca Cortés, T., Jiménez-Reyes, S.J.: 2002, Variation of the low-degree solar acoustic mode parameters over the solar cycle. *Solar Phys.* **209**, 247–263.

Jiménez, A., Pérez Hernandez, F., Claret, A., Pallé, P.L., Régulo, C., Roca Cortés, T.: 1994, The rotation of the solar core. *Astrophys. J.* **435**, 874–880. doi:10.1086/174868.

Jiménez-Reyes, S.J., Corbard, T., Pallé, P.L., Roca Cortés, T., Tomczyk, S.: 2001, Analysis of the solar cycle and core rotation using 15 years of Mark-I observations: 1984–1999. I. The solar cycle. *Astron. Astrophys.* **379**, 622–633. doi:10.1051/0004-6361:20011374.

Korzennik, S.G.: 2005, A mode-fitting methodology optimized for very long helioseismic time series. *Astrophys. J.* **626**, 585–615. doi:10.1086/429748.

Lopes, I., Turck-Chièze, S.: 1994, The second order asymptotic theory for the solar and stellar low degree acoustic mode predictions. *Astron. Astrophys.* **290**, 845–860.

Mathur, S., Eff-Darwich, A.M., García, R.A., Turck-Chièze, S.: 2008, Sensitivity of helioseismic gravity modes to the dynamics of the solar core. *Astron. Astrophys.* in press.

Nigam, R., Kosovichev, A.G.: 1998, Measuring the sun's eigenfrequencies from velocity and intensity helioseismic spectra: Asymmetrical line profile-fitting formula. *Astrophys. J.* **505**, L51. doi:10.1086/311594.

Provost, J., Berthomieu, G., Morel, P.: 2000, Low-frequency p- and g-mode solar oscillations. *Astron. Astrophys.* **353**, 775–785.

Roca Cortés, T., Lazrek, M., Bertello, L., Thiery, S., Baudin, F., Boumier, P., Gavryusev, V., Garcia, R.A., Regulo, C., Ulrich, R.K., Grec, G., the GOLF Team: 1998, The solar acoustic spectrum as seen by GOLF. II. Noise statistics background and methods of analysis. In: Korzennik, S. (ed.) *Structure and Dynamics of the Interior of the Sun and Sun-like Stars* **SP-418**, ESA, Noordwijk, 323–334.

Scherrer, P.H., Bogart, R.S., Bush, R.I., Hoeksema, J.T., Kosovichev, A.G., Schou, J., Rosenberg, W., Springer, L., Tarbell, T.D., Title, A., Wolfson, C.J., Zayer, I., MDI Engineering Team: 1995, The solar oscillations investigation – Michelson Doppler Imager. *Solar Phys.* **162**, 129–188. doi:10.1007/BF00733429.

Schou, J., Antia, H.M., Basu, S., Bogart, R.S., Bush, R.I., Chitre, S.M., Christensen-Dalsgaard, J., di Mauro, M.P., Dziembowski, W.A., Eff-Darwich, A., Gough, D.O., Haber, D.A., Hoeksema, J.T., Howe, R., Korzennik, S.G., Kosovichev, A.G., Larsen, R.M., Pijpers, F.P., Scherrer, P.H., Sekii, T., Tarbell, T.D., Title, A.M., Thompson, M.J., Toomre, J.: 1998, Helioseismic studies of differential rotation in the solar envelope by the solar oscillations investigation using the Michelson Doppler Imager. *Astrophys. J.* **505**, 390–417. doi:10.1086/306146.

Thompson, M.J., Toomre, J., Anderson, E., Antia, H.M., Berthomieu, G., Burtonclay, D., Chitre, S.M., Christensen-Dalsgaard, J., Corbard, T., Derosa, M., Genovese, C.R., Gough, D.O., Haber, D.A., Harvey, J.W., Hill, F., Howe, R., Korzennik, S.G., Kosovichev, A.G., Leibacher, J.W., Pijpers, F.P., Provost, J., Rhodes, E.J., Schou, J., Sekii, T., Stark, P.B., Wilson, P.: 1996, Differential rotation and dynamics of the solar interior. *Science* **272**, 1300–1305.

Toutain, T., Appourchaux, T.: 1994, Maximum likelihood estimators: An application to the estimation of the precision of helioseismic measurements. *Astron. Astrophys.* **289**, 649–658.

Can We Constrain Solar Interior Physics by Studying the Gravity-Mode Asymptotic Signature?

R.A. García · S. Mathur · J. Ballot

Originally published in the journal Solar Physics, Volume 251, Nos 1–2, 135–147.
DOI: 10.1007/s11207-008-9159-y © Springer Science+Business Media B.V. 2008

Abstract Gravity modes are the best probes to infer the properties of the solar radiative zone, which represents 98% of the Sun's total mass. It is usually assumed that high-frequency g modes give information about the structure of the solar interior whereas low-frequency g modes are more sensitive to the solar dynamics (the internal rotation). In this work, we develop a new methodology, based on the analysis of the almost constant separation of the dipole gravity modes, to introduce new constraints on the solar models. To validate this analysis procedure, several solar models – including different physical processes and either old or new chemical abundances (from, respectively, Grevesse and Noels (*Origin and Evolution of the Elements* **199**, Cambridge University Press, Cambridge, 15, 1993) and Asplund, Grevesse, and Sauval (*Cosmic Abundances as Records of Stellar Evolution and Nucleosynthesis* **CS-336**, Astron. Soc. Pac., San Francisco, 25 – 38, 2005)) – have been compared to another model used as a reference. The analysis clearly shows that this methodology has enough sensitivity to distinguish among some of the models, in particular, among those with different compositions. The comparison of the models with the g-mode asymptotic signature detected in GOLF data favors the ones with old abundances. Therefore, the physics of the core – obtained through the analysis of the g-mode properties – is in agreement with the results obtained in the previous studies based on the acoustic modes, which are mostly sensitive to more external layers of the Sun.

Keywords Helioseismology: Observations · Interior: Radiative zone, core

---

Helioseismology, Asteroseismology, and MHD Connections
Guest Editors: Laurent Gizon and Paul Cally.

R.A. García (✉) · S. Mathur
Laboratoire AIM, CEA/DSM-CNRS, U. Paris Diderot, IRFU/SAp, 91191 Gif-sur-Yvette Cedex, France
e-mail: rgarcia@cea.fr

S. Mathur
e-mail: smathur@cea.fr

J. Ballot
Max-Planck-Institut für Astrophysik, Karl-Schwarzschild-Strasse 1, 85748 Garching, Germany
e-mail: jballot@mpa-Garching.mpg.de

1. Introduction

The combination of detailed modeling and helioseismic observations has enabled us to accurately describe the solar interior from the surface to its center where the nuclear reactions take place. The precise characterization of the resonant acoustic (pressure or p) modes has allowed us to constrain the theories describing the solar interior. The profiles of sound speed, density (*e.g.*, Basu *et al.*, 1997; Couvidat, Turck-Chièze, and Kosovichev, 2003), and differential rotation (Thompson *et al.*, 1996; Howe *et al.*, 2000; Chaplin *et al.*, 2001; Couvidat *et al.*, 2003), as well as the position of the base of the convective zone (*e.g.*, Christensen-Dalsgaard, Gough, and Thompson, 1991) and the prediction of the neutrino fluxes (*e.g.*, Turck-Chièze *et al.*, 2001) are some examples of the constraints provided by the p-mode analyses.

Unfortunately, only a few p modes propagate inside the inner core – below 0.25 solar radius ($R_\odot$) – which contains about 50% of the solar mass. To make progress in knowledge of these layers, another type of mode is needed: the gravity (g) modes. These modes have not yet been unambiguously detected individually in the Sun (for example, see Appourchaux *et al.*, 2000; Gabriel *et al.*, 2002; Elsworth *et al.*, 2006) because they become evanescent in the convection zone, reaching the surface of the Sun with very small amplitudes (Andersen, 1996; Kumar, Quataert, and Bahcall, 1996). Some candidates have been detected (Turck-Chièze *et al.*, 2004b; Mathur *et al.*, 2007) by analyzing helioseismic data from the Global Oscillations at Low Frequencies (GOLF; Gabriel *et al.*, 1995) instrument onboard the ESA/NASA *Solar and Heliospheric Observatory* (SOHO; Domingo, Fleck, and Poland, 1995) spacecraft. These candidates have provided some scenarios about the possible dynamics of the solar core but it has been impossible to correctly label these peaks in terms of their orders and degrees (n, ℓ, m) and to give the central frequencies of the candidates. Recently, García *et al.* (2007) were able to uncover the signature of the asymptotic properties of the dipole ($\ell = 1$) g modes by measuring their constant spacing ΔP_1. Indeed, for a given degree ℓ, the difference ΔP_ℓ of the periods of g modes with consecutive radial order n is almost constant when $n \gg \ell$ (Tassoul, 1980; Provost and Berthomieu, 1986). The comparison of the observations with some solar models – including different rotation of the core – showed the great sensitivity of this analysis to the dynamics of the solar core and it implied a rotation rate in the core faster than in the rest of the radiative region.

During the past few decades, many new physical processes have been studied and included in the solar models, whereas others have been refined to better match the helioseismic observations. Today, these models include better descriptions of the microscopic physics such as opacities, equation of state, nuclear reaction rates (*e.g.*, Morel, Provost, and Berthomieu, 1997 and references therein), microscopic diffusion (Michaud and Proffitt, 1993) as well as other processes, for example turbulence mixing in the tachocline (Brun, Turck-Chièze, and Zahn, 1999).

New studies of the solar spectrum using the most recent advances in 3D modeling of the solar atmosphere and taking into account nonlocal thermodynamic equilibrium (NLTE) effects have provided new estimations of the solar photospheric composition (Asplund, Grevesse, and Sauval, 2005). In comparison with the older estimations (*e.g.*, Grevesse and Noels, 1993) based on 1D models and the LTE approximation, this new measurement leads to a smaller metallicity and to a substantial modification of the abundances of C, N, and O, which are crucial in solar modeling. The standard models computed with these new abundances yield a larger discrepancy from the helioseismic observations. They disagree with sound-speed and density profiles inferred from acoustic modes (for example, see Turck-Chièze *et al.*, 2004a; Bahcall, Serenelli, and Basu, 2005; Guzik, Watson, and Cox, 2005) and we need, for example, to modify some opacities to reduce the discrepancy. It has also been shown that the models computed with these new abundances modify the frequencies of the resonant low-degree p and g modes (Basu *et al.*, 2007; Chaplin *et al.*, 2007; Zaatri *et al.*, 2007; Mathur *et al.*, 2007).

In this paper we use a set of solar models (briefly described in Section 2) – including different physical processes combined with the old and new abundances of the Sun and with different core rotation rates – that we utilize as a tool to study whether the methodology used to detect the dipole gravity modes in the Sun by García *et al.* (2007) (and summarized in Section 3) can distinguish among these models. To do so, in Section 4 we compare all of them to our reference model: the Saclay seismic model (Turck-Chièze *et al.*, 2001). Finally, once the method has been demonstrated, we compare, in Section 5, the solar models with the GOLF data.

2. Description of the Solar Models

Several solar models, which differ from each other by the physical processes that they include, have been computed with the Code d'Évolution Stellaire Adaptatif et Modulaire (CESAM) (Morel, 1997). They have been developed to study the influence of different physical processes and the impact of the new abundances on the g-mode frequencies. These models are fully explained in Mathur *et al.* (2007). Here we recall briefly their main characteristics.

First let us describe the solar model that has been taken as a reference: the so-called seismic model (corresponding to the model seismic of Couvidat, Turck-Chièze, and Kosovichev, 2003). It is important to note that this model is produced by an evolutionary code and is not directly deduced from a seismic inversion. It is "seismic" because some physical quantities, such as opacities or nuclear reaction rates, have been tuned to better reproduce the sound-speed profile in the radiative zone deduced from the analysis of p-mode observations. One of its aims was to improve the prediction of the neutrino flux. This model includes turbulent diffusion in the tachocline and the chemical composition of Grevesse and Noels (1993).

As we want to compare the impact of different physical processes in our work, five other solar models have been used. They mainly differ in three physical inputs: microscopic diffusion (including the gravitational settling), horizontal turbulent diffusion in the tachocline, and abundances of chemical elements. The model called *no diffusion* does not include any diffusion processes, whereas all of the following models include diffusion through the prescription of Michaud and Proffitt (1993). Although today this model is obsolete, it is a good test for the detection methodology, since the g-mode frequencies are very different as compared to the seismic ones and therefore the analysis should be able to clearly separate them.

The second process, tachocline mixing, is introduced in the code following the prescription of Spiegel and Zahn (1992). The models including this process are called *tacho* whereas the ones without this turbulent effect are named *std*. This turbulent horizontal motion contributes to the thinness of the tachocline responsible for the mixing of chemical elements. Finally, the chemical composition is either taken from Grevesse and Noels (1993) (GN93 hereafter) or from the recent release of Asplund, Grevesse, and Sauval (2005) (As05 hereafter). The difference between these two chemical abundances is a decrease of the heavy element by $\approx 30\%$. Such differences substantially affect, for instance, the Rossland opacities and lead to noticeably different thermal stratifications. The models using the old abundances have the suffix *93* whereas the models including the new abundances have the suffix *05*. Table 1 summarizes the main characteristics of all of the models used. Finally, to use a model calculated with another stellar evolution code, the following study is also done with model S

Table 1 The main characteristics of the solar evolution models used to compute the g-mode frequencies.

Solar model	Microscopic diffusion	Diffusion in tachocline	GN93	As05
no diffusion			×	
std93	×		×	
std05	×			×
tacho93	×	×	×	
tacho05	×	×		×

from Christensen-Dalsgaard et al. (1996), which is a standard model including microscopic diffusion and using GN93.

From each stellar structure, eigenfrequencies have been computed with the adiabatic pulsation package (adipack, provided by J. Christensen-Dalsgaard and available on http://www.phys.au.dk/~jcd/adipack.n/). The computation of g-mode frequencies can become very sensitive to the numerical approach, especially to the treatment of the boundary condition at the center. Therefore, the numerical uncertainties on the computed g-mode periods become large at very low frequency and we should be extremely cautious. However, in the range used for this study (between 2 and 8.5 hours; see Section 3), it has been shown that, for a given model, frequencies are modified by less than 50 nHz by changing the numerics (see Mathur, 2007), leading to a maximum uncertainty of around 50 seconds on the mode periods. Such errors are smaller than the observed deviations among the different models used here. In fact, they are too small (by more than an order of magnitude) to be correctly distinguished by the methodology developed in this paper.

It is usually considered that the high-frequency g modes (above 150 µHz) are very sensitive to the physical processes included in the models (the structure) giving up to 5 µHz of difference among the models, whereas at lower frequencies – inside the asymptotic region – the frequency differences are smaller than a fraction of a microhertz and they could be more useful for studying the dynamics of the solar interior (for example, see Mathur et al., 2007). Indeed, we have already seen that the difference of the periods of two consecutive g modes with the same degree (ℓ) and successive radial order (n) is almost constant. In Figure 1 the differences (ΔP_1) of the central frequencies ($m = 0$ components) of the g modes have been plotted for all the models described in Table 1. The asymptotic regime is reached for periods above $\approx$ five hours and the differences among all of the models considered here are within one minute (from 24 to 25 minutes). Thus, it is going to be very difficult to distinguish among all these models.

Moreover, until now we have only considered the effect of the structure on the g-mode frequencies but we cannot neglect the effect of the rotation of the solar interior on the other m-components of the modes. Indeed, in Figure 1 we have also drawn the $m \pm 1$ components for one of the solar models, the *std05*, by considering a core rotating as a rigid body, like the rest of the radiative region. The dispersion in ΔP_1 introduced by the dynamics (without even considering a rapid rotation in the core, just a rigid rotation profile) is already of the same order of the structural effects owing to the different physical processes and abundances considered. That means that the two effects cannot be easily disentangled and we have to study both at the same time.

At low frequency, it seems more appropriate to study the g modes in period, and not in frequency. Depending on the physical processes taken into account in the solar models, there can be important differences among the periods of the g modes. Thus, we could expect to

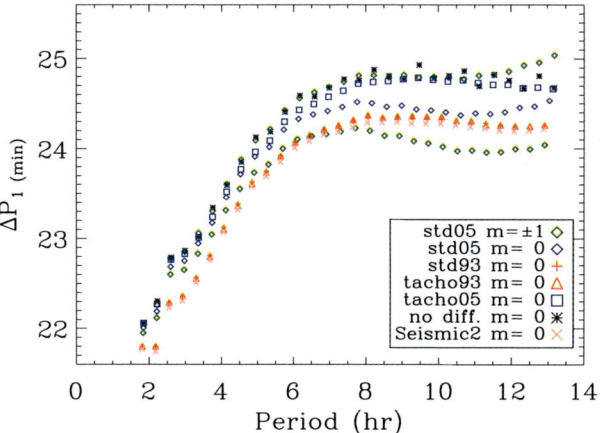

Figure 1 Separations in period (ΔP_1) of the $m = 0$ components of the g modes computed by using the models described in Table 1. For clarity, we have plotted the $m \pm 1$ components only for one model: the *std05* (with a solid rotation of $\Omega_{rz} = 433$ nHz in the radiative zone and a rotation in the convection zone that reproduces the inferred profile deduced from helioseismology data).

use these low-frequency g modes to diagnose some physical processes or abundances taken into account in the models.

To complete this study we need to model the dynamics – the rotation rate – of the solar interior, in particular, in the regions where the g modes are trapped (*i.e.* the radiative region). Today, the solar rotation rate is known well down to the solar core (Thompson *et al.*, 2003) but it is still uncertain in the deepest layers. We have thus computed the g-mode splittings corresponding to ten artificial rotation profiles $\Omega(r, \theta)$. These artificial profiles have a differential rotation in the convection zone (that mimics the real behavior) and a rigid rotation from 0.7 down to a given fractional radius $r_i = r/R_\odot$ (0.1, 0.15, and 0.2), which is equal to $\Omega_{\rm rad} = 433$ nHz. From each of these internal fractional radii (r_i) down to the center we have a step-like profile having a rate of three, five, and ten times larger than the rest of the radiative zone. Finally, we have also computed a rigid rotation rate down to the center of the Sun, which is considered as our reference rotation profile.

3. Methodology

To compare the signature of the g-mode asymptotic properties between real and modeled spectra or even between two different synthetic spectra, we have used the method developed by García *et al.* (2007). Indeed, a numerical simulation of the g-mode power spectrum has been used (see Section S.2 of the on-line material of García *et al.*, 2007, for a full description of the simulation). Briefly, the central frequencies of the g modes have been computed as described in Section 2. Their amplitudes have been considered to all be the same following the calculations done by Kumar, Quataert, and Bahcall (1996), where it has been shown that all the modes in the studied regions have roughly the same amplitudes. We have also simulated the g-mode amplitudes following different laws (*e.g.*, a linear variation or a parabolic decreasing with frequency). In all the cases, the results were qualitatively the same. Finally, one-bin line widths have been used but we have checked that the results do not significantly change when wider modes are simulated. Since we are only interested in the power spectrum, the mode–phase information is lost. It is also important to note that none of the computed simulations contain noise because otherwise we could not quantify those effects coming from the models and those coming from the different noise realizations. Therefore, this study represents what we could obtain in the best conditions. The addition

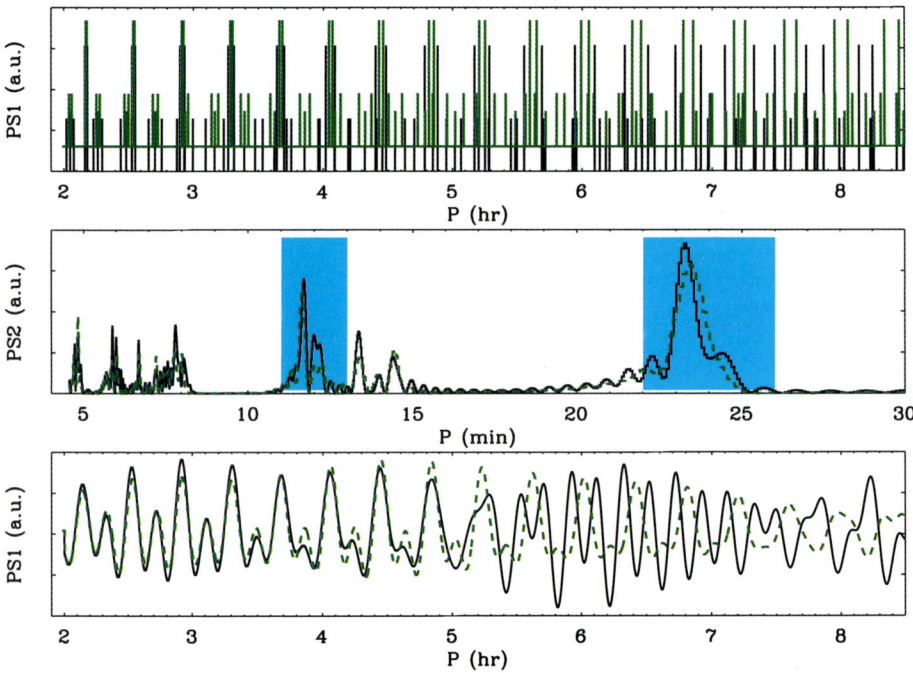

Figure 2 Top panel: Power spectral density, PS1, of $\ell = 1$ and 2 modes computed by using the seismic model and two rotation laws in the core: a rigid core (shifted upward in green) and a core rotating five times faster than the rest of the radiative zone at $r_i = 0.15$ (continuous black lines in all the panels). Middle panel: Power spectral density, PS2, computed from PS1 between 2 and 11 hours. The shaded regions are those used to reconstruct the signal in PS1. Bottom panel: Reconstructed signals in PS1. In the two lower panels, the green dashed lines correspond to the case of a rigid core rotation law.

of noise (both instrumental and solar) would affect the data and could probably degrade the results.

The starting point of this analysis is the oscillation power spectrum density. We will call it hereafter the *first* spectrum (PS1). The real first spectrum is computed from GOLF time series calibrated into velocity (García *et al.*, 2005) with a classical periodogram; the synthetic ones are directly built from the g-mode frequencies and rotational splittings derived from the considered models (see top panel of Figure 2).

PS1 is expressed as a function of the period, since the g modes present a comblike structure in period and not in frequency, as is the case for p modes. We have then performed a spectral analysis of the first spectrum, limited to the range 2–11 hours (*i.e.*, 25–140 µHz), to get the *second* spectrum (hereafter called PS2). It is important to note that the considered PS2 is a complex spectrum, which is required to keep information on both amplitude and phase. For practical purposes, as PS1 is not regularly sampled in period, we cannot use a classical fast Fourier transform (FFT) and we use instead a general method based on sine-wave fittings.

The asymptotic signature of the $\ell = 1$ g modes is detected in $|PS2|^2$ as a high and broad peak around 22 to 25 minutes. Depending on the physics of the solar model and on the rotation rate in the radiative region, the shape and the position of the ΔP_1 pattern is modified. Thus in the middle panel of Figure 2 we show the resultant PS2 of the same seismic model but using two different rotation rates in the core, a rigid rotation in the whole radiative zone and a rotation rate five times faster than the rest of the radiative zone starting at $r_i = 0.15$. The model with the higher rotation rate has more power at 24.5 minutes and widens the ΔP_1 structure.

Once this signature (a pattern of peaks) has been detected in PS2, we can locate in PS1 what contributes to the emergence of the structure in the PS2. To do so, we select in PS2 the vicinity around the pattern at ΔP_1 and around its first harmonic at $\frac{1}{2}\Delta P_1$. We reconstruct the sum of the sine waves corresponding to the amplitudes, frequencies, and phases measured in PS2 in these limited regions. In other words, we perform an inverse Fourier transform of PS2 after having applied a narrow-band filter around what we consider to be the $\ell = 1$ g-mode signature and its first harmonic.

The maxima of these reconstructed waves indicate the location in PS1 corresponding to the positions where we found peaks that have a repetitive pattern. If this signature is indeed the one of the dipolar g modes, the maxima of these reconstructed waves correspond to the m-components – depending on their splittings – of the $\ell = 1$ g modes. The bottom panel of Figure 2 shows the reconstructed signals of the seismic models and the same two rotation rates in the core as in the previous example. At low periods, less than five hours, the reconstructed waves are nearly the same in both cases, but at higher periods, the model with a higher rotation rate in the core follows the separation of the two m-components owing to the higher splittings and the position of the maxima are different compared to the model using the rigid rotation profile.

Finally, we need to quantify the consistency between our models and the GOLF observations. Comparing ΔP_1 is not very informative because, as mentioned by Mathur *et al.* (2007), all of the theoretical computations of ΔP_1 are in agreement to within a minute and, unfortunately, there is not enough resolution to reach such an accuracy in PS2. However, a direct comparison of the periods of the modes shows clear discrepancies between the different models (see bottom panel of Figure 2). Thus, to compare the positions of the modes in PS1, we compute the Pearson's correlation coefficients (*e.g.*, see Press *et al.*, 1992) between the reconstructed waves of the models and the reference one (Section 4) or with the GOLF data (Section 5). If the correlation coefficient is clearly positive, then the maxima of both reconstructed waves coincide generally; that is, in both considered *first* spectra (real and synthetic or both synthetic), the g-mode components are overall at the same position. If the modes in one spectrum are shifted compared to the modes in the other one, the correlation coefficient decreases to zero and even becomes negative. A large negative correlation indicates that each mode component in one spectrum is in between two mode components of the other spectrum.

The range used for the correlation analysis has been restricted to between 2 and 8.5 hours (33 to 140 µHz). The reason for reducing to the low-frequency limit is twofold: On one hand, the dynamical effects on the periods are so important when high core rotation rates are taken into account that the m-components of different modes $\ell = 1$ and consecutive orders (n) are mixed and the interpretation of the results is more complicated; on the other hand, the calculation of the frequencies at such rapid periods (above nine hours) could be less reliable and numerical effects can play an important role, as has been stated in the previous section.

4. Comparison with the Seismic Model

We have seen in the previous sections that the period of the g modes in the low-frequency region (below 150 µHz) is dominated by the physics and the dynamics of the radiative region, and mainly inside the solar core. To study the influence of the dynamics on the correlation coefficients, we apply the methodology described in the previous section to the seismic

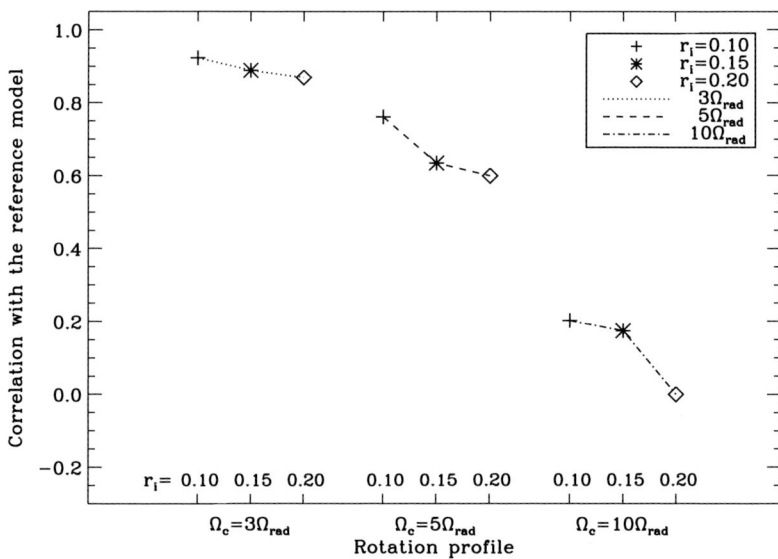

Figure 3 Correlation coefficients between the reconstructed waves for the seismic model with a rigid rotation profile and the ones for the seismic model with other rotation profiles. All of the rotation profiles have a differential rotation in the convective zone and a rigid rotation from 0.7 down to r_i ($= r/R_\odot$) of 0.1 (crosses), 0.15 (asterisks), and 0.2 (diamonds). The rotation rate below r_i is constant: $3\,\Omega_{\rm rad}$ (dotted line), $5\,\Omega_{\rm rad}$ (dashed line), and $10\,\Omega_{\rm rad}$ (dotted-dashed line), where $\Omega_{\rm rad}$ is the rotation rate in the rest of the radiative zone (433 nHz).

model with a rigid rotation profile that is taken as our reference model. Then, we compute the correlation coefficient between the waves reconstructed for this reference model and the ones for the seismic model with different rotation profiles. The results are shown in Figure 3.

We can note that the higher the rotation rate in the core, the lower the value of the correlation will be (*i.e.*, when the difference between the reference and the studied model is bigger, the correlation is smaller). For a given rotation rate in the core, if we shift the value of r_i from $0.1\,R_\odot$ to $0.2\,R_\odot$, the correlation decreases a little, showing that there is low sensitivity to the fractional radius r_i. Then if we had only one reliable model with defined physical processes, we could infer some information on the rotation profile from this correlation. We would be able to say, for instance, that a very rapid rotation in the core is very unlikely to be compatible with our reference model (seismic with rigid rotation profile).

However, another parameter has to be taken into account: the solar structure. In Figure 4, we have plotted the correlation coefficients between the reconstructed waves for the seismic model with a given rotation profile and another model with the same rotation profile. As each point on the plot is calculated with two different models but with the same rotation profile, it indicates the sensitivity of the correlation to the different physical parameters included in the two compared models.

We can see in Figure 4 that there are two clearly separated groups of correlations. The first one contains the models with the old abundances of GN93 that give correlations above 70% whatever rotation profile to which they are attributed. The second group of models, containing the model without diffusion and the models that include the new abundances of As05, gives correlations below 30% and even negative. Thus the *g*-mode frequencies extracted from these two groups are incompatible. Although this result is already known (Zaatri *et al.*, 2007; Mathur *et al.*, 2007), it proves the ability of this methodology to separate

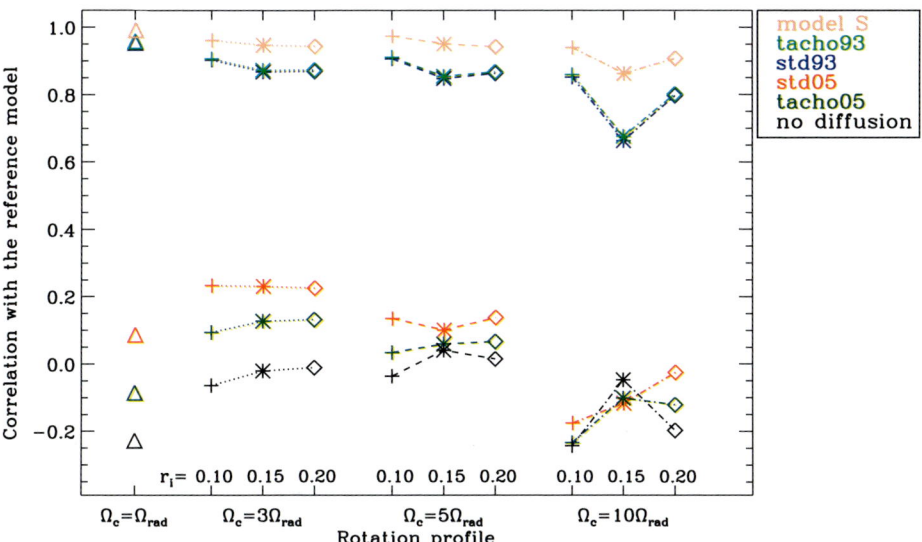

Figure 4 Correlation coefficients between the reference model (seismic) with a given rotation profile depending on the rate in the region below r_i as described in Figure 3 (plus the rigid rotation profile (triangles)) and another model with the same rotation profile. These models (described in Section 2) are *std93* (blue), *std05* (red), *tacho93* (light green), *tacho05* (dark green), *no diffusion* (black), and model S (pink).

the two types of models. Obviously, there are tiny differences among other models such as the *std93* and the *tacho93* models because the frequency differences between the g modes computed from them are very small.

Moreover, we can see other systematics in Figure 4 as the correlation coefficient decreases with higher rotation rates in the core. When higher rotation rates are considered, the reconstructed waves have to match both $m \pm 1$ components of the $\ell = 1$ modes. Therefore, the structure of the waves is more complicated, allowing a higher precision in the calculation of the correlation and more sensitivity to the positions of the modes (the structure parameters of the model).

5. Comparison with the Real Data

To compare our models with the real data, we use the 3481 day-long GOLF time series that was also used in García *et al.* (2007) (see that paper for more details on the data analysis process). Figure 5 shows the results of the correlation between the reconstructed waves obtained from the GOLF data and those from the models. In this figure, instead of having one free parameter (physical processes or rotation), the results depend on both structure and dynamics. However, by mixing them, it is possible to extract some information on the data.

As in the case of the correlation with the reference model, we see a clear separation between the same two groups of models: On one side we have all the models using GN93 opacities and, on the other, the models using As05 opacities and the model without diffusion. This was an expected result. Indeed, we knew that GOLF is well correlated with the seismic model for rotation rates in the core not faster than five times – on average – the rotation rate in the rest of the radiative region whatever the considered fractional radius (r_i). As we have seen in the previous section, the models containing the new abundances and the one

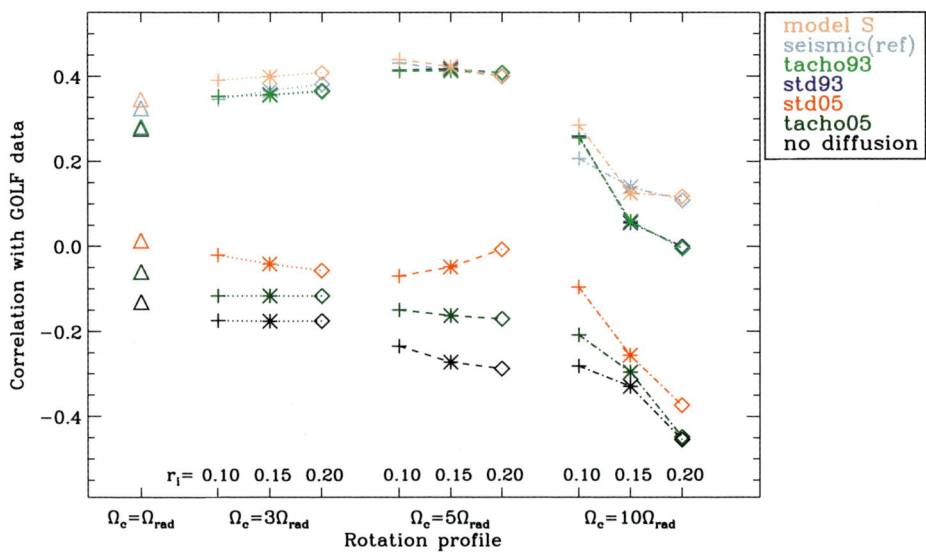

Figure 5 Correlation coefficients between the reconstructed waves from GOLF data and the ones from all of the models with the various rotation profiles. These models (described in Section 2) are *std93* (blue), *std05* (red), *tacho93* (light green), *tacho05* (dark green), *no diffusion* (black), model S (pink), and *seismic* (light blue).

without microscopic diffusion are incompatible with the models including the old chemical abundances. Therefore, they could not be compatible with the GOLF data. Actually, models that are close to the seismic model (that is, with the GN93 composition) associated with a rotation rate in the core $\Omega_c \leq 5\Omega_{rad}$ present the best compatibility with the data. A rotation profile having a rate five times the rotation rate in the rest of the radiative zone gives the highest correlation coefficients.

It is also important to note that the models using As05 and the one without treatment of the microscopic diffusion have, in most cases, negative correlations. The g-mode frequencies of these models are shifted as compared to GOLF and the maxima of the reconstructed waves are placed in between those of the GOLF data.

We know that the largest difference of g-mode frequencies can reach up to 5 µHz between the models with the GN93 abundances and the other group of models. In contrast to that, among the group of models presenting the highest correlations in Figure 5, the frequencies are shifted by less than 1 µHz (see Mathur *et al.*, 2007). Therefore, we can say that using this technique on GOLF data, we are able to distinguish among structural parameters of the solar models (such as the abundances) when the frequency differences induced by these modifications are of the order of 1 µHz or even larger. If solar models that differ by various physical processes produce g-mode frequencies with differences lower than a microhertz, the present methodology would be unable to distinguish among them. To improve our detection capability we should enlarge the analyzed region toward lower frequencies, where we have seen that the g-mode periods are sensitive to the models.

6. Conclusion

In the present paper, we have studied the changes in the signal reconstructed from the period separation ΔP_1, which vary when the physics of the solar models are modified, and we have verified whether it was possible to detect these changes in the g-mode asymptotic signature with the present uncertainties.

By comparing our different models with the reference one, we have learned that the methodology presented here offers enough sensitivity to distinguish between two sets of models: diffusive models using the old GN93 abundances and models using the new As05 abundances and the one not including microscopic diffusion.

This work shows that the g-mode signature detected in GOLF data clearly favors the models including the old abundances from Grevesse and Noels (1993). In this sense, the analysis of the g-mode signature recently detected is in agreement with previous results coming from the p-mode analysis showing a larger discrepancy in the sound-speed profile when we use the new abundances from Asplund, Grevesse, and Sauval (2005).

However, we have to keep in mind that these results do not mean that the new abundances are incorrect. They only tell us that the classic models including them are not compatible with the observations. It could be possible to obtain a better agreement by changing other physical quantities of the models such as the opacities or the reaction rates or by including new processes to compensate for the effects of the reduction in the metallicity.

Acknowledgements The authors want to thank J. Christensen-Dalsgaard who provided us with the model S and the adipack code, J. Provost for having given us her frequencies with which we could compare the ones of our standard model and for having produced the seismic-model frequencies from her own oscillation code, and A. Eff-Darwich and S. Turck-Chièze for useful discussions and comments. The GOLF experiment is based upon a consortium of institutes (IAS, CEA/Saclay, Nice, and Bordeaux Observatories from France and IAC from Spain) involving numerous scientists and engineers, as enumerated in Gabriel *et al.* (1995). SOHO is a mission of international cooperation between ESA and NASA. This work has been partially funded by Grant No. AYA2004-04462 of the Spanish Ministry of Education and Culture and by the European Helio- and Asteroseismology Network (HELAS, http://www.helas-eu.org/), a major international collaboration funded by the European Commission's Sixth Framework Programme.

References

Andersen, B.N.: 1996, Theoretical amplitudes of solar g-modes. *Astron. Astrophys.* **312**, 610–614.
Appourchaux, T., Fröhlich, C., Andersen, B., Berthomieu, G., Chaplin, W.J., Elsworth, Y., Finsterle, W., Gough, D.O., Hoeksema, J.T., Isaak, G.R., Kosovichev, A.G., Provost, J., Scherrer, P.H., Sekii, T., Toutain, T.: 2000, Observational upper limits to low-degree solar g-modes. *Astrophys. J.* **538**, 401–414. doi:10.1086/309124.
Asplund, M., Grevesse, N., Sauval, A.J.: 2005, The solar chemical composition. In: Barnes, T.G. III, Bash, F.N. (eds.) *Cosmic Abundances as Records of Stellar Evolution and Nucleosynthesis* **CS-336**, Astron. Soc. Pac., San Francisko, 25–38.
Bahcall, J.N., Serenelli, A.M., Basu, S.: 2005, New solar opacities, abundances, helioseismology, and neutrino fluxes. *Astrophys. J.* **621**, L85–L88. doi:10.1086/428929.
Basu, S., Christensen-Dalsgaard, J., Chaplin, W.J., Elsworth, Y., Isaak, G.R., New, R., Schou, J., Thompson, M.J., Tomczyk, S.: 1997, Solar internal sound speed as inferred from combined BiSON and LOWL oscillation frequencies. *Mon. Not. Roy. Astron. Soc.* **292**, 243.
Basu, S., Chaplin, W.J., Elsworth, Y., New, R., Serenelli, A.M., Verner, G.A.: 2007, Solar abundances and helioseismology: Fine-structure spacings and separation ratios of low-degree p-modes. *Astrophys. J.* **655**, 660–671. doi:10.1086/509820.
Brun, A.S., Turck-Chièze, S., Zahn, J.P.: 1999, Standard solar models in the light of new helioseismic constraints. II. Mixing below the convective zone. *Astrophys. J.* **525**, 1032–1041. doi:10.1086/307932.
Chaplin, W.J., Elsworth, Y., Isaak, G.R., Marchenkov, K.I., Miller, B.A., New, R.: 2001, Rigid rotation of the solar core? On the reliable extraction of low-l rotational p-mode splittings from full-disc observations of the Sun. *Mon. Not. Roy. Astron. Soc.* **327**, 1127–1136. doi:10.1046/j.1365-8711.2001.04805.x.

Chaplin, W.J., Serenelli, A.M., Basu, S., Elsworth, Y., New, R., Verner, G.A.: 2007, Solar heavy-element abundance: constraints from frequency separation ratios of low-degree p-modes. *Astrophys. J.* **670**, 872 – 884. doi:10.1086/522578.
Christensen-Dalsgaard, J., Gough, D.O., Thompson, M.J.: 1991, The depth of the solar convection zone. *Astrophys. J.* **378**, 413 – 437. doi:10.1086/170441.
Christensen-Dalsgaard, J., Dappen, W., Ajukov, S.V., Anderson, E.R., Antia, H.M., Basu, S., Baturin, V.A., Berthomieu, G., Chaboyer, B., Chitre, S.M., Cox, A.N., Demarque, P., Donatowicz, J., Dziembowski, W.A., Gabriel, M., Gough, D.O., Guenther, D.B., Guzik, J.A., Harvey, J.W., Hill, F., Houdek, G., Iglesias, C.A., Kosovichev, A.G., Leibacher, J.W., Morel, P., Proffitt, C.R., Provost, J., Reiter, J., Rhodes, E.J. Jr., Rogers, F.J., Roxburgh, I.W., Thompson, M.J., Ulrich, R.K.: 1996, The current state of solar modeling. *Science* **272**, 1286 – 1292.
Couvidat, S., Turck-Chièze, S., Kosovichev, A.G.: 2003, Solar seismic models and the neutrino predictions. *Astrophys. J.* **599**, 1434 – 1448. doi:10.1086/379604.
Couvidat, S., García, R.A., Turck-Chièze, S., Corbard, T., Henney, C.J., Jiménez-Reyes, S.: 2003, The rotation of the deep solar layers. *Astrophys. J.* **597**, L77 – L79. doi:10.1086/379698.
Domingo, V., Fleck, B., Poland, A.I.: 1995, The SOHO mission: an overview. *Solar Phys.* **162**, 1 – 37.
Elsworth, Y.P., Baudin, F., Chaplin, W., Andersen, B., Appourchaux, T., Boumier, P., Broomhall, A.M., Corbard, T., Finsterle, W., Fröhlich, C., Gabriel, A., García, R.A., Gough, D.O., Grec, G., Jiménez, A., Kosovichev, A., Provost, J., Sekii, T., Toutain, T., Turck-Chièze, S.: 2006, The internal structure of the Sun inferred from g modes and low-frequency p modes. In: Fletcher, K., Thompson, M. (eds.) *Proceedings of SOHO 18/GONG 2006/HELAS I, Beyond the spherical Sun* **SP-624**, ESA, Noordwijk, 18.
Gabriel, A.H., Grec, G., Charra, J., Robillot, J.M., Cortés, T.R., Turck-Chièze, S., Bocchia, R., Boumier, P., Cantin, M., Céspedes, E., Cougrand, B., Cretolle, J., Dame, L., Decaudin, M., Delache, P., Denis, N., Duc, R., Dzitko, H., Fossat, E., Fourmond, J.J., García, R.A., Gough, D., Grivel, C., Herreros, J.M., Lagardere, H., Moalic, J.P., Pallé, P.L., Petrou, N., Sanchez, M., Ulrich, R., van der Raay, H.B.: 1995, Global oscillations at low frequency from the SOHO mission (GOLF). *Solar Phys.* **162**, 61 – 99.
Gabriel, A.H., Baudin, F., Boumier, P., García, R.A., Turck-Chièze, S., Appourchaux, T., Bertello, L., Berthomieu, G., Charra, J., Gough, D.O., Pallé, P.L., Provost, J., Renaud, C., Robillot, J.M., Roca Cortés, T., Thiery, S., Ulrich, R.K.: 2002, A search for solar g modes in the GOLF data. *Astron. Astrophys.* **390**, 1119 – 1131. doi:10.1051/0004-6361:20020695.
García, R.A., Turck-Chièze, S., Boumier, P., Robillot, J.M., Bertello, L., Charra, J., Dzitko, H., Gabriel, A.H., Jiménez-Reyes, S.J., Pallé, P.L., Renaud, C., Roca Cortés, T., Ulrich, R.K.: 2005, Global solar Doppler velocity determination with the GOLF/SoHO instrument. *Astron. Astrophys.* **442**, 385 – 395. doi:10.1051/0004-6361:20052779.
García, R.A., Turck-Chièze, S., Jiménez-Reyes, S.J., Ballot, J., Pallé, P.L., Eff-Darwich, A., Mathur, S., Provost, J.: 2007, Tracking solar gravity modes: the dynamics of the solar core. *Science* **316**, 1591 – 1593. doi:10.1126/science.1140598.
Grevesse, N., Noels, A.: 1993, Cosmic abundances of the elements. In: Prantzos, N., Vangioni-Flam, E., Casse, M. (eds.) *Origin and Evolution of the Elements* **199**, Cambridge, England, 15 – 25.
Guzik, J.A., Watson, L.S., Cox, A.N.: 2005, Can enhanced diffusion improve helioseismic agreement for solar models with revised abundances? *Astrophys. J.* **627**, 1049 – 1056. doi:10.1086/430438.
Howe, R., Christensen-Dalsgaard, J., Hill, F., Komm, R.W., Larsen, R.M., Schou, J., Thompson, M.J., Toomre, J.: 2000, Dynamic variations at the base of the solar convection zone. *Science* **287**, 2456 – 2460.
Kumar, P., Quataert, E.J., Bahcall, J.N.: 1996, Observational searches for solar g-modes: some theoretical considerations. *Astrophys. J.* **458**, L83 – L85. doi:10.1086/309926.
Mathur, S.: 2007, A la recherche des modes de gravité: étude de la dynamique du coeur solaire. Ph.D. thesis, Université Paris XI Orsay.
Mathur, S., Turck-Chièze, S., Couvidat, S., García, R.A.: 2007, On the characteristics of the solar gravity mode frequencies. *Astrophys. J.* **668**, 594 – 602. doi:10.1086/521187.
Michaud, G., Proffitt, C.R.: 1993, Particle transport processes. In: Weiss, W.W., Baglin, A. (eds.) *IAU Colloq. 137: Inside the Stars* **CS-40**, Astron. Soc. Pac., San Francisco, 246 – 259.
Morel, P.: 1997, CESAM: A code for stellar evolution calculations. *Astron. Astrophys. Suppl.* **124**, 597 – 614.
Morel, P., Provost, J., Berthomieu, G.: 1997, Updated solar models. *Astron. Astrophys.* **327**, 349 – 360.
Press, W.H., Teukolsky, S.A., Vetterling, W.T., Flannery, B.P.: 1992, *Numerical Recipes in FORTRAN. The Art of Scientific Computing*, 2nd edn. Cambridge University Press, Cambridge.
Provost, J., Berthomieu, G.: 1986, Asymptotic properties of low degree solar gravity modes. *Astron. Astrophys.* **165**, 218 – 226.
Spiegel, E.A., Zahn, J.P.: 1992, The solar tachocline. *Astron. Astrophys.* **265**, 106 – 114.
Tassoul, M.: 1980, Asymptotic approximations for stellar nonradial pulsations. *Astrophys. J. Suppl. Ser.* **43**, 469 – 490. doi:10.1086/190678.

Thompson, M.J., Toomre, J., Anderson, E., Antia, H.M., Berthomieu, G., Burtonclay, D., Chitre, S.M., Christensen-Dalsgaard, J., Corbard, T., Derosa, M., Genovese, C.R., Gough, D.O., Haber, D.A., Harvey, J.W., Hill, F., Howe, R., Korzennik, S.G., Kosovichev, A.G., Leibacher, J.W., Pijpers, F.P., Provost, J., Rhodes, E.J., Schou, J., Sekii, T., Stark, P.B., Wilson, P.: 1996, Differential rotation and dynamics of the solar interior. *Science* **272**, 1300–1305.

Thompson, M.J., Christensen-Dalsgaard, J., Miesch, M.S., Toomre, J.: 2003, The internal rotation of the Sun. *Annu. Rev. Astron. Astrophys.* **41**, 599–643. doi:10.1146/annurev.astro.41.011802.094848.

Turck-Chièze, S., Couvidat, S., Kosovichev, A.G., Gabriel, A.H., Berthomieu, G., Brun, A.S., Christensen-Dalsgaard, J., García, R.A., Gough, D.O., Provost, J., Roca-Cortés, T., Roxburgh, I.W., Ulrich, R.K.: 2001, Solar neutrino emission deduced from a seismic model. *Astrophys. J.* **555**, L69–L73. doi:10.1086/321726.

Turck-Chièze, S., Couvidat, S., Piau, L., Ferguson, J., Lambert, P., Ballot, J., García, R.A., Nghiem, P.: 2004a, Surprising Sun: a new step towards a complete picture? *Phys. Rev. Lett.* **93**(21), 211102-1–211102-4. doi:10.1103/PhysRevLett.93.211102.

Turck-Chièze, S., García, R.A., Couvidat, S., Ulrich, R.K., Bertello, L., Varadi, F., Kosovichev, A.G., Gabriel, A.H., Berthomieu, G., Brun, A.S., Lopes, I., Pallé, P., Provost, J., Robillot, J.M., Roca Cortés, T.: 2004b, Looking for gravity-mode multiplets with the GOLF experiment aboard SOHO. *Astrophys. J.* **604**, 455–468. doi:10.1086/381743.

Zaatri, A., Provost, J., Berthomieu, G., Morel, P., Corbard, T.: 2007, Sensitivity of low degree oscillations to the change in solar abundances. *Astron. Astrophys.* **469**, 1145–1149. doi:10.1051/0004-6361:20077212.

Zonal Velocity Bands and the Solar Activity Cycle

K.R. Sivaraman · H.M. Antia · S.M. Chitre · V.V. Makarova

Originally published in the journal Solar Physics, Volume 251, Nos 1–2, 149–156.
DOI: 10.1007/s11207-008-9172-1 © Springer Science+Business Media B.V. 2008

Abstract We compare the zonal-flow pattern in subsurface layers of the Sun with the distribution of surface magnetic features such as sunspots and polar faculae. We demonstrate that, in the activity belt, the butterfly pattern of sunspots coincides with the fast stream of zonal flows, although part of the sunspot distribution does spill over to the slow stream. At high latitudes, the polar faculae and zonal-flow bands have similar distributions in the spatial and temporal domains.

Keywords Solar activity · Helioseismology · Rotation · Magnetic fields

1. Introduction

The pattern of temporal variations in the solar differential-rotation rates discovered by Howard and LaBonte (1980) from the Mt. Wilson full-disc Dopplergram data is known as torsional oscillations. These manifest themselves as alternating latitudinal bands of slightly faster and slower than average rotation velocities migrating from pole to the Equator in about 22 years. The low velocity of $3-5$ m s^{-1} of these zonal flows makes it difficult to detect them from the global rotation signal, which is more than two orders of magnitude stronger, and

---

Helioseismology, Asteroseismology, and MHD Connections
Guest editors: Laurent Gizon and Paul Cally

K.R. Sivaraman
Indian Institute of Astrophysics, Bangalore 560034, India

H.M. Antia (✉)
Tata Institute of Fundamental Research, Homi Bhabha Road, Mumbai 400005, India
e-mail: antia@tifr.res.in

S.M. Chitre
Centre for Basic Sciences, University of Mumbai, Mumbai 400098, India

V.V. Makarova
Kislovodsk Solar Station of the Pulkovo Observatory, Kislovodsk 357700, Russia

it was remarkable that Howard and LaBonte were able to isolate these weak signals. The close correspondence between the torsional oscillation pattern and the surface magnetic-flux distribution in the sunspot latitudes led Howard and LaBonte (1980) to conclude that this velocity field is perhaps the signature of a "large-scale deep-seated phenomenon" and that "this velocity field is associated in some way with the subsurface magnetic fields that are responsible for the solar cycle."

A subsequent investigation by Snodgrass and Howard (1985) used the corrected values for the coefficients A, B, and C of the parabolic fit for the global rotation to show that the full torsional oscillation pattern is not in the form of a continuous wave running from poles to the Equator but rather consisted of a high-latitude branch and a low-latitude branch with a break around $40° - 50°$ with each of the two components consisting of a fast and a slow stream alternating with each other in time. Snodgrass and Howard (1985) also noticed that the low-latitude stream of enhanced shear, which lies in between the fast zone and its poleward adjacent zone, spatially coincides with the centroid of the sunspot distribution. Later, Snodgrass (1987, his Figure 2), in addition to confirming the overlap of the shear enhanced pattern with the sunspot distribution (referred to as the butterfly pattern) for cycles 20 and 21, demonstrated that the zone of diminished shear is located between the two successive sunspot activity zones (*i.e.*, between two successive butterfly patterns). This figure also showed that there are no magnetic features corresponding to the shear increase and decrease zones at latitudes beyond $\pm 60°$. Makarov and Sivaraman (1989), in fact, suggested that these zones of increased or decreased shear at high latitudes correspond spatially with the polar faculae distribution. Subsequent painstaking efforts by Ulrich *et al.* (1988), Snodgrass (1992), and Ulrich (2001 and the references therein) refined the methods of reduction of the Mt. Wilson full-disc velocity maps, leading to a considerable improvement in the visibility of the torsional oscillation signal against the background noise and, indeed, established the reality of the existence of this velocity pattern on the solar surface.

All of these works refer to the rotation rate near the solar surface. With the advent of helioseismology it has become possible to estimate the rotation rate in the solar interior by inversions of the rotational splittings of the solar-oscillation frequencies from the accurately measured helioseismic data obtained by the ground-based Global Oscillation Network Group (GONG) and by the Michelson Doppler Imager (MDI) onboard the SOHO spacecraft. The rotation-rate residuals derived by subtracting the time-averaged rotation rate from the corresponding rotation rate at each depth and latitude show temporal variations with faster and slower rotating bands moving equatorwards with time (Schou, 1999; Howe *et al.*, 2000; Antia and Basu, 2000). We refer to this as the zonal-flow pattern in this paper. This pattern is very similar to the torsional oscillation bands observed on the surface (Howard and LaBonte, 1980; Ulrich *et al.*, 1988; Snodgrass, 1992; Ulrich, 2001) but because of smaller errors in the seismic data, it is better defined and more robust than the latter. Further investigations by Antia and Basu (2001), Vorontsov *et al.* (2002), Basu and Antia (2003), and Howe *et al.* (2006) using more extensive data from GONG and MDI have revealed results sufficient to build a fairly consistent picture of the time-dependent structure and dynamics of these zonal flows at different depths and in different latitudes in the solar interior. The patterns derived from the GONG and MDI data are generally in reasonably good agreement with each other. Finally, the results from the recent study by Basu and Antia (2006) and Howe *et al.* (2006) using data covering almost a complete sunspot cycle (1996 to 2006) have consolidated the properties of the zonal flows and provided a fairly complete picture of these migrating zonal bands.

The zonal-flow pattern in the solar interior has now become available for almost the full sunspot cycle from helioseismic data. It would therefore be of interest to compare the zonal-flow pattern with the distribution of surface magnetic features, which are manifestations of the cyclic magnetic activity, and to look for similarities and differences between them. With this aim, we plot the patterns of distribution of surface magnetic features, namely sunspots and the polar faculae, on zonal-flow contour maps in the subsurface region at $r = 0.98 R_\odot$ derived from the GONG data for the period 1995–2007.

The rest of the paper is organised as follows: In Section 2 we describe the data and the analysis procedure, in Section 3 we present the results, and finally in Section 4 we summarise the conclusions.

2. Data and Analysis

2.1. Helioseismic Data and the Inferred Zonal-Flow Pattern

To infer the rotation rate in the solar interior, we use 120 temporally overlapping data sets from GONG (Hill *et al.*, 1996), each covering a period of 108 days from 7 May 1995 to 15 May 2007, with a spacing of 36 days between successive data sets. Each data set consists of the mean frequencies of different (n, ℓ) multiplets and the splitting coefficients. We use the 2D Regularised Least Squares (2DRLS) inversion technique (Antia, Basu, and Chitre, 1998) for deducing the rotation rate for each of these data sets. We take the temporal average over all of these data sets to find the mean rotation rate at each latitude and depth covered in the study. The temporal mean is subtracted from the rotation rate at each epoch to find the residuals ($\delta\Omega$), which give the temporally varying component of the rotation rate. This is referred to as the zonal flow. Thus we have

$$\delta\Omega(r, \theta, t) = \Omega(r, \theta, t) - \langle\Omega(r, \theta, t)\rangle, \qquad (1)$$

where $\Omega(r, \theta, t)$ is the rotation rate as a function of the radial distance r, latitude θ, and time t. The angular brackets denote the temporal average over the data ensemble. It should be noted that the splitting coefficients are sensitive only to the North–South symmetric component of rotation rate and hence the inferred rotation pattern always looks symmetrical about the Equator. The actual zonal-flow pattern may have some asymmetry that cannot be detected from the seismic data used in this study.

The zonal-flow pattern consisting of fast-rotating streams (represented by contours in red) and slow-rotating streams (represented by contours in blue) is shown in Figure 1. At low latitudes this flow pattern consists of bands of fast- and slow-moving fluid that move towards the Equator with time in both hemispheres, which eventually meet near the Equator. At high latitudes, the behaviour of the zonal-flow pattern is quite different as the bands of fast- and slow-moving streams migrate polewards with time. The slow stream reaches the poles near the time of polar-field reversal around 2000.7. The zonal-flow velocities are rather small around latitudes of $40° - 50°$ (Vorontsov *et al.*, 2002; Basu and Antia, 2003). This region acts as a boundary separating the zonal-flow systems at low and high latitudes. Snodgrass and Howard (1985) also found a break in the surface torsional oscillation pattern around the same latitudes.

2.2. Properties of Polar Faculae

2.2.1. Number Counts and Latitudinal Distribution

Soon after the polar-field reversal, which occurs after sunspot activity has reached its peak, polar faculae (PF) make their appearance in the latitude zones polewards of $40°$ in both

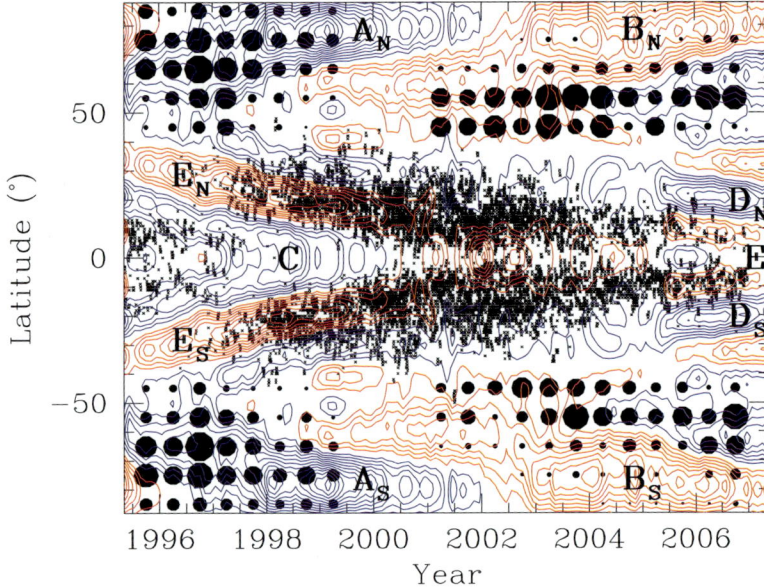

Figure 1 Contours of constant residual rotation velocity, $\delta v_\phi = \delta\Omega r \cos\theta$, at $r = 0.98 R_\odot$ obtained from GONG data shown as a function of time and latitude. The red contours show positive values and the blue contours show negative values. The contour spacing is 1 m s^{-1} and the zero contour is not shown. Different bands of faster (e.g., B_N, B_S, E, E_N, and E_S) and slower (e.g., A_N, A_S, C, D_N, and D_S) than average rotation velocity are marked in the figure. The markings at low latitudes show the position of sunspots. The filled black circles at high latitudes represent the daily mean number of polar faculae averaged over a period of six months in $10°$ latitude bins from $40°$ to $90°$. The area of the circles is proportional to the number of PFs.

hemispheres. They can be identified easily on the broad-band images as tiny bright features of sizes ranging from $3''$ to $10''$ with a contrast $I_{\text{PF}}/I_{\text{photosphere}} = 1.03$ to 1.10 and on the K-line spectroheliograms as bright points located preferentially along the network boundaries (i.e., along the boundaries of the supergranular cells; Makarov and Makarova, 1996). The cyclically varying numbers of PF, which are $180°$ out of phase with the sunspot number, over the entire solar disc are highly correlated with the polar magnetic-field strengths (Sheeley, 1964, 1976, 1991). The total counts and the latitude distribution of PF determined for the period 1940–1985 by Makarov, Makarova, and Sivaraman (1989), Makarov and Sivaraman (1989), and Makarov and Makarova (1996) have been extended to the present day by Makarov, Makarova, and Callebaut (in preparation). The procedure that was adopted for determining the total counts, as well as the latitude distribution of the PF, is described in Makarov and Makarova (1996) and may be summarised as follows: For every month, about 15 high-quality broad-band images were selected from the Kislovodsk Solar Station collections. The gaps in the data were filled by using the Ca II K spectroheliograms of the Kodaikanal Observatory collections. The numbers of PF within the latitude zones starting from $40°$ and reaching to the poles were counted in steps of $10°$ to derive their distribution. This was done plate by plate for the northern and southern hemispheres separately and, from the counts so obtained for every month, the monthly and the six-month averages were worked out. We use the six-month mean values of PF in each of the $10°$ bins for comparison with the zonal-flow pattern in Figure 1. These are represented by filled circles with area proportional to the six-month means plotted at the mean latitude positions of each of the $10°$ bins.

The PFs appear first at latitude zones $\pm 40° - 60°$. The zones of appearance of the PF expand and reach almost to the poles, filling the entire high-latitude regions during the sunspot minimum years. On average, about 900 PF emerge per day during this epoch. The zones of PF are seen to shrink progressively in extent during the rising phase of the then-current sunspot cycle until they finally disappear with the polar-field reversal around the maximum phase (Figure 4 of Makarov and Sivaraman, 1989). Sheeley and Warren (2006) used images and magnetograms from MDI to study PFs during the period 1996 – 2005 and showed that in each hemisphere there is a facula-free zone separating the old-cycle polar field from trailing-polarity flux that is migrating polewards from the sunspot belts. These facula-free zones coincide with the neutral lines of the axisymmetric component of photospheric magnetic field and their arrival at the poles in 2001 marks the reversal of the polar fields.

2.2.2. Magnetic Fields of Polar Faculae as the Source of Polar Fields

PFs have most commonly sizes in the range of $3'' - 10''$ when they occur as individual structures or as bipoles, although a few of them appear at smaller sizes reaching down to $\approx 1''$. A few others showing a complex structure have sizes exceeding $10''$ and at times even as high as $30''$ (Makarov and Makarova, 1996). In the magnetograms they appear in the form of flux knots, either bright or dark depending on their magnetic polarity. Based on the polarimetric measurements by Homann, Kneer, and Makarov (1997), Makarov and Makarova (1998) estimate the magnetic flux per faculae to be $\approx 7 \times 10^{19}$ Mx, whereas Varsik, Wilson, and Li (1999) estimate the flux to be $\approx 10^{19}$ Mx from calibrated low-resolution magnetograms. Thus the PFs possess a range of magnetic-field strengths from 150 to 1700 G depending on both their sizes and the amount of flux they carry, although PF with low field strengths appear to be more common.

The evolution of the integrated flux over the polar regions was traced by Lin, Varsik, and Zirin (1994) using high-resolution magnetograms covering the period from early-1991 to mid-1993 that spans the maximum and declining phases of sunspot cycle 22. According to them, during the solar-maximum phase, the polar regions are populated by magnetic elements of positive and negative polarity of almost equal numbers and of equal field strengths, rendering the net fields at the poles close to zero. This presumably represents the polar-field reversal epoch. With the progress of the sunspot cycle towards the minimum, the elements of one polarity outnumber those of the opposite polarity in terms of the field strengths and numbers, rendering the fields at the poles predominantly of one polarity. A subsequent study using more such high-resolution magnetogram sequences during the sunspot minimum phase by Varsik, Wilson, and Li (1999) confirms that knots of one polarity far exceed those of opposite polarity. It is this excess of one polarity flux elements over the other, itself varying cyclically (from + to − and then to + and so on), that determines the polarity of the field at the poles of the Sun in any given cycle. It is the collective net field of the PF that a magnetograph operating at low resolution measures as the magnetic flux at the poles of the Sun during the sunspot minimum phase.

3. Results

3.1. Latitude Distribution of Sunspots and the Zonal-Flow Pattern

We have shown in Figure 1 the latitudinal distribution of sunspots (the butterfly pattern) extracted from the Greenwich sunspot data superposed on the plot of the zonal velocity band pattern at $r = 0.98 R_\odot$. We find that the dense part of the distribution of spots lies over the zonal bands of the faster than average rotation rate (E_N and E_S) whereas the less dense part of the distribution spills over to the slow streams (D_N and D_S). There is also a sprinkling of sunspots in region C (the slow stream). These might possibly be very small spots or pores, being the last vestiges of solar cycle 22 (1986–1996). Similar spatial coincidence between sunspot distribution and the surface torsional oscillation pattern has been noted earlier by Snodgrass (1987).

3.2. PF Distribution and the High-Latitude Zonal-Flow Pattern

It is evident from Figure 1 that the PF distribution coincides well, both spatially and temporally, with the high-latitude zonal band pattern (A_N in the North and A_S in the South). Both the slow-stream bands A_N and A_S and the PF associated with them that have reached close to the respective poles disappear with the polar-field reversal in 2000.7. Two new zonal bands of fast streams (B_N and B_S) that originated at $\pm 50°$ latitudes about two years prior to the polar-field reversal, have now ascended polewards, filling the latitude regions where A_N and A_S were present before the polar-field reversal. The PF of the new cycle (2002–2006) that appeared soon after the polar-field reversal coinciding with the new fast streams B_N and B_S have also migrated polewards synchronously with the fast streams. The polar-field reversal in 2000.7 that marks the end of one PF cycle and the beginning of the next one in both hemispheres is also the epoch that marks the end of the slower-than-average rotation-rate zonal bands at the poles (A_N and A_S) and the ascendancy of the new faster-than-average rotation-rate zonal bands (B_N and B_S) to their positions. Thus the distribution of polar faculae that provides the magnetic field at the poles shows a distribution in latitude and time similar to the zonal-flow pattern of the subsurface layers at $r = 0.98 R_\odot$ at high latitudes. Although we have used the zonal-flow pattern at $r = 0.98 R_\odot$ for establishing the spatial and temporal coincidence with the PF, similar correspondence should hold for all subsurface layers lying above $r = 0.95 R_\odot$. Below $r = 0.90 R_\odot$ the zonal-flow pattern appears to be smeared out and the phase also changes at low latitudes. Thus at $r = 0.8 R_\odot$ the fast and slow streams can hardly be recognised at the high latitudes (Antia and Basu, 2001, Figure 4; Basu and Antia, 2006, Figure 1). Nonetheless, it appears that the correspondence between zonal-flow pattern and the PF seems to be confined to the layers above $r = 0.95 R_\odot$.

4. Discussion and Conclusions

We have derived the residual rotation rates by subtracting the time-averaged rotation rate from that at each epoch from helioseismic data from the GONG project for a full solar cycle (1996–2007) at $r = 0.98 R_\odot$ in the solar interior. These show a set of zonal velocity bands moving with faster- and slower-than-average rotation rate. The zonal velocity bands have two components per hemisphere (Figure 1): *i*) the high-latitude component of alternating slow (blue contours) and fast (red contours) streams (above $\approx 50°$) that move polewards (A_N or A_S and B_N or B_S) and *ii*) the low-latitude component of fast (E_N or E_S) and slow (D_N or D_S) streams in sunspot latitudes that move towards the Equator. E_N and E_S later merge to form a single fast stream E. From earlier studies (Vorontsov *et al.*, 2002; Basu and Antia, 2003) it is known that the zonal flows in the high-latitude as well as in the sunspot-latitude belts persist throughout much of the convection zone and are quite stable. The $\approx 50°$ latitude region in the two hemispheres seems to be the boundary that separates the high-latitude streams from the low-latitude streams. Interestingly, this is also the latitude region where the rotation-rate residual is close to zero throughout almost the entire convection zone (Vorontsov *et al.*, 2002; Basu and Antia, 2006).

We have established that polar faculae and the zonal-flow bands in the region above $\pm 40° - 50°$ latitudes have very similar distribution in the spatial and temporal domains, irrespective of whether the zonal velocity band is a fast stream or a slow stream (Figure 1). The switch from fast to slow happens around sunspot minimum (around 1996) when there is no reversal in polarity of the polar magnetic field, whereas the switch from slow to fast occurs around the sunspot maximum, which coincides with field polarity reversal at the poles around 2000.7. Thus there is one pair of streams (one fast and one slow) during the period of two successive polar-field reversals and both components of the pair are associated with polar fields of the same polarity, either positive or negative as the case might be. We have also established that, in the sunspot latitudes, the butterfly pattern coincides with the fast streams (E_N, E_S or E) although part of the sunspot distribution spills over to the slow streams (D_N and D_S) too (Figure 1). Of course, our study is restricted to one solar cycle for which the seismic data are available and this association needs to be confirmed in subsequent cycles.

We have further established that there is a striking similarity in spatial and temporal organisation between the zonal-flow streams in the interior and the surface magnetic fields. This would imply a close coupling between the periodic components of flows in the interior and the surface magnetic-field structures, which are visible manifestations of the cyclic magnetic activity. This close similarity raises interesting possibilities about their connection (*cf.* Snodgrass, 1987). Clearly, the velocity changes present in the zonal flow alone are intrinsically too weak to drive the global solar-activity cycle. It is, therefore, possible that both the zonal-flow pattern and the overall solar magnetic activity are manifestations of a common coherently driven global mechanism that remains to be identified and properly understood. There could be other possibilities: Two mechanisms operating on disparate scales at two different depths in the convection zone could conceivably result in mutually coherent velocity and magnetic-field patterns. It is commonly accepted that the shear zone below the base of the convection zone is the seat of the dynamo that amplifies and produces the strong magnetic field that gives rise to sunspots. Likewise, there is an amplification of the magnetic field by the shear in the subsurface layers between $r = 0.98 R_\odot$ and $0.95 R_\odot$ that could produce weaker fields like those in the polar faculae (Dikpati *et al.*, 2002). The possible role of the near-surface shear layer in small-scale amplification of magnetic field has been envisaged earlier by Gilman (2000). More sophisticated theoretical models supported by simulations to explore the mechanisms that can generate and sustain the zonal-band systems in the interior and also organise the magnetic fields in a mutually coherent way are clearly needed. It is gratifying to note that initial efforts in this direction have already produced migrating patterns (*e.g.*, Covas *et al.*, 2000; Covas, Tavakol, and Moss, 2001; Covas, Moss, and Tavakol, 2004; Lanza, 2007). It would be equally important to explore the role of the subsurface shear region in small-scale amplification of magnetic fields and to explain the intimate relationship between the family of zonal-band systems and the distribution of magnetic flux elements.

The helioseismic data from GONG and MDI accumulated over the past solar cycle 23 enabled Antia, Chitre, and Gough (2008) to study temporal variations in the solar rotational kinetic energy. It was demonstrated that at high latitudes ($> 45°$) variation in the kinetic energy through the convection zone is correlated with solar activity, whereas in the equatorial latitudes ($< 45°$) it is anticorrelated, except for the upper 10% of the solar radius where both are in phase. The amplitude of temporal variation of the rotational kinetic energy integrated over the entire convection zone turns out to be $\approx 3 \times 10^{38}$ ergs, implying a rate of variation of about 5×10^{30} ergs s^{-1} over the solar cycle. From energy conservation it is expected that the torsional kinetic energy variation is comparable with that in the magnetic energy but with opposite phase. It thus seems that the temporal variation in rotational kinetic energy in the convection zone is related to the solar cycle with its tantalising similarity with the magnetic activity cycle.

Acknowledgements This work utilises data obtained by the Global Oscillation Network Group (GONG) project, managed by the National Solar Observatory, which is operated by AURA, Inc., under a cooperative agreement with the National Science Foundation. The data were acquired by instruments operated by the Big Bear Solar Observatory, High Altitude Observatory, Learmonth Solar Observatory, Udaipur Solar Observatory, Instituto de Astrofísico de Canarias, and Cerro Tololo Inter-American Observatory. We thank Baba Varghese for his valuable help in formulating the figure. S.M.C. thanks the Indian National Science Academy for support under the INSA Honorary Scientist programme.

References

Antia, H.M., Basu, S.: 2000, *Astrophys. J.* **541**, 442.
Antia, H.M., Basu, S.: 2001, *Astrophys. J.* **559**, L67.
Antia, H.M., Basu, S., Chitre, S.M.: 1998, *Mon. Not. Roy. Astron. Soc.* **298**, 543.
Antia, H.M., Chitre, S.M., Gough, D.O.: 2008, *Astron. Astrophys.* **477**, 657.
Basu, S., Antia, H.M.: 2003, *Astrophys. J.* **585**, 553.
Basu, S., Antia, H.M.: 2006, In: Fletcher, K., Thompson, M.J. (eds.) *SOHO 18/GONG 2006/HELAS I, Beyond the Spherical Sun*, SP-**624**, ESA Publ. Div., Noordwijk, 128.
Covas, E., Moss, D., Tavakol, R.: 2004, *Astron. Astrophys.* **416**, 775.
Covas, E., Tavakol, R., Moss, D.: 2001, *Astron. Astrophys.* **371**, 718.
Covas, E., Tavakol, R., Moss, D., Tworkowski, A.: 2000, *Astron. Astrophys.* **360**, L21.
Dikpati, M., Corbard, T., Thompson, M.J., Gilman, P.A.: 2002, *Astrophys. J.* **575**, L41.
Gilman, P.A.: 2000, *Solar Phys.* **192**, 27.
Hill, F., et al.: 1996, *Science* **272**, 1292.
Homann, T., Kneer, F., Makarov, V.I.: 1997, *Solar Phys.* **175**, 81.
Howard, R.F., LaBonte, B.J.: 1980, *Astrophys. J.* **239**, L33.
Howe, R., Christensen-Dalsgaard, J., Hill, F., Komm, R.W., Larsen, R.M., Schou, J., Thompson, M.J., Toomre, J.: 2000, *Astrophys. J.* **533**, L163.
Howe, R., Rempel, M., Christensen-Dalsgaard, J., Hill, F., Komm, R.W., Larsen, R.M., Schou, J., Thompson, M.J.: 2006, *Astrophys. J.* **649**, 1155.
Lanza, A.F.: 2007, *Astron. Astrophys.* **471**, 1011.
Lin, H., Varsik, J., Zirin, H.: 1994, *Solar Phys.* **155**, 243.
Makarov, V.I., Makarova, V.V.: 1996, *Solar Phys.* **163**, 267.
Makarov, V.I., Makarova, V.V.: 1998, In: Balasubramaniam, K.S., Harvey, J.W., Rabin, D.M. (eds.) *Synoptic Solar Physics*, CS-**140**, Astron. Soc. Pac., San Francisco, 347.
Makarov, V.I., Sivaraman, K.R.: 1989, *Solar Phys.* **123**, 367.
Makarov, V.I., Makarova, V.V., Sivaraman, K.R.: 1989, *Solar Phys.* **119**, 45.
Schou, J.: 1999, *Astrophys. J.* **523**, L181.
Sheeley, N.R. Jr.: 1964, *Astrophys. J.* **140**, 731.
Sheeley, N.R. Jr.: 1976, *J. Geophys. Res.* **81**, 3462.
Sheeley, N.R. Jr.: 1991, *Astrophys. J.* **374**, 386.
Sheeley, N.R. Jr., Warren, H.P.: 2006, *Astrophys. J.* **641**, 611.
Snodgrass, H.B., Howard, R.F.: 1985, *Science* **228**, 945.
Snodgrass, H.B.: 1987, *Solar Phys.* **110**, 35.
Snodgrass, H.B.: 1992, In: Harvey, K.L. (ed.) *Solar Cycle Workshop*, CS-**27**, Astron. Soc. Pac., San Francisco, 205.
Ulrich, R.K.: 2001, *Astrophys. J.* **560**, 466.
Ulrich, R.K., Boyden, J.E., Webster, L., Snodgrass, H.B., Padilla, S.P., Gilman, P., Shieber, T.: 1988, *Solar Phys.* **117**, 291.
Varsik, J.R., Wilson, P.R., Li, Y.: 1999, *Solar Phys.* **184**, 223.
Vorontsov, S.V., Christensen-Dalsgaard, J., Schou, J., Strakhov, V.N., Thompson, M.J.: 2002, *Science* **296**, 101.

Acoustic Radius Measurements from MDI and GONG

S. Kholikov · F. Hill

Originally published in the journal Solar Physics, Volume 251, Nos 1–2, 157–161.
DOI: 10.1007/s11207-008-9205-9 © Springer Science+Business Media B.V. 2008

Abstract We study the temporal autocorrelation function (ACF) of global solar oscillations. It is well known that the "large frequency separation" is proportional to the solar acoustic radius. We analyze the ACF of MDI and GONG spherical-harmonic-coefficient time series for degrees $\ell = 0 - 3$. Acoustic radius measurements obtained from the first dominant peak locations of the ACF show a significant anticorrelation with solar cycle. This technique can be a useful tool to search for stellar activity.

Keywords Sun: helioseismology · Solar radius · Oscillations

1. Introduction

Solar radius changes over the solar cycle during the past ten years have been investigated with different techniques. However, the results were often controversial (Laclare *et al.*, 1996; Noël, 2004; Kuhn *et al.*, 2004). In contrast, acoustic-radius measurements provided by helioseismic methods (Schou *et al.*, 1997) based on f-mode frequencies are quite consistent (Antia, 1998; Dziembowski and Goode, 2005). The properties of the autocorrelation function (ACF) of global solar oscillations have been analyzed by several authors. Estimates of p-mode lifetimes have been obtained from the ACF of 70 days of GOLF data (Grec *et al.*, 1997). A long sequence of 500 days of observations was studied in detail by Gabriel *et al.* (1998). They found that variations of the lag of the ACF peaks are not related to the damping time but are a consequence of the nonconstancy of the so-called large frequency separation, $\Delta \nu = \nu(n+1, \ell) - \nu(n, \ell)$, where ν is the frequency, ℓ is the spherical harmonic degree, and n is the radial order of the mode. An application of the ACF was developed by Fossat *et al.* (1999), who used the quasi-periodicity of the oscillation frequencies to fill gaps in helioseismic time series. Measuring the large and small frequency separations [where the

Helioseismology, Asteroseismology, and MHD Connections
Guest Editors: Laurent Gizon and Paul Cally.

S. Kholikov (✉) · F. Hill
National Solar Observatory, 950 N. Cherry Ave., Tucson, AZ 85719, USA
e-mail: kholikov@nso.edu

small frequency separation is defined as $\delta\nu_\ell = \nu(n,\ell) - \nu(n-1,\ell+2)]$ from the ACF and its modulation has been proposed by Kholikov, Roxburgh, and Vorontsov (2004). Using the ACF as a diagnostic tool for stellar oscillations has been proposed by Roxburgh and Vorontsov (2006), who also developed a method of measuring the strength of acoustic-wave refraction in the stellar core from the modulation of the ACF (Roxburgh and Vorontsov, 2007).

Here, we study the determination of $\Delta\nu$ for the Sun from the ACF. The advantage of this method is that it allows us to use short time series to measure $\Delta\nu$ with much smaller error bars than can be achieved from the individual frequencies determined from fits to the acoustic power spectrum. This advantage is very useful for asteroseismology, where it is extremely difficult to obtain exact frequencies because of the higher noise levels and much shorter observational times for individual stars.

2. Data and Method

We have used GONG and MDI monthly spherical-harmonic (SH) coefficient time series for $\ell = 0 - 3$. These time series are computed by projecting individual full-disk Doppler velocity images onto the spherical-harmonic functions (Y_ℓ^m), typically defined as

$$Y_\ell^m = \sqrt{\frac{2\ell+1}{2\pi}\frac{(\ell-m)!}{(\ell+m)!}} P_\ell^m(\cos\theta)e^{im\phi}, \qquad (1)$$

where ℓ is the spherical harmonic degree, m is the azimuthal degree, θ is the colatitude, ϕ is the longitude, and P_ℓ^m is the associated Legendre function. The decomposition of an image, acquired every minute, provides a set of numbers that represent the relative strengths of the SH with different (ℓ, m) values for that image. These numbers, known as spherical-harmonic coefficients, are then rearranged to form a set of time series for each (ℓ, m). Typical SH time series lengths are 36 days for GONG and 72 days for MDI. Our data set covers the time period from 1995 to 2007.

The time series were filtered with a Gaussian filter of FWHM = 2.0 mHz centered at $\nu = 3.3$ mHz. The filtering was done by applying a Fourier transform to the SH time series, applying the filter in the frequency domain, and then transforming the filtered power spectrum back into the time domain to give the ACF. The ACF was computed separately for each ℓ and m. Figure 1 shows an example of the ACF for the $\ell = 0$ time series of 36 days duration. The most dominant peak, with a lag of about four hours, arises from the solar value of $\Delta\nu$ of ≈ 135 µHz. This value is the inverse travel time of sound from the solar surface to the center and back. Since the acoustic wave undergoes multiple reflections from the solar surface, the peak appears at delays of four, eight, ... hours. Other features of the ACF, such as the decreasing peak amplitudes and the modulation between odd and even reflections, have been discussed in detail elsewhere (Roxburgh and Vorontsov, 2006).

The location of the ACF peak is an estimate of the acoustic radius (T) of the Sun multiplied by four owing to the propagation of the wave through the Sun from the observation point to the far side and back, traveling a distance of four radii. The units of T are seconds, as it is expressed as the sound travel time. To obtain the location of the peak we use a technique from time–distance helioseismology, where it has been shown that the correlations of the waves can be approximately represented in the form of a Gabor function. Briefly, the correlation signal is described in terms of a superposition of stochastically excited normal

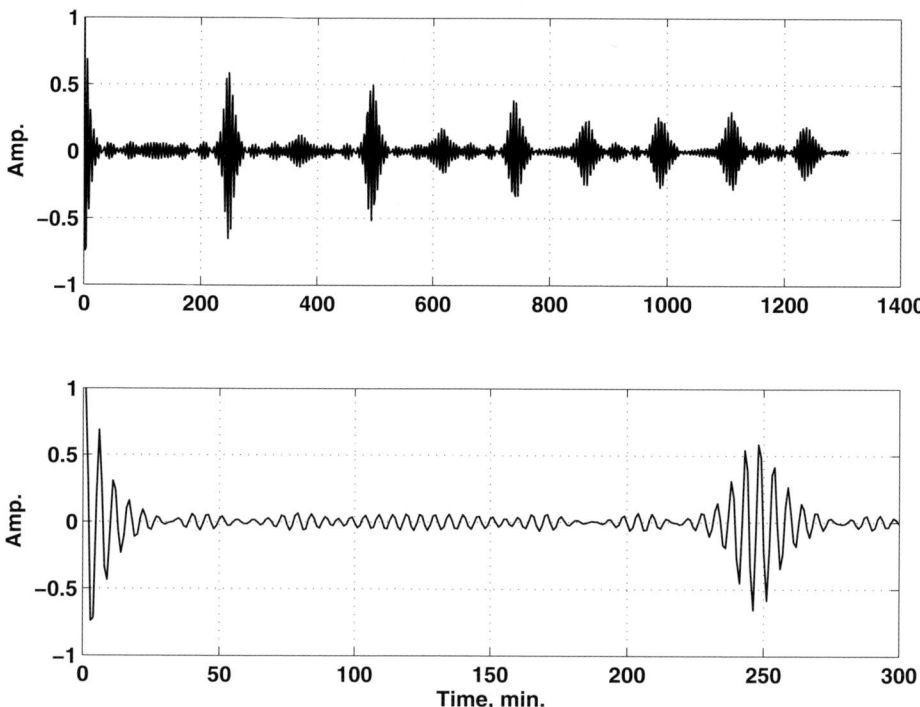

Figure 1 The autocorrelation function computed from a GONG 36-day $\ell = 0$ time series. The dominant peak corresponds to the large frequency separation ($\Delta \nu$)

modes, which are combined by using the spherical harmonic summation theorem. The signals are then grouped in angular-phase-velocity bins, and the method of stationary phase is applied to derive the Gabor representation. Further details can be found in Kosovichev and Duvall (1997) and Kosovichev, Duvall, and Scherrer (1997). We thus fit the peak around its maximum to the Gabor wavelet

$$G(\tau) = A\cos[2\pi\nu(\tau - \tau_{ph})]\exp\left[-\frac{(\tau - \tau_{en})^2}{2\sigma^2}\right], \qquad (2)$$

where A, ν, and σ are the amplitude, frequency, and width of a Gaussian envelope, respectively. The parameters τ_{ph} and τ_{en} correspond to the phase and group travel times. The phase travel time can be measured more accurately than the envelope travel time. Figure 2 shows phase travel times obtained from $\ell = 0$ and $\ell = 3$ time series for a set of five-day-long time series.

The measurements are smoothed by a 1.5-year running window by using the Savitzky–Golay method.

3. Results and Discussion

The large separations obtained from the ACF have very small error bars. In Figure 2 the typical error bars are about 0.1 seconds. These errors are estimated from the scatter of the

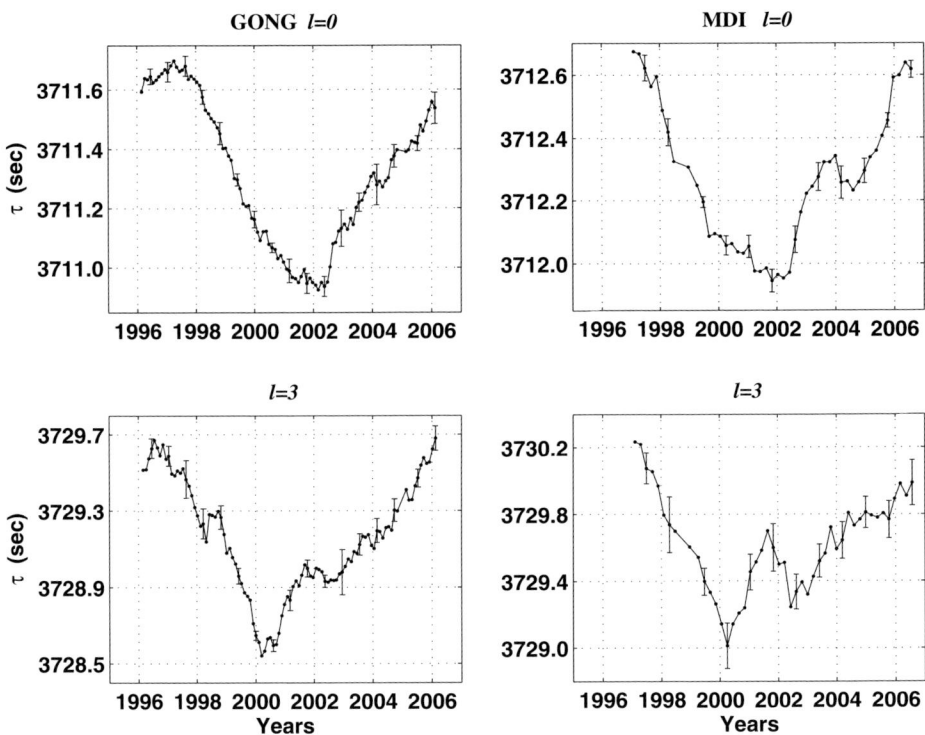

Figure 2 The acoustic radius of the Sun as a function of time from GONG (left) and MDI (right). The individual autocorrelation functions were obtained from five-day time series. The measurements are smoothed by a 1.5-year running window. A significant anticorrelation with the solar activity level can be clearly seen.

averaged measurements computed from the five-day time series, not from formal estimations of the Gabor fitting procedure. The origin of the one-second difference in acoustic radius between the GONG and MDI results is unknown.

We also analyzed measurements of $\Delta \nu$ from GONG and BiSON frequency tables. Time series of T obtained from fitted frequencies do not show any changes with solar activity cycle because of the much larger errors resulting from the peak-fitting analysis. The typical errors in frequency differences from peak fitting are typically about 0.5 µHz, corresponding to an error in T of approximately 25 seconds.

The most important features of Figure 2 are the clearly visible variations of the acoustic radius, which are anticorrelated with the solar activity cycle. For all low-degree modes ($\ell = 0-3$), the magnitude of the change between the minimum and maximum activity phases is less than one second. These small changes cannot be seen from individual frequency separations, where the precision of the determination is poorer by more than a factor of 100.

Variations in the solar acoustic radius arise either from changes in the density scale height, which affects the depth of the upper reflecting point, or from variations in the internal sound speed. However, cycle variations of the solar radius increase near the surface because of the effects of turbulence and magnetic fields in these layers (Lefebvre, Kosovichev, and Rozelot, 2007; Sofia et al., 2005). From our measurements we cannot infer the depth dependence of radius changes, but it clearly is anticorrelated with the solar activity cycle, as seen in inversions of f-mode frequencies (Lefebvre et al., 2006).

According to Equation (8) in Roxburgh and Vorontsov (2006), the relationship between the large separation, $\Delta\omega$ in radians per second, and the stellar acoustic radius (T) in seconds is

$$\Delta\omega = 2\pi(2T)^{-1}. \qquad (3)$$

Thus, a variation in the acoustic radius (δT) is related to a change in the large frequency separation $(\delta\Delta\omega)$ by

$$\delta\Delta\omega = \frac{-\pi}{T^2}\delta T. \qquad (4)$$

In terms of $\nu = \omega/2\pi$, $\delta\Delta\nu = -\delta T/2T^2$. For $\delta T = 1$ second and $T \approx 3710$ seconds, $\delta\Delta\nu \approx 36$ nHz. This is consistent with the frequency dependence of the activity-related shift in the frequency of five-minute low-degree modes separated by 135 µHz. Using plots in Libbrecht and Woodard (1990) and Toutain and Kosovichev (2005), we graphically estimate the change in the large separation between minimum and maximum activity phases to be about 20–40 nHz.

Obtaining precise frequencies in asteroseismology is difficult and requires very long time series The ACF technique is thus an alternative tool to search for stellar activity.

Acknowledgements This work utilizes data obtained by the Global Oscillation Network Group (GONG) program, managed by the National Solar Observatory, which is operated by AURA, Inc., under a cooperative agreement with the National Science Foundation. The data were acquired by instruments operated by the Big Bear Solar Observatory, High Altitude Observatory, Learmonth Solar Observatory, Udaipur Solar Observatory, Instituto de Astrofísica de Canarias, and Cerro Tololo Interamerican Observatory. SOHO is a mission of international cooperation between ESA and NASA.

References

Antia, H.M.: 1998, *Astron. Astrophys.* **330**, 336.
Dziembowski, W.A., Goode, P.R.: 2005, *Astrophys. J.* **625**, 548.
Fossat, E., Kholikov, S., Gelly, B., Schmider, F.X., Fierry-Fraillon, D., Grec, G., Palle, P., Cacciani, A., Ehgamberdiev, S., Hoeksema, J.T., Lazrek, M.: 1999, *Astron. Astrophys.* **343**, 608.
Gabriel, M., Grec, G., Renaud, C., Gabriel, A.H., Robillot, J.M., Roca Cortes, T., Turck-Chieze, S., Ulrich, R.K.: 1998, *Astron. Astrophys.* **338**, 1109.
Grec, G., Turck-Chieze, S., Lazrek, M., Roca Cortes, T., Bertello, L., Baudin, F., Boumier, P., Charra, J., Fierry-Fraillon, D., Fossat, E., Gabriel, A., Garcia, R., Gelly, B., Gouiffes, C., Regulo, C., Renaud, C., Robillot, J., Ulrich, R.: 1997, In: Provost, J., Schmider, F.-X. (eds.) *Sounding Solar and Stellar Interiors*, Kluwer, Dordrecht, 91.
Kholikov, Sh., Roxburgh, I.W., Vorontsov, S.: 2004, In: Favata, F., Aigrain, S., Wilson, A. (eds.) *Second Eddington Workshop, Stellar structure and habitable planet finding* **SP-538**, ESA, Noordwijk, 331.
Kosovichev, A.G., Duvall, T.L., Jr.: 1997, In: Pijpers, F.P., Christensen-Dalsgaard, J., Rosenthal, C.S. (eds.) *SCORe'96: Solar Convection and Oscillations and Their Relationship*, Kluwer, Dordrecht, 241.
Kosovichev, A.G., Duvall, T.L., Jr., Scherrer, P.H.: 1997, *Solar Phys.* **192**, 159.
Kuhn, J.R., Bush, R.I., Emilio, M., Scherrer, P.H.: 2004, *Astrophys. J.* **613**, 1241.
Laclare, F., Delas, C., Coin, J.-P., Irbah, A.: 1996, *Solar Phys.* **166**, 211.
Lefebvre, S., Kosovichev, A., Rozelot, J.P.: 2007, *Astrophys. J.* **658**, L135.
Lefebvre, S., Kosovichev, A., Nghiem, P., Turck-Chieze, S., Rozelot, J.P.: 2006, In: Fletcher, K., Thompson, M. (eds.) *Proceedings of SOHO 18/GONG 2006/HELAS I, Beyond the Spherical Sun*, 624, Published on CDROM.
Libbrecht, K.G., Woodard, M.F.: 1990, *Nature* **308**, 413.
Noël, F.: 2004, *Astron. Astrophys.* **413**, 725.
Roxburgh, I.W., Vorontsov, S.: 2006, *Mon. Not. Roy. Astron. Soc.* **369**, 1491.
Roxburgh, I.W., Vorontsov, S.: 2007, *Mon. Not. Roy. Astron. Soc.* **379**, 801.
Schou, J., Kosovichev, A.G., Goode, P.R., Dziembowski, W.A.: 1997, *Astrophys. J.* **489**, L197.
Sofia, S., Basu, S., Demarque, P., Li, L., Thuillier, G.: 2005, *Astrophys. J.* **632**, L147.
Toutain, T., Kosovichev, A.G.: 2005, *Astrophys. J.* **622**, 1314.

Spatio-Temporal Analysis of Photospheric Turbulent Velocity Fields Using the Proper Orthogonal Decomposition

A. Vecchio · V. Carbone · F. Lepreti · L. Primavera ·
L. Sorriso-Valvo · T. Straus · P. Veltri

Originally published in the journal Solar Physics, Volume 251, Nos 1–2, 163–178.
DOI: 10.1007/s11207-008-9141-8 © Springer Science+Business Media B.V. 2008

Abstract The spatio-temporal dynamics of the solar photosphere are studied by performing a proper orthogonal decomposition (POD) of line-of-sight velocity fields computed from high-resolution data coming from the SOHO/MDI instrument. Using this technique, we are able to identify and characterize the different dynamical regimes acting in the system. All of the POD modes are characterized by two well-separated peaks in the frequency spectra. In particular, low-frequency oscillations, with frequencies in the range 20 – 130 μHz, dominate the most energetic POD modes (excluding solar rotation) and are characterized by spatial patterns with typical scales of about 3 Mm. Patterns with larger typical scales, of about 10 Mm, are dominated by p-mode oscillations at frequencies of about 3000 μHz. The p-mode properties found by POD are in agreement with those obtained with the classical Fourier analysis. The spatial properties of high-energy POD modes suggest the presence of a strong coupling between low-frequency modes and turbulent convection.

Keywords Velocity fields · Granulation · Oscillations · Solar: turbulence

1. Introduction

The solar photosphere provides an important opportunity to observe and study a large number of phenomena related to solar and plasma physics. Among others, the complex

---

Helioseismology, Asteroseismology, and MHD Connections
Guest Editors: Laurent Gizon and Paul Cally

A. Vecchio (✉) · V. Carbone · F. Lepreti · L. Primavera · P. Veltri
Dipartimento di Fisica, Università della Calabria, Rende (CS), Italy
e-mail: vecchio@fis.unical.it

L. Sorriso-Valvo
LICRYL, INFM/CNR, Ponte P. Bucci 31/C, 87036 Rende (CS), Italy

T. Straus
INAF—Osservatorio Astronomico di Capodimonte, Napoli, Italy

dynamics of the plasma motions on the solar surface have been studied in detail in recent years, by taking advantage of the great improvements of the quality and quantity of both ground- and space-based observations (see, *e.g.*, Stix, 2002). Moreover, the photosphere is an interesting example of a system exhibiting complex spatio-temporal behavior, a quite common feature in a wide range of systems far from equilibrium. The early research in pattern formation focused on the presence of simple periodic structures, but the main questions currently addressed concern regimes characterized by higher complexity, that is, patterns that are more irregular in space and time. This is often related to the occurrence of intermediate states between order and turbulence (Akhromeyeva *et al.*, 1989; Rabinovich, Ezersky, and Weidman, 2000). Moreover, the complex photospheric convective dynamics is further complicated by the presence of solar global oscillations.

The new generation of instruments have opened new possibilities in the study of photospheric dynamics. Because of the available high spatio-temporal resolution, many phenomena can be characterized in much more detail. In particular, high spatial resolution is fundamental in studying the coupling between turbulent convective motions and solar oscillations. As an example, solar convection is commonly believed to be responsible for the p-mode excitation (Goode *et al.*, 1998; Strous, Goode, and Rimmele, 2000).

However, the appearance of new features makes the understanding of the phenomena under study rather difficult. Thus with technological improvements, the development of new analysis techniques is needed to better investigate solar atmosphere observations.

The study of photospheric dynamics has so far usually been performed through Fourier analysis. In this framework, different dynamical phenomena are identified as structures in the $k - \omega$ diagram. However, this kind of analysis suffers some important limitations;

1. An exact separation of the different phenomena in the $k - \omega$ diagram is not possible when nonlinear couplings are at work.
2. The Fourier basis functions $\exp[i(kx - \omega t)]$, given *a priori*, are not proper eigenfunctions of the convective signal. The convective phenomenon is complex in both space and time and is known to involve the frequency region of p modes (Straus *et al.*, 1999).
3. Fourier analysis is exact only in the presence of periodic boundary conditions and is optimal when a dispersion relation $\omega(k)$ exists.

In the present paper we introduce the Proper Orthogonal Decomposition (POD) technique, also known as the Karhunen–Loeve expansion, for the modal decomposition of photospheric velocity fields. POD was introduced some time ago in the context of turbulence by Lumley (see Holmes, Lumley, and Berkooz, 1996, and references therein), and it is a powerful technique to extract basis functions that represent ensemble-averaged structures, such as coherent structures in turbulent flows. It provides a basis that optimally represents a flow in the energy norm. Convective structures, which are usually observed in the photosphere, can be seen as a kind of coherent structure within a stochastic field, and this idea represents the basis for the present paper. In the next section we briefly describe POD, in Section 3 we present the results of POD applied to velocity fields acquired by the Michelson Doppler Imager (MDI) instrument onboard SOHO, and in the following sections we discuss the results obtained.

2. The Proper Orthogonal Decomposition

The POD technique is designed to yield a complete set of eigenfunctions that are optimal in energy compared to any other basis (Holmes, Lumley, and Berkooz, 1996). The basic idea is straightforward. Suppose we have an ensemble of fields $\{u^k\}$, in our case the line-of-sight (LOS) photospheric velocity fields $u(x, y, t)$ at different times, where x, y are the coordinates on the solar surface. When searching for a representation of members of $\{u^k\}$ in terms of a series expansion, we need to project each $\{u^k\}$ onto candidate basis functions. We assume that each $\{u^k\}$ belongs to an inner product space $L^2(\Omega)$, where Ω denotes the spatial domain of the fields. The goal is to find an orthogonal basis $[\Psi_j(x, y)]$ for $L^2(\Omega)$ that is optimal for the data set in the sense that a finite N-dimensional representation in the form

$$u(x, y, t) = \sum_{j=0}^{N} a_j(t) \Psi_j(x, y) \qquad (1)$$

[where the $a_j(t)$ are time-dependent modal coefficients] describes typical members of the ensemble better than N-dimensional representations in any other basis. There is no reason, in principle, to distinguish between space and time in deriving the basis functions. However, being mainly interested in detecting coherent structures and constructing low-dimensional models for their dynamics, we can look for basis functions (Ψ_j) depending only on the spatial variables and subsequently determine the time-dependent modal coefficients $[a_j(t)]$ by projecting the original fields onto the basis functions. Thus, we assume the space-time decomposition

$$u(x, y, t) = \sum_{j} a_j(t) \Psi_j(x, y). \qquad (2)$$

Each member of the ensemble $\{u^k\}$ can be interpreted as a different realization of the field described in Equation (2). This allows us to define an averaging operation, which will be denoted by $\langle \cdot \rangle$ and assumed to commute with the spatial integral of the L^2 inner product. The operation $\langle \cdot \rangle$ may be thought of as an ensemble average over a number of separate experiments forming $\{u\}$, or, if ergodicity is assumed, as a time average over an ensemble of observations $u^k(x, y) = u(x, y, t_k)$ obtained from different measurements during a single experimental run. Now we search for an "optimal" basis on which to project the fields. The optimality condition is that a basis element (Ψ) has to be chosen so as to maximize the average projection of u onto Ψ. In mathematical terms, this condition reads

$$\max_{\Psi \in L^2(\Omega)} \frac{\langle |(u, \Psi)|^2 \rangle}{\|\Psi\|^2}, \qquad (3)$$

where $|\cdot|$ denotes the modulus, $(\cdot, \cdot)$ denotes the inner product, and $\|\cdot\|$ is the L^2 norm $[\|f\| = (f, f)^{1/2}]$. Solution of Equation (3) would give only the best approximation to the ensemble members by a single function. Through a variational method, we look for extrema in $\langle |(u, \Psi)|^2 \rangle$ subject to the normalization constraint $\|\Psi\|^2 = 1$. This procedure allows us to obtain from Equation (3) a Fredholm integral equation

$$\int_{\Omega_{x,y}} dx' \, dy' \langle u(x, y, t) u^*(x', y', t) \rangle \Psi(x', y') = \lambda \Psi(x, y). \qquad (4)$$

The unique optimal basis is thus given by the eigenfunctions $\{\Psi_j\}$ of this Fredholm equation, whose kernel is the averaged autocorrelation function $R(x, y, x', y') = \langle u(x, y) u^*(x', y') \rangle$. For experimental data sets the operator $\boldsymbol{R}$ is self-adjoint, compact, and non-negative definite. Thus the Hilbert–Schmidt theory guarantees that there is a countable infinity of real eigenvalues $\lambda_j \geq 0$ and eigenfunctions $\{\Psi_j\}$ [mutually orthogonal in $L^2(\Omega)$], given by solutions of Equation (4) that can be normalized so that $\|\Psi_j\| = 1$. The eigenvalues are ordered so that $\lambda_j \geq \lambda_{j+1}$. The POD coefficients are characterized by the following property:

$$\langle a_j a_k^* \rangle = \delta_{jk} \lambda_j, \tag{5}$$

which means that the modal coefficients are uncorrelated on average.

If $u(x, y, t)$ is a turbulent velocity field, the eigenvalues $\{\lambda_j\}$ are twice the average kinetic energy per unit mass contained in each mode j. This means that if we consider the subspace spanned by the modes corresponding to the first (i.e., largest) N eigenvalues, the representation in Equation (1) captures the most energetic structures in the field. In this way, we form a subspace spanned by the first N eigenfunctions

$$u_N(x, y, t) = \sum_{j=0}^{N} a_j(t) \Psi_j(x, y). \tag{6}$$

These eigenfunctions can be used either to build up a reduced model of the dynamical behavior of the system or, in an indirect way, to quantify the number of relevant modes for the dynamics.

POD is an "optimal" field expansion; that is, a truncated POD expansion given by Equation (6) captures, on average, the most kinetic energy possible for a projection on a given number N of modes (Holmes, Lumley, and Berkooz, 1996). The basis given in Equation (2) is built directly from the experimental data set and is obtained by maximizing the averaged projection of $u(x, y, t)$ onto each $\Psi_j(x, y)$ constrained to a unitary norm. When dealing with a periodic field, the empirical eigenfunctions would just be the Fourier basis (Holmes, Lumley, and Berkooz, 1996). Being extracted directly from experiments, POD eigenfunctions $\Psi_j(x, y)$ provide the proper functional shape of the phenomenon under study and the associated temporal part $a_j(t)$ represents the time evolution of the jth mode associated to that eigenfunction.

It must be realized that POD numerical codes need massive computer resources to solve the Fredholm integral equation, so it is difficult to work with large data sets. The algorithm used in this work follows the snapshot method (Holmes, Lumley, and Berkooz, 1996). The allocated memory depends on the temporal points of the data set (T) and is equal to $T^2 \times b$, where $b = 4$ bytes in single precision. In this way, the calculation can be performed on a personal computer with suitable calculation resources.

3. Data Analysis and Results

3.1. The Data Set

In this work, POD has been applied to high-resolution, photospheric, line-of-sight velocity fields. The analyzed data set has been obtained from images acquired by the MDI instrument onboard the SOHO spacecraft. The image size is 695×695 pixels, with a pixel size of about 0.6 arcsec. The time series spans a time interval $T = 886$ minutes, with images being sampled every minute. Figure 1 shows one of the snapshots of the data set, together with its spatial Fourier spectrum. These data have already been studied with Fourier techniques (Straus et al., 1999). As can be seen from the image, many spatial scales are present and the spectrum appears to be very broad with no characteristic lengths (Roudier et al., 1991).

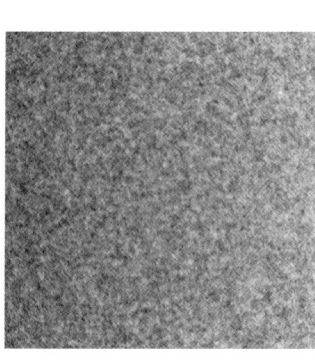
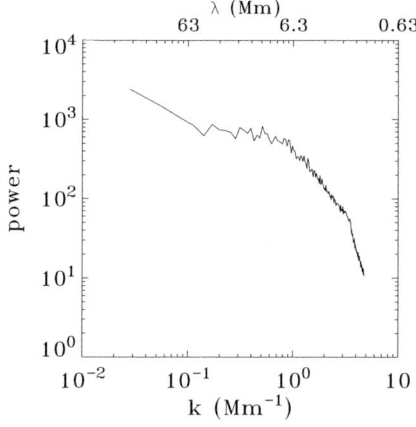

Figure 1 Snapshot of the data set (left) together with the corresponding wavevector spectrum (right). The spatial extent of the image is 417 × 417 arcsec.

The POD of the velocity field $u(x, y, t)$ (x, y being the coordinates on the surface of the Sun) yields a set of eigenfunctions $[\Psi_j(x, y)]$ and corresponding coefficients $[a_j(t)]$, as well as the sequence of eigenvalues λ_j ($j = 0, 1, \ldots, 886$, sorted in decreasing energetic content).

3.2. The POD Results

In Figure 2 we present the energy of the POD modes with $j = 1, \ldots, 400$. The majority of the energy (87%) is associated with the first POD mode $j = 0$ (not shown in the figure) accounting for the line-of-sight velocity component owing to solar rotation. The rest of the energy (see Figure 2) is decreasingly shared by the following 885 modes. A total of 140 modes (excluding the mode $j = 0$ associated with the nonturbulent pattern of solar rotation) are needed to capture 90% of the energy, indicating the absence of dominating coherent structures with a well-defined spatial scale and the presence of turbulent dynamics related to nonlinear interactions among different modes and structures at all scales. For comparison, when applied to laboratory turbulent flows, POD attributes almost 90% of the total energy to the typical large-scale coherent structures confined to $j \leq 2$ (Alfonsi and Primavera, 2002). The energy plot of the POD modes shows an evident change (Figure 2) passing from modes with $j \leq 11$ to modes with $j > 11$. This indicates that modes $j \leq 11$ and $j > 11$ describe contributions characterized by different spatio-temporal behavior.

In general, if a data set has a mean field, the most energetic POD mode will characterize this contribution. This is the case of our data set, where the $j = 0$ mode is dominated by solar rotation. The corresponding eigenfunction is characterized by two large dark and bright regions in the left and right boundaries of the image, respectively (Figure 3). A contribution associated with supergranulation is also present in Ψ_0 near the boundaries of the field of view. The temporal behavior of this mode shows the linear variation of the solar rotation owing to the relative velocity between the Sun and the observer (in this case the satellite). The eigenfunction of the $j = 1$ mode presents an evident pattern because of supergranulation (Figure 3). The corresponding temporal behavior is dominated by a linear trend with a sort of saturation at the boundaries of the temporal interval. The mode $j = 1$ could be associated

Figure 2 Energy of the first 400 POD modes as a function of the index j excluding the solar rotation $j = 0$ mode.

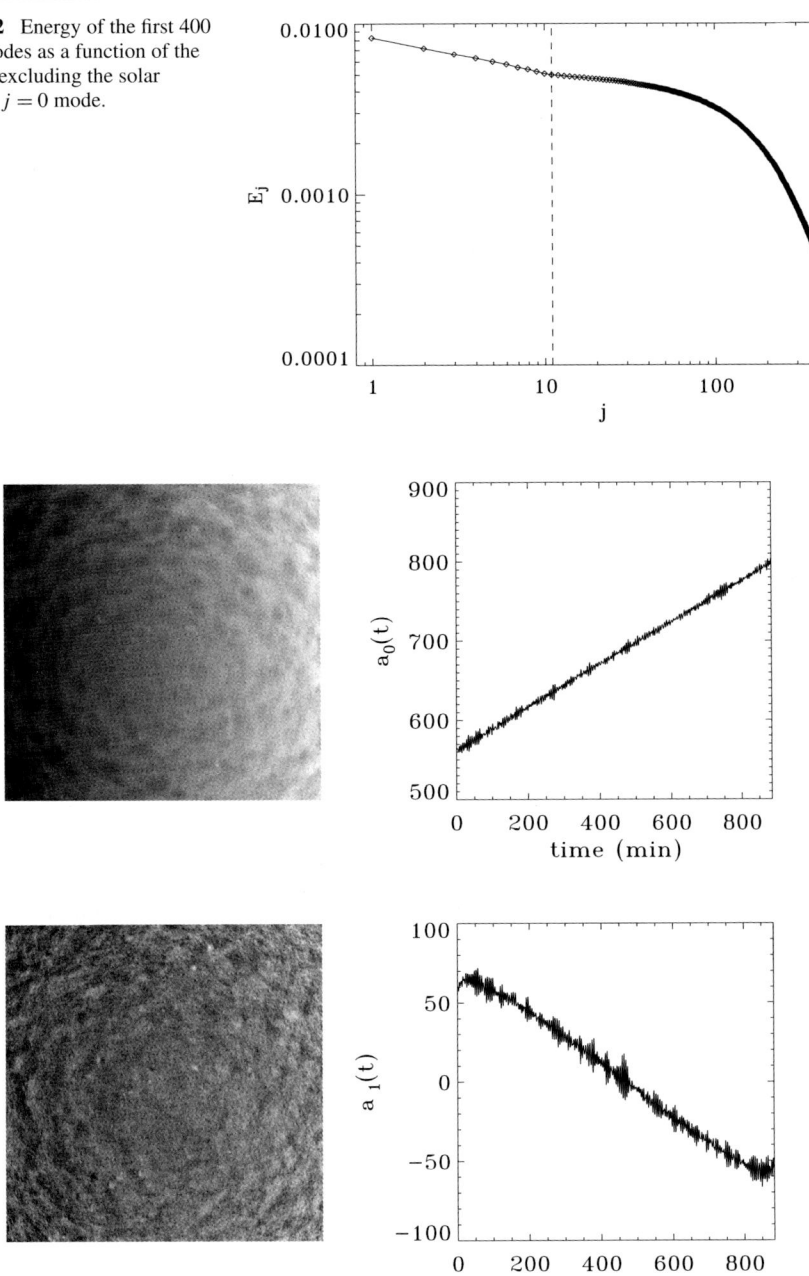

Figure 3 Upper panels: Eigenfunction Ψ_0 (left) together with the corresponding temporal coefficient $a_0(t)$ (right). Lower panels: eigenfunction Ψ_1 (left) together with the corresponding temporal coefficient $a_1(t)$ (right). The spatial extent of the image is 417×417 arcsec.

with the slow temporal variation of the supergranulation pattern. In fact this phenomenon is characterized by a lifetime of $\approx$ one day, longer than our data set. Beyond these dominant trends, coefficients $a_1(t)$ and $a_2(t)$ show a low-amplitude, high-frequency contribution of the five-minute oscillations.

We now turn our attention to the other modes found by POD, starting with a consideration of the spatial properties of the eigenfunctions. As an example, three POD eigenfunctions are shown in Figure 4 for the modes $j = 8$, 12, and 49. The power spectra $[|\Psi_j(k)^2|]$ obtained from the Fourier transform of the eigenfunctions $[\Psi_j(x, y)]$ are also presented in Figure 4 as functions of the wavevector $k = (k_x^2 + k_y^2)^{1/2}$. All of the spatial patterns appear rather irregular and are characterized by a broad spectrum. Moreover, all of the POD modes can be divided, according to the typical size of the structures, into three different groups. The $2 \leq j \leq 11$ modes display a pattern with very fine structures, recalling the photospheric granulation with center-limb modulation. Their broad wavelength spectrum indicates the complex nature of the POD modes. Conversely, eigenfunctions of the modes $j = 12$ and $j = 13$ present a coarser pattern, and their (still broad) spectra are characterized by the presence of some peaks. The further lower-energy modes cannot be precisely classified and seem to present a mixture of the previous characteristics when both structure size and spectral shape are considered. The spatial pattern associated with each mode, although complicated, can be quantitatively characterized by computing the integral scale length $L_j = \int_0^\infty |\Psi_j(k)|^2 k^{-1} dk / \int_0^\infty |\Psi_j(k)|^2 dk$, which represents the typical energy-containing scale of a turbulent field (Pope, 2000). This allows us to estimate the typical scale for the $j = 2, \ldots, 11$ modes as $L \approx 3$ Mm, close to the typical scales of granulation, whereas for the $j = 12$ and 13 modes the typical scale is $L \approx 10$ Mm. For the $j > 13$ modes it is found that L varies within the interval $5 < L < 9$ Mm. The values of L_j as a function of j are shown in Figure 7.

The eigenfunctions described here are associated with the corresponding coefficients accounting for the temporal evolution of each mode. In Figure 5 we report, for the same modes represented in Figure 4, the time behavior of the coefficients $a_j(t)$, together with their Fourier spectra $|a_j(\nu)|^2$. For all of the eigenfunctions, the coefficients are characterized by the presence of a broad range of frequencies. However, in the power spectra one generally observes two peaks, located in two well-defined and separated frequency ranges. The $j = 2, \ldots, 11$ eigenfunctions are associated with time coefficients dominated by very low frequency oscillations in the range 20–130 µHz. For the $j = 12$ and 13 modes the high-frequency peak around 3300 µHz becomes of the same order, and even slightly higher than the low-frequency one. For $j > 13$ we found either modes dominated by low frequencies and small spatial scales (e.g., $j = 15$ and $j = 28$) or modes with intermediate behavior (e.g., $j = 49$), for which the amplitudes of the two peaks are of the same order. The existence of modes with intermediate behavior, with associated non-negligible energy, confirms the coupling among the different photospheric phenomena: A simple separation of the phenomena according to a linear theory does not seem to be possible. The high-frequency $\nu_j^{(p)}$ peaks lie in the p-mode range 3250–3550 µHz. The low frequencies ($\nu_j^{(l)}$) can be measured by applying a low-pass filter $[a_j^F(t)]$ to the temporal coefficients $[a_j(t)]$ and performing a minimum χ^2 fit to a sinusoidal behavior $A_j \sin(\omega_j^{(l)} t + \beta_j)$. Some examples are shown in Figure 6. In Figure 8 we report amplitude versus frequency for the low-frequency oscillations, which dominate the $j = 2, \ldots, 11$ modes (red symbols), and for both low- and high-frequency oscillations present in the $j = 12, \ldots, 49$ modes (blue symbols and light blue symbols).

The POD applied to a time series of line-of-sight velocity maps of the solar photosphere reveals for every mode the presence of two well-separated frequencies. The corresponding spatial patterns are related to the values of the dominating frequency. Looking at Figures 4, 5, and 7, we can see that for small integral lengths, $L \approx 3$ Mm ($j = 2, \ldots, 11$), the low-frequency oscillations ($\nu_j^{(l)}$) dominate, whereas when the integral length is large, $L \approx 10$ Mm

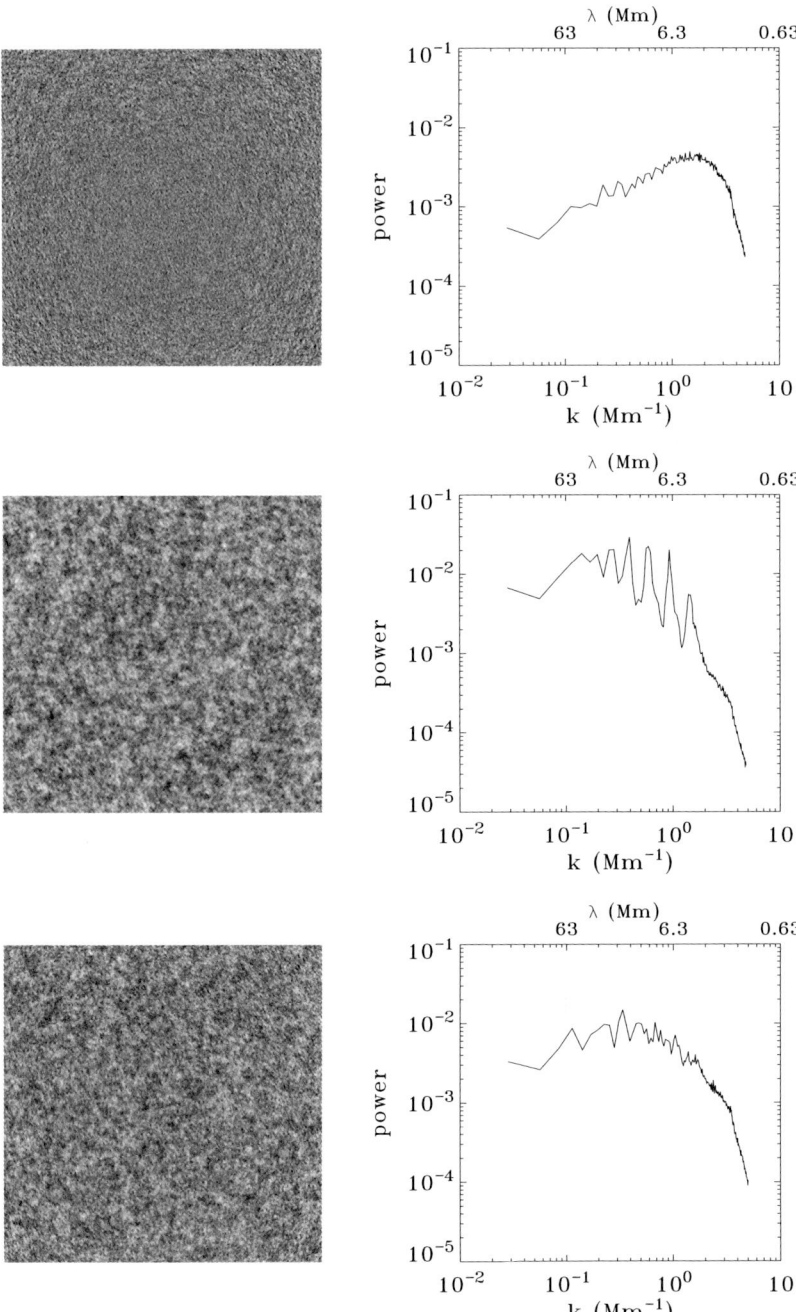

Figure 4 Left: From top to bottom the POD eigenfunctions $\Psi_j(x, y)$ for the three modes $j = 8$, $j = 12$, and $j = 49$. Right: The wavevector spectra $|\Psi_j(k)|^2$ of the corresponding images. The spatial extent of the image is 417×417 arcsec.

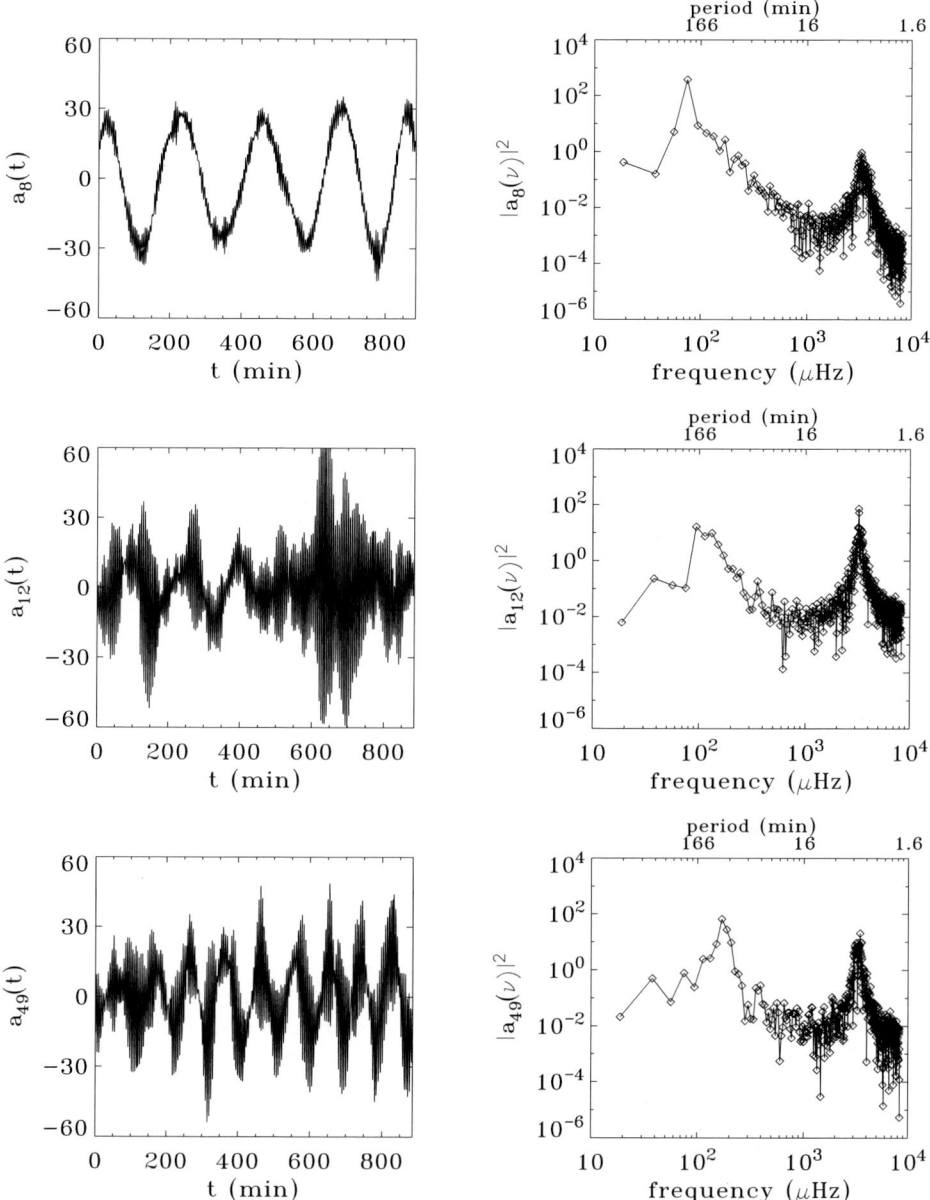

Figure 5 Left: Time evolution of POD coefficients $a_j(t)$ (m s^{-1}) for the same three modes as in Figure 4. Right: The corresponding frequency spectra $|a_j(\nu)|^2$ (m^2 s^{-2}).

($j = 12$ and 13), the amplitude of the high-frequency oscillations ($\nu_j^{(p)}$) becomes larger then that of the low-frequency ones. The modes for which the amplitudes of $\nu_j^{(l)}$ and $\nu_j^{(p)}$ are comparable show intermediate integral scales between 3 and 10 Mm (*e.g.*, $j = 49$).

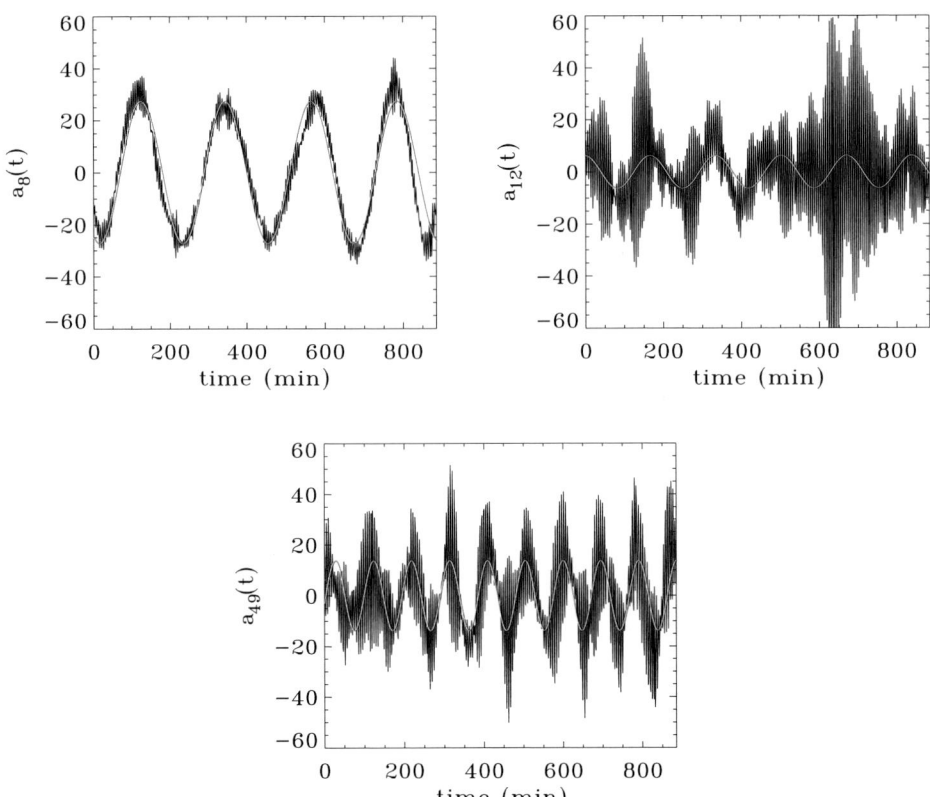

Figure 6 Fits of sinusoidal functions superimposed to POD coefficients $j = 8$, $j = 12$, and $j = 49$.

3.3. Comparison with the $k - \omega$ Analysis

The results shown in the previous subsection indicate that solar p-mode contributions are identified by POD. The measured frequencies (for instance, $\nu = 3254$ µHz for $j = 12$ and $\nu = 3405$ µHz for $j = 13$) are compatible with the p-mode frequencies predicted by helioseismological models (Cristensen Dalsgaard, 2002) and observed using Fourier techniques (Deubner, 1974, 1975; Stein and Leibacher, 1974; Stix, 2002). The spatial scale of about 10 Mm, associated in the POD technique with high-frequency modes, is also in agreement with the horizontal coherence length attributed to solar p modes.

The results obtained by using the POD can be compared with the results of $k - \omega$ analysis. We compare the wavevectors and frequency spectra obtained from POD modes with the $k - \omega$ wavevector spectra. For example, performing a cut of the $k - \omega$ spectrum at the p-mode frequency $\nu^{(p)} \approx 3254$ µHz, we obtain a wavevector spectrum (see Figure 9). The $k - \omega$ cut can be compared with the wavevector spectrum for the eigenfunction $j = 12$ dominated in time by a frequency of ≈ 3254 µHz. The spikes observed in the POD spectrum qualitatively correspond to the typical ridge structures of the $k - \omega$ spectrum (Figure 9). The integral scale length calculated from $k - \omega$ cut $L = 8.241$ Mm is not far from the corresponding $L_{12} = 9.91$ Mm for Ψ_{12}. The frequency spectra can be compared by performing a cut of the $k - \omega$ spectrum at fixed k. The values of k_{12} can be obtained by the corresponding turbulence integral length L_{12}. Obviously, since the wavevector spectra extend over a wide range of k,

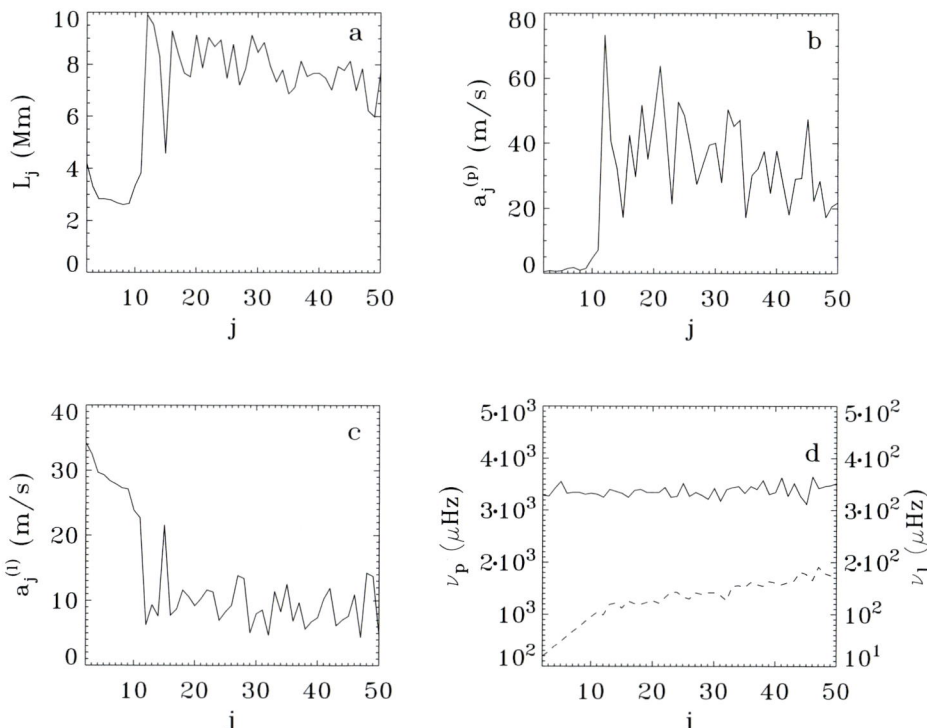

Figure 7 (a) Integral scale length L_j, (b) p-mode amplitudes calculated from Fourier spectra $a_j^{(p)}$, (c) low-frequency oscillation amplitudes $a_j^{(l)}$ from the fits, and (d) high (p modes) and low frequencies (dashed line) calculated from POD coefficients as functions of POD index j.

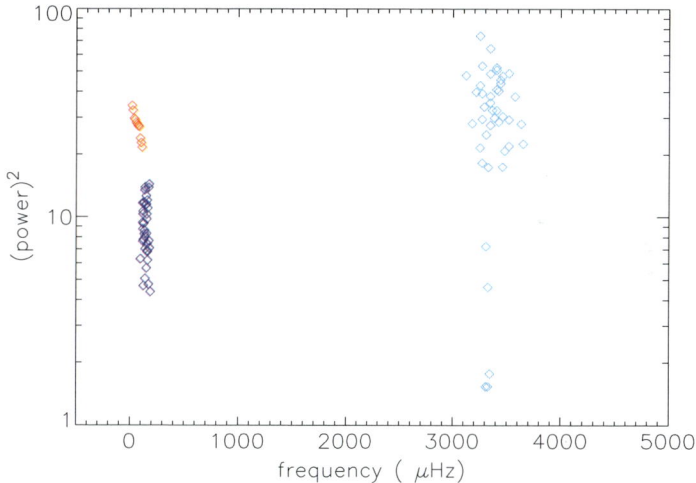

Figure 8 Amplitude–frequency diagram of the POD coefficients. Red symbols: $2 \leq j \leq 11$ low frequencies calculated through fitting. Light blue symbols: $12 \leq j \leq 49$ high frequencies measured from the power spectrum of coefficients. Blue symbols: $12 \leq j \leq 49$ low frequencies calculated through fitting after low-pass filtering.

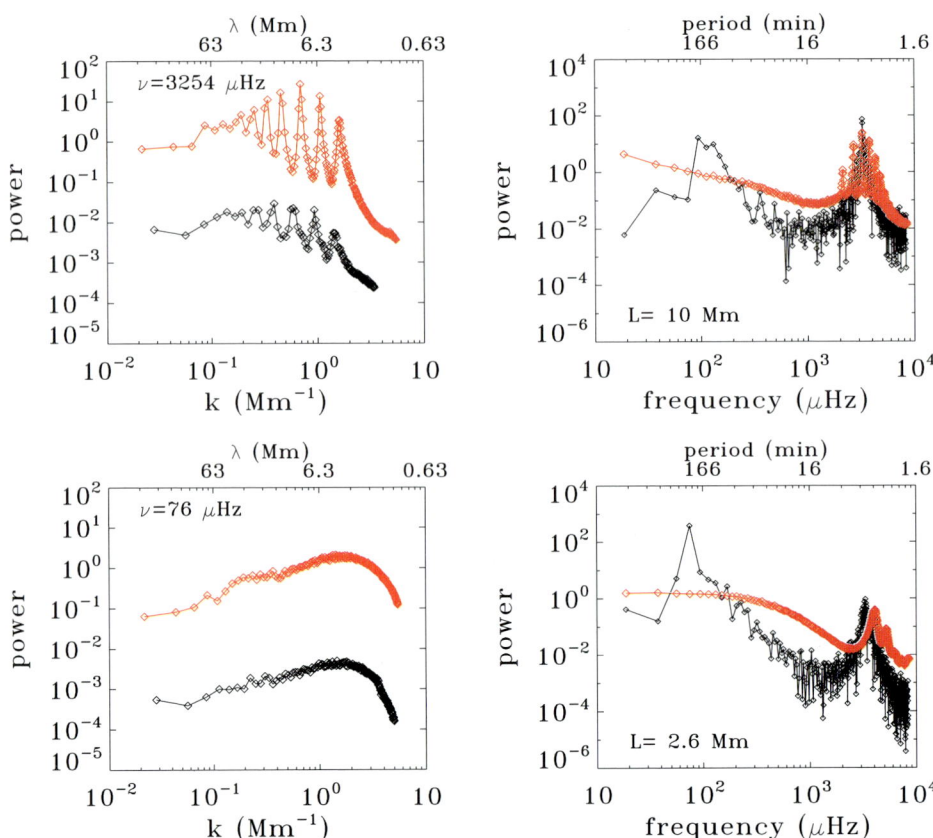

Figure 9 Wavevector spectrum (red) obtained from the $k-\omega$ cut at $\nu = 3254$ µHz and wavevector spectrum (black) calculated from Ψ_{12} (upper left panel). Frequency spectrum obtained from the $k-\omega$ cut at $L = 9.842$ Mm (red) and the spectrum (black) for the coefficient a_{12} (upper right panel). Comparison between the $k-\omega$ cut (red) at $\nu \approx 76$ µHz and the wavevector spectra (black) of Ψ_8 (lower left panel). Comparison between the $k-\omega$ cut (red) at $L_8 = 2592$ Mm and the frequency spectrum (black) of a_8 (lower right panel).

k_{12} is only an indicative scale, useful to compare the spectra. For the mode $j = 12$, the high-frequency peaks, observed in the spectra, nearly coincide. In contrast, at low frequencies the two spectra are different. A peak is found in the POD spectrum in the range where the $k-\omega$ frequency spectrum shows a continuum. The same kind of comparison can be performed for the mode $j = 8$ in which the low frequency is dominant. The cut in the $k-\omega$ spectrum at the low frequency of 75 µHz reveals a qualitative agreement between the two techniques (Figure 9), as the wavevector spectra show a similar structure. The characteristic integral scale computed for the $k-\omega$ cut, $L = 2.74$, is also in agreement with the POD integral scale L_8. The results change when the frequency spectra are compared. Indeed the $|a_8(\nu)|^2$ is dominated by the low-frequency peak where the $k-\omega$ spectrum, obtained by performing a cut at k calculated from $L_8 = 2.61$ Mm, shows a high power continuum. At high frequencies the behavior of the $|a_8(\nu)|^2$ and the $k-\omega$ spectrum is nearly the same. The high-frequency peaks overlap only partially since the k value, used to extract the spectrum, is deduced from

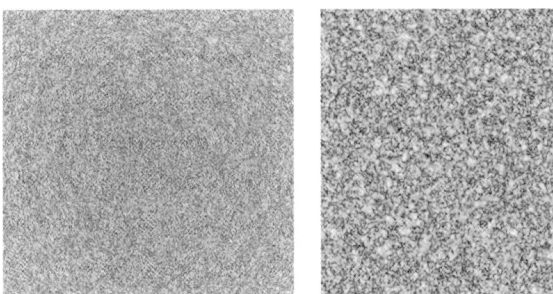

Figure 10 $p(x, y, \nu)$ with $\nu \approx 75$ µHz (left panel) and $\nu = 3254$ µHz (right panel). Images are in arbitrary units. The spatial extent of the image is 417×417 arcsec.

the integral length L_8 so it is not exactly coincident with the corresponding k of the Fourier analysis.

The pattern of the POD eigenfunction can be compared with the power $p(x, y, \nu)$ obtained by performing a Fourier transform only in time:

$$f(x, y, \nu) = \int v(x, y, t) e^{-i\nu t} \, dt \tag{7}$$

and

$$p(x, y, \nu) = f(x, y, \nu) \times f^*(x, y, \nu). \tag{8}$$

By looking at $p(x, y, \nu)$ it is possible to visualize the scales containing the energy for each selected frequency. In Figure 10 the spatially resolved Fourier power, $p(x, y, \nu)$, is shown for the frequencies of $\nu^{(l)} \approx 75$ µHz and $\nu^{(p)} \approx 3254$ µHz. The spatial power distribution and the POD eigenfunction show substantially the same features. In particular, at low frequencies the energy is distributed on small scales, as for $\Psi_8(x, y)$, and the center-limb modulation can be recognized. In the same way the spatial pattern $p(x, y, \nu)$, for $\nu = 3254$ µHz, is characterized by the same structures observed in $\Psi_{12}(x, y)$ and associate with the p-mode coherence pattern.

4. Discussion

It seems evident that although the modes $j = 12$ and $j = 13$ are associated with the solar p modes, the dynamics described by the low-j POD modes seems to be highly complex and may be driven by nonlinear couplings between different phenomena. In this section we discuss, in a qualitative way, some characteristics of the high-energy $j = 2, \ldots, 11$ POD modes.

4.1. Some Dimensional Considerations

In the turbulence framework an *eddy turnover time* (τ) can be introduced to quantify the characteristic time of nonlinear evolution of a vortex in a turbulent fluid. For each POD mode we can simply define an eddy turnover time using the integral scale length of the eigenfunction and the velocity amplitude of the coefficients. For example, for the modes $j = 8$ and $j = 11$ we have

$$\tau_8 = L_8/a_8 \approx 120 \text{ minutes},$$

$$\tau_{11} = L_{11}/a_{11} \approx 36 \text{ hours}.$$

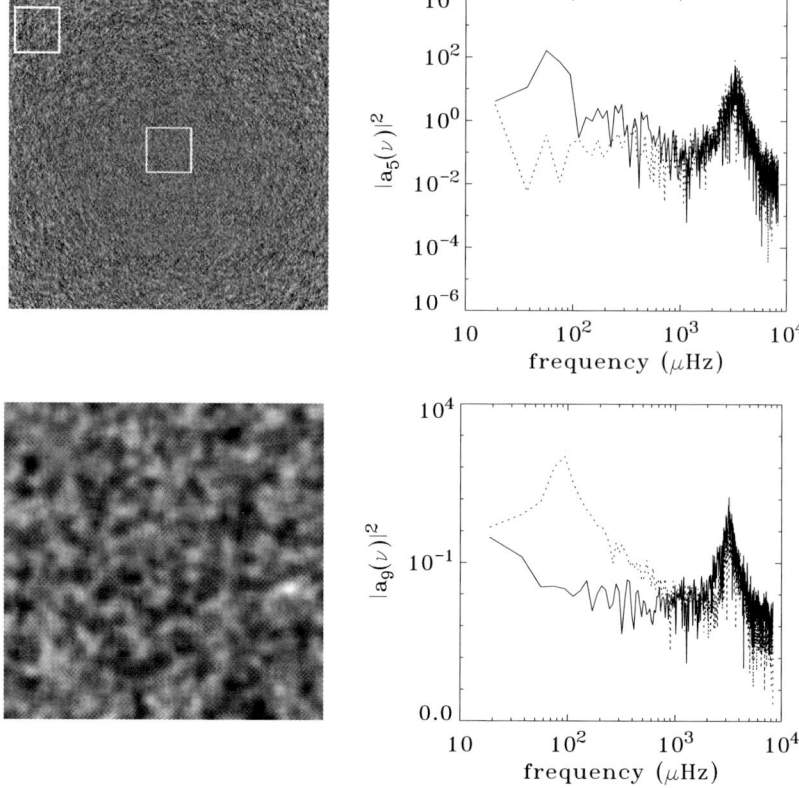

Figure 11 Upper panels: The POD eigenfunction $\Psi_5(x, y)$ (left; the center-limb modulation is evident) and the Fourier spectra (right) of the modal coefficient $a_5(t)$ obtained from two reduced square boxes of size 100×100 pixels, taken at the center (dotted curve) and at the border (solid curve) of the image (see left panel). Lower panels: 50×50 pixel running averaged data at a given time (left) and the $j = 9$ Fourier spectrum of the modal coefficient $a_9(t)$ (right) for both the smoothed field (solid curve) and the original field (dotted curve). The spatial extent of the image is 417×417 arcsec.

This value of τ_{11} is in agreement with the typical coherence time of the p modes (Deubner, 1981). However, the value of τ_8, associated with a mode characterized by granular scale, is very different from the granulation lifetime of about five minutes (Harvey, 1985). From this simple dimensional analysis we can conclude that another phenomenon, characterized by long time scales and strongly coupled to the turbulent convection, is at work in the solar photosphere.

4.2. Some Properties of the Low-j Modes

4.2.1. Center-Limb Modulation

The POD eigenfunctions with $2 < j < 11$ show a typical feature of center-limb modulation. The signal at the disk center is significantly lower than at the boundary of the domain (Figure 4).

The increasing constrast toward the limb suggests that such modes are mainly associated with horizontal velocities. This could be the signature of either the fine structure of supergranular flows (see, *e.g.*, Getling, 2006) or low-frequency gravity waves characterized by horizontal polarization.

To quantify the center-limb modulation of the eigenfunctions, we apply the POD on reduced squares of about 100×100 pixels taken at different positions in the field of view. We used a square around the center of the field of view, where the horizontal components of the velocity are weaker, and around a corner of each image, where the contribution of horizontal velocities is dominant. For the disk-center square, low-frequency oscillations and small spatial scales are not observed in the most energetic POD modes. As an example in Figure 11 it is shown that the peak corresponding to low-frequency oscillations in the $j = 5$ mode disappears. This is strong evidence that the low-frequency oscillations are associated with a physical process phenomenon manifesting itself in horizontal velocities.

4.2.2. Spatial Averages

The association between small scales and low frequencies is different from the standard phenomenological expectations in turbulence. If the low frequencies are really tied to the granular pattern, spatial integration or averages, before applying POD, should remove the contribution of these modes. The effects of averages on our data set can be verified by a simple test. Data have been smoothed by applying a spatial 50×50 pixel running average before applying POD. In this way we completely lose not only small scales but also low-frequency oscillations in the most energetic POD modes. As can be seen in Figure 11, the low-frequency peaks in the coefficient spectra are completely lost. As a consequence, observations performing a spatial average or disk integration, commonly used in helioseismology, cannot reveal the contribution to the dynamics of the low-frequency processes detected by POD.

5. Conclusions

In this paper, we presented the application of POD on high-resolution solar photospheric velocity fields (Vecchio *et al.*, 2005). Solar convection presents an example of convective turbulence in a high-Rayleigh-number natural fluid. POD is able to capture the main energetic and spatial features of the solar photosphere. The dynamical processes at work are separated according to their energy. In addition, the most energetic modes, capturing solar rotation and supergranulation contributions, two main oscillatory processes, well separated in frequency, are detected in all of the other modes. High-frequency oscillations are detected in the range 3250–3550 µHz and low-frequency oscillations in the range 20–130 µHz. The high-frequency waves are the well-known acoustic *p* modes and their properties, as obtained from POD, are in agreement with previous results based on Fourier techniques. However, low-frequency oscillations, in the frequency range where different processes are expected to overlap [*e.g.*, *g* modes (García *et al.*, 2007) and convection], prevail in the most energetic POD modes, which are characterized by spatial eigenfunctions with typical scales of ≈ 3 Mm. Another interesting feature of the most energetic modes is a clear center-limb modulation. The same properties can be found in the spatially-resolved Fourier power map at low frequencies. This is a clear indication that horizontal velocities are dominant at low frequencies and at high energy. The clear association between low-frequency oscillations and small spatial scales, which are close to the solar granulation, is the most interesting and most surprising result provided by our analysis, as large spatial scales might be the more obvious candidates for such an association. However, it should be noted that also in $k - \omega$ spectra obtained through classical Fourier analysis there is significant power in the "low-frequency, small scale" region. This point needs to be investigated in future theoretical studies to understand whether the origin of these modes is purely convective or due to nonlinear interactions with low-frequency oscillations.

A further improvement of the analysis presented here can be expected from the application of POD to longer time series. This will allow one to extend our study to lower frequencies and to better understand the role of the low-frequency-dominated POD modes found in this work.

References

Akhromeyeva, T.S., Kurdyumov, S.P., Malinetskii, G.G., *et al.*: 1989, *Phys. Rep.* **176**, 189.
Alfonsi, G., Primavera, L.: 2002, *J. Flow Vis. Image Process.* **9**, 89.
Cristensen Dalsgaard, J.: 2002, *Rev. Mod. Phys.* **74**, 1073.
Deubner, F.-L.: 1974, *Astron. Astrophys.* **39**, 31.
Deubner, F.-L.: 1975, *Astron. Astrophys.* **44**, 371.
Deubner, F.-L.: 1981, *Nature* **290**, 682.
García, R.A., Turck-Chièze, S., Jimènez-Reyes, S.J., Ballot, J., Pallè, P.L., Eff-Darwich, A., Mathur, S., Provost, J.: 2007, *Science* **316**, 1591.
Getling, A.V.: 2006, *Solar Phys.* **239**, 93, 2006.
Goode, P.R., Strous, L.H., Rimmele, T.R., Stebbins, R.T.: 1998, *Astrophys. J.* **516**, 939 ,1998.
Harvey, J.: 1985, *Future Missions in Solar, Heliospheric & Space Plasma Physics* **SP 235**, ESA, Nordwijk, 199.
Holmes, P., Lumley, J.L., Berkooz, G.: 1996, *Turbulence, Coherent Structures, Dynamical Systems and Symmetry*, Cambridge University Press, Cambridge.
Pope, S.B.: 2000, *Turbulent Flows*, Cambridge University Press, Cambridge.
Rabinovich, M.I., Ezersky, A.B., Weidman, P.D.: 2000, *The Dynamics of Patterns*, World Scientific, Singapore.
Roudier, T., Muller, R., Mein, P., Vigneau, J., Malherbe, J.M., Espagnet, O.: 1991, *Astron. Astrophys.* **248**, 245.
Stein, R.F., Leibacher, J.W.: 1974, *Ann. Rev. Astron. Astrophys.* **12**, 470S.
Stix, M.: 2002, *The Sun: An Introduction*, Springer, Berlin.
Straus, T., Severino, G., Deubner, F.-L., Fleck, B., Jefferies, S.M., Tarbell, T.: 1999, *Astrophys. J.* **516**, 939.
Strous, L.H., Goode, P.R., Rimmele, T.R.: 2000, *Astrophys. J.* **535**, 1000S.
Vecchio, A., Carbone, V., Lepreti, F., Primavera, L., Sorriso-Valvo, L., Veltri, P., Alfonsi, G., Straus, Th.: 2005, *Phys. Rev. Lett.* **95**, 061102.

Calculation of Spectral Darkening and Visibility Functions for Solar Oscillations

C. Nutto · M. Roth · Y. Zhugzhda · J. Bruls ·
O. von der Lühe

Originally published in the journal Solar Physics, Volume 251, Nos 1–2, 179–188.
DOI: 10.1007/s11207-008-9132-9 © Springer Science+Business Media B.V. 2008

Abstract Calculations of spectral darkening and visibility functions for the brightness oscillations of the Sun resulting from global solar oscillations are presented. This has been done for a broad range of the visible and infrared continuum spectrum. The procedure for the calculations of these functions includes the numerical computation of depth-dependent derivatives of the opacity caused by p modes in the photosphere. A radiative-transport code was used for this purpose to get the disturbances of the opacities from temperature and density fluctuations. The visibility and darkening functions are obtained for adiabatic oscillations under the assumption that the temperature disturbances are proportional to the undisturbed temperature of the photosphere. The latter assumption is the only way to explore any opacity effects since the eigenfunctions of p-mode oscillations have not been obtained so far. This investigation reveals that opacity effects have to be taken into account because they dominate the violet and infrared part of the spectrum. Because of this dominance, the visibility functions are negative for those parts of the spectrum. Furthermore, the darkening functions show a wavelength-dependent change of sign for some wavelengths owing to these opacity effects. However, the visibility and darkening functions under the assumptions used contradict the observations of global p-mode oscillations, but it is beyond doubt that the opacity effects influence the brightness fluctuations of the Sun resulting from global oscillations.

Helioseismology, Asteroseismology, and MHD Connections
Guest Editors: Laurent Gizon and Paul Cally.

C. Nutto (✉) · Y. Zhugzhda · J. Bruls · O. von der Lühe
Kiepenheuer-Institut für Sonnenphysik, Schöneckstraße 6, 79104, Freiburg, Germany
e-mail: nutto@kis.uni-freiburg.de

M. Roth
Max-Planck-Institut für Sonnensystemforschung, 37191, Katlenburg-Lindau, Germany

Y. Zhugzhda
Institute of Terrestrial Magnetism, Ionosphere and Radio Wave Propagation of the Russian Academy of Sciences, Troitsk, Moscow Region 142092, Russia

Keywords Oscillations: Solar · Waves: Acoustic · Integrated Sun observations · Helioseismology: Theory · Spectrum: Continuum

1. Introduction

Since the ACRIM experiment (Woodard, 1984) it has been well known that solar oscillations lead to brightness fluctuations of the Sun as a star. However, the spatial integration of the flux disturbances over the whole solar disk suppresses high-frequency modes, and photometric observations of the Sun as a star can only resolve spatial oscillation modes with low harmonic degree ($\ell < 3$). But if the visibility of these low-degree modes is suppressed at some positions on the solar disk through opacity effects, then it might be possible to observe oscillation modes with harmonic degree $\ell \geq 3$.

The space experiment IPHIR onboard the *Phobos* spacecraft (Fröhlich *et al.*, 1988) showed that the amplitudes of the brightness fluctuations resulting from p modes are different for different optical wavelengths. Zhugzhda, Dzhalilov, and Staude (1993) showed with the consideration of nonadiabatic waves in a uniform, nongray atmosphere that this problem cannot be solved with the introduction of the Rosseland mean opacity. With the calculation of visibility functions for low-degree nonradial oscillations and a comparison with IPHIR data, Toutain and Gouttebroze (1993) suggested two limiting cases for the explanation of the origin of the brightness fluctuations:

- Intensity and flux perturbations, caused by adiabatic oscillations, are driven by opacity effects and thus have to be taken into account for calculations of visibility functions.
- The visibility functions depend on nonadiabatic effects, and opacity disturbances can be neglected (blackbody approximation).

The direct way to solve the problem is to solve the eigenproblem for nonadiabatic oscillations for a solar model that includes a standard model of the photosphere. To our knowledge nobody has done this so far since the set of integro-differential equations of radiative hydrodynamics has to be solved. Staude, Dzhalilov, and Zhugzhda (1994) and Zhugzhda, Staude, and Bartling (1996) proposed using another approach in which the temperature and density fluctuations resulting from the p modes are assumed to be given. Under some special assumptions, this makes it possible to obtain so-called darkening functions, which show the dependence of brightness fluctuations with respect to the position on the solar disk. The intention was to develop a qualitative explanation of brightness oscillations to improve observations of solar oscillations. The different wavelengths under consideration indicated that the brightness fluctuations of the most dominant oscillation mode ($\ell = 0$) vanishes for certain positions on the solar disk (*e.g.*, the darkening function changes its sign). With the assumption that these calculations can be easily extended to modes with harmonic degree $0 < \ell < 3$ and that those functions will also show a change of sign, modes with higher harmonic degree ($\ell \geq 3$) might be observable.

Zhugzhda, Staude, and Bartling (1996) carried out calculations of darkening functions for a few wavelengths only. However, the visibility functions were not calculated. In this paper the calculations of darkening and visibility functions are performed for a wide range of wavelengths, from the near UV to the infrared, to see whether the opacity effects are essential for this range of the spectrum. This investigation will help us to analyze the observations from photometric instruments such as DIFOS onboard *CORONAS-F* (Lebedev *et al.*, 2004).

2. Intensity and Flux Fluctuations from p Modes

To avoid the full solution of the equations of radiation hydrodynamics (Zhugzhda, Staude, and Bartling, 1996) we expand the general expression for the intensity in a series around the equilibrium state up to the first order of small disturbances of the opacity. Under the assumption of local thermodynamic equilibrium (LTE), the fluctuations of the emergent intensity (δI_ν) normalized to the specific intensity ($I_{0\nu}$) at the center of the solar disk are (Zhugzhda, Staude, and Bartling, 1996)

$$\frac{\delta I_\nu}{I_{0\nu}} = \int_0^\infty e^{-\tau_\nu/\mu} \left[\frac{dB_\nu(\tau_\nu)}{d \ln T_0} \frac{\delta T}{T_0} + \frac{\delta \kappa_\nu}{\kappa_{0\nu}} (B_\nu(\tau_\nu) - I_\nu(\tau_\nu, \mu)) \right] \frac{d\tau_\nu}{\mu}$$
$$\times \left[\int_0^\infty e^{-\tau_\nu} B_\nu(\tau_\nu) \, d\tau_\nu \right]^{-1}, \tag{1}$$

where τ_ν is the optical depth at the frequency ν, B_ν is the Planck function, $\mu = \cos\theta$, and θ is the polar angle. The unperturbed, outgoing intensity, originating at optical depth τ_ν, is

$$I_\nu(\tau_\nu, \mu) = \int_{\tau_\nu}^\infty e^{-(\tau'_\nu - \tau_\nu)/\mu} B_\nu(\tau'_\nu) \frac{d\tau'_\nu}{\mu}. \tag{2}$$

The fluctuation of the opacity $[\delta \kappa_\nu(\rho, T)]$ is given by the fluctuation of the density ($\delta\rho$) and the temperature (δT),

$$\frac{\delta \kappa_\nu}{\kappa_{0\nu}} = \frac{d \ln \kappa_{0\nu}}{d \ln T} \frac{\delta T}{T_0} + \frac{d \ln \kappa_{0\nu}}{d \ln \rho} \frac{\delta \rho}{\rho_0}. \tag{3}$$

Equation (1) considers the various contributions to the intensity fluctuations. The first term appears because of the temperature fluctuation in the source function, which can be described by the Planck function in the case of LTE. The second term appears because of the perturbation of the opacity from disturbances in density and temperature. An approximation for the temperature and the density perturbations will be used. The solution yields the intensity fluctuation of a nongray atmosphere caused by small disturbances. This can be used for the disturbances caused by p modes since they are small enough and the linear approximation works perfectly.

To simplify Equation (1) it is assumed that disturbances of the equilibrium state are adiabatic. In this special case the temperature and density disturbances are connected by the simple relation

$$\frac{\delta T}{T} = (\Gamma_3 - 1) \frac{\delta \rho}{\rho}, \tag{4}$$

where Γ_3 is the third adiabatic exponent. For a neutral or fully ionized gas it is constant and has the value $\Gamma_3 - 1 = 2/3$, whereas for a partially ionized gas it can reach values of $\Gamma_3 - 1 \leq 0.1$. The first case is given for the photosphere, which is completely dominated by neutral hydrogen. The second case applies for the deeper photosphere and the chromosphere, where the gas is partially ionized.

In the case of adiabatic disturbances, the intensity fluctuations depend only on temperature fluctuations and on parameters of the quiet undisturbed photosphere. The dependence of the intensity fluctuations on μ is a darkening function for the adiabatic disturbances. To obtain the darkening function for p modes, the dependence of the temperature disturbances on depth $[\delta T(\tau_\nu)]$ has to be substituted into Equation (1). But this function is not known, even for adiabatic disturbances. To make a qualitative analysis of the opacity effects we simplify the problem by assuming that the disturbances are proportional to their undisturbed values (Zhugzhda, Staude, and Bartling, 1996),

$$\frac{\delta T}{T_0} = \text{constant}, \tag{5}$$

which is a rather crude approximation. It holds only for continuum radiation that originates from a thin layer of the solar atmosphere. With this assumption the relative disturbances $\delta T/T_0$ and $\delta \rho/\rho_0$ are independent of the optical depth (τ_ν) and can be factored out of the integral. Now, with relationship (4) the relative intensity variation can be described in units of $\delta \rho/\rho$ or $\delta T/T$. This leads to the darkening function

$$\Delta_\nu = \frac{\delta I_\nu/I_{0\nu}}{\delta T/T_0}, \tag{6}$$

which in our case is expressed in units of $\delta T/T_0$.

To obtain the visibility function, the darkening function has to be integrated over the entire solar disk. This yields the following expression for the flux perturbation for the most dominant mode $\ell = 0$:

$$\frac{\delta F_\nu/F_{0\nu}}{\delta T/T_0} = \left\{ \int_0^\infty \left[\frac{\partial B_\nu(\tau_\nu)}{\partial \ln T} E_2(\tau_\nu) \right. \right.$$
$$- \left(\frac{\partial \ln \kappa_\nu}{\partial \ln T}\right)_\rho \int_{\tau_\nu}^\infty \frac{dB_\nu(\tau'_\nu)}{d\tau'_\nu} E_2(\tau'_\nu) d\tau'_\nu$$
$$\left. \left. - \frac{1}{\Gamma_3(\tau_\nu) - 1} \left(\frac{\partial \ln \kappa_\nu}{\partial \ln \rho}\right)_T \int_{\tau_\nu}^\infty \frac{dB_\nu(\tau'_\nu)}{d\tau'_\nu} E_2(\tau'_\nu) d\tau'_\nu \right] d\tau_\nu \right\}$$
$$\left/ \left\{ \int_0^\infty B_\nu(\tau_\nu) E_2(\tau_\nu) d\tau_\nu \right\}. \right. \tag{7}$$

3. Numerical Calculation of Darkening and Visibility Functions for Solar Oscillations with Degree $\ell = 0$

For the evaluation of the darkening functions one needs to know radiative transport quantities for the solar atmosphere as functions of the optical depth τ_ν. For this purpose the radiative transport code RH[1] was used (Uitenbroek, 2001). The code is based on the MALI (Multi-level Approximate Lambda Iteration) formalism by Rybicki and Hummer (1991). It solves the combined equations of statistical equilibrium and radiative transport under the general assumption of non-LTE conditions for a multilevel atom in a given plasma. In our case we choose hydrogen to be the considered atom. In addition to opacities and emissivities from the transitions in the hydrogen atom, the code accounts for background radiation sinks and sources from other atoms and molecules and all relevant continuum processes such as H^- bound-bound and bound-free processes, Rayleigh scattering off neutral hydrogen, Thompson scattering off free electrons, helium, and H_2, and hydrogen free-free processes. The atoms and molecules considered for the background radiation and opacities are listed

---

[1] http://www.nso.edu/staff/uitenbr/rh.html.

Table 1 Atoms and molecules that are considered as background elements by the radiative transport code.	Atoms	He, C, N, O, Na, Mg, Al, Si, S, Ca, Fe
	Molecules	H_2, H_2^+, C_2, N_2, O_2, CH, CO, CN, NH, NO, OH, H_2O

in Table 1. In addition, it is necessary to provide the radiative transport code with a model of the solar atmosphere. We choose the semi-empirical, one-dimensional, quiet-Sun model FAL-C by Fontenla *et al.* (1999).

For the calculation of the darkening function we are interested not only in the radiative quantities alone but also in their disturbances. Considering Equation (1), we need the disturbances of the Planck function from temperature variations and the disturbances of the opacity from variation of temperature and density.
To calculate the derivatives

$$\left(\frac{\partial B_\nu}{\partial \ln T}\right), \quad \left(\frac{\partial \ln \kappa_\nu}{\partial \ln T}\right)_\rho, \quad \text{and} \quad \left(\frac{\partial \ln \kappa_\nu}{\partial \ln \rho}\right)_T, \quad (8)$$

we disturb the quantities T and ρ of the atmospheric model; these are then used as input for the radiative transport code. We use temperature disturbances of -5 K and $+5$ K, which is less then 1% of the undisturbed temperature of the atmosphere, and the derivatives behave linearly under these small temperature disturbances. With Equation (4) for the adiabatic relation, it can be assumed that the relative disturbances of the density are around 1%. Thus, the hydrogen and electron populations are modified accordingly in the model of the atmosphere. While one modification is applied, all other values are kept constant. This procedure allows us to calculate the partial derivatives of the Planck function and the opacity using the centered-difference approximation (Pozrikidis, 1998).

Our interest is in the calculation of the darkening and visibility functions for a broad range of the visible and infrared spectrum. This has been done for 79 wavelengths between 250 and 2100 nm. We have to avoid any lines since we assume a temperature disturbance that is constant (see Equation (5)). This is only justified as long as the radiation originates from a thin layer. This applies for the continuum radiation, but spectral lines form in higher layers and are not constrained to a small part of the solar atmosphere.

4. Results

4.1. Derivatives

First, the derivatives of the Planck function $(\partial B_\nu/\partial \ln T)$ were calculated. The Planck function has the advantage that the derivative can be evaluated analytically and the numerical values can be compared with the analytical function. A comparison showed that the numerical calculation of the derivatives coincides with the analytical values within a few tenths of a percent.

In Figure 1 the derivative of the opacity with respect to temperature is shown as a function of wavelength λ and monochromatic optical depth τ_ν. Although the model atmosphere extends from the photosphere to the transition region, we restrict ourselves to the photosphere since we are only interested in continuum radiation. Certain discontinuities in the plot are noticeable. The corresponding wavelengths coincide with the ionization wavelengths of neutral hydrogen, namely the Balmer continuum ($\lambda = 364.6$ nm). Just above the Balmer continuum there is another visible ionization edge that belongs to magnesium with $\lambda = 375.6$ nm.

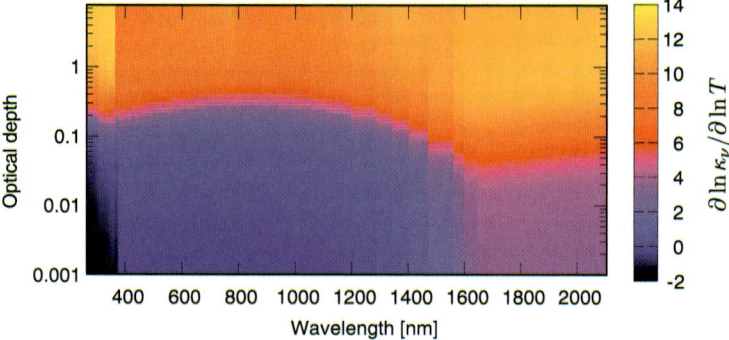

Figure 1 Derivative of opacity with respect to temperature as a function of wavelength and the monochromatic optical depth τ_ν.

These discontinuities can be explained with the ionization setting in abruptly at wavelengths just smaller than the ionization edges. If we assume positive temperature disturbances, then higher temperatures give rise to a higher ionization rate, which causes a greater increase of the opacity. Thus, the derivative of the opacity is greater for wavelengths shorter than the ionization edges. The effect diminishes with smaller optical depth, since the influence of neutral hydrogen on the opacity vanishes for higher parts of the atmosphere (Vernazza, Avrett, and Loeser, 1976). The influence of neutral hydrogen causes the steep rise of the derivative for the deeper layers. For higher layers, the derivative is dominated by the influence of H^- ions (Vernazza, Avrett, and Loeser, 1976) and the opacity is less sensitive to temperature disturbances. This can be explained by the combined effect of the destruction of H^- ions from higher temperatures and the higher ionization of neutral hydrogen, which results in a higher electron density. Investigations showed that if those higher electron densities are not taken into account the derivatives turn out to be negative for those parts of the spectrum where the opacity is dominated by H^-. In conclusion, the creation of H^- resulting from a higher electron density wins over the destruction of H^- resulting from higher temperatures and thus the opacity increases. There is also a change of the behavior of the derivative for $\lambda = 1645$ nm. This can be explained by the fact that above this wavelength the free-free processes of H^- start to dominate the contribution to the opacity.

The results for the derivative of the opacity with respect to the density are shown in Figure 2. First, we can point out the discontinuities at the continuum edges again, where the same explanation as for Figure 1 applies. For deeper layers, the discontinuities for the Paschen continuum ($\lambda = 820.4$ nm) and the Brackett continuum ($\lambda = 1458.4$ nm) of neutral hydrogen are more pronounced than for the derivative of the opacity with respect to temperature.

Knowing the derivatives of opacity, we can now calculate the darkening and visibility functions for the most dominant solar oscillation mode with $\ell = 0$.

4.2. Darkening and Visibility Functions

In Figure 3 the darkening functions (Δ_ν) (see Section 2) are shown for selected wavelengths. Some of the these wavelengths coincide with the wavelengths examined by Zhugzhda, Staude, and Bartling (1996) and a comparison[2] of the results for those wavelengths shows

---

[2] In Zhugzhda, Staude, and Bartling (1996) the graphs for $\lambda = 275$ nm and $\lambda = 1645$ nm should be switched.

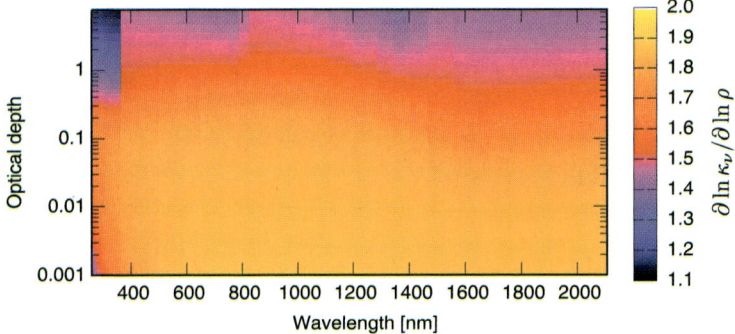

Figure 2 Derivative of opacity with respect to density as a function of wavelength and optical depth.

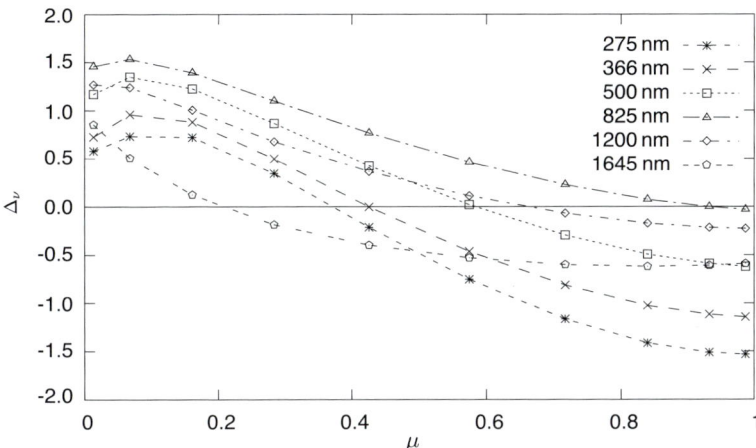

Figure 3 The relative darkening functions (Δ_ν) for selected wavelengths versus position on the solar disk ($\mu = \cos\theta$). The marks show the sampling points on the solar disk.

that they are similar, although there are some small differences in the position of the change of sign. This might be caused by a small difference in the derivatives of the opacity derived in this paper and the derivatives used by Zhugzhda, Staude, and Bartling (1996) for higher layers of the atmosphere. The wavelengths displayed show that the position of the change of sign on the solar disk depends strongly on the wavelength. For the wavelengths whose opacity is mostly dominated by the H^- ions (*e.g.*, 825 nm), the position of the change of sign of the darkening function is almost at the center of the solar disk whereas the function increases toward the limb of the Sun. Thus, for this particular wavelength the oscillation mode $\ell = 0$ would be best observed at the limb while it almost vanishes at the center of the disk. To get the visibility of global oscillation modes for each wavelength, the darkening functions have to be integrated over the solar disk. This can be done by either numerical integration of the darkening function for each wavelength or by direct evaluation of Equation (7). We choose the latter option for the calculation of the monochromatic visibility function.

The result of this integration can be seen in Figure 4, where the absolute visibility function $[\delta F_\nu/(\delta T/T_0)]$ is plotted together with its different contributions from the Planck function and opacity. This plot clearly indicates that the visibility functions are dominated by

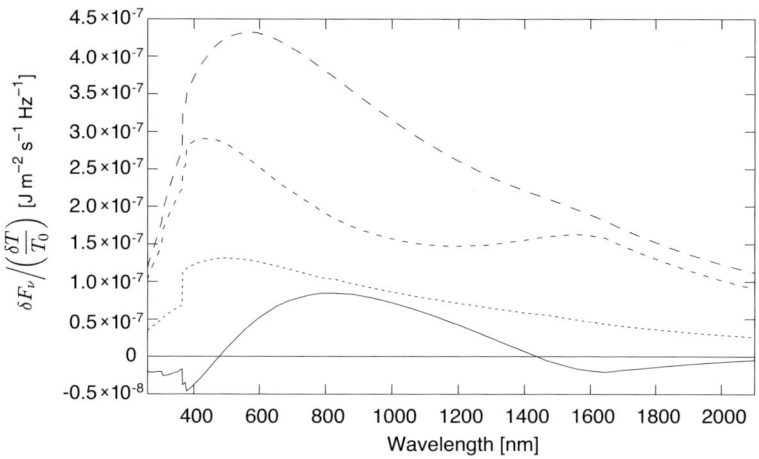

Figure 4 The absolute visibility function versus wavelength (solid line). In addition, the different contributions $\partial B_\nu / \partial \ln T$ (long-dashed line), $\partial \ln \kappa_\nu / \partial \ln T$ (medium-dashed line), and $\partial \ln \kappa_\nu / \partial \ln \rho$ (short-dashed line) originating from the Planck function and the opacity are shown.

the effects of the opacity. This is due to the assumption of adiabatic oscillations for which the temperature and density disturbances are in phase. The effect of the oscillation on the opacity is big enough to overcome the disturbances of the Planck function and for shorter and longer wavelengths positive disturbances actually result in a darkening of the solar flux. Considering the visibility function, one sees that the oscillation mode $\ell = 0$ has the biggest influence on the disturbance of the flux at wavelengths around $\lambda = 800$ nm. Since the visibility functions also show a change of sign for two wavelengths of the considered spectrum, it is interesting to note that for $\lambda = 480$ nm and $\lambda = 1450$ nm the global oscillation of mode $\ell = 0$ should not be visible. This can actually be used as a test for the assumption of adiabatic oscillations, as will be discussed in the following.

5. Conclusions

There is an essential distinction between visibility functions for the global solar and stellar oscillations observed by measurements of Doppler shifts and continuum brightness fluctuations. In the case of Doppler shifts the visibility is defined by the projection of the velocity vector of the oscillation modes onto the line of sight. Thus, it is not complicated to calculate the visibility function for every spherical harmonic.

The derivation of the visibility functions for brightness fluctuations resulting from global oscillations is more difficult. We have succeeded in obtaining the spectral darkening and visibility functions over a broad range of the visible and infrared spectrum of the Sun. Two simplifications of the problem make this possible: the adiabatic approximation and the condition $\delta T / T_0 = $ constant.

We found that the brightness fluctuations are not necessarily proportional to temperature fluctuations. The darkening function shows a change of sign and the visibility function can be negative (*e.g.*, the brightness can decrease with the increase of the temperature) owing to radiative flux blocking.

However, there are some problems considering the visibility function. In accordance with Figure 4, the brightness fluctuations should be absent for 480 and 1450 nm. But this is not the case. Brightness fluctuations are observed at those wavelengths by SOHO/VIRGO and *CORONAS*/DIFOS. This clearly indicates that one of the assumptions is not correct, if not both of them. The condition with constant temperature disturbances should be relaxed in general. It is necessary to know how the temperature fluctuations depend on the depth in the solar atmosphere.

The second assumption in use is the adiabatic approximation. Our calculations confirm one of the results of Toutain and Gouttebroze (1993). If adiabatic conditions are assumed, the opacity perturbations show a great influence on the visibility functions. But, as already mentioned, in this case the brightness fluctuations should not be visible for 480 and 1450 nm, which is not in accordance with observations. Thus, this approximation is violated in the solar photosphere. Zhugzhda (2006) discovered that there are phase shifts between brightness fluctuations in different spectral channels of the DIFOS photometric instrument. This is only possible if oscillations are nonadiabatic since in this case there is a phase shift between the temperature and density fluctuations. Moreover, the nonadiabatic p modes are coupled with temperature waves (Zhugzhda, 1983). Consequently, the darkening and visibility functions for nonadiabatic oscillations are complex functions. Multichannel observations of *CORONAS* together with SOHO observations make it possible to explore these nonadiabatic oscillations.

The visibility and darkening functions are needed for the interpretation of the observations of solar and stellar observations. The main result of the current exploration is that the effect of the opacity fluctuations on brightness oscillations dominates. However, it is a rather complicated problem to find the correct functions for adiabatic and nonadiabatic oscillations. The current paper just presents the first step on this road.

Acknowledgements YZ is grateful to RFFI (Grant No. 06-02-16359). CN, MR, YZ and OvdL acknowledge support from the European Helio- and Asteroseismology Network–HELAS. HELAS is funded by the European Union's Sixth Framework programme. We also thank J. Staude for his helpful comments on our research. We thank H. Uitenbroek for the provision of his radiative transport code RH. Furthermore, we gratefully thank the anonymous referee for the substantial revision and valuable advice that helped to improve this paper.

References

Fontenla, J., White, O.R., Fox, P.A., Avrett, E.H., Kurucz, R.L.: 1999, Calculation of Solar irradiances. I. Synthesis of the Solar spectrum. *Astrophys. J.* **518**, 480–499.
Fröhlich, C., Bonnet, R.M., Bruns, A.V., Delaboudinière, J.P., Domingo, V., Kotov, V.A., Kollath, Z., Rashkovsky, D.N., Toutain, T., Vial, J.C.: 1988, IPHIR: The helioseismology experiment on the PHOBOS mission. In: Rolfe, E.J. (ed.) *Seismology of the Sun and Sun-Like Stars*, ESA-SP **286**, European Space Agency, Noordwijk, 359–362.
Lebedev, N.I., Kuznetsov, V.D., Oraevskii, V.N., Staude, J., Kostyk, R.I.: 2004, The helioseismological CORONAS-F DIFOS experiment. *Astron. Rep.* **48**, 871–875. doi:10.1134/1.1809399.
Pozrikidis, C.: 1998, *Numerical Computation in Science and Engineering*, Oxford University Press, New York.
Rybicki, G.B., Hummer, D.G.: 1991, An accelerated lambda iteration method for multilevel radiative transfer. I. Non-overlapping lines with background continuum. *Astron. Astrophys.* **245**, 171–181.
Staude, J., Dzhalilov, N.S., Zhugzhda, Y.D.: 1994, Radiation-hydrodynamic waves and global solar oscillations. *Solar Phys.* **152**, 227–239.
Toutain, T., Gouttebroze, P.: 1993, Visibility of solar p-modes. *Astron. Astrophys.* **268**, 309–318.
Uitenbroek, H.: 2001, Multilevel radiative transfer with partial frequency redistribution. *Astrophys. J.* **557**, 389–398. doi:10.1086/321659.
Vernazza, J.E., Avrett, E.H., Loeser, R.: 1976, Structure of the solar chromosphere. II. The underlying photosphere and temperature-minimum region. *Astrophys. J. Suppl.* **30**, 1–60.

Woodard, M.: 1984, Observations of low-degree modes from the solar maximum mission (extended abstract). In: Ulrich, R.K., Harvey, J., Rhodes, E.J. Jr., Toomre, J. (eds.) *Solar Seismology from Space*, JPL, Pasadena, 195–197.

Zhugzhda, Y.D.: 1983, Non-adiabatic oscillations in an isothermal atmosphere. *Astrophys. Space Sci.* **95**, 255–275.

Zhugzhda, Y.D.: 2006, Analytical signal as a tool for studying solar p-modes. *Astron. Lett.* **32**, 329–343.

Zhugzhda, Y.D., Dzhalilov, N.S., Staude, J.: 1993, Radiation-hydrodynamic waves in an optically non-grey atmosphere. *Astron. Astrophys.* **278**, L9–L12.

Zhugzhda, Y.D., Staude, J., Bartling, G.: 1996, Spectral darkening functions of solar p-modes – an effective tool for helioseismology. *Astron. Astrophys.* **305**, L33–L36.

Uncovering the Bias in Low-Degree p-Mode Linewidth Fitting

W.J. Chaplin · Y. Elsworth · B.A. Miller · R. New · G.A. Verner

Originally published in the journal Solar Physics, Volume 251, Nos 1–2, 189–196.
DOI: 10.1007/s11207-008-9215-7 © Springer Science+Business Media B.V. 2008

Abstract Obtaining reliable estimates of linewidths in the power spectra of low-degree p modes is problematic at low frequency. In this regime, the mode coherence time increases with decreasing frequency, often causing the modes to be unresolved in relatively long duration spectra. The signal-to-noise ratio is also less favourable at low frequency, resulting in fits to power spectra underestimating the true linewidth of the p modes owing to the tails of the Lorentzian peaks becoming dominated by the background noise. We use a numerical simulation approach to assess the effect of this bias on the fitted widths of p-mode peaks and calculate observational duration limits required to obtain an unbiased estimate of the p-mode linewidth as a function of frequency. This is done in four different cases, where the precision of the artificial data is set at 0.25, 0.50, 0.75, and 1.00 m s^{-1} by adding random scatter to increase the sample standard deviation per 40-second measurement. In all cases, the observational duration required to accurately obtain width estimates increases beyond that required for sufficient spectral resolution below a certain threshold frequency. For modes at ≈ 1500 µHz, with an amplitude of approximately ten times the background, obser-

---

Helioseismology, Asteroseismology, and MHD Connections
Guest Editors: Laurent Gizon and Paul Cally.

W.J. Chaplin · Y. Elsworth · B.A. Miller · G.A. Verner
School of Physics and Astronomy, University of Birmingham, Edgbaston, Birmingham, UK

W.J. Chaplin
e-mail: w.j.chaplin@bham.ac.uk

Y. Elsworth
e-mail: y.p.elsworth@bham.ac.uk

B.A. Miller
e-mail: b.a.miller@bham.ac.uk

R. New
Faculty of Arts, Computing, Engineering and Sciences, Sheffield Hallam University, Sheffield, UK
e-mail: R.New@shu.ac.uk

G.A. Verner (✉)
Astronomy Unit, School of Mathematical Sciences, Queen Mary University of London, London, UK
e-mail: g.verner@qmul.ac.uk

vations of up to 972 days are required to obtain an unbiased estimate of the linewidth. This is equivalent to ≈ 18 times the coherence time of the corresponding p modes.

Keywords Helioseismology: Observations · Oscillations: Solar

1. Introduction

Solar p-mode oscillations are thought to be stochastically excited by turbulence within the upper convection zone, with characteristic damping and excitation properties. The expected mode profile in the frequency domain of a stochastically driven, damped harmonic oscillator is essentially Lorentzian, with a degree of asymmetry in the solar case reflecting the correlated background noise (Roxburgh and Vorontsov, 1997; Nigam and Kosovichev, 1998) and the properties of the excitation source (Gabriel, 1992, 1993). The linewidths of the p-mode profiles, once sufficient resolution is available to resolve the peaks, reflect the mode damping rates. These damping rates can be used to constrain convection parameters and test the stochastic excitation theory (Rabello-Soares, Houdek, and Christensen-Dalsgaard, 1999).

The linewidths of p modes have a well-known dependence on frequency and decrease rapidly below 2000 µHz (see, for example, Chaplin et al., 1997, 1998). The mode coherence time, proportional to the inverse of the linewidth, is therefore longest for low-frequency modes. These modes are of particular interest because of their long lifetimes and thus well-constrained frequencies. The mode amplitudes and signal-to-noise ratios also decrease rapidly at low frequency. Obtaining reliable linewidths in this low-frequency domain is then complicated by fitting narrow profiles to modes that are not clearly prominent against the background noise.

It should be noted that the linewidth is not the only parameter that becomes problematic to obtain at low frequency. Models that fit the resonant peaks in helioseismic power spectra also have the mode amplitudes, frequencies, and asymmetries as free parameters, in addition to a background-noise component. While the mode frequency is a very robust fitted parameter, the fitted mode amplitude often exhibits an anticorrelation with the fitted linewidth; thus an underestimated mode width is usually accompanied by an overestimated amplitude. Determining the asymmetry of the peaks at low frequency is also affected by the decreasing signal-to-noise ratio. Chaplin et al. (1999) showed that the significance of the asymmetries obtained from fitting a 32-month power spectrum is very low for modes below 1800 µHz. The spectral background component contains contributions from a number of solar convection and instrumental effects (Harvey, 1985; Chaplin et al., 2005). When fitting the peaks in the power spectrum, this background can be modelled and fitted while iteratively fitting the mode peaks over a number of spectral windows (see, for example, Gelly et al., 2002). However, in the analysis carried out here, the fitting window was sufficiently narrow (54 µHz) such that a constant fitted background parameter for each mode pair provided enough freedom to the model without adding further complexity.

In previous work, Chaplin et al. (2002) set limits on the length of time series for which p modes are resolved or unresolved based on their coherence times. Here we use numerical simulations with artificial helioseismic time series to show that the fitted linewidths of modes that are properly resolved may be underestimated owing to the effect of decreasing signal-to-noise ratio at low frequencies.

2. Simulations with Artificial Data

The numerical simulations were based on 100 "solar Fitting at Low Angular Degree Group" (solarFLAG) artificial time series (Chaplin *et al.*, 2006). Each time series was created with a duration of 6048 days, a 100% duty cycle, and a cadence of 40 seconds, in agreement with the low-degree Doppler velocity observations of the Birmingham Solar Oscillations Network (BiSON). The artificial time series contained modes of degrees $0 \leq \ell \leq 3$ and frequencies $1100 \leq \nu \leq 5000$ µHz and also modes of degrees $4 \leq \ell \leq 5$ and frequencies $1500 \leq \nu \leq 4100$ µHz. The low-amplitude modes of degree $4 \leq \ell \leq 5$ were included to simulate a Sun-like low-degree power spectrum, with possible contamination from higher-degree modes, but were not fitted as part of the analysis. The mode parameters and correlated solar noise were set in accordance to the known parameters for the Sun. Solar-cycle effects on the mode parameters were not included in the simulations.

In real observations, many of the fitted parameters display systematic changes that are correlated with solar activity. The frequencies of *p* modes increase by as much as 40% of the mode linewidth (at 3000 µHz) from typical solar minimum to maximum activity. Meanwhile, the linewidths themselves increase by up to 20%, and the mode amplitudes decrease by up to 50% (Gelly *et al.*, 2002). The variation in each of these parameters is also a strong function of the mode frequency itself. Chaplin *et al.* (2008) carried out an analysis into whether the distortion introduced into the mode profiles by the variation of frequency with solar activity introduces bias into the fitted parameters if one assumes an asymmetric Lorentzian limit spectrum. A small bias was found to be introduced into the fitted linewidth and amplitude. Although not including temporal changes in the mode parameters removes some of the Sun-like accuracy from the artificial data, this was deemed necessary to infer that any bias observed comes from the fits themselves rather than from activity-driven distortion of the mode profiles.

The stochastic-excitation noise was different for each of the time series. Additional normally distributed random scatter was also added to each time series to simulate limited observational precision. Random noise data with sample standard deviations per 40-second sample of 0.25, 0.50, 0.75, and 1.00 m s^{-1} were added to each artificial time series to give four different observational scenarios. The lowest noise levels were indicative of solar Doppler velocity observational precision, whereas the higher noise levels were of the order of the precision expected for spectroscopic Doppler velocity observations of oscillations on nearby bright stars (Grundahl *et al.*, 2006). Adding normally distributed scatter to the time series data has the effect of whitening the noise background, which otherwise has a red noise profile. Examples of 432-day solarFLAG power spectra with added white noise background are shown in Figure 1. Adding white noise to the solarFLAG time series caused the amount of reduction in signal-to-noise ratio in each of the scenarios to vary as a function of frequency. The average ratio of the mode amplitudes to the fitted backgrounds, for modes of degree $\ell = 0$, is shown in Figure 2 for each of the noise scenarios over the low-frequency range.

Each artificial time series was divided into two sets of shorter series of increasing duration. One of the sets used 5940 days of data from each series to create ten series of evenly spaced durations from 108 to 1080 days. This set was specifically for investigating the fitted widths of low-frequency modes, which require extended observations to resolve. The second set used 486 days of data to create evenly spaced subsets from 13.5 to 108 days in duration. This set was used to investigate the widths of higher frequency modes, which can be resolved in short-duration observations. These are referred to as the "long" and "short" sets, respectively. Each subseries within the long and short sets is independent; however,

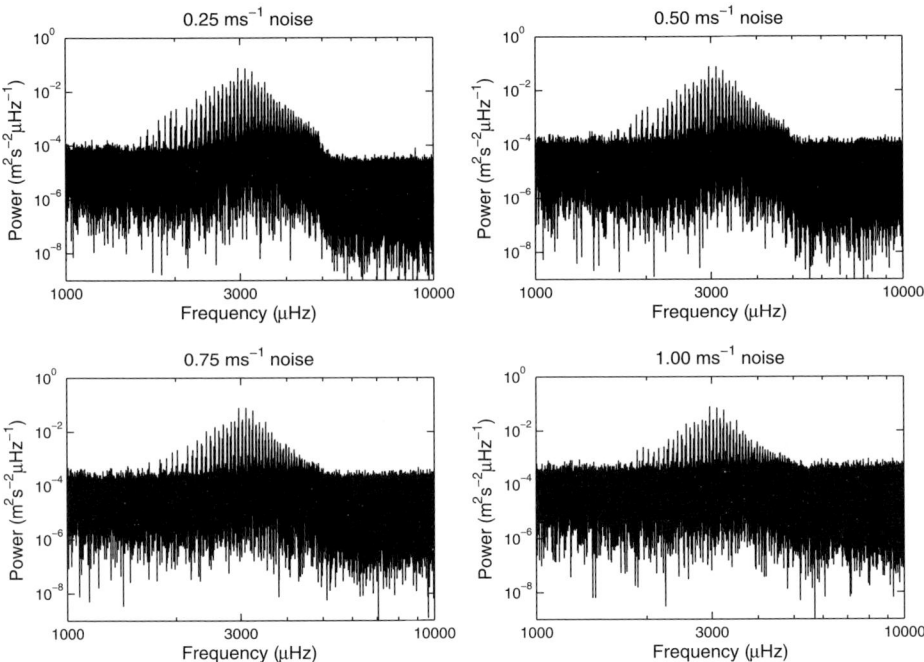

Figure 1 Power spectra of 432-day solarFLAG artificial time series with increasing amounts of added white noise. Noise levels are given in the title of each frame.

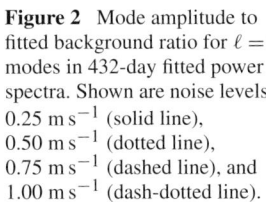

Figure 2 Mode amplitude to fitted background ratio for $\ell = 0$ modes in 432-day fitted power spectra. Shown are noise levels of 0.25 m s^{-1} (solid line), 0.50 m s^{-1} (dotted line), 0.75 m s^{-1} (dashed line), and 1.00 m s^{-1} (dash-dotted line).

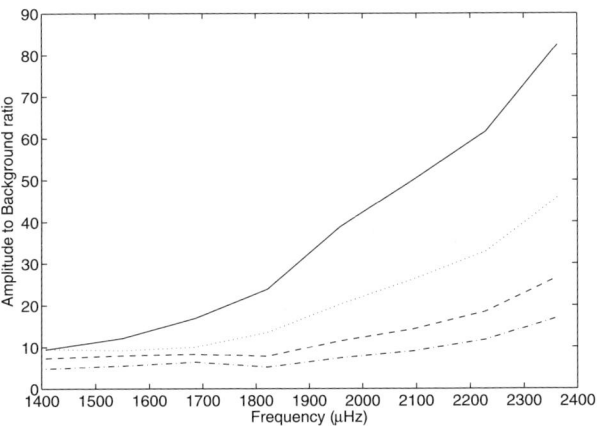

the short subsets overlap with some of the data from the 1080-day subset. Additional short sets of observations were required because modes with frequencies above ≈ 2000 µHz displayed no fitting bias when using 108-day data. The shorter sets allowed the determination of any bias present at higher frequencies. For low-frequency modes, extended observations are required; therefore a duration interval of 108 days was chosen as a compromise between resolution and the processing time needed to generate the artificial time series.

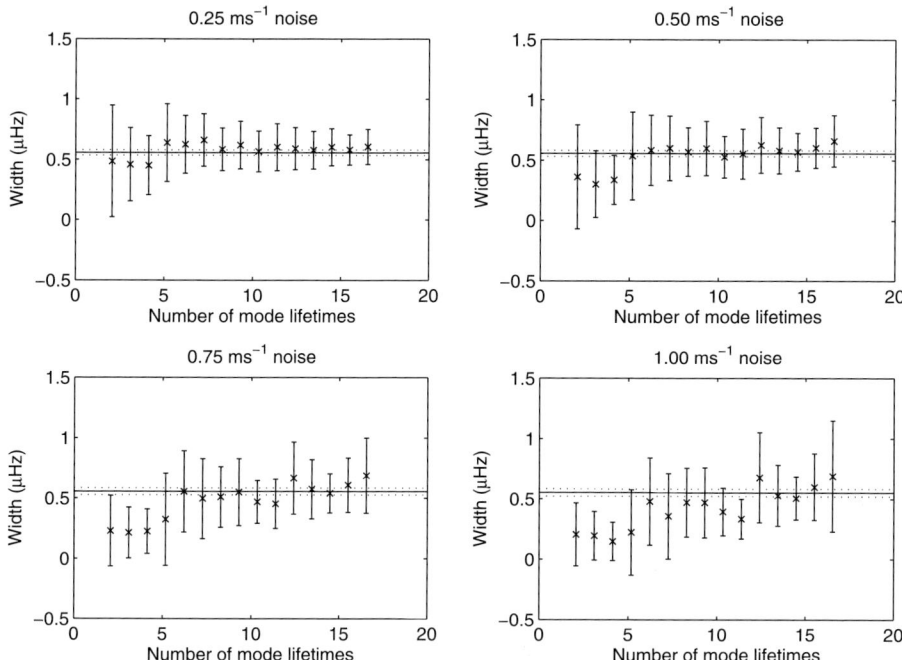

Figure 3 Fitted widths for the $(\ell, n) = (0, 14)/(2, 13)$ mode pair in the short-duration set as a function of the number of realisations of these modes. The noise levels are given in the title of each plot. Each point is the mean of the fits to 100 simulations, with the error bars showing the one-σ standard deviation. The solid lines show the mean fitted width to the complete 6048-day time series, with the dotted lines showing the corresponding one-sigma standard deviation.

The resonant peaks in the power spectra of each of the time series in the short and long sets, along with those from the complete 6048-day series, were fitted in pairs to asymmetric Lorentzian profiles for modes of degrees $0 \leq \ell \leq 3$ by using the usual maximum likelihood technique (Chaplin et al., 1999). In the mode-pair fitting, adjacent modes and individual m-components in modes of degree $\ell > 0$ were assumed to have the same linewidth.

3. Bias in Fitted Widths

The simulations often showed an underestimation of the true width of the p-mode peaks when the observational duration was less than four times the mode lifetime. This bias was more pronounced when the additional noise level in the time series was increased. An example of this can be seen in Figure 3 for the $(\ell, n) = (0, 14)/(2, 13)$ mode pair at ≈ 2093 µHz.

To find the minimum time series length required to obtain an unbiased estimate of the mode width, the fitted mode widths from the simulations were used to determine the duration at which one sigma (68.27%) of the simulation fits were within their one-sigma formal error of the input value for the width of the corresponding mode. This was carried out separately for each noise level, and for the long- and short-duration sets. The minimum duration required was compared with the time required to observe the mode for three mode lifetimes,

Figure 4 Observational duration required to obtain an unbiased measure of the fitted mode width in the cases where the added noise level is 0.25 m s^{-1} (crosses), 0.50 m s^{-1} (squares), 0.75 m s^{-1} (diamonds), and 1.00 m s^{-1} (triangles). The solid line shows the duration required to observe for three complete coherence times. Error bars indicate the difference in duration between subsequent time series in the short ($\leq$108 days) and long ($>$ 8 days) sets.

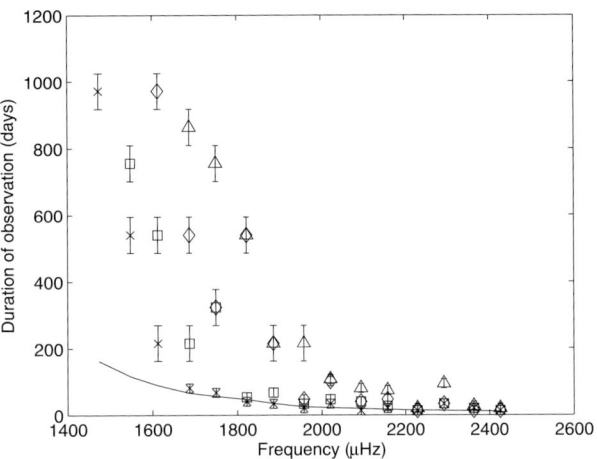

where the lifetime was defined as

$$\tau = \frac{1}{\pi \Delta},\qquad(1)$$

where Δ is the mode width. This definition of mode lifetime depends on an exponentially damped model of solar oscillations (Kennedy, 1998). Because of the factor of π in Equation (1), a time series of length corresponding to three mode lifetimes is equivalent to a frequency resolution in the power spectrum of the order of the width of the mode. This provides a useful reference and indicates the length of time series required if the only limitation on fitting an unbiased width is sufficient resolution in the power spectrum. The minimum observational durations, as a function of frequency, and a line showing three times the corresponding mode lifetimes are shown in Figure 4.

For the cases where the additional noise is low (*i.e.*, 0.25 and 0.50 m s^{-1}) at frequencies above 1800 µHz the minimum observational duration required to obtain an unbiased estimate of the mode linewidth is in agreement with the duration required to resolve the mode. For the simulations with higher noise levels, this frequency threshold increases to 2200 µHz for a noise level of 0.75 m s^{-1} and 2400 µHz when the additional noise is 1.00 m s^{-1}.

Poor results at low frequency are demonstrated visually in Figure 4 by the fact that the symbols depart significantly from the solid line in this regime. Some of the low-frequency modes could not be reliably fitted in the higher-noise sets in time series of less than 1080 days. In the 0.25 m s^{-1} noise case, the linewidth of the $(\ell, n) = (1, 9)/(3, 8)$ mode pair at ≈ 1472 µHz was underestimated in spectra of less than 972 days, equivalent to 18 times the coherence time of these modes.

4. Discussion

We have shown that the linewidths of low-frequency p modes may be underestimated despite the time series containing many realisations of the modes. At higher frequencies, this underestimation bias is not present and the limitation on obtaining reliable linewidths depends on the frequency resolution of the power spectrum relative to the linewidth of the mode.

For spectra of sufficient duration to resolve the low-frequency modes, this bias is introduced as a result of the decreasing signal-to-noise ratio present in the data. At low frequency, the mode amplitudes decrease while the noise background increases. In this regime, the tails of the fitted Lorentzian functions become poorly determined owing to the amplitude of the background relative to the mode peaks. A maximum likelihood minimisation effectively fits part of the tail of the mode to the background, reducing the fitted mode linewidth. For modes at higher frequencies, the tails of the mode profiles remain significantly above the background noise and can be fitted appropriately. The mode peak amplitude-to-background ratio at low frequency may be sufficient to obtain a reliable estimate of the centroid frequency; however, it is the mode tail amplitude-to-background ratio that is also important when obtaining a nonbiased estimate of the linewidth. As the tails of the mode peaks become poorly defined, the mode linewidths become underestimated.

The obvious implication of this bias is where low-frequency mode linewidths are obtained from spectra that sample a small number of mode lifetimes. It should be noted that the lowest frequency linewidths of Chaplin *et al.* (1997), which were determined from 32 months of solar data, are shown here to be unbiased owing to the extended observations.

Accurate determination of linewidths at low frequency can be used to test models of the physical contributions to the damping of solar p modes. Goldreich and Murray (1994) suggest that scattering should dominate the damping over absorption at low frequency. The low-degree linewidths of Chaplin *et al.* (1997), providing linewidths down to $\nu = 1472.8$ µHz, could not discriminate between scattering and absorption models over the low-frequency range. With increased resolution from longer-duration observations, further low-frequency mode detections can be fitted to asymmetric Lorentzian profiles to obtain unbiased estimates of the mode damping at lower frequencies, which may facilitate the discrimination between these models.

In the case where the observational precision is lower than that obtained for solar data, in forthcoming asteroseismic observations on Sun-like stars for example, the width bias is more pronounced owing to the increased background noise underlying p-mode peaks at higher frequency. Despite this bias at low frequency, many of the modes at higher frequency can be fitted to obtain reliable linewidths in short-duration sets with low signal-to-noise ratio. For solar observations of 34 days, one can expect to obtain unbiased linewidths for p modes with frequencies as low as 1885 µHz by using time series data with 0.25 m s^{-1} precision, 2157 µHz with 0.50 m s^{-1} precision, 2229 µHz with 0.75 m s^{-1} precision, and 2363 µHz with 1.00 m s^{-1} precision. The characteristics of power spectra from asteroseismic data will be different from that of the Sun, particularly from those data obtained from intensity measurements that have a significantly higher noise component. However, it is reassuring that unbiased mode lifetimes can be obtained from relatively short observations with limited signal-to-noise ratio.

Acknowledgements We acknowledge the solar Fitting at Low Angular degree Group (solarFLAG) for use of their artificial helioseismic data. The authors acknowledge the support of STFC.

References

Chaplin, W.J., Elsworth, Y., Isaak, G.R., McLeod, C.P., Miller, B.A., New, R.: 1997, *Mon. Not. Roy. Astron. Soc.* **288**, 623.
Chaplin, W.J., Elsworth, Y., Isaak, G.R., Lines, R., McLeod, C.P., Miller, B.A., New, R.: 1998, *Mon. Not. Roy. Astron. Soc.* **298**, L7.
Chaplin, W.J., Elsworth, Y., Isaak, G.R., Miller, B.A., New, R.: 1999, *Mon. Not. Roy. Astron. Soc.* **308**, 424.
Chaplin, W.J., Elsworth, Y., Isaak, G.R., Miller, B.A., New, R.: 2002, *Mon. Not. Roy. Astron. Soc.* **330**, 731.

Chaplin, W.J., Elsworth, Y., Isaak, G.R., Miller, B.A., New, R., Pintér, B.: 2005, *Mon. Not. Roy. Astron. Soc.* **359**, 607.
Chaplin, W.J., Appourchaux, T., Baudin, F., Boumier, P., Elsworth, Y., Fletcher, S.T., Fossat, E., García, R.A., Isaak, G.R., Jiménez, A., *et al.*: 2006, *Mon. Not. Roy. Astron. Soc.* **369**, 985.
Chaplin, W.J., Elsworth, Y., New, R., Toutain, T.: 2008, *Mon. Not. Roy. Astron. Soc.* **384**, 1668.
Gabriel, M.: 1992, *Astron. Astrophys.* **265**, 771.
Gabriel, M.: 1993, *Astron. Astrophys.* **274**, 935.
Gelly, B., Lazrek, M., Grec, G., Ayad, A., Schmider, F.X., Renaud, C., Salabert, D., Fossat, E.: 2002, *Astron. Astrophys.* **394**, 285.
Goldreich, P., Murray, N.: 1994, *Astrophys. J.* **424**, 480.
Grundahl, F., Kjeldsen, H., Christensen-Dalsgaard, J., Arentoft, T., Frandsen, S.: 2006, *Commun. Asteroseismol.* **150**, 1.
Harvey, J.: 1985, In: Rolfe, E., Battrick, B. (eds.) *Future Missions in Solar, Heliospheric and Space Plasma Physics* **SP-235**, ESA, Noordwijk, 199.
Kennedy, J.R.: 1998, *Solar Phys.* **181**, 265.
Nigam, R., Kosovichev, A.G.: 1998, *Astrophys. J.* **505**, L51.
Rabello-Soares, M.C., Houdek, G., Christensen-Dalsgaard, J.: 1999, In: Guinan, E.F., Montesinos, B. (eds.) *Theory and Tests of Convective Energy Transport* **CS-173**, Astron. Soc. Pac., San Francisco, 301.
Roxburgh, I.W., Vorontsov, S.V.: 1997, *Mon. Not. Roy. Astron. Soc.* **292**, L33.

Analysis of MDI High-Degree Mode Frequencies and their Rotational Splittings

M.C. Rabello-Soares · S.G. Korzennik · J. Schou

Originally published in the journal Solar Physics, Volume 251, Nos 1–2, 197–224.
DOI: 10.1007/s11207-008-9231-7 © Springer Science+Business Media B.V. 2008

Abstract We present a detailed analysis of solar acoustic mode frequencies and their rotational splittings for modes with degree up to 900. They were obtained by applying spherical harmonic decomposition to full-disk solar images observed by the Michelson Doppler Imager onboard the *Solar and Heliospheric Observatory* spacecraft. Global helioseismology analysis of high-degree modes is complicated by the fact that the individual modes cannot be isolated, which has limited so far the use of high-degree data for structure inversion of the near-surface layers ($r > 0.97 R_\odot$). In this work, we took great care to recover the actual mode characteristics using a physically motivated model which included a complete leakage matrix. We included in our analysis the following instrumental characteristics: the correct instantaneous image scale, the radial and non-radial image distortions, the effective position angle of the solar rotation axis, and a correction to the Carrington elements. We also present variations of the mode frequencies caused by the solar activity cycle. We have analyzed seven observational periods from 1999 to 2005 and correlated their frequency shift with four different solar indices. The frequency shift scaled by the relative mode inertia is a function of frequency alone and follows a simple power law, where the exponent obtained for the p modes is twice the value obtained for the f modes. The different solar indices present the same result.

Keywords Helioseismology, observations · Instrumental effects · Oscillations, solar · Solar cycle, observations

---

Helioseismology, Asteroseismology, and MHD Connections
Guest Editors: Laurent Gizon and Paul Cally

M.C. Rabello-Soares (✉) · J. Schou
W.W. Hansen Experimental Physics Laboratory, Stanford University, 455 Via Palou, Stanford, CA 94305, USA
e-mail: csoares@sun.stanford.edu

J. Schou
e-mail: schou@sun.stanford.edu

S.G. Korzennik
Harvard-Smithsonian Center for Astrophysics, 60 Garden St, Cambridge, MA 02138, USA
e-mail: sylvain@cfa.harvard.edu

1. Introduction

The central frequencies of solar acoustic modes, which are obtained using spherical harmonic decomposition, have been successfully used to determine the solar interior structure to as close as 21 Mm to the solar surface ($r < 0.97 R_\odot$) using modes with angular degrees $\ell \leq 300$ (*e.g.*, Gough *et al.*, 1996). The inclusion of high-degree modes (*i.e.*, up to $\ell = 1000$) has the potential to improve dramatically the inference of the sound speed and the adiabatic exponent (Γ_1) in the outermost 2 to 3% of the solar radius, allowing construction of localized kernels as close to the solar surface as 1.75 Mm (Rabello-Soares *et al.*, 2000). The effects of the equation of state, through the ionization of hydrogen and helium, are felt most strongly in the outer layers of the Sun, making this shallow region of particular interest. Furthermore, dynamical effects of convection, and the processes that excite and damp the solar oscillations, are predominantly concentrated in this region. Although the spatial resolution of modern helioseismic instruments allows us to observe oscillation modes up to $\ell = 1000$ and higher, only a small fraction of them are currently used ($\ell \leq 300$). Unfortunately, analysis of high-degree data is complicated by the fact that the individual modes cannot be isolated (*e.g.*, Rabello-Soares, Korzennik, and Schou, 2001).

Solar structure is not static, but changes over the solar cycle. It is well known that the mode frequencies change with solar activity. It seems that the responsible mechanism is restricted to the outer layers of the Sun (Libbrecht and Woodard, 1990), where the high-degree modes are confined. However, at the moment, there is no general agreement as to the precise physical mechanism that gives rise to the frequency variation. It is likely a product of the change in the subphotospheric, small-scale, magnetic field strength with the solar activity cycle (*e.g.*, Goldreich *et al.*, 1991). Accordingly to Dziembowski and Goode (2004), the frequency shift is easily explained in terms of a variation in the turbulent velocities associated with the magnetic field variation rather than the direct effect of the magnetic field itself. Li *et al.* (2003), using models of the structure and evolution of the Sun, found that turbulence near the surface of the Sun plays a major role in solar variability, and only a model that includes a magnetically modulated turbulent mechanism can agree with the observed correlation between the frequency shift and the solar cycle. In such a dynamic model, the evolution of the subsurface layers of the Sun through the activity cycle plays an important role.

The frequencies of the global modes give the radius and latitude (r, θ) part of the structure. While, local helioseismic techniques such as ring-diagram analysis (Hill, 1988) allow the determination of the three-dimensional structure of the Sun, allowing the study of localized areas in the solar surface, such as those in active regions. Large variations of the mode frequencies observed in and near sunspots, in comparison to magnetically quiet regions, are well known to be correlated with variations in the average surface magnetic field between the corresponding regions (*e.g.*, Rajaguru, Basu, and Antia, 2001 and Rabello-Soares, Bogart, and Basu, 2008). Whether the frequencies are changed directly by the magnetic field or indirectly through an associated change in the solar structure (such as a pressure change) is still a matter of debate. A detailed analysis of the frequency-shift characteristics will hopefully help understand their physical origin.

Basu, Antia, and Bogart (2004), using ring-diagram analysis, found that the sound speed is lower in the immediate subsurface layers of an active region than of a magnetically quiet region, while the opposite is true for depths below about 7 Mm. However, Basu and Antia (2003), using global analysis, have found no observable structural changes in the inner layers of the Sun below a depth of 21 Mm associated with the magnetic-activity induced frequency shifts. They were, however, unable to get closer to the solar surface due to the lack of high-degree modes in their mode set. High-degree global analysis is important to complete the picture of the near-surface layers. Besides, the determination of high-degree frequencies using different methods allows us to check the results against each other giving confidence in the results and avoiding systematic errors. We should point out that, although the high-degree modes have short lifetimes (one to ten hours for $100 \leq \ell \leq 600$ according to Burtseva *et al.*, 2007) propagating only locally, they are averaged over most of the solar surface using spherical harmonic decomposition (over a relatively long time series) and thus can still be called global analysis.

In the traditional global helioseismology data-analysis methodology, a time series of full-disk Doppler solar images is decomposed into spherical harmonic coefficients, characterized by its degree (ℓ) and its azimuthal order (m). Each coefficient time series is Fourier transformed, and the order of the radial wave function (n) gets separated in the frequency domain. However, a spherical harmonic decomposition is not orthonormal over less than the full sphere – *i.e.*, the solar surface that can be observed from a single view point – resulting in what is referred to as spatial leakage. At low and intermediate degrees, most of these leaks are separated in the frequency domain from the target mode (except for some m leaks) and individual modes can be identified and fitted. However, at high degrees, the spatial leaks lie closer in frequency (due to a smaller mode separation) and, at high frequency, the modes become wider (as the mode lifetimes get smaller), resulting in the overlap of the target mode with the spatial leaks that merges individual peaks into ridges (see Figure 1 in Rabello-Soares, Korzennik, and Schou, 2001). The characteristics of the resulting ridge (central frequency, amplitude, *etc.*) do not correspond to those of the underlying target mode. This has so far hindered the estimation of unbiased mode parameters at high degrees.

To recover the actual mode characteristics, we need a very good estimation of the relative amplitude of the spatial leaks present in a given (ℓ, m) power spectrum, also known as the leakage matrix, which in turn requires a very good knowledge of the instrumental properties (Rabello-Soares, Korzennik, and Schou, 2001). In our previous papers (Rabello-Soares, Korzennik, and Schou, 2001 and Korzennik, Rabello-Soares, and Schou, 2004, hereafter KRS), we described in detail the large influence of the instrumental properties on the amplitude of the leaks and as a consequence in the determination of unbiased high-degree mode parameters.

In the following, we will first describe the data used in this analysis and the ridge-to-mode correction applied to them (Sections 2 and 3). In Section 4, we will discuss the influence on the mode parameters of each of the instrumental properties that were included in the spherical harmonic decomposition of the solar images. We then analyze in Section 5 the characteristics of the high-degree mode frequencies and their rotational splittings obtained in this work. Finally, in Section 6, we analyze the frequency variation induced by the solar cycle.

2. Observations and Ridge-Parameter Extraction

The data used in this work consist of full-disk Dopplergrams obtained at a one-minute cadence by the Michelson Doppler Imager (MDI) onboard the *Solar and Heliospheric Observatory* (SOHO). We have used two distinct sets of data. One while MDI was operating in its 4″ resolution mode, allowing the detection of oscillation modes up to $\ell \approx 1500$, which we will call from now on the high-ℓ data set. This is the so-called *Dynamics Program* observing mode, which is available every year for two to three months. The second one (hereafter referred to as the medium-ℓ data set) using MDI *Structure Program* that provides almost continuous coverage year round. In this observing mode, the original Dopplergrams are convolved, onboard the MDI instrument, with a Gaussian and subsampled on a 200×200 grid,

Table 1 Details of the analyzed time series. The corresponding relative mean values of solar UV spectral irradiance and their standard deviation are also listed as an indication of the solar activity level.

Year	Starting date high-ℓ set	Duration high-ℓ set	Solar index rel. to max. (in %)	Starting date medium-ℓ set
1999	Mar. 13	77 days	40 ± 13	Feb. 03
2000	May 27	45 days	69 ± 11	Apr. 10
2001	Feb. 28	90 days	61 ± 14	Jan. 23
2002	Feb. 23	72 days	80 ± 8	Mar. 31
2003	Oct. 18	38 days	51 ± 17	Oct. 28
2004	Jul. 04	65 days	36 ± 11	Aug. 11
2005	Jun. 25	67 days	30 ± 9	May 26

thus reducing the telemetry requirements but also limiting the spatial resolution to modes with degree $\ell \leq 300$.

For the high-ℓ set, we computed the spherical harmonic decomposition of the MDI images for modes with $100 \leq \ell \leq 900$. The resulting time series were Fourier transformed in small segments (4096 minutes) whose spectra were averaged to produce an averaged power spectrum with a low but adequate frequency resolution to fit the ridge while reducing the realization noise. The number of averaged spectra varies with the year, from 12 segments in 2003, to 30 in 2001. Most of the known instrumental effects relevant to the high-degree analysis were included in the spatial decomposition, and they will be described in Section 4. The peaks in each (ℓ, m) spectrum were then fitted using an asymmetric Lorentzian profile with an additive background term, given by ten to the power of a second-degree polynomial in frequency (Equation (5) in KRS). Since the number of segments used in the average of each (ℓ, m) spectrum is large enough, its χ^2 distribution can be approximate by a Gaussian distribution and a least-square fitting was used. The fitted Lorentzian profile is characterized by the following parameters: frequency ($\nu_{n,\ell,m}$), amplitude ($A_{n,\ell,m}$), width ($\gamma_{n,\ell,m}$), and asymmetry ($\alpha_{n,\ell,m}$). The asymmetric profile used is equivalent to the one defined by Nigam and Kosovichev (1998) where their asymmetric parameter is equal to $\alpha/(2 - \alpha)$. The frequency splittings were parametrized in terms of Clebsch–Gordan coefficients up to a_6 (Ritzwoller and Lavely, 1991). The number of modes analysed was reduced without affecting the results of this paper and hence was easy to handle. The fitting was carried out only for every tenth ℓ and only for some 50 equally spaced m values at each ℓ. The central frequency, i.e., the frequency free of splitting effects, is taken to be the frequency given by $m = 0$ in the splitting parametrization. Since the even splitting coefficients are zero on average, it is the same as the mean frequency (averaged over m). Notice that the fitted parameters are ridge parameters and do not correspond to the associated mode parameters as discussed in the introduction. In this study, high-ℓ time series available from 1999 until 2005 were used and their properties are listed in Table 1.

Modes with $\ell > 300$ are the focus of this work, and therefore it is centered on the analysis of the high-ℓ set. We used however the results obtained by one of us analyzing the medium-ℓ time series (Schou, 1999) to compare with and complement our analysis. Using 72-day long time intervals, the central frequency and splitting coefficients for a given (n, ℓ) mode were determined directly by fitting symmetric Lorentzian profiles to its power spectra (Schou, 1999). Every p mode up to $\ell \approx 200$ (up to $\ell \approx 300$ for the f modes) and every m were fitted. The 72-day time series that best overlap in time with the high-ℓ time series were selected and are listed in Table 1. Only modes with $\ell \geq 20$ were used. The mode coverage of both analyses is illustrated in Figure 1.

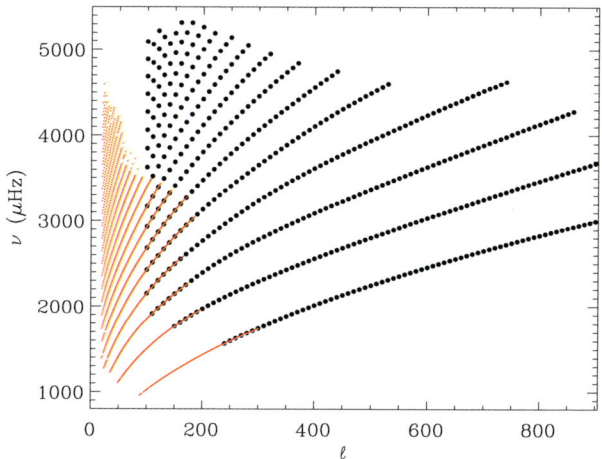

Figure 1 Coverage, in an $\ell - \nu$ diagram, of the medium-ℓ (red) and high-ℓ (black) mode parameters for 2005. The coverage is very similar for all epochs.

In the medium-ℓ analysis, the instrumental effects described in Section 4 were not taken into account, and neither were the distortion of the eigenfunctions by the solar differential rotation nor the horizontal component of the oscillation (both described in Section 3). However, in the frequency and degree ranges of the medium-ℓ analysis, the spatial leaks are well separated from the target mode and individual modes can be identified and fitted, making the above-mentioned effects not as crucial as they are for the high-ℓ analysis. Another relevant difference between the medium- and high-ℓ analysis is that the medium-ℓ power spectra were fitted using symmetric profiles, which is well known to lead to systematic errors in the frequency measurements (*e.g.*, Toutain *et al.*, 1998; Basu and Antia, 2000). Larson and Schou (2008) have recently reanalyzed the medium-ℓ time series and reported that several of these physical effects result in highly significant changes in the mode parameters. Their Figure 1 shows the total correction to be applied to the medium-ℓ frequency and splitting coefficients, a_1 and a_2, used here. Their Figure 2 illustrates the frequency changes due to each of these effects. The first two panels correspond to the distortion of the eigenfunctions by the solar differential rotation and the horizontal component of the leakage matrix respectively. The third, fourth, and fifth panels correspond to the instrumental effects described here in Sections 4.1, 4.2, and 4.3 respectively. The panels at the bottom show the difference obtained between fitting symmetric and asymmetric Lorentzians. Although these corrections are significant, they correspond to very small variations in the results presented in this paper and do not affect our conclusions, as it will be discussed later (Section 5).

3. Ridge-to-Mode Correction

Our methodology to recover the mode characteristics from the ridges observed at high-degree and high-frequency power spectra consists in generating and fitting a sophisticated model of the underlying modes that contribute to the ridge power distribution and deduce the offset ($\Delta°$) between the ridge properties and the target mode (Korzennik, 1998). A synthetic power spectrum is computed for each (ℓ, m) mode consisting of several asymmetric Lorentzians for each n: one for the target mode (ℓ, m) and one for each spatial leak (ℓ', m') with a relative amplitude given by the leakage matrix. It has the same frequency resolution as the high-ℓ set power spectra and it is generated for the same set of (ℓ, m) modes.

We used the complete leakage matrix (*i.e.*, radial and horizontal components) where the horizontal-to-vertical displacement ratio is given by Christensen-Dalsgaard (2003): $GM_\odot L/(R_\odot^3 \omega_{n,\ell}^2)$, where G is the gravitational constant, $M_\odot$ is the solar mass, $R_\odot$ is the solar radius, ω is the cyclic frequency ($\omega = 2\pi\nu$), and $L^2 = \ell(\ell+1)$. For each (ℓ, m) synthetic power spectrum, we have taken into account the contribution of the spurious modes (ℓ', m') that obey the following expressions: $|m' - m| \leq 10$ and $|\ell' - \ell| \leq \ell_d$, where $\ell_d = 12$ for $\ell \leq 600$ and ℓ_d is equal to $0.02\,\ell$ rounded to the closest integer for $\ell > 600$. Spurious modes with (ℓ', m') values that differ from those of the power spectrum (ℓ, m) by more than the amount specified in the equations above have a very small relative amplitude and their contribution to the synthetic power spectrum profile is negligible, *i.e.*, their inclusion or not does not affect the fitted parameters of the peaks in the power spectrum (Rabello-Soares, Korzennik, and Schou, 2005). We also included in our model the distortion of the eigenfunctions by the solar differential rotation as described in Woodard (1989), using the values given by Schou *et al.* (1998) for the solar rotation. The coefficients in this superposition become negligible when $|\ell - \ell'| \geq \ell_c$ where $\ell_c = 10$ for $\ell \leq 400$ and the next even integer to $3 + 0.02\,\ell$ for $\ell > 400$ (Rabello-Soares, Korzennik, and Schou, 2005).

The profile resulting from the overlap of several profiles of nearby spatial leaks is reasonably well modeled by a single profile when the ratio of the mode width to their separation (given by $\partial\nu/\partial\ell$) becomes large. The difference is smaller than the rms of the observed residuals to the fit, 5% – 7% at all degrees, which are dominated by the realization noise (KRS). The synthetic power spectrum is then fitted following the methodology applied to the high-ℓ set (Section 2), providing the ridge central frequency $\nu_{n,\ell}^{\text{model}}$ and its splittings $a_{i_{n,\ell}}^{\text{model}}$ ($i = 1, 6$). The offset is given by the difference between the modeled ridge and the target mode parameter given by the input value of the parameter used to generate the synthetic power spectra. For a given mode (n, ℓ), the offsets in central frequency can be then written as

$$\Delta^\circ \nu_{n,\ell} = \nu_{n,\ell}^{\text{model}} - \nu_{n,\ell}^{\text{input}}. \tag{1}$$

The offsets in each of the fitted a-coefficients are obtained in the same manner. Realistic input values based on observed mode parameters were used (as described in KRS). The input linewidth is given by the square root of $\gamma_{\text{mode}}^2 + \gamma_{\text{WF}}^2$, where γ_{mode} is an estimate of the mode linewidth and γ_{WF} is the width of the window function in our case given by the length of the high-ℓ set time string (*i.e.*, 4096 minutes). This expression was obtained by Korzennik (1990) assuming that the observed power spectrum is a convolution of the "true" power spectrum (*i.e.*, the one that would have been obtained with a infinite time series), with the power spectrum of the window function both represented by Gaussian profiles, which is adequate for the purpose at hand.

If the leakage matrix is correct and complete, our simulations should adequately estimate the parameter offsets (Δ°) and we would be able to obtain corrected mode parameters from the observed ridge. Thus, in the case of the central frequency, we would have

$$\nu_{n,\ell}^{\text{corrected}} = \nu_{n,\ell}^{\text{observed}} - \Delta^\circ \nu_{n,\ell}. \tag{2}$$

The corrected mode a-coefficients are obtained as in Equation (2).

Figure 2 shows the estimated offsets for the central frequency and the splitting coefficients a_1, a_2, and a_3. They are primarily a function of frequency. Except for a_2, the offsets are quite large in comparison with the observed fitting uncertainties of the corresponding parameter which, for the central frequency, a_1 and a_3 coefficients, are usually smaller than 0.5 µHz, 1 nHz, and 0.8 nHz, respectively. The width and amplitude offsets will be described

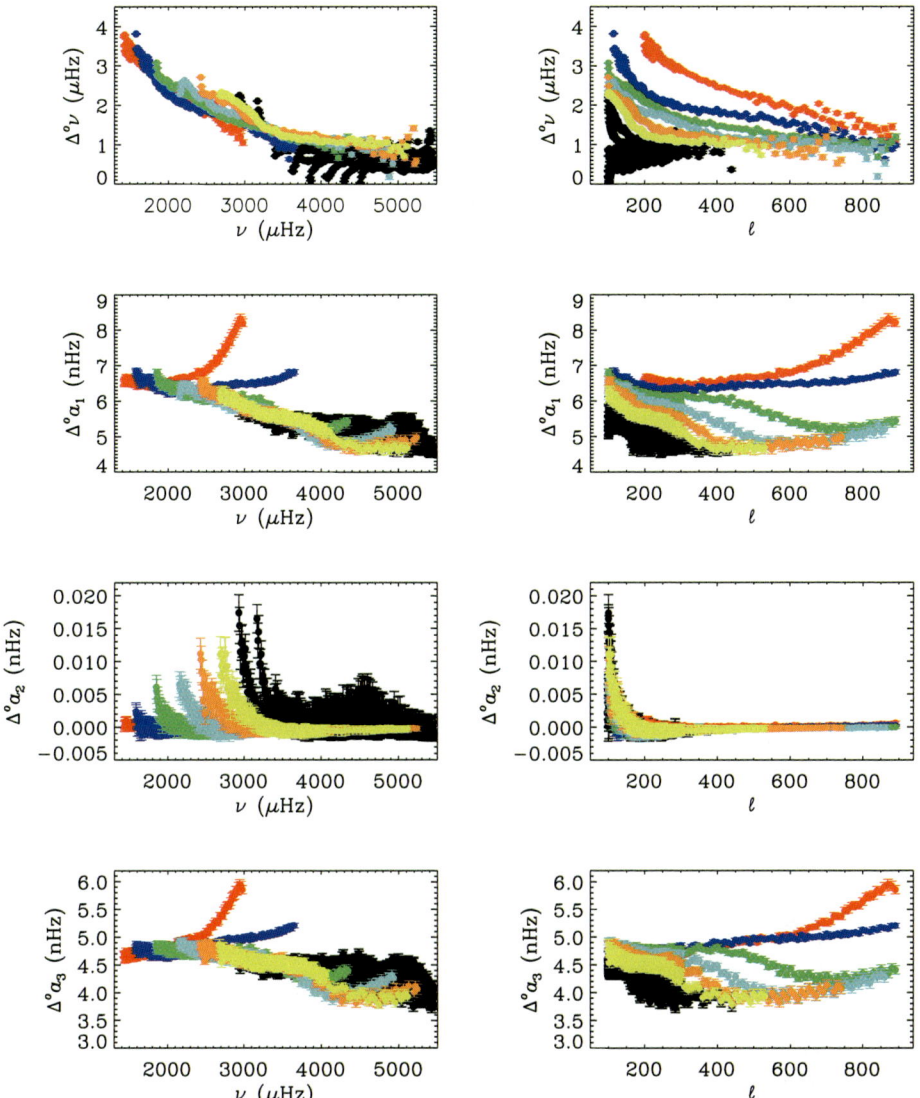

Figure 2 Estimated offsets for the central frequency (ν) and splitting coefficients (a_1, a_2, and a_3). The error bars correspond to the uncertainties when fitting the synthetic power spectra. Modes with $n \leq 5$ are in color with the f modes in red.

in a future paper. Accordingly to Korzennik (1998), the asymmetry of the ridge seems to be the same as the asymmetry of the underlying mode (see also KRS).

Another method to recover the mode characteristics from the observed ridges is to fit a sum of Lorentzians or asymmetric profiles to a given ridge: one for the target mode and one for each spatial leak that has a significant amplitude (Reiter *et al.*, 2002, 2003, 2004). Such an approach is likely to be particularly appropriate in the transition region between well-resolved modes and ridges, where the spatial leaks start to overlap with the target mode but do not yet blend fully into a ridge. The reasoning behind the method used here is that, when the spatial leaks are completely blended into a ridge, there is not enough information in the power spectra to justify modeling the individual spatial leaks as part of the fitting of a given observed spectrum. The difference in the profile of a sum of overlapped asymmetric profiles and a single asymmetric profile is much smaller than the observed fitting residuals. This method is significantly less computationally demanding since the profiles fitted are much simpler. As a result we can more easily check its reliability, by using different leakage matrices or time series produced with a different spatial decomposition, and quantitatively validating the corrections, as described in Section 4 and KRS. This validation step ensures the reliability of the method and allows us to estimate a quantitative upper limit on any residual bias, a crucial step in the intricate high-ℓ analysis. Note that the method described by Reiter *et al.* also relies on a very good estimation of the leakage matrix to obtain unbiased mode parameters. Indeed, the same leakage matrices have been used for both methods and it is likely that both the random and systematic errors will be quite similar where the modes are fully blended into ridges.

4. Influence of Instrumental Properties on the High-Degree Power Spectra

In the determination of unbiased high-degree mode parameters, it is necessary to have a good knowledge of the properties of the instrument used to collect the data. The instrumental characteristics must be taken into account either in the image spatial decomposition or in the leakage matrix calculation to obtain a correct estimation of the amplitude of the spatial leaks. In a continuing effort to infer unbiased estimates of high-degree mode parameters using the high-resolution observations from the MDI *Dynamics Program*, the following instrumental properties were included in the spatial decomposition of the high-ℓ data set one at a time and their effect on the observed power spectra analyzed: *i*) the correct instantaneous image scale, *ii*) the radial image distortion, *iii*) the non-radial image distortion, *iv*) the effective *P* angle, and *v*) a correction to the Carrington elements. Once a given instrumental property is analyzed, it is incorporated in the analysis from then on in the paper. For example, in the analysis of the radial image distortion (Section 4.2), the correct image scale (Section 4.1) was used and, in the non-radial image distortion analysis (Section 4.3), the correct image scale and the radial image distortion were included.

4.1. Image Scale

Variations in the amount of defocus of the instrument have a direct influence in the image scale at the detector. (Image scale is the ratio between the size of the solar image at the detector and its actual size.) Although the MDI instrument has been very stable over the more than 11 years that it has been observing the Sun, continuous exposure to solar radiation has increased the instrument front window absorption resulting in a continuous small increase of the instrument defocus. Moreover the change of the front-window temperature due to the satellite orbit around the Sun also adds a small annual variation in the image defocus. The instrument has however an adjustable focus with nine possible positions chosen to best suit various science needs, resulting unfortunately in abrupt jumps in the image defocus every time that a new position is chosen and which are responsible for the largest variations (see Figure 5 in Rabello-Soares, Korzennik, and Schou, 2001). The average size change (at the solar limb) per focus step is 0.529 ± 0.002 pixels (Kuhn *et al.*, 2004). Due to these different time-varying focus variation, the image scale must be continuously estimated and the correct value used in the spatial decomposition. The image scale is obtained by measuring the observed image radius, which is defined as the inflection point in the radial limb-darkening function. The FNDLMB routine in the GONG Reduction and Analysis Software Package (GRASP) was used. It is available from the National Solar Observatory, Tucson, AZ, USA.

To show the influence of the image scale on the high-ℓ mode parameters, we compared the observed ridge parameters obtained using two different spatial analyses of the 1999 time series. In one of the analyses, the time-varying image scale was obtained for the actual observational period (*i.e.*, the correct image scale) and used in the spatial decomposition. In the other one, a constant value obtained from observations taken in early 1996 (at the beginning of the mission), *i.e.*, the wrong value for the 1999 time series, was used. It is 0.27% larger on average than the actual 1999 time-varying image scale. A variation in the observed ridge parameter due to a change in the data analysis corresponds to an identical variation in its offset ($\Delta°$) in order to obtain the correct mode parameter (see Equation (2)). Figure 3 shows the corresponding offset variation. The frequency offsets obtained using the correct image scale are systematically larger than the ones using a slightly larger image scale and their difference increases with frequency (Figure 3). There is no indication of a degree dependence. The variation in the frequency offset is larger for the f modes than for p modes (respectively upper and lower branches seen at frequencies smaller than 3 mHz in Figure 3). This suggests that the image-scale correction has a strong effect on the horizontal component of the leakage matrix. The a_1 coefficients obtained using the correct image scale are systematically larger by 1.44 ± 0.02 nHz than those using an image scale 0.27% larger on average. Their difference is plotted in Figure 3 (bottom panel) arbitrarily against degree instead of frequency. The effect on the other parameters is very small and their mean difference is listed in Table 2. The 2000 data set was used to calculate the values in Table 2, except for the variations due to the effect of the image scale where the 1999 epoch was used. The values in the table should not depend significantly on the observational period used since we are comparing variations in the analysis of the same time series. The variations in the frequency offset (Figure 3 top) are quite large in comparison with their absolute values shown in Figure 2. The theoretical offset estimation in Figure 2 corresponds to the case where the correct values for all instrumental effects were taken into account in the spherical harmonic decomposition.

The red points in Figure 3 were calculated taking into account the image-scale error in the leakage matrix calculation instead of in the image spatial decomposition. The leakage matrix is calculated assuming that a constant and 0.27% larger image scale than the actual value was used in the image spatial decomposition. The variation in the parameter offsets estimated from the synthetic power spectra using Equation (1) matches very well with the observations in most cases. For modes above 4 mHz, the frequency changes were underestimated by this method. This could be because we did not take into account the image-scale temporal variation, only the average difference for the entire observing period, while the spatial decomposition was carried out using a instantaneous image scale estimation.

Note that the smearing of the image represented by the point-spread function (PSF) is not taken into account nor is its variation with the amount of defocus. Our preliminary analysis indicated that including a simple model of the azimuthally averaged estimation of the PSF in the leakage matrix calculation has a very small effect on the offsets. It affected mostly the frequency offset and only by less than 0.2 µHz (Rabello-Soares, Korzennik, and Schou, 2006). Recently we found that the observed ridge frequencies obtained for the observational periods where MDI was set to a large defocus, *i.e.*, 1996 to 1998, are larger than the ones obtained for the other periods where the instrument was nearly in focus, after correcting for the solar-cycle variation; their maximum difference is of the order of a few µHz (Rabello-Soares, Korzennik, and Schou, 2008). A possible explanation is that an azimuthally averaged

Figure 3 Difference in the frequency and the a_1-coefficient offsets using, in the spatial decomposition for the 1999 high-ℓ set, the correct time-varying image scale ("corr") and a constant value, which is 0.27% larger on average than the actual values ("wrong"). The errors are given by the fitting uncertainties. The estimation of this effect when it is included in the leakage matrix calculation instead (see text for details) is shown in red.

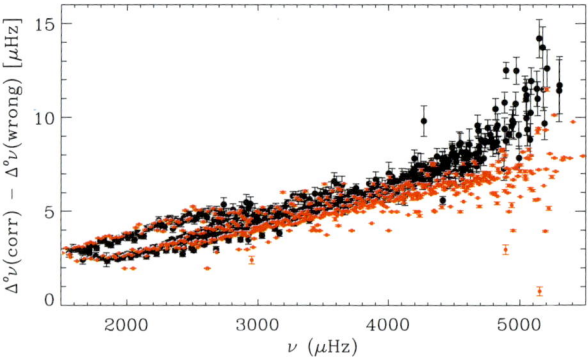

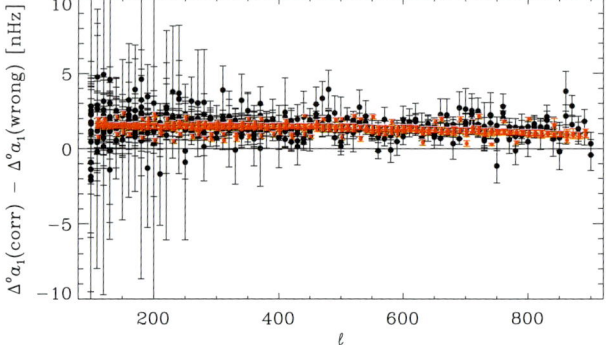

estimation of the PSF is not a good approximation to the true PSF of the instrument, which is known to depend on the azimuth angle (Schou and Bogart, 1998) with a phase that changes with focus position (*e.g.*, KRS). Unfortunately, there is not a good estimate of the PSF for the MDI *Dynamics Program* at the moment. We plan to use an approximation to the azimuthally-varying PSF and analyze its influence in the mode parameter determination in the future.

4.2. Radial Image Distortion

The ray-trace model of the MDI optical configuration predicts a radial distortion (Δr) which depends on the distance from the CCD center (r):

$$\frac{\Delta r}{r} = b \left(r^2 - \langle r \rangle^2 \right), \tag{3}$$

where $b = 7 \times 10^{-9}$ pixels^{-2} and $\langle r \rangle$ is the observed image mean radius (Kuhn *et al.*, 2004). The distortion causes the apparent solar radius to be larger by $\approx 0.17\%$ (≈ 0.8 pixels or 17 μm). Thus, the second term in Equation (3) was added to ensure that the distorted and undistorted images have the same mean radius.

As in Figure 3, Figure 4 shows the offset variation corresponding to the difference in the observed ridge frequencies when including this distortion in the spatial decomposition. Similarly to the image scale, the frequency offset increases with frequency. The similarity is expected, since the radial distortion changes the image scale by an amount that is a function

Table 2 Mean differences in the observed ridge parameters in the sense improved minus unimproved. The weighted average and its standard error were calculated over all observed modes (500 in all cases). The mode range is shown in Figure 1 (black circles). The weight of each determination was taken as inversely proportional to the square of the uncertainty. The uncertainties of the central frequency and the splitting coefficients are given by the uncertainties in the Clebsch–Gordan expansion. The uncertainties in the width, amplitude, and asymmetry are given by the standard deviation of the m-average. The first line ("Uncertainty") gives the minimum and maximum values of the uncertainties, which are very similar in all the analysis presented in the table. The other lines, from top to bottom, are the differences in the image scale, radial image distortion, non-radial image distortion, P-angle and Carrington elements respectively. The amplitude values are given in arbitrary units and the asymmetry parameter is dimensionless.

Parameter	Frequency (µHz)	a_1 (nHz)	$a_2 \times 10^5$ (nHz)	a_3 (nHz)
Uncertainty	0.05 – 1.4	0.07 – 2.0	70 – 20000	0.2 – 8
Scale	4.54 ± 0.06^a	1.44 ± 0.02	-25 ± 9	0.03 ± 0.02
Radial	0.95 ± 0.01^a	0.42 ± 0.02	-95 ± 7	0.03 ± 0.02
Non-rad.	0.133 ± 0.008	-0.11 ± 0.02	240 ± 10	0.03 ± 0.01
P angle	-0.0012 ± 0.006	0.04 ± 0.02	-0.5 ± 8	0.025 ± 0.016
Carr. el.	0.00020 ± 0.005	-0.018 ± 0.02	-1 ± 6	-0.013 ± 0.01

Parameter	Width (µHz)	Amplitude	Asymmetry
Uncertainty	0.3 – 6	50 – 10^5	0.003 – 0.12
Scale	0.047 ± 0.007	-4 ± 2	$(-1 \pm 2) \times 10^{-4}$
Radial	-0.018 ± 0.005^a	0.4 ± 1	$(-7.5 \pm 0.5) \times 10^{-4}$
Non-rad.	-0.044 ± 0.005	2 ± 1	$(1 \pm 0.4) \times 10^{-4}$
P angle	-0.00026 ± 0.005	-2 ± 2	$(9 \pm 4) \times 10^{-7}$
Carr. el.	-0.0050 ± 0.004	0.9 ± 1	$(-0.4 \pm 4) \times 10^{-5}$

[a] In these cases, the difference depends on the frequency (see text).

of the distance from the CCD center (r; Equation (3)). The difference in the ridge width also increases with frequency from -0.064 ± 0.004 µHz at $\nu < 2.5$ mHz to 0.28 ± 0.04 µHz at $\nu > 4.5$ mHz, in the sense "including" minus "not including" the distortion, but it is very small in comparison with the fitting uncertainties and barely significant. The effect on the other parameters is small and is listed in Table 2. The ridge modeling of this effect, introducing the radial distortion in the leakage matrix, agrees well with the observations (red circles in Figure 4).

4.3. Non-Radial Image Distortion

Solar images as observed by MDI have a nearly elliptical shape with a difference between the semi-major and the semi-minor axis of about 0.6 pixels. This is consistent with a small tilt ($\approx 2°$) in the detector around an axis that is rotated 56° from the detector's horizontal x-axis and it introduces a non-radial image distortion. In KRS, we estimated this distortion using different observational methods. Unfortunately, the distortion varies by as much as 35% depending on the data used, with the correspondent CCD tilt varying from 1.71° to 2.6°. Kuhn *et al.* (2004) also estimated the non-radial distortion using yet another observational method – the Mercury transit across the Sun on 7 May 2003 – and found a 3.3° tilt. Although their equations to estimate the distortion have the same general form as ours (KRS), there are

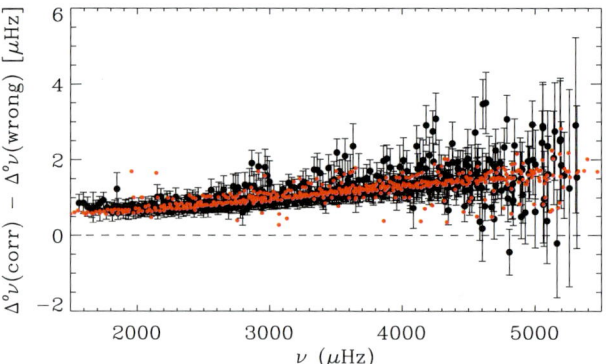

Figure 4 Difference in the frequency offset obtained when including or not the radial distortion in the spherical harmonic decomposition using the 2000 high-ℓ data set. The errors are given by the fitting uncertainties. The estimation of this effect using our model is shown in red.

other differences besides the tilt angle between the two calculations. Their estimation of the distortion varies by 14% in relation to our estimation using a 2.6° tilt. These discrepancies in the estimated distortion using different observational methods might be attributed to a number of reasons such as the inaccuracy of our simple model for the non-circular shape of the solar images or the influence of an optical aberration (an asymmetric PSF, for example).

To analyze the effect of the non-radial distortion on the power spectra, we included in the spherical harmonic decomposition our estimation that has the largest tilt angle (2.6°), which corresponds to the distortion that better reproduces the solar limb shape. Although, to the moment, we have been unable to precisely determine the non-radial distortion pattern, our estimation provides an improvement to the analysis and it will be incorporated from now on in the paper. Fortunately, overall it has a small effect on the observed spectra, as can be inferred from Table 2, and we can safely extrapolate that small variations from this distortion pattern can only correspond to variations in the parameters smaller than the differences shown in the table which were obtained comparing with using no correction for the non-radial distortion, and are most likely negligible.

4.4. Position Angle (P)

The roll angle of the SOHO spacecraft is maintained such that the effective position angle (P_{eff}) of the MDI images should always be zero. (It is the position angle of the northern extremity of the solar rotation axis with respect to MDI detector y-axis.) However, a 0.2° difference has been measured by intercomparing MDI and GONG images obtained in 1999 and 2000 (Cliff Toner, private communication). Beck and Giles (2005) estimated a 0.07° difference, assuming that there is no Equator-crossing flow, after regaining contact with the SOHO spacecraft (in 1999) and 0.1° before losing contact. No noticeable effect is seen in the observed ridge parameters after including a 0.2° correction (Table 2). We do not see the 3 nHz variation in the a_1-coefficient offset that was predicted by our ridge model using a slightly higher correction of 0.25° (KRS). This is probably because we did not accurately model the P-angle correction in the leakage matrix calculation, but added a very simple approximation of its effect.

4.5. Carrington Elements

Accordingly to Beck and Giles (2005), the standard values used for the two angles specifying the orientation of the solar rotation axis (i, Ω) (i is the angle between the plane of the ecliptic and the solar Equator and Ω is the angle between the crossing point of the solar Equator with the Ecliptic and the Vernal Equinox) known as the Carrington elements, are off by $\Delta i = 0.095° \pm 0.002°$ and $\Delta \Omega = 0.17° \pm 0.1°$. This introduces a time-varying correction in the calculation of the rotation axis projection in the plane of the sky, i.e., the position angle (P) and the roll angle (B_0 is the heliographic latitude of the central point of the disk and presents an annual variation). This correction in the P angle will be on top of the one mentioned in the previous section. Introducing a correction of $\Delta i = 0.1°$ and $\Delta \Omega = 0.1°$ in the image spatial decomposition has no significant effect on the observed ridge parameters (Table 2).

5. The High-Degree Mode Parameters

Here we analyze the frequencies and splitting coefficients obtained using the high-ℓ data set. The mode parameters were obtained after including the five known instrumental effects in the spatial decomposition of the high-ℓ data sets described in Section 4, fitting their observed power spectra using an asymmetric Lorentzian profile (Section 2), and applying the ridge-to-mode correction to the fitted ridge parameters (Section 3).

First, in Section 5.1, the high-ℓ data set mode frequencies and splitting coefficients are compared with the values obtained by the medium-ℓ analysis. Figure 1 shows the region of common modes between the two sets. This comparison is done to check the quality of the estimation of the mode parameters from the observed ridge in the high-ℓ data set at these high medium-ℓ common modes. Then, the estimated high-ℓ frequencies and splitting coefficients for $\ell \geq 100$ are analyzed (Sections 5.2 and 5.3).

5.1. Medium- and High-ℓ Set Comparison

Figure 5 shows the differences between the mode frequency and the splitting coefficients obtained using the 72-day medium-ℓ data set described in Section 2 and using the high-ℓ set both observed during 2004 (Table 1). The high-ℓ set used in this comparison was analyzed as described in Sections 2 and 3, and including all instrumental effects analyzed in the Section 4. In the medium-ℓ mode range, the spatial leaks are well separated from the target mode and individual modes can be identified and fitted. By decreasing the observed high-ℓ set frequency resolution using a short time string (4096 minutes), and thus increasing the width of the window function, we force the width of the spatial leaks to increase. (An even shorter time series, 2048 minutes, was used to check the results at the lowest high-ℓ modes.) The spatial leaks in the high-ℓ power spectra now overlap the target mode forming a ridge at ℓ as low as 100 and the ridge-to-mode correction described in Section 3 is applied. To increase the number of common modes, the high-ℓ power spectra used in the comparison were fitted for every ℓ (and not every tenth ℓ). The set of modes used in this comparison consists of 420 p modes with $100 \leq \ell < 200$ and $1.7 < \nu < 3.5$ mHz and 60 f modes with $230 < \ell < 300$ and $1.5 < \nu < 1.8$ mHz observed during 2004.

The mode frequencies agree within 1 μHz (Figure 5 top panel). Their difference varies with frequency having a maximum at 2 mHz and decreasing to zero (or near zero) at 1.6 and 3 mHz. There is no indication of a degree dependence. Although 1 μHz is a small value, it is almost one order of magnitude larger than the fitting uncertainty of these modes (0.12 μHz on average) and hence undesirable. This seems to indicate that the frequency offsets $\Delta°\nu$ (Figure 2) are underestimated by $\approx 35\%$ for modes with frequencies in the range 1.7 to 2.3 mHz, this amount decreases to 10% around 1.6 mHz and zero in the interval between 2.6

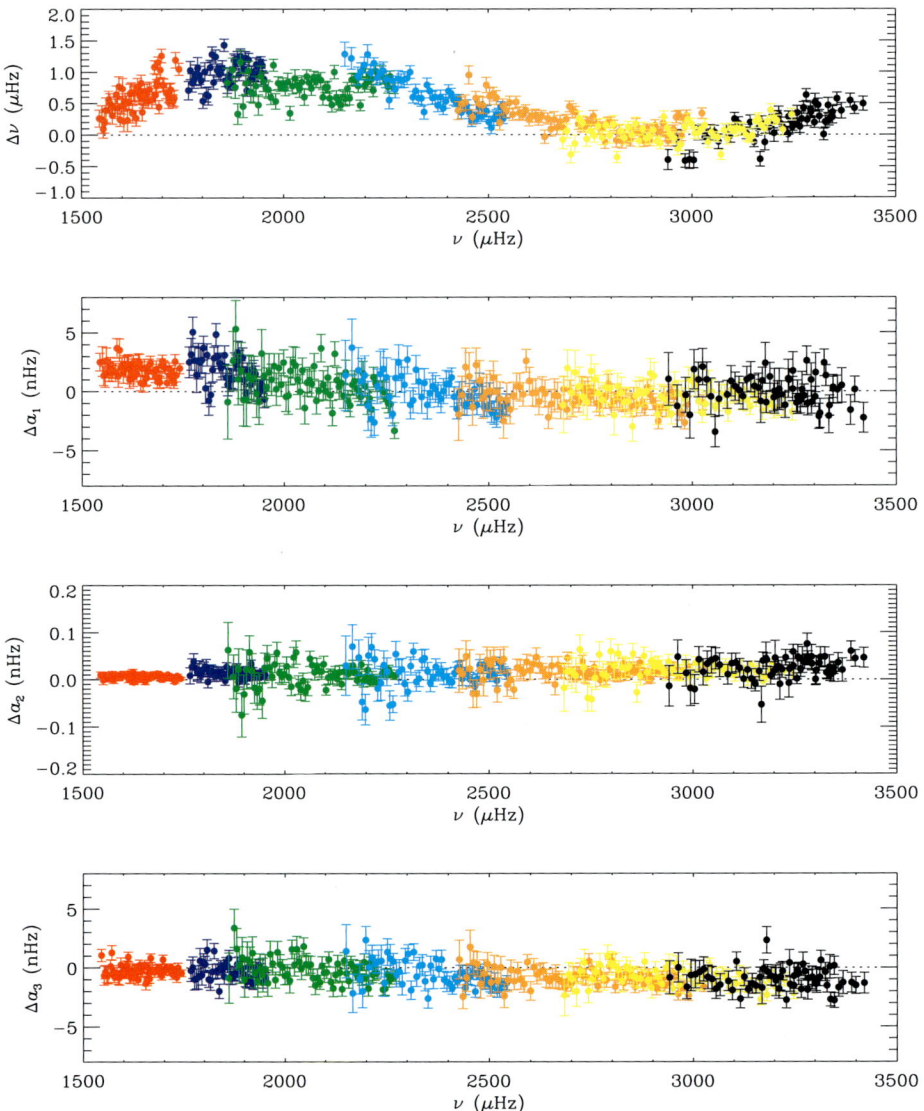

Figure 5 Central frequency and splitting coefficients difference between the medium- and high-ℓ sets observed during 2004 (Table 1). The differences are in the sense high- minus medium-ℓ set. Modes with $n \leq 5$ are in color where the f modes are in red.

and 3.2 mHz where the estimation is correct. There is some indication that at 3.4 mHz the offsets are again underestimated (by $\approx 30\%$). Unfortunately, there are no common modes at higher frequencies.

A variation in the mode frequency that is purely a function of frequency does not affect the outcome of the solar structure inversion, since it is removed together with the near-surface errors in the physics of the solar model (see Section 5.2). The instrumental effect that is probably causing the frequency offset to be underestimated at certain frequencies might be the instrumental PSF which is not included in the image spatial decomposition or in the leakage matrix calculation (see Section 4.1). Besides the frequency dependence, there is a puzzling frequency difference (≈ 0.5 µHz) in the p_2 modes (green circles in Figure 5 top panel) with respect to the adjacent n values.

The difference in the a_1-coefficients between the medium and high-ℓ set is negligible except for $n = 0, 1$ modes (second panel in Figure 5). Their mean difference normalized by the fitting uncertainties is, in units of σ, -0.2 for $n \geq 2$, 2 for $n = 1$, and 3 for the f modes. The horizontal-to-vertical displacement ratio decreases exponentially with n. Thus, the differences observed only for $n = 0, 1$ modes could be an indication of an unaccounted for instrumental effect that has a stronger effect on the horizontal component of the leakage matrix than on its vertical component. Note also that the f modes in the medium-ℓ set are fitted using a different frequency interval around the peak in the power spectrum than the p modes (Schou, 1999).

There are small differences between the medium- and high-ℓ set for the splitting coefficients a_2 and a_3 (third and bottom panels in Figure 5). Their mean differences normalized by the fitting uncertainties of the high-ℓ power spectra are $a_2 = 1$ and $a_3 = -1$ in units of σ.

In conclusion, the estimation of the mode frequency and splitting coefficients from the observed ridge in the high-ℓ data set at these moderate-degree values is, in general, quite good. However, there is still room for improvement, specially for the central frequency and the f-mode a_1-coefficient.

The corrections in the medium-ℓ mode parameters calculated by Larson and Schou (2008), described in Section 2, do not have a large influence in our results. Their effect in Figure 5, including the variation between fitting symmetric and asymmetric profiles, is small in comparison with the difference between the medium- and high-ℓ sets. The mean total correction in the medium-ℓ frequency and splitting coefficients a_1, a_2, and a_3 to be applied to the values used here are: 0.130 ± 0.004 µHz, -0.244 ± 0.009 nHz, $(1.78 \pm 0.08) \times 10^{-3}$ nHz, and -0.137 ± 0.006 nHz, respectively (Larson and Schou, 2008). The average was calculated over the 420 medium- and high-ℓ common modes. The largest correction is in the central frequency. The mean frequency variation between fitting symmetric and asymmetric profiles is 0.043 ± 0.002 µHz for the common modes (Larson and Schou, 2008). The improved medium-ℓ analysis will decrease slightly the differences between medium- and high-ℓ frequencies showed in the top panel of Figure 5. However, it does not change the overall behavior of the frequency differences.

5.2. The Central Frequency

Figure 6 shows the difference between the high-ℓ set mode frequencies ($100 \leq \ell \leq 900$) and their theoretical value calculated from Christensen-Dalsgaard's model S (Christensen-Dalsgaard *et al.*, 1996) as a function of degree for different n values. The high-ℓ set used in this comparison was analyzed as described in Sections 2, 3, and includes all instrumental effects analyzed in Section 4. The mode range is shown in Figure 1. The medium-ℓ set frequencies are also plotted as a reference. In the absence of any acceptable theory to describe the physics of the layers near the solar photosphere, the difference between the observed and theoretical frequencies due to the near-surface errors in the model are well known to be quite large (*e.g.*, Christensen-Dalsgaard and Berthomieu, 1991). The general trend is such that the observed p-mode frequencies are smaller than their theoretical prediction and, at a high enough degree, the differences increase with degree and with frequency. Accordingly to the results obtained by the high-ℓ set, this difference can be as large as 60 µHz. In fact, for a given n, the frequency differences increase almost linearly with degree with a slope

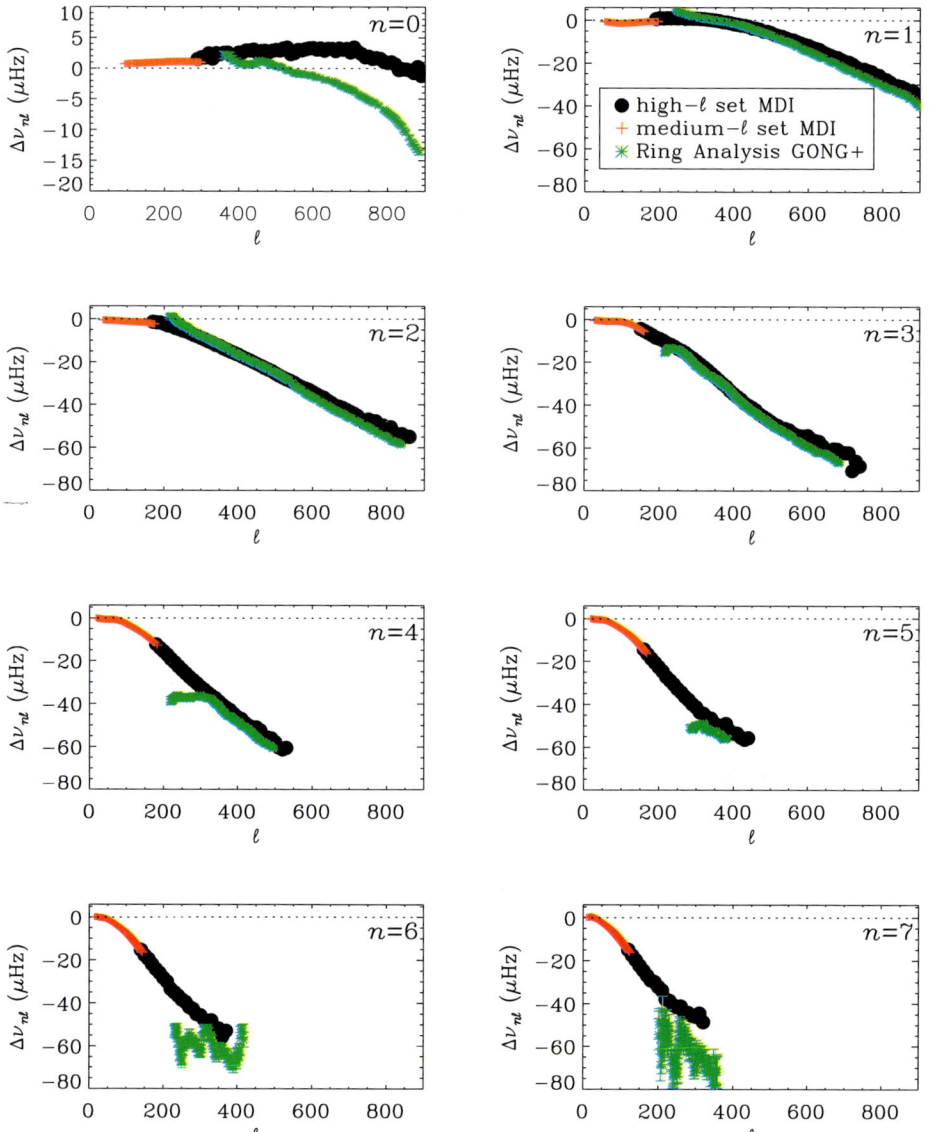

Figure 6 Difference between observed frequencies and their theoretical value as a function of degree for different n values using the high-ℓ set data obtained in 2004 (in black). The differences are in the sense observed minus theoretical. For comparison, the results for the medium-ℓ set (red crosses) and from ring-diagram analysis (green stars) are also plotted. The fitting uncertainties are given by the error bars. Note the different scale for the f modes.

that also increases linearly with n. The high-ℓ set analysis gives consistent frequencies for all observational periods listed in Table 1, where the differences in the frequencies obtained at the different observational periods can be explained by their well known variation with the solar activity cycle, and it is described in detail in Section 6.

In order to compensate for the frequency shifts due to the near-surface errors in the model, an unknown function (the so-called surface term: F_{surf}) is usually added to the equation governing helioseismic inversions [$F_{\text{surf}} = Q_{n,\ell} \times (\delta \nu_{n,\ell}/\nu_{n,\ell})_{\text{surf}}$] where $Q_{n,\ell}$ is the mode inertia normalized by the inertia of a radial mode of the same frequency. Christensen-Dalsgaard, Thompson, and Gough (1989), from the asymptotic theory of solar p modes, pointed out that low- and moderate-degree modes propagate nearly vertically near the surface; thus, their behavior in this region is essentially independent of degree and depends only on frequency. This however does not hold for high-degree modes as can be inferred from Figure 6. A second-order asymptotic approximation, where the surface term is a function not only of frequency but also of $L/\nu_{n,\ell}$ (Brodsky and Vorontsov, 1993), must be used as shown by Di Mauro et al. (2002).

Figure 6 also shows for comparison the frequency difference applying the ring-analysis technique obtained by Rabello-Soares, Bogart, and Basu (2008). A magnetically quiet region (15° in diameter) observed by the GONG++ network on 4 July 2005 during Carrington rotation 2031 centered at 115° longitude and 3° South in latitude was tracked for 8192 minutes at the appropriate photospheric rotation rate. The region crossed the central meridian at the middle of its tracking interval. The tracked region was mapped to a plane using Postel's projection and its power spectra, given by the three-dimensional Fourier transform of the temporal series of images, were fitted using the 13-parameter model of Basu and Antia (1999). The wavenumber (k) can be identified with the degree of a spherical harmonic mode of global oscillations by $L = kR_\odot$. As the oscillations in a plane-parallel geometry are only discrete in radial order, ℓ does not need to be an integer. Each "mode" is obtained by fitting a region of power spectrum that has significant overlap with those covered by neighboring "modes", making them not strictly independent. To confirm that the frequencies obtained for this particular solar region correspond to typical values, they were compared with the frequencies obtained for 13 additional quiet regions, with latitudes ranging from $-12°$ to $+12°$ (Rabello-Soares, Bogart, and Basu, 2008). Their difference is smaller than 3 µHz for $n \leq 5$. Part of this variation is due to the fact that the projection of the spherical solar surface onto a flat detector introduces some foreshortening that depends on the distance of the region from disk center, which can introduce systematic errors in determining the mode characteristics.

The frequency differences (in relation to the theoretical value) obtained using ring analysis present the same trend as using spherical harmonic decomposition, except for the f modes. In most cases, the ring-analysis frequencies are smaller than the ones obtained by the high-ℓ set. For p modes, their difference in the sense global minus ring analysis varies between -4 and 6 µHz for $n \leq 5$ and it could be as large as 20 µHz for $n > 5$. Due to the small size of the region analyzed, the ring-analysis power spectra have low spatial resolution doing poorly at medium degree and at high n, as can be seen for $n = 6, 7$ in Figure 6. Note also that for $n = 4$ and $\ell \leq 300$ modes in Figure 6, the difference between the ring-analysis and theoretical frequencies is constant and does not continue to decrease as ℓ decreases. This is probably an indication of a poor frequency determination by the ring analysis at these medium-ℓ values. In conclusion, for the p modes, the ring-analysis frequencies agree with the global-analysis values within ± 6 µHz, which is much smaller than the difference between either of them and the frequencies obtained from the solar model.

For the f modes, the frequency differences (in relation to the theoretical value) obtained using ring and global analysis are different. The medium-ℓ f-mode frequency is on average 0.9 ± 0.1 µHz larger than the theoretical values. The f-mode frequency obtained by ring analysis for $\ell \leq 450$ is on average 1.1 ± 0.6 µHz larger than the theoretical frequency. For $\ell > 450$, it decreases sharply with degree and it can be described by a fifth-degree polynomial. (The polynomial in ℓ has the following coefficients in units of µHz: $c_0 = 350 \pm 80$,

$c_1 = -2.7 \pm 0.7$, $c_2 = 0.008 \pm 0.002$, $c_3 = (-1.3 \pm 0.3) \times 10^{-5}$, $c_4 = (1 \pm 0.2) \times 10^{-8}$ and $c_5 = (-3.2 \pm 0.7) \times 10^{-12}$.) At $\ell = 880$, it is 13 µHz smaller than the theoretical values. Such a sharp decrease in the f-mode frequency in relation to the theoretical values similar to that observed for the p-modes was already reported by other authors (Duvall, Kosovichev, and Murawski, 1998 and references within). The frequencies of the f modes, contrary to the p modes, depend only weakly on the hydrostatic structure of the model (*e.g.*, Gough, 1993) and a possible explanation is that the f-mode frequencies are reduced by granulation (Murawski and Roberts, 1993). However, the f-mode frequency obtained by global analysis presents a different trend in relation to the theoretical frequency. For $\ell < 710$, it is only 2.7 ± 0.5 µHz on average larger than the theoretical frequency. At larger ℓ, it decreases linearly with degree with a slope of -0.021 ± 0.001 µHz; it is zero at $\ell = 850$. The global-analysis results suggest a much weaker interaction between the f modes and the granulation than predicted by the ring analysis.

The error bars in Figure 6, given by the fitting uncertainties, are too small to be seen, except for the ring-analysis frequencies at $n \geq 6$. The observed frequency errors obtained using the medium-ℓ and high-ℓ sets and ring analysis are in the range 0.007 – 0.4 µHz, 0.06 – 0.6 µHz and 0.2 – 15 µHz, respectively. The ratio between the global- and ring-analysis fitting uncertainties varies from 0.9 to 33. The ring-analysis uncertainties are similar to those of global analysis at low n ($n \leq 2$) and high ℓ ($\ell > 700$). They become much higher than the global-analysis uncertainties as n increases and as ℓ decreases. The frequency is not a parameter in the model used to fit the ring-analysis power spectra (see Basu and Antia, 1999), the observational error estimated for the fitted width is used instead, following Rajaguru, Basu, and Antia (2001).

5.3. The Rotational Splitting Coefficients

High-ℓ splittings can be used to infer the solar rotation rate in the outermost layers of the solar convection zone. In these layers, there is a steep gradient of the rotation, making it a very interesting region to study. This was first suggested by the fact that different surface markers such as sunspots, faculae, Hα filaments, and supergranular network present different rotation rates (*e.g.*, Snodgrass, 1992) which has been interpreted by assuming that the markers are anchored at different depths (Foukal, 1972). The solar rotation rate determined by helioseismology (using modes with $\ell \leq 300$) has a local maximum at about $0.95 R_\odot$ (35 Mm below the solar surface) and decreases rapidly towards the surface (Howe *et al.*, 2007). Using f-mode splittings with degree between 117 and 300, Corbard and Thompson (2002) estimated a constant radial gradient of solar angular velocity of around -400 nHz $R_\odot^{-1}$ in the outer 15 Mm between the Equator and 30° latitude. Above 30° it decreases in absolute magnitude to zero around 50°. The addition of high-degree splittings to this analysis will help to constrain the determination of the solar-rotation profile both closer to the solar surface and at higher latitudes.

Figure 7 shows a smooth variation with degree of the high-ℓ a_1-coefficients as expected, implying a good estimation of the mode splittings. For a given n, the p-mode a_1 coefficients decrease with degree until $\ell \approx 300$ by 1 – 2% (which corresponds to 4 – 6 nHz) where it becomes approximately constant. And the f-mode a_1-coefficients decrease almost linearly with degree from $\ell \approx 300$ to 900 by 2.5% (or 9 nHz). For a given n, the a_1 variation with ℓ shown in Figure 7 presents the same trend if plotted against the location of the mode lower turning point (given by L/ν), indicating the expected decrease of the solar rotation with solar radius near the solar surface. The p-mode a_1-coefficients, especially for $2 \leq n \leq 5$, present a sharp decrease near $\ell = 200$ corresponding to a turning point of depth of ≈ 30 Mm, which is at the beginning of the solar rotation decrease.

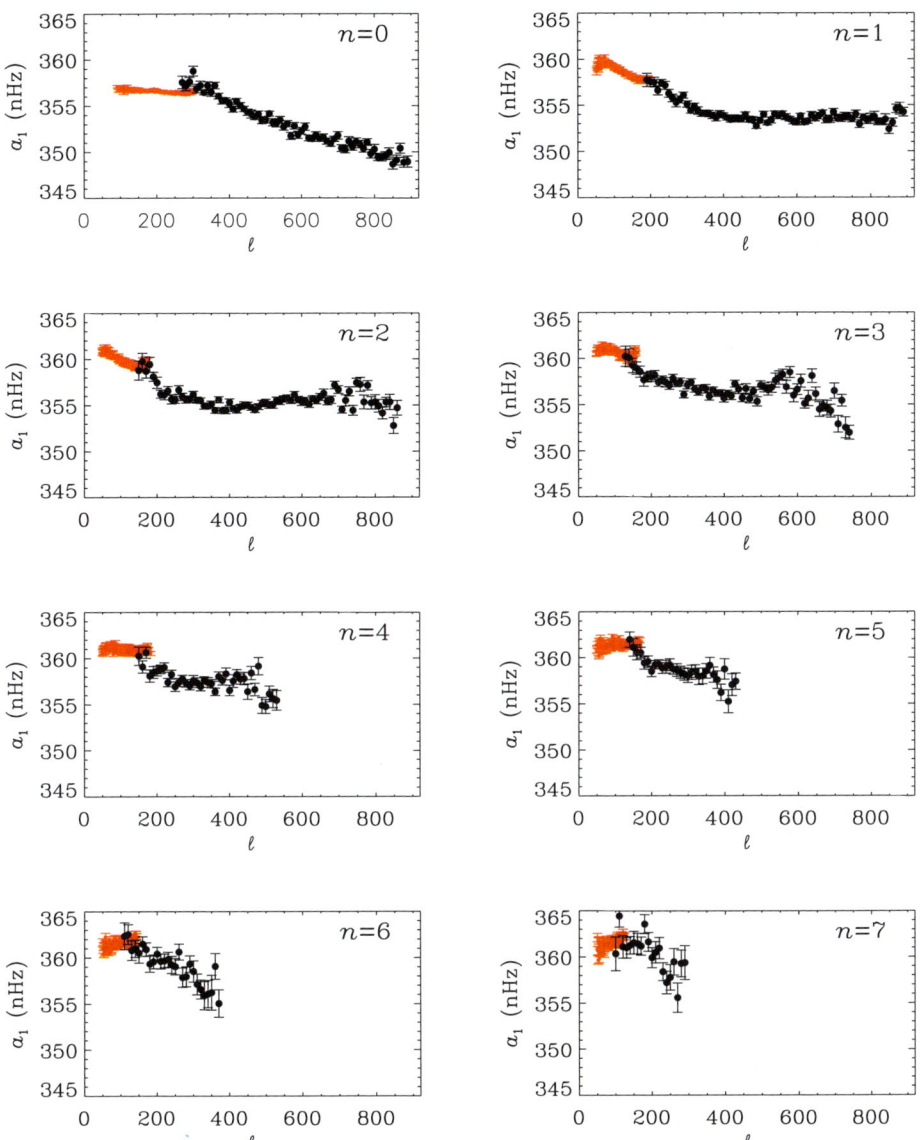

Figure 7 The a_1 rotational-splitting coefficients as a function of degree for the 2004 data set, where the ones obtained with the medium-ℓ set are in red and with the high-ℓ set in black. The fitting uncertainties are given by the error bars.

Although the medium-ℓ a_3-coefficient decreases slightly in absolute value with degree (from -7.8 nHz at $\ell \approx 50$ to -7.3 nHz at $\ell \approx 300$), the high-ℓ a_3-coefficient mean value (-8.6 ± 1.2 nHz) is larger in absolute value than the mean medium-ℓ value (-7.5 ± 0.3 nHz). The high-ℓ a_3-coefficient does not show any variation with degree. However, it has a small absolute value at low frequency ($\nu \leq 2.2$ mHz), which is similar to the medium-ℓ mean: -7.4 ± 0.5 nHz.

While the odd coefficients provide information about the internal solar rotation rate, the even coefficients arise from latitudinal structural variation, centrifugal distortion, and magnetic fields. The a_2-coefficients obtained with the high-ℓ set are on average zero ($1500 < \nu < 5200$ μHz). The a_2-coefficients obtained with the medium-ℓ set are on average zero for $\nu < 2$ mHz. At higher frequencies, they start to decrease with frequency, where for $\nu = 3550 \pm 50$ μHz, its mean value is -0.08 ± 0.03 nHz.

6. Frequency Variation Over the Solar Cycle

The correlation between solar acoustic mode frequencies and the magnetic activity cycle is well established and has been substantially studied during the last and current solar cycles (see for example Chaplin *et al.*, 2007; Dziembowski and Goode, 2005 and references therein). However, its physical origin is still a matter of debate and the detailed analysis of the frequency shift characteristics are likely to contribute to solving this problem. The data sets used in this work and listed in Table 1 cover a considerable part of solar cycle 23. They also cover a large degree range ($20 \leq \ell \leq 900$) when combining the medium- and high-ℓ sets. In a previous work (Rabello-Soares, Korzennik, and Schou, 2008), we used these two data sets to analyze the characteristics of the frequency variation along the solar cycle. Here we improved our previous analysis by calculating the probability that a linear relationship indeed exists between the frequency shift of a given (n, ℓ) mode and the solar-activity index. Although the medium- and high-ℓ sets are fitted with symmetric and asymmetric profiles respectively, the resulting difference in frequency does not change on average over the solar cycle (Larson and Schou, 2008).

The frequency variation with the solar cycle is given by the frequency difference between 1999–2004 epochs with respect to the 2005 epoch, the one with the lowest activity index in our data set. For each (n, ℓ) mode, their frequency shifts were fitted assuming a linear relationship with the solar-activity index with a zero intercept and using a weighted least-squares minimization. For a detailed description, see Rabello-Soares, Korzennik, and Schou (2008). Four different solar-activity indices commonly used in the literature and that are available in the period 1999–2005 were used in this analysis: the solar UV spectral irradiance (given by the NOAA Mg II core-to-wing ratio (http://www.ngdc.noaa.gov/stp/SOLAR/ftpsolaruv.html)), the Magnetic Plage Strength Index (MPSI, Mt. Wilson Observatory (http://www.astro.ucla.edu/~obs/intro.html)), the solar-radio 10.7-cm flux (National Research Council of Canada (http://www.ngdc.noaa.gov/stp/SOLAR/ftpsolarradio.html.)) and the sunspot number (SSN, SIDC, RWC, Belgium (http://sidc.oma.be)). The minimum-to-maximum solar cycle frequency shift ($\delta \nu_{n,\ell}^e$) was estimated by multiplying the slope of the linear fit by the corresponding solar-index variation between the maximum and minimum of cycle 23. The mean activity level for each epoch in relation to the maximum reached in January 2002 is listed in Table 1, given by the solar UV spectral irradiance.

In an attempt to include in the analysis of the solar-cycle-induced frequency shifts only modes that are, in fact, correlated with the solar activity, Rabello-Soares, Korzennik, and Schou (2008) rejected modes whose Pearson correlation coefficient is smaller than 0.8, or whose slope uncertainty is larger than 20% of its absolute value. However, the correlation coefficient cannot be used directly to indicate the probability that a linear relationship exists between two observed quantities, *i.e.*, if, indeed, there exists a physical linear relation between them. A small correlation coefficient might indicate only a small slope in their linear relation. A better approach is to calculate the probability that the data points represent a sample derived from an uncorrelated parent population. We estimated the probability

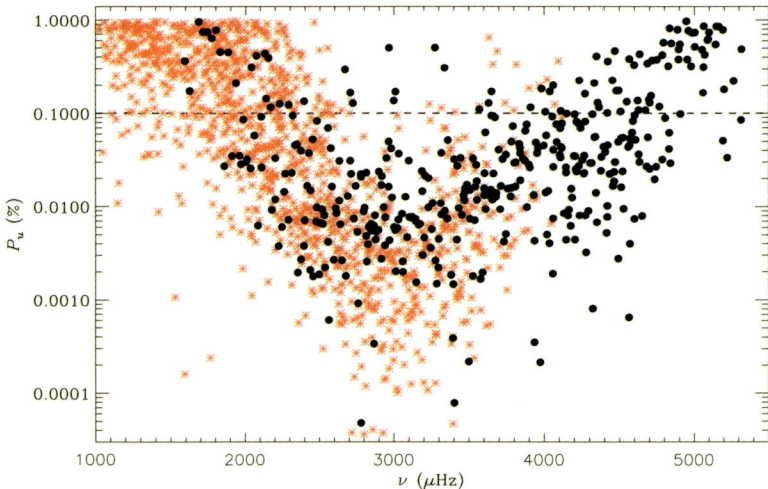

Figure 8 Probability $P_u(n, \ell; i)$ as a function of the mode frequency for the medium (red) and high-ℓ (black) sets, using the sunspot number as the solar activity index i.

$[P_u(r)]$ that a random sample of N uncorrelated data points would yield a linear-correlation coefficient as large as or larger than the observed absolute value of r (Bevington and Robinson, 1992). If $P_u(r)$ is very small, it would indicate that it is very improbable that they are linearly uncorrelated. Thus, the probability is high that the frequency shift of a particular (n, ℓ) mode and the solar-activity index used are correlated and the linear fit is justified.

For each (n, ℓ) mode and solar-activity index i, we calculated the Pearson correlation coefficient $r(n, \ell; i)$ and the probability $P_u(r) \equiv P_u(n, \ell; i)$. Figure 8 shows the probability $P_u(n, \ell; i)$ as function of the mode frequency. It is very similar for all four solar-activity indices. Modes with frequency around 3 mHz have the highest probability of being correlated. It decreases for modes with frequency smaller than ≈ 2.5 mHz or larger than ≈ 4.5 mHz. There is no noticeable dependence of P_u with ℓ or ν/ℓ.

Libbrecht and Woodard (1990) were the first to suggest that the solar-cycle frequency variation ($\delta\nu_{n,\ell}$) is linearly proportional to the inverse mode inertia ($I_{n,\ell}$) using observations obtained at the Big Bear Solar Observatory. The observed frequency shift is larger for higher-frequency and for higher-ℓ modes. As pointed out by Libbrecht and Woodard (1990), these modes are more sensitive to surface perturbations, because they have, respectively, higher upper- and lower-reflection points in the Sun. This indicates that the dominant structural changes during the solar cycle, inasmuch as they affect p-mode frequencies, occur near the solar surface. Chaplin *et al.* (2001) obtained a similar shift at low frequencies ($\nu \leq 2.5$ mHz): $\delta\nu_{n,\ell} \propto 1/I_{n,\ell}$. However, at higher frequencies, they found: $\delta\nu_{n,\ell} \propto \nu^\alpha/I_{n,\ell}$, where $\alpha = 1.91 \pm 0.03$. They used data sets with a much higher duty cycle obtained by the ground-based networks, GONG and BiSON, and the SOHO satellite (VIRGO/LOI).

The value of α depends upon the physical mechanism responsible for affecting the acoustic mode frequencies during the solar cycle. Several authors (Gough, 1990; Libbrecht and Woodard, 1990; Goldreich *et al.*, 1991) estimated $\alpha = 3$ assuming modifications in the thermal structure of the Sun in a thin layer at the photosphere. On the other hand, Gough (1990) pointed out that if the frequency shifts are caused by variations in the efficacy of the convection during the solar cycle, $\alpha \approx -1$.

Rabello-Soares, Korzennik, and Schou (2008) showed that the scaled frequency shift can be described with a simple power law at all frequencies:

$$\delta v_{n,\ell}^e = C_\gamma \frac{(v_{n,\ell})^\gamma}{Q_{n,\ell}}, \qquad (4)$$

where $Q_{n,\ell}$ is the mode inertia, normalized by the inertia of a radial mode of the same frequency, $I_{\ell=0}(v_{n,\ell})$, calculated from Christensen-Dalsgaard's model S (see Christensen-Dalsgaard et al., 1996). The f and p modes were fitted independently of one another. Chaplin et al. (2001) also plotted the frequency shift scaled by the normalized mode inertia (upper right-hand panel of their Figure 1). However, they chose to fit the frequency shift scaled by the mode inertia instead. The multiplication of the frequency shift by $Q_{n,\ell}$ normalizes the shift to its expected radial equivalent.

As in Rabello-Soares, Korzennik, and Schou (2008), Equation (4) was fitted to estimate γ_p (for the p modes) and γ_f (f modes), using a weighted least-squares minimization (thin green line and black dashed line in Figure 9, respectively). Only modes that have a probability of 0.1% or less that its variation is linearly uncorrelated with the solar index were included in the present analysis. They represent 80% of the high-ℓ set modes and 65% of the medium-ℓ set modes. The frequency ranges used in the fitting of the medium- and high-ℓ p modes and high-ℓ f modes are: 2.5 – 4.1 mHz, 2.8 – 4.9 mHz and 1.8 – 3.0 mHz, respectively. It seems that there is a step in the p-mode frequency shift around 2.3 mHz and only modes with $v \geq 2.5$ mHz were included in the fitting of the medium-ℓ set. p modes with $v \leq 2.3$ mHz seem to have a similar slope to the fitted high-frequency modes, but a different y-intercept. This is illustrated by the thick short green line in Figure 9. The fact that the f modes are affected by the solar cycle in a different way than the p modes is expected since they have very different properties. The f mode is essentially a surface wave and its frequency is less likely to be influenced by the solar stratification than the p-mode frequency. We repeated the analysis for all four solar indices. The results are plotted in Figure 10. The exponents obtained using the different solar indices agree within their fitting uncertainty. The weighted mean from the four solar indices is $\bar{\gamma}_f = 1.29 \pm 0.07$ for the f modes and $\bar{\gamma}_p = 3.60 \pm 0.01$ for the p modes. $\bar{\gamma}_p$ was calculated after averaging the values obtained for the medium- and high-ℓ sets (given by the circles in Figure 10). Only f modes obtained with the high-ℓ set were used to estimate γ_f. There are only a few modes (seven) in the medium-ℓ set with a probability of 0.1% or less of being uncorrelated; they have a very small frequency shift (0.05 ± 0.01 μHz on average), and as a consequence they were excluded from the fit. The f modes in the medium-ℓ set have very small frequencies ($v < 1.5$ mHz) and, accordingly to Figure 8, low-frequency modes are not well correlated with the solar cycle. High-ℓ f modes with $v \leq 1.8$ mHz are also not well correlated with the solar cycle and were excluded from the analysis. Their frequency shifts show a steep increase from values similar to those of the well-correlated medium-ℓ f modes at 1.7 mHz to values similar to the diamonds in Figure 9 at 1.8 mHz.

In Rabello-Soares, Korzennik, and Schou (2008) only the solar UV spectral irradiance was used as a proxy for the solar-cycle index to calculate the exponent γ. The value obtained here for the p modes, using the same solar index but a better criteria for selecting the modes included in the analysis, is the same as before: $\gamma_p = 3.63 \pm 0.02$. However, the value obtained for the f modes ($\gamma_f = 1.33 \pm 0.2$) is 20% smaller than our previous result. The distinct mode selection criteria account for this difference. As the number of f modes is small (45), the fitting of γ_f is more sensitive to the mode selection.

To compare the above-mentioned values of α obtained by Chaplin et al. (2001) with our results, $v_{n,\ell}^\alpha$ was divided by $I_{\ell=0}(v_{n,\ell})$ and then fitted by Equation (4) to estimate the

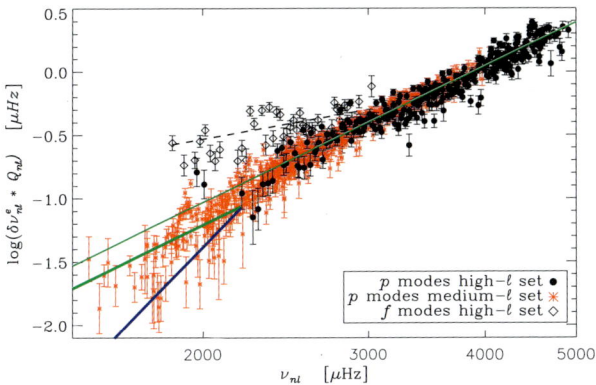

Figure 9 Frequency shift multiplied by the normalized mode inertia as a function of frequency obtained, in this case, using the sunspot number as a proxy of solar activity. The long green continuous line is the fit to all p modes with $\nu \geq 2.5$ mHz and the dashed black line is to the f modes in the high-ℓ set. The thick short green line is the fit to p modes with $\nu \leq 2.3$ mHz using the same slope as the long green line but a different y-intercept. The blue line corresponds to a frequency shift given by $\alpha = 0$ with an arbitrarily chosen y-intercept.

corresponding γ exponent. The α values obtained at the two frequency intervals $1.6 < \nu < 2.5$ mHz ($\alpha = 0$) and $2.5 < \nu < 3.9$ mHz ($\alpha = 1.91$) correspond to $\gamma = 7.59 \pm 0.18$ and $\gamma = 3.58 \pm 0.03$, respectively. Accordingly to the authors, the division of the data into two parts has been made on a purely "artificial" basis. Rabello-Soares, Korzennik, and Schou (2008) showed that this imposed "breakpoint" can be explained by the fact that $\log(\delta \nu^e_{n,\ell} I_{n,\ell})$ has a quadratic dependence on $\log(\nu_{n,\ell})$, with an inflection point at 2.59 mHz (see Figure 6 in Rabello-Soares, Korzennik, and Schou, 2008). Chaplin *et al.* (2001) used 10.7-cm radio flux as the solar activity index. Their results at the high-frequency interval agrees well with ours (represented by a square in Figure 10 bottom panel). However, at low frequency, their result is twice as large as the one obtained here (blue line in Figure 9). The difference could be due to the mode selection used to obtain our results. Dziembowski and Goode (2005) also used a simple frequency difference including all modes. They fitted $\delta \nu^e_{n,\ell} I_{n,\ell}$ using truncated Legendre polynomial series and fitted the f and p modes independently of one another. For p modes, their fitting agrees with Chaplin *et al.* (2001) thus disagreeing with ours at low frequency. For f modes, in the frequency range 1.37–1.60 mHz, the corresponding γ_f (inferring from their Figure 2) is five times larger than our determination at $\nu > 1.8$ mHz, and it is similar to the low-frequency γ_p obtained by Chaplin *et al.* (2001). Dziembowski and Goode (2005) noted that the frequency shifts normalized by $I_{n,\ell}$ present opposite trends for the f and p modes. They increase with increasing frequency for p modes and decrease for f modes. This opposite trend disappears when the frequency shifts are normalized by $Q_{n,\ell}$ instead.

Several authors have reported an abrupt decrease of the frequency shifts at high frequency. The positive shifts suddenly drop to zero and become negative, reaching absolute values much larger than the positive shifts at moderate frequency. The falloff was reported to happen around 3.7 mHz by Jefferies (1998) – using $100 \leq \ell \leq 250$ modes – and Salabert *et al.* (2004) – using $\ell \leq 3$ modes, which is supported by Libbrecht and Woodard (1990) determinations ($5 \leq \ell \leq 60$). However, Howe, Komm, and Hill (2002) do not see any falloff for $\nu \leq 4$ mHz modes ($\ell \leq 300$) and Rhodes, Reiter, and Schou (2002) see the decrease at 5 mHz (using $\ell \leq 1000$ modes). The exact frequencies where the frequency shifts become

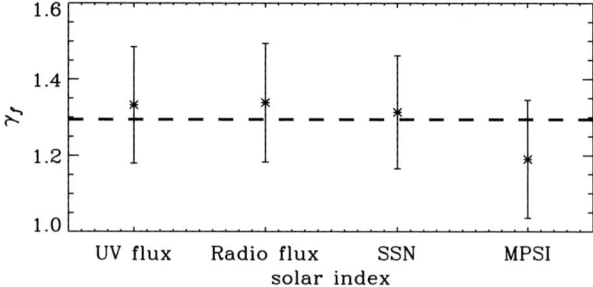
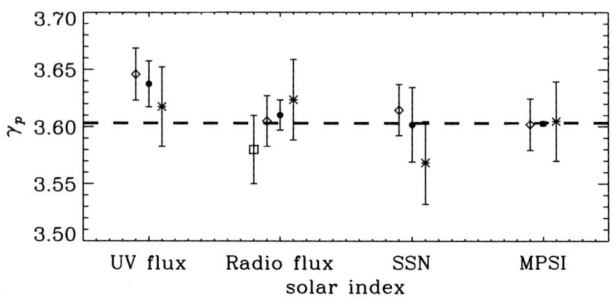

Figure 10 Exponent γ obtained fitting Equation (4) to the f (top) and p (bottom) modes for four different solar indices. In the top panel, the exponents were obtained using only the high-ℓ set. In the bottom panel, the diamonds and stars were obtained using the medium- and high-ℓ sets respectively and their weighted average is given by the circles. The square is the γ corresponding to the α value quoted in Chaplin *et al.* (2001) for the high-frequency interval. The fitting uncertainties are given by the error bars. The dashed lines are the weighted average of the four solar indices.

zero, negative, or reach a maximum negative value also vary widely between the publications. This sharp decrease in the frequency shift has been associated with an increase in the chromospheric temperature (*e.g.*, Goldreich *et al.*, 1991; Jain and Roberts, 1996). From Figure 8, high-frequency modes have a large probability P_u of being uncorrelated with the solar cycle, at least until 5.2 mHz. Note that $P_u(r)$ is calculated for the absolute value of r, estimating the probability of being correlated or anti-correlated with solar cycle. In our analysis, the frequency shift drops sharply around 4.6 mHz and there are a few modes (≈ 10) with a negative frequency shift with frequencies between 5 and 5.35 mHz. However, only modes with frequency smaller than 4.9 mHz obey the mode-selection criteria (*i.e.*, $P_u \leq 0.1\%$) and were included in Figure 9. The frequency shift decrease from 4.6 to 4.9 mHz is not seen in Figure 9 because of the mode-inertia normalization.

As first suggested by Libbrecht and Woodard (1990), the most significant sources of frequency shift must be localized near the solar surface since the frequency shifts seem to be independent of ℓ. Consider a small perturbation in the solar equilibrium model localized near the solar surface and the corresponding changes in the mode frequency, $\delta \nu_{n,\ell}$. For modes extending substantially more deeply than the region of the perturbation, the eigenfunctions are nearly independent of ℓ at fixed frequency in that region (see Figure 8 and the associated discussion in Christensen-Dalsgaard and Berthomieu, 1991). From this, it can be inferred that if the quantity $\delta \nu_{n,\ell} Q_{n,\ell}$ is independent of ℓ at fixed ν for a given set of modes, then the perturbation is probably largely localized outside the radius given by the maximum lower turning point of the set of modes considered (Christensen-Dalsgaard and Berthomieu, 1991). In the case of the frequency shift induced by the solar cycle, $\delta \nu_{n,\ell} Q_{n,\ell}$ is independent of ℓ at fixed ν for all modes analyzed here, where $\ell \leq 900$ and $\nu/L > 4.2$ μHz. Thus, the perturbation causing the frequency shift is probably localized in the region 4 Mm below the solar surface or less. Using the same reasoning, the thickness of the near-surface region where the uncertainties in the physics of the model are confined can be inferred. In this case, the set of modes where $\delta \nu_{n,\ell} Q_{n,\ell}$ is independent of ℓ at fixed ν is such that $\nu/L > 12$ μHz and

$\ell \leq 370$ using the high-ℓ frequency determination (Section 5.2). Including higher-degree modes, $\delta\nu_{n,\ell}Q_{n,\ell}$ is not a function of frequency alone anymore (see, for example, Figure 3 in Rabello-Soares *et al.*, 2000). As a result, the frequency uncertainties in the model are probably largely restricted to the layers 18 Mm below the surface.

7. Conclusion

In the determination of unbiased high-degree mode parameters, the instrumental characteristics must be taken into account in the image spatial decomposition, or in the leakage matrix calculation itself, to obtain a correct estimation of the relative amplitude of the spatial leaks. Among the instrumental characteristics analyzed here, the image scale is the one that affects the parameter determination the most. The image scale is the ratio of the image dimensions observed on the CCD detector and the dimensions in the actual Sun. An error in the image scale introduces an error in the estimated central frequency which increases with the mode frequency. A 0.27% error in the image scale would shift the estimated central frequency by as much as 11 µHz at 5 mHz. The radial distortion also has an important effect, which is expected since it is similar to an image scale error. An instrumental property not taken into account here (due to a lack of a good estimation) that could have an important effect on the measured parameters is an azimuthally-varying PSF.

The applied ridge-to-mode correction recovers frequencies at moderate degree that differ from the assumed corrected values by 1 µHz or less depending on the mode frequency. The fitting uncertainty of the recovered frequencies is in the range 0.07–0.18 µHz. The agreement for the a_1, a_2, and a_3-coefficients is very good, except maybe for f- and p_1-mode a_1-coefficients, their mean difference with respect to the assumed correct values is, respectively, three and two normalized by the ridge fitting uncertainties.

At high degree, the differences between our frequency determination and theoretical frequencies for the p modes shows the same general variation with degree as the results obtained with ring analysis. For $n \leq 5$, the global and ring analysis agree within 6 µHz. The high-degree f-mode frequencies obtained using ring analysis, like previous observations (Duvall, Kosovichev, and Murawski, 1998 and references within), are substantially lower than the theoretical frequencies. Surprisingly, the f-mode high-ℓ set frequencies agree well with the model frequencies (within 3 µHz) whereas the ring-analysis frequency differences can be as large as 13 µHz for $\ell > 700$ modes. The implications of the high-ℓ frequencies and splitting coefficients on the solar structure and rotation will be addressed in a future paper.

As noted by other authors for low- and moderate-degree modes (*e.g.*, Libbrecht and Woodard, 1990), the frequency shift induced by the solar cycle scales well with the mode inertia. We extended this analysis to high-degree modes and found out that scaling with the mode inertia normalized by the inertia of a radial mode of the same frequency follows a simple power law (given by Equation (4)) with one exponent for all frequency ranges, where the f and p modes are fitted independently of one another. The exponents obtained using four different solar indices agree within their fitting uncertainty, where: $\bar{\gamma}_f = 1.29 \pm 0.07$ and $\bar{\gamma}_p = 3.60 \pm 0.01$. The f-mode exponent is less than half of the p-mode value. The fundamental mode of solar oscillations has essentially the character of surface gravity waves and, contrary to the p modes, it is essentially incompressible and independent of the hydrostatic structure of the Sun. Hence, it is not a surprise that these different types of modes have different exponents. Due to their different properties, it is also very likely that different physical effects are responsible for their frequency variation. Accordingly to Dziembowski and Goode (2005), for the f modes, the dominant cause of frequency shift is the variation of the subphotospheric magnetic field. For the p modes, it is the decrease in the radial component of the turbulent velocity in the outer layers during the increase in solar activity, which is accompanied by a decrease in temperature (due to a decrease in the efficiency of convective transport). At low frequency ($\nu < 2.3$ mHz), the p-mode frequency shifts have a different behavior than at high ν: a step (with the same exponent γ_p) or, as found by other authors, an exponent twice as large as the one at high ν. Low-frequency modes have a large probability of being uncorrelated with the solar cycle, which was not taken into account in the case where a large exponent was estimated. The f-mode frequency shifts also have a different behavior around 1.7 mHz: they increase abruptly by an order of magnitude.

Modes with frequency around 3 mHz have the smallest probability that their variation is linearly uncorrelated with the solar index, while modes with $\nu < 2.5$ mHz or $\nu > 4.5$ mHz have the largest probability of being uncorrelated. A large probability (P_u) of being uncorrelated does not necessarily means that a given mode is not physically correlated with the solar cycle, instead it could be due to uncertainties in the measurements, that is a low signal-to-noise ratio. The logarithm of P_u is well correlated with the logarithm of the relative uncertainty of the estimated frequency-shift $\delta\nu^e$, the Pearson correlation coefficient is 0.71 for medium-ℓ modes and 0.54 for high-ℓ modes. If a given mode has a large probability of being linearly uncorrelated, it is expected that the linear fitting of its frequency shifts will have a large uncertainty, hence the high correlation coefficient between P_u and the estimated frequency-shift uncertainty. However, it raises the question of what could be the physical process that would make those modes less sensitive to solar activity. For a given ℓ, the upper reflection point for lower-frequency modes is deeper in the Sun than for high-frequency modes. If the perturbation layer causing the frequency shift is above the upper turning point of the mode, it would not be affected by the solar cycle. Accordingly to model S of Christensen-Dalsgaard, the depth of the upper turning point increases sharply with decreasing frequency below 2.3 mHz. The upper turning point for a radial mode with $\nu = 2$ mHz is 0.5 Mm deeper in the Sun than a 3-mHz mode (from Figure 2 in Chaplin *et al.*, 2001). At high-frequency, the observed frequency shift seems to suddenly drop to zero. However, there is no agreement on the exact frequency that this happens, the observed values range from 3.7 to 5 mHz. The frequency-shift falloff is explained by an increase of chromospheric temperature and magnetic field at solar maximum (Jain and Roberts, 1996 and references within). In the presence of an inclined magnetic field, high-frequency modes tunnel through the temperature minimum and are particularly sensitive to changes in the chromosphere, which are expected to be well correlated with solar activity.

Acknowledgements   We are grateful to Tim Larson of Stanford University for discussing with us the results of his improved analysis of MDI medium-ℓ data. The Solar Oscillations Investigation (SOI) involving MDI is supported by NASA grant NNG05GH14G at Stanford University. SOHO is a mission of international cooperation between ESA and NASA. SGK is supported by NASA grant NNG05GD58G. NOAA Mg II Core-to-wing ratio data are provided by Dr. R. Viereck, NOAA Space Environment Center. The solar radio 10.7 cm daily flux (2800 MHz) have been made by the National Research Council of Canada at the Dominion Radio Astrophysical Observatory, British Columbia. The International Sunspot Number was provided by SIDC, RWC Belgium, World Data Center for the Sunspot Index, Royal Observatory of Belgium. This study includes data from the synoptic program at the 150-Foot Solar Tower of the Mt. Wilson Observatory. The Mt. Wilson 150-Foot Solar Tower is operated by UCLA, with funding from NASA, ONR, and NSF, under agreement with the Mt. Wilson Institute. This work utilizes data obtained by the Global Oscillation Network Group (GONG) program, managed by the National Solar Observatory, which is operated by AURA, Inc. under a cooperative agreement with the National Science Foundation. The data were acquired by instruments operated by the Big Bear Solar Observatory, High Altitude Observatory, Learmonth Solar Observatory, Udaipur Solar Observatory, Instituto de Astrofísica de Canarias, and Cerro Tololo Interamerican Observatory.

References

Basu, S., Antia, H.M.: 1999, *Astrophys. J.* **525**, 517.

Basu, S., Antia, H.M.: 2000, *Astrophys. J.* **531**, 1088.
Basu, S., Antia, H.M.: 2003, In: Sawaya-Lacoste, H. (ed.) *Local and Global Helioseismology: The Present and Future* **SP-517**, ESA, Noordwijk, 231.
Basu, S., Antia, H.M., Bogart, R.S.: 2004, *Astrophys. J.* **610**, 1157.
Beck, J.G., Giles, P.: 2005, *Astrophys. J.* **621**, L153.
Bevington, P.R., Robinson, D.K.: 1992, In: *Data Reduction and Error Analysis for the Physical Sciences*, McGraw-Hill, New York.
Brodsky, M., Vorontsov, S.V.: 1993, *Astrophys. J.* **409**, 455.
Burtseva, O., Kholikov, S., Serebryanskiy, A., Chou, D.-Y.: 2007, *Solar Phys.* **241**, 17.
Chaplin, W.J., Appourchaux, T., Elsworth, Y., Isaak, G.R., New, R.: 2001, *Mon. Not. Roy. Astron. Soc.* **324**, 910.
Chaplin, W.J., Elsworth, Y., Miller, B.A., Verner, G.A., New, R.: 2007, *Astrophys. J.* **659**, 1749.
Christensen-Dalsgaard, J.: 2003, In: *Lecture Notes on Stellar Oscillations*, Aarhus University Press, Aarhus.
Christensen-Dalsgaard, J., Berthomieu, G.: 1991, In: Cox, A.N., Livingston, W.C., Matthews, M. (eds.) *Solar Interior and Atmosphere*, University of Arizona Press, Tucson, 401.
Christensen-Dalsgaard, J., Thompson, M.J., Gough, D.O.: 1989, *Mon. Not. Roy. Astron. Soc.* **238**, 481.
Christensen-Dalsgaard, J., Dappen, W., Ajukov, S.V., Anderson, E.R., Antia, H.M., Basu, S., Baturin, V.A., Berthomieu, G., Chaboyer, B., Chitre, S.M., *et al.*: 1996, *Science* **272**, 1286.
Corbard, T., Thompson, M.J.: 2002, *Solar Phys.* **205**, 211.
Di Mauro, M.P., Christensen-Dalsgaard, J., Rabello-Soares, M.C., Basu, S.: 2002, *Astron. Astrophys.* **384**, 666.
Duvall, T.L. Jr., Kosovichev, A.G., Murawski, K.: 1998, *Astrophys. J.* **505**, L55.
Dziembowski, W.A., Goode, P.R.: 2004, *Astrophys. J.* **600**, 464.
Dziembowski, W.A., Goode, P.R.: 2005, *Astrophys. J.* **625**, 548.
Foukal, P.: 1972, *Astrophys. J.* **173**, 439.
Goldreich, P., Murray, N., Willette, G., Kumar, P.: 1991, *Astrophys. J.* **370**, 752.
Gough, D.O.: 1990, In: Osaki, Y., Shibahashi, H. (eds.) *Progress of Seismology of the Sun and Stars*, Lecture Notes in Physics **367**, Springer, New York, 283.
Gough, D.O.: 1993, In: Weiss, W.W., Baglin, A. (eds.) *Inside the Stars* **CS-40**, Astron. Soc. Pac., San Francisco, 767.
Gough, D.O., Kosovichev, A.G., Toomre, J., Anderson, E., Antia, H.M., Basu, S., Chaboyer, B., Chitre, S.M., Christensen-Dalsgaard, J., Dziembowski, W.A., *et al.*: 1996, *Science* **272**, 1296.
Hill, F.: 1988, *Astrophys. J.* **333**, 996.
Howe, R., Komm, R.W., Hill, F.: 2002, *Astrophys. J.* **580**, 1172.
Howe, R., Christensen-Dalsgaard, J., Hill, F., Komm, R., Schou, J., Thompson, M.J., Toomre, J.: 2007, *Adv. Space Res.* **40**, 7, 915.
Jain, R., Roberts, B.: 1996, *Astrophys. J.* **456**, 399.
Jefferies, S.M.: 1998, In: Deubner, F.L., Christensen-Dalsgaard, J., Kurtz, D. (eds.) *New Eyes to See Inside the Sun and Stars*, IAU Symposium **185**, Kyoto, 415.
Korzennik, S.G.: 1990, Ph.D. Thesis, Univ. California, Los Angeles.
Korzennik, S.G.: 1998, In: Korzennik, S.G., Wilson, A. (eds.) *Structure and Dynamics of the Interior of the Sun and Sun-like Stars* **SP-418**, ESA, Noordwijk, 933.
Korzennik, S.G., Rabello-Soares, M.C., Schou, J.: 2004, *Astrophys. J.* **602**, 481. (KRS).
Kuhn, J.R., Bush, R.I., Emilio, M., Scherrer, P.H.: 2004, *Astrophys. J.* **613**, 1241.
Larson, T.P., Schou, J.: 2008, In: Gizon, L. (ed.) *Helioseismology, Asteroseismology and MHD Connections*, J. Phys.: Conf. Ser. in press.
Li, L.H., Basu, S., Sofia, S., Robinson, F.J., Demarque, P., Guenther, D.B.: 2003, *Astrophys. J.* **591**, 1267.
Libbrecht, K.G., Woodard, M.F.: 1990, *Nature* **345**, 779.
Murawski, K., Roberts, B.: 1993, *Astron. Astrophys.* **272**, 595.
Nigam, R., Kosovichev, A.G.: 1998, *Astrophys. J.* **505**, L51.
Rabello-Soares, M.C., Korzennik, S.G., Schou, J.: 2001, In: Wilson, A., Pallé, P.L. (eds.) *Helio- and Asteroseismology at the Dawn of the Millennium* **SP-464**, ESA, Noordwijk, 129.
Rabello-Soares, M.C., Korzennik, S.G., Schou, J.: 2005, *AGU Spring Meeting*, SP11B-08.
Rabello-Soares, M.C., Korzennik, S.G., Schou, J.: 2006, In: Fletcher, K., Thompson, M. (eds.) *Beyond the Spherical Sun* **SP-624**, ESA, Noordwijk, 71.1.
Rabello-Soares, M.C., Bogart, R.S., Basu, S.: 2008, In: Gizon, L. (ed.) *Helioseismology, Asteroseismology and MHD Connections*, J. Phys.: Conf. Ser. in press.
Rabello-Soares, M.C., Korzennik, S.G., Schou, J.: 2008, *Adv. Space Res.* **41**, 861.
Rabello-Soares, M.C., Basu, S., Christensen-Dalsgaard, J., Di Mauro, M.P.: 2000, *Solar Phys.* **193**, 345.
Rajaguru, S.P., Basu, S., Antia, H.M.: 2001, *Astrophys. J.* **563**, 410.

Reiter, J., Rhodes, E.J. Jr., Kosovichev, A.G., Schou, J., Scherrer, P.H.: 2002, In: Wilson, A. (ed.) *From Solar Min to Max: Half a Solar Cycle with SOHO* **SP-508**, ESA, Noordwijk, 91.
Reiter, J., Kosovichev, A.G., Rhodes, E.J. Jr., Schou, J.: 2003, In: Sawaya-Lacoste, H. (ed.) *Local and Global Helioseismology: The Present and Future* **SP-517**, ESA, Noordwijk, 369.
Reiter, J., Rhodes, E.J. Jr., Kosovichev, A.G., Schou, J.: 2004, In: Danesy, D. (ed.) *Helio- and Asteroseismology: Towards a Golden Future* **SP-559**, ESA, Noordwijk, 61.
Rhodes, E.J. Jr., Reiter, J., Schou, J.: 2002, In: Wilson, A. (ed.) *From Solar Min to Max: Half a Solar Cycle with SOHO* **SP-508**, ESA, Noordwijk, 37.
Ritzwoller, M.H., Lavely, E.M.: 1991, *Astrophys. J.* **369**, 557.
Salabert, D., Fossat, E., Gelly, B., Kholikov, S., Grec, G., Lazrek, M., Schmider, F.X.: 2004, *Astron. Astrophys.* **413**, 1135.
Schou, J.: 1999, *Astrophys. J.* **523**, L181.
Schou, J., Bogart, R.S.: 1998, *Astrophys. J.* **504**, L131.
Schou, J., Antia, H.M., Basu, S., Bogart, R.S., Bush, R.I., Chitre, S.M., Christensen-Dalsgaard, J., di Mauro, M.P., Dziembowski, W.A., Eff-Darwich, A., *et al.*: 1998, *Astrophys. J.* **505**, 390.
Snodgrass, H.B.: 1992, In: Harvey, K.L. (ed.) *The Solar Cycle* **CS-27**, Astron. Soc. Pac., San Francisco, 250.
Toutain, T., Appourchaux, T., Fröhlich, C., Kosovichev, A.G., Nigam, R., Scherrer, P.H.: 1998, *Astrophys. J.* **506**, L147.
Woodard, M.F.: 1989, *Astrophys. J.* **347**, 1176.

Recent Developments in Local Helioseismology

M.J. Thompson · S. Zharkov

Originally published in the journal Solar Physics, Volume 251, Nos 1–2, 225–240.
DOI: 10.1007/s11207-008-9143-6 © Springer Science+Business Media B.V. 2008

Abstract Local helioseismology is providing new views of subphotospheric flows from supergranulation to global-scale meridional circulation and for studying structures and dynamics in the quiet Sun and active regions. In this short review we focus on recent developments, and in particular on a number of current issues, including the sensitivity of different measures of travel time and testing the forward modelling used in local helioseismology. We discuss observational and theoretical concerns regarding the adequacy of current analyses of waves in sunspots and active regions, and we report on recent progress in the use of numerical simulations to test local helioseismic methods.

Keywords Sun: local helioseismology · Sun: time-distance analysis · Sun: forward problem · Sun: inversion

1. Introduction

Helioseismology provides a remarkable capability to image beneath the surface of the Sun and thus to test our understanding of the physics of solar and stellar interiors. Through the study of properties of global modes, and in particular their frequencies, global-mode helioseismology has enabled us to image the radial variation of the Sun's hydrostatic structure through most of the interior as well as the radial and latitudinal variation of its rotation rate. However, the global-mode frequencies do not discriminate in the longitudinal direction, nor do they distinguish between northern and southern hemispheres. The global eigenfunctions can in principle be used to probe such variation, but the global modes are not the natural tool to explore such aspects. Rather, with the advent of high-spatial-resolution helioseismic observations of the solar surface, a number of different techniques have been developed, going

Invited Review.

Helioseismology, Asteroseismology, and MHD Connections
Guest Editors: Laurent Gizon and Paul Cally.

M.J. Thompson (✉) · S. Zharkov
Department of Applied Mathematics, School of Mathematics and Statistics, University of Sheffield, Hounsfield Road, Sheffield S3 7RH, UK
e-mail: michael.thompson@sheffield.ac.uk

under the generic name of local helioseismology. Local helioseismic techniques are much better suited to the task of studying horizontally localised features than is global-mode helioseismology. These techniques are providing us today with subphotospheric maps of supergranulation (Duvall *et al.*, 1997) and of flow patterns surrounding active regions (Gizon, Duvall, and Larsen, 2001; Haber *et al.*, 2001; Zhao and Kosovichev, 2004). Local helioseismological methods help us detect large-scale azimuthal and meridional flows (Giles *et al.*, 1997; Basu, Antia, and Tripathy, 1999) and to monitor active regions on the Sun's far surface (Lindsey and Braun, 2000). They have also allowed us to take a first look into internal structures beneath sunspots and active regions (*e.g.*, Kosovichev, Duvall, and Scherrer, 2000).

The principal local helioseismic techniques in common use are time–distance helioseismology (Duvall *et al.*, 1993), ring-diagram analysis (Hill, 1988), and helioseismic holography (Lindsey and Braun, 1997) and the closely related method of acoustic imaging (Chou, 2000). Each has a different conceptual basis. Ring-diagram analysis is closest to global-mode helioseismology: what is determined is the local dispersion relation of waves observed in a localised region of solar surface. Time–distance helioseismology works in terms of travel times of wave packets travelling between pairs of points on the solar surface; the wave packets are made manifest by cross-correlating the oscillatory signals at the surface points. Helioseismic holography reconstructs subsurface wave fields from surface observations, in particular the ingoing and outgoing waves with respect to a pupil, from which correlations can be measured and phase shifts derived. Other methods of local helioseismology, which shall not however be mentioned further in this paper, are Hankel spectral analysis (Braun, Duvall, and Labonte, 1987) and statistical waveform analysis (Woodard, 2002).

Local helioseismology has been excellently reviewed by Gizon and Birch (2005). Here we consider some of the developments in the field since then. We do not presume to a complete review of all such developments. Indeed we focus largely on time–distance helioseismology. For a recent short review of local helioseismology focussing on the techniques, see Birch (2006). A discussion of some outstanding issues in time–distance helioseismology is presented by Gizon and Thompson (2007).

2. The Fundamental Observable of Time–Distance Helioseismology

Time–distance helioseismology proceeds by cross-correlating wave-field measurements (typically Doppler velocities) at different locations on the solar surface. These cross-correlations are then compared with a reference cross-correlation, obtained from a model.

The equations of solar oscillations may be combined in the form

$$L[\xi] = \mathbf{s}, \qquad (1)$$

where $L = \rho \partial_t^2 + \cdots$ is a linear differential operator governing adiabatic oscillations and $\mathbf{s}$ is a stochastic source function. Operator L depends on the solar interior parameters such as sound speed (c), density (ρ), bulk flow ($\mathbf{u}$), magnetic field ($\mathbf{B}$), and damping. A seismic model is one that fits the observed data and is consistent with the estimated properties of the data errors. Given a set of observations, one task of helioseismology is to construct, if possible, a seismic model, that is, to obtain such a distribution of the underlying parameters that fits the totality of the observed data. The common approach to this task in local helioseismology is to make some basic assumptions about wave sources and then to consider the linearised perturbations to some initial model of the solar interior. The only perturbations commonly considered to date have been to introduce a flow and to allow a modification of the sound-speed distribution [*i.e.*, the operator is perturbed from $L(c, \ldots; \mathbf{u} = 0)$ to $L(c + \delta c, \ldots; \mathbf{u})$].

2.1. Filtering

By ignoring systematic errors such as instrumental and geometric artifacts (*e.g.*, foreshortening), the line-of-sight wave velocity observed at some height $z = z_{\text{obs}}$ in the solar atmosphere can be written as

$$\phi = \text{PSF} * \left[\hat{\mathbf{n}} \cdot \partial_t \boldsymbol{\xi}(\mathbf{r}, z_{\text{obs}}, t)\right], \tag{2}$$

where $\mathbf{r}$ is a horizontal position vector on the solar surface defined by the height z_{obs}, t is time, $\hat{\mathbf{n}}$ is a unit vector pointing in the direction of the observer, and $\boldsymbol{\xi}$ is the wave displacement vector. Equation (2) also contains the convolution (indicated with the star operator) with the instrument point-spread function (PSF). In local helioseismology, a reference frame that is co-rotating with the Sun is normally used to remove the main component of the rotation.

Further, the oscillation signal may be filtered in the Fourier domain to obtain

$$\psi(\mathbf{k}, \omega) = F(\mathbf{k}, \omega)\phi(\mathbf{k}, \omega), \tag{3}$$

where $\mathbf{k}$ is the horizontal wave vector, ω is the angular frequency, and F is a filter function chosen by the observer to remove granulation noise and to select parts of the wave-propagation diagram. These may include, but are not limited to, a filter selecting data on the f-mode ridge for working with f modes, or removal of the f-mode ridge to leave acoustic modes only, combined with phase-speed filters (Duvall *et al.*, 1997).

2.2. Cross-Correlations and Travel Times

The *fundamental observable* of time–distance analysis is the cross-correlation function between the filtered Doppler velocities at two surface locations $\mathbf{r}_1$ and $\mathbf{r}_2$:

$$C(\mathbf{r}_1, \mathbf{r}_2, t) = \frac{1}{T - |t|} \int \psi(\mathbf{r}_1, t')\psi(\mathbf{r}_2, t' + t)\,dt', \tag{4}$$

where T is the duration of the observation. Positive time lags ($t > 0$) provide information about waves propagating from $\mathbf{r}_1$ to $\mathbf{r}_2$, whereas negative time lags ($t < 0$) give information about waves travelling in the opposite direction.

A seismic model in a strong sense would be a three-dimensional model of the solar interior that was consistent with the large set of observed cross-correlations. Full waveform modelling to fit the cross-correlations has not yet been attempted. Instead, only a few parameters are chosen to describe the behaviour of the cross-correlation. Indeed, commonly a single functional measure obtained from the cross-correlation is used: Travel time is the widely used parameter in time–distance helioseismology. We note in passing that the phase perturbations of helioseismic holography are also commonly expressed in terms of travel-time perturbations (see, *e.g.*, Braun *et al.*, 2007).

By expressing the wave field as a superposition of global normal-mode solutions for standing waves of a spherically symmetric Sun, Kosovichev and Duvall (1997) and Giles (1999) modelled the measured cross-correlation function as a Gabor wavelet, a five-parameter analytic function incorporating the values of the central frequency (ω_0) of the wave packet, its width ($\Delta\omega$), group travel time (t_g), and phase travel time (t_p), as well as its amplitude (A):

$$G(\mathbf{p}, t) = A \exp\left[-\frac{\Delta\omega^2}{4}(t - t_g)^2\right]\cos\left[\omega_0(t - t_p)\right]. \tag{5}$$

By **p** we denote the vector of parameters $\{A, \Delta\omega, t_g, \omega_0, t_p\}$. Travel times are then obtained by fitting expression (5) to the measured cross-correlation function. Generally the phase travel-time parameter (t_p) is taken as the measured travel time and we shall do that here. A travel-time perturbation is then computed as the difference between the travel time so obtained and the corresponding travel time in an unperturbed medium, for example obtained from a Gabor-wavelet fit to the cross-correlation signal averaged over a region of quiet Sun. Nigam, Kosovichev, and Scherrer (2007) have recently provided a generalised formula for the Gabor wavelet, taking into account both the radial and horizontal components of the oscillation displacement signal as well as phase-speed filtering used in measuring the observed cross-correlation signal.

Motivated by studies in geophysics, Gizon and Birch (2002, 2004) have developed a definition of travel time that consists of linearising the difference between the observed cross-correlation signal and a sliding reference cross-correlation signal (C^{ref}):

$$X_{\pm}(\mathbf{r}_1, \mathbf{r}_2, \tau) = \int_{-\infty}^{\infty} dt' f(\pm t') \left(C(\mathbf{r}_1, \mathbf{r}_2, t') - C^{\text{ref}}(\mathbf{r}_1, \mathbf{r}_2, t' \mp \tau) \right)^2. \quad (6)$$

Here $f(\pm t)$ is a window function designed to separate waves travelling in different directions (*i.e.*, positive and negative branches of the cross-correlation function) and also different bounces for the p modes. The travel-time perturbation for waves travelling between $\mathbf{r}_1$ and $\mathbf{r}_2$ is then defined as the value of τ that minimises $X_{\pm}$, which leads to the following formula for computing travel-time perturbations:

$$\tau_{\pm} = \frac{\sum_i \mp f(\pm t_i) \dot{C}^{\text{ref}}(t_i) [C(t_i) - C^{\text{ref}}(t_i)]}{\sum_i f(\pm t_i) [\dot{C}^{\text{ref}}(t_i)]^2}, \quad (7)$$

where the dots denote derivatives with respect to time. The travel-time perturbation can be expressed as a summation over the values at the discrete (presumed uniformly spaced) times (t_i) at which the observations were made.

The reference cross-correlation function (C^{ref}) may be calculated by solving the hydrodynamic equations in a solar model with stochastically distributed sources (Gizon and Birch, 2002; Birch, Kosovichev, and Duvall, 2004). Indeed, one attractive feature of the Gizon–Birch definition of travel time is that it allows an immediate interpretation in terms of perturbations to a reference solar model. Alternatively, as for some of the results presented later in Section 4, the reference cross-correlation can be obtained from observation as an average cross-correlation over a region of quiet Sun.

To gain some insight into what the Gizon–Birch method's travel-time measure is actually sensitive to, one may consider applying it in the case where a simulated cross-correlation function and the reference cross-correlation both take the form of Gabor wavelets. Then, neglecting the effect of the window function, and using Equation (5) and the continuous version of Equation (7), one can show that

$$\tau_{+} \approx -\int_{-\infty}^{\infty} \frac{\partial G}{\partial \mathbf{p}} \cdot \delta \mathbf{p} \frac{\partial G}{\partial t} dt \bigg/ \int_{-\infty}^{\infty} \left(\frac{\partial G}{\partial t} \right)^2 dt, \quad (8)$$

where G and **p** are the same as in Equation (5). By considering the response of τ_{+} to individual perturbations of the parameters contained in **p**, it can be shown that to first order the perturbations to amplitude (A) and to wave packet width ($\Delta\omega$) have no effect on the travel-time perturbation obtained via the Gizon–Birch definition, whereas the response to a perturbation in phase or group travel times would be

$$\tau_+ = \delta t_p + (\delta t_p - \delta t_g)\frac{\Delta\omega^2}{\Delta\omega^2 + 4\omega_0^2}. \tag{9}$$

Note that in practice the second term is small ($\Delta\omega$ generally being rather smaller than ω_0), so that the measured travel-time perturbation is approximately equal to the perturbation in the phase travel time. Finally, a perturbation ($\delta\omega_0$) to the central frequency (ω_0) of the Gabor wave packet would also introduce changes to the travel-time measured by the Gizon–Birch method if the waves were dispersive:

$$\tau_+ \approx (t_p - t_g)\frac{4\omega_0^2}{\Delta\omega^2 + 4\omega_0^2}\frac{\delta\omega_0}{\omega_0}. \tag{10}$$

An observational comparison between the results of applying the two travel-time definitions – Gabor-wavelet fitting and Gizon–Birch – has recently been made (Roth, Gizon, and Beck, 2007). Using a SOHO/MDI Structure Program 72-hour times series of Dopplergrams for the quiet Sun, the study shows good agreement between the inferred travel times for the two methods of measurement.

3. Modelling

Forward modelling is the process of computing the wave field or some of its observable properties, based on a prescribed model of the solar structure and, for example, the background flow, magnetic field, or mode physics (excitation, damping, etc.). We shall refer to all of these as background properties. In general, changes to the background properties will result in changes to the wave field and its observable properties. An outcome of the modelling is often a linear-response kernel, which quantifies how a small change in background properties will change some observable property. The change in observable property is then expressible as a convolution of the kernel and the changes in background properties. The formulation of helioseismic linear-response kernels using the Born approximation was first presented by Birch and Kosovichev (2000) and using the Rytov approximation by Jensen and Pijpers (2003).

Gizon and Birch had the insight to relate the changes in background properties to changes in the fundamental observable of time–distance helioseismology, namely the cross-correlation function. From there, one can compute as required the kernel relating those changes to the change in, say, the measured travel time. Following such a procedure, the linearised relationship between changes in the background properties and changes in, for example, the travel time (τ) can be written as

$$\tau = \sum_\alpha \int K_\alpha \delta q_\alpha \, dV, \tag{11}$$

where α labels different background properties q_α whose change may affect the observable. Thus the contribution that each perturbation (δ_q) makes to the travel-time perturbation (τ) is a convolution of that δq_α with a kernel K_α.

3.1. Empirical Test of Kernel Calculations

The linear-response kernels have been computed from theoretical considerations. Recently, however, Duvall, Birch, and Gizon (2006) have devised a means of estimating empirically the linear-response kernel in a particular circumstance. They have measured the mean travel-time perturbations for f modes in the vicinity of isolated, small magnetic features. A small magnetic feature may perturb the travel times through various physical effects, but we assume that all of the contributions can be parameterised by a single parameter,

$$\delta q_\alpha = c_\alpha \delta q, \qquad (12)$$

where the vector of coefficients c_α is the same for all magnetic elements. Hence

$$\tau = \int K \delta q \, dV, \qquad (13)$$

where $K = \sum_\alpha c_\alpha K_\alpha$. This convolution may be written conveniently in Fourier space, where it simply becomes a product:

$$\tau(\mathbf{k}) = K(\mathbf{k}) \delta q(\mathbf{k}). \qquad (14)$$

We have used the same symbols for quantities and their Fourier transforms: The latter are distinguished by being written explicitly as functions of the horizontal wavenumber ($\mathbf{k}$).

In the analysis of Duvall, Birch, and Gizon (2006) the single perturbative parameter (δq) is the absolute photospheric magnetic field strength. Since the travel-time perturbations can be measured in the vicinity of many such isolated magnetic features, the Fourier-transformed kernel can then be estimated by making a fit of expression (14) to the observed magnetograms (δq) and travel-time perturbations (τ), for each $\mathbf{k}$ separately.

Of course the resulting kernel corresponds to whatever combination of physical effects is present in the ensemble of magnetic features. The resulting kernels obtained by Duvall, Birch, and Gizon show close similarity to the theoretical kernels for wave damping computed by Gizon and Birch (2002).

The results from this new technique are fascinating as they are the first empirical verification of the kernels produced from theory. The f-mode kernels are essentially two dimensional, because the f modes are confined close to the surface. By contrast, p-mode kernels are intrinsically three dimensional. The technique devised by Duvall, Birch, and Gizon (2006) can be extended to p modes. However, because the perturber is confined to the near-surface layers (the ratio of magnetic to gas pressure presumably falling off very quickly with increasing depth), the results will only provide the kernels evaluated at the surface.

3.2. Validity of Time–Distance Kernels for Flows

Jackiewicz et al. (2007) have made a careful study of flow-speed kernels for interpreting f-mode travel-time data. One question that they address is the range of flow speeds in which the linear approximation is valid. By considering a uniform flow velocity ($\mathbf{u}$), and making an expansion in powers of the small parameter ($\mathbf{u} \cdot \mathbf{k}$), they find that the linearisation of travel times with the flow amplitude is valid to within 10% for flow amplitude less than 250 m s^{-1} and travel distances less than 25 Mm. For larger distances or flow speeds, nonlinear effects become important. The same authors further deduce that the main effect on travel times of the flow is its advection of the waves: Advection of the sources may be neglected to a very good approximation when calculating the kernels.

Jackiewicz *et al.* also show that, at small spatial scales, the kernels for Doppler velocity measurements depend significantly on position on the solar disk and on the angle between the observation points and the center-to-limb direction. This means that the kernels are not translationally invariant and so the inverse problem is not simply a deconvolution. Close to disk centre one may hope that translational invariance is still a reasonable approximation.

3.3. Ring-Analysis Kernels

It is assumed in inversions of ring-analysis data that the measured ring parameters are sensitive only to conditions directly beneath the region in which the observations were made. Hindman *et al.* (2005) have sought to test this assumption by a series of forward-modelling experiments. The regions of observation are typically rectangular "tiles," and within each tile a circular apodization is normally applied to the data before the power spectrum is calculated. Hindman *et al.* indeed find that the sensitivity of the ring parameters to various flows is essentially zero outside the tile and is proportional to the apodization function within the tile. These results have been confirmed by the recent work of Birch *et al.* (2007), who have computed the linear-response kernels of ring parameters to flows. Specifically, they calculate the linear sensitivity to small changes in the local power spectrum and then compute the sensitivity of the power spectrum to time-independent weak local flows. By combining these two results, they obtain the three-dimensional kernels that give the linear sensitivity of ring parameters to both vertical and horizontal flows. The sensitivity to flows is essentially limited to the region where the apodization is nonzero, and the depth dependence of the kernels is very close to the mode's kinetic-energy density.

4. Helioseismology of Sunspots and Active Regions

Local helioseismology has provided intriguing views of the structures and flows under sunspots and active regions. However, there is a growing body of evidence that there are surface effects and other inadequacies of the modelling of wave propagation in these regions that may introduce systematic errors into the present local helioseismic inferences.

The forward problem as applied to the analysis of helioseismic data is currently almost universally based on solving nonmagnetic hydrodynamical equations. Sunspots and active regions are then considered as local perturbations. Most of the analysis to date considers that the sunspot or active region introduces only an isotropic perturbation to the propagation speed of the waves.

Cally and collaborators (see Cally, 2007, and references therein) have emphasised the need for more sophisticated modelling and interpretation of wave propagation in strongly magnetised regions. Their analytical and numerical work highlights several effects on seismic waves, including wave transmission and conversion at the layer where the values of the Alfvén speed and sound speed are equal, and a directional filtering of acoustic waves entering the overlying atmosphere. Work is currently being undertaken by several groups to include MHD effects in numerical simulations of wave propagation suitable for testing local helioseismic techniques.

Using helioseismic holography, Schunker, Braun, and Cally (2007) have found that the measured phase shifts and equivalent travel-time perturbations in sunspots depend on the line-of-sight angle in the plane containing the magnetic field and the local vertical. This appears to be a surface effect. The existence of the effect in time–distance Dopplergram data has been confirmed by Zhao and Kosovichev (2006), who also find however that the effect does not exist for continuum-intensity data.

Couvidat and Rajaguru (2007) have investigated travel times and associated sound-speed perturbations associated with sunspots. They detect rings of negative mean travel-time perturbations in the travel-time maps of most of the sunspots that they have studied, for travel distances between 8.7 and 11.6 Mm. These rings (see also the top panels in Figure 1) produce significant arclike or ringlike structures in the inversion results, mimicking regions of increased sound speed. The apparent structures are located beneath the sunspot penumbrae and are sensitive to the frequency filtering applied. Couvidat and Rajaguru conclude that the rings are most probably artifacts caused, in a way not yet accounted for, by near-surface interaction between the waves and the magnetic field.

The strong reduction of p-mode power in regions of strong surface magnetic field is well known (*e.g.,* Hindman and Brown, 1998; Nicholas, Thompson, and Rajaguru, 2004; and references therein). Possible causes of this reduction include mode conversion and suppression of wave sources within the active region. Rajaguru *et al.* (2006) have shown that the reduction in acoustic power interacts with a phase-speed filter, if one is applied in the data reduction, to introduce spurious travel-time shifts. An *ad hoc* solution to reduce this systematic error is to artificially restore the level of the acoustic power signal in the magnetised region before proceeding with the rest of the data analysis. In helioseismic holography, Lindsey and Braun (2003) have coined the term "showerglass effect" for the modulation of amplitudes and phases in regions of strong surface magnetic field.

Jensen, Pijpers, and Thompson (2006) have reported strong asymmetries between cross-correlation functions corresponding to waves propagating towards or away from regions of strong magnetic field.

Whilst the Gizon–Birch and Gabor wavelet-fitting methods for travel times give results that are in good agreement for quiet-Sun measurements (Roth, Gizon, and Beck, 2007), there may be significant differences when measurements are made in and around sunspots. The wave-packet central frequency obtained from Gabor wavelet fitting is seen to be lower in a sunspot than in the quiet Sun (Figure 2). According to Equation (10), the Gizon–Birch travel times can be sensitive to a shift in central frequency, and hence this may introduce a significant bias compared to measured Gabor-wavelet phase travel times. Evidently, it is important for any inversion in active regions that the kernels used are computed in a manner that is consistent with the definition of the measured travel times. Also Nigam, Kosovichev, and Scherrer (2007) have shown that although the phase-speed filtering procedure does not change the functional form of the basic time–distance Gabor fitting formula (5), it systematically shifts the travel times, if the central phase speed of the filter is different from the actual phase or group speeds for a given distance. Thus, if a change in wave-packet phase speed occurs in a sunspot region, perhaps because of a shift in the frequency of the wave packet (as seen in Figure 2), this might introduce systematic artifacts in the travel times measured using Gabor-wavelet fitting.

One might expect, at least to a first approximation, that any effect of subsurface anomalies on measured travel times would depend on the phase speed of the waves but not on their frequency at fixed phase speed. Braun and Birch (2006) have used helioseismic holography to measure travel-time perturbations within active regions after filtering both according to phase speed and frequency. At fixed phase speed, they find that the inferred travel-time perturbations do depend on frequency. In fact, the functional dependence of the travel-time perturbations on the modes in the region left by the filtering is roughly inversely proportional to mode mass. This is strongly indicative that a significant part of the observed negative travel-time perturbation originates near the surface. In global-mode inversions and some ring-analysis inversions for structure, such a component of the data is explicitly accounted for (Dziembowski, Pamyatnykh, and Sienkiewicz, 1990; Simmons and Basu, 2003). This

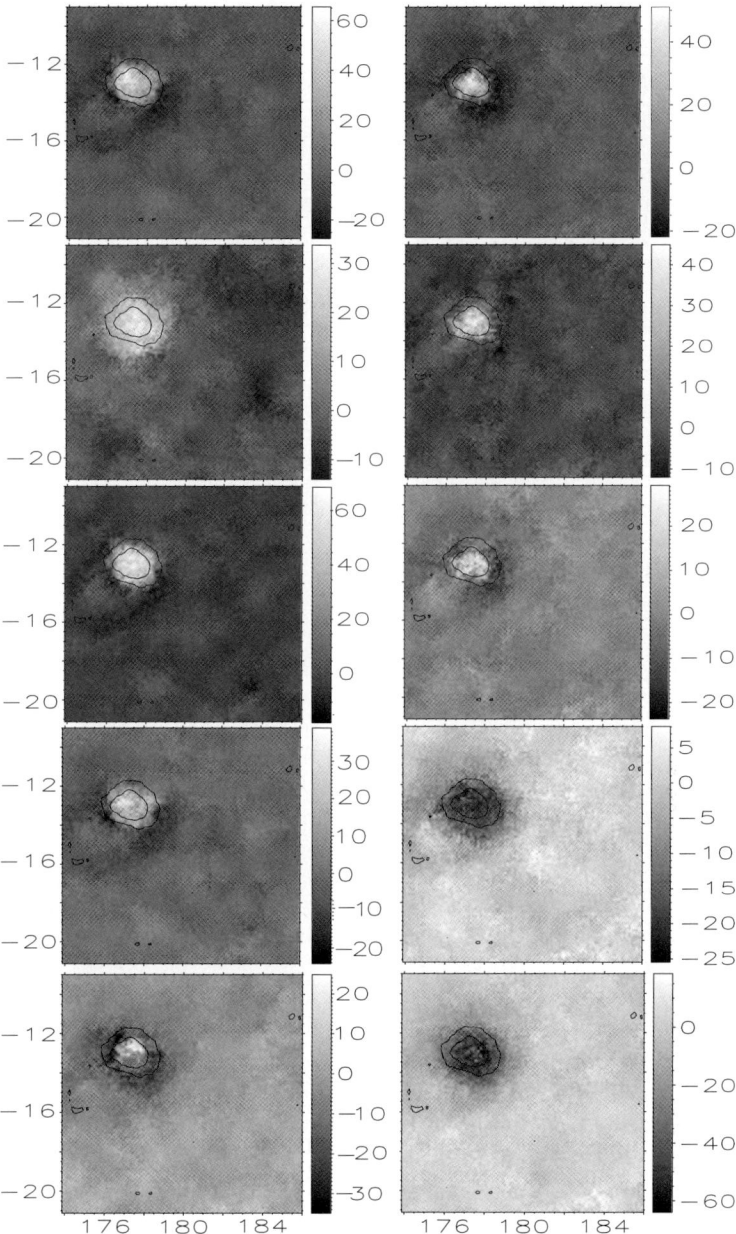

Figure 1 Mean travel times obtained from NOAA 9056 data via Gabor wavelet fitting for skip distances of 7.29 Mm (left column) and 11.664 Mm (right column). The grey scale is in seconds, Carrington longitude is plotted along the x-axis, and latitude is along the y-axis. The phase-speed filter values were as in Couvidat, Birch, and Kosovichev (2006). Top row: Travel times obtained without frequency bandpass filter. From the second row to the bottom row are the mean travel times obtained by applying additional frequency filters centred at 3.0, 3.5, 4.0, and 4.5 mHz. The contours are deduced from the magnetograms and correspond to -150 and -700 gauss.

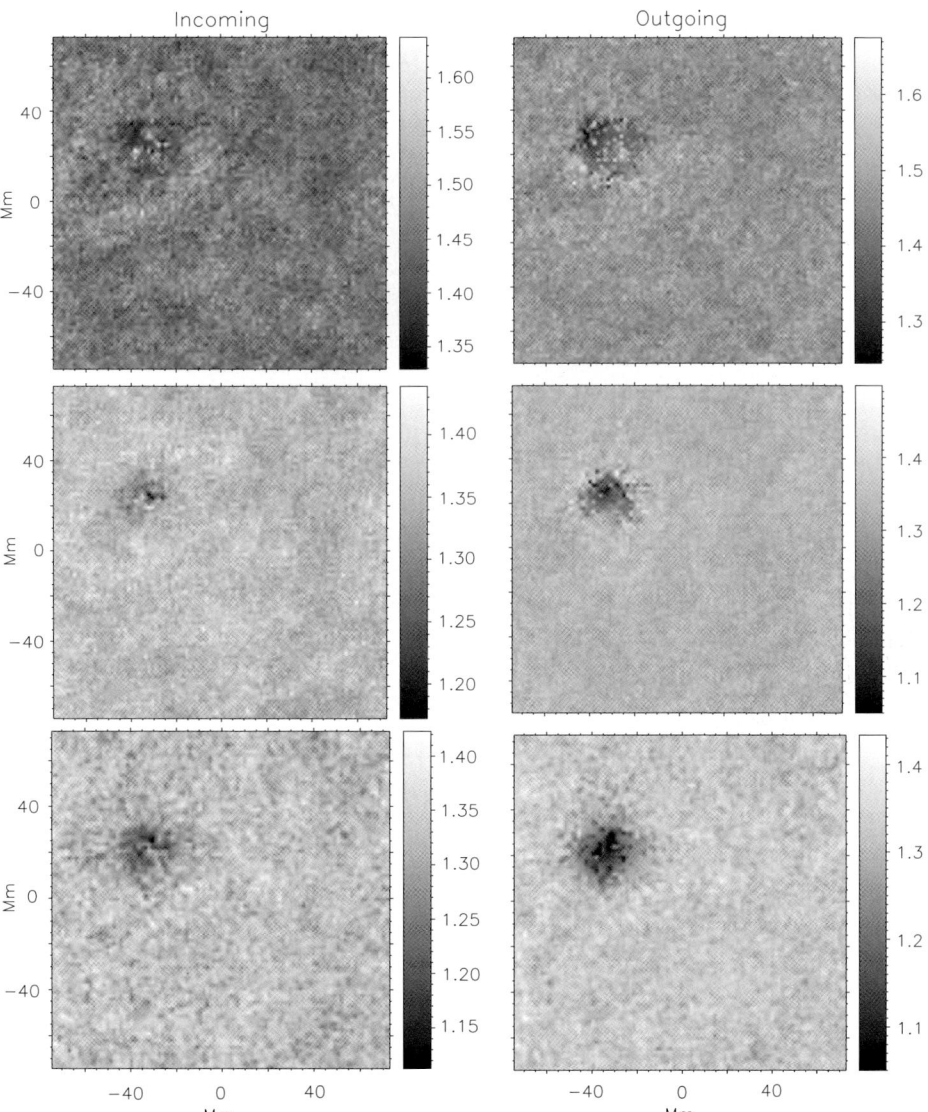

Figure 2 Results of Gabor-wavelet fitting for the central frequency (ω_0) for incoming (left) and outgoing (right) waves extracted from NOAA 9056 SOHO/MDI data. The skip distances used are (top to bottom rows) 7.29, 11.664, and 16.038 Mm. The grey scale in all images is in min^{-1}.

suggests a similar inversion strategy for travel-time perturbations, such that these might be fitted as a sum of a contribution from the interior plus a function $F(\nu)$ of frequency multiplied by a horizontal weighting function Φ and inversely weighted by mode mass $\mathcal{M}$:

$$\tau = \sum_\alpha \int K_\alpha \delta q_\alpha \, dV + \Phi \mathcal{M}^{-1} F(\nu). \tag{15}$$

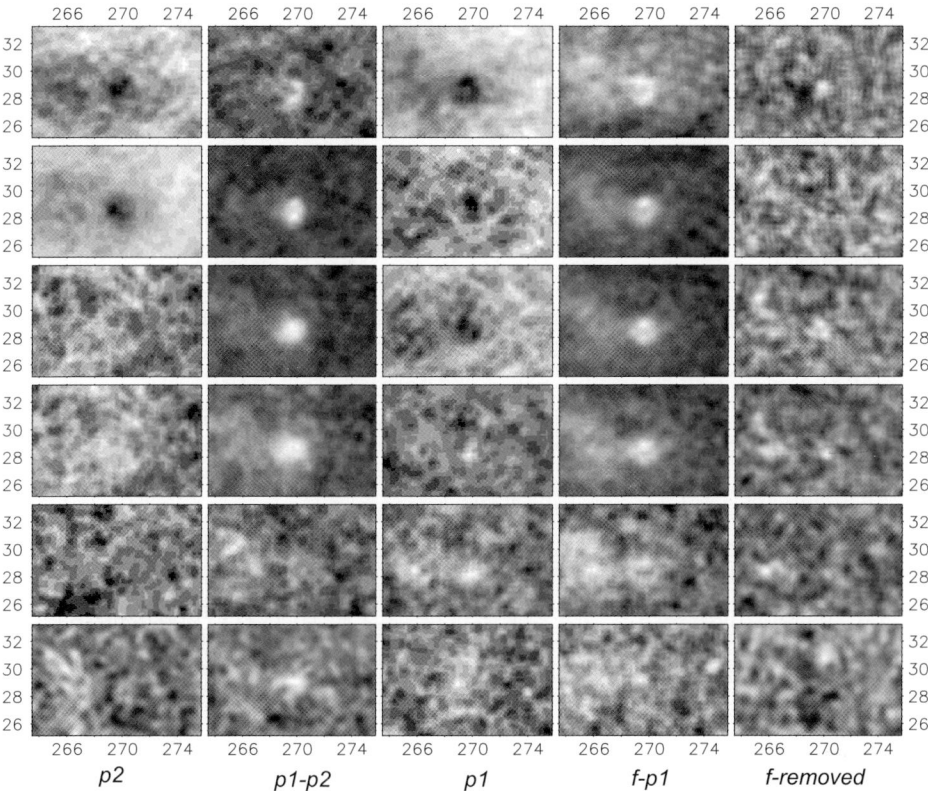

Figure 3 Gizon–Birch travel-time perturbations for isolated sunspot NOAA 9779 obtained for various filtering schemes: Dark tones represent negative perturbations; light tones represent positive perturbations. The skip distance is equal to 5.89 Mm. From the right, the columns, respectively, have filters applied as follows: filter to remove the f mode and lower frequencies; pass-filter centred between the f and p_1 ridges; pass-filter centred on the p_1 ridge; pass-filter centred between the p_1 and p_2 ridges; pass-filter centred on the p_2 ridge. From the bottom, the rows have bandpass filters centred on 2.0, 2.5, 3.0, 3.5, 4.0, and 4.5 mHz, respectively. More details are given in the text.

A further very interesting result has recently been presented by Braun and Birch (2008). This relates to the positive travel-time perturbations obtained for short skip distances in sunspots and widely interpreted as indicating a region of slower wave propagation in the shallow subsurface layers beneath the spot. By investigating where in the space of frequency and phase speed the positive travel-time perturbations show up, and also by applying ridge filters rather than the more usual phase-speed filtering, Braun and Birch find evidence that the positive perturbations arise from the p_1 ridge or beneath it and are not seen in the higher order p-mode data. A similar finding is illustrated in Figures 3 and 4. Here we present travel-time perturbations measured in the vicinity of isolated sunspot NOAA 9779 relative to the surrounding quiet Sun, using a centre-to-annulus geometry and skip distances of 5.83 Mm (Figure 3) and 16.04 Mm (Figure 4). For each row, a bandpass filter was used to select data within a 1-mHz frequency band with a 0.2-mHz Gaussian roll-off, centred (from bottom to top) at 2.0, 2.5, 3.0, 3.5, 4.0, and 4.5 mHz. These were combined for each column (left to right) with filters selecting the data from the p_2 ridge, in between the p_1 and p_2 ridges, the p_1 ridge, in between f and p_1 ridges, and finally removing only the f-ridge

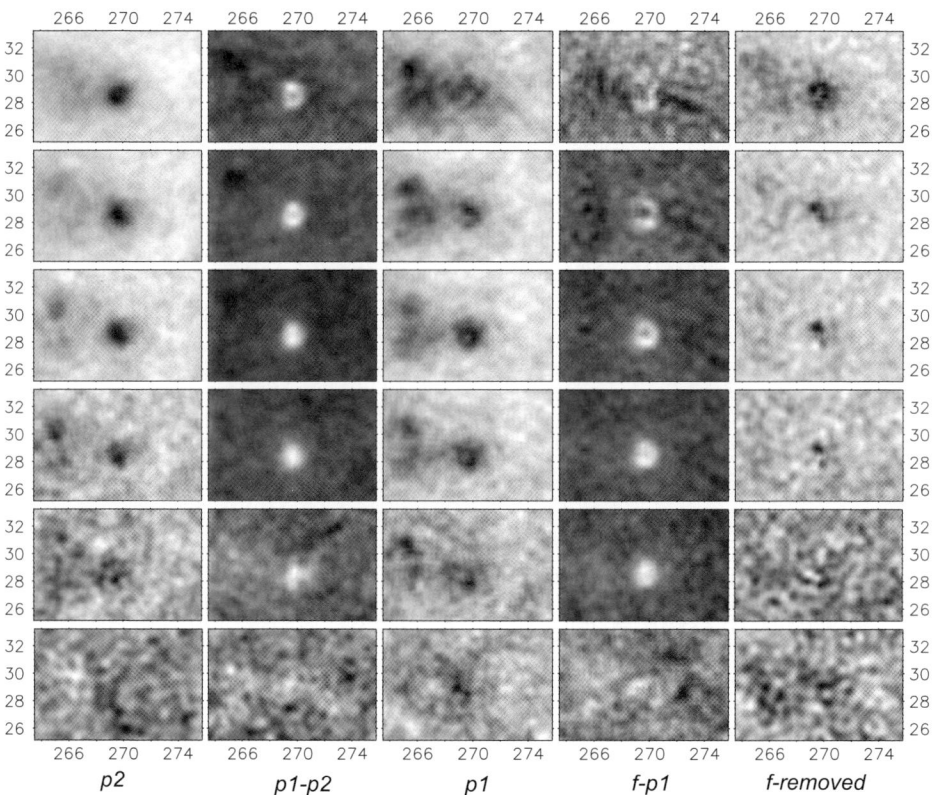

Figure 4 Same as Figure 3, but for a skip distance of 16.038 Mm.

data. The filters were constructed as follows: At constant frequency we apply a filter that takes the value of unity at the horizontal wavenumber corresponding to either a particular ridge, for example, p_1 (a "ridge filter"), or a midpoint between the adjacent ridges, for example, $p_1 - p_2$ (an "off-ridge filter"). On either side of this centre line the filter has a Gaussian roll-off, with half width at half maximum (HWHM) equal to 0.3175 times the distance to the neighbouring ridge on that side for the ridge filter, and with HWHM equal to 0.625 times the distance to the adjacent ridge in the case of the off-ridge filter. The "f-removed" column corresponds to applying simply a standard filter to remove f-ridge data as described by Giles (1999). No phase-speed filter was applied. To illustrate the ridge and off-ridge filters, Figure 5 shows the filtered power spectrum for the data, prior to application of the bandpass filters. In agreement with Braun and Birch (2008), as illustrated in Figures 3 and 4, we find a positive travel-time perturbation in the region beneath the p_1 ridge, but we also find such a signal between the p_1 and p_2 ridges and find that on the p_1 ridge the positive perturbation is absent. Our tentative conclusion is that the positive travel-time perturbation signal arises only in the regions between the p-mode ridges and that the travel-time perturbations associated with the data on the ridges themselves are all consistently negative. We do not at present have an explanation for the off-ridge behaviour.

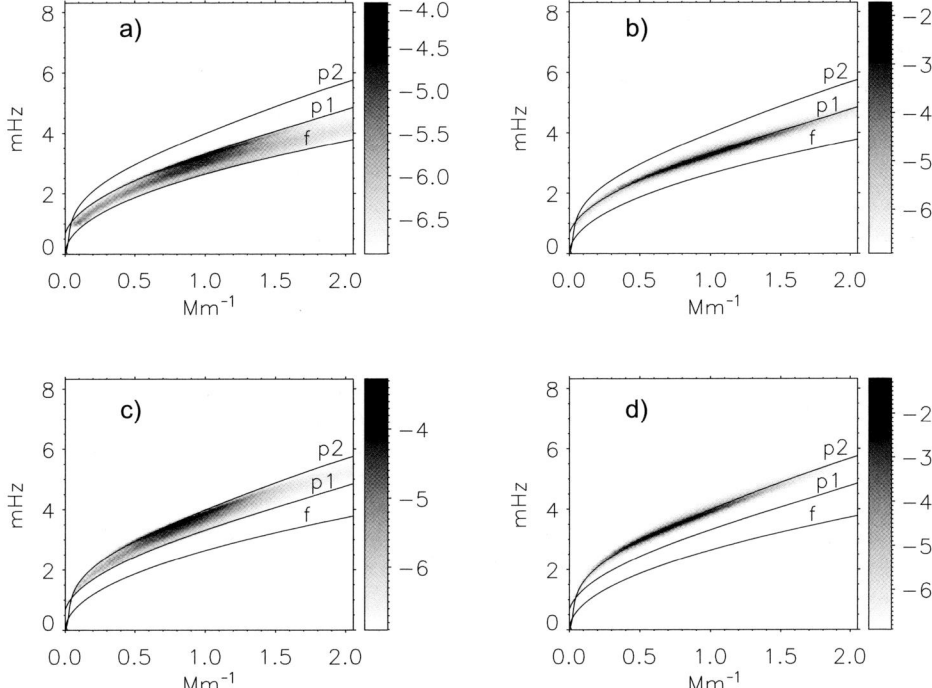

Figure 5 Examples of ridge and off-ridge filters showing the power spectrum after application of, respectively, (a) the $f - p_1$ off-ridge filter, (b) the p_1-ridge filter, (c) the $p_1 - p_2$ off-ridge filter, and (d) the p_2-ridge filter. The locations of the f, p_1, and p_2 ridges, for modes of a theoretical solar model, are indicated by the solid curves. The grey scale is logarithmic with power in arbitrary units.

5. Simulations

Recent progress in realistic simulation of the solar interior, in particular of the upper convection zone, is providing opportunities to test the accuracy and robustness of local helioseismic techniques.

Zhao *et al.* (2007) have used the three-dimensional, radiative-hydrodynamic simulations of the upper convection zone and photosphere of Benson, Stein, and Nordlund (2006) to evaluate the results of time–distance inversion in quiet-Sun regions. The waves are generated consistently within the simulations. The horizontal-flow fields inferred from inverting f-mode travel times are in good agreement with the actual flows in the simulations, down to 3 Mm depth.

Other work has introduced flows into numerical simulations of wave propagation to see how reliably the flows can be recovered. Shelyag, Erdélyi, and Thompson (2007) tested ray-theory application to the flows in the quiet Sun. They find that most of the error in a ray-theoretic inversion result can be understood to arise from the sensitivity of the waves to the flows over a greater depth than is accounted for by the ray paths.

Braun *et al.* (2007) have applied helioseismic holography to the simulations by Benson, Stein, and Nordlund (2006). They obtain phase shifts, which they then converted to travel-time perturbations. The holography recovers the model travel times well, consistent with the noise in the simulated measurements. Moreover, the travel times have been measured in different frequency bands and compared with the model predictions from kernels computed in the Born approximation: The level of agreement gives a fair degree of confidence in the kernels for the kinds of flows in the study. Also in the context of helioseismic holography, Birch, Braun, and Hanasoge (2008) have conducted simulations to investigate the sensitivity of measured travel times to choice of filter parameters (as discussed for observational data in Section 4).

In the previous section it was remarked that the acoustic power is observed to be reduced in regions of strong magnetic field. This reduction might be caused by a lack of granulation-related wave sources in the region. This has been demonstrated in numerical simulation of wave propagation, in which reduction of the amplitudes of random wave sources in a circular region corresponding to a sunspot gives rise to a reduction in the oscillation amplitude in the region by a factor of two to four (Parchevsky and Kosovichev, 2007). Hanasoge *et al.* (2008) have explored the effects on travel times: By suppressing the sources, these authors have shown in their simulations that meaningful travel times depend strongly on the homogeneity of sources in the medium, with asymmetric ingoing and outgoing wave travel-time perturbations being obtained. The results of Hanasoge *et al.* appear to be in very good agreement with the cross-correlation asymmetries observed around an active region by Jensen, Pijpers, and Thompson (2006).

It is essential for testing inferences in strongly magnetic regions that MHD simulations are developed. Such a simulation code has now been developed for studying the interaction of linear waves with magnetic structures in an inhomogeneous atmosphere (Cameron, Gizon, and Daiffallah, 2007) and has already been used, for example, to compare observed cross-correlations on the Sun with the interaction of a simulated plane wave with a magnetic flux cylinder.

6. Prospects

Local helioseismology has enormous potential and has already produced many intriguing insights into the subsurface layers of the Sun. We have touched on several areas of application. Although we have not discussed it here, there have also been a number of recent papers resulting from the holographic study of the seismic response of the solar interior to flares, which should improve our understanding of flare physics (Donea, 2006; Kosovichev, 2006; Kosovichev, 2007; Martinez-Oliveros *et al.*, 2007; Moradi *et al.*, 2007; Zharkova and Zharkov, 2007).

In regions of the quiet Sun, comparative studies using more than one method and studies using numerical simulations are leading to increasing confidence in the reliability of the helioseismic findings.

In regions of strong surface magnetic field, however, there is evidence from observation and from theory that our forward modelling and data analysis techniques are not yet developed well enough to allow robust conclusions about the subsurface structure, nor possibly about the flows. Nonetheless, we are optimistic that we shall see substantial advances in the next year or two. The analytical and simulations work is progressing rapidly and needs now to inform the forward modelling used in interpreting the data. The observations are giving a strong lead to the theorists as to the areas where improved understanding is needed and may also suggest data analysis strategies that are less sensitive to uncertain aspects of the modelling.

Soon a wealth of high-resolution data for local helioseismology will be available from the HMI instrument onboard the *Solar Dynamics Observatory*, covering the entire solar disk. The SOT instrument onboard *Hinode*, with high resolution and a pointing capability, will enable seismic studies (Mitra-Kraev, Kosovichev, and Sekii, 2008) of small regions including at high latitudes. No doubt these will provide new challenges as well as insights both for our techniques and for our understanding of the Sun.

Acknowledgements We thank Aaron Birch, Doug Braun, Laurent Gizon, and the referee for helpful comments that have improved this paper. This work was supported by the UK Science and Technology Facilities Council through Grant Nos. PP/C502914/1 and PP/B501512/1 and by the European Helio- and Asteroseismology Network (HELAS), a major international collaboration funded by the European Commission's Sixth Framework Programme.

References

Basu, S., Antia, H.M., Tripathy, S.C.: 1999, *Astrophys. J.* **512**, 458.
Benson, D., Stein, R., Nordlund, Å: 2006, In: Uitenbroek, H., Leibacher, J., Stein, R.F. (eds.) *Solar MHD Theory and Observations: A High Spatial Resolution Perspective* **CS-354**, Astron. Soc. Pac., San Francisco, 92.
Birch, A.C.: 2006, In: Fletcher, K., Thompson, M. (eds.) *Beyond The Spherical Sun, Proc SOHO 18/GONG 2006/HELAS I* **SP-624**, ESA, Noordwijk, 2.1, publ on CDROM.
Birch, A.C., Kosovichev, A.G.: 2000, *Solar Phys.* **192**, 193.
Birch, A.C., Kosovichev, A.G., Duvall, T.L. Jr.: 2004, *Astrophys. J.* **608**, 580.
Birch, A.C., Braun, D.C., Hanasoge, S.M.: 2008, *Solar Phys.* in press.
Birch, A.C., Gizon, L., Hindman, B.W., Haber, D.A.: 2007, *Astrophys. J.* **662**, 730.
Braun, D.C., Birch, A.C.: 2006, *Astrophys. J.* **647**, 187.
Braun, D.C., Birch, A.C.: 2008, *Solar Phys.* accepted.
Braun, D.C., Duvall, T.L. Jr., Labonte, B.J.: 1987, *Astrophys. J.* **319**, L27.
Braun, D.C., Birch, A.C., Benson, D., Stein, R.F., Nordlund, Å.: 2007, *Astrophys. J.* **669**, 1395.
Cally, P.S.: 2007, *Astron. Nachr.* **328**, 286.
Cameron, R., Gizon, L., Daiffallah, K.: 2007, *Astron. Nachr.* **328**, 313.
Chou, D.-Y.: 2000, *Solar Phys.* **192**, 241.
Couvidat, S., Rajaguru, S.P.: 2007, *Astrophys. J.* **661**, 558.
Couvidat, S., Birch, A.C., Kosovichev, A.G.: 2006, *Astrophys. J.* **640**, 516.
Donea, A.: 2006, *Solar Phys.* **239**, 113.
Duvall, T.L. Jr., Birch, A.C., Gizon, L.: 2006, *Astrophys. J.* **646**, 553.
Duvall, T.L. Jr., Jefferies, S.M., Harvey, J.W., Pomerantz, M.A.: 1993, *Nature* **362**, 430.
Duvall, T.L. Jr., Scherrer, P.H., Bogart, R.S., Bush, R.I., De Forest, C., Hoeksema, J.T., Schou, J., Saba, J.L.R., Tarbell, T.D., Title, A.M., Wolfson, C.J., Milford, P.N.: 1997, *Solar Phys.* **170**, 63.
Dziembowski, W.A., Pamyatnykh, A.A., Sienkiewicz, R.: 1990, *Mon. Not. Roy. Astron. Soc.* **244**, 542.
Giles, P.M.: 1999, Ph.D. dissertation, Stanford University.
Giles, P.M., Duvall, T.L. Jr., Scherrer, P.H., Bogart, R.S.: 1997, *Nature* **390**, 52.
Gizon, L., Birch, A.C.: 2002, *Astrophys. J.* **571**, 966.
Gizon, L., Birch, A.C.: 2004, *Astrophys. J.* **614**, 472.
Gizon, L., Birch, A.C.: 2005, *Living Rev. Solar Phys.* **2**. http://solarphysics.livingreviews.org/Articles/lrsp-2005-6/.
Gizon, L., Thompson, M.J.: 2007, *Astron. Nachr.* **328**, 204.
Gizon, L., Duvall, T.L. Jr., Larsen, R.M.: 2001, In: Brekke, P., Fleck, B., Gurman, J. (eds.) *Proc. 23rd IAU General Assembly*, **203**, Astron. Soc. Pac., San Francisco, 189.
Haber, D.A., Hindman, B.W., Toomre, J., Bogart, R.S., Hill, F.: 2001, In: Wilson, A. (ed.) *Helio- and Asteroseismology at the Dawn of the Millennium, Proc. SOHO 10/GONG 2000* **SP-464**, ESA, Noordwijk, 209.
Hanasoge, S.M., Couvidat, S., Rajaguru, S.P., Birch, A.C.: 2008, *Astrophys. J.* submitted. arXiv:astro-ph/0707.1369v3.
Hill, F.: 1988, *Astrophys. J.* **333**, 996.
Hindman, B.W., Brown, T.M.: 1998, *Astrophys. J.* **504**, 1029.
Hindman, B.W., Gough, D., Thompson, M.J., Toomre, J.: 2005, *Astrophys. J.* **621**, 512.
Jackiewicz, J., Gizon, L., Birch, A.C., Duvall, T.L. Jr.: 2007, *Astrophys. J.* **671**, 1051–1064.
Jensen, J.M., Pijpers, F.P.: 2003, *Astron. Astrophys.* **412**, 257.
Jensen, J.M., Pijpers, F.P., Thompson, M.J.: 2006, *Astrophys. J.* **648**, 75.
Kosovichev, A.G.: 2006, *Solar Phys.* **238**, 1.

Kosovichev, A.G.: 2007, *Astrophys. J.* **670**, L65.
Kosovichev, A.G., Duvall, T.L.: 1997, In: Pijpers, F.P., Christensen-Dalsgaard, J., Rosenthal, C.S. (eds.) *SCORe'96: Solar Convection and Oscillations and Their Relationship*, ASSL **225**, Kluwer Academic, Dordrecht, 241.
Kosovichev, A.G., Duvall, T.L. Jr., Scherrer, P.H.: 2000, *Solar Phys.* **192**, 159.
Lindsey, C., Braun, D.C.: 1997, *Astrophys. J.* **485**, 895.
Lindsey, C., Braun, D.C.: 2000, *Science* **287**, 1799.
Lindsey, C., Braun, D.C.: 2003, In: Sawaya-Lacoste, H. (ed.) *Local and Global Helioseismology: The Present and Future, Proc SOHO 12/GONG+ 2002* **SP-517**, ESA, Noordwijk, 23.
Martinez-Oliveros, J.C., Moradi, H., Besliu-Ionescu, D., Donea, A.-C., Cally, P.S., Lindsey, C.: 2007, *Solar Phys.* **245**, 121.
Mitra-Kraev, U., Kosovichev, A.G., Sekii, T.: 2008, *Astron. Astrophys. Lett.* accepted. arXiv:astro-ph/0711.2210v1.
Moradi, H., Donea, A.-C., Lindsey, C., Besliu-Ionescu, D., Cally, P.S.: 2007, *Mon. Not. Roy. Astron. Soc.* **374**, 1155.
Nicholas, C.J., Thompson, M.J., Rajaguru, S.P.: 2004, *Solar Phys.* **225**, 213.
Nigam, R., Kosovichev, A.G., Scherrer, P.H.: 2007, *Astrophys. J.* **659**, 1736.
Parchevsky, K., Kosovichev, A.G.: 2007, *Astrophys. J.* **666**, L53.
Rajaguru, S.P., Birch, A.C., Duvall, T.L. Jr., Thompson, M.J., Zhao, J.: 2006, *Astrophys. J.* **646**, 543.
Roth, M., Gizon, L., Beck, J.G.: 2007, *Astron. Nachr.* **328**, 215.
Schunker, H., Braun, D.C., Cally, P.S.: 2007, *Astron. Nachr.* **328**, 292.
Shelyag, S., Erdélyi, R., Thompson, M.J.: 2007, *Astron. Astrophys.* **469**, 1101.
Simmons, B., Basu, S.: 2003, In: Sawaya-Lacoste, H. (ed.) *Local and Global Helioseismology: The Present and Future, Proc. SOHO 12/GONG+ 2002* **SP-517**, ESA, Noordwijk, 393.
Woodard, M.F.: 2002, *Astrophys. J.* **565**, 634.
Zhao, J., Kosovichev, A.G.: 2004, *Astrophys. J.* **603**, 776.
Zhao, J., Kosovichev, A.G.: 2006, *Astrophys. J.* **643**, 1317.
Zhao, J., Georgobiani, D., Kosovichev, A.G., Benson, D., Stein, R.F., Nordlund, Å.: 2007, *Astrophys. J.* **659**, 848.
Zharkova, V.V., Zharkov, S.I.: 2007, *Astrophys. J.* **664**, 573.

Observation and Modeling of the Solar-Cycle Variation of the Meridional Flow

Laurent Gizon · Matthias Rempel

Originally published in the journal Solar Physics, Volume 251, Nos 1–2, 241–250.
DOI: 10.1007/s11207-008-9162-3 © The Author(s) 2008

Abstract We present independent observations of the solar-cycle variation of flows near the solar surface and at a depth of about 60 Mm, in the latitude range $\pm 45°$. We show that the time-varying components of the meridional flow at these two depths have opposite sign, whereas the time-varying components of the zonal flow are in phase. This is in agreement with previous results. We then investigate whether the observations are consistent with a theoretical model of solar-cycle-dependent meridional circulation based on a flux-transport dynamo combined with a geostrophic flow caused by increased radiative loss in the active region belt (the only existing quantitative model). We find that the model and the data are in qualitative agreement, although the amplitude of the solar-cycle variation of the meridional flow at 60 Mm is underestimated by the model.

Keywords Solar cycle: Models · Solar cycle: Observations · Velocity fields: Interior · Interior: Convective zone · Helioseismology: Observations · Magnetic fields: Models · Oscillations: Solar · Active regions · Supergranulation

1. Introduction

Solar oscillations are a unique tool to infer conditions inside the Sun. They have been recorded with great precision since 1996 with the Michelson Doppler Imager (MDI, Scherrer *et al.*, 1995) onboard the *Solar and Heliospheric Observatory* (SOHO). Large-scale rotation inside the Sun can be estimated by inversion of the frequencies of millions of global

---

Helioseismology, Asteroseismology, and MHD Connections
Guest Editors: Laurent Gizon and Paul Cally.

L. Gizon (✉)
Max-Planck-Institut für Sonnensystemforschung, 37191 Katlenburg-Lindau, Germany
e-mail: gizon@mps.mpg.de

M. Rempel
High Altitude Observatory, National Center for Atmospheric Research, P.O. Box 3000, Boulder, CO 80307, USA

modes of oscillation (*e.g.*, Schou *et al.*, 1998). Rotation is known to vary with time in the solar interior at the level of about ± 10 m s^{-1} (*e.g.*, Schou, 1999; Howe *et al.*, 2000, 2006a, 2006b; Vorontsov *et al.*, 2002). These variations, known as torsional oscillations, consist of bands of faster and slower rotation that migrate in latitude as the 11-year solar magnetic cycle develops. Torsional oscillations may be driven by the Lorentz force owing to a dynamo wave (Schüssler, 1981; Yoshimura, 1981; Covas *et al.*, 2000). Other explanations have been proposed (see Shibahashi, 2004, and references therein), including the suggestion by Spruit (2003) that torsional oscillations are driven by horizontal pressure gradients caused by photospheric magnetic activity.

Long time averages of surface Doppler measurements have shown the existence of a flow from the equator to the poles with an amplitude of $10-20$ m s^{-1}. An introduction to the theory of solar meridional circulation is provided by Shibahashi (2007). Temporal variations in the meridional flow have been reported by several authors. By tracking the small photospheric magnetic features, Komm, Howard, and Harvey (1993b), Komm (1994), and Meunier (1999) found a significant change in the meridional flow near sunspot latitudes, implying a solar-cycle variation. Variations in the surface Doppler meridional velocity have been detected by Ulrich and Boyden (2005), in particular at latitudes above 60°.

Local helioseismology (see, *e.g.*, Gizon and Birch, 2005) also provides reasonable measurements of the meridional circulation for latitudes below about 50° (Giles *et al.*, 1997). The time-varying component of the meridional flow with respect to a long-term average does not exceed ± 5 m s^{-1} and is consistent with a small near-surface inflow toward active latitudes (Basu and Antia, 2003; Gizon, 2004; Zhao and Kosovichev, 2004; González Hernández *et al.*, 2006; Komm, Howe, and Hill, 2006) and an outflow from active latitudes at depths greater than 20 Mm (Chou and Dai, 2001; Beck, Gizon, and Duvall, 2002; Chou and Ladenkov, 2005). As discussed by Gizon (2004), these variations would appear to be caused by localized flows around localized regions of magnetic activity (Gizon, Duvall, and Larsen, 2001; Haber *et al.*, 2001; Haber *et al.*, 2004).

The purpose of this paper is twofold. First, we provide independent measurements of the temporal variations of the meridional circulation near the solar surface and at a depth of about 60 Mm. For the near-surface meridional flow, we use an original technique, which consists of measuring the advection of the supergranulation pattern (Gizon, Duvall, and Schou, 2003). The deeper meridional flow is calibrated from an earlier time–distance helioseismology observation by Beck, Gizon, and Duvall (2002). The meridional circulation measurements at these two depths are compared, in particular by looking at the 11-year periodicity of the flows (Section 2). Our study of meridional circulation confirms previous observations (as listed earlier) and is complementary to the analysis of zonal flows by Howe *et al.* (2006a).

Second, we wish to compare the observations with a theoretical model (Rempel, 2005; Rempel, Dikpati, and MacGregor, 2005; Rempel, 2006) based on a flux-transport dynamo combined with a geostrophic flow caused by increased radiative loss in the active region belt, according to Spruit's (2003) original idea. Section 3 provides a description of the model with a focus on meridional flows, since this aspect of the model has not been discussed elsewhere in detail. To our knowledge, this model is the only existing quantitative model of the solar-cycle dependence of internal flows: It is natural to ask whether it is consistent with the observations. The comparison between the observations and the model (Section 4) is encouraging, although some inconsistencies cannot be ignored.

2. Observations of the Meridional Flow

2.1. Near-Surface Layers

The method we employ to infer flows near the solar surface is based on the analysis of Gizon, Duvall, and Schou (2003), which was originally applied to a single data set from 1996. Here we use a series of MDI full-disk Doppler images covering the period 1996–2002. Each year, two to three months of continuous Dopplergrams are available for analysis (MDI Dynamics runs). The MDI data after 2002 were not used simply because we are analyzing an existing preprocessed data set.

Dopplergrams were tracked at the Carrington rotational velocity to remove the main component of rotation. We used f-mode time–distance helioseismology (Duvall and Gizon, 2000) to obtain every 12 hours a $120° \times 120°$ map of the horizontal divergence of the flow field 1 Mm below the photosphere. The main component of the divergence signal is due to supergranulation. For any given target latitude (λ) we considered a longitudinal section of the data 10° wide in latitude. Using a local plane-parallel approximation in the neighborhood of latitude λ, we interpolated the divergence signal onto a Cartesian grid sampled at 2.92 Mm in the x (prograde) and y (northward) coordinates. The divergence signal was decomposed into its harmonic components $\exp[i(k_x x + k_y y - \omega t)]$ to obtain a local power spectrum $P(\mathbf{k}, \omega; \lambda)$, where $\mathbf{k} = (k_x, k_y)$ is the horizontal wavevector and ω is the angular frequency. At fixed $kR_\odot = 120$, we fit for two functions (f and g) and a horizontal vector ($\mathbf{v}$) such that $P(\mathbf{k}, \omega; \lambda) = f(\mathbf{k}) g[\omega - \mathbf{k} \cdot \mathbf{v}(\lambda)]$. This representation fits the data adequately. As was done by Gizon, Duvall, and Schou (2003), we interpret $\mathbf{v}(\lambda)$ to be a horizontal flow causing a Doppler shift $\Delta\omega = \mathbf{k} \cdot \mathbf{v}$. This flow is likely to be an average over the supergranulation layer, which has been estimated to reach depths greater than 10 Mm by Zhao and Kosovichev (2003).

Figure 1 is a plot of $v_x(\lambda)$ and $v_y(\lambda)$ for each full-disk MDI run as a function of latitude in the range $|\lambda| < 50°$. To reduce random noise, the North–South symmetric component of v_x and the antisymmetric component of v_y are extracted. Over the period 1996–2002, v_x varies by 12 m s^{-1} peak-to-peak at the equator (Figure 1(a)). The meridional flow is poleward with a mean amplitude of 10 m s^{-1} at latitude 20° (Figure 1(b)). The peak-to-peak variation of the meridional flow is 7 m s^{-1} at $\lambda = 30°$ (*i.e.*, a significant fraction of the time-average value). We estimate that the standard deviation of the noise at a particular latitude (5° bin) for any given year is less than 1 m s^{-1}. The systematic errors that depend on position on the solar disk have been measured to be very low (less than 5 m s^{-1} over the $120° \times 120°$ region of analysis).

2.2. Deeper Inside the Sun

To probe deeper layers in the solar convection zone, we used acoustic waves and time–distance helioseismology. For each three-month period, travel times were measured by cross-correlation of the Doppler oscillation signal recorded during the MDI structure program (offering nearly continuous coverage but lower spatial resolution) according to the procedure described by Giles (1999). Using a mean travel distance of 17° enables us to probe layers about 60 Mm below the surface. The full details of this analysis can be found in Beck, Gizon, and Duvall (2002). Waves that propagate in the North–South direction are used to infer the meridional flow, whereas waves that propagate East–West are used to infer the zonal flows. To convert travel-time shifts into flows in units of meters per second, we use a simple calibration of v_x at a depth based on the observation by Howe *et al.* (2006a)

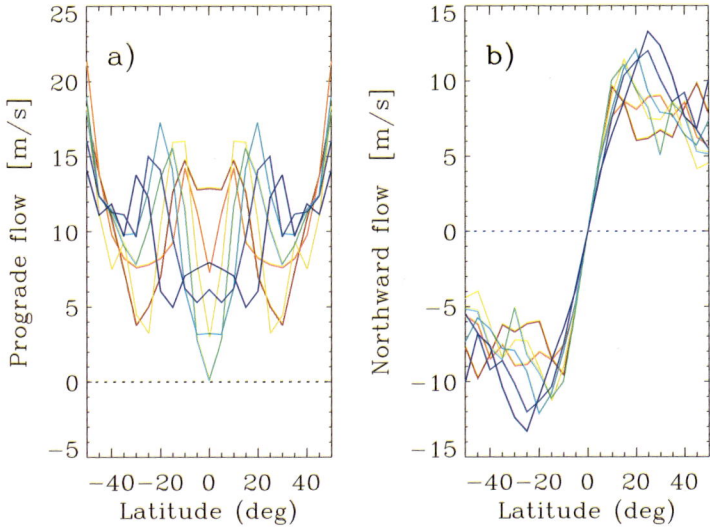

Figure 1 (a) Rotational velocity (v_x) and (b) meridional flow (v_y) near the solar surface as a function of latitude (λ). Each MDI dynamics run is plotted with a different color from blue in 1996 to red in 2002. The rotational velocity is given with respect to the rotational velocity of the small magnetic features (Komm, Howard, and Harvey, 1993a).

(global-mode helioseismology) that the amplitude of the time-varying component of the zonal flow is nearly independent of depth. We choose the near-surface zonal-flow measurements of Section 2.1 as a reference. The calibration of v_x is then used to calibrate v_y. We find that the meridional flow at a depth of 60 Mm is poleward at all latitudes and has a maximum value of 6 m s^{-1} at latitude 25°. For a particular year and at fixed latitude (5° bin), the standard deviation of the noise is about 2 m s^{-1}, significantly more than for the near-surface measurements.

2.3. Solar-Cycle Variations

To discuss the solar-cycle dependence of the flows and to study the phase relationship between the flows measured at the two different depths, we extract the 11-year periodic component from the data, as was done by Vorontsov *et al.* (2002) and Howe *et al.* (2006a) for zonal flows. At each latitude λ and for each depth, we fit a function of the form

$$\tilde{v}_i(\lambda, t) = \overline{v}_i(\lambda) + v'_i(\lambda) \cos\left(\frac{2\pi t}{11\,\mathrm{yr}} + \phi_i(\lambda)\right) \tag{1}$$

to the observed velocity $v_i(\lambda, t)$, where the index i refers to either the x or the y component of the flow. The long-term average is given by $\overline{v}_i$, and the amplitude and the phase of the time-varying component are denoted by v'_i and ϕ_i, respectively. We extract an 11-year periodicity from the data, since it is known from other observations that this is the dominant mode. (We do not determine the 11-year periodicity based on the data set itself.) Shorter and longer periodicities are certainly present in the data; however, the length of the data set does not allow for a determination of the full spectrum of modes.

The 11-year periodic components of the meridional and zonal flows are shown in Figure 2. The torsional-oscillation pattern is clearly seen at both depths with an amplitude and a

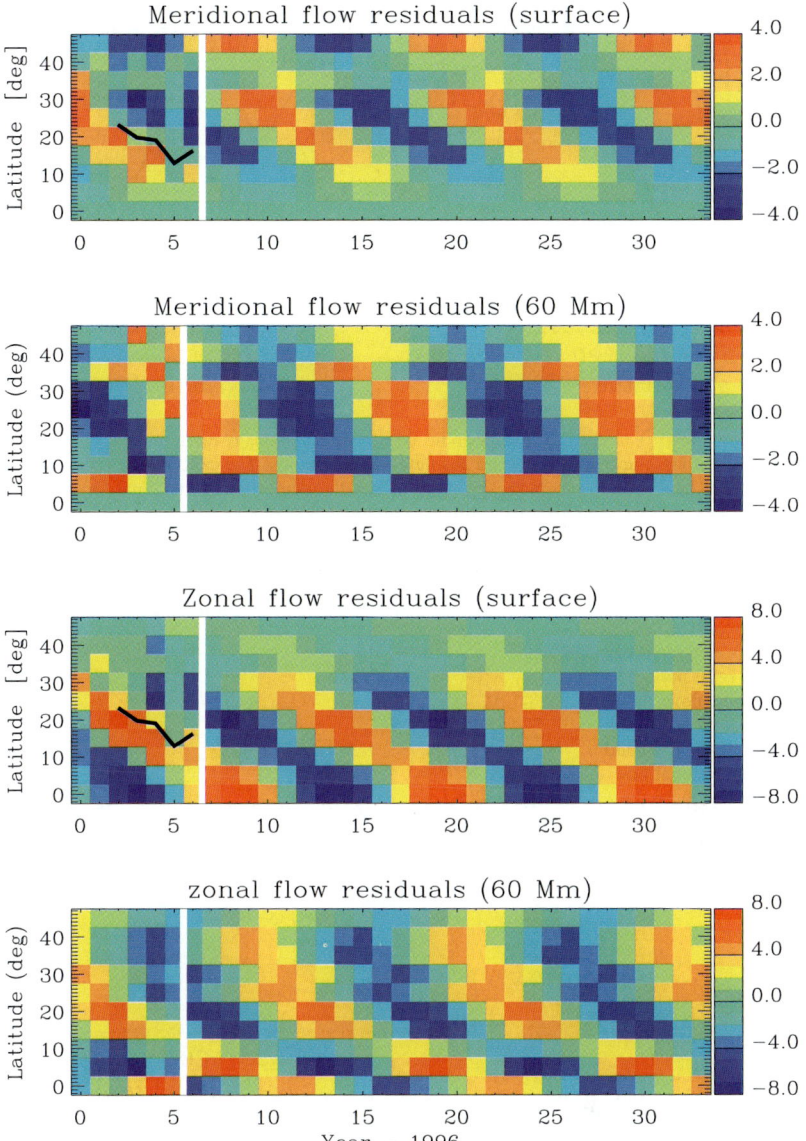

Figure 2 Eleven-year periodic component of the meridional (top two panels) and zonal (bottom two panels) flows as function of time and latitude at two different depths in the solar interior: near the surface (top and third panels) and 60 Mm deep (second and bottom panel). The color bar is in units of m s^{-1}. A positive value indicates a poleward meridional or prograde zonal flow. The observations ($v_i - \bar{v}_i$) cover the first six years, whereas the purely sinusoidal component ($\tilde{v}_i - \bar{v}_i$) is extrapolated in time (beyond the white vertical white line). The black curves indicate the mean latitude of magnetic activity.

phase comparable to previous measurements (*e.g.*, Howe *et al.*, 2006a). The meridional flow also contains a significant 11-year periodic component. Near the solar surface, the residuals indicate the presence of a North–South inflow toward the mean latitude of activity (*e.g.*, Zhao and Kosovichev, 2004; Komm, Howe, and Hill, 2006), but the data are consistent with

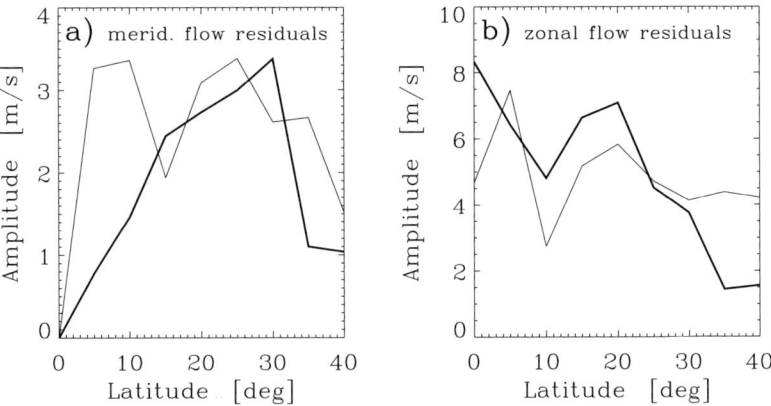

Figure 3 Amplitude (v_i') of the 11-year periodic component of the (a) meridional and (b) zonal flows. The near-surface values (thick solid lines) are absolute measurements. The calibration of the observations at 60 Mm depth (thin lines) follows from the assumption that the amplitude of the zonal torsional oscillation (panel (b)) is independent of depth over the latitude range $|\lambda| < 45°$.

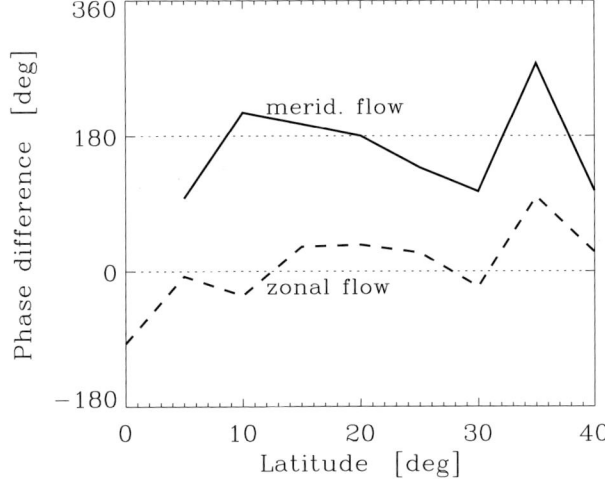

Figure 4 Phase difference $[\Delta \phi = \phi(\text{deep}) - \phi(\text{surface})]$ between the 11-year periodic components of the flows measured at a depth of 60 Mm and near the surface. The solid line is for the meridional flow and the dashed line is for the zonal flow.

a horizontal outflow from the mean latitude of activity deeper into the convection zone (*e.g.*, Chou and Dai, 2001; Beck, Gizon, and Duvall, 2002).

Figure 3 gives the amplitudes of the time-varying components of the flows (v_i'). Under the assumption (Section 2.2) that v_x' does not vary appreciably with depth (5 m s^{-1} latitudinal average), then the amplitude of the time-varying meridional flow (v_y') is also found to be approximately independent of depth ($v_y' \approx 3$ m s^{-1} at 20° latitude). The evidence that the time-varying components of the meridional flow near the surface and deeper in the interior are anticorrelated is given in Figure 4, which shows the difference in ϕ_y at the two depths. In contrast, there is no significant phase variation with depth for the zonal flow.

3. Theoretical Model of Time-Varying Flows

The model results presented here are based on a nonkinematic flux-transport dynamo model developed recently by Rempel. This model combines the differential rotation and meridional flow model of Rempel (2005) with a flux-transport dynamo similar to the models of Dikpati and Charbonneau (1999) and Dikpati and Gilman (2001). We emphasize that this model is intended to give a fundamental understanding of the basic cycle properties and their relation to observable variations of zonal and meridional flows. Therefore we focus here only on axisymmetric and North–South averaged quantities. Details of the model can be found in Rempel (2006). Since a detailed comparison with observed torsional oscillations can be found in Rempel (2006) and Howe *et al.* (2006b), we focus here on the meridional flow variations.

The differential rotation model utilizes a mean-field Reynolds-stress approach that parametrizes the turbulent angular momentum transport (Kitchatinov and Rüdiger, 1993; Λ-effect) leading to the observed equatorial acceleration. In this model the tachocline is forced through a uniform rotation boundary condition at the lower boundary of the computational domain. A meridional circulation, as required for a flux-transport dynamo, follows self-consistently through the Coriolis force resulting from the differential rotation.

The computed differential rotation and meridional flow are used to advance the magnetic field in the flux-transport dynamo model, while the magnetic field is allowed to feed back through the mean-field Lorentz force $\langle \mathbf{J} \rangle \times \langle \mathbf{B} \rangle$ (where the contribution of the fluctuating part $\langle \mathbf{J}' \times \mathbf{B}' \rangle$ is not well known and so is neglected here).

We find in our model that the Lorentz-force feedback can only account for the poleward propagating branch of the torsional oscillations, whereas the equatorward propagating branch in latitudes beneath 30° requires additional physics. Parametrizing the idea proposed by Spruit (2003) that the low-latitude torsional oscillation is a geostrophic flow caused by increased radiative loss in the active region belt (owing to small-scale magnetic flux) leads in our model to a surface oscillations pattern in good agreement with observations. To force a torsional oscillation with around 1 nHz amplitude, a temperature variation of around 0.2 K is required. As a side effect, close to the surface (in our model at $r = 0.985 R_\odot$) the cooling produces an inflow into the active region belt of around 2.3 m s^{-1}.

We incorporated this process by adding a surface-cooling term that is dependent on the toroidal-field strength at the base of the convection zone, which is assumed to be the source for active region magnetic field (the small-scale flux required for the surface-cooling being a consequence of the decay of active regions). Observations show that the low-latitude branch of torsional oscillations starts around one to two years before the sunspots of the new cycle appear. It is possible that magnetic flux rises toward the surface without forming sunspots in the beginning of a cycle, providing enough small-scale magnetic field; this is however currently neither confirmed nor ruled out by observations. Alternative explanations for the low-latitude branch of torsional oscillations such as the models of Schüssler (1981), Yoshimura (1981), and Covas *et al.* (2000) are based on the longitudinal component of the Lorentz force. Recently, Rempel (2007) showed that torsional oscillations forced that way are close to the Taylor–Proudman state (alignment of phase with the axis of rotation), which contradicts observations. In addition, the resulting meridional surface flow has the wrong sign (active region belt outflow). Despite some shortcomings, the model of Spruit (2003) is currently the only proposed explanation that is consistent with the observed meridional and zonal-flow variations close to the solar surface.

Figure 5 summarizes the results of the model in latitudes below 45°. Figure 5(a) shows the temperature fluctuation (color shades) caused by increased surface cooling in the active region belt. The contour lines indicate the magnetic butterfly diagram computed from

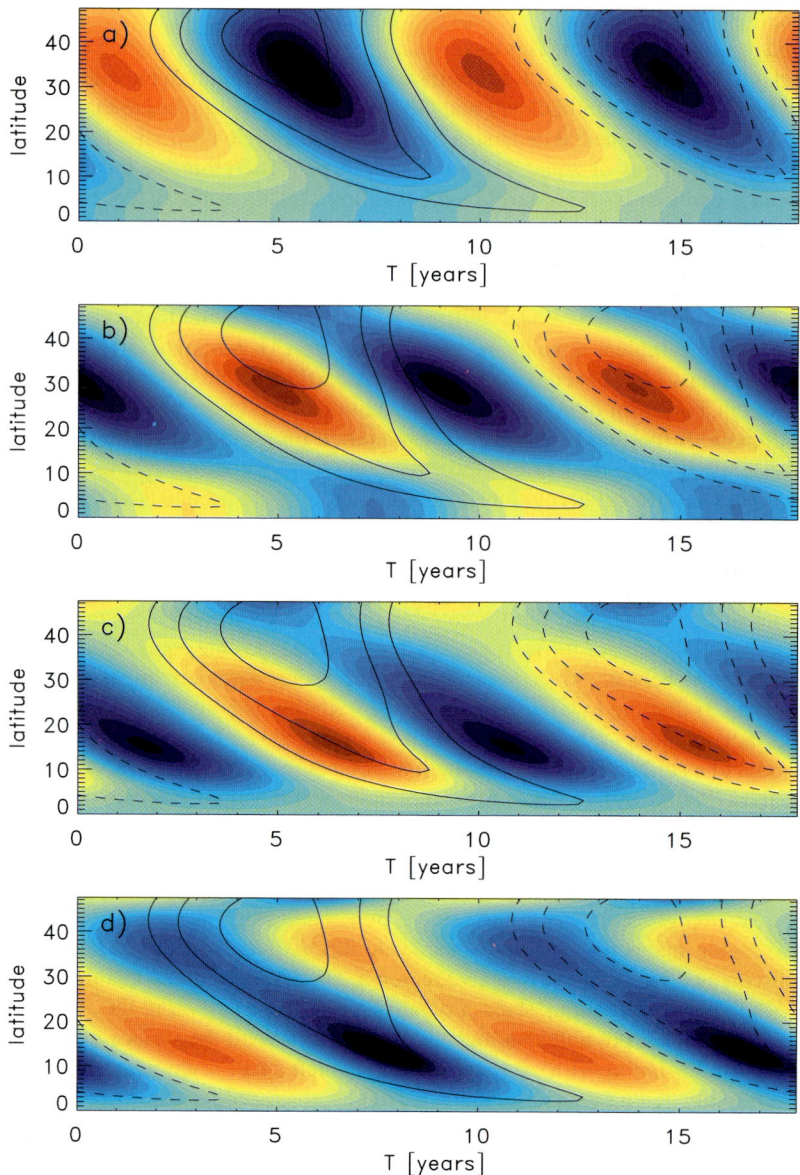

Figure 5 Model results. (a) Surface temperature variation (blue: cold, red: hot, amplitude: 0.2 K). (b) Torsional oscillations (blue: slower rotation, red: faster rotation, amplitude: 1.35 nHz). (c) Meridional flow variation at $r = 0.985 R_\odot$ (blue: equatorward motion, red: poleward motion, amplitude: 2.3 m s^{-1}). (d) Meridional flow variation at $r = 0.93 R_\odot$ (blue: equatorward motion, red: poleward motion, amplitude: 0.22 m s^{-1}). The variation of the meridional flow pattern at $r = 0.985 R_\odot$ is almost in anticorrelation with the flow at $r = 0.93 R_\odot$ ($\approx$ 50 Mm depth). In all four panels the contour lines indicate the butterfly diagram computed from the toroidal field at the base of the convection zone.

2.1. Equations

In Section 3, a broadly realistic solar-interior model will be coupled to an overlying isothermal atmosphere. A uniform, inclined magnetic field,

$$\mathbf{B}_0 = B_0(\sin\theta\cos\phi, \sin\theta\sin\phi, \cos\theta), \tag{1}$$

will be assumed throughout, where $\theta < 90°$ is the field inclination from the vertical, and ϕ is the angle by which it is rotated from the $x-z$ plane. The isothermal top adequately represents the chromosphere for our purposes and conveniently allows us to take advantage of known exact series solutions when selecting top boundary conditions (see Section 2.2 and the Appendix). A horizontal and time dependence $\exp[i(kx - \omega t)]$ is assumed, with z-dependence to be determined, corresponding to wave propagation in the $x-z$ plane.

The linearised, adiabatic MHD equations for this scenario may be expressed in terms of the components of the displacement vector $\boldsymbol{\xi} = (\xi, \eta, \zeta)$, the sound speed c, Alfvén speed a, gravitational acceleration g, and density scale height H as

$$a^2\big[\cos\phi\sin^2\theta\sin\phi\eta\, k^2 - (\cos^2\theta + \sin^2\theta\sin^2\phi)\xi k^2$$
$$+ \sin\theta\big(ik\sin\phi(\sin\theta\sin\phi\zeta' - \cos\theta\eta') - \cos\theta\cos\phi\zeta''\big) + \cos^2\theta\xi''\big]$$
$$+ k\big(a^2 k\cos\theta\cos\phi\sin\theta - ig\big)\zeta + \big(\omega^2 - c^2 k^2\big)\xi + ic^2 k\zeta' = 0, \tag{2}$$

$$-\big(ik\xi + \zeta'\big)\frac{c^2}{H} - ik\cos\phi\sin^2\theta\sin\phi\eta' a^2 + ik\sin^2\theta\sin^2\phi\xi' a^2 + \big(a^2\sin^2\theta + c^2\big)\zeta''$$
$$- \cos\theta\sin\theta\sin\phi\eta'' a^2 - \cos\theta\cos\phi\sin\theta\xi'' a^2 + \big(\omega^2 - a^2 k^2\cos^2\phi\sin^2\theta\big)\zeta$$
$$+ k\xi\big[k\cos\theta\cos\phi\sin\theta a^2 + i(g + (c^2)')\big] + (c^2)'\zeta' + ic^2 k\xi' = 0, \tag{3}$$

and

$$ik\cos\theta\xi' c^2 + \cos\theta\zeta'' c^2 + \big(\omega^2\cos\theta - igk\cos\phi\sin\theta\big)\zeta + \omega^2\sin\theta\sin\phi\eta$$
$$+ \xi\left(-\frac{ik\cos\theta c^2}{H} + igk\cos\theta + \big(\omega^2 - c^2 k^2\big)\cos\phi\sin\theta + ik\cos\theta c^{2'}\right)$$
$$+ \left(ik\cos\phi\sin\theta c^2 - \frac{\cos\theta c^2}{H} + \cos\theta c^{2'}\right)\zeta' = 0, \tag{4}$$

where primes indicate derivatives with respect to z.

2.2. Fast, Slow, and Alfvén Waves and Radiation Boundary Conditions

The top boundary conditions for the fast and slow waves in the superposed isothermal atmosphere are straightforward. The asymptotic controlling factors for the displacements as $z \to \infty$ are, respectively (see the Appendix),

$$\exp[\pm kz] \quad \text{and} \quad \exp\left[\left(1 - 2ikH\tan\theta\cos\phi \pm i\sqrt{\frac{\omega^2}{\omega_c^2}\sec^2\theta - 1}\right)\frac{z}{2H}\right], \tag{5}$$

where $\omega_c = c/2H$ is the acoustic cutoff frequency.

The fast wave is clearly evanescent, with $\exp[-kz]$ the appropriate solution. The term $2ikH\tan\theta\cos\phi$ in the slow-wave controlling factor is purely geometric, accounting for the change in x along an inclined field line: $x - z\tan\theta\cos\phi = $ constant along a field line so the $\exp[ikx]$ dependence cancels this term. (Recall that the slow wave is rigidly channelled along **B** in the high-altitude limit.) The $\exp[z/2H]$ term is extinguished by the $\rho^{1/2}$ factors in the kinetic-energy density $\frac{1}{2}|\rho^{1/2}\dot{\boldsymbol{\xi}}|^2$, and similarly in the acoustic energy. The square-root term however is more important. If $\omega > \omega_c \cos\theta$, the slow wave can propagate, and the "+" sign is the appropriate choice for it to be upgoing at infinity. However, for $\omega < \omega_c \cos\theta$ (i.e., below the ramp-modified acoustic cutoff frequency), the slow wave is evanescent. In that case too, the "+" sign is the correct choice.

The selection of appropriate top boundary condition for Alfvén waves is more subtle. The difficulty is best illustrated by reference to the well-known exact solution for the perpendicular (to both **B** and **g**) velocity η in terms of Bessel functions in the 2D isothermal case $\phi = 0$:

$$\eta = s^{2i\kappa \tan\theta}\left[A J_0(2s\sec\theta) + B Y_0(2s\sec\theta)\right], \tag{6}$$

where $s = \omega H/a = (\omega H/a_0)\exp(-z/2H)$, $\kappa = kH$ is a dimensionless horizontal wavenumber, and A and B are integration constants. If the atmosphere is deemed to extend to $z = \infty$ (i.e., $s = 0$), the common practice is to set $B = 0$, resulting in a standing wave: the J_0 solution alone (Ferraro and Plumpton, 1958; An et al., 1989). With this choice, the Alfvén wave may not take away energy, contrary to expectations. However, although the Y_0 Bessel function diverges as $s \to 0^+$ ($z \to \infty$), its energy density is bounded, and so there is no need to dispense with it. The correct choice is actually $B = -iA$, whence

$$\eta = A s^{2i\kappa \tan\theta} H_0^{(2)}(2s\sec\theta) \tag{7}$$

(Schwartz, Cally, and Bel, 1984), where $H_0^{(2)} = J_0 - iY_0$ is the second Hankel function of order zero, which is well known to represent a wave travelling in the negative s (positive z) direction. Although the Alfvén wave [travelling at speed $a = a_0 \exp(z/2H)$] actually reaches $z = \infty$ in a finite time ($2H/a_0$ from $z = 0$), it is what it does there that is important: if it is assumed that it reflects totally, a standing wave quickly results (the J_0 solution), but if, however, it is assumed lost there, the $H_0^{(2)}$ solution develops. Of course, in reality the isothermal atmosphere (representing the chromosphere) does not extend to infinity. Nevertheless, if we imagine that the wave is lost once it reaches a great height above the solar surface, which is the more physical scenario, the Hankel solution is appropriate. Because we are focused on the conversion process per se, we choose not to complicate issues by including a multilayer (chromosphere and corona) atmosphere and the resulting resonant partial reflection from the transition region (Cally, 1983; Schwartz, Cally, and Bel, 1984).

The full 3D case is more complicated, because it inherently involves coupling among all three MHD modes: Alfvén, fast, and slow. Closed-form solutions in terms of special functions are not available. Instead, a Frobenius series expansion is required. This is carried out in the Appendix, where the construction of the outward and inward travelling wave solutions is explained.

With all of this in place, the three top boundary conditions in the numerical solution of the full 3D sixth-order coupled wave equations consist of a matching onto the three physical solutions in the isothermal atmosphere: the evanescent fast wave, the outgoing or evanescent slow wave, and the outgoing Alfvén wave.

The bottom magnetic boundary conditions are applied deep enough in the interior that $a \ll c$, and so both the slow and Alfvén waves are rigidly field-guided and highly oscillatory with respect to the density scale height. This suggests the radiation condition

$$\left(\frac{\partial}{\partial t} - \mathbf{a}\cdot\nabla\right)\nabla\times\boldsymbol{\xi} = \mathbf{0}, \tag{8}$$

where $\mathbf{a} = a\widehat{\mathbf{B}}_0$ is the Alfvén velocity. The term in brackets is the field-directed upgoing wave operator, and it is applied to $\nabla\times\boldsymbol{\xi}$ to suppress the acoustic wave (which is irrotational to a high degree of accuracy) and accentuate the highly oscillatory magnetic waves.

3. Numerical Model and Solutions

We perform a "scattering experiment" (in the spirit of Spruit and Bogdan, 1992) by placing a plane acoustic source at $z_b < 0$ beneath the solar surface. The Magnetically Modified Model S (MMMS), a variant on the well-known nonmagnetic Model S of Christensen-Dalsgaard et al. (1996), is used below $z = 0.5$ Mm, and an isothermal slab is used above. This model, which reduces the gas pressure (and consequently the sound speed) near the surface to compensate for the introduction of a magnetic pressure, is detailed in the appendix of Schunker and Cally (2006). In practice, we use $z_b = -4$ Mm, although this is not important, providing it is well below the acoustic–Alfvénic equipartition level $z_{eq} \approx -297$ km (for a 2-kG field) where $a = c$. At z_b, an acoustic wave of frequency ω is generated; this wave has a natural cavity depth $z_1 < z_b$. This determines its horizontal wavenumber (k) or alternately spherical harmonic degree (ℓ). Since $a \ll c$ at z_b, the slow mode is predominantly magnetic and transverse, much like the Alfvén wave. Radiation conditions are applied on both: no slow or Alfvén waves enter the computational region from below. Together with the amplitude of the acoustic driver (set by normalising the total acoustic energy per unit horizontal area in $z_b < z < 0$ to unity), these constitute the three lower boundary conditions. The upper radiation–evanescence conditions are as prescribed in Section 2.2. Subject to these six boundary conditions (one of which is nonhomogeneous), Equations (2)–(4) are solved numerically in $z_b < z < z_t$. We are then free to adjust ω, z_1, θ, ϕ, B_0, etc. and to observe the vertical components of acoustic and magnetic wave energy fluxes F_{ac} and F_{mag} reaching the computational top z_t, typically set at 2 Mm.

For reference, we note that the vector wave energy flux is

$$\mathbf{F} = \mathbf{F}_{ac} + \mathbf{F}_{mag} = \mathrm{Re}\left[p_1\mathbf{v}^* + \mathbf{e}\times\mathbf{b}^*\right] \tag{9}$$

(Bray and Loughhead, 1974), where $p_1 = -\rho c^2 \nabla\cdot\boldsymbol{\xi} + \rho g\zeta$ is the Eulerian gas pressure perturbation, $\mathbf{v} = -i\omega\boldsymbol{\xi}$ is the plasma velocity, $\mathbf{b} = \nabla\times(\boldsymbol{\xi}\times\mathbf{B}_0)$ is the magnetic-field perturbation, and $\mathbf{e} = -\mathbf{v}\times\mathbf{B}_0$ is the electric-field perturbation. The energy density is made up, respectively, of kinetic, acoustic, gravitational, and magnetic parts,

$$E = \frac{\rho}{2}|\mathbf{v}|^2 + \frac{|p_1|^2}{2\rho c^2} + \frac{\rho}{2}N^2|\zeta|^2 + \frac{|\mathbf{b}|^2}{2\mu}, \tag{10}$$

where N is the Brunt–Väisälä frequency. E and $\mathbf{F}$ satisfy the expected conservation equation $\partial E/\partial t + \nabla\cdot\mathbf{F} = 0$ by construction.

Figure 1 displays the vertical fluxes as functions of z for various cases with a 2-kG magnetic field inclined at 30° to the vertical. The top-left panel corresponds to a frequency

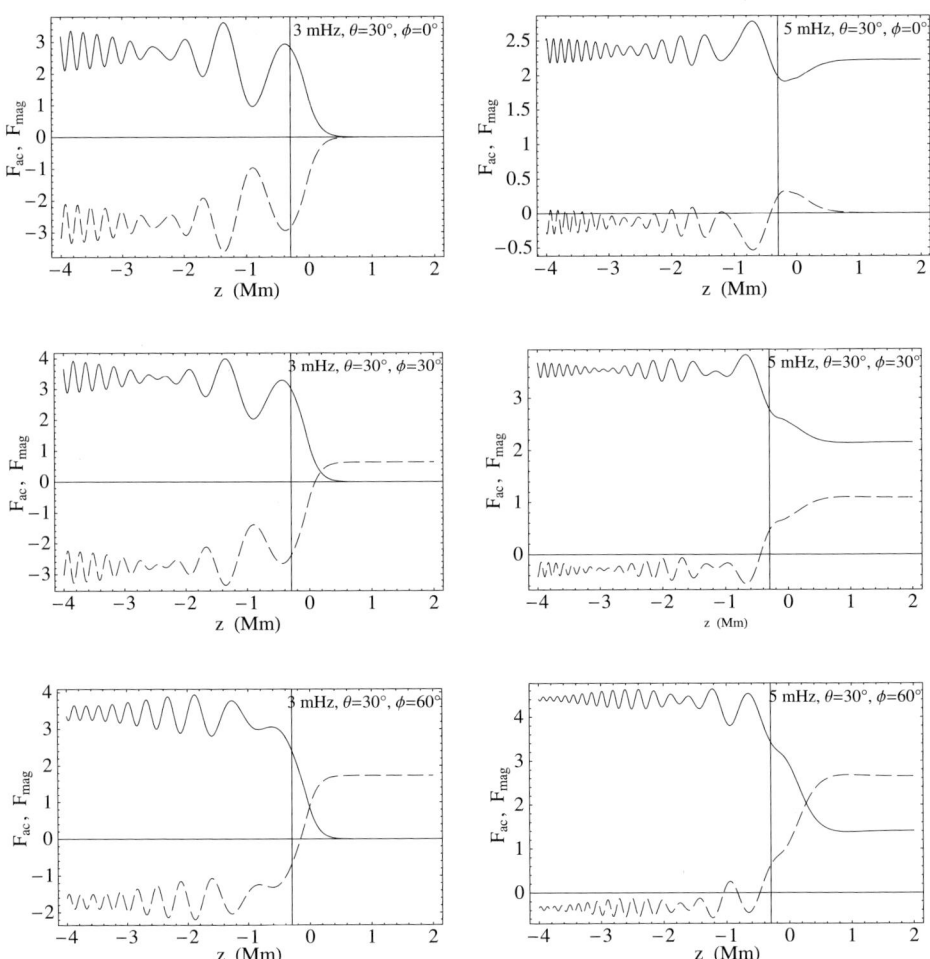

Figure 1 Acoustic (solid) and magnetic (dashed) wave energy fluxes as functions of height (z) over the full computational domain, for two different frequencies: 3 mHz (left panels) and 5 mHz (right panels). In all cases, the 2-kG magnetic field is inclined at $\theta = 30°$ to the vertical. The top panels refer to the 2D case ($\phi = 0°$), the middle panels to $\phi = 30°$, and the bottom panels to $\phi = 60°$. The acoustic cavity base is at $z_1 = -5$ Mm throughout, corresponding to $k = 1.37$ Mm^{-1}, or $\ell \approx 955$. The vertical axis at $z = -0.297$ Mm indicates the position of the $a = c$ equipartition.

of 3 mHz with $\phi = 0°$, so the Alfvén wave is entirely decoupled and cannot be excited by the acoustic driver at $z_b = -4$ Mm. All magnetic flux in this case is therefore associated with the magnetoacoustic waves, predominantly the slow wave below z_{eq}, where it is negative, in accord with the lower radiation boundary condition. Since the (magnetically dominated) fast wave in $z \gg z_{eq}$ is evanescent in all cases, as is the (acoustic) slow wave here since $\omega < \omega_c \cos\theta$ (the acoustic cutoff frequency being 5.2 mHz in the isothermal slab), both magnetic and acoustic fluxes vanish quickly as z increases. In the top-right panel, the frequency is increased to 5 mHz, with other parameters unaltered. In this case, the ramp effect is sufficient to reduce $\omega_c \cos\theta$ below the wave frequency, and so the slow (acoustic) wave propagates

Zhugzhda and Dzhalilov (1984) did not pursue the logarithmic case. We therefore develop the necessary solution here.

We adopt the same dimensionless variables as in the case of the decoupled Alfvén wave, $v = \omega H/c$, $s = \omega H/a = (\omega H/a_0)\exp(-z/2H)$, and $\kappa = kH$. The MHD wave equations can be represented as a set of three coupled second-order ODEs in the displacements $\xi(s)$, $\eta(s)$, and $\zeta(s)$, derived from Equations (2)–(4). However, computationally, we find it more convenient to write them as a single sixth-order matrix equation[1]

$$s\mathbf{U}' = \mathbf{A}\mathbf{U}, \tag{12}$$

where $\mathbf{U}(s) = (\xi, \eta, \zeta, s\xi', s\eta', s\zeta')^\mathrm{T} = (\boldsymbol{\xi}, s\boldsymbol{\xi}')^\mathrm{T}$ and

$$\mathbf{A} = \begin{pmatrix} \mathbf{0} & \mathbf{I} \\ \mathbf{P} & \mathbf{Q} \end{pmatrix}, \tag{13}$$

and where each of the four constituent blocks is 3×3. Specifically,

$$\mathbf{Q} = \begin{pmatrix} 2\mathrm{i}\kappa\cos\phi\tan\theta & -2\mathrm{i}\kappa\sin\phi\tan\theta & 2\mathrm{i}\kappa\left(\frac{s^2\sec^2\theta}{v^2} + \tan^2\theta\right) - 2\cos\phi\tan\theta \\ 0 & 4\mathrm{i}\kappa\cos\phi\tan\theta & -2\sin\phi\tan\theta \\ 2\mathrm{i}\kappa & 0 & 2\mathrm{i}\kappa\cos\phi\tan\theta - 2 \end{pmatrix}, \tag{14}$$

and the components of $\mathbf{P}$ are

$$P_{11} = -2(\cos 2\phi\tan^2\theta - 1)v^2 + \left(\left(\frac{4\kappa^2}{v^2} - 4\right)s^2 + 4\kappa^2 - 2v^2\right)\sec^2\theta$$
$$+ 4\mathrm{i}(1 - \gamma^{-1})\kappa\cos\phi\tan\theta, \tag{15}$$

$$P_{12} = -2(\kappa^2 + v^2)\sin 2\phi\tan^2\theta, \tag{16}$$

$$P_{13} = \frac{4\mathrm{i}(\cos\phi\tan\theta(\mathrm{i}\gamma(\kappa^2 + v^2) + \kappa\cos\phi\tan\theta)v^2 + s^2\kappa\sec^2\theta)}{\gamma v^2}, \tag{17}$$

$$P_{21} = \frac{4\mathrm{i}\sin\phi\tan\theta(\mathrm{i}\gamma\cos\phi\tan\theta v^2 + (\gamma - 1)\kappa)}{\gamma}, \tag{18}$$

$$P_{22} = -4(s^2\sec^2\theta + (v^2\sin^2\phi - \kappa^2\cos^2\phi)\tan^2\theta), \tag{19}$$

$$P_{23} = -\frac{4\sin\phi\tan\theta(\gamma v^2 - \mathrm{i}\kappa\cos\phi\tan\theta)}{\gamma}, \tag{20}$$

$$P_{31} = \frac{4\mathrm{i}(\gamma - 1)\kappa}{\gamma} + 4(\kappa^2 - v^2)\cos\phi\tan\theta, \tag{21}$$

$$P_{32} = -4v^2\sin\phi\tan\theta, \tag{22}$$

$$P_{33} = \frac{4\mathrm{i}\kappa\cos\phi\tan\theta}{\gamma} - 4v^2, \tag{23}$$

where γ is the ratio of specific heats. Note that $\mathbf{A}(s) = \mathbf{A}_0 + \mathbf{A}_2 s^2$.

---

[1] A partial theory of Frobenius expansion for matrix equations is set out in Rubinstein (1969).

By adopting the standard Frobenius expansion about the regular singular point $s = 0$ ($z = \infty$),

$$\mathbf{U}(s) = \sum_{n=0}^{\infty} \mathbf{u}_n s^{n+\mu}, \qquad (24)$$

the lowest order balance yields the indicial equation

$$\mathbf{A}_0 \mathbf{u}_0 = \mu \mathbf{u}_0, \qquad (25)$$

indicating that μ is an eigenvalue of $\mathbf{A}_0$ and $\mathbf{u}_0$ is the corresponding eigenvector. The spectrum is

$$\mu \in \{ -2\kappa, 2\kappa, -i\sqrt{4\nu^2 - \cos^2\theta}\sec\theta + 2i\kappa\cos\phi\tan\theta - 1,$$

$$i\sqrt{4\nu^2 - \cos^2\theta}\sec\theta + 2i\kappa\cos\phi\tan\theta - 1, 2i\kappa\cos\phi\tan\theta, 2i\kappa\cos\phi\tan\theta \}, \qquad (26)$$

corresponding, respectively, to the growing (unphysical) fast mode, the evanescent fast mode, the outgoing slow mode (with the assumption $\nu > \frac{1}{2}\cos\theta$), the incoming slow mode, and the Alfvén mode. As previously mentioned, the Alfvén eigenvalue has algebraic multiplicity 2, although its geometric multiplicity (dimensionality of its eigenspace) is only 1.

Beyond the first order, we find that $\mathbf{u}_1 = \mathbf{0}$ in all cases, and we derive the recurrence relation

$$\mathbf{u}_n = -\bigl(\mathbf{A}_0 - (n+\mu)\mathbf{I}\bigr)^{-1} \mathbf{A}_2 \mathbf{u}_{n-2}, \qquad (27)$$

from which it is apparent that all odd coefficients vanish. This completes the description of the first five solutions.

To discover the sixth solution, we must replace Equation (24) with

$$\mathbf{U}_6(s) = \frac{2}{\pi}\left[\mathbf{U}_5(s)\ln s + \sum_{n=0}^{\infty} \mathbf{v}_n s^{n+\mu_6} \right]. \qquad (28)$$

The factor $2/\pi$ is just a convenient normalisation. Substituting this into (12) and equating coefficients we find

$$(\mathbf{A}_0 - \mu_6 \mathbf{I})\mathbf{v}_0 = \mathbf{u}_0, \qquad (29)$$

$$\mathbf{v}_1 = \mathbf{0}, \qquad (30)$$

$$\bigl(\mathbf{A}_0 - (n+\mu_6)\mathbf{I}\bigr)\mathbf{v}_n = \mathbf{u}_n - \mathbf{A}_2 \mathbf{v}_{n-2}, \qquad (31)$$

where the $\mathbf{u}_n$ are the coefficients in the first Alfvénic solution $\mathbf{U}_5$. Equation (29) yields a generalised eigenvector, in the sense that $\mathbf{A}_0 - \mu_6\mathbf{I}$ is singular by definition, with $\mathbf{u}_0$ in its null space [i.e., $(\mathbf{A}_0 - \mu_6\mathbf{I})^2 \mathbf{v}_0 = \mathbf{0}$]. We find

$$\mathbf{v}_0 = \left(\frac{i\cos^3\theta\sin\theta\sin\phi}{2\kappa(1-\sin^2\theta\sin^2\phi)}, \frac{\cos^2\theta}{2\gamma\nu^2}, \frac{i\cos^2\theta\cos\phi\sin^2\theta\sin\phi}{2\kappa(\sin^2\theta\sin^2\phi - 1)}, \right.$$

$$\frac{\sin^2\theta((\cos 2\phi - 3)\sin^2\theta + 4)\sin 2\phi}{4(\sin^2\theta\sin^2\phi - 1)}, 1 - \sin^2\theta\sin^2\phi + \frac{i\kappa\cos\theta\cos\phi\sin\theta}{\gamma\nu^2},$$

$$\left. \frac{\cos^3\theta\sin\theta\sin\phi}{\sin^2\theta\sin^2\phi - 1} \right)^{\mathrm{T}}. \qquad (32)$$

Naturally, this is determined only up to an arbitrary multiple of the Alfvénic eigenvector

$$\mathbf{u}_0 = \bigl(-\cos\phi \sin^2\theta \sin\phi,\, 1 - \sin^2\theta \sin^2\phi,\, -\cos\theta \sin\theta \sin\phi,$$
$$\qquad -2i\kappa \cos^2\phi \sin^2\theta \sin\phi \tan\theta,\, 2i\kappa \cos\phi\bigl(1 - \sin^2\theta \sin^2\phi\bigr) \tan\theta,$$
$$\qquad -2i\kappa \cos\phi \sin^2\theta \sin\phi\bigr)^{\mathrm{T}}, \tag{33}$$

although Equation (32) has been engineered to yield a solution $\mathbf{U}_6$ with zero net energy flux (i.e., it is a standing wave), as we shall see shortly.

The remaining task is to determine which combinations of $\mathbf{U}_5$ and $\mathbf{U}_6$ represent incoming and outgoing Alfvén waves at $z = \infty$. To do this, we evaluate the field-aligned component of the Pointing flux, $F_\parallel = \widehat{\mathbf{B}}_0 \cdot \mathbf{F}_{\mathrm{mag}}$, where $\mathbf{F}_{\mathrm{mag}} = \mathrm{Re}[i\omega(\boldsymbol{\xi} \times \mathbf{B}_0) \times \nabla \times (\boldsymbol{\xi}^* \times \mathbf{B}_0)]$, as $s \to 0^+$, which only involves the $n = 0$ coefficients. Letting $\mathbf{U} = C_5 \mathbf{U}_5 + iC_6 \mathbf{U}_6$, we find

$$F_\parallel = -\tfrac{1}{2} F_0 \cos\theta \bigl(C_5^* C_6 + C_5 C_6^*\bigr)\bigl(1 - \sin^2\theta \sin^2\phi\bigr), \tag{34}$$

where $F_0 = B_0^2 c v/(\pi H^2)$. Notice that $F_\parallel = 0$ if either $C_5 = 0$ or $C_6 = 0$, indicating that $\mathbf{U}_5$ and $\mathbf{U}_6$ are both standing waves. Also note that $F_\parallel$ is invariant under the transformation $C_5 \to C_5 + i\alpha C_6$, $\alpha \in \mathbb{R}$, and similarly for C_6.

Setting $C_5 = A_+ + A_-$ and $C_6 = A_- - A_+$, we may alternatively represent the general Alfvén wave as a linear combination of upgoing and downcoming modes, $\mathbf{U} = A_+ \mathbf{U}_+ + A_- \mathbf{U}_-$, where $\mathbf{U}_\pm = (1 \mp i\alpha)\mathbf{U}_5 \mp i\mathbf{U}_6$, with total flux

$$F_\parallel = F_0\bigl[|A_+|^2 - |A_-|^2\bigr] \cos\theta \bigl(1 - \sin^2\theta \sin^2\phi\bigr). \tag{35}$$

The vertical flux is $F_z = F_\parallel \cos\theta$. The real coefficient α does not affect the energy flux, but it does contribute to the energy density and the solution matchings. To determine it, we look for guidance in the two-dimensional case.

2D Case

Consider the 2D case $\phi = 0$, where $\xi = \zeta = 0$ and only the transverse displacement η remains:

$$\eta_5 = s^{2i\kappa \tan\theta} J_0(2s \sec\theta), \tag{36}$$

$$\eta_6 = s^{2i\kappa \tan\theta} \left[Y_0(2s \sec\theta) + \left(\frac{\cos^2\theta}{\pi \gamma v^2} - \frac{2\ln \sec\theta}{\pi} - \frac{2\mathcal{C}}{\pi} \right) J_0(2s \sec\theta) \right], \tag{37}$$

where $\mathcal{C} = 0.577216\ldots$ is Euler's constant. Clearly, $J_0(2s \sec\theta)$ and $Y_0(2s \sec\theta)$ are linearly independent solutions, as expected. The corresponding outgoing wave solution then takes the familiar Hankel function form

$$\eta_+ = s^{2i\kappa \tan\theta} \bigl[(1 - i\beta) J_0(2s \sec\theta) - i Y_0(2s \sec\theta)\bigr] \tag{38}$$
$$= s^{2i\kappa \tan\theta} \bigl[H_0^{(2)}(2s \sec\theta) - i\beta J_0(2s \sec\theta) \bigr], \tag{39}$$

where $\beta = \alpha + \bigl(\frac{\cos^2\theta}{\pi \gamma v^2} - \frac{2\ln \sec\theta}{\pi} - \frac{2\mathcal{C}}{\pi}\bigr)$. It is well known[2] that the $H_0^{(2)}$ Hankel function represents a "pure" wave travelling in the negative s (positive z) direction, so it is apparent that we must set $\beta = 0$. This determines α in the 2D case.

---

[2] This is verified using the large-argument asymptotic behaviour (Abramowitz and Stegun, 1965), $H_0^{(2)}(x) \sim \sqrt{2/(\pi x)} \exp[-i(x - \pi/4)]$.

3D Case

In three dimensions, we do not have the luxury of closed-form solutions from which the large-s asymptotic behaviour may be determined. In fact, this would not even be appropriate, as the Alfvén wave undergoes coupling to the magnetoacoustic waves around $s = \mathcal{O}(1)$. We are trying to find the real α that delivers a pure outgoing wave *at infinity*, despite it not exhibiting sinusoidal behaviour there. But how do we identify a "pure" wave in this regime? To do this, we can again use the Hankel function, or at least its asymptotic behaviour as $s \to 0^+$,

$$H_0^{(2)}(2s\sec\theta) \sim 1 - \frac{2i}{\pi}(\ln s + \ln\sec\theta + C) + \mathcal{O}(s^2 \ln s). \tag{40}$$

(Because $s\sec\theta$ is distance along a field line, independent of ϕ, it is the appropriate spatial coordinate for the field-guided Alfvén wave.) A pure outgoing wave should have this asymptotic structure.

Now, it is easily confirmed that the asymptotic polarisation direction of $\boldsymbol{\xi}_+$ as $s \to 0$ is $\mathbf{d} = (-\cos\phi\sin^2\theta\sin\phi,\, 1 - \sin^2\theta\sin^2\phi,\, -\cos\theta\sin\theta\sin\phi)$, which is perpendicular to $\mathbf{B}_0$ as expected, and that the small-s behaviour of displacement in this direction is

$$\mathbf{d}\cdot\boldsymbol{\xi}_+ \sim 1 - i\alpha - \frac{2i}{\pi}\ln s - \frac{i\cos^2\theta}{\pi\gamma v^2}, \tag{41}$$

where an arbitrary normalisation has been suppressed. Comparing this with Equation (40), we again infer that

$$\alpha = \frac{2\ln\sec\theta}{\pi} + \frac{2C}{\pi} - \frac{\cos^2\theta}{\pi\gamma v^2}, \tag{42}$$

as in two dimensions.

In summary, the three required radiation solutions at large z are $\mathbf{U}_2$, $\mathbf{U}_3$, and $\mathbf{U}_+ = (1 - i\alpha)\mathbf{U}_5 - i\mathbf{U}_6$. Physical solutions must match to a linear combination of these. In practice, we apply the boundary conditions at $z = 2$ Mm, where $s = \omega H/a \sim \mathcal{O}(10^{-4})$ or smaller, so convergence of the series is very rapid.

Acknowledgements PSC wishes to acknowledge the generous support of the research council of the K. U. Leuven through the award of a visiting senior postdoctoral fellowship (F/05/088) and the hospitality of the Centre for Plasma Astrophysics, where this work was begun. He also wishes to express his gratitude to Charlie Lindsey, Ashley Crouch, and Ineke De Moortel for very useful discussions during and after the week of the SOHO 19/GONG 2007 meeting in Melbourne.

References

Abramowitz, M., Stegun, I.A.: 1965, *Handbook of Mathematical Functions*, Dover, New York.
An, C.-H., Musielak, Z.E., Moore, R.L., Suess, S.T.: 1989, *Astrophys. J.* **345**, 597.
Bray, R.J., Loughhead, R.E.: 1974, *The Solar Chromosphere 252*, Chapman and Hall, London.
Bogdan, T.J., Cally, P.S.: 1997, *Proc. Roy. Soc. Lond. A* **453**, 943.
Cally, P.S.: 1983, *Solar Phys.* **88**, 77.
Cally, P.S.: 2000, *Solar Phys.* **192**, 395.
Cally, P.S.: 2001, *Astrophys. J.* **548**, 473.
Cally, P.S.: 2007, *Astron. Nachr.* **328**, 286.
Cally, P.S., Bogdan, T.J.: 1993, *Astrophys. J.* **402**, 732.
Cally, P.S., Bogdan, T.J.: 1997, *Astrophys. J.* **486**, L67.
Cally, P.S., Bogdan, T.J., Zweibel, E.G.: 1994, *Astrophys. J.* **437**, 505.

Christensen-Dalsgaard, J., Däppen, W., Ajukov, S.V., *et al.*: 1996, *Science* **272**, 1286.
Crouch, A.D., Cally, P.S.: 2003, *Solar Phys.* **214**, 201.
Crouch, A.D., Cally, P.S.: 2005, *Solar Phys.* **227**, 1.
Ferraro, V.C.A., Plumpton, C.: 1958, *Astrophys. J.* **127**, 459.
Melrose, D.B.: 1977, *Aust. J. Phys.* **27**, 43.
Melrose, D.B., Simpson, M.A.: 1977, *Aust. J. Phys.* **30**, 495.
Rubinstein, Z.: 1969, *A Course in Ordinary and Partial Differential Equations*, Academic, New York.
Schunker, H., Cally, P.S.: 2006, *Mon. Not. Roy. Astron. Soc.* **372**, 551.
Schwartz, S.J., Cally, P.S., Bel, N.: 1984, *Solar Phys.* **92**, 81.
Spruit, H.C., Bogdan, T.J.: 1992, *Astrophys. J.* **391**, L109.
Tracy, E.R., Kaufman, A.N., Brizard, A.J.: 2003, *Phys. Plasmas* **10**, 2147.
Zhugzhda, Y.D., Dzhalilov, N.S.: 1984, *Astron. Astrophys.* **132**, 45.

Surface-Focused Seismic Holography of Sunspots: I. Observations

D.C. Braun · A.C. Birch

Originally published in the journal Solar Physics, Volume 251, Nos 1–2, 267–289.
DOI: 10.1007/s11207-008-9152-5 © Springer Science+Business Media B.V. 2008

Abstract We present a comprehensive set of observations of the interaction of p-mode oscillations with sunspots using surface-focused seismic holography. Maps of travel-time shifts, relative to quiet-Sun travel times, are shown for incoming and outgoing p modes as well as their mean and difference. We compare results using phase-speed filters with results obtained with filters that isolate single p-mode ridges, and we further divide the data into multiple temporal frequency bandpasses. The f mode is removed from the data. The variations of the resulting travel-time shifts with magnetic-field strength and with the filter parameters are explored. We find that spatial averages of these shifts within sunspot umbrae, penumbrae, and surrounding plage often show strong frequency variations at fixed phase speed. In addition, we find that positive values of the mean and difference travel-time shifts appear exclusively in waves observed with phase-speed filters that are dominated by power in the low-frequency wing of the p_1 ridge. We assess the ratio of incoming to outgoing p-mode power using the ridge filters and compare surface-focused holography measurements with the results of earlier published p-mode scattering measurements using Fourier–Hankel decomposition.

Keywords Active regions · Magnetic fields · Helioseismology · Observations

1. Introduction

The use of solar acoustic (p-mode) waves to probe the subsurface structure of active regions (ARs) was first proposed by Thomas, Cram, and Nye (1982). Although efforts have been made to deduce properties of sunspots from interpretations of oscillations observed within

Helioseismology, Asteroseismology, and MHD Connections
Guest Editors: Laurent Gizon and Paul Cally

D.C. Braun (✉) · A.C. Birch
NWRA, CoRA Division, Boulder, CO, USA
e-mail: dbraun@cora.nwra.com

A.C. Birch
e-mail: aaronb@cora.nwra.com

sunspots (see reviews by Lites, 1992, and Bogdan, 2000), recent advances in sunspot seismology have been largely driven by the observations of the strong influences of sunspots (and ARs in general) on externally impinging p modes. This includes both absorption (*e.g.*, Braun, Duvall, and LaBonte, 1988; Bogdan *et al.*, 1993) and changes in phase (often characterized in terms of a change in travel time; *e.g.*, Braun *et al.*, 1992; Braun, 1995; Duvall *et al.*, 1996). A prevalent, largely phenomenological, approach to exploiting the travel-time shifts (relative to travel times in the quiet Sun) to model the subsurface properties of sunspots has been the characterization of the spot as a perturbation in the background sound speed. These types of models have been constructed by using observations from a variety of local-helioseismic techniques, including Fourier–Hankel decomposition (*e.g.*, Fan, Braun, and Chou, 1995), time–distance analysis (*e.g.*, Kosovichev, 1996; Kosovichev, Duvall, and Scherrer, 2000; Jensen *et al.*, 2001), ring diagrams (*e.g.*, Basu, Antia, and Bogart, 2004), and holography (*e.g.*, Lindsey and Braun, 2005b). The discovery of travel-time asymmetries between waves propagating toward and away from sunspots (Duvall *et al.*, 1996) have led to the inclusion of subsurface flows in many of these efforts. Travel times inferred from time–distance (TD) helioseismology have in particular been inverted to model flows and sound-speed perturbations by using a variety of assumptions including Fermat's principle and the ray approximation (*e.g.*, Kosovichev and Duvall, 1997; Kosovichev, Duvall, and Scherrer, 2000; Zhao, Kosovichev, and Duvall, 2001; Hughes, Rajaguru, and Thompson, 2005), the Fresnel-zone approximation (*e.g.*, Jensen *et al.*, 2001; Couvidat *et al.*, 2004), and the Born approximation (Couvidat, Birch, and Kosovichev, 2006). A consensus of many of these 3D inversions has emerged consisting of a relatively shallow (approximately 3 Mm deep) "slower" sound-speed perturbation above a "faster" sound-speed layer extending 10 Mm or more below the photosphere (see the review by Gizon and Birch, 2005).

These phenomenological models have been useful as foundations for developing both forward and inverse methods under a variety of approximations and assumptions (Gizon and Birch, 2005). At the same time, uncertainties about the degree to which the magnetic fields may contribute (in ways other than through associated thermal perturbations and flows) to phase or travel-time shifts, particularly in the near-surface layers, have persisted. Most local-helioseismic models of travel-time shifts, to date, do not include provisions for contributions from unresolved near-surface layers. [Near the photosphere, the typical vertical resolution provided by observed p modes is around 1 Mm (*e.g.*, Couvidat, Birch, and Kosovichev, 2006).] Notable exceptions include some 1D (horizontally invariant) structural inversions obtained from ring-diagram analyses (*e.g.*, Simmons and Basu, 2003; Basu, Antia, and Bogart, 2004).

Some observations and inferred sound-speed models may show direct evidence of strong near-surface contributions to the helioseismic signatures associated with ARs. An early example of this is the predominantly near-surface sound-speed perturbation consistent with Fourier–Hankel analysis (Fan, Braun, and Chou, 1995). Birch, Braun, and Hanasoge (2008; Paper 2) examine the relevance of this particular result to a more recent modeling effort. Lindsey and Braun (2005b) have shown that helioseismic signatures beneath ARs obtained by using holography largely vanish when a surface ("showerglass") phase shift, empirically related to photospheric magnetic flux density, is removed from the data. Some peculiar properties of the 3D time–distance inversions have also been presented as evidence for surface "contamination." Korzennik (2006) demonstrated that an inferred subsurface sound-speed "plume" structure depends critically on the inclusion of observations made within a sunspot penumbra and umbra. A test of inversions for flows performed by masking only the umbra showed little effect of the mask (Zhao and Kosovichev, 2003). Ringlike regions of enhanced sound speed in TD inversions of sunspots have also been examined as possible artifacts arising from the surface (Couvidat and Rajaguru, 2007). Surface effects in magnetic fields also include changes in the upper turning points (*e.g.*, Kosovichev and Duvall, 1997; Braun and Lindsey, 2000; Barnes and Cally, 2001). The observed reduction of p-mode amplitudes in spots has been shown to contribute to travel-time shifts independent of actual structural changes (Rajaguru *et al.*, 2006) as has reduced wave excitation (Hanasoge *et al.*, 2007; Parchevsky, Zhao, and Kosovichev, 2008).

Schunker *et al.* (2005) and Schunker, Braun, and Cally (2007) found that travel-time shifts obtained from seismic holography in sunspot penumbrae vary with the line-of-sight angle projected onto the plane containing the magnetic field and the vertical direction. A similar effect has also been noted by Zhao and Kosovichev (2006) with time–distance measurements. A satisfactory theory explaining these observations remains to be constructed, but some preliminary suggestions include mode conversion (Schunker and Cally, 2006) or radiative transfer effects in combination with mode propagation asymmetries (Rajaguru *et al.*, 2007). Whatever the cause, the observed line-of-sight dependence of travel-time shifts implies that a significant component of the shifts, at least in sunspot penumbrae, must be photospheric in origin.

In 1D inversions in global helioseismology (*e.g.*, Christensen-Dalsgaard, Gough, and Perez Hernandez, 1988) and ring-diagram analyses (*e.g.*, Basu, Antia, and Bogart, 2004), surface effects are largely characterized by their frequency-dependent contribution to the helioseismic signatures (mode or ridge frequencies). In contrast, the observations used in 3D travel-time inversions are typically made over a single wide-frequency bandpass and do not easily allow the assessment of possible frequency-dependent surface terms. However, there is increasing evidence for frequency variations in the travel-time shifts observed in ARs (Braun and Lindsey, 2000; Chou, 2000; Lindsey and Braun, 2004b; Braun and Birch, 2006; Couvidat and Rajaguru, 2007). Braun and Birch (2006) found evidence for a frequency variation, at fixed phase speed, of the travel times measured in active regions using helioseismic holography. This variation exceeds the smaller frequency variation expected from travel-time shifts computed from a proxy sound-speed model, with properties similar to recent two-component 3D inversions, by using the ray approximation.

The observed travel-time asymmetries in sunspots (differences in travel times between the incoming and outgoing propagating waves) have been interpreted and modeled as due to flows (Duvall *et al.*, 1996; Zhao, Kosovichev, and Duvall, 2001; Zhao and Kosovichev, 2003). The shallow inflows, within the first 3 Mm below the surface, characteristic of some of these p-mode-based TD models appear to be inconsistent with outflows inferred from other methods including f-mode time–distance analysis (Gizon, Duvall, and Larsen, 2000) and holography (Braun, Birch, and Lindsey, 2004). Some questions have been raised whether travel-time asymmetries may arise from other mechanisms, including the suppression of acoustic sources (Gizon and Birch, 2002; Hanasoge *et al.*, 2007) or absorption (Woodard, 1997; Lindsey, Schunker, and Cally, 2007).

Including magnetic fields in helioseismic models of sunspots appears to be a substantially more formidable task than constructing models that include only thermal perturbations. Some progress has been made with MHD models of the absorption in sunspots observed from Fourier–Hankel decomposition (*e.g.*, see the review by Bogdan and Braun, 1995). More recent efforts have also addressed the observed phase shifts (*e.g.*, Cally, Crouch, and Braun, 2003; Crouch *et al.*, 2005; Gordovskyy and Jain, 2007). It is expected that considerable advances in modeling helioseismic data will follow from the current development and application of hydrodynamic (HD) and magnetohydrodynamic (MHD) simulations (*e.g.*, Jensen, Duvall, and Jacobsen, 2003; Tong *et al.*, 2003; Mansour *et al.*, 2004; Werne, Birch, and Julien, 2004; Benson, Stein, and Nordlund, 2006; Hanasoge and Duvall, 2007; Hanasoge, Duvall, and Couvidat, 2007; Khomenko and Collados, 2006; Shelyag, Erdélyi, and

Thompson, 2006, 2007; Cameron, Gizon, and Daiffallah, 2007; Cameron, Gizon, and Duvall, 2008).

Our primary motivation in this paper is to expand the measurements of Braun and Birch (2006). We hope that a comprehensive exposition of helioseismic observations of travel-time shifts, and their dependence on p-mode properties, will promote and support improved modeling efforts, including the use of numerical simulations. As we are specifically interested in the importance of near-surface effects we also examine the relationship between the observed travel-time shifts and the photospheric magnetic field. Of particular importance is the measurement of frequency variations of the travel-time shifts, by using methods similar to Braun and Birch (2006). However, we extend those measurements to include both mean travel-time shifts and travel-time asymmetries, and we determine spatial averages of these quantities over sunspot umbrae, penumbrae, and other magnetic regions. Our principal tool is surface-focused helioseismic holography (*e.g.*, Braun and Lindsey, 2000; Braun and Birch, 2006), for which the travel-time shifts are expected to have the most sensitivity to near-surface perturbations. The use of surface-focus holography (described in Section 2) contrasts this work with other recent ("lateral-vantage") holographic studies of ARs (*e.g.*, Lindsey and Braun, 2005a, 2005b). In addition, to ensure a meaningful comparisons of our results (described in Section 3) to TD observations we use narrow annular pupils and corresponding phase-speed filters as discussed in Section 3.1. An overriding theme in our findings is a strong sensitivity of the results to the choice of filter and frequency bandwidth employed. To investigate this further, we also employ filters centered on the p-mode ridges (Section 3.2). The ridge-based filters allow a detailed comparison of surface-focused holography measurements of both travel-time shifts and absorption with published results of Fourier–Hankel analysis (Section 3.3).

2. Analysis

Helioseismic holography (HH) is a method based on the phase-coherent imaging of the solar interior acoustic field. In HH one computationally extrapolates the surface acoustic field into the solar interior (Lindsey and Braun, 1997, 2000) to estimate the amplitudes of the waves propagating into and out of a focus point at a chosen depth and position in the solar interior. These amplitudes, called the ingression (H_-) and egression (H_+) are estimated by a convolution of the surface oscillation signal (ψ) (typically the line-of-sight component of velocity observed from Dopplergrams) with appropriate Green's functions (Lindsey and Braun, 2000). For this work, the Green's functions are computed in the eikonal formulation (Lindsey and Braun, 1997, 2000). For surface-focused HH, the Green's functions represent propagators that evolve the acoustic field forward or backward in time from a position on the solar surface into the solar interior, and back up to the surface focus. To select a particular set of p modes, these functions are evaluated over a chosen annular pupil. A dispersion correction, empirically determined from statistics obtained from measurements in the quiet Sun, is applied to the computation of the Green's functions (see Lindsey and Braun, 2000).

The basis of our analysis consists of what are termed *local control correlations* (Lindsey and Braun, 2004a, 2005a). These are directly comparable to center-annulus TD correlations (*e.g.*, Duvall *et al.*, 1996; Braun, 1997). In the space-frequency domain, the correlation

$$C_+(\mathbf{r}) = \langle H_+(\mathbf{r}, \nu)\psi^*(\mathbf{r}, \nu)\rangle_{\Delta\nu} \tag{1}$$

describes the egression control correlation, and

$$C_-(\mathbf{r}) = \langle \psi(\mathbf{r}, \nu) H_-^*(\mathbf{r}, \nu)\rangle_{\Delta\nu} \tag{2}$$

describes the ingression control correlation. Here, $\psi(\mathbf{r}, \nu)$ represents the temporal Fourier transform of the surface wave field, ν is the temporal frequency, $\mathbf{r}$ is the horizontal position on the solar surface, and $H_-(\mathbf{r}, \nu)$ and $H_+(\mathbf{r}, \nu)$ represent the temporal Fourier transforms of the ingression and egression, respectively. The asterisk denotes complex conjugation, and the angle brackets indicate an average over a chosen positive frequency range $\Delta \nu$.

The primary quantities of interest are the travel-time shifts, which are related to the phase of the correlations,

$$\delta\tau_\pm = \arg\bigl[C_\pm(\mathbf{r})\bigr]/2\pi\nu_0, \qquad (3)$$

where ν_0 is the central frequency of the bandpass $\Delta\nu$. These represent travel-time shifts of the observed incoming (τ_-) or outgoing (τ_+) waves, as sampled by a chosen filter, relative to the travel times expected for the same ensemble of waves propagating in the solar model used to compute the Green's functions. Small systematic deviations of the quiet-Sun values from zero, which vary with pupil size and filter and are likely caused by imperfections in the Green's functions and dispersion correction, are removed by subtracting averaged quiet-Sun values from the observed control correlation phases. Of interest are the mean travel-time shift, $\delta\tau_{\text{mean}} = (\delta\tau_+ + \delta\tau_-)/2$, and the difference (or travel-time asymmetry), $\delta\tau_{\text{diff}} = \delta\tau_+ - \delta\tau_-$.

A 27-hour sequence of full-disk Dopplergrams with one-minute cadence, obtained from the Michelson Doppler Imager (MDI; Scherrer *et al.*, 1995) onboard the *Solar and Heliospheric Observatory* (SOHO), was used in this study. The data set starts on 1 April 2002, 21:01 UT, and includes several sunspot groups (NOAA groups 9885, 9886, 9887, and 9888) within a $60° \times 60°$ Postel-projected region. This area was tracked at the Carrington rotation rate and includes four sunspots with penumbral radii greater than 15 Mm as well as other smaller spots. The three largest sunspots are very similar in size, with mean umbral and penumbral radii of 7 and 18 Mm, respectively.

The following steps summarize the general data reduction: (1) projection of the desired region from full-disk Dopplergrams to a Postel projection that rotates with a fixed Carrington rate, (2) temporal detrending by subtraction of a linear fit to each pixel signal in time, (3) removal of poor quality images, identified by a five-σ deviation of any pixel from the linear trend, (4) Fourier transform of the data in time, (5) (optional) correction for the amplitude suppression in magnetic regions (Rajaguru *et al.*, 2006), (6) spatial Fourier transform of the data and multiplication by a chosen filter, (7) extraction of the desired frequency bandpass, (8) computation of Green's functions over the appropriate pupil, (9) computation of ingression and egression amplitudes by a 3D convolution of the data with the Green's functions, and (10) computation of the travel-time shift maps by Equations (1)–(3).

The optional correction for amplitude suppression (step 5) involves dividing the amplitude of each pixel in the data by its root-mean-square value over the frequency bandpass. In step 6 we have used two sets of filters: phase-speed filters and ridge filters. Their description and the results obtained from each set are described in Section 3.1 and Section 3.2. For the phase-speed filters (Section 3.1) we compare results with and without the amplitude-suppression correction. The difference is relatively small and, for the ridge-filtering (Section 3.2), we use only uncorrected data.

3. Observations

3.1. Phase-Speed Filters

The phase-speed filters used are of the type specified by Couvidat, Birch, and Kosovichev (2006); namely, the three-dimensional Fourier transform of the data (step 6) is multiplied by a function

$$F(\mathbf{k}, \nu) = \exp[-(2\pi \nu/|\mathbf{k}| - w)^2/2\delta w^2], \qquad (4)$$

where w and δw are the mean phase speed and filter width, respectively. We use the same set of ten filters (denoted A through J) of Braun and Birch (2006) with parameters listed in Table 1 of that paper. These filters are of the same type, but are somewhat narrower in width, than the eleven common filters often employed in TD analyses (*e.g.*, Couvidat, Birch, and Kosovichev, 2006; Zhao and Kosovichev, 2006). Each filter is used with a corresponding pupil, over which the ingression and egression are evaluated. This pupil is a complete annulus defined so that acoustic rays at a frequency of $\nu = 3.5$ mHz reaching the inner and outer radii span the full width at half maximum (FWHM) of the squared filter. The parameter δw is related to the FWHM by $\delta w = \text{FWHM}/[2(\ln 2)^{1/2}]$. The filters were chosen such that the sets of FWHM and corresponding pupils span a continuous range of phase speed and radius, respectively.

All of the phase-speed filters used also remove the contribution of the f mode, a practice first used by Giles (2000). Our f-mode cutoff consists of a high-pass filter with Gaussian roll-off in temporal frequency. The position and rate of the roll-off varies with spatial wavenumber such that full transmission occurs at a frequency midway between the f mode and p_1 ridges and 10^{-7} transmission occurs at the frequency of the f mode.

The filters are applied to, and the travel-time shift maps computed from, portions of the data extracted in 1-mHz-wide frequency bandpasses centered at 2, 3, 4, and 5 mHz. We also compute travel-time shifts over a wider bandpass (2.5–5.5 mHz) typical of common TD measurements. The use of narrow frequency bandpasses, although frequently employed in HH, differs from typical applications of TD correlations. Consequently, the wide bandpass measurements provide a useful check and basis for comparison. It should be noted, however, that the power spectra of solar acoustic oscillations are naturally limited in bandwidth. Power spectra computed by using typical phase-speed filters applied to MDI data, without any additional temporal filtering, show a concentration of between 60% and 90% of the power within a 1-mHz bandwidth, depending on the choice of phase-speed filter.

Some maps of travel-time shifts computed with phase-speed filters are shown in Figures 1 and 2. Figure 1 shows maps of $\delta\tau_-$ (top panels) and $\delta\tau_+$ (bottom panels); Figure 2 shows the corresponding maps of $\delta\tau_{\text{mean}}$ (top) and $\delta\tau_{\text{diff}}$ (bottom). The maps are stacked into columns of increasing phase speed (left to right) and rows representing increasing frequency (bottom to top). For simplicity we will refer to each map by a number–letter combination denoting filter and frequency combination; for example, "4B" refers to filter B applied to the frequency bandpass centered at 4 mHz. Because of the filter masking the f mode, and the decrease in acoustic wave amplitudes at frequencies below the p_1 ridge, only a subset of possible frequency–filter combinations produce meaningful correlations. The maps analyzed here are limited to filters 2D–2J, 3B–3J, 4A–4J, and 5A–5J.

Individual maps exhibit spatial relationships between the travel-time shifts and the surface magnetic flux explored below. The stacking of maps in Figures 1 and 2 reveals several striking properties of the travel-time shifts, including frequency dependencies of $\delta\tau_+$, $\delta\tau_-$, $\delta\tau_{\text{mean}}$, and $\delta\tau_{\text{diff}}$ at all phase-speed filters, and a surprising connection between the sign of the shifts and the value of the central frequency of the filter with respect to the frequency of the p_1 ridge. In particular, positive travel-time shifts (for both incoming and outgoing waves) are observed exclusively in frequency bandwidths that are centered below the p_1 ridge, shown by the solid line in Figures 1 and 2. Maps of $\delta\tau_+$, $\delta\tau_-$, $\delta\tau_{\text{mean}}$, and $\delta\tau_{\text{diff}}$ for filters with frequencies closest to the p_1 ridge (*i.e.*, 4B, 3C, and 2D – not shown) show positive

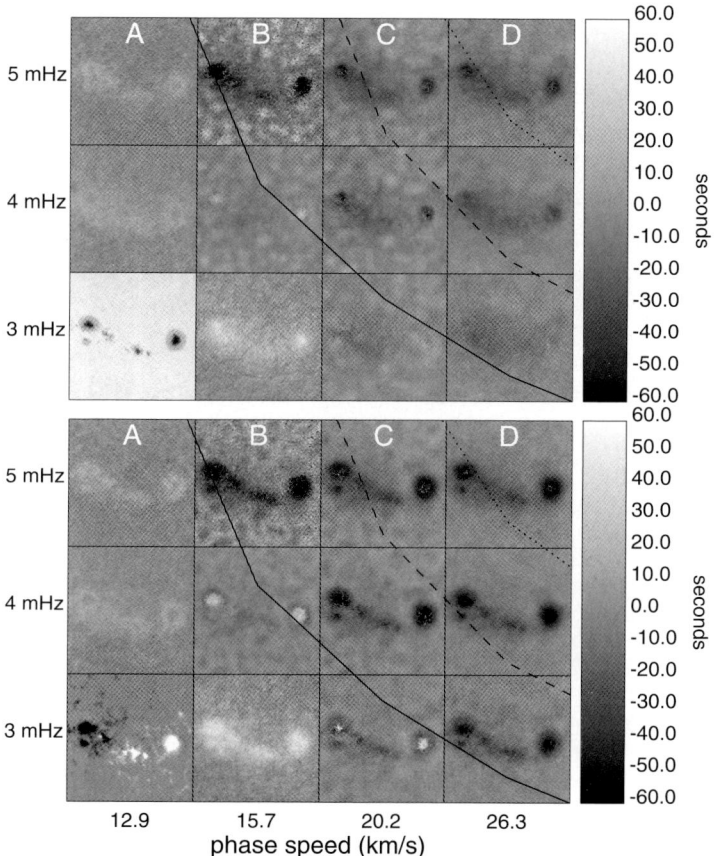

Figure 1 Maps of travel-time shifts $\delta\tau_-$ (top panels) and $\delta\tau_+$ (bottom panels) covering a portion of the region studied and showing sunspot group 9885. The columns of maps labeled A through D indicate the phase-speed filter used; the rows indicate the frequency bandpass. The solid jagged line running diagonally through the panels connects the location of the p_1 ridge in the $\nu - w$ domain for each filter, with the centers of the maps assigned to values of frequency and phase speed as indicated on the left and bottom edges of the plot. The dashed and dotted lines indicate the locations of the p_2 and p_3 ridges, respectively. The map in the lowest left position of the top set of panels shows an MDI continuum intensity image; the map in the same position in the bottom set shows a line-of-sight magnetogram.

shifts near sunspot umbra and negative travel-time shifts elsewhere in the ARs. At frequencies above (below) these values, the filters yield exclusively negative (positive) travel-time shifts. Filters E–J (not shown) exhibit trends similar to filter D. For these filters, negative values for $\delta\tau_+$ and $\delta\tau_-$ are observed throughout the active region. Both incoming and outgoing time shifts measured with the larger phase-speed filters increase with increasing frequency, with the values of the outgoing shifts exceeding the incoming shifts. These filters (E–J) show AR travel-time differences ($\delta\tau_{\text{diff}}$) that decrease with increasing frequency.

As noted earlier by Braun and Birch (2006), the travel times are nonlinearly related to the surface magnetic-flux density. The quantity B_{tot} is derived from an MDI line-of-sight magnetogram by assuming the magnetic field is the gradient of a potential and is used as a proxy for the total flux density. Figures 3 and 4 show plots of travel-time shifts against B_{tot} for phase-speed filters B and E, respectively. For clarity, the scatter of individual pixel

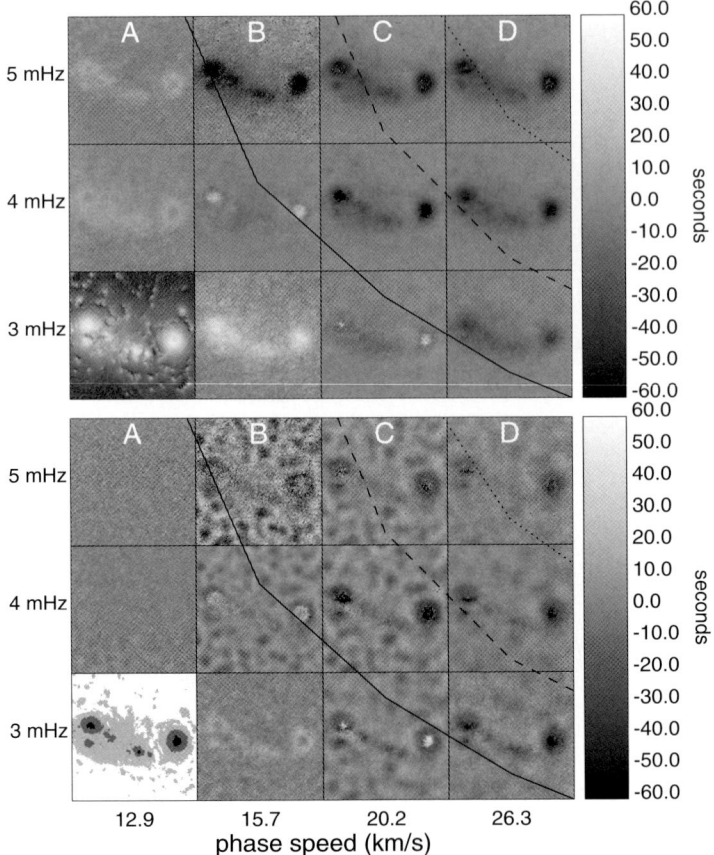

Figure 2 Maps of the travel-time shifts $\delta\tau_{\mathrm{mean}}$ (top panels) and $\delta\tau_{\mathrm{diff}}$ (bottom panels) covering a portion of the region studied and showing sunspot group 9885. The columns of maps labeled A through D indicate the phase-speed filter used; the rows indicate the frequency bandpass. The solid jagged line running diagonally through the panels connects the location of the p_1 ridge in the $\nu - w$ domain for each filter, with the centers of the maps assigned to values of frequency and phase speed as indicated on the left and bottom edges of the plot. The dashed and dotted lines indicate the locations of the p_2 and p_3 ridges, respectively. The map in the lowest left position of the top set of panels shows B_{tot} (see text); the image in the same position in the bottom set isolates in shades of gray different portions of the AR for study (see text).

values is not shown (although see Figure 1 of Braun and Birch, 2006, for some examples). Instead, Figures 3 and 4 show the average of the shifts derived from bins equally spaced in the logarithm of the flux density. Solid (dotted) lines indicate time shifts computed with (without) the amplitude-suppression correction discussed in Section 2. The effect of the correction is to decrease the travel-time shifts, especially at lower phase speeds (and smaller pupils) and in regions of high flux densities typical of the sunspot umbrae and penumbrae, by amounts on the order of one to ten seconds. Vertical bars in Figures 3 and 4, and in most of the other plots shown in this paper, indicate the total deviation (maxima minus minima) of the averaged value as determined within three independent subregions containing the three largest sunspots. Thus, the bars include contributions not only from sources of random error but also from potentially systematic differences between sunspots. Typically, however, these deviations as a whole are very small.

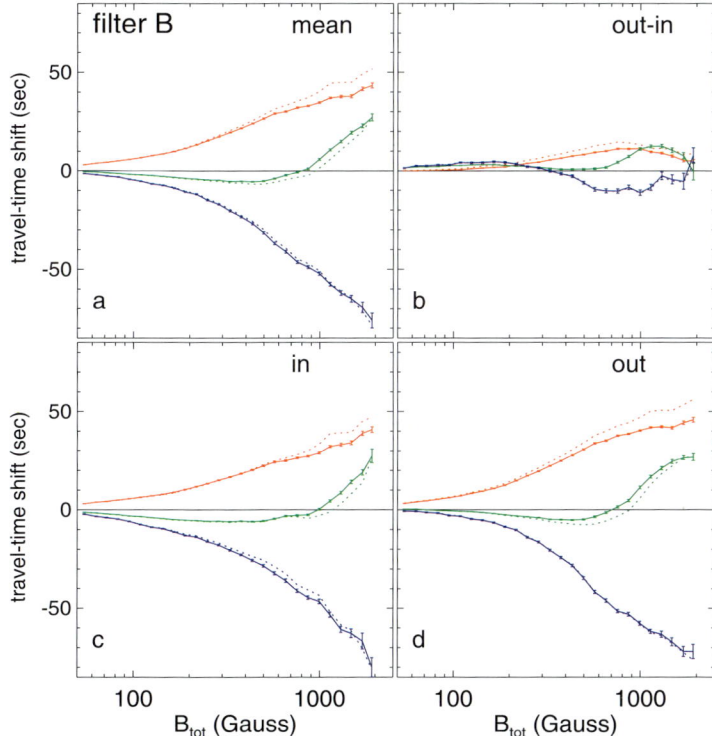

Figure 3 Travel-time shifts for filter B: (a) the mean travel-time shift, (b) the travel-time difference, (c) the incoming travel-time shift, and (d) the outgoing travel-time shift as functions of the magnetic field B_{tot}. Solid (dotted) lines connect averaged travel-time shifts, averaged over equally spaced bins in the logarithm of the flux density, computed with (without) an amplitude-suppression correction (see text). Red, green, and blue lines indicate frequencies of 3, 4, and 5 mHz, respectively. Vertical bars indicate the total deviation (maxima minus minima) of the averages of three independent subregions containing the three largest sunspots.

The frequency variation shown for filter B (Figure 3) is typical of the results with smaller phase speeds (and smaller pupils) that undergo the transition from positive time shifts (at sub-p_1 frequencies) to negative time shifts (at frequencies higher than the p_1 ridge). The different frequencies shown in Figure 3a exhibit the three types of dependence on flux density of the mean travel-time shift described by Braun and Birch (2006). The travel-time differences (Figure 3b) show similar trends except at the highest flux densities, where the results for all frequencies approach zero. For filter E (Figure 4), it is observed that the frequency variations are in general larger for $\delta\tau_-$ shifts than for $\delta\tau_+$. As noted earlier, the differences $\delta\tau_{\text{diff}}$ for these larger phase speeds show larger shifts at lower frequencies. This trend is opposite to that observed with the mean shifts $\delta\tau_{\text{mean}}$.

As Braun and Birch (2006) note, the close relationship between $\delta\tau_{\text{mean}}$ and B_{tot} is consistent with predominately near-surface perturbations, but it does not rule out subsurface perturbations that may very well correlate with surface flux. As discussed in Section 1, the variation of travel-time shifts with both temporal frequency and phase speed is of critical importance in understanding the depth variations of the underlying perturbations. In particular, Braun and Birch (2006) have suggested that the variation with frequency, at fixed phase speeds, of travel-time shifts may be a signature of surface effects (see also Paper 2).

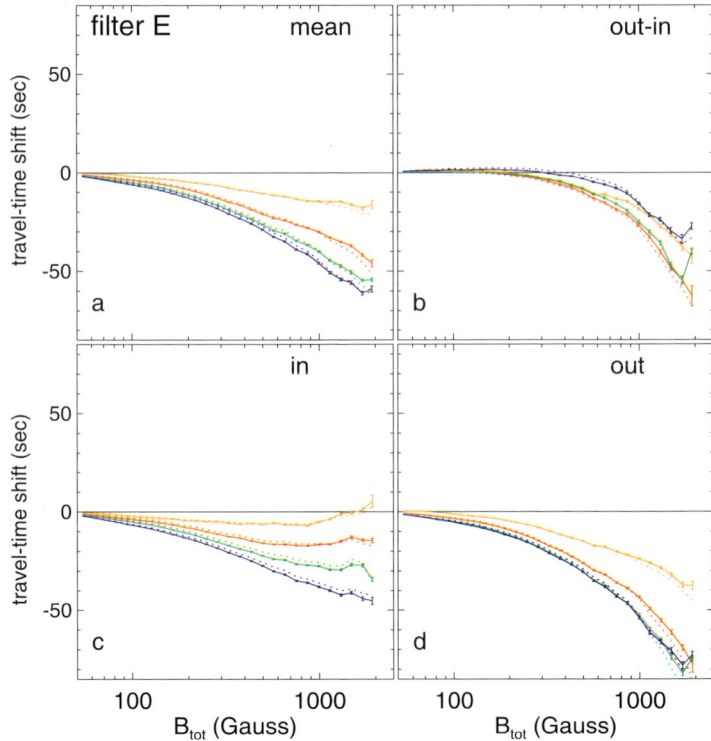

Figure 4 Travel-time shifts for filter E: (a) the mean travel-time shift, (b) the travel-time difference, (c) the incoming travel-time shift, and (d) the outgoing travel-time shift as functions of the magnetic field B_{tot}. Solid (dotted) lines connect averaged travel-time shifts, averaged over equally spaced bins in the logarithm of the flux density, computed with (without) an amplitude-suppression correction (see text). Orange, red, green, and blue lines indicate frequencies of 2, 3, 4, and 5 mHz, respectively. Vertical bars indicate the total deviation (maxima minus minima) of the averages of three independent subregions containing the three largest sunspots.

To quantify these variations, we compute spatial averages of the travel-time shifts over three types of regions characteristic of the sunspot groups. The first two types are sunspot umbrae and penumbrae, identified by brightness values of less than 50% and 92% of the mean MDI continuum values, respectively. The third region of interest (which we simply call "plage") is identified by values of B_{tot} above 100 gauss and excluding areas previously marked as umbrae or penumbrae. The panel in the lowest left corner of Figure 2 illustrates in increasingly darker shades of gray the three regions (plage, penumbrae, and umbrae) identified in this manner around NOAA 9885. The umbral and penumbral averages are of particular importance since they represent time shifts experienced by waves propagating through the immediate subsurface layers of a sunspot (*e.g.*, within 10 Mm depth below a 30-Mm-diameter spot). Figures 5, 6, and 7 show the spatially averaged time shifts for the umbrae, penumbrae, and plage, respectively. These figures quantify many of the properties already noted in the travel-time shift maps, including the strong frequency variations at each fixed phase speed and the changes of sign at smaller phase speeds. Figures 5–7 also include the measurements made with the wide temporal bandpass (2.5–3.5 mHz). These values, as expected from the relative contributions of modes in the power spectra, fall largely between the results obtained in the 3- and 4-mHz bandpasses. Also noteworthy is that, for the plage,

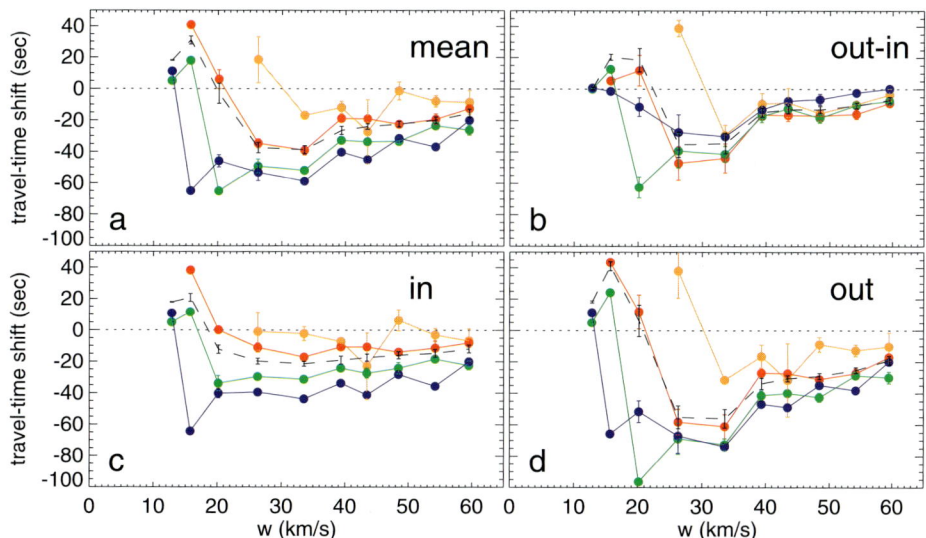

Figure 5 (a) The mean travel-time shift, (b) the travel-time difference, (c) the incoming travel-time shift, and (d) the outgoing travel-time shift, averaged over the sunspot umbrae, as functions of the phase speed w. Yellow, red, green, and blue lines indicate frequencies of 2, 3, 4, and 5 mHz, respectively. The black dashed line indicates the use of the wide (2.5 – 5.5 mHz) frequency bandpass. Vertical bars indicate the total deviation (maxima minus minima) of the averages of three independent subregions containing the three largest sunspots.

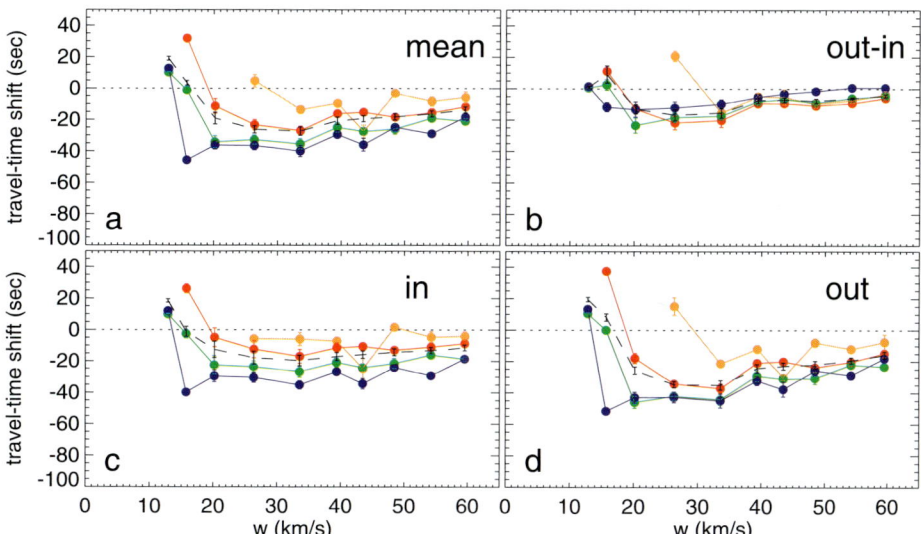

Figure 6 (a) The mean travel-time shift, (b) the travel-time difference, (c) the incoming travel-time shift, and (d) the outgoing travel-time shift, averaged over the sunspot penumbrae, as functions of the phase speed w. Yellow, red, green, and blue lines indicate frequencies of 2, 3, 4, and 5 mHz, respectively. The black dashed line indicates the use of the wide (2.5 – 5.5 mHz) frequency bandpass. Vertical bars indicate the total deviation (maxima minus minima) of the averages of three independent subregions containing the three largest sunspots.

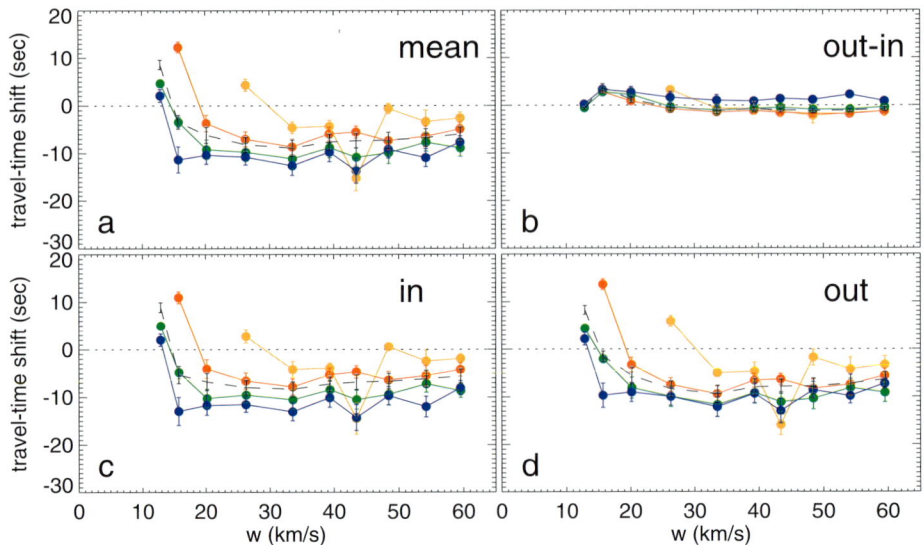

Figure 7 (a) The mean travel-time shift, (b) the travel-time difference, (c) the incoming travel-time shift, and (d) the outgoing travel-time shift, averaged over the plage, as functions of the phase speed w. Yellow, red, green, and blue lines indicate frequencies of 2, 3, 4, and 5 mHz, respectively. The black dashed line indicates the use of the wide (2.5 – 5.5 mHz) frequency bandpass. Vertical bars indicate the total deviation (maxima minus minima) of the averages of three independent subregions containing the three largest sunspots. Note that the vertical scale differs from those in the figures for the umbrae and penumbrae.

$\delta\tau_{\text{diff}}$ varies from slightly negative at low frequencies to slightly positive at high frequencies and is mostly independent of phase speed.

It is noteworthy that essentially all of the travel-time shifts observed in Figures 5 – 7 show significant frequency variations. At smaller phase speeds, the variations of both mean travel-time shifts and travel-time asymmetries show strong variations, which often include a change of sign of the shifts. At higher phase speeds, the mean travel-time shifts also show large systematic frequency variations. The mean shifts observed at 5 mHz, for example, are typically twice the value at 3 mHz, with the difference being about 15 – 20 seconds.

The effects of dispersion, including changes of ray paths as a function of temporal frequency (*e.g.*, Barnes and Cally, 2001), may cause frequency variations of travel-time shifts. However, the frequency variations in the mean travel-time shifts observed here may be much larger than are expected for sound-speed perturbations inferred from recent inversions of travel times. Using the ray approximation, Braun and Birch (2006) computed differences of the mean travel-time shifts between 3 and 5 mHz, for a sound-speed perturbation similar to that of Kosovichev, Duvall, and Scherrer (2000), of about five to ten seconds for phase speeds less than 30 km s^{-1}, and less than one to two seconds for higher phase speeds. The implications of these variations are best explored in the context of modeling (and some initial efforts are addressed in Paper 2). However, it is expected that observations such as shown in Figures 5 – 7 may lead to methods for identifying and removing surface effects. There are no precedents in either global or local helioseismic inversions that offer any hope of including the contribution of unresolved near-surface structure without making use of the temporal frequency dependencies of the observables.

3.2. Ridge Filters

In this section, we present results obtained by using filters that isolate individual p-mode ridges. The use of ridge filters allows us to judge the sensitivity of the travel-time shifts (especially those experienced by waves near the p_1 ridge) to the choice of filter. Ridge-based filters also facilitate a more direct comparison with results obtained with Fourier–Hankel decomposition (see Section 3.3). Ridge filters have been used previously in time–distance helioseismology for f-mode studies (*e.g.*, Duvall and Gizon, 2000; Gizon and Birch, 2002; Jackiewicz *et al.*, 2007a, 2007b), and recently for p modes (Jackiewicz, Gizon, and Birch, 2008). Here we employ filters for HH that isolate the p_1, p_2, p_3, and p_4 ridges. At each temporal frequency, the filters have full transmission for wavenumbers spanning the midpoints between the desired ridge and the neighboring ridges. Sharp Gaussian roll-offs (similar to those used to remove the f mode in combination with the phase-speed filters in Section 3.1) remove the contributions above and below these wavenumbers.

Unlike the common use of ray theory to define the radii of annuli (TD) or pupils (HH), there is no "standard" procedure for adopting a pupil geometry for ridge filters. After some trial and error, we settled on a fixed pupil for each ridge. We found that the results were not largely dependent on the outer pupil radius, but they did change significantly with the choice of inner pupil radius. A choice of inner pupil radius smaller than roughly the horizontal p-mode wavelength of the highest wavenumber present in the power spectra apparently produces undesired leakage of the oscillatory signal at the focus directly into the egression and ingression regressions. With these considerations, we adopted a set of pupils with radii 9–42 Mm (p_1), 12–90 Mm (p_2), 14–167 Mm (p_3), and 17–195 Mm (p_4). We have also experimented with varying the width of the frequency bandpass. The travel-time shift maps analyzed here were made with $\Delta \nu = 0.26$ mHz and were critically sampled with a frequency spacing of $\Delta \nu/2$.

The results for the mean travel-time shifts, averaged over umbrae, penumbrae, and plage, are shown in Figure 8, whereas the results for the travel-time differences are shown in Figure 9. In light of the results obtained by using phase-speed filters (Section 3.1), what is most striking from Figures 8 and 9 is that both the mean and difference travel-time shifts obtained by using ridge filters are essentially (with a few noisy exceptions) always negative. This is also true along the p_1 ridge. Comparisons between the ridge-filtered and phase-filtered results are facilitated by overlaying the nearest phase-speed-determined values on Figures 8 and 9. By "nearest" we mean that for a given 1-mHz-frequency bandpass and radial order, the nearest phase-speed filter is that closest to the phase speed of the ridge at the central frequency. For p_1, the nearest filter combinations are 2F, 3C, 4B, and 5B. For p_2, these combinations are 2J, 3E, 4D, and 5C. For p_3, they are 3I, 4E, and 5D, and for p_4, they are 4H and 5E. For p_2, p_3, and p_4 there is very good agreement between phase-speed and ridge-filtered results for the mean travel-time shifts averaged in umbrae or penumbrae, and reasonable agreement for the time differences. The largest discrepancies are clearly in the p_1 ridge and (especially in the umbra) involve a change in sign in the measurements of τ_{mean} and τ_{diff} between the two types of filters.

It is noteworthy that the discrepancies between ridge-filtered and phase-speed-filtered results are largest for phase-speed filters with frequency bandwidths that are centered below p_1. These are the same phase-speed and frequency combinations that produce the positive travel-time shifts seen in Figures 1 and 2. In contrast, adjacent phase-speed filters centered above the p_1 ridge apparently produce travel-time shifts in sunspot umbrae and penumbrae (*i.e.*, the black and magenta squares in the top panels of Figures 8 and 9), which are very close to that observed with the p_1 ridge filter. Repeating the ridge-filtered measurements for some of the cases where these sign changes occur (*e.g.*, near filter 3C) with the

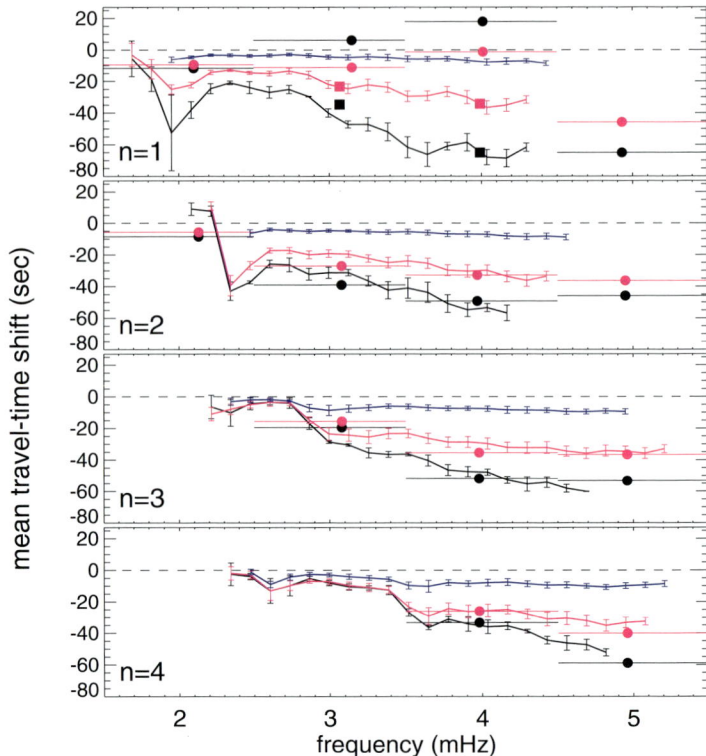

Figure 8 Mean travel-time shifts determined by using ridge filters and averaged over umbrae (black lines), penumbrae (magenta lines), and plage (blue lines), as functions of frequency along the $p_1 - p_4$ ridges. Vertical bars indicate the total deviation (maxima minus minima) of the averages of three independent subregions containing the three largest sunspots. The filled black and magenta circles (with 1-mHz-wide horizontal bars) represent comparisons of umbral and penumbral time shifts determined from travel-time maps made with "nearby" phase-speed filters (see text). The circles are placed at the power-weighted average frequency for the given filter and frequency combination. The squares shown near 3 and 4 mHz in the top panel indicate umbral (black) and penumbral (magenta) mean travel-time shifts for the phase-speed filters (3D and 4C), which have higher values of phase speed compared to those denoted by the filled circles (3C and 4B).

same pupil as used with the phase-speed filter yields deviations in both the mean shifts and travel-time asymmetry of only a few seconds (out of a total of 30–40 seconds) from results obtained by using the fixed pupil range stated earlier. Thus, the discrepancy in the sign of travel-time perturbations between the two types of filters is not the result of using different pupils.

Figure 10 illustrates the extreme sensitivity of the sign of the travel-time shifts in sunspots near the p_1 ridge on the choice of filters. This figure compares the results of travel-time shifts computed with a commonly used TD phase-speed filter (filter 1 of Couvidat, Birch, and Kosovichev, 2006, and Zhao and Kosovichev, 2006; hereafter TD1) with shifts computed with a p_1 ridge filter over a frequency bandpass between 3.5 and 5.5 mHz. Despite the gross similarity of the filtered power included in the measurements, the resulting maps of τ_{mean} and τ_{diff} are drastically different. The travel-time shift maps made with the TD1 filter have larger positive values (*e.g.*, by about a factor of two in the mean travel-time shift and a factor of four in the travel-time asymmetry in the penumbrae) than results obtained

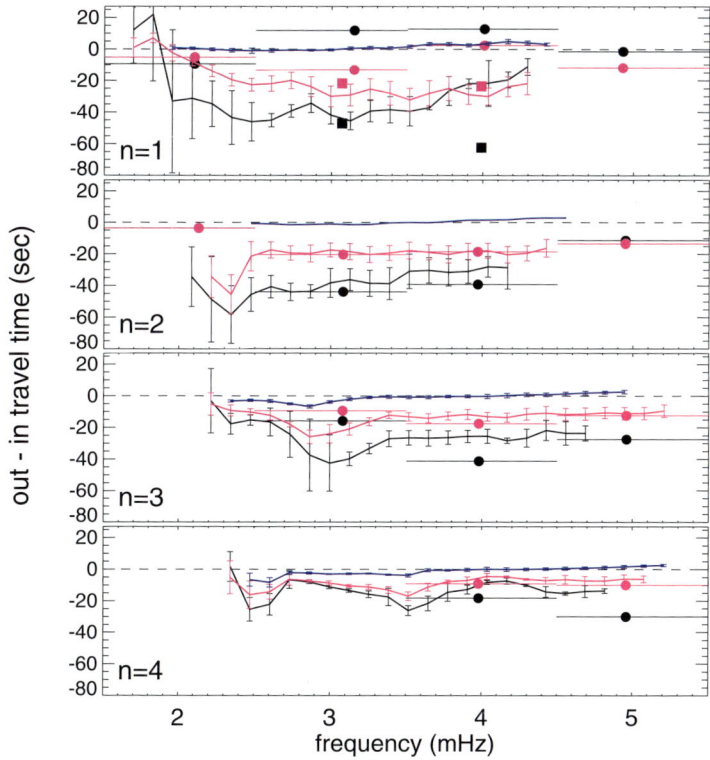

Figure 9 Difference travel-time shifts determined by using ridge filters in the same format as Figure 8.

with the filter 5A shown in Figure 1, even though the mean phase speed of these filters are both approximately 13 km s^{-1}. We have found that incrementally increasing the width of a phase-speed filter, centered at 12.8 km s^{-1}, produces maps with incrementally stronger positive travel-time shifts in sunspots. Based on our experience with a variety of filters, we find in general that the requirement for producing positive travel-time shifts appears to be a disproportionate contribution to the correlations of wave power from the low-frequency wing of the p_1 ridge relative to the high-frequency wing. This is illustrated in Figure 11. This asymmetry apparently results from the fact that the mean phase speed of the filter (12.8 km s^{-1}; shown by the dashed line in Figure 10a) falls significantly below the p_1 ridge.

The reasons for the strong sensitivity, including sign changes, of the travel-time shifts to details of the filter (*e.g.*, width) and the relative weighting of the low-frequency wing of p_1 are not fully understood at this time. Thompson and Zharkov (2008) have presented evidence that a sign switch in mean travel-time shifts also apparently occurs with the application of filters centered half-way (*i.e.*, in the trough) between the p_1 and p_2 ridges.

3.3. Comparison with Fourier–Hankel Analysis

Here we compare the ridge-filtered (mean) travel-time shifts with the phase shifts observed in sunspots obtained by using Fourier–Hankel (FH) analysis (Braun, 1995). To do this, we use the published values of phase shifts determined from FH decomposition of waves around two sunspot groups, NOAA 5229 and 5254, observed with Ca II intensity images made at

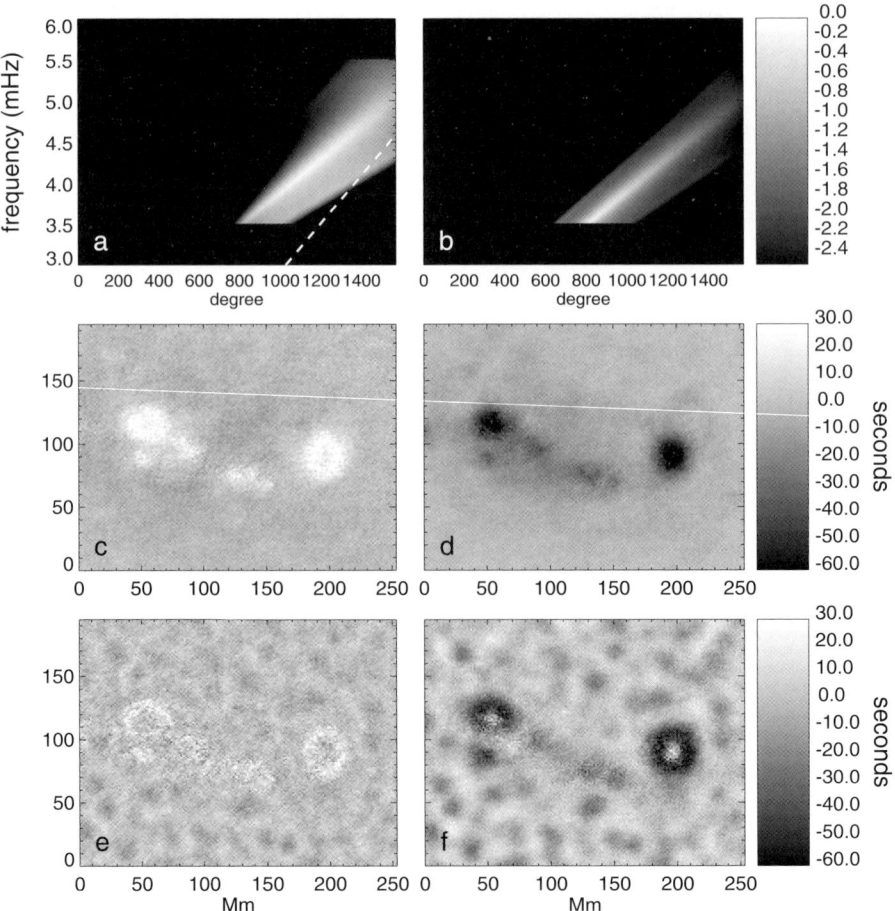

Figure 10 (a) Azimuthally averaged MDI power spectrum multiplied by phase-speed filter "1" of Couvidat, Birch, and Kosovichev (2006). The gray scale indicates the logarithm of the power, normalized to the peak power. The dashed line indicates the mean phase speed of the filter ($w = 12.8$ km s^{-1}). (b) The power spectrum at frequencies between 3.5 and 5.5 mHz multiplied by a p_1 ridge filter, normalized to the peak power. (c) Mean travel-time shifts determined by using the phase-speed filter. (d) Mean shifts determined by using the ridge filter. (e) Travel-time differences determined by using the phase-speed filter. (f) Travel-time differences determined by using the ridge filter.

the geographic South Pole (Braun, 1995). Fortunately, the sunspots studied by Braun (1995) are similar in size to those included in this work. As noted in an earlier comparison with TD measurements (Braun, 1997), the phase shifts in FH analysis are divided by twice the angular frequency for comparison with travel-time shifts (with a switch in the sign of the Hankel results also being needed owing to different Fourier transform conventions). The results are shown in Figure 12. The agreement between the two sets of measurements is very respectable, despite differences in methods and data sets. In particular, the agreement is significantly better than an earlier comparison between FH phase shifts and TD travel-time shifts of Braun (1997), which may be due to the poorer spatial resolution and lack of mode discrimination (*i.e.*, lack of filtering) in that study. The agreement is especially good for the p_3 and p_4 ridges, and it is fair for the p_2 ridge. There are clear, systematic differences for

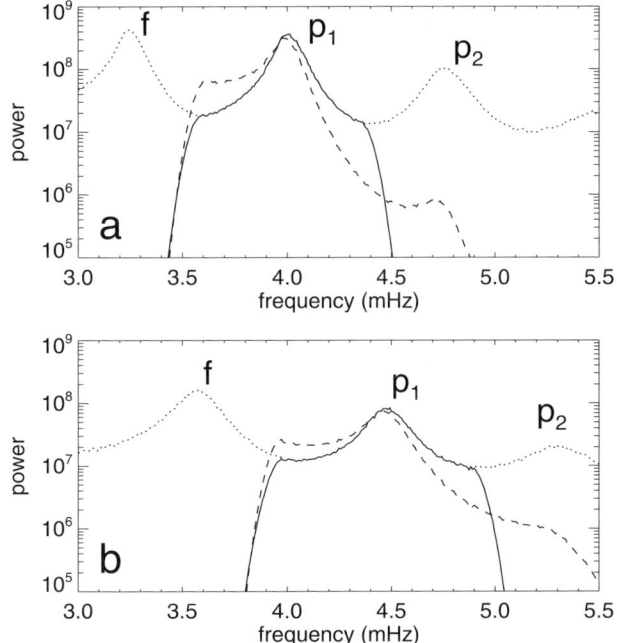

Figure 11 Cuts of the power as a function of frequency at two fixed wavenumbers illustrating the effect of the filters shown in Figure 10. (a) Power at a constant spatial wavenumber corresponding to a spherical harmonic degree of 1055. The dotted line indicates the unfiltered power, with the f, p_1, and p_2 ridges labeled. The solid line shows the power multiplied by the ridge filter for p_1. The dashed line indicates the power multiplied by a commonly used time–distance phase speed filter (TD1; see text). For comparison with the ridge-filtered power spectra, the phase-speed-filtered power spectrum is multiplied by a factor so that the integrated power over the frequency bandwidth is the same as the ridge-filtered spectrum. (b) Power as wavenumber corresponding to a spherical harmonic degree of 1230. For both cases, it is clear that the principle difference between the two filter types is the relative weighting of the two wings of the p_1 ridge, such that the phase-speed filter enhances the contribution of the low-frequency wing relative to the high-frequency wing.

p_1, however, with the FH results, indicating stronger travel-time shifts at higher frequencies than the HH results.

Even under ideal circumstances, the two methods (FH decomposition and surface-focused HH) may be expected to yield systematic differences. For example, surface-focused HH (like TD) is primarily sensitive to the set of wave components that propagate to the surface at chosen locations, unlike FH decomposition, which provides no such discrimination. At low phase speeds, for example, phase shifts and absorption coefficients determined from FH analysis include contributions from wave components passing under the sunspot, and at high phase speeds include contributions from waves that refract to the surface multiple times in the sampling annulus. The expectation is that the results of FH analysis may be less sensitive to perturbations at the target than those obtained by using TD or HH. In light of these considerations, the agreement shown in Figure 12 is remarkable.

Maps of the ratio of egression to ingression power ($|H_+|^2/|H_-|^2$) can be used in surface-focused HH to probe local emission and absorption properties. Representative maps of this quantity, computed from several frequency bandpasses by using the p_1 (top panels) and p_3 (bottom panels) ridge filters, are shown in Figure 13. The quantity $\alpha_{\mathrm{HH}} = 1 - |H_+|^2/|H_-|^2$ (which we denote as the HH absorption parameter) can be directly compared with the

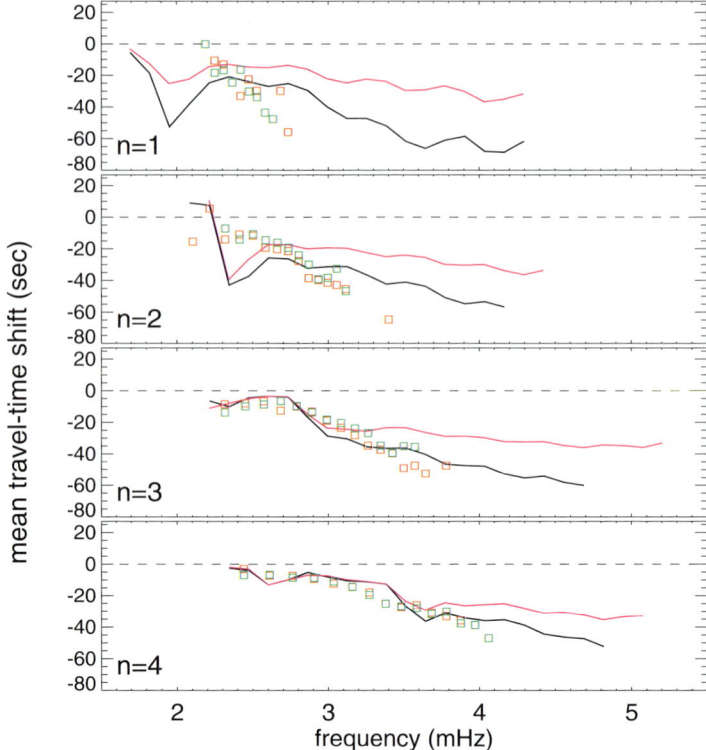

Figure 12 The mean travel-time shift determined by using ridge filters and averaged over umbrae (black lines) and penumbrae (magenta lines) as shown in Figure 8 as a function of frequency along the $p_1 - p_4$ ridges. The squares represent equivalent travel-time shifts determined by previously published FH decomposition methods applied to two sunspots, NOAA 5229 (red) and 5254 (green).

absorption coefficients determined by using FH decomposition measurements, α_{FH} (Figure 14). The FH absorption coefficients are typically smaller than the surface-focused HH absorption parameter in either the umbrae or penumbrae. Peak values of α_{HH} of around 0.7 are observed in sunspot umbrae as compared to typical peak values of α_{FH} of around 0.5. Also striking is the difference in behavior of α_{FH} and α_{HH} at high frequencies, with the absorption parameter from FH methods decreasing toward zero at much lower frequencies than the absorption parameter determined from HH. Some prior HH measurements have shown evidence for this behavior (Lindsey and Braun, 1999), and it has been speculated that the presence of surrounding emission (called "acoustic glories"), which may not be readily resolved by FH decomposition methods, may be at least partly responsible for the lower observed values of α_{FH}. It is highly likely that the high-frequency (e.g., $\nu \geq 5$ mHz) properties of the measured absorption parameters are determined not only by actual absorption mechanisms in the sunspots but by local emission properties as well. For example, a decrease in the local emission can in principle decrease the egression power, and hence the HH absorption parameter. What is known from measurements of p-mode lifetimes suggests that such a mechanism, as an explanation for all apparent absorption, is not viable for waves at lower frequencies, where the contribution of acoustic flux actually originating at a given target is expected to be only a very small fraction of the total observed egression (or outgoing) power

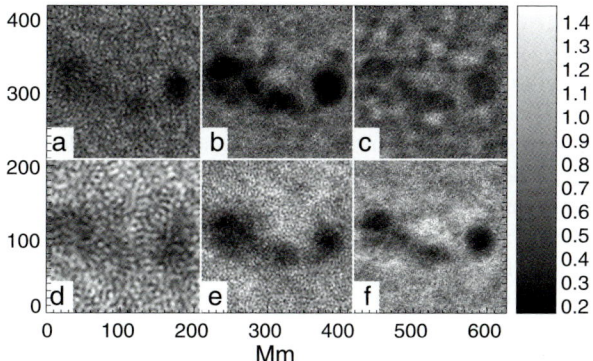

Figure 13 Maps of the ratio of egression to ingression power determined by using ridge filters and 0.26-mHz-wide frequency bandpasses centered at (a) 2.47 mHz, (b) 3.52 mHz, and (c) 4.55 mHz along the p_1 ridge and (d) 2.99 mHz, (e) 4.04 mHz, and (f) 5.08 mHz along the p_3 ridge. At all frequencies, the areas containing sunspots (and other magnetic flux) are dark, which corresponds to $\alpha_{HH} > 1$. At high frequencies (panels c and f), however, note the additional presence of brighter regions.

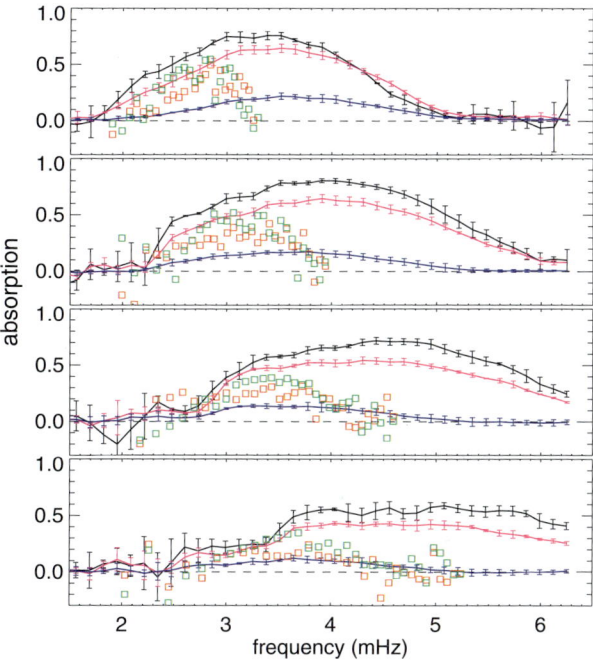

Figure 14 The surface-focus HH absorption parameter α_{HH} determined by using ridge filters and averaged over umbrae (black lines) and penumbrae (magenta lines) as shown in Figure 8 and plage (blue lines) as a function of frequency along the $p_1 - p_4$ ridges. The *squares* represent values of the absorption coefficient determined by previously published Fourier–Hankel decomposition methods α_{FH} applied to two sunspots, NOAA 5229 (red) and 5254 (green).

(Braun, Duvall, and LaBonte, 1987). Finally, we urge caution regarding any interpretation of the observed fall-off of α_{HH} shown in Figure 14. No attempt has been made to assess (and remove) any "background" contributions to the measured egression and ingression (*e.g.*, contributions from locally generated oscillatory motion occurring within the pupils over which the ingression and egression amplitudes are evaluated). Thus, the observed decrease toward zero of the HH absorption measurements may be unphysical.

4. Discussion

It is our preference to let most of the results, as presented through Figures 1 – 14, speak for themselves. A summary of these observations would likely be either too lengthy or otherwise incomplete in that potentially important relationships not directly addressed the text may be neglected. It ought to be fairly clear from observations such as these, however, that the interaction between solar magnetic regions and acoustic waves is highly complex and that no existing model is sufficient in explaining or predicting the complete range of observed behavior. In addition, it should be kept in mind that our observations represent measurements of only three active regions, with only one observable, using one spectral line, and along (essentially) one line-of-sight. It is known, from other observations, that dependencies of phase or travel-times shifts on these and other variables are now recognized, if not fully understood. Deep-focus methods (as applied to TD or HH analyses) are expected to provide further important constraints on models. We also note that we have only briefly touched on the observations relevant to p-mode absorption in magnetic fields, with the expectation that we will return to this in further publications.

An open question is the degree to which "surface effects" – unknown or unaccounted for physical influences of magnetic fields on acoustic waves – are important in the modeling of subsurface structure of sunspots and active regions. As the set of observations shown here confirms and expands upon those presented by Braun and Birch (2006), it is worth restating the general conclusions derived there. Namely, the strong frequency variation of the measured travel-time shifts cannot be explained by using standard assumptions (*i.e.*, standard ray-approximation-based modeling applied to sound-speed models that are typical of published 3D inversion results). To these unexplained frequency variations must now be added apparently "anomalous" positive travel-time shifts (both the mean and travel-time asymmetry) in sunspots, so called here because the conditions (*i.e.*, p-mode properties and choice of filter) under which they appear defy the expectations of standard assumptions and models. We will return to both of these issues, in the context of models of sound-speed perturbations, in Paper 2.

It has been argued (*e.g.*, Zhao and Kosovichev, 2006) that the sign change of travel-time shifts with varying phase speed provides evidence for the relative lack of importance of surface effects for standard inferences from 3D inversions. Certainly, a change of sign is not a typical property of known "surface terms" in models of frequency shifts in structural inversions in global helioseismology and ring-diagram analyses. However, rather than identify the sensitivity of the sign of τ_{mean} or τ_{diff} to the choice of filter with a magnetic surface effect, these observations seriously raise the possibility that the positive values of these shifts represent an artifact, by which we mean a property that is more sensitive to the methods of the analysis than to actual physical conditions within or below sunspots.

Acknowledgements We thank an anonymous referee for useful suggestions. This work is supported by funding through NASA Contract Nos. NNH04CC05C, NNH05CC76C, and NNG07E151C, NSF Grant No. AST-0406225, and a subcontract through the HMI project at Stanford University awarded to NWRA.

References

Barnes, G., Cally, P.S.: 2001, Frequency dependent ray paths in local helioseismology. *Publ. Astron. Soc. Aust.* **18**, 243 – 251.
Basu, S., Antia, H.M., Bogart, R.S.: 2004, Ring-diagram analysis of the structure of solar active regions. *Astrophys. J.* **610**, 1157 – 1168.

Benson, D., Stein, R., Nordlund, Å.: 2006, Supergranulation scale convection simulations. In: Uitenbroek, H., Leibacher, J., Stein, R.F. (eds.) *Solar MHD Theory and Observations: A High Spatial Resolution Perspective*, Astron. Soc. Pac. Conf. Ser. **354**, Astron. Soc. Pac., San Francisco, 92–96.
Birch, A.C., Braun, D.C., Hanasoge, S.M.: 2008, Surface-focused seismic holography of sunspots: II. Expectations from numerical simulations using sound-speed perturbations. *Solar Phys.*, submitted (Paper 2).
Bogdan, T.J.: 2000, Sunspot oscillations: A review. *Solar Phys.* **192**, 373–394.
Bogdan, T.J., Braun, D.C.: 1995, Active region seismology. In: Hoeksema, J.T., Domingo, V., Fleck, B., Battrick, B. (eds.) *Helioseismology*, SP-**376**, ESA, Noordwijk, 31–45.
Bogdan, T.J., Brown, T.M., Lites, B.W., Thomas, J.H.: 1993, The absorption of p modes by sunspots – variations with degree and order. *Astrophys. J.* **406**, 723–734.
Braun, D.C.: 1995, Scattering of p modes by sunspots. I. Observations. *Astrophys. J.* **451**, 859–876.
Braun, D.C.: 1997, Time–distance sunspot seismology with GONG data. *Astrophys. J.* **487**, 447–456.
Braun, D.C., Birch, A.C.: 2006, Observed frequency variations of solar p-mode travel times as evidence for surface effects in sunspot seismology. *Astrophys. J.* **647**, L187–L190.
Braun, D.C., Lindsey, C.: 2000, Phase-sensitive holography of solar activity. *Solar Phys.* **192**, 307–319.
Braun, D.C., Birch, A.C., Lindsey, C.: 2004, Local helioseismology of near-surface flows. In: Danesy, D. (ed.) *SOHO 14 Helio- and Asteroseismology: Towards a Golden Future*, SP-**559**, ESA, Noordwijk, 337–340.
Braun, D.C., Duvall, T.L. Jr., LaBonte, B.J.: 1987, Acoustic absorption by sunspots. *Astrophys. J.* **319**, L27–L31.
Braun, D.C., Duvall, T.L. Jr., LaBonte, B.J.: 1988, The absorption of high-degree p-mode oscillations in and around sunspots. *Astrophys. J.* **335**, 1015–1025.
Braun, D.C., Duvall, T.L. Jr., Labonte, B.J., Jefferies, S.M., Harvey, J.W., Pomerantz, M.A.: 1992, Scattering of p modes by a sunspot. *Astrophys. J.* **391**, L113–L116.
Cally, P.S., Crouch, A.D., Braun, D.C.: 2003, Probing sunspot magnetic fields with p-mode absorption and phase shift data. *Mon. Not. Roy. Astron. Soc.* **346**, 381–389.
Cameron, R., Gizon, L., Daiffallah, K.: 2007, SLiM: A code for the simulation of wave propagation through an inhomogeneous, magnetised solar atmosphere. *Astron. Nachr.* **328**, 313–318.
Cameron, R., Gizon, L., Duvall, T. Jr.: 2008, Helioseismology of sunspots: Confronting observations with three-dimensional MHD simulations of wave propagation. *Solar Phys.*, submitted.
Chou, D.Y.: 2000, Acoustic imaging of solar active regions. *Solar Phys.* **192**, 241–259.
Christensen-Dalsgaard, J., Gough, D.O., Perez Hernandez, F.: 1988, Stellar disharmony. *Mon. Not. Roy. Astron. Soc.* **235**, 875–880.
Couvidat, S., Rajaguru, S.P.: 2007, Contamination by surface effects of time–distance helioseismic inversions for sound speed beneath sunspots. *Astrophys. J.* **661**, 558–567.
Couvidat, S., Birch, A.C., Kosovichev, A.G.: 2006, Three-dimensional inversion of sound speed below a sunspot in the Born approximation. *Astrophys. J.* **640**, 516–524.
Couvidat, S., Birch, A.C., Kosovichev, A.G., Zhao, J.: 2004, Three-dimensional inversion of time–distance helioseismology data: Ray-path and Fresnel-zone approximations. *Astrophys. J.* **607**, 554–563.
Crouch, A.D., Cally, P.S., Charbonneau, P., Braun, D.C., Desjardins, M.: 2005, Genetic magnetohelioseismology with Hankel analysis data. *Mon. Not. Roy. Astron. Soc.* **363**, 1188–1204.
Duvall, T.L. Jr., Gizon, L.: 2000, Time–distance helioseismology with f modes as a method for measurement of near-surface flows. *Solar Phys.* **192**, 177–191.
Duvall, T.L. Jr., D'Silva, S., Jefferies, S.M., Harvey, J.W., Schou, J.: 1996, Downflows under sunspots detected by helioseismic tomography. *Nature* **379**, 235–237.
Fan, Y., Braun, D.C., Chou, D.Y.: 1995, Scattering of p modes by sunspots. II. Calculations of phase shifts from a phenomenological model. *Astrophys. J.* **451**, 877–888.
Giles, P.M.: 2000, Time–distance measurements of large-scale flows in the solar convection zone. Ph.D. thesis, Stanford University.
Gizon, L., Birch, A.C.: 2002, Time–distance helioseismology: The forward problem for random distributed sources. *Astrophys. J.* **571**, 966–986.
Gizon, L., Birch, A.C.: 2005, Local helioseismology. *Liv. Rev. Solar Phys.* **2**. http://www.livingreviews.org/lrsp-2005-6 (cited on 14 February 2008).
Gizon, L., Duvall, T.L. Jr., Larsen, R.M.: 2000, Seismic tomography of the near solar surface. *J. Astrophys. Astron.* **21**, 339–342.
Gordovskyy, M., Jain, R.: 2007, Scattering of p modes by a thin magnetic flux tube. *Astrophys. J.* **661**, 586–592.
Hanasoge, S.M., Duvall, T.L. Jr.: 2007, The solar acoustic simulator: Applications and results. *Astron. Nachr.* **328**, 319–322.
Hanasoge, S.M., Duvall, T.L. Jr., Couvidat, S.: 2007, Validation of helioseismology through forward modeling: Realization noise subtraction and kernels. *Astrophys. J.* **664**, 1234–1243.

Hanasoge, S.M., Couvidat, S., Rajaguru, S.P., Birch, A.C.: 2007, Impact of locally suppressed wave sources on helioseismic travel times. *Astrophys. J.*, submitted. http://arxiv.org/abs/0707.1369v3.
Hughes, S.J., Rajaguru, S.P., Thompson, M.J.: 2005, Comparison of GONG and MDI: Sound-speed anomalies beneath two active regions. *Astrophys. J.* **627**, 1040–1048.
Jackiewicz, J., Gizon, L., Birch, A.C.: 2008, High-resolution mapping of flows in the solar interior: Fully consistent OLA inversion of helioseismic travel times. *Solar Phys.*, submitted.
Jackiewicz, J., Gizon, L., Birch, A.C., Duvall, T.L. Jr.: 2007a, Time-distance helioseismology: Sensitivity of f-mode travel times to flows. *Astrophys. J.* **671**, 1051–1064.
Jackiewicz, J., Gizon, L., Birch, A.C., Thompson, M.J.: 2007b, A procedure for the inversion of f-mode travel times for solar flows. *Astron. Nach.* **328**, 234–239.
Jensen, J.M., Duvall, T.L. Jr., Jacobsen, B.H.: 2003, Noise propagation in inversion of helioseismic time–distance data. In: Sawaya-Lacoste, H. (ed.) *GONG+ 2002. Local and Global Helioseismology: The Present and Future*, SP-**517**, ESA, Noordwijk, 315–318.
Jensen, J.M., Duvall, T.L. Jr., Jacobsen, B.H., Christensen-Dalsgaard, J.: 2001, Imaging an emerging active region with helioseismic tomography. *Astrophys. J.* **553**, L193–L196.
Khomenko, E., Collados, M.: 2006, Numerical modeling of magnetohydrodynamic wave propagation and refraction in sunspots. *Astrophys. J.* **653**, 739–755.
Korzennik, S.G.: 2006, The cookie cutter test for time–distance tomography of active regions. In: *Proceedings of SOHO 18/GONG 2006/HELAS I, Beyond the Spherical Sun*, SP-**624**, ESA, Noordwijk, 60.
Kosovichev, A.G.: 1996, Tomographic imaging of the Sun's interior. *Astrophys. J.* **461**, L55–L57.
Kosovichev, A.G., Duvall, T.L. Jr.: 1997, Acoustic tomography of solar convective flows and structures. In: Pijpers, F.P., Christensen-Dalsgaard, J., Rosenthal, C.S. (eds.) *SCORe'96: Solar Convection and Oscillations and their Relationship*, *Astrophys. Spa. Science* **225**, 241–260.
Kosovichev, A.G., Duvall, T.L. Jr., Scherrer, P.H.: 2000, Time–distance inversion methods and results. *Solar Phys.* **192**, 159–176.
Lindsey, C., Braun, D.C.: 1997, Helioseismic holography. *Astrophys. J.* **485**, 895–903.
Lindsey, C., Braun, D.C.: 1999, Chromatic holography of the sunspot acoustic environment. *Astrophys. J.* **510**, 494–504.
Lindsey, C., Braun, D.C.: 2000, Basic principles of solar acoustic holography. *Solar Phys.* **192**, 261–284.
Lindsey, C., Braun, D.C.: 2004a, Principles of seismic holography for diagnostics of the shallow subphotosphere. *Astrophys. J. Suppl.* **155**, 209–225.
Lindsey, C., Braun, D.C.: 2004b, The penumbral acoustic anomaly. In: Danesy, D. (ed.) *SOHO 14 Helio- and Asteroseismology: Towards a Golden Future*, SP-**559**, ESA, Noordwijk, 552.
Lindsey, C., Braun, D.C.: 2005a, The acoustic showerglass. I. Seismic diagnostics of photospheric magnetic fields. *Astrophys. J.* **620**, 1107–1117.
Lindsey, C., Braun, D.C.: 2005b, The acoustic showerglass. II. Imaging active region subphotospheres. *Astrophys. J.* **620**, 1118–1131.
Lindsey, C., Schunker, H., Cally, P.S.: 2007, Magnetoseismic signatures and flow diagnostics beneath magnetic regions. *Astron. Nachr.* **328**, 298–304.
Lites, B.W.: 1992, Sunspot oscillations–observations and implications. In: Thomas, J.H., Weiss, N.O. (eds.) *NATO ASIC Proc. 375: Sunspots. Theory and Observations*, Kluwer, Dordrecht, 261–302.
Mansour, N.N., Kosovichev, A.G., Georgobiani, D., Wray, A., Miesch, M.: 2004, Turbulence convection and oscillations in the Sun. In: Danesy, D. (ed.) *SOHO 14 Helio- and Asteroseismology: Towards a Golden Future*, SP-**559**, ESA, Noordwijk, 164–171.
Parchevsky, K.V., Zhao, J., Kosovichev, A.G.: 2008, Influence of non-uniform distribution of acoustic wavefield strength on time–distance helioseismology measurements. *Astrophys. J.*, submitted.
Rajaguru, S.P., Birch, A.C., Duvall, T.L. Jr., Thompson, M.J., Zhao, J.: 2006, Sensitivity of time–distance helioseismic measurements to spatial variation of oscillation amplitudes. I. Observations and a numerical model. *Astrophys. J.* **646**, 543–552.
Rajaguru, S.P., Sankarasubramanian, K., Wachter, R., Scherrer, P.H.: 2007, Radiative transfer effects on Doppler measurements as sources of surface effects in sunspot seismology. *Astrophys. J.* **654**, L175–L178.
Scherrer, P.H., Bogart, R.S., Bush, R.I., Hoeksema, J.T., Kosovichev, A.G., Schou, J., Rosenberg, W., Springer, L., Tarbell, T.D., Title, A., Wolfson, C.J., Zayer, I., MDI Engineering Team: 1995, The solar oscillations investigation–Michelson Doppler imager. *Solar Phys.* **162**, 129–188.
Schunker, H., Cally, P.S.: 2006, Magnetic field inclination and atmospheric oscillations above solar active regions. *Mon. Not. Roy. Astron. Soc.* **372**, 551–564.
Schunker, H., Braun, D.C., Cally, P.S.: 2007, Surface magnetic field effects in local helioseismology. *Astron. Nachr.* **328**, 292–297.
Schunker, H., Braun, D.C., Cally, P.S., Lindsey, C.: 2005, The local helioseismology of inclined magnetic fields and the showerglass effect. *Astrophys. J.* **621**, L149–L152.

Shelyag, S., Erdélyi, R., Thompson, M.J.: 2006, Forward modeling of acoustic wave propagation in the quiet solar subphotosphere. *Astrophys. J.* **651**, 576–583.
Shelyag, S., Erdélyi, R., Thompson, M.J.: 2007, Forward modelling of sub-photospheric flows for time–distance helioseismology. *Astron. Astrophys.* **469**, 1101–1107.
Simmons, B., Basu, S.: 2003, A method for inverting high-degree modes. In: Sawaya-Lacoste, H. (ed.) *GONG+ 2002. Local and Global Helioseismology: The Present and Future*, SP-**517**, ESA, Noordwijk, 393–396.
Thomas, J.H., Cram, L.E., Nye, A.H.: 1982, Five-minute oscillations as a subsurface probe of sunspot structure. *Nature* **297**, 485–487.
Thompson, M.J., Zharkov, S.: 2008, Recent developments in local helioseismology. *Solar Phys.* doi:10.1007/s11207-008-9143-6.
Tong, C.H., Thompson, M.J., Warner, M.R., Pain, C.C.: 2003, Helioseismic signals and wave field helioseismology. *Astrophys. J.* **593**, 1242–1248.
Werne, J., Birch, A., Julien, K.: 2004, The need for control experiments in local helioseismology. In: Danesy, D. (ed.) *SOHO 14 Helio- and Asteroseismology: Towards a Golden Future*, SP-**559**, ESA, Noordwijk, 172–181.
Woodard, M.F.: 1997, Implications of localized, acoustic absorption for heliotomographic analysis of sunspots. *Astrophys. J.* **485**, 890–894.
Zhao, J., Kosovichev, A.G.: 2003, Helioseismic observation of the structure and dynamics of a rotating sunspot beneath the solar surface. *Astrophys. J.* **591**, 446–453.
Zhao, J., Kosovichev, A.G.: 2006, Surface magnetism effects in time–distance helioseismology. *Astrophys. J.* **643**, 1317–1324.
Zhao, J., Kosovichev, A.G., Duvall, T.L. Jr.: 2001, Investigation of mass flows beneath a sunspot by time–distance helioseismology. *Astrophys. J.* **557**, 384–388.

Helioseismology of Sunspots: Confronting Observations with Three-Dimensional MHD Simulations of Wave Propagation

R. Cameron · L. Gizon · T.L. Duvall Jr.

Originally published in the journal Solar Physics, Volume 251, Nos 1–2, 291–308.
DOI: 10.1007/s11207-008-9148-1 © The Author(s) 2008

Abstract The propagation of solar waves through the sunspot of AR 9787 is observed by using temporal cross-correlations of SOHO/MDI Dopplergrams. We then use three-dimensional MHD numerical simulations to compute the propagation of wave packets through self-similar magnetohydrostatic sunspot models. The simulations are set up in such a way as to allow a comparison with observed cross-covariances (except in the immediate vicinity of the sunspot). We find that the simulation and the f-mode observations are in good agreement when the model sunspot has a peak field strength of 3 kG at the photosphere and less so for lower field strengths. Constraining the sunspot model with helioseismology is only possible because the direct effect of the magnetic field on the waves has been fully taken into account. Our work shows that the full-waveform modeling of sunspots is feasible.

Keywords Sun: helioseismology · Sun: sunspots · Sun: magnetic fields

1. Introduction

The subsurface magnetic structure of sunspots is poorly known. The theoretical picture is that sunspots are either monolithic or not (Parker, 1979), or change from being monolithic to "disconnected" over the course of the first few days of their lives (Schüssler and Rempel, 2005). The observational picture is limited because the subsurface structure is inherently difficult to infer. The most promising possibility by far is local helioseismology. Local helioseismology includes several techniques of data analysis, such as Fourier–Hankel analysis (*e.g.*, Braun, 1995), time–distance analysis (*e.g.*, Duvall *et al.*, 1993), and helioseismic

---

Helioseismology, Asteroseismology, and MHD Connections
Guest Editors: Laurent Gizon and Paul Cally.

R. Cameron (✉) · L. Gizon
Max-Planck-Institut für Sonnensystemforschung, 37191 Katlenburg-Lindau, Germany
e-mail: cameron@mps.mpg.de

T.L. Duvall Jr.
Solar Physics Laboratory, NASA Goddard Space Flight Center, Greenbelt, MD 20771, USA

holography (*e.g.*, Lindsey and Braun, 1997). For example, the time–distance approach (Duvall *et al.*, 1993) has been applied to determine wave-speed variations and flows associated with sunspots (*e.g.*, Kosovichev, Duvall, and Scherrer, 2000; Zhao, Kosovichev, and Duvall, 2001; Couvidat, Birch, and Kosovichev, 2006). These inversions did not take into account the direct effects of the magnetic field. We do not mean to give the impression that such effects have not been investigated, but rather that they are only beginning to be incorporated into helioseismic inversions. Fourier–Hankel decompositions of the acoustic wave field (p modes) near sunspots by Braun, Duvall, and LaBronte (1987) showed that incoming p modes are phase-shifted and "absorbed" by sunspots. This triggered many studies of the effects of the magnetic field on solar waves. For example, Spruit (1991) suggested that sunspot magnetic fields are responsible for the observed acoustic wave "absorption" by partially converting incoming p modes into slow magnetoacoustic waves. This idea was followed up in detail by Spruit and Bogdan (1992), Cally and Bogdan (1993, 1997), Cally, Bogdan, and Zweibel (1994), Hindman, Zweibel, and Cally (1996), Bogdan and Cally (1997), and Rosenthal and Julien (2000). An important result was the realization that nonuniform and nonvertical magnetic fields are required to explain the observations. In particular, Cally (2000) reported on numerical calculations, showing that inclined magnetic fields are able to achieve levels of absorption compatible with the observations. This aspect of the problem was followed up by Cally, Crouch, and Braun (2003) and Crouch *et al.* (2005), who placed constraints on the strength of the sunspot's magnetic field. More recently, attention has been placed on upward propagating magnetoacoustic waves and their possible observational signatures (Schunker and Cally, 2006; Khomenko and Collados, 2006; Cally, 2007; Cally and Goossens, 2008). The use of direct numerical simulations in helioseismology, with or without magnetic fields, has also undergone rapid development. The aims of these simulations include validating helioseismic techniques (*e.g.*, Zhao *et al.*, 2007) and, more importantly, increasing our understanding of what the seismic observations reveal about the solar interior. Here we restrict our focus to simulations of linearized wave propagation through an inhomogeneous solar atmosphere. Examples of these linear wave propagation studies, most of which do not include the direct effects of magnetic fields, include Birch *et al.* (2001), Khomenko and Collados (2006), Hanasoge, Duvall, and Couvidat (2007), Hanasoge *et al.* (2006), and Hanasoge and Duvall (2007). Some results of these studies clearly have a large bearing on the correct interpretation of the helioseismic signatures of sunspots.

In this paper we first design a technique to image the interaction of waves with sunspots, using appropriate averaging of the cross-correlations of the random seismic wave field (SOHO/MDI data). We proceed further by using numerical simulations to do full-waveform forward modeling of the passage of solar waves through a sunspot. This is done with the SliM code (Cameron, Gizon, and Daiffallah, 2007), specifically developed for this purpose. The ultimate goal is to understand the observed cross-correlations in terms of the properties of our parametric sunspot models.

This paper is organized as follows. The SOHO/MDI observations of sunspot AR 9787 are presented in Section 2 and their helioseismic analysis in Section 3. We describe our numerical simulation code in Section 4. We compare the observations (cross-correlations) against the vertical velocity on a horizontal cut taken from the simulation at the height of the quiet-Sun photosphere. This, among other reasons, means the comparison is only meaningful outside the sunspot. Except in the immediate vicinity of the sunspot, the comparison between the observations and the simulations is very encouraging, as shown in Section 5. We place a helioseismic constraint on the sunspot magnetic field in Section 6. Our work strongly suggests that we will be able to use observations and simulations in combination to constrain the subsurface structure of sunspots by taking direct and indirect magnetic effects into account.

2. SOHO/MDI Observations of Active Region 9787

The helioseismic analysis of a sunspot is more or less difficult depending on which sunspot is being studied. We have been searching the SOHO/MDI database for the ideal sunspot, that is, a sunspot that is isolated (does not belong to a complex sunspot group), has a simple geometry (circular shape), and evolves slowly in time as it moves across the solar disk. To find such a "theorist's sunspot," we searched the MDI Dynamics Program, which combines the advantage of good spatial resolution ($0.12°$ at disk center) with a complete view of the solar disk. The full-disk Dopplergrams are available each minute during two to three months each year since 1996. The sunspot we have selected is that of Active Region 9787, continuously observed by MDI during nine days: 20–28 January 2002. The Dopplergrams were remapped by using Postel's azimuthal equidistant projection almost – but not exactly – centered on the sunspot, using a tracking angular velocity of $-0.1102°$ day^{-1} (in the Carrington frame). In addition to the Dopplergrams, we also used the line-of-sight magnetograms (each minute) and all the intensity images (one per six hours). The daily averages of these three quantities are shown in Figure 1. Apart from some plage, there is no other active region in the vicinity of the sunspot during the entire observation sequence. The sunspot of AR 9787 is large and quite stable over the nine days of the observations, even though it starts decaying from 27 January onward.

Using the MDI intensity images, we measured the center of the sunspot's umbra. As shown in Figure 2, the sunspot has a significant amount of proper motion. Thus we chose to split the time series into six-hour subsets and to analyze each subset separately. The images belonging to each particular subset were remapped into a new Postel map, with the center of the projection corresponding exactly to the center of the sunspot's umbra in the middle of the six-hour time interval. These 36 six-hour time series of Dopplergrams, centered on the sunspot's position, are the basic data that we used for the helioseismic analysis, which we discuss in the next section.

The average intensity image of the sunspot, corrected for the proper motion of the sunspot, is shown in Figure 3. From this average image, which is still sharp, we determine the average umbral and penumbral radii to be 9 and 20 Mm, respectively.

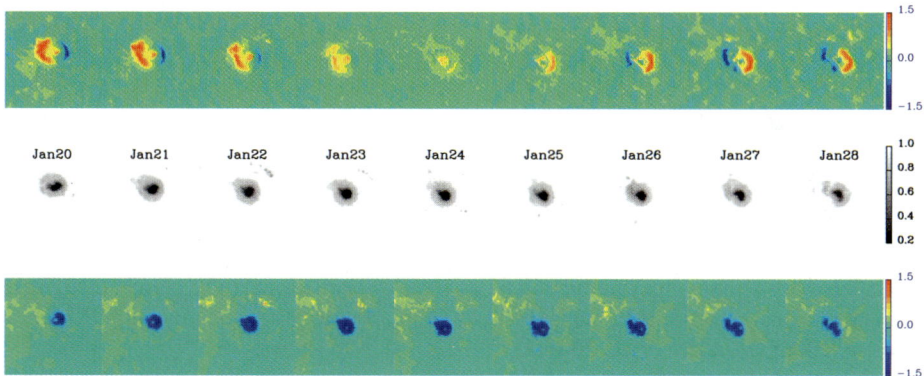

Figure 1 SOHO/MDI observations of the sunspot of Active Region 9787 during the period 20–28 January 2002. Shown are the daily averages of the line-of-sight Doppler velocity (top row), the intensity (middle row), and the line-of-sight magnetic field (bottom row). The color bars are in units of kilometers per second, relative intensity, and kilogauss, from top to bottom.

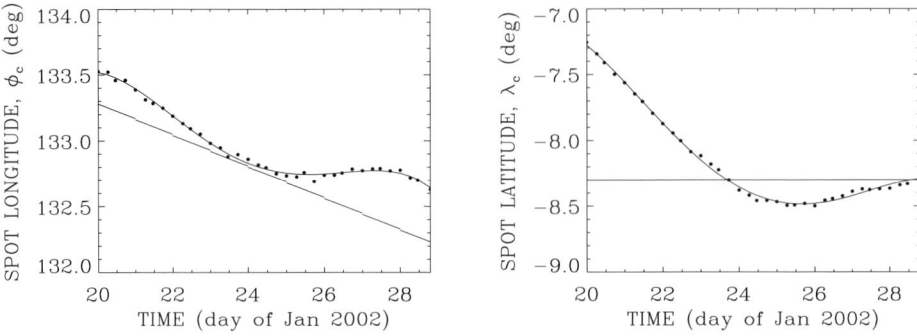

Figure 2 Coordinates of the center of the sunspot's umbra as a function of time. Left: Carrington longitude of the umbra (dots) and Carrington longitude at the center of the Postel projection (line segments). Right: Latitude of the umbra (dots) and latitude at the center of the Postel projection (horizontal line).

Figure 3 SOHO/MDI intensity image of AR 9787 averaged over nine days, after correcting for the proper motion of the sunspot. The intensity is measured relatively to the quiet-Sun value.

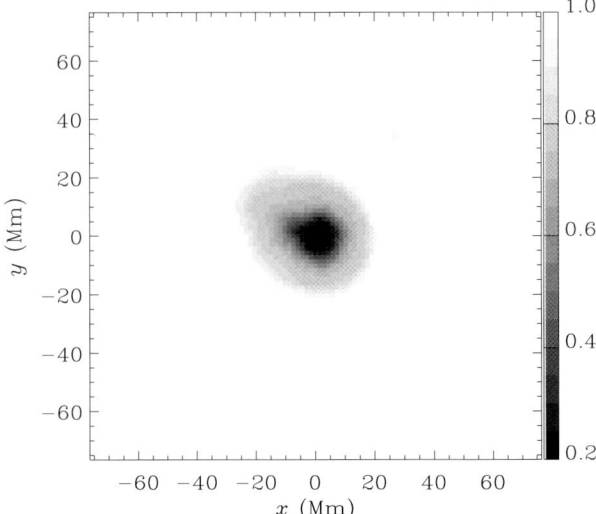

3. Observed f-Mode Cross-Covariance Function

We study sunspot AR 9787 using cross-covariances of the random wavefield, as is done in time–distance helioseismology (Duvall *et al.*, 1993). The temporal cross-covariance between two points on the solar surface provides information about the Green's function between these two points. This interpretation has recently been shown to be correct in the case of homogeneously distributed random sources in an arbitrarily complex medium (Colin de Verdière, 2006; Gouédard *et al.*, 2008).

In this paper, for the sake of simplicity, we wish to observe the propagation of f modes through the sunspot. Thus we filter the Dopplergrams in 3D Fourier space to only keep the f modes. Let us denote by $\phi(\mathbf{r}, t)$ the filtered Doppler velocity, where $\mathbf{r}$ is a position vector and t is time. Rather than studying the cross-covariance of ϕ between two spatial points, we consider the cross-covariance between ϕ averaged over a great circle γ, denoted by $\overline{\phi}(\gamma, t)$, and ϕ measured at any other point **r**:

$$C(\mathbf{r}, t) = \int_0^T \overline{\phi}(\gamma, t')\phi(\mathbf{r}, t+t')\,\mathrm{d}t', \quad (1)$$

where the effective integration time, T, is nine days, or 36 times six hours. Since averaging ϕ over γ is equivalent to selecting only the horizontal wavevectors that are perpendicular to γ, this cross-covariance function gives us information about plane wave packets that propagate away from γ. In simple words, the cross-covariance at time lag t tells us about the position of a wave packet, a time t after it has left γ. In the rest of the paper we fix the distance between γ and the center of the sunspot at $\Delta = 40$ Mm. Since $\Delta \gg \lambda$, where $\lambda \approx 5$ Mm is the dominant wavelength, the sunspot is in the far field of γ.

For any particular choice of orientation of γ, the computed cross-covariance is very noisy, thus explaining the need for some spatial averaging. Let us pick a reference great circle coincident with the meridian at a distance 40 Mm eastward of the center of the sunspot. Because sunspot AR 9787 is almost rotationally invariant around its center, we can compute many equivalent cross-covariance functions corresponding to many different directions of the incoming wave packet, derotate these about the center of the sunspot so that they match the reference cross-covariance, and average them to reduce the noise. We have performed this averaging over all the directions of the incoming wavevectors, with a fine sampling of 1°. As seen in Figure 4, this enables us to reach a very good signal-to-noise level.

4. Three-Dimensional MHD Simulation of Wave Propagation through a Sunspot

4.1. The Equations

We use the ideal MHD equations, linearized about an arbitrary, inhomogeneous, magnetized atmosphere. We assume a local Cartesian geometry defined by horizontal coordinates x and y, and the vertical coordinate z increases upward. The level $z = 0$ is assumed to correspond to the photosphere, as defined in model S (Christensen-Dalsgaard *et al.*, 1996). Under the assumptions that gravitational acceleration (g) is constant and that there is no background steady flow, the equation governing the wave-induced displacement vector $\boldsymbol{\xi}$ is (*e.g.*, Cameron, Gizon, and Daiffallah, 2007)

$$\rho_0 \partial_t^2 \boldsymbol{\xi} = \mathbf{F}', \quad (2)$$

where

$$\mathbf{F}' = -\nabla P' + \rho' g \hat{\mathbf{z}} + \frac{1}{4\pi}(\mathbf{J}' \times \mathbf{B}_0 + \mathbf{J}_0 \times \mathbf{B}') \quad (3)$$

is the linearized force acting on a fluid element (first order in $\boldsymbol{\xi}$), with the density, pressure, magnetic field, and electric current denoted by the symbols ρ, P, $\mathbf{B}$, and $\mathbf{J}$, respectively. In this paper the subscript 0 variables represent the steady inhomogeneous background atmosphere and the primed quantities represent the wave-induced perturbations. The system is closed by the following relations that define $\mathbf{F}'$ in terms of $\boldsymbol{\xi}$:

$$\rho' = -\nabla \cdot (\rho_0 \boldsymbol{\xi}), \quad (4)$$

$$P' = c_0^2(\rho' + \boldsymbol{\xi} \cdot \nabla \rho_0) - \boldsymbol{\xi} \cdot \nabla P_0, \quad (5)$$

$$\mathbf{B}' = \nabla \times (\boldsymbol{\xi} \times \mathbf{B}_0), \quad (6)$$

$$\mathbf{J}' = \nabla \times \mathbf{B}', \quad (7)$$

Figure 4 Upper panel: The observed f-mode MDI cross-covariance function at zero time lag. The sunspot (black circle) is at the center of the Postel map, a distance $\Delta = 40$ Mm from the meridian γ (white line). Lower panel: The same cross-covariance at time lag $t = 130$ minutes. To increase the signal-to-noise ratio, the cross-covariance was averaged over all wave directions. It is easy to see, by eye, that the waves are affected by their passage through the sunspot. The white rectangles show the size of the simulation box.

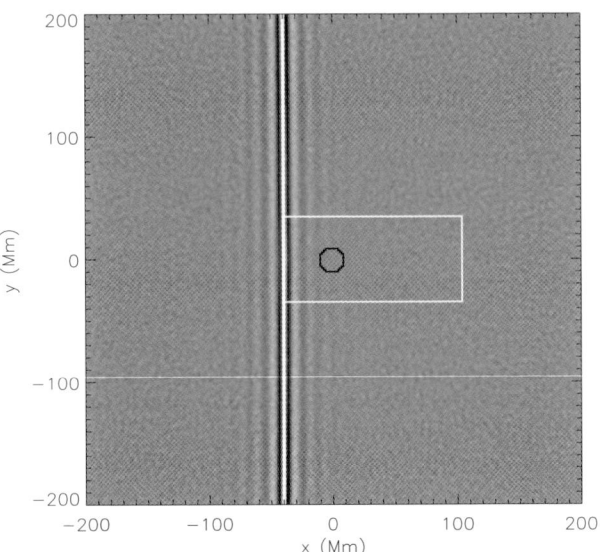

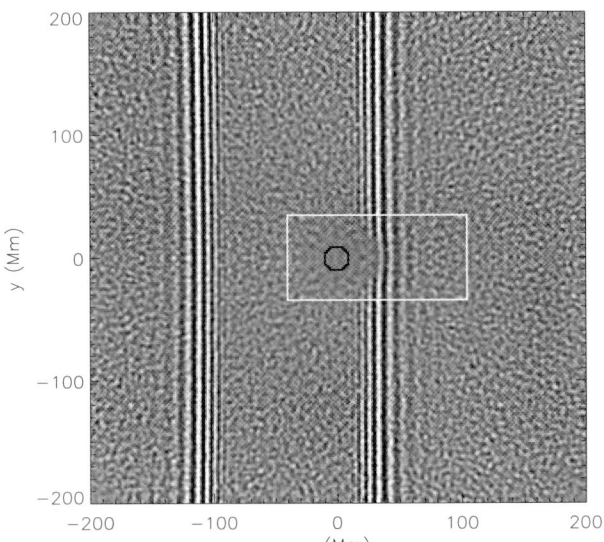

where c_0 is the background sound speed.

Although we have employed ideal MHD, the waves on the Sun are strongly attenuated, presumably as a result of scattering off the time-dependent granulation. Empirically, a solar mode with horizontal wavevector $\mathbf{k}$ decays as $e^{-\gamma_k t}$, where $1/\gamma_k$ is the e-folding lifetime at wavenumber $k = \|\mathbf{k}\|$. Introducing $\mathbf{v}'$ as the wave velocity, for each horizontal Fourier mode we write

$$\rho_0(\partial_t + \gamma_k)\mathbf{v}'(\mathbf{k}, z, t) = \mathbf{F}'(\mathbf{k}, z, t), \qquad (8)$$

$$(\partial_t + \gamma_k)\boldsymbol{\xi}(\mathbf{k}, z, t) = \mathbf{v}'(\mathbf{k}, z, t). \qquad (9)$$

In these equations, any function $f(\mathbf{k}, z, t)$ refers to the spatial Fourier transform of $f(x, y, z, t)$. For $\gamma_k > 0$ the system is damped. The eigenstates, pairs of $(\boldsymbol{\xi}, \mathbf{v}')$, are independent of γ_k, although the eigenvalues are naturally affected. This is the advantage of this approach. The attenuation that we have implemented acts only in the time domain and is suitable for our purpose, although in the Sun attenuation is more complicated. An alternate way of introducing this phenomenological time attenuation would have been to perform the ideal calculation and then impose the decay in the time domain *post facto* after the calculation has ended.

In this paper we focus on the f modes (solar-surface gravity waves), for which the attenuation has been measured to be

$$\gamma_k = \gamma_* (k/k_*)^{2.2}, \tag{10}$$

where $\gamma_*/\pi = 100$ μHz and $k_* = 902/R_\odot$ is a reference wavenumber (Duvall, Kosovichev, and Murawski, 1998; Gizon and Birch, 2002).

4.2. The Code

We use a modified version of the SLiM code, which is, apart from the modifications discussed in the following, described in Cameron, Gizon, and Daiffallah (2007). The code has been tested against analytic solutions; some of these tests are also described in detail in Cameron, Gizon, and Daiffallah (2007).

The present code includes two absorbing layers at the top and the bottom of the box. In the top layer above the temperature minimum we heavily damp the waves and systematically reduce the effect of the Lorentz force. This layer only affects waves that have escaped through the photosphere. Likewise the bottom layer damps the waves that propagate downward. The purpose of both layers is to minimize the effects of the boundaries, which would otherwise artificially reflect the waves.

An additional change has been to introduce the mode attenuation described in Section 4.1. Since we use a semispectral scheme the implementation was straightforward.

The scheme uses finite differences in the vertical direction, with 558 uniformly spaced grid points sampled at $\Delta z = 25$ km. In the horizontal direction we use a Fourier decomposition with 200 modes in the x-direction and only 50 in the y-direction. The spatial sampling is $\Delta x = 0.725$ Mm and $\Delta y = 1.45$ Mm. This seemingly low resolution is satisfactory because neither the initial wave packet nor the sunspot has any significant power at short wavelengths. The size of the simulation box is 145 Mm long (x-coordinate), 72.5 Mm across (y-coordinate), and 14 Mm in depth (12.5 Mm below the photosphere). A typical run, such as the one from Section 5.3, takes approximately 14 days on a single-CPU core.

4.3. Stabilizing the Quiet-Sun Atmosphere

Our aim is to model the propagation of waves through the solar atmosphere in such a way as to allow a direct comparison with the observations. The most direct way of proceeding would be to use an existing solar-like atmosphere such as that of model S (Christensen-Dalsgaard *et al.*, 1996). This most direct approach is however unavailable when considering the full evolution of wave packets using numerical codes because both the solar and model S convection zones are convectively unstable.

Wave propagation through unstable atmospheres is difficult to study numerically because any numerical noise in the unstable modes grows exponentially and eventually dominates the numerical solution. The problem is then to stabilize the atmosphere while leaving it as solar-like as possible, at least in terms of the nature of the waves propagating through it.

The convective instability is easily understood by reference to Equation (5). We imagine a small vertical displacement of a blob of plasma and assume it is evolving slowly enough that it is in pressure balance with its surroundings. This implies $P' = 0$ and hence $c_0^2 \rho' = \boldsymbol{\xi} \cdot \nabla P_0 - c_0^2 \boldsymbol{\xi} \cdot \nabla \rho_0$. For a vertically stratified atmosphere, this becomes $\rho' = \xi_z (\partial_z P_0 - c_0^2 \partial_z \rho_0)/c_0^2$. The atmosphere is convectively unstable if an upward displacement ($\xi_z > 0$) corresponds to a region of lowered density ($\rho' < 0$); in such a case the fluid parcel is buoyant and accelerates upward. The atmosphere is convectively stable when ξ_z and ρ' have the same sign, requiring $(\partial_z P_0 - c_0^2 \partial_z \rho_0)/c_0^2 > 0$, or equivalently

$$\partial_z P_0 > c_0^2 \partial_z \rho_0. \tag{11}$$

We have the freedom to modify any combination of P_0, c_0, or ρ_0 to satisfy Equation (11). It is somewhat natural to regard these three quantities as being related through an equation of state or through the constraint that the atmosphere be hydrostatic; however, neither of these relationships is necessary in terms of the properties of the propagating waves. We choose P_0, c_0, and ρ_0 so that the wave speed is solar-like. Changing the sound speed would obviously have a major impact on the propagation of sound waves, and hence we have chosen to keep c_0 unchanged. Varying ρ_0 would have the effect of varying the distribution of the kinetic energy density of the different modes as a function of height, which would significantly change the sensitivity of wave packets to inhomogeneities; so we have also kept ρ_0 fixed. Thus we choose P_0 so that

$$\partial_z P_0 = \max\{0.9 c_0^2 \partial_z \rho_0, \partial_z P_u\}, \tag{12}$$

where P_u was the pressure distribution of the unstable atmosphere. Notice that $\partial_z P_0$ is negative so that the factor of 0.9 does indeed mean that Equation (11) is satisfied. The factor of 0.9 is possibly unnecessary, but it does little harm: Setting this factor to 1 should create stationary eigenmodes, and setting it to 0.9 introduces internal gravity modes, which propagate very slowly.

4.4. An Example Oscillation Power Spectrum

Since we have modified the atmosphere, we need to check whether it supports oscillations that are those of the Sun and model S in the absence of a sunspot. We chose initial conditions consisting of a "line source" of the form

$$\xi_z = e^{-x^2/2s^2} e^{-(z-z_0)^2/2s^2} \quad \text{at } t = 0, \tag{13}$$

where $s = 0.7$ Mm and $z_0 = -250$ km is an arbitrary height below the photosphere. All other wave perturbations, ξ_x, ξ_y, and $\mathbf{v}'$, are zero at $t = 0$. This disturbance encompasses many different modes allowing several ridges of the dispersion diagram to emerge. Figure 5 directly compares these ridges with the eigenfrequencies of model S (A.C. Birch, private communication). The f modes and the acoustic modes p_1, p_2, and p_3 lie where they are expected for $kR_\odot > 300$. It is also to be noted that there is a very low amplitude ridge (not visible on the plot) at frequencies below 1 mHz that corresponds to internal gravity modes, which are not present in the Sun. They are an artifact of having made the system convectively stable as already explained, but they can be safely ignored.

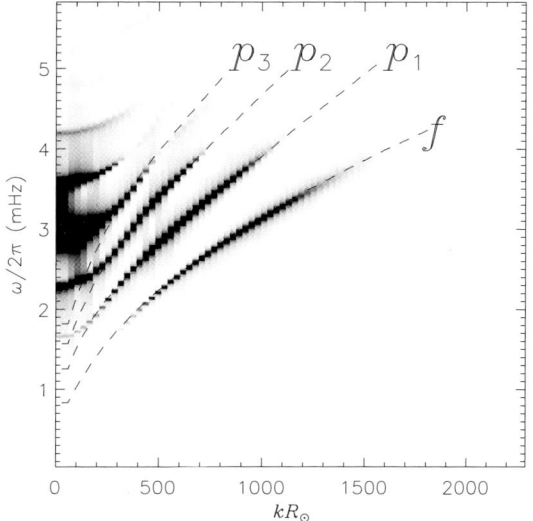

Figure 5 Power as a function of wavenumber and frequency for the example simulation described in Section 4.4 (no sunspot). The dashed lines correspond to the eigenfrequencies of the model S atmosphere.

The simulations and model S eigenfrequencies differ strongly at wavenumbers less than $300/R_\odot$. This difference occurs because our computational domain extends to only 12 Mm below the solar surface. We have not tested a deeper box because the f, p_1, and p_2 modes look satisfactory in the range of wavelengths in which we are interested ($k > 300/R_\odot$).

4.5. A Parametric Sunspot Model

In this section we will describe the sunspots that we have embedded in our atmosphere. We begin by noting that we embed the sunspot in the atmosphere before we stabilize it, to get the correct sound speed and density structures. We use P_u to denote the pressure of the atmosphere *before* it has been stabilized as previously described. The sunspot is modeled by an axisymmetric magnetic field $\mathbf{B}_0 = \mathbf{B}_0(r, z)$, where r is the horizontal radial distance from the sunspot axis. Within the part of the atmosphere described by model S, the magnetic field is made hydrostatic in a standard way, by calculating the Lorentz force $\mathbf{L} = (\mathbf{J}_0 \times \mathbf{B}_0)/4\pi$ and noting that the horizontal force balance then requires a horizontal pressure gradient: $\partial_r P_u = \hat{\mathbf{r}} \cdot \mathbf{L}$. Since P_u is unaffected by the spot at large distances, we can integrate from infinity toward the center of the spot to find P_0. Having thus found P_u, we can find ρ_0 by the constraint of vertical force balance. In this case the gravitational force needs to balance both the vertical component of the Lorentz force as well as that of the pressure gradient. In principle given ρ_0 and P_0 the Saha–Boltzmann equations need to be solved to obtain the first adiabatic exponent (Γ_1) and the sound speed (c_0). However, for the purposes of this paper we have assumed that Γ_1 is a function of z only and is unaffected by the sunspot. This assumption will be relaxed in future studies.

The procedure thus outlined can always be applied; however, in some cases the results will involve negative pressures or densities. Such solutions are of course unphysical and indicate that no hydrostatic solution exists for the given magnetic configuration and quiet-Sun pressure stratification. The problem typically arises in the very upper layers of the box when the magnetic pressure is large. For example, a purely vertical flux tube with a magnetic pressure P_m cannot be in hydrostatic balance in an atmosphere with external pressure $P_{ext} < P_m$. In practice the Sun rapidly evolves to almost force-free field configurations above the solar surface. The role of the Lorentz force in structuring the atmosphere in the low-β region is thus artificial as well as being responsible for the lack of equilibrium. It is also dynamically of minor importance to the waves since it is above both the acoustic cutoff and the layer where the acoustic and Alfvén wave speeds become equal. We have therefore adopted the approach (in the context of realistic photospheric magnetoconvection simulations of active regions; Å. Nordlund, private communication) of scaling the Lorentz force in this region when constructing the background atmosphere. The scaling factor was chosen to be $1/[1 + B_0^2/(8\pi P_{qs})]$, where P_{qs} is the quiet-Sun pressure in the absence of the spot. This scaling factor works, although a more conservative scaling will be tested in the future.

Above the model S atmosphere, the density and pressure fall rapidly and hydrostatic balance requires an extreme scaling of the Lorentz force. However, since we are not aiming to realistically model waves that reach these heights (we only want to damp them) this is acceptable. Instead of requiring force balance in this purely artificial region, we have chosen to put $P_0(x, y, z) = P_{qs}(z)P_0(x, y, z_0)/P_{qs}(z_0)$, where z_0 is the top of the model S atmosphere and $P_{qs}(z)$ is the pressure stratification of the system in the absence of the spot (the quiet-Sun value). The density was treated similarly.

In this paper we follow Schlüter and Temesvary (1958), Deinzer (1965), and Schüssler and Rempel (2005) among others in concentrating on axisymmetric self-similar solutions. For this class of models, the vertical component of the magnetic field is assumed to satisfy

$$B_{0z}(r, z) = \mathcal{B}_0 \, Q\bigl(r\sqrt{H(z)}\bigr), H(z), \tag{14}$$

where the functions Q and H satisfy $Q(0) = H(0) = 1$ but are otherwise arbitrary functions, and $\mathcal{B}_0$ is a scalar measure of the vertical-field strength at the surface. The usual practice (Solanki, 2003), which we adopt in this paper, is to assume Q is a Gaussian,

$$Q(r) = e^{-(\ln 2)r^2/R_0^2}, \tag{15}$$

where R_0 is the half-width at half-maximum. We chose $R_0 = 10$ Mm to correspond to the observed value for sunspot AR 9787, since R_0 is the half-width at half-maximum of the model sunspot at the surface. The value of R_0 is fixed throughout this paper. For the function H we chose the exponential function

$$H(z) = e^{z/\alpha}, \tag{16}$$

with $\alpha = 6.25$ Mm. This choice is rather arbitrary and will certainly be varied in the near future. In this paper we concentrate on varying $\mathcal{B}_0$, the peak magnetic field at the surface. Having thus prescribed the formula for $B_{0z}(r, z)$, we determine $B_{0r}(r, z)$ by the requirement that $\nabla \cdot \mathbf{B}_0 = 0$. Once we have such a hydrostatic solution, we adjust the pressure everywhere to make it convectively stable in the manner described earlier.

5. Comparison between Simulations and Observations

In this section we want to compare the observed f-mode cross-covariance from AR 9787 and the simulations. The computational domain is $-40 < x < 105$ Mm, $-36.25 < y < 36.25$ Mm, and $-12.5 < z < 1.5$ Mm. The sunspot axis is $x = y = 0$. The relationship of the computational box to the observations is shown by the white rectangle in Figure 4.

5.1. The Initial Conditions for the Simulation

In this section we discuss the choice of the initial conditions for the simulation. The aim is to allow a direct comparison between the simulated wave field (Section 4) and the observed cross-covariance function (Section 3). Since the observed cross-covariance uses the z-component of the wave velocity as input data, we choose to use v'_z from the simulations for the comparison (which is also in accordance with a deeper interpretation of the cross-covariance; see Campillo and Paul, 2003). The vertical component of velocity of any f-mode wave packet propagating in the $+\hat{\mathbf{x}}$-direction in a horizontally homogeneous atmosphere is of the form

$$v'_z(x, y, z, t) = \operatorname{Re} \sum_k A_k e^{kz} e^{ik(x-x_0) - i\omega_k t - \gamma_k t}, \qquad (17)$$

where A_k are complex amplitudes, and $\exp(kz)$ and $\omega_k = \sqrt{gk}$ are the f-mode eigenfunction and eigenfrequency, respectively, at wavenumber k. We introduced the reference coordinate $x_0 = -\Delta$, where $\Delta = 40$ Mm is defined in the previous section as the distance between Γ and the sunspot. Given the evolution (Equations (8) and (9)), this wave packet is uniquely determined by the initial conditions

$$\mathbf{v}' = \operatorname{Re} \sum_k \left(i\hat{\mathbf{x}} + \hat{\mathbf{z}}\right) A_k e^{kz + ik(x-x_0)}, \qquad (18)$$

$$\boldsymbol{\xi} = \operatorname{Re} \sum_k \left(-\hat{\mathbf{x}} + i\hat{\mathbf{z}}\right) \omega_k^{-1} A_k e^{kz + ik(x-x_0)}. \qquad (19)$$

Thus our problem is reduced to fixing the amplitudes A_k. In this study, the amplitudes A_k are real (all waves are in phase at $x = x_0$ and $t = 0$) and are shown in Figure 6. This choice was simply the result of requiring that the simulated $v'_z(x, y, z = 0, t)$ and the cross-covariance look approximately the same, in the far field and in the absence of the sunspot. As will be shown in the next section, this spectrum of wavenumbers is sufficiently accurate for the present study. A more systematic analysis is planned. One important difficulty that arises is the existence of background solar noise, which is not easy to model and has been ignored in Equation (17).

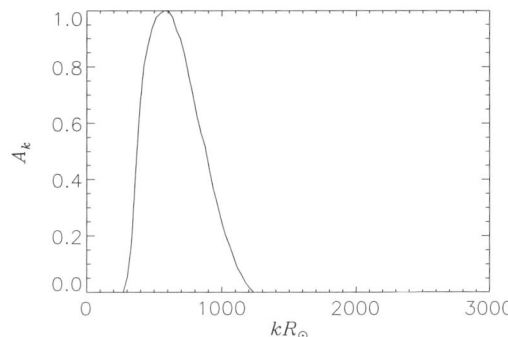

Figure 6 The initial distribution of f-mode amplitudes used in the simulations, A_k, as a function of $kR_\odot$.

5.2. Magnetoacoustic Waves

First we consider a sunspot model with $\mathcal{B}_0 = 3$ kG. The simulation enables us to understand what happens below the solar surface. The theory of the interaction of solar waves with sunspots had been developed by using ray theory and two-dimensional numerical simulations (*e.g.*, Cally, 2007; Bogdan *et al.*, 1996). We find what appears to be very similar physics in our three-dimensional simulations: The strong "absorption" of the f modes is consistent with partial conversion of the incoming waves into slow magnetoacoustic waves, which propagate along the field lines. This can be seen in Figure 7. The downward propagating slow magnetoacoustic modes experience a decrease in their speed as they propagate downward and are therefore shifted to increasingly short wavelengths. Mode conversion is a very robust feature of this type of simulation. We note that the slow magnetoacoustic waves are much easier to see in the x-component of wave velocity than in v'_z, which is why Figure 7 shows the former.

Since we are strongly damping short wavelengths, these magnetoacoustic modes rapidly decay. Any upward propagating wave encounters the damping buffer situated above the photosphere and is also damped. In principle this decay is unphysical in both instances; however, because neither the downward nor upward propagating waves return to the surface this is not undesirable.

5.3. Waveforms

In this section we compare the simulated v'_z at $z = 0$ with the observed cross-covariance. We emphasize that at the moment this comparison is not expected to be appropriate in the immediate vicinity of the sunspot.

Figure 8 shows such a comparison between simulations and observations for times $t = 40$, 100, and 130 minutes. The peak field strength used in this simulation was $\mathcal{B}_0 = 3$ kG and the radius was $R_0 = 10$ Mm. The comparison appears to be very good at time $t = 130$ minutes, at which time the wave packet has completely traversed the sunspot. At this time, all aspects of the waveform – amplitudes (including "absorption"), phases, and spatial spectrum – would appear to have been approximately reproduced by the simulation. The match appears to be poorer for $t = 40$ minutes; this will be discussed in the following.

To be more precise, however, the match between the simulations and observations must be better quantified. To reduce the observational noise, we have averaged the cross-covariance function in the y-direction over two bands. Both bands, shown in Figure 9, are 14.5 Mm wide in the y-direction. The bands are labeled A and B and centered around $y = 0$ and $y = \pm 29$ Mm. Band A is centered on the spot, and band B acts as a reference. The simulations are averaged in the same manner for comparison with the observations.

In Figure 10 we have plotted both the simulation (v'_z, thick red lines) and the observed cross-covariance (C, thin blue lines). In the top panels we compare the wave propagation at the edge of the computational box (*i.e.*, in band B). This part of the wave is little affected by the sunspot and the match between the simulation and observation is quite good since the initial amplitude spectrum A_k was chosen accordingly. In the lower panels of Figure 10 we see the results for the waves passing through the spot (*i.e.*, band A). First we notice that the wave amplitude is remarkably well reproduced in the simulations, meaning that the observed wave "absorption" is consistent with mode conversion. For $t = 130$ minutes, after the waves have crossed the sunspot, there is no appreciable phase shift between the simulation and the observations. The match, however, is somewhat worse at $t = 40$ minutes and there is an obvious phase shift between the two waveforms with the observations lagging behind the

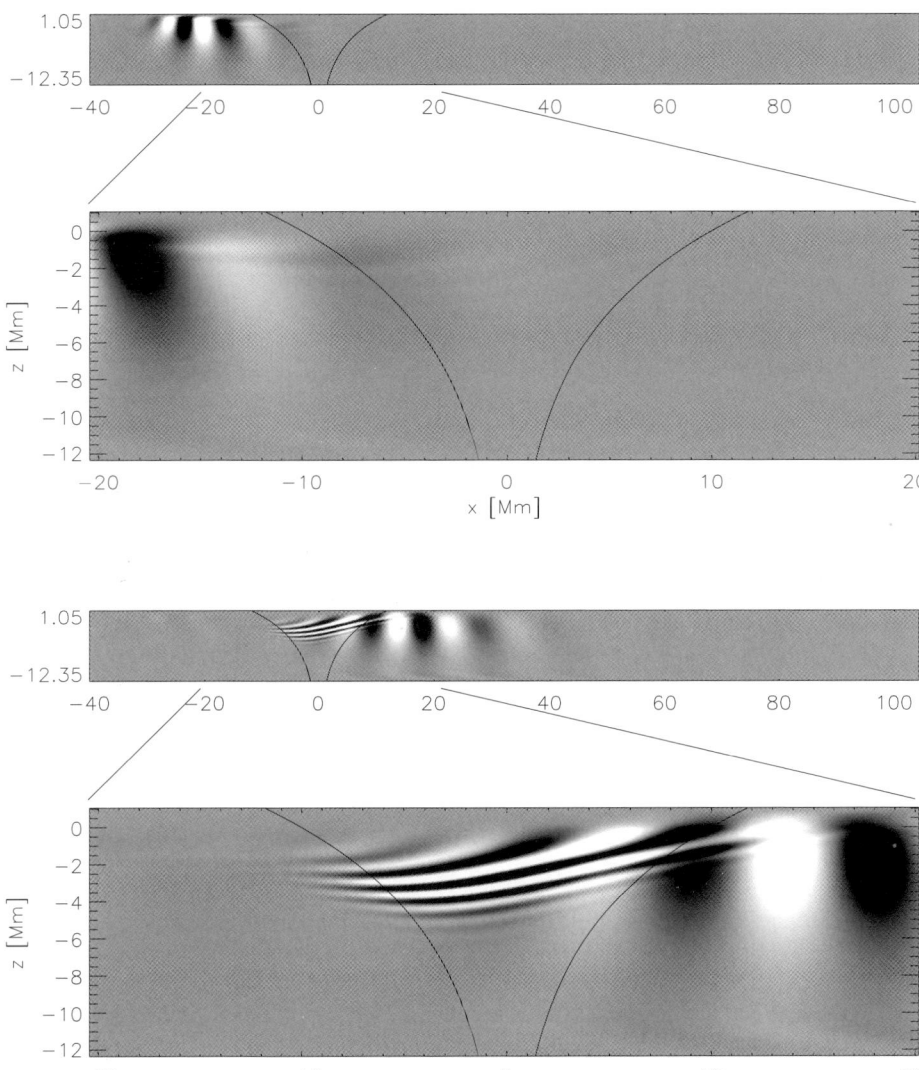

Figure 7 Plot showing $\rho_0 v'_x$ in the $x-z$ plane, through the sunspot axis. The upper frames are for time $t = 34$ minutes (before the f-mode wavepacket crosses the sunspot) and the lower frames are for $t = 84$ minutes (afterward). The black curves with equation $B_z(r, z) = B_z(r = 0, z)/2$ give an estimate of the "width" of the sunspot. The conversion of the incoming f modes into slow magnetoacoustic modes is evident in the lower frames.

simulations. Given the value of the phase shift, we suspect that it is due to the effect of the moat flow. The moat flow is a horizontal outflow from the sunspot, which we have measured by tracking the small moving magnetic features. The observed moat velocity (averaged over nine days) has a peak value of 230 m s^{-1} at a distance of 25 Mm from the center of the sunspot and vanishes at a distance of 45 Mm. The solar waves moving through the flow are first slowed down (against the flow) and later sped up again (with the flow). We have not

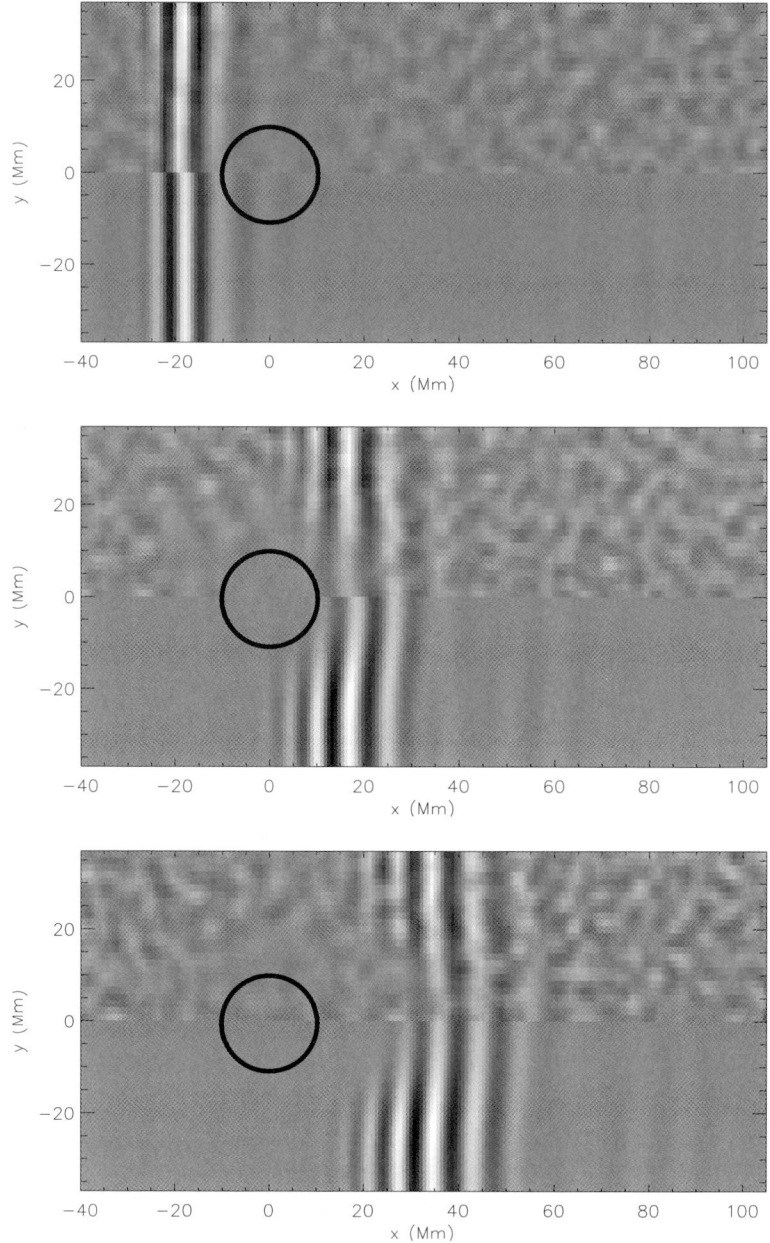

Figure 8 Comparison of the simulated vertical velocity and the observed cross-covariance. In each panel the upper frame shows the observed cross-covariance and the bottom panel the simulated wavepacket. The panels correspond to $t = 40$ minutes, $t = 100$ minutes, and $t = 130$ minutes, from top to bottom. The black circles of radius $R_0 = 10$ Mm indicate the location of the sunspot.

modeled this effect yet. It is reassuring, however, to see that the Doppler shift caused by the moat appears to have disappeared at $t = 130$ minutes.

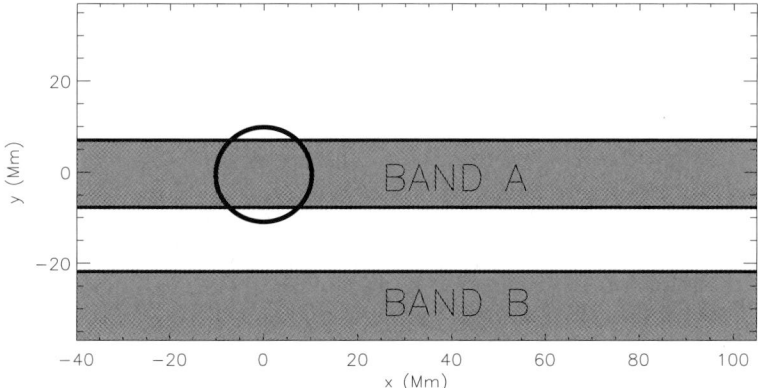

Figure 9 Sketch of the bands over which the data have been averaged in the y-direction in later plots. Band A is centered on the spot, and band B acts as a reference.

Reiterating, we find that, by using numerical simulations, it appears that seismic signatures can be followed through their passage across the spot.

6. Constraining B

The solutions shown thus far have been for a model sunspot with a peak vertical-field strength $B_0 = 3$ kG at the surface $z = 0$. The match is good, which raises the question of whether other field strengths would match as well. The answer to this question can be seen in Figure 11, where we show the comparison between the simulations and observations for different field strengths, $B_0 = 2$ kG and $B_0 = 2.5$ kG. The comparison is made some time after the wave packet has passed through the sunspot, so the issue of the moat flow does not arise. At this stage we restrict ourselves to commenting that qualitatively the match is best, in phase and amplitude, for $B_0 = 3$ kG; for the other values of B_0 there is an apparent phase mismatch along $y = 0$ and the amplitude of that part of the wave passing through the spot is not sufficiently damped. The quantification of the "goodness of fit" between the simulation and observations, which will allow a more accurate determination of the field strength, will be the subject of a future study.

7. Discussion

We have performed three-dimensional MHD simulations of waves propagating through sunspot models. The computations are set up in such a way as to allow comparing observed cross-covariances (except in the immediate vicinity of the sunspot). The parameters of the sunspot model can be chosen in such a way that its helioseismic signature is in good agreement with helioseismic observations of sunspot AR 9787.

A qualitative study using f modes has enabled us to place a constraint, $B_0 \geq 3$ kG, on the sunspot's near-surface field strength. The remaining differences reflect real differences between the model and the observed sunspot, such as the moat flow.

The model atmosphere that we have constructed also supports p modes with a dispersion relation that is very close to that of the Sun. Obviously, the p modes in combination with the

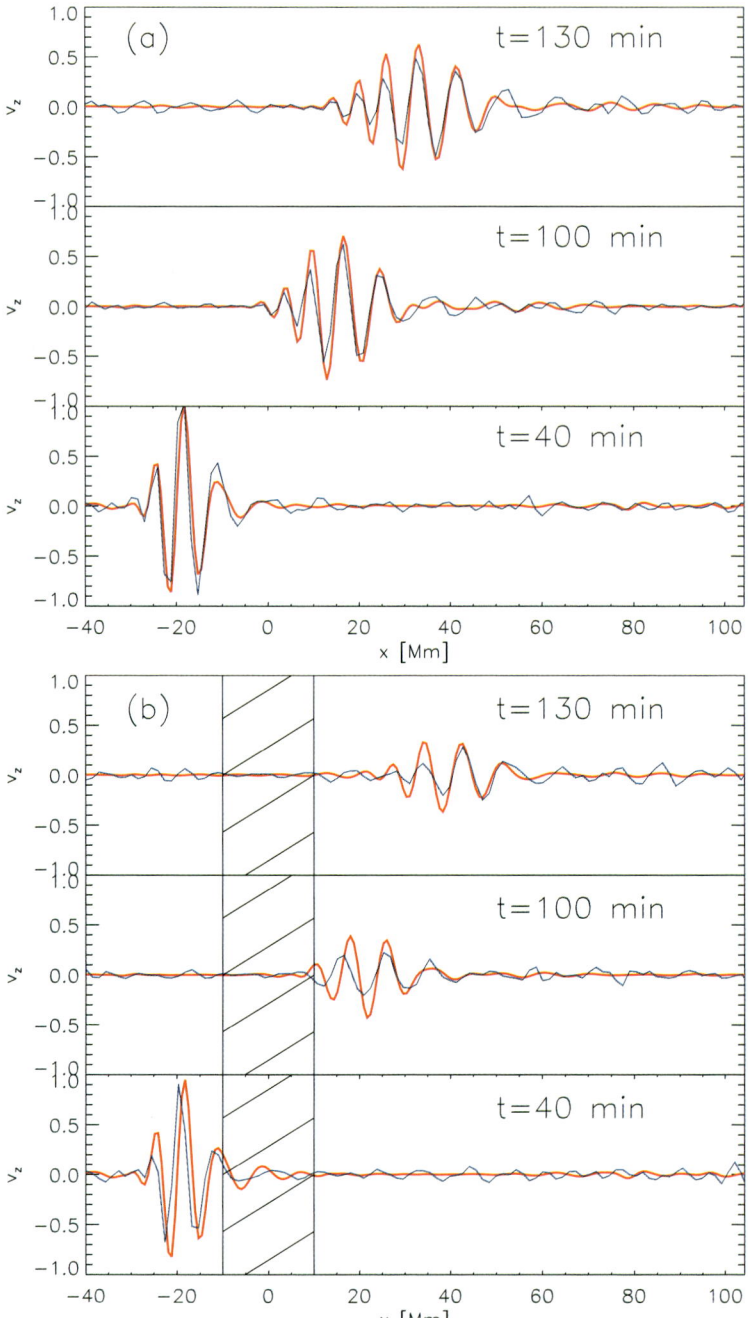

Figure 10 Top panels: The simulation (v'_z, thick red lines) and the observed cross-covariance (C, thin blue lines) averaged in the y-direction across band B (reference), at three different times (t). Bottom panels: The simulation (v'_z, $\mathcal{B}_0 = 3$ kG, thick red lines) and observed cross-covariance (C, thin blue lines) averaged in the y-direction across band A, at three different times (t). The effect of the sunspot is very easily seen, by comparing with band B. The stripes indicate the location of the sunspot umbra.

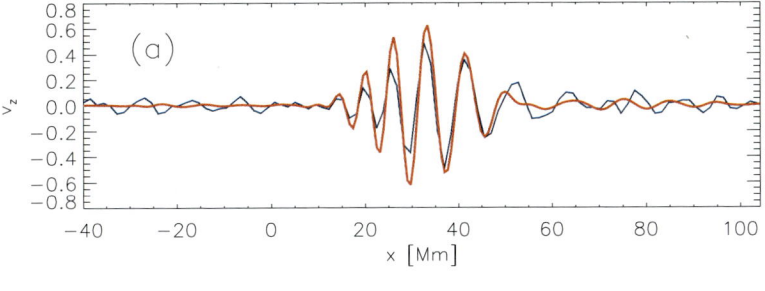
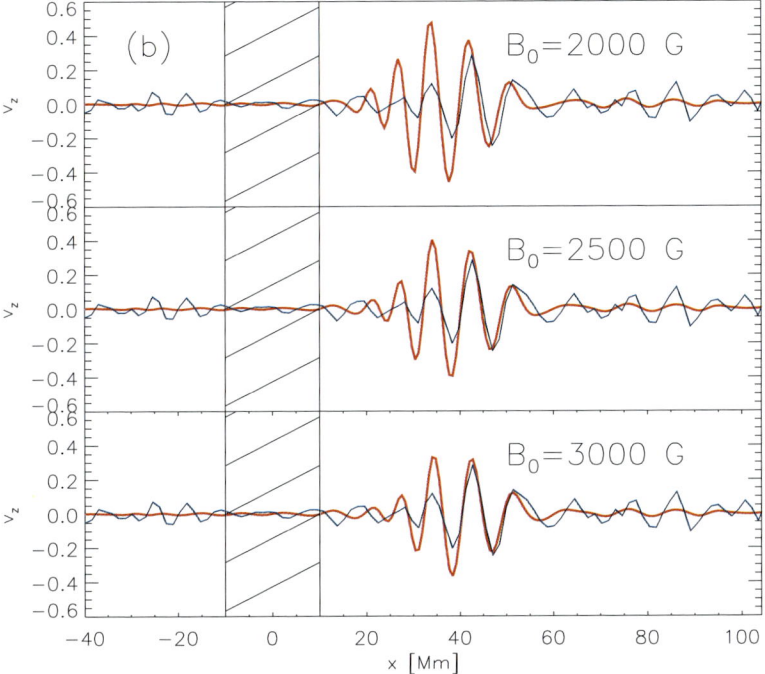

Figure 11 Simulated vertical velocity (thick red lines) and observed cross-covariance (thin blue lines) at $t = 130$ minutes (a) at the edge of the domain (large y) and (b) passing through the sunspot ($y = 0$). The three panels correspond to simulations using $\mathcal{B}_0 = 2$, 2.5, and 3 kG. It is seen that the simulation with $\mathcal{B}_0 = 3$ kG provides the best match in terms of both amplitude and phase. The stripes indicate the location of the sunspot umbra.

f modes should enable us to place more substantial constraints on the subsurface structure of the sunspot. In particular, it should be possible to simultaneously constrain the magnetic field strength $\mathcal{B}_0$, the sunspot radius R_0, and the magnetic field inclination (controlled by the parameter α). Of course, the parametric sunspot model that we have considered in this paper is just one particular model, we plan to consider other types of sunspot models in the future.

In summary, we believe that we have shown that the full-waveform modeling of sunspots is feasible.

Acknowledgements SOHO is a project of international collaboration between ESA and NASA. We are grateful to Manfred Schüssler for insightful discussions. This work was supported in part by the European Helio- and Asteroseismology Network (HELAS) funded by the European Union.

References

Birch, A., Kosovichev, A.G., Price, G.H., Schlottmann, R.B.: 2001, *Astrophys. J.* **561**, L229.
Bogdan, T.J., Cally, P.S.: 1997, *Proc. Roy. Soc. London, Ser. A* **453**, 919.
Bogdan, T.J., Hindman, B.W., Cally, P.S., Charbonneau, P.: 1996, *Astrophys. J.* **465**, 406.
Braun, D.C.: 1995, *Astrophys. J.* **451**, 859.
Braun, D.C., Duvall, T.L. Jr., LaBonte, B.J.: 1987, *Astrophys. J.* **319**, L27.
Cally, P.: 2000, *Solar Phys.* **192**, 395.
Cally, P.: 2007, *Astron. Nachr.* **328**, 286.
Cally, P., Bogdan, T.: 1993, *Astrophys. J.* **402**, 721.
Cally, P., Bogdan, T.: 1997, *Astrophys. J.* **486**, L67.
Cally, P., Goossens, M.: 2008, *Solar Phys.* DOI: 10.1007/s11207-007-9086-3.
Cally, P., Bogdan, T., Zweibel, E.G.: 1994, *Astrophys. J.* **437**, 505.
Cally, P., Crouch, A., Braun, D.C.: 2003, *Mon. Not. Roy. Astron. Soc.* **346**, 381.
Cameron, R., Gizon, L., Daiffallah, K.: 2007, *Astron. Nachr.* **328**, 313.
Campillo, M., Paul, A.: 2003, *Science* **299**, 547.
Colin de Verdière, Y.: 2006, http://fr.arxiv.org/abs/math-ph/0610043/.
Christensen-Dalsgaard, J., Däppen, W., Ajukov, S.V., Anderson, E.R., Antia, H.M., Basu, S., Baturin, V.A., Berthomieu, G., Chaboyer, B., Chitre, S.M., Cox, A.N., Demarque, P., Donatowicz, J., Dziembowski, W.A., Gabriel, M., Gough, D.O., Guenther, D.B., Guzik, J.A., Harvey, J.W., Hill, F., Houdek, G., Iglesias, C.A., Kosovichev, A.G., Leibacher, J.W., Morel, P., Proffitt, C.R., Provost, J., Reiter, J., Rhodes, E.J. Jr., Rogers, F.J., Roxburgh, I.W., Thompson, M.J., Ulrich, R.K.: 1996, *Science* **272**, 1286.
Couvidat, S., Birch, A.C., Kosovichev, A.G.: 2006, *Astrophys. J.* **640**, 516.
Crouch, A.D., Cally, P.S., Charbonneau, P., Braun, D.C., Desjardins, M.: 2005, *Mon. Not. Roy. Astron. Soc.* **363**, 1188.
Deinzer, W.: 1965, *Astrophys. J.* **141**, 548.
Duvall, T.L. Jr., Kosovichev, A.G., Murawski, K.: 1998, *Astrophys. J.* **505**, L55.
Duvall, T.L. Jr., Jefferies, S.M., Harvey, J.W., Pomerantz, M.A.: 1993, *Nature* **362**, 430.
Gizon, L., Birch, A.C.: 2002, *Astrophys. J.* **571**, 966.
Gouédard, P., Stehly, L., Brenguier, F., Campillo, M., Colin de Verdière, Y., Larose, E., Margerin, L., Roux, Ph., Sánchez-Sesma, F.J., Shapiro, N., Weaver, R.: 2008, *Geophys. Prospect.* accepted.
Hanasoge, S.M., Duvall, T.L. Jr.: 2007, *Astron. Nachr.* **328**, 319.
Hanasoge, S.M., Duvall, T.L. Jr., Couvidat, S.: 2007, *Astrophys. J.* **664**, 1243.
Hanasoge, S.M., Larsen, R.M., Duvall, T.L. Jr., DeRosa, M.L., Hurlburt, N.E., Schou, J., Roth, M., Christensen-Dalsgaard, J., Lele, S.K.: 2006, *Astrophys. J.* **648**, 1268.
Hindman, B.W., Zweibel, E.G., Cally, P.: 1996, *Astrophys. J.* **459**, 760.
Khomenko, E., Collados, M.: 2006, *Astrophys. J.* **653**, 739.
Kosovichev, A.G., Duvall, T.L. Jr., Scherrer, P.H.: 2000, *Solar Phys.* **192**, 159.
Lindsey, C., Braun, D.C.: 1997, *Astrophys. J.* **485**, 895.
Parker, E.: 1979, *Astrophys. J.* **230**, 905.
Rosenthal, C.S., Julien, K.A.: 2000, *Astrophys. J.* **532**, 1230.
Schüssler, M., Rempel, M.: 2005, *Astron. Astrophys.* **441**, 337.
Schlüter, A., Temesvary, S.: 1958. In: Lehnert, B. (ed.) *Electromagnetic Phenomena in Cosmical Physics, IAU Symp.* **6**, Cambridge University Press, Cambridge, 263.
Schunker, H., Cally, P.S.: 2006, *Mon. Not. Roy. Astron. Soc.* **372**, 551.
Solanki, S.: 2003, *Astron. Astrophys. Rev.* **11**, 153.
Spruit, H.C.: 1991. In: Toomre, J., Gough, D.O. (eds.) *Challenges to Theories of the Structure of Moderate Mass Stars, Lecture Notes in Physics* **388**, Springer, Berlin, 121.
Spruit, H.C., Bogdan, T.: 1992, *Astrophys. J.* **391**, L109.
Zhao, J., Kosovichev, A.G., Duvall, T.L. Jr.: 2001, *Astrophys. J.* **557**, 384.
Zhao, J., Georgobiani, D., Kosovichev, A.G., Benson, D., Stein, R.F., Nordlund, Å.: 2007, *Astrophys. J.* **659**, 848

Time–Distance Modelling in a Simulated Sunspot Atmosphere

H. Moradi · P.S. Cally

Originally published in the journal Solar Physics, Volume 251, Nos 1–2, 309–327.
DOI: 10.1007/s11207-008-9190-z © Springer Science+Business Media B.V. 2008

Abstract In time–distance helioseismology, wave travel times are measured from the cross-correlation between Doppler velocities recorded at any two locations on the solar surface. However, one of the main uncertainties associated with such measurements is how to interpret observations made in regions of strong magnetic field. Isolating the effects of the magnetic field from thermal or sound-speed perturbations has proved to be quite complex and has yet to yield reliable results when extracting travel times from the cross-correlation function. One possible way to decouple these effects is by using a 3D sunspot model based on observed surface magnetic-field profiles, with a surrounding stratified, quiet-Sun atmosphere to model the magneto-acoustic ray propagation, and analyse the resulting ray travel-time perturbations that will directly account for wave-speed variations produced by the magnetic field. These artificial travel-time perturbation profiles provide us with several related but distinct observations: *i*) that strong surface magnetic fields have a dual effect on helioseismic rays – increasing their skip distance while at the same time speeding them up considerably compared to their quiet-Sun counterparts, *ii*) there is a clear and significant frequency dependence of both skip-distance and travel-time perturbations across the simulated sunspot radius, *iii*) the negative sign and magnitude of these perturbations appears to be directly related to the sunspot magnetic-field strength and inclination, *iv*) by "switching off" the magnetic field inside the sunspot, we are able to completely isolate the thermal component of the travel-time perturbations observed, which is seen to be both opposite in sign and much smaller in magnitude than those measured when the magnetic field is present. These results tend to suggest that purely thermal perturbations are unlikely to be the main effect seen in travel times through sunspots, and that strong, near-surface magnetic fields may be directly and significantly altering the magnitude and lateral extent of sound-speed inversions of sunspots made by time–distance helioseismology.

Helioseismology, Asteroseismology, and MHD Connections
Guest Editors: Laurent Gizon and Paul Cally

H. Moradi (✉) · P.S. Cally
Centre for Stellar and Planetary Astrophysics, School of Mathematical Sciences, Monash University, Clayton, Victoria 3800, Australia
e-mail: hamed.moradi@sci.monash.edu.au

Keywords Helioseismology, direct modelling · Sunspots, magnetic fields · Magnetic fields, models

1. Introduction

Time–distance helioseismology is a powerful diagnostic tool used in local helioseismology to probe the subsurface structure and dynamics of the solar interior, in particular in and around solar active regions. To date however, results obtained by time–distance helioseismology have not directly accounted for the effects of the magnetic field on the wave-speed in travel-time perturbation maps, forward modelling or inversions, but have indirectly included magnetic effects only through their influence on the acoustic properties of the medium (*e.g.*, the sound speed). Standard forward-modelling is based on a number of assumptions including, but not limited to, Fermat's Principle and the ray approximation (*e.g.*, Kosovichev, Duvall, and Scherrer, 2000; Zhao, Kosovichev, and Duvall, 2001; Hughes, Rajaguru, and Thompson, 2005), the Fresnel-Zone approximation (*e.g.*, Jensen *et al.*, 2001; Couvidat *et al.*, 2004) and the Born approximation (*e.g.*, Couvidat, Birch, and Kosovichev, 2006). These models do not include any provision for surface effects. In fact, no standard local-helioseismic method includes provisions for contributions from near-surface magnetic fields.

Recent work in sunspot seismology has pointed to the significant influence of near-surface magnetic fields and possible contamination due to their effects in helioseismic inversions for sound speed beneath sunspots (Couvidat and Rajaguru, 2007). Prior to this, a number of other very important results have highlighted the complications of interpreting helioseismic observations (in particular, the interaction of p modes) in the near-surface regions of sunspots (see, *e.g.*, Fan, Braun, and Chou, 1995; Cally, Crouch, and Braun, 2003; Lindsey and Braun, 2005; Schunker *et al.*, 2005; Schunker and Cally, 2006; Braun and Birch, 2006).

The key issues are *i*) how to successfully model the effects of wave-speed inhomogeneities thought to be produced by the magnetic field in solar active regions, *ii*) how to isolate such effects from those thought to be associated with temperature, flow perturbations, and other observational constraints and effects, and finally *iii*) how will inferences made about subsurface structure change as a result of incorporating these effects into the modelling process? Efforts to address these issues both observationally and computationally have been largely unsuccessful, mainly because of a general lack of understanding of the process involved. But there is some light at the end of the tunnel, as there are currently under development a number of robust magnetohydrodynamical (MHD) simulations modelling helioseismic data and wave propagation that may aid our understanding considerably in the near future (*e.g.*, Cameron, Gizon, and Duvall, 2008; Hanasoge and Duvall, 2007). In this work, we shall attempt to address some of these outstanding issues by using helioseismic ray theory to perform forward modelling of helioseismic rays in a simulated sunspot atmosphere with the aim of modelling the magneto-acoustic ray propagation and analysing the resulting artificial ray travel-time perturbations that will directly account for wave-speed variations produced by the magnetic field. We shall also address the problem of trying to isolate and analyse the thermal contributions to the observed travel-time perturbations using our simulations.

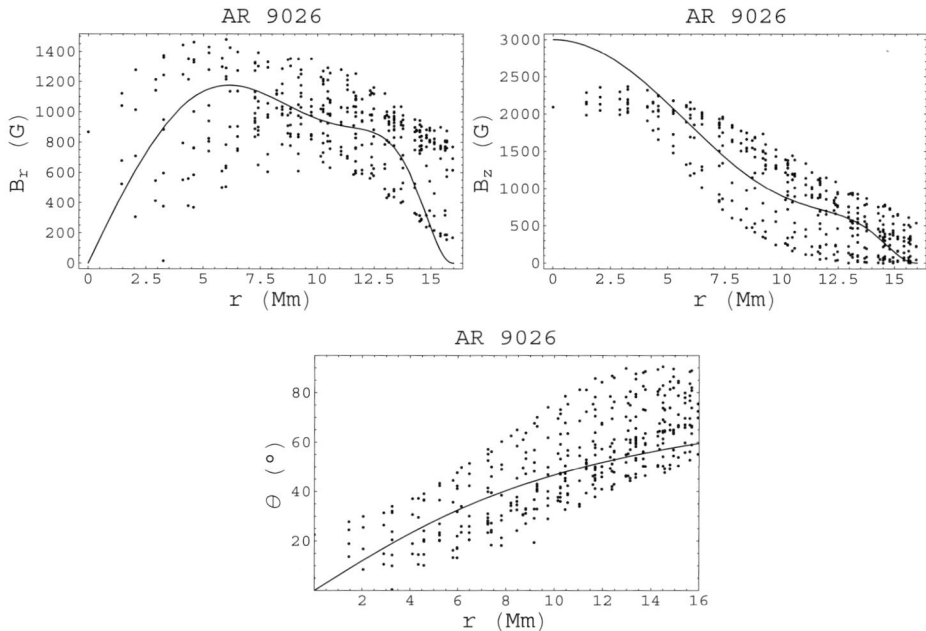

Figure 1 Plots of the radial (B_r, left), vertical components of the observed magnetic field (B_z, right), and magnetic-field inclination from the vertical ($\theta°$) as derived from IVM surface magnetic field profiles of Active Region (AR) 9026 on 5 June 2000, shown as a function of sunspot radius (r, Mm). Solid lines indicate constrained polynomial fits. Values of B are shown in Gauss (G).

2. The Sunspot Model

The axisymmetric sunspot model chosen for our analysis consists of a non-potential, untwisted, magnetohydrostatic sunspot model constrained to fit observed surface magnetic field profiles. The surface field is therefore quite realistic, which is important because there is evidence (Schunker and Cally, 2006) that magnetic effects on helioseismology are dominated by the top Mm.

The sunspot also needs to be surrounded by an unperturbed, stratified atmosphere. The background model employed consists of a Global Oscillation Network Group (GONG) Model S atmosphere (Christensen-Dalsgaard *et al.*, 1996) (obtained from the {L5BI.D.15C.PRES.960126.AARHUS} Model S package). The preferred surface field configuration of the flux tube was derived from constrained polynomial fits to the observed scatter plots of the radial (B_r) and vertical (B_z) surface magnetic field profiles (see Figure 1) of AR 9026 on 5 June 2000 – a fairly symmetrical sunspot near disk-centre, ideal for helioseismic analysis – obtained from Imaging Vector Magnetograph (IVM) vector magnetograms (see Mickey *et al.* (1996) for more details regarding the observations). We note that in the B_z profile of AR 9026, the vertical-field strength tends to *decrease* to around 2 kG as it approaches $r = 0$. We find this highly improbable for a sunspot, so we extrapolate to a peak field of 3 kG for our model at $r = 0$. (A separate analysis was conducted for the model with the (unrealistic) peak field of 2 kG. As expected, the only difference we observed was the magnitude of the perturbations produced being slightly smaller than the ones we report in Section 4. All other results appeared to be identical.) The fits of B_r and B_z are then used to

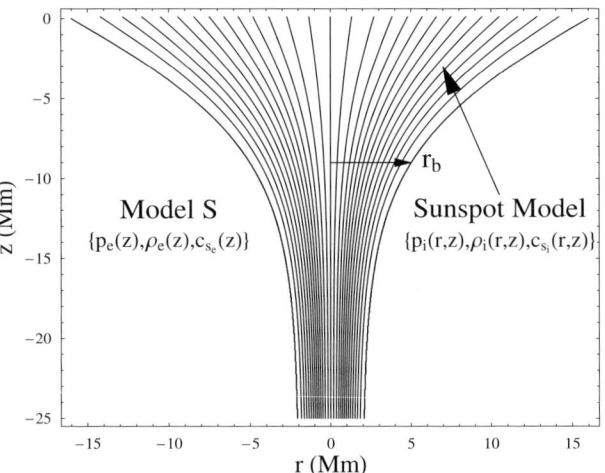

Figure 2 The magnetic-field configuration for the sunspot model. The field lines plotted indicate equidistant magnetic-flux values. Internal and external (Model S) variables are indicated for reference. r_b represents the radius of the outermost field line, which varies with depth (z) along the sunspot radius.

derive an analytical form for the potential function,

$$\Psi(r, z) = \psi_0\left(\frac{R_0 r}{r_b(z)}\right), \quad (1)$$

where ψ_0 is the derived surface field at the surface ($z = Z_0$), the radius of the sunspot at the surface ($r = R_0$) is fixed at $R_0 = 16$ Mm. Instead of a current sheet along the boundary, we prescribe an analytical form for the outermost field line,

$$r_b(z) = \frac{R_0 - R_m}{(1-c)e^{-(z-Z_0)/\lambda} + c} + R_m, \quad (2)$$

where the field strength drops to zero and R_m and c are free parameters. We ensure that all calculations (*e.g.*, change in pressure, density, *etc.*) made across the boundary layer/transition region between the sunspot atmosphere and the external environment are both consistent and continuous along r_b. The magnetic field configuration of the sunspot model is presented in Fig. 2.

The next step essentially involves solving the standard equations of magnetohydrostatics (MHS), using the Model S atmosphere and its variables as the quiet-Sun environment. The magnetic pressure and tension resulting from the Lorentz force,

$$\mathbf{f}_L = \mathbf{J} \times \mathbf{B}, \quad (3)$$

are confined within the simulated sunspot atmosphere, where μ is the magnetic permeability and $\mathbf{J} = \frac{1}{\mu}(\nabla \times \mathbf{B})$. The gas pressure $p(r, z)$ is calculated using horizontal force balance,

$$p_i(r, z) = p_e(z) + \Delta p(r, z), \quad (4)$$

where $p_i(r, z)$ and $p_e(z)$ denote internal and external (*i.e.*, Model S) pressure, respectively, and the change in pressure is therefore

$$\Delta p(r, z) = \int_{r_b}^{r} f_{L_r} \, dr \quad (5)$$

which drops to zero as we approach r_b. Once the pressure inside the sunspot and along the boundary are known, the density $\rho(r, z)$, can similarly be calculated using vertical force balance,

$$\rho_i(r, z) = \rho_e(z) + \Delta\rho(r, z), \tag{6}$$

where the change in density is given by

$$\Delta\rho(r, z) = \frac{1}{g}\left[f_{L_z} - \frac{\partial \Delta p(r, z)}{\partial z}\right]. \tag{7}$$

This is essentially all that is required to then compute the modified sound speed or thermal profile of the sunspot atmosphere,

$$c_{s_i}^2(r, z) = c_{s_e}^2(z) + \Gamma_1(z)\left[\frac{p_i(r, z)}{\rho_i(r, z)} - \frac{p_e(z)}{\rho_e(z)}\right], \tag{8}$$

while for the sake of simplicity, assuming the ratio of specific heat (Γ_1) that appears in the sound speed is the same function of height as it is in the external atmosphere. Finally, all that is left is to calculate the Alfvén speed,

$$a^2(r, z) = \frac{1}{\mu\rho_i(r, z)}\left[B_r^2 + B_z^2\right]. \tag{9}$$

Some of the important internal properties of the resulting sunspot model (*e.g.*, pressure, density, sound and Alfvén speeds) are shown in Figure 3. The external (Model S) profiles for each variable are also shown for reference. The near-surface thermal structure of the sunspot and the ($a = c_s$) equipartition depth is also shown for reference in Figure 4. We can clearly see the modified sound-speed structure (c_s^2) as a result of the magnetic field in this image. It is interesting to note that in our (simple) model the region of decreased sound-speed does not appear to extend as deep as 3D time–distance inversions of the real Sun have suggested. Estimates for the lateral extent of the decreased sound-speed region using tomographic imaging of the sub-surface layers of sunspots have ranged from depths of approximately $z = -2.4$ to $z = -3.5$ Mm using the Born and ray approximations, respectively (Couvidat, Birch, and Kosovichev, 2006). Nevertheless, the sunspot model exhibits the broad features expected of a real sunspot, and presents a useful test case.

3. Ray Path Calculations

The ray paths are calculated in Cartesian geometry, in the realm of frequency dependent ray paths described by Barnes and Cally (2001), with the complete form of the three-dimensional dispersion relation:

$$\begin{aligned}\mathcal{D} &= \omega^2\omega_c^2 a_y^2 k_h^2 + \left(\omega^2 - a^2 k_\parallel^2\right) \times \left[\omega^4 - \left(a^2 + c^2\right)\omega^2 k^2 \right.\\ &\quad \left. + a^2 c^2 k^2 k_\parallel^2 + c^2 N^2 k_h^2 - \left(\omega^2 - a_z^2 k^2\right)\omega_c^2\right] = 0,\end{aligned} \tag{10}$$

where k_h and $k_\parallel$ are the horizontal and parallel components of the wave-vector (**k**) and

$$N^2 = \frac{g}{H_\rho} - \frac{g^2}{c^2} \tag{11}$$

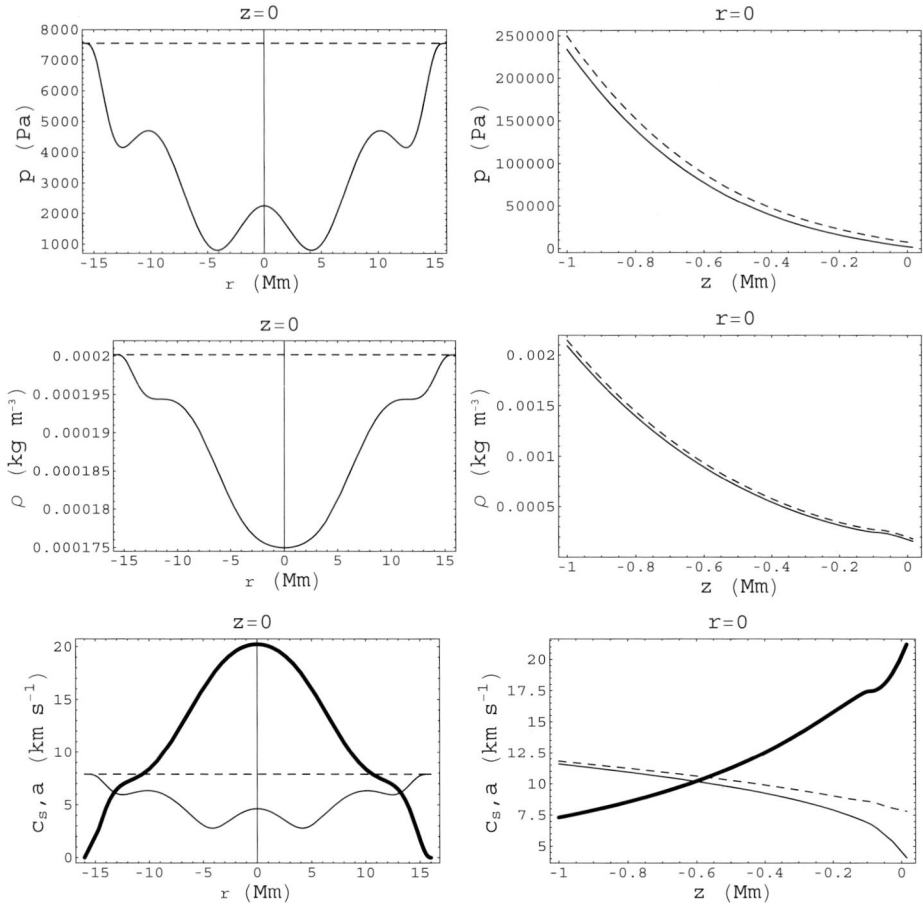

Figure 3 Internal pressure (p), density (ρ), sound (c_s), and Alfvén speed (a) profiles of the sunspot model with an external GONG Model S atmosphere. Left-hand column profiles are calculated along the surface of the sunspot ($z = 0$), while right-hand column profiles are calculated along the axis of the sunspot ($r = 0$). Internal profiles are indicated by solid lines in all plots. The thick solid line in the bottom two panels indicate Alfvén speeds. The dashed lines represent GONG Model S values in all plots.

is the squared Brunt–Väisälä frequency, with g being the gravitational acceleration, $H_\rho(z)$ the density scale height, and $H'_\rho = dH_\rho/dz$ and ω_c^2 is the square of the acoustic-cutoff frequency. For completeness, we calculate the ray paths using two forms of ω_c. The most commonly used form,

$$\omega_c^2 = \frac{c^2}{4H_\rho^2}(1 - 2H'_\rho), \qquad (12)$$

exhibits an extended sharp spike around $z = -100$ km (see Figure 5). This form of ω_c is often used by helioseismologists. However, as Cally (2007) points out, this sharp spike in the cutoff frequency is inconsistent with the WKB assumption of slowly varying coefficients on which $\mathcal{D}$ is based. A much smoother isothermal form,

$$\omega_{c_i} = c/2H, \qquad (13)$$

Figure 4 The thermal profile (c_s^2) in the top 1 Mm of the sunspot. Lighter coloured contours (*i.e.*, cyan/green) indicate regions of decreased sound speed (cooler regions) under the sunspot surface, while darker (hotter) regions (*i.e.*, orange/red) are indicative of areas of enhanced sound speed. The dashed line marks the position of the $a = c_s$ layer. Field lines are over-plotted.

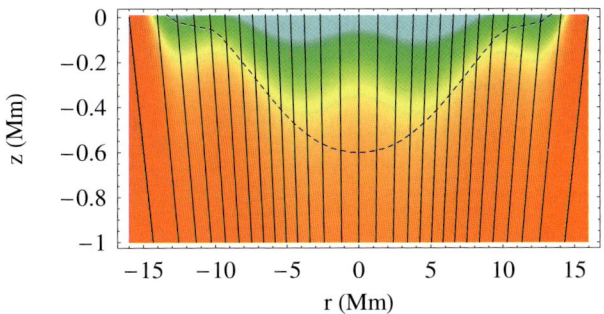

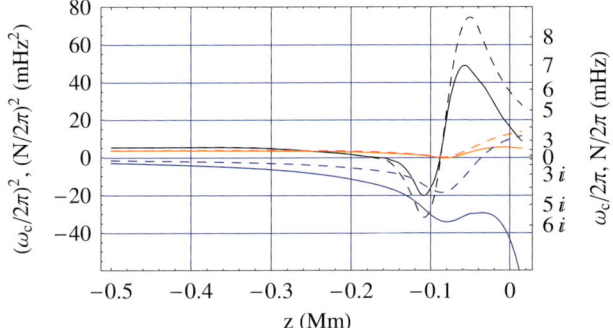

Figure 5 Plots of the various forms of the acoustic cutoff (ω_c) and Brunt–Väisälä (N) frequencies. The later is indicated by a blue, solid line inside the sunspot atmosphere and dashed blue line indicating Model S values. The solid, black line indicates the acoustic cutoff frequency ω_c for the sunspot atmosphere, while the dashed, black line indicates Model S values. The isothermal form, ω_{c_i}, is indicated by the solid, red line for the sunspot atmosphere, dashed, red line indicates Model S values.

is consistent with the derivation of $\mathcal{D}$, and does not suffer from the spike (see Figure 5). Unless otherwise stated, all results shown here utilise ω_{c_i}. (Simulations using the form of ω_c in Equation (12) were also conducted, the results being very similar to those reported in Section 4, expect for a certain amount of unsmoothness being present in the travel-time perturbation profiles (mainly affecting shallow rays which are more sensitive to the reflecting boundary near the surface) as a result of using the more rigid form of ω_c.) Naturally, the magnetic field slightly modifies both ω_c and ω_{c_i}, the results of which can be seen in Figure 5.

Following Weinberg (1962), the construction of **k** is completed by specifying the governing equations of the ray paths

$$\frac{d\mathbf{x}}{d\tau} = \frac{\partial \mathcal{D}}{\partial \mathbf{k}}, \tag{14}$$

$$\frac{d\mathbf{k}}{d\tau} = -\frac{\partial \mathcal{D}}{\partial \mathbf{x}}, \tag{15}$$

$$\frac{dt}{d\tau} = -\frac{\partial \mathcal{D}}{\partial \omega}, \tag{16}$$

$$\frac{d\omega}{d\tau} = -\frac{\partial \mathcal{D}}{\partial \tau}, \tag{17}$$

where τ parameterises the progress of a disturbance along the ray path. For a time-independent medium, for which $\partial \mathcal{D}/\partial t = 0$ and ω is constant, the phase function $S(\mathbf{x})$ evolves according to

$$\frac{dS}{dt} = \mathbf{k} \cdot \frac{d\mathbf{x}}{dt} - \omega. \tag{18}$$

Hence,

$$S(\mathbf{x}) = \int \mathbf{k} \cdot d\mathbf{x} - \omega t, \tag{19}$$

where the first term (integral) represents the contribution to the phase due to motion along the ray path, and the second term represents the Eulerian part. Since we are only going to be concerned about the change in phase due to motion along the ray path, we can essentially ignore the Eulerian part for the rest of our analysis.

We iteratively find the initial wave vector ($\mathbf{k}_{\text{init}}$) by using an initial guess which comes from solving $\mathcal{D} = 0$ for the wave number, assuming that the wave vector is in the directions α, β – where α and β are angles from the vertical and the $x-z$ plane respectively of the initial shot. Initially, we initiated the rays from the top of the ray path, adjusting the initial shooting angle (α) to obtain the desired range of skip distances. However, given the very sensitive nature of the near-surface region of the sunspot atmosphere, we used a much finer computational grid in the top 1.5 Mm. As a result, we encountered many instances of rays initiated inside evanescent regions (which should obviously be avoided) and also obtaining very shallow rays with little or no helioseismic value. So, in order to reduce computation time and also have greater flexibility in choosing the desired range of ray skip distances, we initialised the rays from the minima of their trajectories (essentially the lower turning point of the ray, z_{bot}). Hence, the value of α was fixed at $\alpha = 90°$, allowing us to adjust the initial shooting depth z_{bot} to obtain the desired range of skip distances.

A number of other important points regarding the simulations should also be noted. Firstly, in this paper we only examine the 2D case ($\beta = 0$) where rays are confined to the $x-z$ plane. Furthermore, by ensuring that the rays remain on the fast-wave branch at all times, we avoid any mode-conversion effects as rays pass through the $a = c_s$ layer (where fast/slow conversion occurs, see Figure 4). Of course, as numerous works exploring MHD mode conversion in local helioseismology have shown (*e.g.*, Spruit and Bogdan, 1992; Cally and Bogdan, 1993, 1997; Cally, Bogdan, and Zweibel, 1994; Bogdan and Cally, 1997; Cally, 2000, 2007; Crouch and Cally, 2003, 2005; Schunker and Cally, 2006), mode transmission and conversion between fast and slow magneto-acoustic waves indeed occurs as rays of helioseismic interest pass through the $a = c_s$ equipartition level and have distinct effects on helioseismic waves that should not be ignored. But in our current analysis (and as with actual time–distance inversions) we do not directly account for these effects. As a result the complexities of the ray-path calculations are greatly reduced. We also note that we ignore any finite-wavelength effects and filtering of observations in our simulations.

The computational ray propagation grid extends across the 16 Mm radius of the sunspot model in regular 1 Mm spatial increments in the horizontal x-direction and down to a depth of 25 Mm in the vertical z-direction, employing a much finer grid spacing in the top 1.5 Mm, followed by 1 Mm increments down to a depth of 25 Mm. The cutoff height (depth) for all rays propagated in the grid was fixed at $z = -0.1$ Mm, regardless of frequency. This computational grid, although not exhaustive, allows us to obtain the desired range of skip distances required to replicate the "centre-to-annulus" skip distance geometry (*i.e.*, averaging rays from a central point/pixel to a surrounding annulus of different sizes to probe varying depths beneath the solar surface) often employed in time–distance helioseismology for the

Table 1 The annuli geometry (or skip-distances) used to bin the ray travel-time measurements.

Δ	Pupil size (Mm)
1	3.7– 8.7
2	6.2–11.2
3	8.7–14.5
4	14.5–19.4
5	19.4–29.3
6	26.0–35.1
7	31.8–41.7
8	38.4–47.5
9	44.2–54.1
10	50.8–59.9
11	56.6–66.7

derivation of mean travel-time perturbation maps (see Gizon and Birch (2005) for a more comprehensive description of this process). The 11 standard skip distance bin/annuli (Δ) sizes usually used for these calculations are detailed in Table 1.

4. Results

4.1. Travel-Time and Skip-Distance Perturbations

The ray-propagation grids were computed for three frequencies: $\omega = 3.5$, 4, and 5 mHz. Both the phase (t_p, associated with the *phase* velocity) and group (t_g, associated with the envelope peak of a wave packet as it travels at the *group* velocity) ray travel times were calculated along each ray path for every radial grid position (r_{spot}, which is essentially the radial position of the lower turning point of the ray) along the sunspot model. In time–distance helioseismology, centre-to-annulus travel times are extracted from Gaussian wavelet fits – usually represented by a function of the form

$$W_{\pm}(t) = A e^{-\gamma^2 (t \mp t_g)^2} \cos[\omega_0 (t \mp t_p)] \quad (20)$$

(where all parameters are free) – to both the positive and negative time parts of the observed cross-correlations (Gizon and Birch, 2005). However, t_p is more often used in time–distance literature, primarily as a result of difficulties (mainly observational noise) associated with fitting to the envelope peak. Furthermore, because t_p is much more independent of the shape of the wave packet than t_g (as the shape of the wave packet depends on (unmodelled) mode conversion), we shall also limit our analysis to t_p calculations in this paper. We identify the phase travel time as

$$t_p = \frac{S(\mathbf{x})}{\omega}, \quad (21)$$

which is consistent with the form of t_p described by the Gaussian wavelet. These travel times are then subtracted from similar ray travel times calculated using the quiet-Sun atmosphere to produce *travel-time perturbation* (δt_p) profiles. In general, travel-time *differences* are sensitive to sub-surface flows, while *mean* travel times are sensitive to wave-speed perturbations. However, as our model does not contain flows, we do not need to distinguish directions along ray paths.

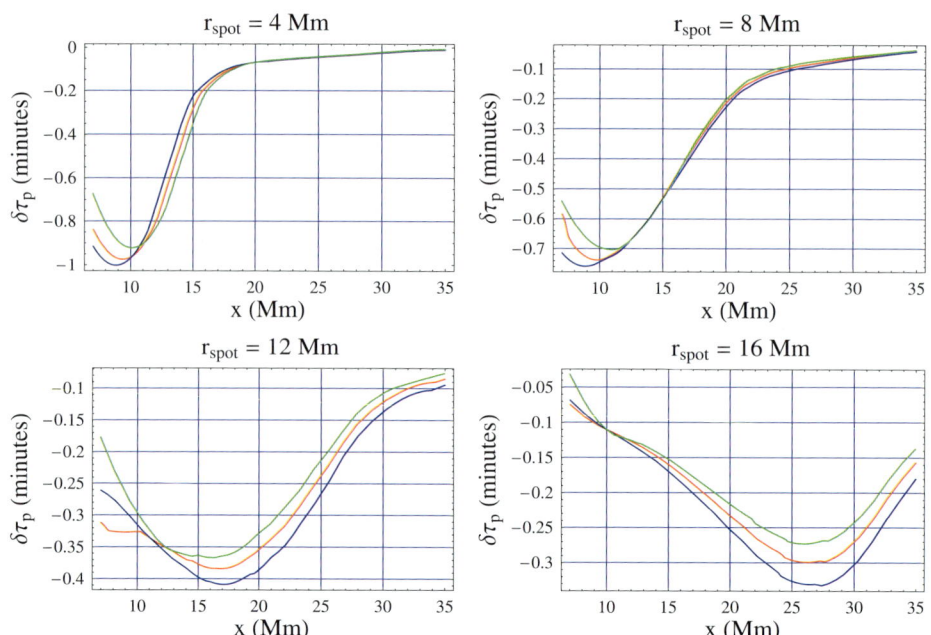

Figure 6 Travel-time perturbations ($\delta\tau_p$) as a function of skip distance (x) for $r_{spot} = 4, 8, 12$, and 16 Mm on the sunspot (where r_{spot} is the radial position of the lower turning point of the ray), as calculated for three frequencies: $\omega = 3.5$ (green), $\omega = 4$ (red) and $\omega = 5$ mHz (blue).

In Figure 6 we see some sample $\delta\tau_p$ profiles for $r_{spot} = 4, 8, 12$, and 16 Mm shown as a function of ray skip distance (x) for $\omega = 3.5$ (green), 4 (red), and 5 mHz (blue). By and large, there are significant perturbations as we approach the centre of the sunspot (*i.e.*, regions associated with stronger surface magnetic-field strength). The sign of the perturbations appears to remain exclusively negative, regardless of position on the sunspot. This means that all rays propagated within the simulated sunspot atmosphere are significantly sped up when compared to their Model S counterparts.

Furthermore, in Figure 7 we can see that there are also significant skip-distance perturbations (δx) associated with rays that are propagated through the sunspot atmosphere. These calculations are for similar positions and frequencies as in Figure 6. The exclusively positive values of δx that we can see along the sunspot radius indicates that at the same time that these rays are being sped up, they are also undertaking a longer journey than their Model S counterparts in the process, and, as with $\delta\tau_p$, the magnitude of the calculated δx appears to be closely related to surface magnetic-field strength. For both $\delta\tau_p$ and δx we also observe a particular pattern of perturbation associated with each position along the sunspot. Whereas the perturbations appear to mainly decrease when we are close to spot centre (*e.g.*, $r_{spot} = 4, 8$ Mm), they appear to increase when further away (*e.g.*, $r_{spot} = 12, 16$ Mm) from spot centre. This is clearly a byproduct of both varying field strength and inclination angle of field lines (see Figure 1) as we move across the sunspot. Field strength tends to decrease, while field lines become more significantly inclined, as we move away from centre of the sunspot.

Also clearly obvious from both Figures 6 and 7 is the presence of a significant frequency dependence of both $\delta\tau_p$ and δx measurements in the sunspot, with the magnitudes of the

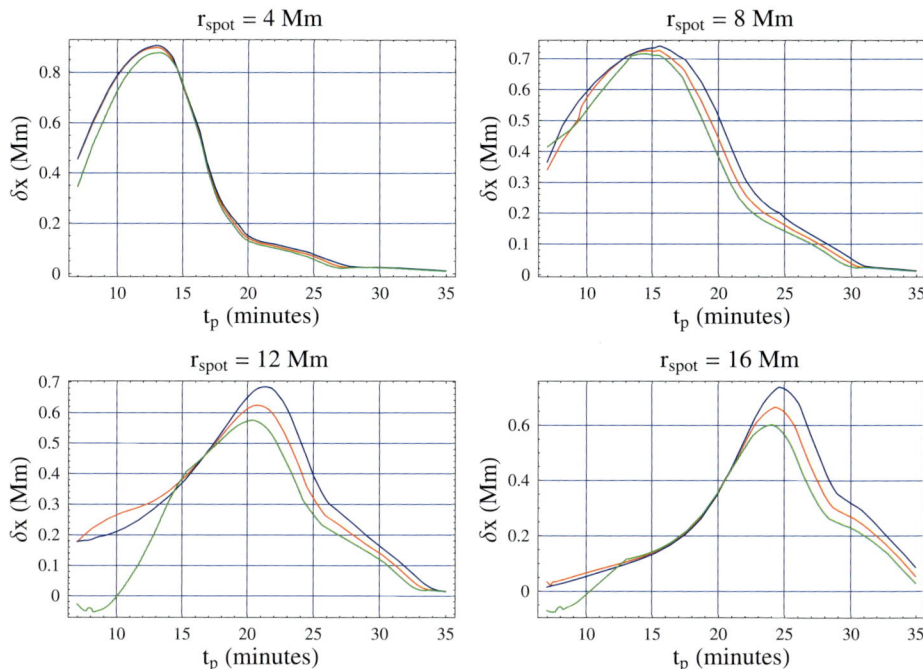

Figure 7 Skip distance perturbations (δ_x) as a function of phase travel time (t_p) for $r_{spot} = 4, 8, 12$, and 16 Mm on the sunspot, calculated for three frequencies $\omega = 3.5$ (green), $\omega = 4$ (red), and $\omega = 5$ mHz (blue).

perturbations increasing as the frequency is increased from 3.5 to 5 mHz. This is particularly evident for rays with short skip distances (*i.e.*, surface skimmers with very shallow lower turning points). Frequency dependence of travel-time perturbations in active regions has also been observed by both helioseismic holography (Braun and Birch, 2006) and time–distance helioseismology (Couvidat and Rajaguru, 2007). We shall discuss the importance of these observations in greater detail in the upcoming sections. Cally (2007) also observed a similar behaviour when modelling rays in inclined fields and described several related but distinct effects that strong magnetic fields appear to have on seismic waves, with an important "dual effect" that the magnetic field has on individual ray paths (that is, increasing their skip distances while at the same time, speeding them up considerably) being one of these effects.

A comparison between rays propagated inside the sunspot model with rays propagated in the quiet-Sun clearly reveals these effects to the naked eye. All rays shown in Figure 8 are initialised at a depth of $z_{bot} = -2$ Mm, with the rays inside the sunspot model (solid rays, colours identify frequencies) also being initialised at varying positions along the sunspot ($r_{spot} = 0, 4, 8, 12$, and 16 Mm). While the rays propagated inside the Model S atmosphere (dashed rays) are symmetrical about their turning points (as expected), strong asymmetries (at both turning points) are associated with the same rays when initiated inside the sunspot. We can clearly see that the rays inside the sunspot (at all three frequencies) appear to have undergone a longer skip distance, in a slightly shorter amount of time (dots along ray paths indicate one-minute t_g intervals), confirming the perturbation profiles of Figures 6 and 7. Of course Figure 8 shows a very small sample of rays initialised at a given depth, but even so, they are quite clearly indicative of the large-scale effects of the magnetic field on ray

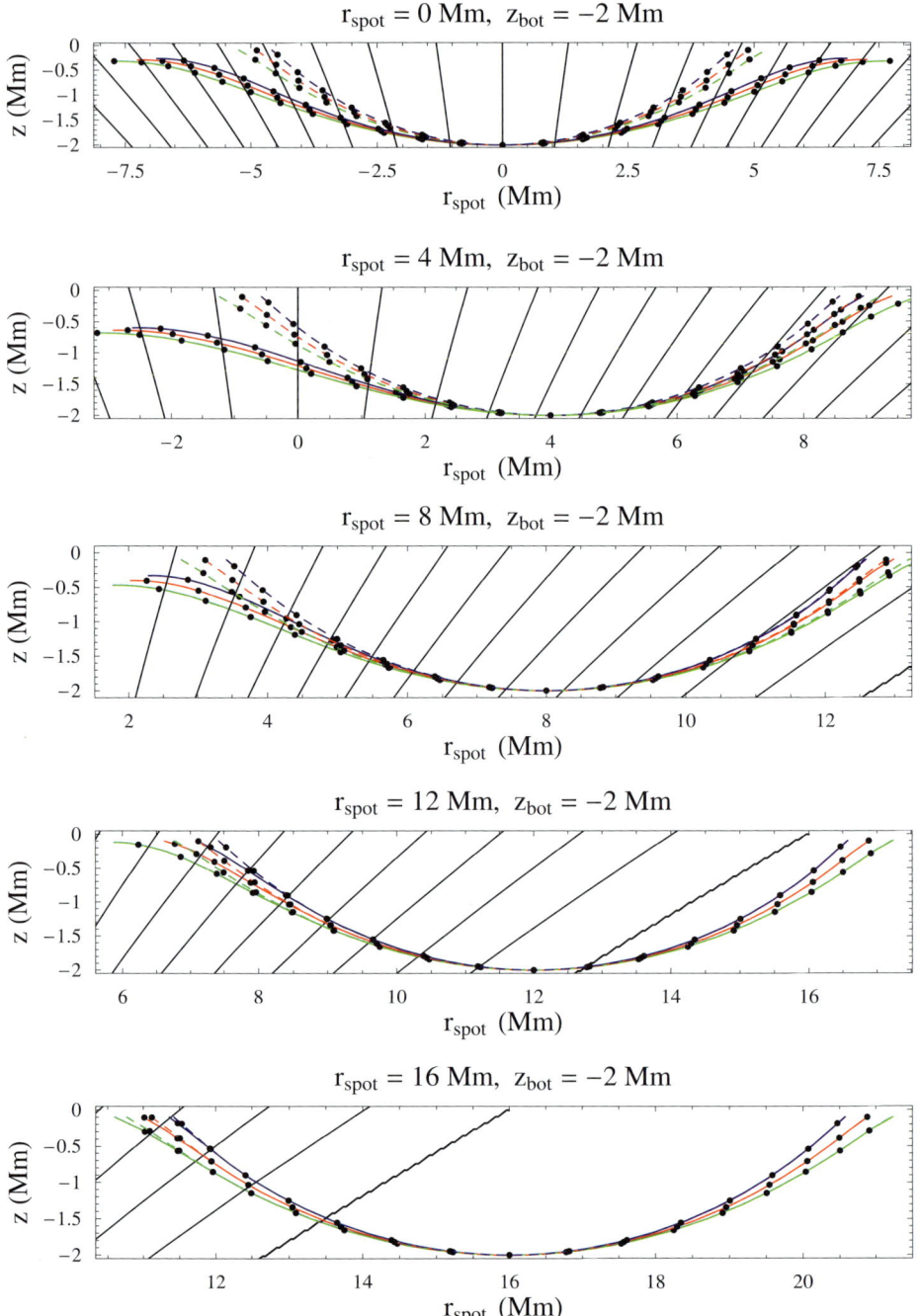

Figure 8 Individual rays propagated through the simulated sunspot (solid rays) and Model S (dashed rays) atmospheres, calculated for three frequencies: $\omega = 3.5$ (green), $\omega = 4$ (red), and $\omega = 5$ mHz (blue). The top of each frame indicates the initial depth ($z_{\rm bot}$, Mm) and radial grid position of the lower turning point of the ray ($r_{\rm spot}$, Mm).

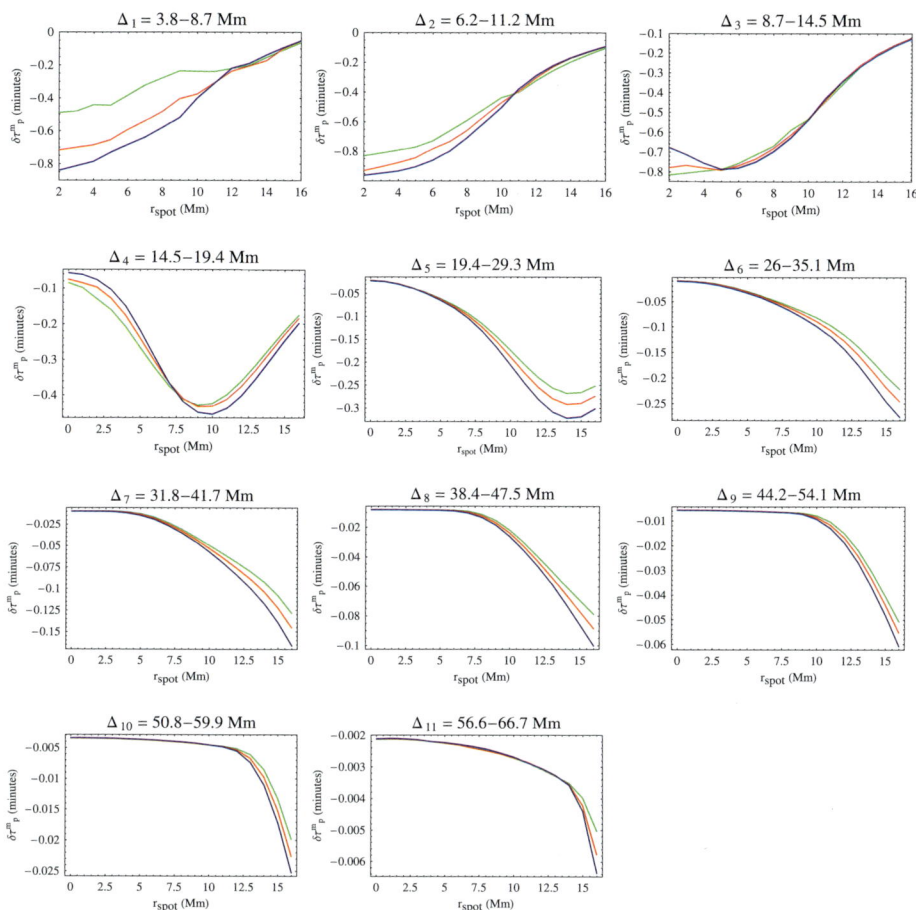

Figure 9 Binned (mean) travel-time perturbation ($\delta\tau_p^m$, minutes) profiles as a function of position (r_{spot}, Mm) on the sunspot, calculated for three frequencies: $\omega = 3.5$ (green), $\omega = 4$ (red), and $\omega = 5$ mHz (blue). Annuli number and sizes are indicated on the top of the frame of each bin.

propagation – effects which are more pronounced as we approach the spot centre and in regions of significantly inclined magnetic fields.

4.2. Binned Travel-Time Perturbation Profiles

The mean ray travel-time perturbations ($\delta\tau_p^m$) for each frequency and grid position were calculated and binned into 11 annuli ($\Delta_1 - \Delta_{11}$) of various sizes (outlined in Table 1). The $\delta\tau_p^m$ profiles of the bins are shown in Figure 9. Once again, we can see the clear frequency dependence of travel-time perturbations evident in all bins, with perturbations increasing with increasing frequency as before. Also, all $\delta\tau_p^m$ bins contain negative perturbations as we saw before in Figure 6. We also observe that the magnitude of $\delta\tau_p^m$ decreases as we move away from the centre of the sunspot (*i.e.*, decreasing field strength) for the smaller bins (*e.g.*, $\Delta_1 - \Delta_3$).

These smaller bins are representative of shallow rays that spend a considerable proportion of their journey inside the magnetic field, consistent with the larger magnitude of the perturbations seen in these bins. Larger bins (*e.g.*, $\Delta_4 - \Delta_{11}$) sample rays with much deeper lower turning points, hence a considerable amount of the journey undertaken by these rays would be spent in the quiet-Sun Model S atmosphere. Therefore the magnitude of the perturbations tends to be smaller than that for the smaller bins. However, they are found to increase in magnitude as we move away from the centre of the sunspot as rays sample larger areas of the magnetic field throughout their journey across the sunspot radius.

It should be noted that for the smaller bins (particularly for $\Delta_1 - \Delta_3$), it becomes quite difficult to obtain a sufficient sampling of rays to average near the centre of the flux tube, even with a very fine grid spacing of $\Delta z = -0.025$ Mm in the very sensitive top 1.5 Mm of the computational grid. As such, we get a certain level of rigidity in the $\delta\tau_p^m$ profiles of these bins. No such restriction is encountered when using a pure Model S atmosphere, which tends to suggest that strong near-surface magnetic fields are severely restricting the propagation of helioseismic rays with short skip distances (or very shallow lower turning points).

Although our sunspot model has many of the qualitative features that we might expect in a real spot, it is nonetheless rather *ad hoc*, and consequently our time–distance results do not warrant detailed comparison with solar observations. Nevertheless, it is of interest to qualitatively compare the $\delta\tau_p^m$ results obtained from our simulations to those reported for AR 8243 (18 June 1998) by Couvidat, Birch, and Kosovichev (2006). S. Couvidat kindly provided us with the actual set of travel-time maps used in their analysis.

To compare the $\delta\tau_p^m$ profiles as closely as possible, we first compute the azimuthal average of the four $\delta\tau_p^m$ maps presented in Figure 3 of Couvidat, Birch, and Kosovichev (2006) (corresponding to Δ_1, Δ_3, Δ_6, and Δ_9, noting that the travel-times were obtained without a frequency bandpass filter), to obtain $\delta\tau_p^m$ profiles of AR 8243, akin to our artificial $\delta\tau_p^m$ profiles contained in Figure 9. We observe peak (positive) travel-time perturbations of ≈ 0.29 and ≈ 0.16 minutes respectively for Δ_1 and Δ_3 in the sunspot umbra, while the sign of $\delta\tau_p^m$ in the sunspot changes for the larger bins, Δ_6 and Δ_9, with $\delta\tau_p^m$ ranging from ≈ -0.38 to ≈ -0.31 minutes, respectively. The perturbations for all four bins also appear to decrease in the penumbra relative to the umbra. In comparison, if we assume a central frequency of 3.5 mHz, the artificial $\delta\tau_p^m$ profiles for the bins produced by our simulations (Figure 9, 3.5 mHz profiles indicated by solid green lines) show opposite-in-sign and larger-in-magnitude $\delta\tau_p^m$ for both Δ_1 (≈ -0.7 minutes) and Δ_3 (≈ -0.82 minutes), while similar-in-sign yet smaller-in-magnitude $\delta\tau_p^m$ profiles were observed for Δ_6 (≈ -0.22 minutes) and Δ_9 (≈ -0.05 minutes). When we consider higher frequencies, the magnitude of the artificial $\delta\tau_p^m$ increases with frequency for all four bins, with all perturbations being negative in sign. However, the general pattern of the artificial $\delta\tau_p^m$ profiles for all frequencies appears to be similar to the observations of Couvidat, Birch, and Kosovichev (2006), with perturbations decreasing with increasing radius from the centre of the sunspot.

While the differences in the magnitudes of $\delta\tau_p^m$ between our simulations and those of Couvidat, Birch, and Kosovichev (2006) (at a given fixed central frequency) can be explained, to some extent, by magnetic and thermal differences between our model and their sunspot, the frequency dependence of $\delta\tau_p^m$ and the sign change of the smaller bins in particular (*i.e.*, positive $\delta\tau_p^m$ resulting from actual time–distance observations, negative $\delta\tau_p^m$ from the simulations) cannot be dismissed as easily. Traditionally, positive $\delta\tau_p^m$ obtained for short skip distances in sunspots have been interpreted as representing a region of slower wave-speed propagation in the shallow sub-surface layers of the sunspot. However, as we briefly noted in the previous section, Braun and Birch (2006) (using helioseismic holography) found that, at a given fixed phase speed, travel-time perturbations within active regions exhibit a strong frequency dependence. Couvidat and Rajaguru (2007) confirmed these results using time–distance helioseismology, applying additional frequency bandpass filters (centred at 3, 4, and 4.5 mHz) to the standard phase-speed filters used in Couvidat, Birch, and Kosovichev (2006) in order to determine the cause of the dark rings of negative $\delta\tau_p^m$ they detected in the travel-time maps (mainly associated with the Δ_2 and Δ_3 skip-distance bins) of a majority of the sunspots they studied. These rings, which are sensitive to the frequency filtering applied, are found to produce significant ring-like structures in the inversion results, mimicking regions of increased sound speed. The authors conclude that the rings are most likely to be artifacts caused by surface effects, probably of magnetic origin.

In addition to these results, the very recent work undertaken by Braun and Birch (2008) (using ridge filters, in addition to the standard phase-speed filters) provide strong evidence that the positive perturbations observed arise from the p_1 ridge or beneath it. These positive travel-time shifts were not seen in the higher order p-mode data. These results, when considered in conjunction with our artificial $\delta\tau_p^m$ profiles (and the results contained in the next section), provide further concrete evidence that positive travel-time perturbations obtained for short skip distances are likely to be artifacts or bi-products of the data reduction or analysis method used, rather than some actual physical sub-surface anomaly below the sunspot.

4.3. Isolating the Thermal Component of Travel Time Perturbations

One of the keys to understanding the role played by near-surface magnetic fields in local helioseismology is to be able to isolate it from effects thought to be produced by thermal or flow perturbations. The simplest way to isolate such effects is to essentially "switch off" the magnetic field when calculating the ray paths in the simulations – that is, set $a = 0$ in the simulated sunspot atmosphere, but maintain the modified sound-speed profile obtained (seen in Figure 4).

The external atmosphere, ray-path simulations and computational grid remain identical to those described previously. The only difference is the resulting thermal travel-time perturbations ($\delta\tau_p^{mt}$) which would then be purely a result of what can be referred to as "thermal variations" along the ray path. One can then compare the resulting perturbation profiles to those obtained when the magnetic field is included in the simulations (*i.e.*, Figure 9) to better understand the role of the thermal contributions to the observed $\delta\tau_p^m$ profiles. Figure 10 shows the resulting bins of the thermal component of $\delta\tau_p^{mt}$.

In general, the resulting $\delta\tau_p^{mt}$ profiles are relatively smooth and all bins clearly show exclusively positive travel-time perturbations (compared to exclusively negative travel-time perturbations observed in Figure 9), this implies that rays are travelling considerably slower than in the Model S atmosphere – a clear contrast with simulations where the magnetic field is present. The magnitude of $\delta\tau_p^{mt}$ is also decreasing with increasing radius for the smaller bins ($\Delta_1 - \Delta_4$) and *vice versa* for the larger bins ($\Delta_5 - \Delta_{11}$), a similar behaviour to what is observed in Figure 9. However, when considering the magnitude of the perturbations between Figures 9 and 10, it is clear that thermal perturbations appear to be much smaller for a majority of the bins – in fact up to 400% smaller for some frequencies when comparing the perturbations in the near-surface regions $\Delta_1 - \Delta_3$. The magnitude of the perturbations become much more comparable when looking at the larger bins (Δ_7 onwards), and from Δ_8 onwards $\delta\tau_p^{mt}$ becomes ever slightly larger than the ones we see in Figure 9 for the same bins. Frequency dependence of $\delta\tau_p^{mt}$ is also evident, but only clearly discernible for the first six bins ($\Delta_1 - \Delta_6$).

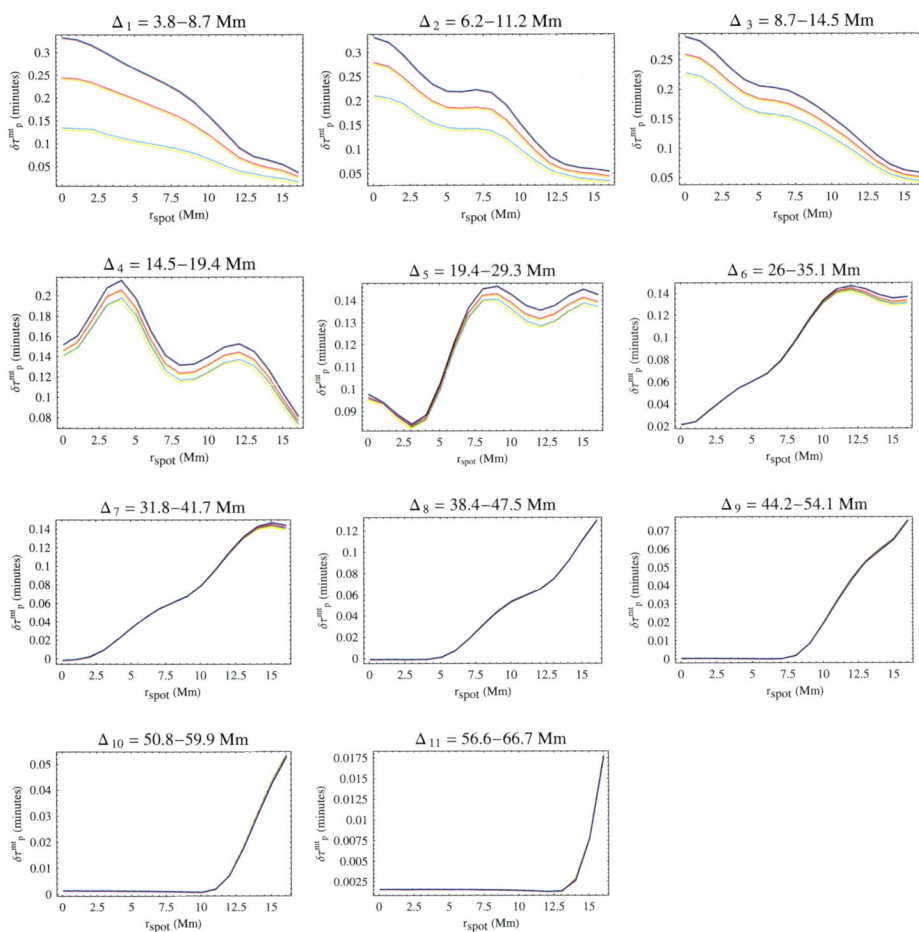

Figure 10 Binned (mean) thermal travel-time perturbation ($\delta\tau_p^{mt}$, minutes) profiles as a function of position (r_{spot}, Mm) on the sunspot, calculated for three frequencies: $\omega = 3.5$ (green), $\omega = 4$ (red), and $\omega = 5$ mHz (blue). Annuli number and sizes are indicated on the top of the frame of each bin.

5. Summary and Discussion

Whether it be through direct observations, forward modelling, or inversions, in order to be able to confidently interpret helioseismic observations and inferences made in regions of strong magnetic field, the actual physical effects of near-surface magnetic fields on ray propagation must be better understood and taken into account when analysing or modelling active region sub-photospheres. Our approach here is akin to forward modelling of rays, but in a simulated sunspot atmosphere based on IVM surface magnetic-field profiles with a peak field strength of 3 kG and an external field-free Model S atmosphere used as the background or unperturbed medium. The main aim of these simulations was to isolate and understand the effects of the wave-speed inhomogeneities produced by the magnetic field from those thought to be produced from thermal or flow perturbations.

The magneto-acoustic rays were propagated across the sunspot radius for a range of depths to produce a skip-distance geometry similar to centre-to-annulus cross-covariances used in time–distance helioseismology. The perturbations from the Model S atmosphere were calculated for each radial grid position and range of frequencies (3.5–5 mHz), then binned into 11 different skip-distance geometries of increasing size. A separate, yet similar, set of simulations was then produced to isolate the role played by thermal variations inside the sunspot atmosphere on the ray skip-distance and travel-time perturbation profiles. This was achieved by having the magnetic field switched off in the sunspot model, thus essentially maintaining a modified sound-speed structure, but with no calculations of the Alfvén speed.

These *artificial* skip-distance and travel-time perturbation profiles, which directly account for the effects near-surface magnetic fields and thermal variations separately, have provided us with a number of very distinct and interesting observations:

1. The sunspot magnetic field has a clear and distinct "dual effect" on helioseismic rays – increasing their skip distances, while at the same time, shortening their travel time (compared to similar rays in a Model S atmosphere). Higher frequency rays propagated within the magnetic field also tend to undergo a more substantial speed up than their non-magnetic counterparts.
2. There is a clear and significant frequency dependence of both ray skip-distance and travel-time perturbations across the simulated sunspot atmosphere. This frequency dependence of perturbations was prevalent for all skip-distance bins, but particularly so for shallow rays, which sample the near-surface layers of the sunspot.
3. The negative sign of travel-time shifts, along with the general pattern and magnitude of these perturbations (*i.e.*, tending to increase with increasing magnetic-field strength and inclination) points to more evidence of the significant role played by the sunspot magnetic field. Rays with shorter skip distances were seen to experience greater perturbations as a result of spending a considerable proportion of their journey within the confines of the magnetic field.
4. With the magnetic field switched off, the simulated travel-time perturbation profiles changed sign for all bins (*i.e.*, only positive perturbations were observed across the sunspot radius, meaning that rays in the thermal model are actually slower than their Model S counterparts), and the magnitude of these perturbations appeared to be significantly smaller in magnitude (300–400% at times) than when the magnetic field is included in the model. This was particularly evident for the bins that sample rays in the near-surface layers, whereas bins of larger skip distances produce slightly larger perturbations than the magnetic model. Frequency dependence of travel-time perturbations were also observed, but only for half of the bins. A majority of bins sampling larger skip distances did not exhibit this behaviour.

These observations as a whole tend to suggest that active-region magnetic fields play a direct and significant role in sunspot seismology, and it is the interaction of the near-surface magnetic field with solar oscillations, rather than purely thermal (or sound-speed) perturbations, that is the major cause of observed travel-time perturbations in sunspots. (We note here that we are only commenting on the interpretation of time–distance results in terms of *thermal/sound-speed* perturbations, and not, for example, in terms of *wave-speed* perturbations.)

The frequency dependence of these perturbations is one of the strongest indications that the magnetic field is a significant contributor to the travel-time shifts. When isolating the thermal component of $\delta\tau_p$ we did observe some frequency dependence in a limited number of bins/skip-distance geometries, certainly not to the extent that we saw when the magnetic field was included. Of course in the absence of any perturbations, rays propagated at different frequencies will naturally have slightly different upper turning points, this could certainly explain a part of a frequency dependence, but this effect combined with the (negative) sign and magnitude of the simulated $\delta\tau_p$ profiles, along with the relatively small (positive) thermal component extracted from the perturbations, makes it very difficult for one to argue that what we are seeing in these travel-time perturbation profiles is a result of a sub-surface flow or sound-speed perturbation, as has been traditionally interpreted in time–distance literature.

Instead, these observations indicate that strong near-surface magnetic fields may be seriously altering the magnitude and lateral extent of sound-speed inversions made by time–distance helioseismology. This is because standard time–distance observations (*e.g.*, Couvidat, Birch, and Kosovichev (2006), see Section 4.2) show $\delta\tau_p^m$ maps derived from the averaged cross-correlations shifting from positive values for the first couple of bins (usually $\Delta_1 - \Delta_3$), to negative ones for the remainder of the bins. Traditionally, positive perturbations result in regions of decreased sound speed in inversions, while negative perturbations result in regions of enhanced sound speed. But we have clearly seen from our forward modelling that the inclusion of the magnetic field in the near-surface layers consistently results in negative values for all bins of $\delta\tau_p^m$. This implies that any inversion of time–distance data that does not account for surface magnetic field effects will be significantly contaminated in the shallower layers of the sunspot (*i.e.*, down to a depth of a few Mm below the surface), in strong agreement with the conclusions of Couvidat and Rajaguru (2007). Hence it is almost certain from these simulations that the two-structure sunspot sound-speed profile, *i.e.*, region of decreased sound speed immediately below the sunspot (corresponding to positive $\delta\tau_p^m$), is most likely an artifact due to surface effects, instead of thermal perturbations. Deeper sound-speed profiles do not appear to be affected as much, given that the sign and magnitude of the simulated $\delta\tau_p^m$ for the larger bins are comparable to actual time–distance calculations – which is to be expected, as the flux tube becomes gas-pressure dominated at such depths.

Of course, we must bear in mind that some of our assumptions outlined earlier (*e.g.*, 2D treatment of rays, our choice of ray cutoff height in the atmosphere, the fact that we are not directly accounting for mode conversion, even the form of the surface magnetic field and background model in general, *etc.*), can certainly alter our results quantitatively in one manner or another. Indeed it would certainly be interesting and worthwhile to conduct a full 3D simulation (*i.e.*, vary the shooting angle β around the sunspot) and also test the ray propagation code with other sunspot and quiet-Sun models in the future. But in any case, it would be surprising, given the self-consistency of our current results, if our qualitative conclusions were changed as a result.

Acknowledgements The authors are very grateful to Hannah Schunker for providing the IVM observations of AR 9026, and also to Sébastien Couvidat for providing us with the travel-time maps of AR 8243. We also thank the anonymous referee for useful comments that helped improve this paper.

References

Barnes, G., Cally, P.S.: 2001, *Publ. Astron. Soc. Aust.* **18**, 243.
Bogdan, T.J., Cally, P.S.: 1997, *Proc. Roy. Soc. Lond. A* **453**, 943.
Braun, D.C., Birch, A.C.: 2006, *Astrophys. J.* **647**, L187.
Braun, D.C., Birch, A.C.: 2008, *Solar Phys.* doi:10.1007/s11207-008-9152-5.
Cally, P.S.: 2000, *Solar Phys.* **192**, 395.
Cally, P.S.: 2007, *Astron. Nachr.* **328**, 286.
Cally, P.S., Bogdan, T.J.: 1993, *Astrophys. J.* **402**, 732.
Cally, P.S., Bogdan, T.J.: 1997, *Astrophys. J.* **486**, L67.
Cally, P.S., Bogdan, T.J., Zweibel, E.G.: 1994, *Astrophys. J.* **437**, 505.
Cally, P.S., Crouch, A.D., Braun, D.C.: 2003, *Mon. Not. Roy. Astron. Soc.* **346**, 381.

Cameron, R., Gizon, L., Duvall, T.L. Jr.: 2008, *Solar Phys.* doi:10.1007/s11207-008-9148-1.
Christensen-Dalsgaard, J., Dappen, W., Ajukov, S.V., *et al.*: 1996, *Science* **272**, 1286.
Couvidat, S., Rajaguru, S.P.: 2007, *Astrophys. J.* **661**, 558.
Couvidat, S., Birch, A.C., Kosovichev, A.G.: 2006, *Astrophys. J.* **640**, 516.
Couvidat, S., Birch, A.C., Kosovichev, A.G., Zhao, J.: 2004, *Astrophys. J.* **607**, 554.
Crouch, A.D., Cally, P.S.: 2003, *Solar Phys.* **214**, 201.
Crouch, A.D., Cally, P.S.: 2005, *Solar Phys.* **227**, 1.
Fan, Y., Braun, D.C., Chou, D.-Y.: 1995, *Astrophys. J.* **451**, 877.
Gizon, L., Birch, A.C.: 2005, *Living Rev. Solar Phys.* **2**, 6. http://www.livingreviews.org/lrsp-2005-6.
Hanasoge, S.M., Duvall, T.L. Jr.: 2007, *Astron. Nachr.* **328**, 319.
Hughes, S.J., Rajaguru, S.P., Thompson, M.J.: 2005, *Astrophys. J.* **627**, 1040.
Jensen, J.M., Duvall, T.L. Jr., Jacobsen, B.H., Christensen-Dalsgaard, J.: 2001, *Astrophys. J.* **553**, L193.
Kosovichev, A.G., Duvall, T.L. Jr., Scherrer, P.H.: 2000, *Solar Phys.* **192**, 159.
Lindsey, C., Braun, D.C.: 2005, *Astrophys. J.* **620**, 1107.
Mickey, D.L., Canfield, R.C., Labonte, B.J., Leka, K.D., Waterson, M.F., Weber, H.M.: 1996, *Solar Phys.* **168**, 229.
Schunker, H., Cally, P.S.: 2006, *Mon. Not. Roy. Astron. Soc.* **372**, 551.
Schunker, H., Braun, D.C., Cally, P.S., Lindsey, C.: 2005, *Astrophys. J.* **621**, L149.
Spruit, H.C., Bogdan, T.J.: 1992, *Astrophys. J.* **391**, L109.
Weinberg, S.: 1962, *Phys. Rev.* **126**, 1899.
Zhao, J., Kosovichev, A.G., Duvall, T.L. Jr.: 2001, *Astrophys. J.* **557**, 384.

Modelling the Coupling Role of Magnetic Fields in Helioseismology

B. Pintér

Originally published in the journal Solar Physics, Volume 251, Nos 1–2, 329–340.
DOI: 10.1007/s11207-008-9128-5 © Springer Science+Business Media B.V. 2008

Abstract Helioseismic global modes change in time, in particular on time scales of the solar cycle. These changes show, in fact, strong correlation with the magnetic activity cycle of the Sun, indicating that a most likely cause of the variation of the mode characteristics, such as frequency, is the magnetic field. In the present paper I attempt to find out in what ways and to what degree the magnetic atmosphere of the Sun can influence the f and p modes of helioseismology. Frequency shifts of the order of a microhertz, line widths of the order of a nanohertz, and penetration depths of the order of a megameter are obtained.

Keywords Solar physics · Helioseismology · Solar interior · Solar atmosphere · Solar cycle · Magnetic fields · MHD modelling

1. Introduction

Helioseismology is an effective way of discovering the hidden physical properties beneath the photosphere, which is shielding the solar interior from our eyes and telescopes. The power of this field of solar physics relies on observing oscillations continually and inverting these observations by using some predefined models. More and more details can be seen from the solar interior by monitoring the continual surface pulsations and analysing the expanding data archive.

So far, probably the most severe handicap of this approach is the lack of detailed modeling (*e.g.*, the missing ubiquitous magnetic field). Recent attempts that address the relevance of magnetism have been made by Erdélyi (2006a, 2006b) and Thompson (2006).

The f and p modes are present above the photosphere with small amplitudes as they are evanescent there. Hence only weak coupling is expected between the f and p modes and the lower atmosphere. However, helioseismic observations confirm that variations in the

---

Helioseismology, Asteroseismology, and MHD Connections
Guest Editors: Laurent Gizon and Paul Cally

B. Pintér (✉)
Institute of Mathematics and Physics, Aberystwyth University, Aberystwyth, Wales SY23 3BZ, UK
e-mail: b.pinter@aber.ac.uk

magnetic atmosphere are reflected in their frequencies (Chaplin *et al.*, 2004; Dziembowski and Goode, 2005) and in their line widths (Chaplin *et al.*, 2000; Komm, Howe, and Hill, 2000), which indicates that the solar atmospheric effects on the f and p modes cannot be ignored completely.

Campbell and Roberts (1989) were the first to investigate the possible effects of an atmospheric magnetic field on global solar acoustic oscillations. They used a magnetohydrodynamic (MHD) model in Cartesian geometry with a polytropic interior and an overlying atmosphere with constant sound and Alfvén speeds. According to their results, the f-mode frequency and the p-mode frequency for radial order $n = 1$ increases owing to the presence of an atmospheric horizontal magnetic field, whereas the p-mode frequencies for $n > 1$ are reduced by the magnetic field.

Evans and Roberts (1990) replace the chromospheric (atmospheric) magnetic profile of constant Alfvén speed by a constant magnetic field. The frequency shifts of the f and p modes obtained are positive, and their magnitude is larger than for the constant-Alfvén-speed model. Evans and Roberts (1991) introduced an isothermal, nonmagnetic intermediate layer between a polytropic interior and an isothermal atmosphere with constant horizontal magnetic field and found that a change in the thickness of the intermediate layer significantly modifies the frequency shifts caused by the atmospheric canopy magnetic field. Jain and Roberts (1994a) extended the analysis to f and p modes for which the horizontal wave vector is not parallel to the canopy magnetic field lines. Jain and Roberts (1994b) used the same three-layer model as Evans and Roberts (1991) but the magnetic field decays with height in the uppermost layer so that the Alfvén speed is constant. The influence of the temperature profile was investigated by Vanlommel and Čadež (1998). Erdélyi and Taroyan (2001) and Taroyan (2003) addressed the coupling of global oscillations into the magnetised solar atmosphere.

Global oscillations can be coupled resonantly to local slow and Alfvén waves, which can increase the amplitudes of the oscillations significantly. This resonant coupling was studied by Zhukov (1997), Tirry *et al.* (1998), and Pintér and Goossens (1999). The mathematical treatment of the MHD equations in the vicinity of the highly dissipative resonant position used in these and follow-up papers (see, *e.g.*, Pintér, Erdélyi, and New, 2001, and Pintér, New, and Erdélyi, 2001) was worked out previously in the linear approximation by Sakurai, Goossens, and Hollweg (1991), Erdélyi, Goossens, and Ruderman (1995), and Erdélyi (1997) and in nonlinear theory by Ruderman, Hollweg, and Goossens (1997) and Ballai, Ruderman, and Erdélyi (1998). A main effect of a resonant interaction between global modes and slow or Alfvén MHD waves is that the global f or p mode is damped because of dissipative effects where the resonance occurs. A comprehensive review of the observational and theoretical results on the interaction between the solar interior and atmosphere is given by Erdélyi (2006a).

In the present paper, magnetic effects on the frequency spectrum and penetration depth of f and p modes are studied in a three-layer model with slab geometry, where resonant interaction between global oscillations and slow waves can occur in the intermediate layer. The model has a horizontal, canopy-like magnetic field in the atmosphere, although the orientation of the solar atmospheric magnetic fields changes continuously in time and space. For modelling effects of stochastic magnetic fields see Erdélyi, Kerekes, and Mole (2005) and Kerekes, Erdélyi, and Mole (2008a, 2008b).

2. Model

The Sun is modelled as follows. The solar interior and atmosphere with the photosphere between them is represented by a one-dimensional, plane-parallel structure of three layers in Cartesian geometry. The lower semi-infinite layer ($z < 0$) is the polytropic interior of the Sun with an adiabatic temperature profile. The upper semi-infinite layer ($z > L$) is an isothermal atmosphere embedded in a horizontal magnetic field with constant Alfvén speed. The magnetic field strength increases from zero to its maximum [$B_L \equiv B_0(z = L)$] in the intermediate layer between the two semi-infinite layers, which we call the transition layer. Here the squares of the sound speed and of the Alfvén speed increase linearly. In equilibrium, the total pressure (plasma pressure plus magnetic pressure) is balanced against gravity. The gravitational acceleration (g) is considered constant, and we take its photospheric value ($g \approx 274$ m s^{-2}) as the perturbations studied in the present paper oscillate with the largest amplitudes and thus are most sensitive, right beneath the photosphere. The equilibrium density, plasma pressure, sound speed, and Alfvén speed have the following profiles, respectively:

$$\rho_0(z) = \begin{cases} \rho_{\text{ph}}\left(1 - \frac{z}{H_{\text{in}}}\right)^{1/(\gamma-1)}, & z \leq 0, \\ \frac{\rho_{\text{ph}}}{(1+\frac{z}{H_{\text{tr2}}})^{\alpha}}, & 0 \leq z \leq L, \\ \rho_L \exp\left(-\frac{z-L}{H_{\text{co}}}\right), & L \leq z, \end{cases} \quad (1)$$

$$p_0(z) = \begin{cases} p_{\text{ph}}\left(1 - \frac{z}{H_{\text{in}}}\right)^{\gamma/(\gamma-1)}, & z \leq 0, \\ p_{\text{ph}}\frac{(1+\frac{z}{H_{\text{tr1}}})}{(1+\frac{z}{H_{\text{tr2}}})^{\alpha}}, & 0 \leq z \leq L, \\ p_L \exp\left(-\frac{z-L}{H_{\text{co}}}\right), & L \leq z, \end{cases} \quad (2)$$

$$v_s(z) = \begin{cases} v_{s,\text{ph}}\sqrt{1 - \frac{z}{H_{\text{in}}}}, & z \leq 0, \\ v_{s,\text{ph}}\sqrt{1 + \frac{z}{H_{\text{tr1}}}}, & 0 \leq z \leq L, \\ v_{s,\text{co}}, & L \leq z, \end{cases} \quad (3)$$

$$v_A(z) = \begin{cases} 0, & z \leq 0, \\ v_{A,\text{co}}\sqrt{\frac{z}{L}}, & 0 \leq z \leq L, \\ v_{A,\text{co}}, & L \leq z, \end{cases} \quad (4)$$

with characteristic constants

$$p_{\text{ph}} = \frac{1}{\gamma}\rho_{\text{ph}}v_{s,\text{ph}}^2, \qquad v_{s,\text{co}}^2 = \frac{T_{\text{co}}}{T_{\text{ph}}}v_{s,\text{ph}}^2, \quad (5)$$

$$\rho_L = \rho_0(z = L), \qquad p_L = p_0(z = L), \quad (6)$$

$$H_{\text{in}} = \frac{v_{s,\text{ph}}^2}{(\gamma - 1)g}, \qquad H_{\text{co}} = \frac{1+\beta}{\beta}\frac{v_{s,\text{co}}^2}{\gamma g}, \quad (7)$$

$$H_{\text{tr1}} = \frac{v_{s,\text{ph}}^2 L}{v_{s,\text{co}}^2 - v_{s,\text{ph}}^2}, \qquad H_{\text{tr2}} = \frac{2v_{s,\text{ph}}^2 L}{2v_{s,\text{co}}^2 - 2v_{s,\text{ph}}^2 + \gamma v_{A,\text{co}}^2}, \quad (8)$$

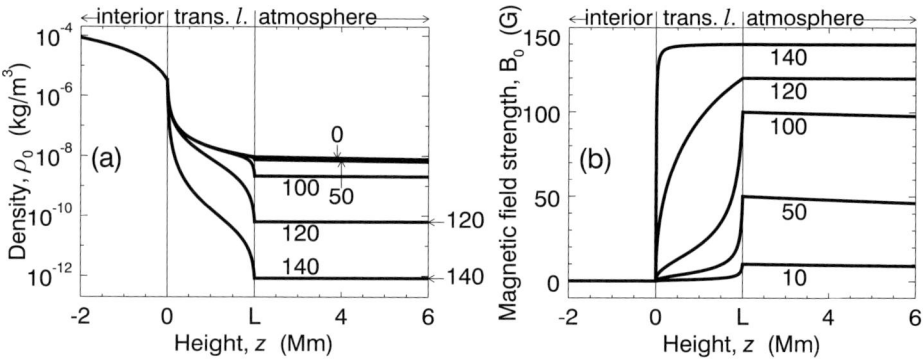

Figure 1 (a) Plasma density (ρ_0) and (b) magnetic field strength (B_0) as a function of height (z) in equilibrium. The graphs are labelled by the value of B_L in gauss.

$$\alpha = 1 + \frac{2\gamma g L}{2v_{s,co}^2 - 2v_{s,ph}^2 + \gamma v_{A,co}^2}, \quad \beta = \frac{2v_{s,co}^2}{\gamma v_{A,co}^2}. \tag{9}$$

The indices "ph", "L", and "co" refer to equilibrium quantities taken at the photosphere ($z = 0$), at the top of the transition layer ($z = L$), and in the corona ($z \geq L$), respectively. The definition of the sound speed square is $v_s^2(z) \equiv \gamma p_0(z)/\rho_0(z)$, and P_0 is the total pressure at equilibrium: $P_0(z) \equiv p_0(z) + B_0^2(z)/2\mu$.

The radial profiles of the equilibrium density [$\rho_0(z)$] and magnetic field strength [$B_0(z)$] are shown in Figures 1(a) and (b), respectively, for $B_L = 0$, 10, 50, 100, 120, and 140 G. The plasma density drops at the photosphere and slowly but still exponentially decays in the corona with a large value of scale height (H_{co}). The interior is free of magnetic field. The magnetic-field strength increases in the transition layer from zero. The combination of a constant Alfvén speed and an exponentially decreasing plasma density results in a magnetic field strength decreasing exponentially in the corona with scale height $H_{co}/2$. Dissipation is taken into account near resonance. Far from resonance, the ideal MHD equations are used. The amplitudes of the helioseismic oscillations are generally small, hence we can use the linearized MHD equations. The model is essentially one-dimensional, as all of the physical quantities vary only with height (z). Hence the perturbed physical quantities in the MHD equations can be replaced by their Fourier-transformed forms

$$f_1(x, y, z, t) = f(z; \omega, k_x, k_y) e^{i(k_x x + k_y y - \omega t)}. \tag{10}$$

This analysis is restricted to global oscillations for which the horizontal component of the wave vector is parallel to the atmospheric magnetic field lines. (For oblique propagation, see Jain and Roberts (1994a), and Pintér, Erdélyi, and Goossens (2007).) By these considerations, the governing MHD equations for parallel propagation can be reduced to two first-order differential equations for the vertical (*i.e.*, radial) component of the Lagrangian displacement vector [($\xi_z(z)$)] and the Eulerian total-pressure perturbation [$P(z)$]:

$$D(z)\frac{d\xi_z}{dz} = C_1(z)\xi_z - C_2(z)P, \quad D(z)\frac{dP}{dz} = C_3(z)\xi_z - C_1(z)P. \tag{11}$$

The linear Fourier-transformed, ideal MHD equations for parallel propagation then can be reduced to two ordinary differential equations of the first order for the vertical component of the Lagrangian displacement [$\xi_z(z)$] and for the Eulerian perturbation of total pressure [$P(z)$]: The coefficient functions $D(z)$, $C_1(z)$, $C_2(z)$, and $C_3(z)$ in Equations (11) are given by

$$D(z) = \rho_0(v_s^2 + v_A^2)(\omega^2 - \omega_c^2), \quad C_1(z) = g\rho_0\omega^2, \quad C_2(z) = \omega^2 - \omega_s^2,$$

$$C_3(z) = \left(\rho_0(\omega^2 - \omega_A^2) + g\frac{d\rho_0}{dz}\right)D + g^2\rho_0^2(\omega^2 - \omega_A^2). \quad (12)$$

The characteristic sound, slow, and Alfvén frequencies, respectively, are defined by

$$\omega_s^2 = k^2 v_s^2, \quad \omega_c^2 = k^2 v_c^2, \quad \omega_A^2 = k^2 v_A^2.$$

According to Equation (10), the quantities $\xi_z(z)$ and $P(z)$ in Equation (12) are the z-dependent magnitudes of the oscillating perturbations around the equilibrium state of the plasma, where at equilibrium gravity and the gradient of the total (plasma and magnetic) pressure are in balance. The linear oscillations in an inhomogeneous magnetic plasma are governed by Equations (11). The interior, transition layer, and corona are joined by the physical requirements that the Lagrangian displacement [$\xi_z(z)$] and the Lagrangian perturbation of total pressure [$\delta P(z) \equiv P(z) + g\rho_0(z)\xi_z(z)$] must be continuous functions of height. The boundary conditions far from the transition layer are that the kinetic and magnetic energy of the eigenoscillations must tend to zero for $z \to -\infty$ (towards the solar centre) and for $z \to \infty$ (towards the outer corona). Equations (11) together with these boundary conditions define an eigenvalue problem for the global frequency (ω).

3. Results

For numerical results, we fix the harmonic degree at $\ell = 200$ and study the frequencies as the magnetic field strength is varied. The ratio of the coronal temperature to that of the photosphere (see Equation (5)) is fixed at $T_{co}/T_{ph} = 200$. The magnetic-field strength is given by its maximum, which is taken at the top of the transition layer, that is, at $B_L \equiv B_0(z = L)$. The results are presented in terms of $\nu (\equiv \omega/2\pi)$, which is more commonly used. The horizontal wavenumber (k) is also replaced by the harmonic degree (ℓ), which can be obtained from $k \equiv \sqrt{\ell(\ell+1)}/R_{Sun}$.

Figure 2 shows the frequency spectrum with B_L varying from zero to about 150 G. This upper limit for the magnetic field strength is determined by the requirement in the model that the total (plasma and magnetic) equilibrium pressure is kept constant in the photosphere, at $z = 0$. Stronger magnetic fields would result in negative plasma pressure [$p(z = 0) < 0$] at equilibrium.

The frequency spectrum in Figure 2 is structured by characteristic frequencies. Eigenfrequencies can be found between the lower (ν_I) and upper (ν_{II}) magneto-acoustic cutoff frequencies. Global oscillation modes with frequencies above the upper cutoff frequency (ν_{II}) or between the characteristic slow frequency [$\nu_c \equiv \nu_c(z = L)$] and the lower magneto-acoustic cutoff frequency (ν_I) are leaky modes, having nonzero total (kinetic plus magnetic) energy in the outer atmosphere, as z tends to ∞. Oscillation modes with frequencies below ν_c have complex frequencies, where the imaginary part of frequency is related to the nonzero lifetime of the mode: $\Gamma = -2\,\text{Im}(\nu)$. The region in the spectrum below ν_c is called the slow continuum.

Only the f mode, the first six p modes, and the Lamb mode exist in the field-free model, because of the upper cutoff (ν_{II}). The Lamb mode, labelled a, is an acoustic oscillation.

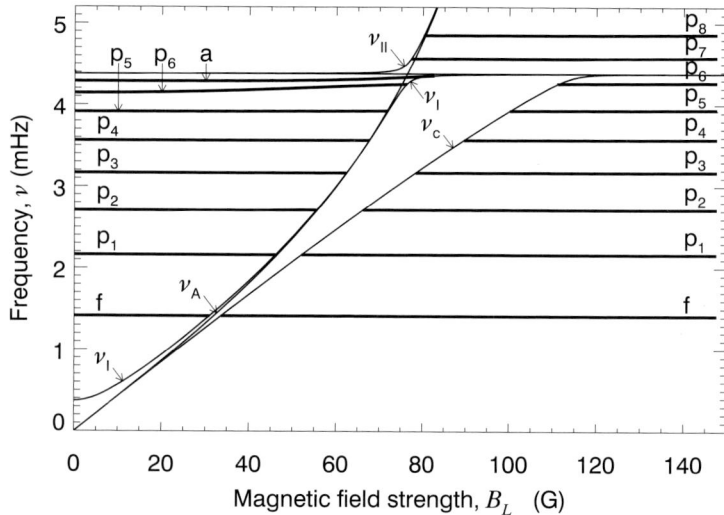

Figure 2 Frequency spectrum of the f, a, and the first six p modes. There are no eigenfrequencies between ν_I and ν_c and above ν_{II}.

It may exist in stratified media, with a frequency near one of the cutoff frequencies (see Lamb, 1932). (Notice the difference between ν_A, which denotes the characteristic Alfvén frequency, and a, which denotes the Lamb mode.)

With increasing B_L, the mode frequencies vary. The eigenmodes cease to exist for threshold values of B_L for which the mode frequencies approach the lower cutoff frequency (ν_I). The modes exist again for strong enough magnetic fields as damped modes, with frequencies in the slow continuum. Pressure modes with order larger than six can be present only for strong magnetic fields. Their frequencies are between the lower and upper cutoff frequencies.

The actual variations in the mode frequencies are hardly perceptible in Figure 2 as they are of the order of microhertz only. We define the frequency shift $\Delta \nu(B_L)$ as the frequency for $B = B_L$ minus the frequency taken for the nonmagnetic case ($B = 0$). If the mode does not exist in a field-free medium (*i.e.*, for p modes with $n > 6$) then the frequency shift is defined as $\nu(B_L)$ minus the frequency taken for the weakest magnetic field for which the mode exists. For example, the p_7 mode exists for $B \geq 77.29$ G; hence its shift is defined as $\Delta \nu(B_L) = \nu(B_L) - \nu(B = 77.29 \text{ G})$.

The frequency shifts ($\Delta \nu$) are plotted in Figure 3 as a function of B_L. The lower part of the diagram ($\Delta \nu < 16$ µHz) is enlarged in Figure 4 for clearer visibility. The frequency shifts are positive except for the f mode, for which the shift is negative between 70 and 103 G. The shift generally increases with increasing B_L, except for the f, p_1, p_2, and p_7 modes for an interval of B_L, where the $\Delta \nu(B_L)$ graphs have negative slopes between a local maximum and minimum. This decrease is the largest for the first p mode, as the shift decreases from 2.83 to 1.96 µHz as B_L changes from 93.8 to 109.3 G. The horizontal dotted lines indicate the gaps of modes in the intervals of magnetic field strength for which the mode frequencies would be between ν_I and ν_c. The a and p_6 modes, for which the frequencies are the closest to the upper cutoff frequency (ν_{II}), have the largest shifts, increasing above 0.1 mHz, which is recognisable even in the frequency spectrum in Figure 2.

Global oscillations with a frequency within the slow continuum interact resonantly with a slow wave in the transition layer, at the height where the local slow frequency matches the

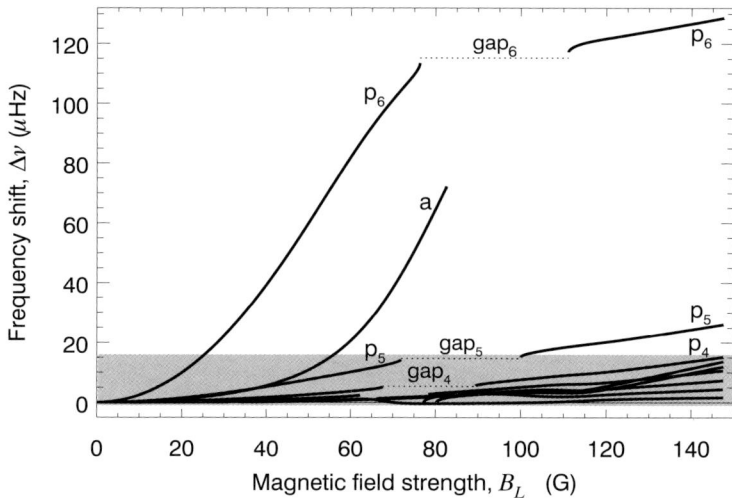

Figure 3 Frequency shift $\Delta \nu$ for the f, a, and the first six p modes with $\ell = 200$ resulting from a varying atmospheric magnetic field strength, measured by B_L.

global frequency. For $\nu_c(z) = \nu$, the coefficient function $D(z)$, defined in Equations (12), becomes zero. In that case Equations (11) are singular. Dissipation makes the global oscillation of high local amplitudes a damped mode. The solutions of the dissipative MHD equations (as an eigenvalue problem) are complex. The frequencies of the f and the first six p modes occur in the slow continuum of the frequency spectrum, as seen in Figure 2. Their line width [$\Gamma \equiv -2\text{Im}(\nu)$] is shown in Figure 5. Γ increases with B_L for the p modes until it reaches its maximum, then the line width decreases to zero as B_L tends to its maximum. The damping rate is larger for smaller radial order (n) and it takes its maximum for weaker magnetic fields. The p_1 mode has the largest damping rate. Its maximum is $\nu \approx 4.4$ nHz, taken at $B_L \approx 102.3$ G. The line width of the f mode has two local maxima, $\nu \approx 3.4$ nHz at $B_L \approx 68.3$ G and $\nu \approx 1.4$ nHz at $B_L \approx 108.1$ G, before it tends to zero. The difference in the behaviour of the f mode from that of the p modes is apparent in Figure 5 and also in Figure 4. The peaks in line width and in frequency shift in the region $60 < B_L < 70$ G may not be expected, as no reason for that can be seen in the spectrum in Figure 2.

A more detailed analysis of the low-frequency region of the spectrum – which is beyond the scope of the present study – shows that another set of global oscillation modes can exist besides the f and p modes, which have frequencies below the f-mode frequency. In contrast to the f and p modes, these global modes are evanescent in the interior ($z < 0$) but oscillate with nodes in a narrow layer right above the photosphere, within the transition layer. Their restoring force is buoyancy. More on these atmospheric gravity oscillations, which appear in the present model, can be read in Pintér and Goossens (1999) and Pintér (2008). The peaks of the frequency shift and line width of the f mode are effects of the atmospheric gravity mode with the highest frequency. Another unexpected feature of the f mode is that its frequency shift is not positive for the whole range of B_L, as found, for example, by Campbell and Roberts (1989), where the two-layer model represents the solar interior and atmosphere. The negative shift between $B_L = 70$ G and $B_L = 103$ G is caused most likely also by the first atmospheric gravity mode. This effect is not present in the model in Campbell and Roberts (1989), which has an interface instead of a transition layer. As the thickness of the transition layer (L) is gradually reduced, the effects of the atmospheric gravity modes

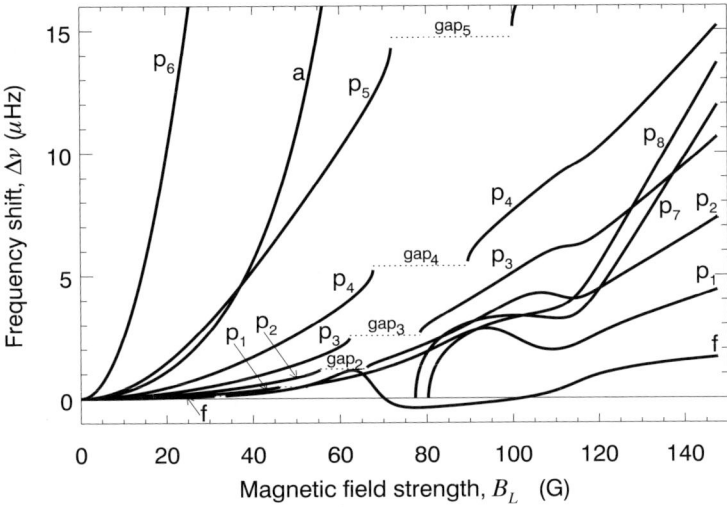

Figure 4 The low-frequency region, the enlarged grey-shaded part from Figure 3.

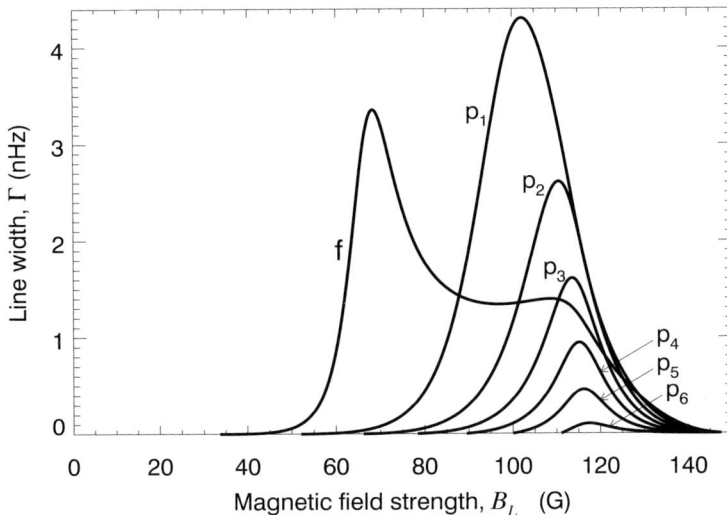

Figure 5 The line width (Γ) of the f and the first six p modes for $\ell = 200$ as a function of the magnetic field strength (B_L).

are also reduced, and thus, the shift of the f-mode frequency becomes positive for every value for B_L, as also obtained by Campbell and Roberts (1989). The results in Figure 5 are for harmonic degree fixed at $\ell = 200$; modes with lower or higher degrees generally have smaller or larger widths, respectively, for the same magnetic field strength (Pintér, Erdélyi, and New, 2001).

The frequency spectrum and the spatial behaviour of the eigenmodes are strongly related to each other, as, in mathematical terms, the spatial functions of perturbations are the eigenvectors – with eigenvalues given by the eigenfrequencies – of the eigenvalue problem, given

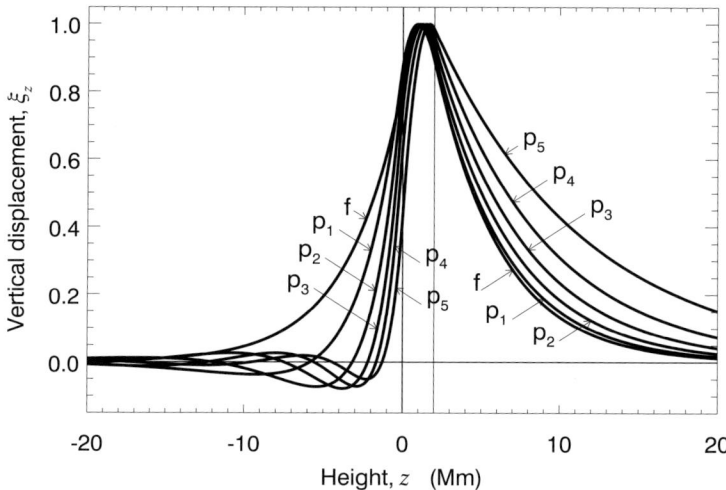

Figure 6 The vertical component of the displacement vector as a function of height for $B = 0$ and $\ell = 200$ for the f and the first five p modes. Each function is normalised by its absolute maximum.

by the governing equations and boundary conditions. Figure 6 displays the oscillation amplitude of the vertical component of the displacement vector (ξ_z) as a function of height (z) of the f and the first five p modes for $B = 0$ and $\ell = 200$ to illustrate the spatial behaviour of the perturbations. The functions are normalised by their absolute maximum. No information is lost by the normalisation, as the linear equations do not hold information about the magnitude of oscillations.

The f mode has the shape of a typical surface mode, being concentrated just above the photosphere and being evanescent towards the solar centre and the outer atmosphere without having a nodal point. The amplitudes of the p-mode perturbations have a similar z-dependence, except that they have n nodes in the solar interior ($z < 0$) as the function tends to zero for $z \to -\infty$. The vertical displacement (ξ_z) decreases as $z \to \infty$ without changing sign for every p mode. Hence the radial order (n) characterises the p modes by their amplitude oscillations in the solar interior but not in the atmosphere. Another feature of the modes can be seen in Figure 6: The higher n is, the more confined the p mode is to the photosphere in the interior (*i.e.*, the slope of the function is steeper before the first node); however, in the atmosphere, the higher n is, the more spread out is the p mode (*i.e.*, the decay rate is smaller – the slope of the function is shallower).

The magnetic field modifies the frequencies of the global oscillations significantly. To see how strong the magnetic effect is on the spatial distribution of the perturbations, we study how confined the modes are for different magnetic field strengths. We determine the confinement by measuring the height where the perturbation amplitude decays to half the maximum amplitude. Figures 7(a) and (b) show the lower (below the height of the maximum amplitude) and the upper (above the height of the maximum amplitude) extent of the f and the first six p modes as a function of B_L. (Note that here we restrict this study to the eigenmodes of which the eigenfrequency is not in the slow continuum.) Both the lower and upper penetration depths vary significantly with the magnetic-field strength. The effect becomes stronger as B_L increases and the lower cutoff frequency (ν_I) approaches the eigenfrequency. This behaviour is evident for the upper extent (Figure 7(b)) as the function of the height where the perturbation magnitude is half of the maximum becomes extremely

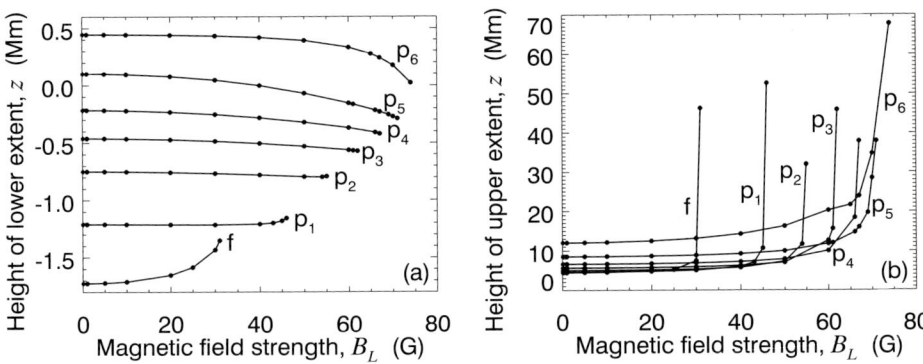

Figure 7 Height where the vertical displacement (ξ_z) is half its maximum (a) below and (b) above the position of the maximum for the f and the first six p modes as a function of B_L with $\ell = 200$.

steep just below B_L for which the mode frequency matches ν_I and the mode ceases to exist (compare to Figure 2).

4. Conclusion

We examined effects of a magnetic atmosphere on global helioseismic modes by using a three-layer slab model. In the transition layer between the interior and atmosphere, the temperature and the magnetic field strength increase continuously with height, resulting in a slow continuum and an Alfvén continuum, where global oscillations may be coupled resonantly to local slow and Alfvén waves. The present study is restricted to global waves propagating horizontally along the magnetic field lines. Parallel propagating modes are decoupled from Alfvén waves, and thus only slow resonance can occur.

The present one-dimensional, static model of the solar interior and atmosphere is a substantial simplification of the real Sun; hence we should not expect that the modelling results agree with observations in detail. However, the model is able to predict the general effects of atmospheric magnetic fields on helioseismic f and p modes. We found significant shifts in the mode frequencies, of the order of a microhertz, which agrees with observations. Frequencies of low-degree p modes were shifted by about 0.5 μHz during the last two or three solar cycles according to Chaplin *et al.* (2004, 2007) and Dziembowski and Goode (2005). The model results suggest that the presence of the magnetic field generally increases the f- and p-mode frequencies, as found in observations. The mode frequencies are maximal during high solar activity, when strong magnetic fields are present in the atmosphere.

Observing the changes in the line width of oscillation modes is a bigger challenge, as the variations of line widths are typically of the order of nanohertz or some tens of nanohertz during a solar cycle. In the model, f and p modes can be coupled resonantly to atmospheric, slow, continuum modes. Besides a frequency shift of the f and p modes, resonant interactions result in a non-zero line width of the oscillation modes, caused by an enhanced dissipation at the resonant coupling. The model is not used to predict *absolute values* of line widths, as we do not assume that only the resonant coupling contributes to the line widths of helioseismic modes. The line widths as a function of the atmospheric magnetic-field strength, obtained from the model, is to predict the *variation* of line widths during a solar cycle, when the atmospheric magnetic field varies between its minimum and maximum. We have found that the line width of global modes caused by magnetic resonant coupling increase from zero to a maximum then decrease as the atmospheric magnetic-field strength increases. The peak heights of $\Gamma(B)$ are of the order of a nanohertz, which suggests that line-width variations between low and high magnetic activities of the solar atmosphere during a solar cycle are of the order of nanohertz. This agrees well with observations (Komm, Howe, and Hill, 2000; Chaplin *et al.*, 2000). The increasing data archive of ground-based networks (see *e.g.* Miller *et al.*, 2004; Chaplin *et al.*, 2007) and observations of the current space missions (see *e.g.* Komm, Howe, and Hill, 2000) will provide observational data of resolution high enough to see the rate of line width variation during the solar cycle. We can conclude that the frequency variation with the solar activity cycle is most likely a consequence of the variation of the atmospheric magnetic field.

The spatial behaviour was also considered in the present paper. It was found that the atmospheric magnetic field also changes the depth of penetration significantly (both into the solar interior and lower atmosphere). The effect is stronger in the lower atmosphere: The presence of the global mode can be extended by the order of tens of megameters. In the interior, the penetration depth is found to be decreased in the case of the f mode and increased for the p modes by a few tens or even hundreds of kilometers. These results indicate that it may be worth trying to observe the f and p modes also in the lower atmosphere, especially during high magnetic activity. The sensitivity of penetration depth to atmospheric magnetic fields, as obtained from the present model, also shows that atmospheric effects when using inversion techniques in helioseismic data analysis should not be ignored.

We have to be aware of the limits of the model and of the fact that no complete parametric analysis was carried out in the present study. For example, we considered only the penetration depth of the vertical displacement, and not the perturbation of other physical quantities, such as density or pressure. Though the results for those properties are certainly different quantitatively, the order of the effects can be expected to be the same.

The present modelling of the behavior of helioseismic modes in a magnetic environment confirms that the atmospheric magnetic field has an important role in coupling the solar interior and atmosphere.

Acknowledgements The author thanks R. Erdélyi and M. Goossens for their useful comments, which substantially improved the article, and also acknowledges the financial support from STFC.

References

Ballai, I., Ruderman, M.S., Erdélyi, R.: 1998, Nonlinear theory of slow dissipative layers in anisotropic plasmas. *Phys. Plasmas* **5**, 252–260.
Campbell, W.R., Roberts, B.: 1989, The influence of a chromospheric magnetic field on the solar p- and f-modes. *Astrophys. J.* **338**, 538–556.
Chaplin, W.J., Elsworth, Y., Isaak, G.R., Miller, B.A., New, R.: 2000, Variations in the excitation and damping of low-ℓ solar p modes over the solar activity cycle. *Mon. Not. Roy. Astron. Soc.* **313**, 32–42.
Chaplin, W.J., Elsworth, Y., Isaak, G.R., Miller, B.A., New, R.: 2004, The solar cycle as seen by low-l p-mode frequencies: comparison with global and decomposed activity proxies. *Mon. Not. Roy. Astron. Soc.* **352**, 1102–1108.
Chaplin, W.J., Elsworth, Y., Miller, B.A., Verner, G.A.: 2007, Solar p-mode frequencies over three solar cycles. *Astrophys. J.* **659**, 1749–1760.
Dziembowski, W.A., Goode, P.R.: 2005, Sources of oscillation frequency increase with rising solar activity. *Astrophys. J.* **625**, 548–555.
Erdélyi, R.: 1997, Analytical solutions for cusp resonance in dissipative MHD. *Solar Phys.* **171**, 1083–1088.
Erdélyi, R.: 2006a, Magnetic coupling of waves and oscillations in the lower solar atmosphere: can the tail wag the dog? *Phil. Trans. Roy. Soc. A* **364**, 351–381.

Erdélyi, R.: 2006b, Magnetic seismology of the lower solar atmosphere. In: Fletcher, K., Thompson, M. (eds.) *Beyond the Spherical Sun* **SP-624**, ESA, Noordwijk 15.1–13.

Erdélyi, R., Taroyan, Y.: 2001, Effect of a steady flow and an atmospheric magnetic field on the solar p- and f-modes. In: Brekke, P., Fleck, B., Gurman, J.B. (eds.) *Recent Insights into the Physics of the Sun and Heliosphere: Highlights from SOHO and Other Space Missions* **IAUS-203**, ASP, San Francisco 208–210.

Erdélyi, R., Goossens, M., Ruderman, M.S.: 1995, Analytic solutions for resonant Alfvén waves in 1D magnetic flux tubes in dissipative stationary MHD. *Solar Phys.* **161**, 123–138.

Erdélyi, R., Kerekes, A., Mole, N.: 2005, Influence of random magnetic field on solar global oscillations: The incompressible f-mode. *Astron. Astrophys.* **431**, 1083–1088.

Evans, D.J., Roberts, B.: 1990, The influence of a chromospheric magnetic field on the solar p- and f-modes. II. Uniform chromospheric field. *Astrophys. J.* **356**, 704–719.

Evans, D.J., Roberts, B.: 1991, The sensitivity of chromospherically induced p- and f-mode frequency shifts to the height of the magnetic canopy. *Astrophys. J.* **371**, 387–395.

Jain, R., Roberts, B.: 1994a, Effects of nonparallel propagation on p- and f-modes. *Astron. Astrophys.* **286**, 243–253.

Jain, R., Roberts, B.: 1994b, Surface effects of a magnetic field on p-modes: two layer atmosphere. *Astron. Astrophys.* **286**, 254–262.

Kerekes, A., Erdélyi, R., Mole, N.: 2008a, Effects of random flows on the Solar f-mode: II. Horizontal and vertical flow. *Solar Phys.* submitted.

Kerekes, A., Erdélyi, R., Mole, N.: 2008b, A novel approach to the solar interior-atmosphere EVP. *Astrophys. J.* submitted.

Komm, R.W., Howe, R., Hill, F.: 2000, Solar-cycle changes in gong p-mode widths and amplitudes 1995–1998. *Astrophys. J.* **531**, 1094–1108.

Lamb, H.: 1932, *Hydrodynamics*. Cambridge University Press, Cambridge.

Miller, B.A., Hale, S.J., Elsworth, Y., Chaplin, W.J., Isaak, G.R., New, R.: 2004, Twenty-eight years of BiSON data. In: Danesy, D. (ed.) *Helio- and Asteroseismology: Towards a Golden Future* **SP-559**, ESA, Noordwijk, 571–573.

Pintér, B.: 2008, Modelling solar atmospheric gravity oscillation modes. *Astron. Nachr.* in press.

Pintér, B., Goossens, M.: 1999, Oscillations in a magnetic solar model. I. Parallel propagation in a chromospheric and coronal magnetic field with constant Alfven speed. *Astron. Astrophys.* **347**, 321–334.

Pintér, B., Erdélyi, R., New, R.: 2001, Damping of helioseismic modes in steady state. *Astron. Astrophys.* **372**, 17–20.

Pintér, B., New, R., Erdélyi, R.: 2001, Rotational splitting of helioseismic modes influenced by a magnetic atmosphere. *Astron. Astrophys.* **378**, 1–4.

Pintér, B., Erdélyi, R., Goossens, M.: 2007, Global oscillations in a magnetic solar model. II. Oblique propagation. *Astron. Astrophys.* **466**, 377–388.

Ruderman, M.S., Hollweg, J.V., Goossens, M.: 1997, Nonlinear theory of resonant slow waves in dissipative layers. *Phys. Plasmas* **4**, 75–90.

Sakurai, T., Goossens, M., Hollweg, J.V.: 1991, Resonant behaviour of MHD waves on magnetic flux tubes. I – Connection formulae at the resonant surfaces. *Solar Phys.* **133**, 227–245.

Taroyan, Y.: 2003, *MHD waves and resonant interactions in steady states*. Ph.D. thesis, University of Sheffield.

Thompson, M.J.: 2006, Magnetohelioseismology. *Phil. Trans. Roy. Soc. A* **364**, 297–311.

Tirry, W.J., Goossens, M., Pintér, B., Čadež, V.M., Vanlommel, P.: 1998, Resonant damping of solar p-modes by the chromospheric magnetic field. *Astrophys. J.* **503**, 422–428.

Vanlommel, P., Čadež, V.I.: 1998, Influence of temperature profile of solar acoustic modes. *Solar Phys.* **182**, 263–281.

Zhukov, V.I.: 1997, Resonant absorption and the spectrum of 5-min oscillations of the Sun. *Astron. Astrophys.* **332**, 302–306.

Physical Properties of Wave Motion in Inclined Magnetic Fields within Sunspot Penumbrae

H. Schunker · D.C. Braun · C. Lindsey · P.S. Cally

Originally published in the journal Solar Physics, Volume 251, Nos 1–2, 341–359.
DOI: 10.1007/s11207-008-9142-7 © The Author(s) 2008

Abstract At the surface of the Sun, acoustic waves appear to be affected by the presence of strong magnetic fields in active regions. We explore the possibility that the inclined magnetic field in sunspot penumbrae may convert primarily vertically-propagating acoustic waves into elliptical motion. We use helioseismic holography to measure the modulus and phase of the correlation between incoming acoustic waves and the local surface motion within two sunspots. These correlations are modeled by assuming the surface motion to be elliptical, and we explore the properties of the elliptical motion on the magnetic-field inclination. We also demonstrate that the phase shift of the outward-propagating waves is opposite to the phase shift of the inward-propagating waves in stronger, more vertical fields, but similar to the inward phase shifts in weaker, more-inclined fields.

Keywords Helioseismology · Observations · Sunspots · Penumbrae

1. Introduction

Helioseismology uses the observed solar-surface acoustic wave field to construct images of the subsurface structure of the Sun. Of particular interest has been the three-dimensional (3D) modeling of time–distance (Duvall *et al.*, 1993) observations of travel-time shifts to

---

Helioseismology, Asteroseismology, and MHD Connections
Guest Editors: Laurent Gizon and Paul Cally.

H. Schunker (✉)
Max Planck Institute for Solar System Research, Max-Planck Strasse 2, Katlenburg-Lindau 37197, Germany
e-mail: schunker@mps.mpg.de

H. Schunker · P.S. Cally
Monash University, Melbourne, VIC 3800, Australia

D.C. Braun · C. Lindsey
Colorado Research Associates Division, NorthWest Research Associates, Inc., 3380 Mitchell Lane, Boulder, CO 80301, USA

deduce the subsurface structure of active regions (Kosovichev, Duvall, and Scherrer, 2000; Zhao and Kosovichev, 2003). By assuming that the travel-time shifts are due to perturbations in the sound speed below the spot, a general consensus in the models has emerged consistent with sound-speed reductions (relative to the surrounding quiet Sun) near the surface ($\lesssim 4$ Mm) and enhancements up to 15 Mm below sunspots (Kosovichev, Duvall, and Scherrer, 2000; Couvidat, Birch, and Kosovichev, 2006). Using ring-diagram analysis, Basu, Antia, and Bogart (2004) also find a lower sound speed immediately below the surface and an increase in the sound speed below 7 Mm.

A comparison between Fourier–Hankel analysis and time–distance results by Braun (1997) first prompted caution in the interpretation of acoustic-oscillation signals within sunspots. The influences of strong surface magnetic fields have not been explicitly included in most helioseismic models of active regions. Lindsey and Braun (2005a, 2005b) have shown that helioseismic phase shifts observed with helioseismic holography vanish below a depth of about 5 Mm, when a surface phase shift ("showerglass") based on photospheric magnetic flux density, is removed from the data.

Other evidence supports the possibility of strong near-surface contributions to the helioseismic phase (or travel-time) shifts. These include the possible contamination of 3D inversions by surface perturbations (*e.g.*, Korzennik, 2006; Couvidat and Rajaguru, 2007). It has also been shown that the reduction of *p*-mode amplitudes in magnetic regions can cause travel-time shifts (Rajaguru *et al.*, 2006). The suppression of sources of wave excitation within sunspots can also produce measurable shifts (Hanasoge *et al.*, 2007).

Schunker *et al.* (2005) and Schunker, Braun, and Cally (2007) have found that phase shifts obtained from seismic holography in sunspot penumbrae vary with the line-of-sight angle (from vertical) as projected onto the plane containing the magnetic field and the vertical direction. A similar effect has also been noted by Zhao and Kosovichev (2006) with time–distance measurements. Schunker, Braun, and Cally (2007) find that the effect is dependent upon the strength and/or inclination of the magnetic field. In the penumbra, the magnetic field strength decreases as the magnetic field angle from vertical increases; hence the two properties of the magnetic field cannot be extricated. The phase variation with line-of-sight viewing angle is demonstrated most substantially at frequencies around 5 mHz, with a strong, almost vertical magnetic field close to the umbra. Schunker, Braun, and Cally (2007) also find that the total variation across all lines of sight increases with temporal frequency, particularly in the stronger fields in the penumbrae.

Mode conversion of the acoustic waves in the near surface has been explored, in two dimensions, as the physical cause of the observed absorption of acoustic waves by sunspots. A fast acoustic wave, propagating toward the surface from the interior, encounters the depth at which the Alfvén speed is equal to the sound speed ($a \approx c$), which is typically close to the surface in a sunspot. Under these conditions, it is able to transmit to a slow acoustic mode and convert to a fast magnetic mode (Cally, 2005). Crouch and Cally (2003) explore the mode conversion in two dimensions with a uniform *inclined* magnetic field relevant to sunspot penumbrae. The inclination of the magnetic field is found to have a significant effect on the likelihood of conversion and fits extremely well with the analysis of Braun (1995), as shown by Cally, Crouch, and Braun (2003). Further work (Cally, 2005; Schunker and Cally, 2006) using ray theory has since established that it is the angle between the acoustic wave path and the magnetic field (the "attack angle") that is the crucial factor inducing conversion. With large attack angle at the $a \approx c$ level there is maximum conversion from a fast-acoustic mode to a fast-magnetic mode. A small attack angle encourages transmission to a slow-acoustic mode, which is guided "up" the magnetic field lines to observational heights in the atmosphere. A consequence of mode conversion may be the observational signature of elliptical motion in regions of inclined magnetic field.

The aim of this paper is to model the observations of two sunspots previously analyzed by Schunker, Braun, and Cally (2007) to determine the properties of velocity ellipses consistent with the data. These models are based on a least-squares fit of the phase and modulus information of the local ingression control correlation. We explore the properties of these ellipses, observed with waves of different temporal frequencies, as functions of the magnetic field inclination angle. We also examine the variation with line-of-sight angle of the phase shifts in the outgoing waves, using the local egression control correlation, to assess the relation of the phase-shift variations between incoming and outgoing waves.

In the following sections we describe the data (Section 2), give an outline of the helioseismic-holography technique used (Section 3), describe the results (Section 4), and discuss our results in the context of mode conversion (Section 5).

2. Observations

As this is a continuation of studies by Schunker *et al.* (2005) and Schunker, Braun, and Cally (2007) we use the same data; however, we provide a short description here to maintain coherence. The Michelson Doppler Imager (MDI) onboard the *Solar and Heliospheric Observatory* (SOHO) (Scherrer *et al.*, 1995) provides the solar-surface Doppler-velocity information. The Dopplergrams are full disk, have a 60-second cadence, and have a resolution of ≈ 1.4 Mm per pixel. These full-disk Dopplergrams are Postel projected and a 512×512 pixel area is extracted centered on the active region. We analyze two sunspots. The first, in AR9026, was observed over ten days from 3 to 12 June 2000 with a Carrington longitude (L_0) of $75°$ and latitude (B_0) of $20°$ and penumbral boundaries defined by inner and outer radii of 7 Mm and 16 Mm, respectively. The second is a sunspot in AR9057 observed over nine days from 24 June to 2 July 2000 with $L_0 = 158°$ and $B_0 = 13°$ and inner and outer penumbral boundaries given by 6 Mm and 13 Mm, respectively. The penumbral boundaries are defined to be between 50% and 85% of the nearby quiet-Sun continuum intensity. The sunspots in these active regions were chosen based on the existence of continuous MDI observation as they traversed the solar disk, and for their relatively simple magnetic structure and evolution.

Magnetograms from the Imaging Vector Magnetograph (IVM) at the University of Hawaii Mees Solar Observatory (Mickey *et al.*, 1996) provide the orientation and strength of the surface magnetic field in the sunspots chosen in AR9026 and AR9057. The IVM observations are made over a 28-minute interval: for AR9026 starting at 18:29 Universal Time (UT) on 5 June 2000 and for AR9057 starting at 16:19 UT on 28 June 2000. The IVM data reveal an azimuthally spreading magnetic-field configuration for both sunspots although they are not entirely symmetric. It is assumed, supported largely by available line-of-sight magnetograms, that there is no significant evolution of the magnetic field in the sunspots during the time of observation. Therefore, using only one vector magnetogram for the duration of the observation is reasonable. Rotation and scaling are applied to align the IVM data to the line-of-sight MDI magnetograms.

Figure 1 shows a strong correlation between the magnetic field inclination (from vertical, γ) and field strength (with the strong field being almost vertical whereas the highly inclined field is relatively weak). In this paper we use the inclination γ as the primary variable, dividing the penumbrae into three regions defined by the values of γ (Figure 1). It is understood in this analysis that the magnetic field strength is implicitly correlated with the inclination through Figure 1 and that independent dependencies of observables with field strength and inclination are not extracted. The dependence of the phase shifts on the line-of-sight viewing angle of the magnetic field is facilitated by knowing the full vector magnetic field.

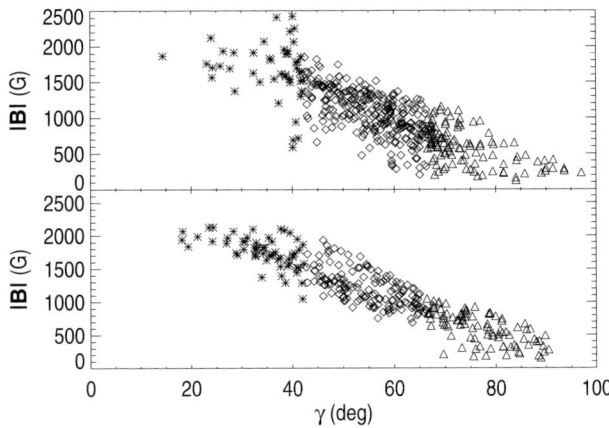

Figure 1 Magnetic field strength ($|B|$) plotted against inclination from vertical (γ) determined from IVM vector magnetograms for AR9026 (5 June 2000) (bottom) and AR9057 (28 June 2000) (top). The different symbols divide the penumbra into roughly equal regions of inclination: $\gamma < 42°$ (asterisks); $42° < \gamma < 66°$ (diamonds); $\gamma > 66°$ (triangles). This corresponds to average magnetic field strengths of 1700, 1000, and 600 G for AR9057 and 1900, 1400, and 600 G for AR9026. In general, the progression from the upper left portion of the distribution to the lower right portion represents increasing distance from the center of the spot.

3. Helioseismic Holography

Helioseismic holography (Lindsey and Braun, 2000; Braun and Lindsey, 2000) is the phase-coherent imaging of the solar subsurface based on photospheric acoustic oscillations. The ingression is an assessment of the observed wave field [$\psi(\mathbf{r}',t)$] converging to a selected focal point ($\mathbf{r}, z, t$) and the egression is the time reverse (an assessment of waves diverging from that point). In this case, we calculate the quantities at the surface, $z = 0$. In practice, the observed wave field used for the calculation is usually an annulus surrounding the chosen focal point. The pupil used here is identical to that described by Schunker *et al.* (2005) and is constructed for the calculations with inner radius $a = 20.7$ Mm and outer radius $b = 43.5$ Mm, designed to be large enough that when the focal point is within the penumbra the area covered by the annulus does not include large areas of strong magnetic field. At a frequency of 5 mHz this selects p modes with spherical harmonic degree and radial degree between $\ell \approx 450$ and $\ell \approx 700$. The ingression at the surface is given by

$$H_-(\mathbf{r},0,t) = \int_{a<|\mathbf{r}-\mathbf{r}'|<b} d^2\mathbf{r}' G_-\big(|\mathbf{r}-\mathbf{r}'|,0,t-t'\big)\psi(\mathbf{r}',t), \qquad (1)$$

where G_- is the ingression Green's function (Lindsey and Braun, 2000). The egression (H_+) is simply the time reverse of Equation (1) (*i.e.*, $t' - t$).

The ingression (1) is correlated with the observed wave field in the space–frequency domain,

$$C_-(\mathbf{r},\nu) = \big\langle \hat{H}_-(\mathbf{r},\nu)\hat{\psi}^*(\mathbf{r},\nu)\big\rangle_{\Delta\nu} = |C_-|e^{-i\delta\phi_-}, \qquad (2)$$

where $\hat{H}_-(\mathbf{r},\nu)$ and $\hat{\psi}(\mathbf{r},\nu)$ are the temporal Fourier transforms of H_- and ψ, respectively, and we have dropped the dependence on depth (z). The correlation has a modulus $|C_-|$ and a phase $\delta\phi_-$. In the frequency spectrum, with the pupil covering mostly quiet Sun, the local ingression control correlation simply characterizes how the local magnetic photosphere responds acoustically to upcoming waves, prescribed by H_-, originating within the pupil.

In Schunker *et al.* (2005) and Schunker, Braun, and Cally (2007) the phase shift, caused by the surface perturbations to the incident wave, is calculated at various frequencies. Here we extend that research and monitor the variation of the correlation amplitude within the sunspot penumbral region, which is then combined with the phase information to form estimates of the surface velocity ellipse. We then go on to examine the phase of the local egression control correlation [$\delta\phi_+ = \text{Arg}[C_+(\mathbf{r}, \nu)]$] with the line-of-sight angle in the penumbra of sunspots in AR9057 and AR9026.

4. Elliptical Representation of Surface Doppler Signals

For better statistics, multiple days of observation of each sunspot are combined and a least-squares fit of the observations allows an estimation of the velocity ellipse. To calculate the velocity ellipse, the ingression correlation (Equation (2)) is used as a proxy for the local velocity. The modulus and phase of the correlation vary with respect to the line-of-sight angle and these variations are modeled to construct an ellipse representing the surface velocity associated with a particular magnetic field element.

A smeared ingression "flat field" takes out undesired contributions to the correlation modulus owing to the temporal or spatial variations of the ingression and the presence of magnetic regions in the pupil. This is achieved by dividing $|C_-|$ by the Gaussian smear of the root-mean square of the ingression to get $|C_-|_\text{flat}$. The resulting correlations are normalized so that the ingression correlation modulus in the nearby quiet Sun has a value of $\cos(\zeta)$, where ζ is the heliocentric angle of the quiet region from disk center. A dependence of the quiet Sun (root-mean-squared) amplitude with $\cos(\zeta)$ is expected for predominantly vertically oscillating p modes. The normalization is achieved by dividing $|C_-|_\text{flat}$ by the quiet-Sun average divided by the cosine of the heliocentric angle, that is,

$$|C_-|_\text{norm} = |C_-|_\text{flat} \frac{\cos(\zeta)}{\langle|C_-|\rangle_\text{QS}}. \tag{3}$$

The normalization corrects for variations in the modulus from duty-cycle variations, foreshortening, and other effects that cause undesired variations in the modulus from day to day. These procedures assume that these factors have the same relative effect on the (desired) correlations in the penumbra as they do to the quiet Sun. It is difficult to assess the validity of this, and so it is used as a reasonable working assumption subject to some caution.

We use the same angle θ_p as defined in Schunker *et al.* (2005). The line-of-sight vector is projected onto the plane containing the local magnetic field vector and the radial vector. We define θ_p as the angle between this projected line-of-sight vector and the radial vector. We now simply refer to $|C_-|_\text{norm}$ as $|C_-|$ and see how it varies with θ_p.

Figures 2, 3, 4 and 6, 7, 8 show the variation of $|C_-|$ with θ_p for AR9026 and AR9057 in the left columns at 3, 4, and 5 mHz in the same three bins of inclination shown in Figure 1. The right column shows $\delta\phi_-$. The variations of $\delta\phi_-$ with θ_p have been the subject of our previous analyses (Schunker *et al.*, 2005; Schunker, Braun, and Cally, 2007). In the absence of magnetic effects, we expect the local oscillatory wave field, as assessed by the ingression correlation, to be consistent with purely vertical motion. This would predict a dependence of $|C_-|$ on $\cos(\theta_\text{p})$ and no dependence of $\delta\phi_-$ on $\cos(\theta_\text{p})$. Departures from these expectation are clearly visible in Figures 2–4 and 6–8. A rough understanding of these results can be gained by noting that the net variation in the phase shift with θ_p is related to the eccentricity of the ellipse, whereas the value of θ_p for maximum $|C_-|$ determines the orientation of the

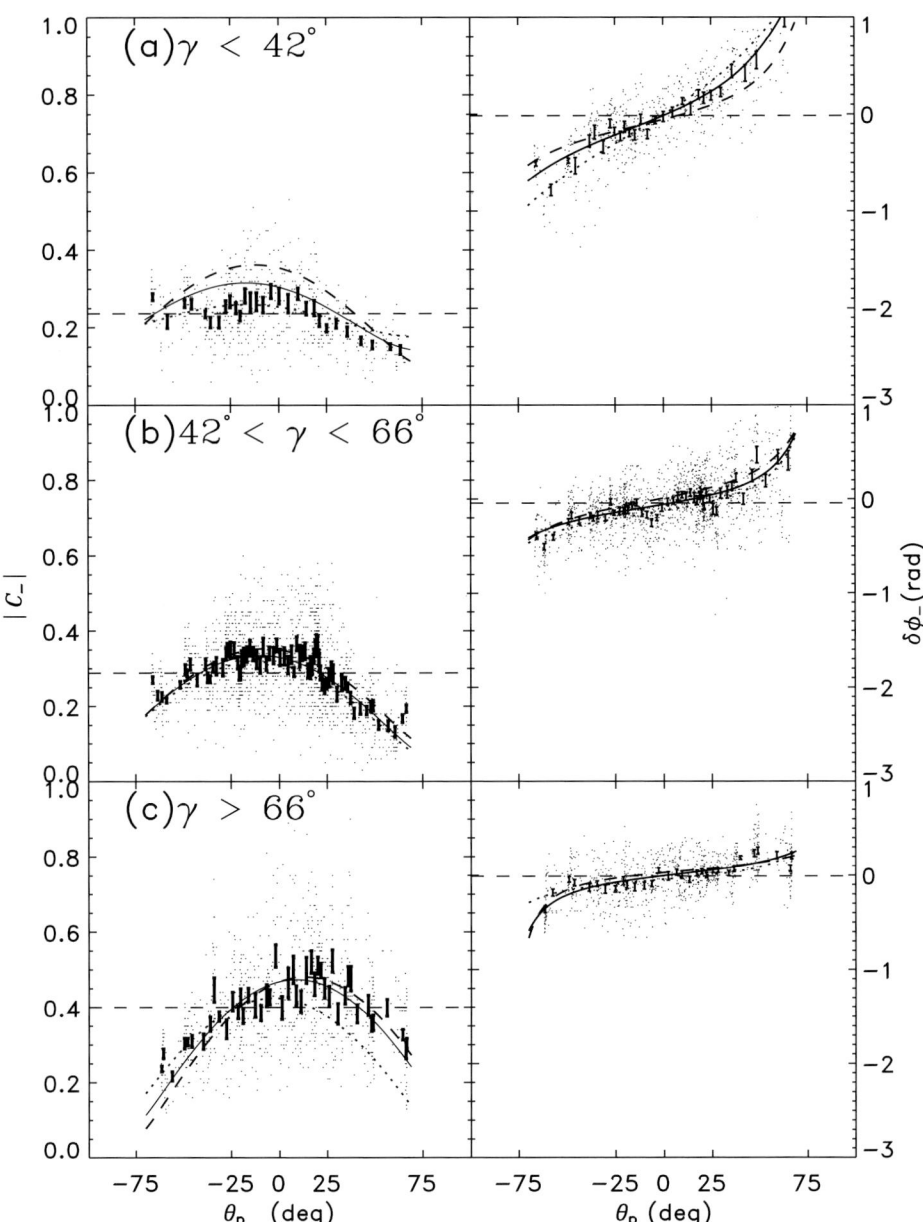

Figure 2 The modulus of the correlation, $|C_-|$ (left column), and the phase $\delta\phi_-$ (right column) in the penumbra of AR9026 at 3 mHz for all days of observation plotted against projected angle θ_p for different values of magnetic field inclinations as indicated: (a) $\gamma < 42°$, where the mean field strength is $\langle\mathbf{B}\rangle = 1900$ G; (b) $42° < \gamma < 66°$, where $\langle\mathbf{B}\rangle = 1400$ G; (c) $\gamma > 66°$, where $\langle\mathbf{B}\rangle = 600$ G. The horizontal dashed lines indicate the mean value of $|C_-|$ for each panel. The error bars indicate the standard deviation of the mean over bins of 20 measurements in θ_p. The solid line is a fit for all the displayed data; the dotted line is a fit for the data from 3 to 7 June 2000; the dashed line is a fit for data from 8 to 12 June 2000.

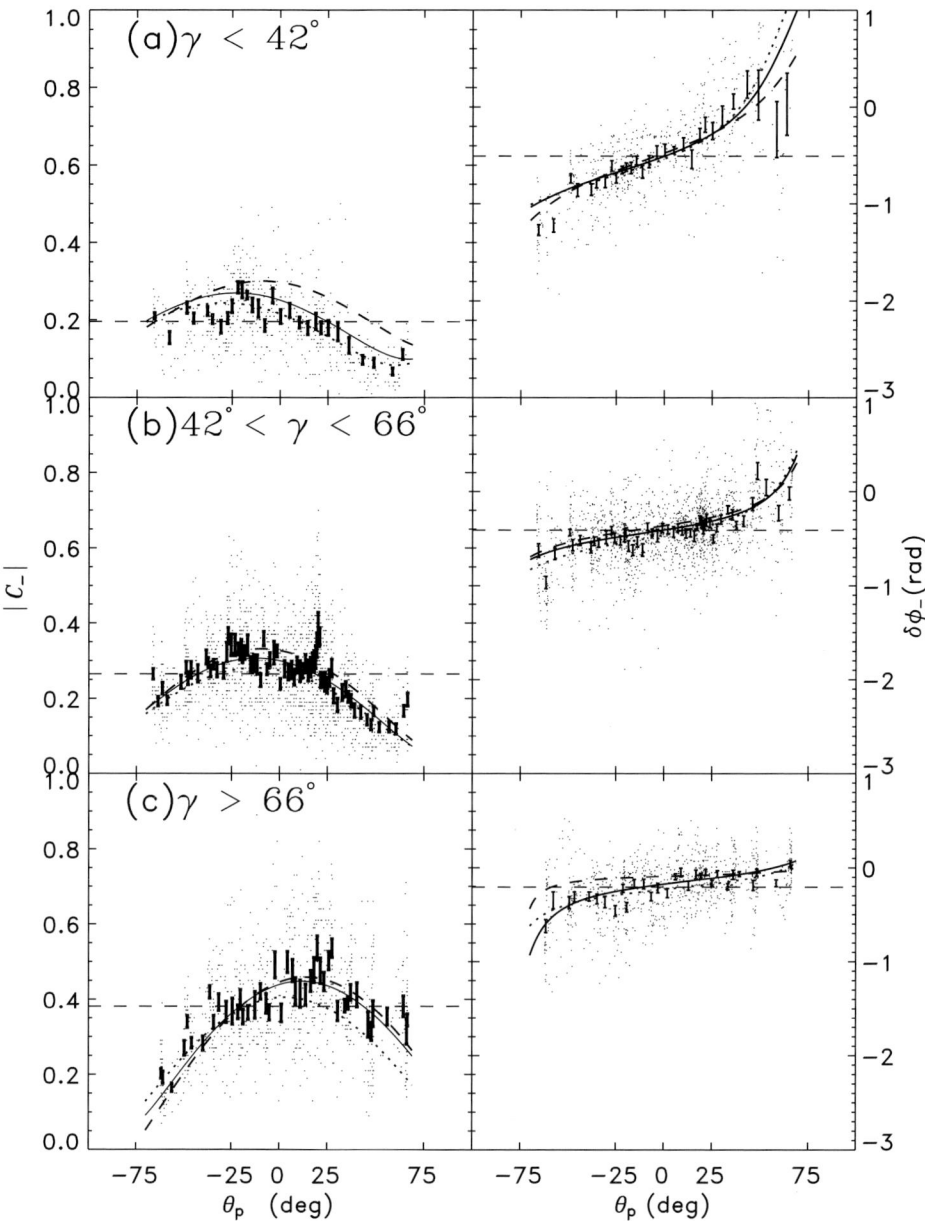

Figure 3 The modulus of the correlation, $|C_-|$ (left column), and the phase $\delta\phi_-$ (right column) in the penumbra of AR9026 at 4 mHz for all days of observation plotted against projected angle θ_p for different values of magnetic field inclinations as indicated: (a) $\gamma < 42°$, where the mean field strength is $\langle B \rangle = 1900$ G; (b) $42° < \gamma < 66°$, where $\langle B \rangle = 1400$ G; (c) $\gamma > 66°$, where $\langle B \rangle = 600$ G. The horizontal dashed lines indicate the mean value of $|C_-|$ for each panel. The error bars indicate the standard deviation of the mean over bins of 20 measurements in θ_p. The solid line is a fit for all the displayed data; the dotted line is a fit for the data from 3 to 7 June 2000; the dashed line is a fit for data from 8 to 12 June 2000.

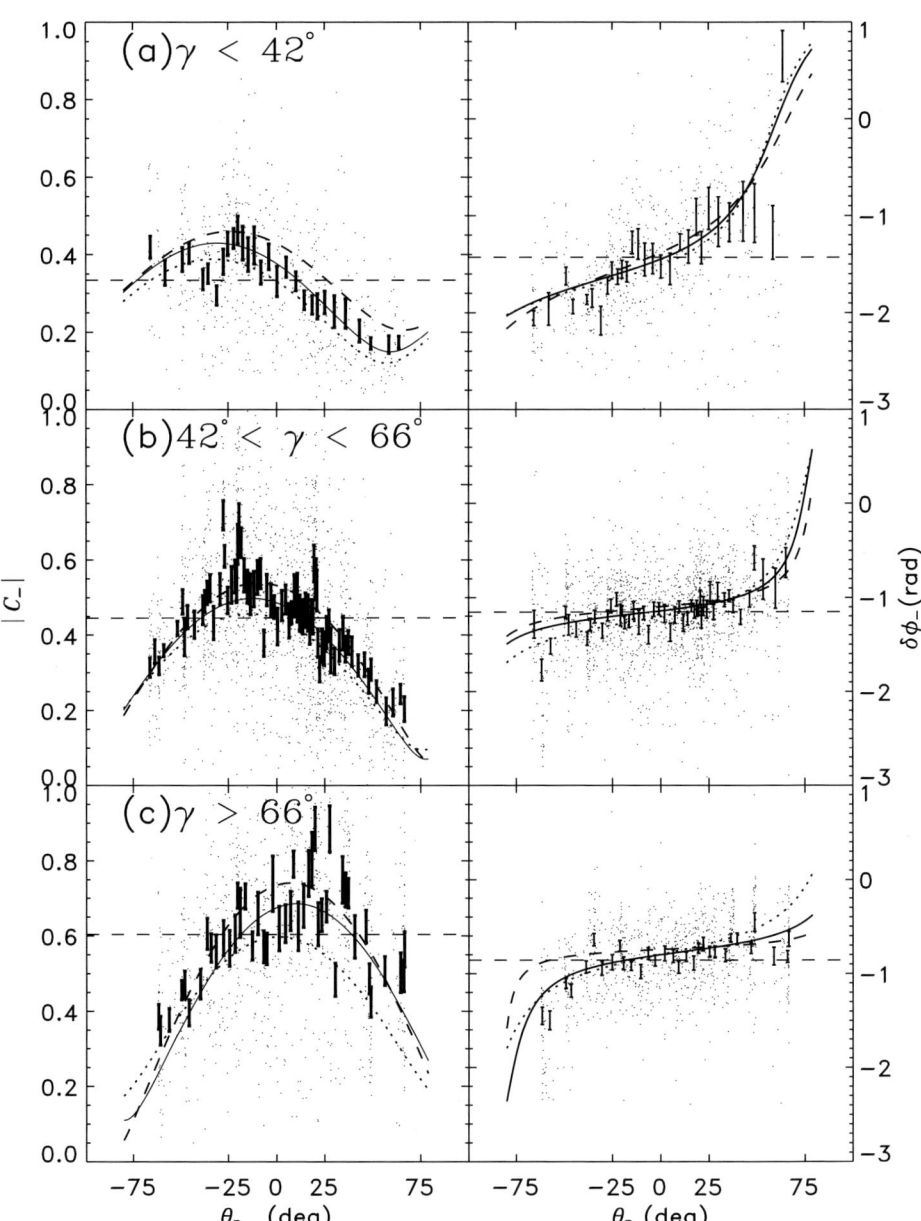

Figure 4 The modulus of the correlation, $|C_-|$ (left column), and the phase $\delta\phi_-$ (right column) in the penumbra of AR9026 at 5 mHz for all days of observation plotted against projected angle θ_p for different values of magnetic field inclinations as indicated: (a) $\gamma < 42°$, where the mean field strength is $\langle \mathbf{B} \rangle = 1900$ G; (b) $42° < \gamma < 66°$, where $\langle \mathbf{B} \rangle = 1400$ G; (c) $\gamma > 66°$, where $\langle \mathbf{B} \rangle = 600$ G. The horizontal dashed lines indicate the mean value of $|C_-|$ for each panel. The error bars indicate the standard deviation of the mean over bins of 20 measurements in θ_p. The solid line is a fit for all the displayed data; the dotted line is a fit for the data from 3 to 7 June 2000; the dashed line is a fit for data from 8 to 12 June 2000.

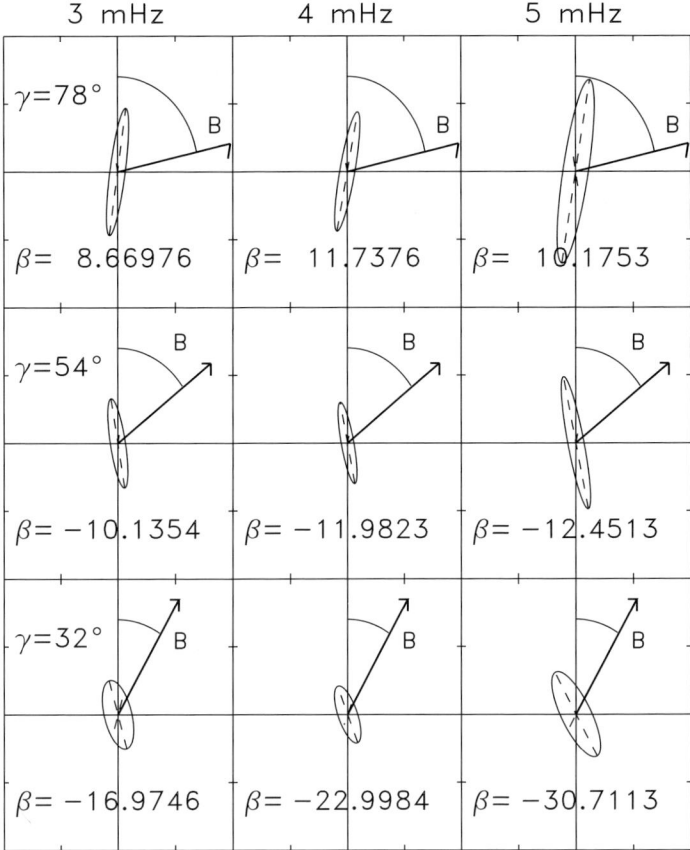

Figure 5 Least-squares-fit surface velocity ellipses as given by the phase and amplitude of the local ingression control correlation in the penumbra of AR9026 for $\gamma > 66°$ (top row), $42° < \gamma < 66°$ (middle row), and $\gamma < 42°$ (bottom row). The γ listed in the plot is the angle at which the magnetic field vector is drawn. β is the inclination angle of the semi-major axis of the velocity ellipse. The left column is for frequencies of 3 mHz, the middle column for 4 mHz, and the right column for 5 mHz.

semi-major axis. We perform a least-squares fit to the observed $|C_-|$ and $\delta\phi_-$ to determine the elliptical motion consistent with the data. The best-fit ellipses are shown in Figures 5 and 9. The moduli and phase of C_- as determined from the fits are plotted (as solid lines) with the data points in Figures 2–4 and 6–8. Separate fits were performed for independent five-day subsets of the data (also shown in the figures by the dotted and dashed lines). There are no obvious, systematic differences in the fits over time. The orientations and eccentricities are highly consistent across all frequency bands in both sunspots (Figures 5 and 9). Some systematic differences between the two spots are evident. For AR9026, the ellipses are aligned slightly toward the magnetic field direction at high field inclination but "swing" over as the inclination becomes smaller (and the magnetic field stronger). For AR9057, the motion is nearly vertical at low field inclination, but it also tilts away from the field in the stronger, more vertical fields. In both spots, the eccentricity of the ellipses increases with decreasing field strength or increasing inclination.

We can define a deviation angle as the angle between the magnetic field vector and surface velocity ellipse semi-major axis ($\delta = \gamma - \beta$, where β is the inclination of the semi-

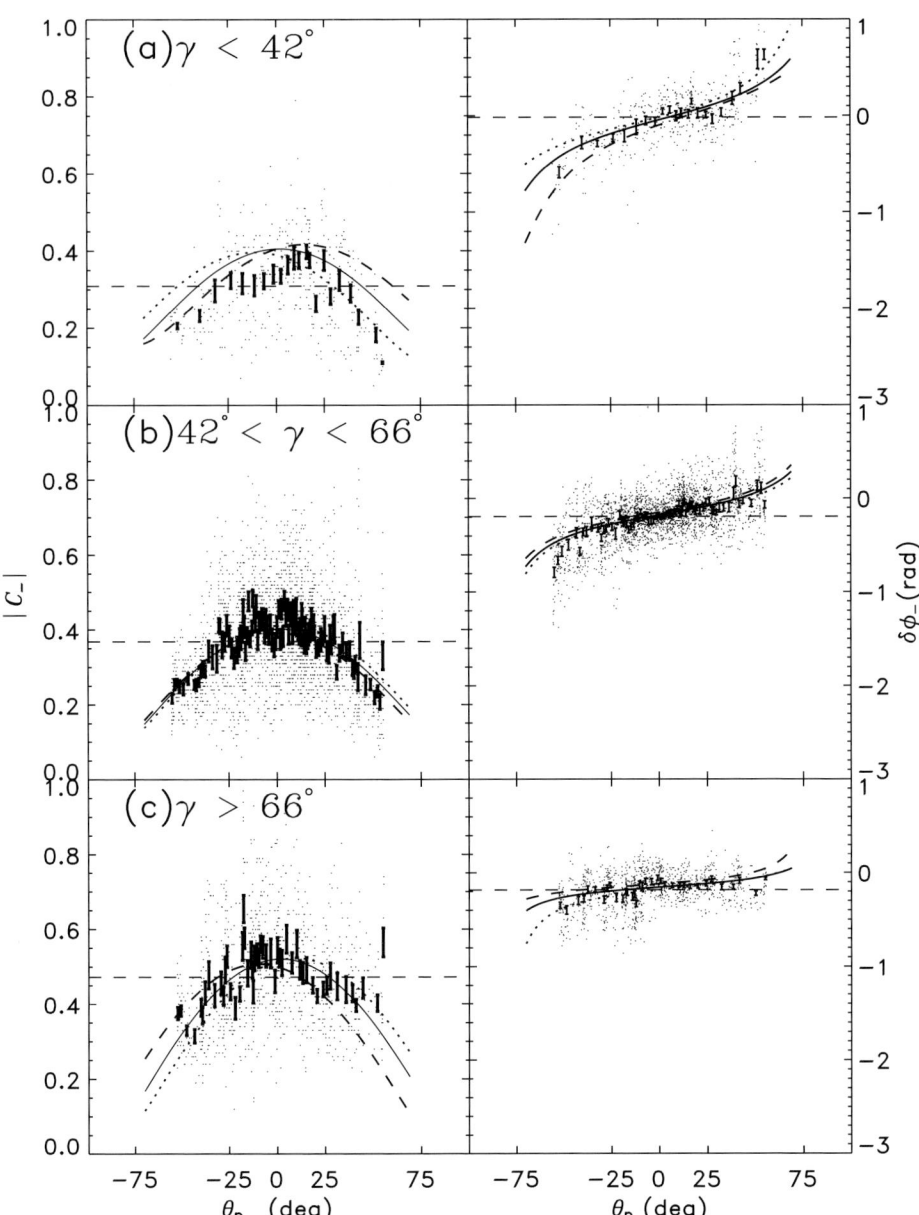

Figure 6 The modulus of the correlation, $|C_-|$ (left column), and phase $\delta\phi_-$ (right column) in the penumbra of AR9057 at 3 mHz for all days of observation plotted against projected angle θ_p for different values of magnetic field inclinations as indicated: (a) $\gamma < 42°$, where the mean field strength is $\langle \mathbf{B} \rangle = 1700$ G; (b) $42° < \gamma < 66°$, where $\langle \mathbf{B} \rangle = 1000$ G; (c) $\gamma > 66°$, where $\langle \mathbf{B} \rangle = 600$ G. The horizontal dashed lines indicate the mean value of $|C_-|$ for each panel. The error bars indicate the standard deviation of the mean over bins of 20 measurements in θ_p. The solid line is a fit for all the displayed data; the dotted line is a fit for the data from 24 to 28 June 2000; the dashed line is a fit for data from 29 June to 2 July 2000.

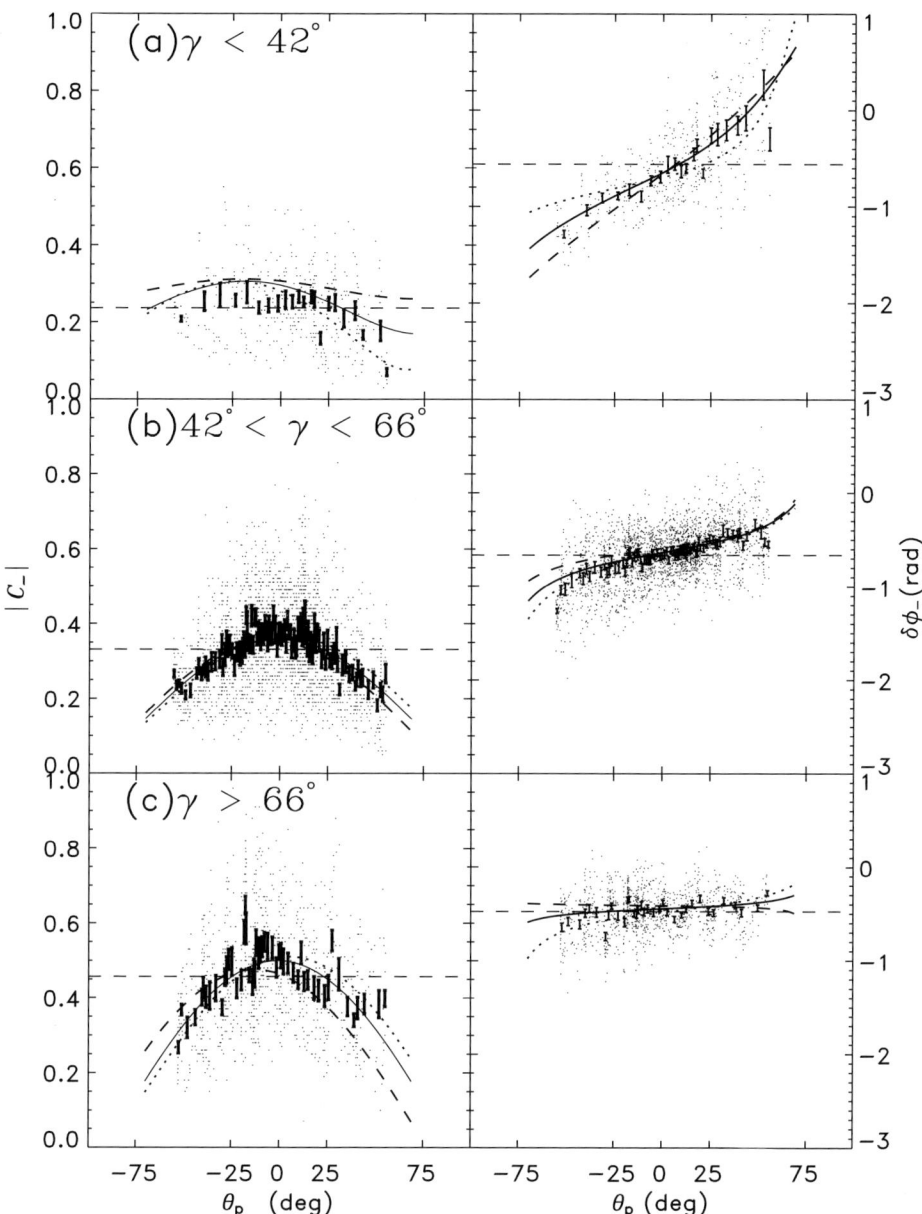

Figure 7 The modulus of the correlation, $|C_-|$ (left column), and phase $\delta\phi_-$ (right column) in the penumbra of AR9057 at 4 mHz for all days of observation plotted against projected angle θ_p for different values of magnetic field inclinations as indicated: (a) $\gamma < 42°$, where the mean field strength is $\langle\mathbf{B}\rangle = 1700$ G; (b) $42° < \gamma < 66°$, where $\langle\mathbf{B}\rangle = 1000$ G; (c) $\gamma > 66°$, where $\langle\mathbf{B}\rangle = 600$ G. The horizontal dashed lines indicate the mean value of $|C_-|$ for each panel. The error bars indicate the standard deviation of the mean over bins of 20 measurements in θ_p. The solid line is a fit for all the displayed data; the dotted line is a fit for the data from 24 to 28 June 2000; the dashed line is a fit for data from 29 June to 2 July 2000.

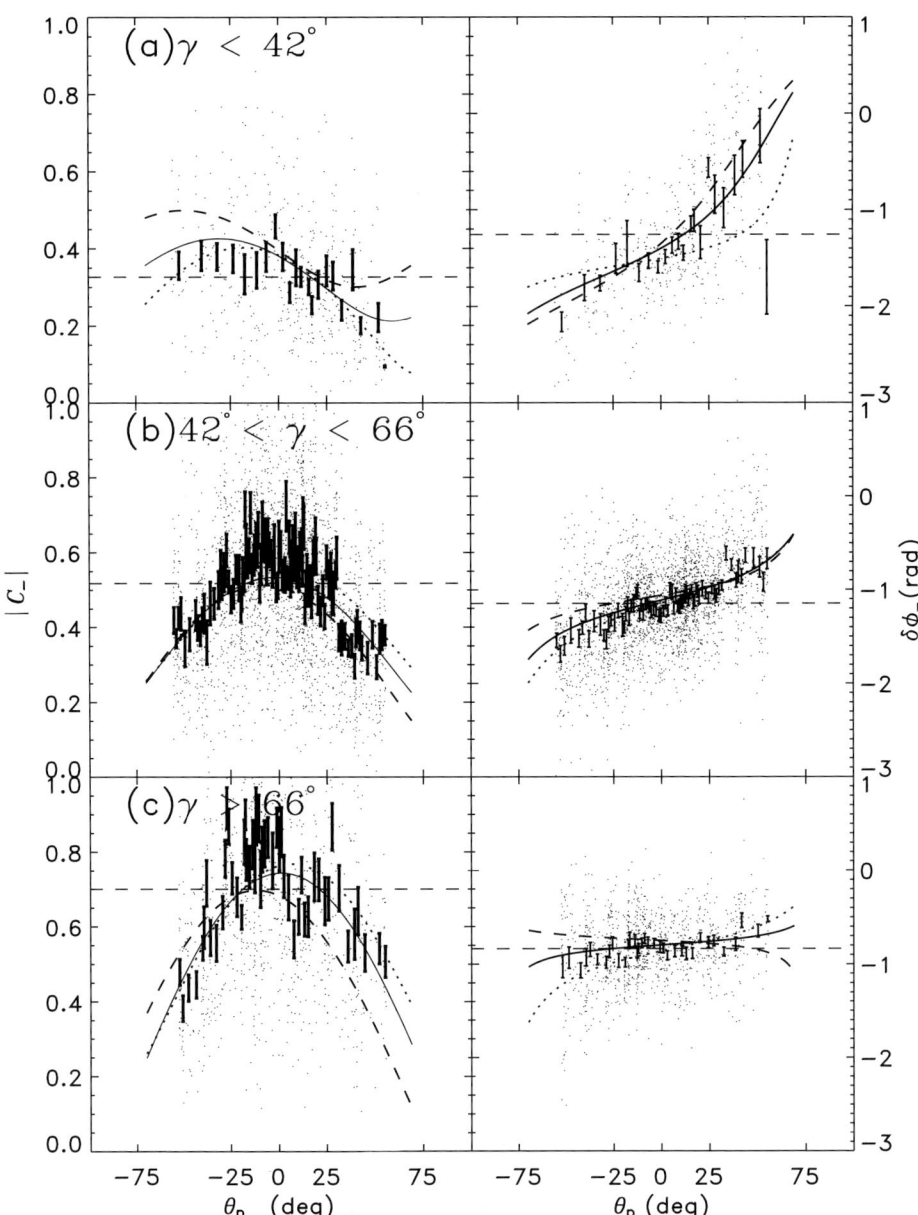

Figure 8 The modulus of the correlation, $|C_-|$ (left column), and phase $\delta\phi_-$ (right column) in the penumbra of AR9057 at 5 mHz for all days of observation plotted against projected angle θ_p for different values of magnetic field inclinations as indicated: (a) $\gamma < 42°$, where the mean field strength is $\langle B \rangle = 1700$ G; (b) $42° < \gamma < 66°$, where $\langle B \rangle = 1000$ G; (c) $\gamma > 66°$, where $\langle B \rangle = 600$ G. The horizontal dashed lines indicate the mean value of $|C_-|$ for each panel. The error bars indicate the standard deviation of the mean over bins of 20 measurements in θ_p. The solid line is a fit for all the displayed data; the dotted line is a fit for the data from 24 to 28 June 2000; the dashed line is a fit for data from 29 June to 2 July 2000.

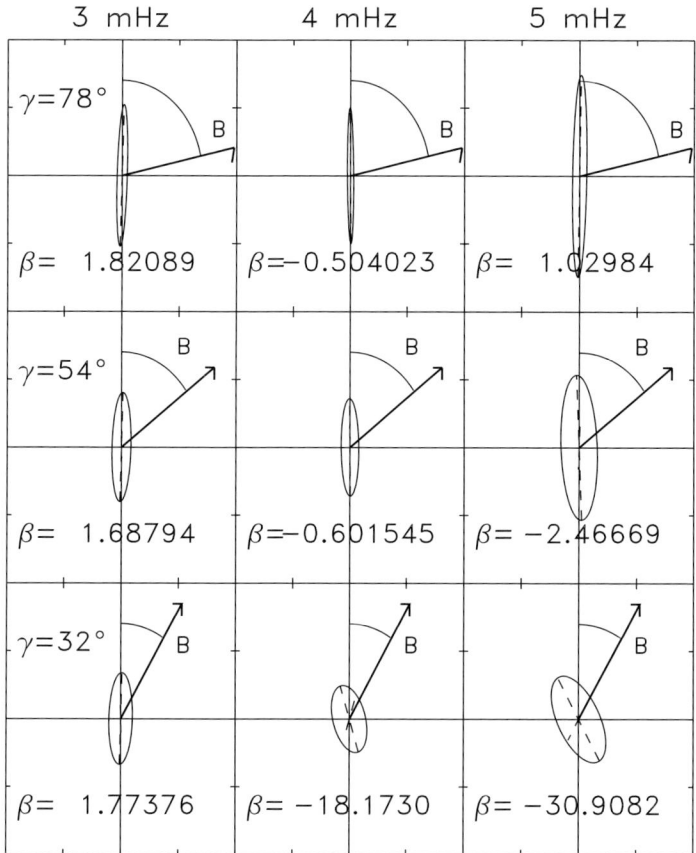

Figure 9 Least-squares-fit surface velocity ellipses as given by the phase and amplitude of the local ingression control correlation in the penumbra of AR9057 for $\gamma > 66°$ (top row), $42° < \gamma < 66°$ (middle row), and $\gamma < 42°$ (bottom row). The γ listed in the plot is the angle at which the magnetic field vector is drawn. β is the inclination angle of the semi-major axis of the velocity ellipse. The left column is for frequencies of 3 mHz, the middle column for 4 mHz, and the right column for 5 mHz.

major axis from the vertical). We show the variation of the deviation angle, length of the semi-major axis, and eccentricity with magnetic-field strength and/or inclination in Figure 10. Also shown is the phase difference between the fits at $\theta_p = -60°$ and $\theta_p = +60°$. This quantity is generally inversely correlated to the ellipse eccentricity, but it is a more direct measure of the total variation of observed phase shift for a specific penumbral region (Schunker *et al.*, 2005). In Figure 10, data from AR9026 are represented by an asterisk and AR9057 by a diamond, and the frequencies are color-coded as follows: 5 mHz is black, 4 mHz is purple, and 3 mHz is red.

In general, the trends shown in Figure 10 – namely an increase in the deviation angle, eccentricity, and semi-major axis length and a decrease in the phase variation – with increasing inclination are observed in both sunspots and at all frequencies. There are some deviations from this. For example, at low inclinations it is observed that the eccentricity (phase variation) decreases (increases) with frequency. Note that the semi-major-axis length is an indication of the total wave amplitude in the magnetic region. The trend observed in

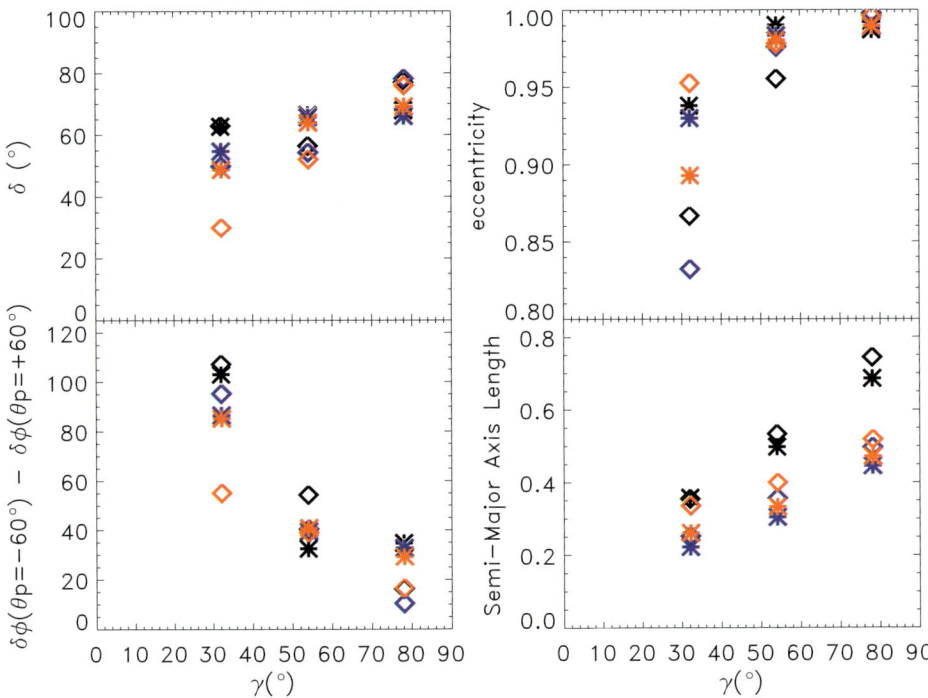

Figure 10 Properties of the ellipses: deviation angle δ (top left); the ellipse eccentricity for both sunspots at all frequencies against the three average magnetic field inclinations (top right); the phase difference between the least-squares-fit correlation phase at $\theta_p = -60°$ and at $\theta_p = +60°$ ($\delta\phi_{(\theta_p=-60°)} - \delta\phi_{(\theta_p=+60°)}$) (bottom left); the length of the semi-major axis of each ellipse in each region for both sunspots at all frequencies against the three average magnetic field inclinations (bottom right). AR9026 is represented by the asterisk and AR9057 by the diamond, and the colors indicate the following frequencies: 5 mHz (black), 4 mHz (purple), and 3 mHz (red).

the lower left panel of Figure 10 is consistent with a reduction in wave amplitude related to the field strength.

5. Local Egression Control Correlation

In this section we examine the phase of the local egression control correlation,

$$C_+(\mathbf{r}, \nu) = \left\langle \hat{H}_+(\mathbf{r}, \nu)\hat{\psi}^*(\mathbf{r}, \nu) \right\rangle_{\Delta\nu} = |C_+|e^{-i\delta\phi_+}. \qquad (4)$$

Since the egression is simply the time reverse of the ingression, we might expect to see a reversal of the phase change compared to the ingression phases.

Using all of the days of data, we plot the egression correlation phase against θ_p in Figures 11 and 12, along with the fits for the phase variation caused by elliptical motion. We see similar trends in both AR9026 and AR9787. Figure 13 shows the phase difference between the least-squares-fit correlation phase at $\theta_p = -60°$ and at $\theta_p = +60°$ of the ingression plotted against that of the egression. The colors and symbols represent the same as before: AR9026 is represented by an asterisk and AR9057 by a diamond, and the frequencies are

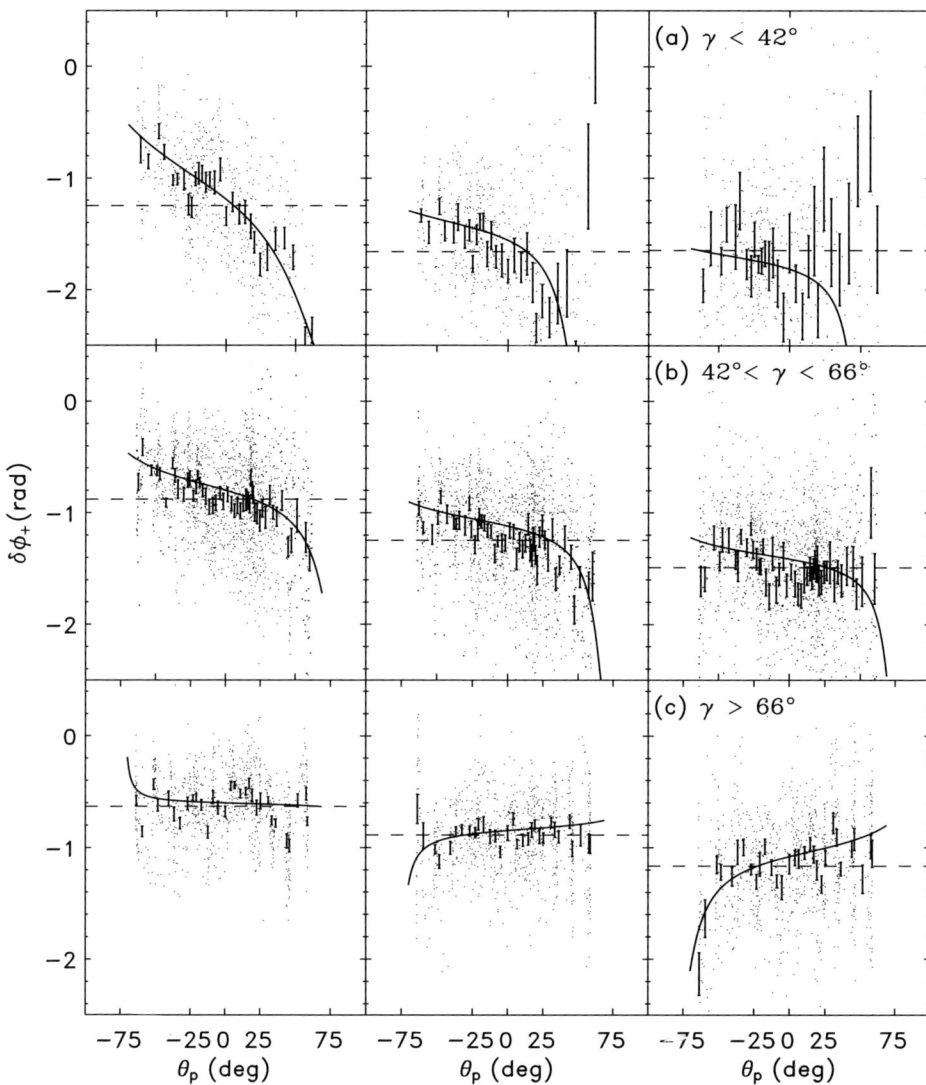

Figure 11 The 3-mHz (left column), 4-mHz (middle column), and 5-mHz (right column) egression correlation phase ($\delta\phi_+$) versus θ_p within the penumbra of sunspot AR9026 for different values of magnetic field inclination as indicated: (a) $\gamma < 42°$, where the mean field strength is $\langle \mathbf{B} \rangle = 1900$ G; (b) $42° < \gamma < 66°$, where $\langle \mathbf{B} \rangle = 1400$ G; (c) $\gamma > 66°$, where $\langle \mathbf{B} \rangle = 600$ G. The three rows represent different portions of the penumbra as shown in Figure 1 (bottom). The horizontal dashed lines indicate the mean value of $\delta\phi_+$ for each panel.

5 mHz (black), 4 mHz (purple), and 3 mHz (red); in addition the size of the symbols represent the average inclination from vertical. The solid line has a slope of -1. There is a reverse behavior of the ingression, compared to the previous egression results, present for all frequencies at most magnetic field inclinations. The exception is when the field is highly inclined, where we observe a trend in the same sense as the ingression. This is unexpected and warrants further study.

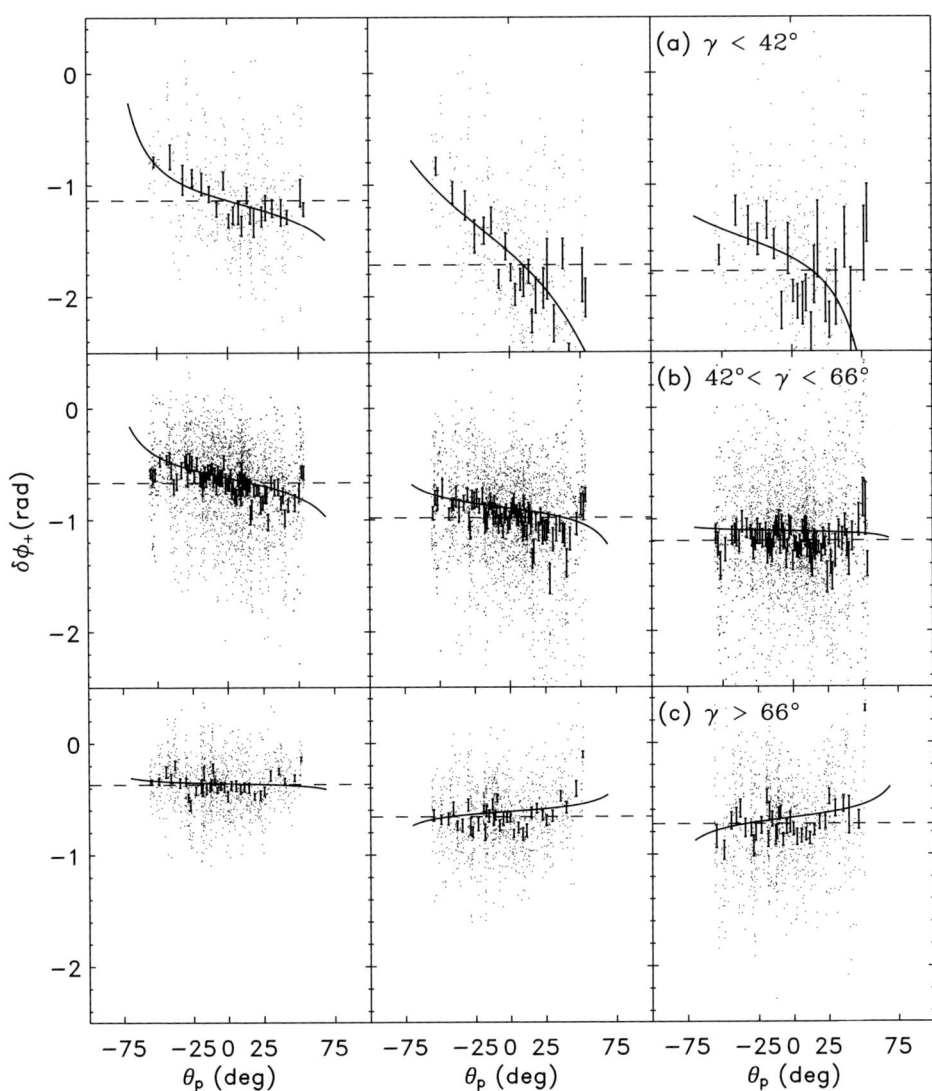

Figure 12 The 3-mHz (left column), 4-mHz (middle column), and 5 mHz (right column) egression correlation phase ($\delta\phi_+$) versus θ_p within the penumbra of sunspot AR9057 for different values of magnetic field strength as indicated: (a) $\gamma < 42°$, where the mean field strength is $\langle B \rangle = 1700$ G; (b) $42° < \gamma < 66°$, where $\langle B \rangle = 1000$ G; (c) $\gamma > 66°$, where $\langle B \rangle = 600$ G. The three different panels represent different portions of the penumbra, similar to Figure 1 (top). The horizontal dashed lines indicate the mean value of $\delta\phi_+$ for each panel.

6. Discussion

Two-dimensional mode conversion predicts (among other things; Cally, 2007) that when the attack angle is small most of the observable energy will be in the slow acoustic mode and when the attack angle is large most of the observable energy will be in the fast magnetic mode. In this case we will be seeing the line-of-sight effect of a combination of waves coming from all directions impinging on magnetic field with a particular orientation. The ob-

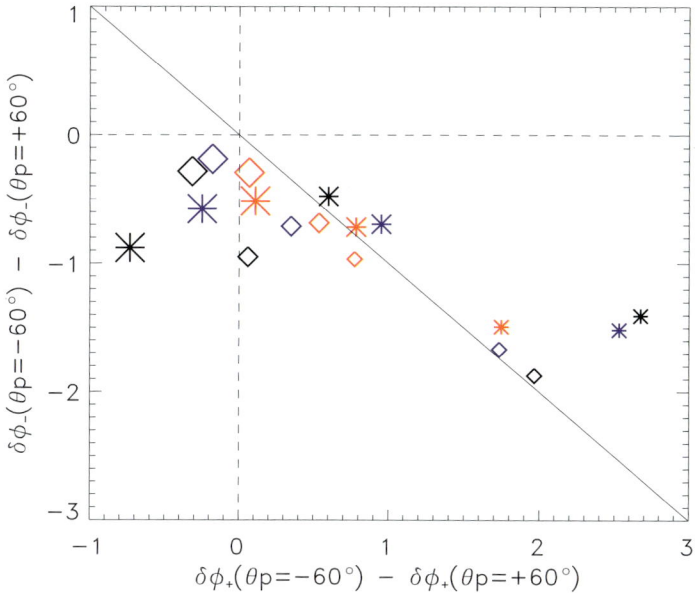

Figure 13 The phase difference between the least-squares-fit correlation phase at $\theta_p = -60°$ and at $\theta_p = +60°$ of the ingression plotted against that of the egression. The solid line is a line of slope -1. AR9026 is represented by an asterisk and AR9057 by a diamond, and the colors indicate the following frequencies: 5 mHz (black), 4 mHz (purple), and 3 mHz (red). The largest symbols represent $\gamma > 66°$ and the smallest symbols $\gamma < 42°$.

servations by MDI consist of line-of-sight Doppler signatures of the surface motion, which are presumably caused mainly by pressure perturbations. This means that we would expect to be observing only the slow acoustic mode; however, it is possible that we are observing a combination of the acoustic and magnetic modes.

The dependence of the phase shift of the observed ingression correlations with azimuthal angle around sunspot penumbrae, as viewed from different observational vantage points, shows that the incoming phase shifts must be (at least partly) photospheric in origin and are influenced by the presence of inclined magnetic fields (Schunker *et al.*, 2005). Analysis of the variation of both the amplitude and phase of the surface velocities provides an opportunity to characterize the magnetically influenced acoustic signature as an ellipse with properties determined by the magnetic fields. We found that the ellipses are either nearly vertical (for weaker, more inclined fields) or generally directed *away* from the magnetic field direction (for stronger, more vertical fields). Largely consistent results for the two active regions AR9026 and AR9057 are found. Some properties of the surface ellipses (*e.g.*, their inclinations) are different for the two sunspots. Some of this variation may be due to differences in the field properties. For example, the field in AR9057 is on average $\approx 15\%$ weaker than that in AR9026.

Fits of the elliptical motion in Figures 5 and 9 depend critically on the correlation modulus, which is prone to systematic uncertainties. But the trend is that a stronger, less inclined magnetic field produces elliptical motion with smaller amplitude, eccentricity, and deviation angle and a larger inclination from the vertical. These trends exist for both spots and, by and large, at all frequencies. The shorter semi-major axis at strong magnetic field strengths is consistent with previous knowledge of surface acoustic amplitude suppression in magnetic fields. It is curious to note, however, that at 4 mHz the amplitude is consistently smaller than at 3 mHz.

These are the first results to estimate the behavior of the surface velocity ellipse at the photosphere within sunspots. The results do not immediately suggest an observation of the slow wave as shown by Cally (2005) or Schunker and Cally (2006), but they are consistent with their expectations of the behavior of slow waves at this height in the atmosphere. Mode conversion theory states that the alignment depends on a^2/c^2, which at the observational heights of ≈ 200 km in the atmosphere may not be large enough to invoke alignment. Since the ray analysis is somewhat unrealistic, we would expect to observe a combination of fast and slow waves at the surface, which will contribute to a spurious view of the surface velocities. In three dimensions, where the wave vector is not in the vertical plane containing the magnetic field lines, Cally and Goossens (2007) have shown theoretically that there should also be substantial conversion to an upgoing Alfvén wave, which further complicates any analysis of oscillation modes in the low solar magnetic atmosphere. Simultaneous multi-height observations would be useful in resolving this issue. Our results are necessary, but not sufficient, for conclusive evidence of mode conversion occurring close to the solar surface. Nevertheless, at this stage, there is nothing in our results that is inconsistent with MHD mode conversion.

Rajaguru et al. (2007) are currently exploring the possibility that these apparent surface effects are due to changes in radiative transfer within active regions and the formation height of the observational Ni 678 nm line. This explanation requires an absorption mechanism or some other means of producing a difference between the amplitudes of upward and downward propagating waves. Thus mode conversion may still be important in this proposed mechanism. A test of the mechanism proposed by Rajaguru et al. (2007) would be to repeat the observations performed here in a magnetically insensitive line, where the proposed radiative transfer effects would not be present. In terms of mode conversion, it is suggested that the main effect occurs along the bright radial filaments of the interlocking comb structure as presented in the penumbral models of Weiss et al. (2004). However, observational helioseismic spatial resolution cannot currently resolve this.

This is also the first time that the variation of the phase of the local egression correlation has been analyzed in the penumbra. It is curious that the egression correlation shows a reverse dependence when the magnetic field is weak and highly inclined. This is evidence of a reverse ingression dependence on the line of sight, but further investigation is required to understand the behavior at high frequencies in the weaker, more inclined fields.

References

Basu, S., Antia, H.M., Bogart, R.S.: 2004, Ring-diagram analysis of the structure of solar active regions. *Astrophys. J.* **610**, 1157.
Braun, D.C.: 1995, Scattering of *p*-modes by sunspots. I. Observations. *Astrophys. J.* **451**, 859.
Braun, D.C.: 1997, Time–distance sunspot seismology with GONG data. *Astrophys. J.* **487**, 447.
Braun, D.C., Lindsey, C.: 2000, Phase-sensitive holography of Solar activity. *Solar Phys.* **192**, 307.
Cally, P.S.: 2005, Local magnetohelioseismology of active regions. *Mon. Not. Roy. Astron. Soc.* **358**, 353.
Cally, P.S.: 2007, What to look for in the seismology of solar active regions. *Astron. Nachr.* **328**, 286.
Cally, P.S., Goossens, M.: 2007, Three-dimensional MHD wave propagation and conversion to Alfvén waves near the Solar surface. I. Direct numerical solution. *Solar Phys.* doi:10.1007/s11207-007-9086-3.
Cally, P.S., Crouch, A.D., Braun, D.C.: 2003, Probing sunspot magnetic fields with *p*-mode absorption and phase shift data. *Mon. Not. Roy. Astron. Soc.* **346**, 381.
Couvidat, S., Rajaguru, S.P.: 2007, Contamination by surface effects of time–distance helioseismic inversions for sound speed beneath sunspots. *Astrophys. J.* **661**, 558.

Couvidat, S., Birch, A.C., Kosovichev, A.G.: 2006, Three-dimensional inversion of sound speed below a sunspot in the Born approximation. *Astrophys. J.* **640**, 516.
Crouch, A., Cally, P.S.: 2003, Mode conversion of solar p modes in non-vertical magnetic fields. I. Two-dimensional model. *Solar Phys.* **214**, 201.
Duvall, T.L., Jr., Jefferies, S.M., Harvey, J.W., Pomerantz, M.A.: 1993, Time–distance helioseismology. *Nature* **362**, 430.
Hanasoge, S.M., Couvidat, S., Rajaguru, S.P., Birch, A.C.: 2007, Impact of locally suppressed wave sources on helioseismic travel times. 707.
Korzennik, S.G.: 2006, The cookie cutter test for time-distance tomography of active regions. In: Fletcher, K., Thompson, M. (eds.) *Proceedings of SOHO 18/GONG 2006/HELAS I, Beyond the Spherical Sun* **624**, ESA, Noordwijk, 60.
Kosovichev, A.G., Duvall, T.L.J., Scherrer, P.H.: 2000, Time–distance inversion methods and results. *Solar Phys.* **192**, 159.
Lindsey, C., Braun, D.C.: 2000, Basic principles of Solar acoustic holography. *Solar Phys.* **192**, 261.
Lindsey, C., Braun, D.C.: 2005a, The acoustic showerglass. I. Seismic diagnostics of photospheric magnetic fields. *Astrophys. J.* **620**, 1107.
Lindsey, C., Braun, D.C.: 2005b, The acoustic showerglass. II. Imaging active region subphotospheres. *Astrophys. J.* **620**, 1118.
Mickey, D.L., Canfield, R.C., Labonte, B.J., Leka, K.D., Waterson, M.F., Weber, H.M.: 1996, The imaging vector magnetograph at Haleakala. *Solar Phys.* **168**, 229.
Rajaguru, S.P., Birch, A.C., Duvall, T.L., Jr., Thompson, M.J., Zhao, J.: 2006, Sensitivity of time–distance helioseismic measurements to spatial variation of oscillation amplitudes. I. Observations and a numerical model. *Astrophys. J.* **646**, 543.
Rajaguru, S.P., Sankarasubramanian, K., Wachter, R., Scherrer, P.H.: 2007, Radiative transfer effects on Doppler measurements as sources of surface effects in sunspot seismology. *Astrophys. J.* **654**, L175.
Scherrer, P.H., Bogart, R.S., Bush, R.I., Hoeksema, J.T., Kosovichev, A.G., Schou, J., Rosenberg, W., Springer, L., Tarbell, T.D., Title, A., Wolfson, C.J., Zayer, I., MDI Engineering Team, 1995, The Solar oscillations investigation – Michelson Doppler imager. *Solar Phys.* **162**, 129.
Schunker, H., Cally, P.S.: 2006, Magnetic field inclination and atmospheric oscillations above solar active regions. *Mon. Not. Roy. Astron. Soc.* **372**, 551.
Schunker, H., Braun, D.C., Cally, P.S.: 2007, Surface magnetic field effects in local helioseismology. *Astron. Nachr.* **328**, 292.
Schunker, H., Braun, D.C., Cally, P.S., Lindsey, C.: 2005, The local helioseismology of inclined magnetic fields and the showerglass effect. *Astrophys. J.* **621**, L149.
Weiss, N.O., Thomas, J.H., Brummell, N.H., Tobias, S.M.: 2004, The origin of penumbral structure in sunspots: Downward pumping of magnetic flux. *Astrophys. J.* **600**, 1073.
Zhao, J., Kosovichev, A.G.: 2003, Helioseismic observation of the structure and dynamics of a rotating sunspot beneath the solar surface. *Astrophys. J.* **591**, 446.
Zhao, J., Kosovichev, A.G.: 2006, Surface magnetism effects in time–distance helioseismology. *Astrophys. J.* **643**, 1317.

On Absorption and Scattering of P Modes by Small-Scale Magnetic Elements

Rekha Jain · M. Gordovskyy

Originally published in the journal Solar Physics, Volume 251, Nos 1–2, 361–368.
DOI: 10.1007/s11207-007-9102-7 © Springer Science+Business Media B.V. 2007

Abstract The solar surface is characterised everywhere by the presence of small-scale magnetic structures. Their collective behaviour in the form of active regions is known to have strong influence on p-mode power. For example, sunspots and plages are strong absorbers of acoustic waves. This paper studies the effects of individual small-scale magnetic elements to understand the details of absorption of p-mode power. For this, we consider a thin magnetic flux tube and calculate the phase shifts and the absorption coefficients by numerically solving the linearised MHD equations. The phase shifts calculated from the Born Approximation are then compared for the same range of degrees. The results are discussed with a view to understanding the physical mechanism.

Keywords Helioseismology: P modes · Magnetic: Magnetic fields, photosphere

1. Introduction

The development of local helioseismic techniques over the last two decades and the progress made in observations of the solar surface have rekindled the efforts of direct theoretical modelling to understand the behaviour of solar oscillations in the presence of magnetic fields. While the absorption of p-mode power by sunspots has long been known observationally (*e.g.* Braun, 1995) and theoretically (Rosenthal, 1992; Cally and Bogdan, 1993; Fan, Braun, and Chou, 1995; Chou *et al.*, 1996), it is now realised that the development of theoretical models to explain the observations and to predict what will be detected at fine scales is essential for further progress (see *e.g.* Duvall, Birch, and Gizon, 2006 for scattering of f modes by small-scale magnetic elements).

Helioseismology, Asteroseismology, and MHD Connections
Guest Editors: Laurent Gizon and Paul Cally.

R. Jain (✉) · M. Gordovskyy
Applied Mathematics Department, University of Sheffield, Sheffield S3 7RH, UK
e-mail: r.jain@sheffield.ac.uk

M. Gordovskyy
e-mail: m.gordovskyy@sheffield.ac.uk

In this paper, we develop a theoretical model of a thin magnetic flux tube and calculate the phase shifts and the absorption coefficients by numerically solving the linearised MHD equations. Section 2 gives a general description of the model that we use with some governing equations. Section 3 gives brief results followed by a discussion. Our conclusions are given in Section 4.

2. The Model

In the solar photosphere, the pressure and density change with depth (or height) and latitude but we are interested in the effects of small-scale magnetic fields. Therefore, in the absence of magnetic field we assume a plane-parallel, stratified ambient atmosphere where the equilibrium density and pressure are adiabatic and uniform in the horizontal direction but varying with depth. The basic state of the fluid is considered to be at rest ($\mathbf{v} = 0$). Thus, the linearized equation of ideal MHD may be written as

$$\frac{\partial \rho'}{\partial t} + \nabla \cdot \rho \mathbf{v}' = 0, \tag{1}$$

$$\rho \frac{\partial \mathbf{v}'}{\partial t} + \nabla p' - \frac{1}{\mu} \{ [(\nabla \times \mathbf{B}) \times \mathbf{B}'] + [(\nabla \times \mathbf{B}') \times \mathbf{B}] \} = \rho' \mathbf{g}, \tag{2}$$

$$\left(\frac{\partial p'}{\partial t} + (\mathbf{v}' \cdot \nabla) p \right) - \gamma \frac{p}{\rho} \left(\frac{\partial \rho'}{\partial t} + (\mathbf{v}' \cdot \nabla) \rho \right) = 0, \tag{3}$$

$$\frac{\partial \mathbf{B}'}{\partial t} - \nabla \times (\mathbf{v}' \times \mathbf{B}) = 0; \qquad \nabla \cdot \mathbf{B}' = 0, \tag{4}$$

where p', ρ', $\mathbf{v}'$, and $\mathbf{B}'$ are the wave perturbations in pressure, density, velocity, and magnetic induction.

In this study we use cylindrical coordinates (r, ϕ, z), where r is the horizontal radius, z is the vertical depth, and ϕ is the azimuthal angle. We assume the inhomogeneity to be axisymmetric and that gravity ($\mathbf{g}$) is uniform *i.e.* $\mathbf{g} = g\hat{\mathbf{z}}$ (see Gordovskyy and Jain, 2007a, 2007b for the details of the model). The equilibrium pressure, density and sound speed are decomposed into their background components (*i.e.* their magnitudes in the absence of magnetic field) as $p_a(z)$, $\rho_a(z)$, and $c_a^2(z)$, respectively, and their perturbed components $p_m(r, z)$, $\rho_m(r, z)$, and $c_m^2(r, z)$ due to the presence of magnetic field $\mathbf{B}(r, z)$:

$$p(r, z) = p_a(z) + p_m(r, z), \tag{5}$$

$$\rho(r, z) = \rho_a(z) + \rho_m(r, z), \tag{6}$$

$$c^2(r, z) = c_a^2(z) + c_m^2(r, z). \tag{7}$$

We assume that in the absence of a magnetic field, the gas pressure, density, and hence the sound speed vary only in the vertical direction as

$$p_a(z) = p_0 \left(1 + \frac{z}{L} \right)^{\alpha+1}; \qquad \rho_a(z) = (\alpha + 1) \frac{p_0}{gL} \left(1 + \frac{z}{L} \right)^{\alpha};$$

$$c_a^2 = \frac{gL}{\alpha} \left(1 + \frac{z}{L} \right), \tag{8}$$

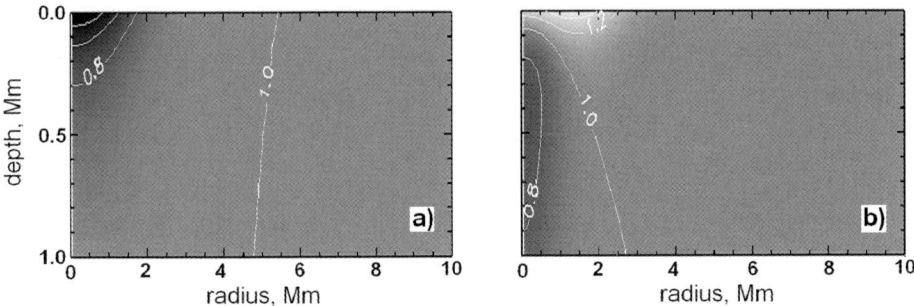

Figure 1 The pressure distribution inside and outside the thin flux tube normalised with respect to the background pressure p_a. Panels (a) and (b) are for $D = 16$ Mm and 0.5 Mm, respectively. Here $B_0 = 1.4$ kG and $R = 2$ Mm.

where α is the polytropic index which is related to the ratio of specific heats as $\gamma = (\alpha + 1)/\alpha$ and L is a characteristic length scale. For calculations, we use $p_0 = 1.4 \times 10^5$ Pa, $g = 2.76 \times 10^2$ m s^{-2}, $\alpha = 2.5$ and $L = 0.5$ Mm.

The perturbation in pressure and density is assumed to be caused by the nonpotential magnetic field of the form $\mathbf{B} = \{B_r(r,z), 0, B_z(r,z)\}$, where

$$B_z(r,z) = B_0\left(1 + \frac{z}{D}\right) e^{-\frac{r^2}{R^2}(1+\frac{z}{D})}; \qquad B_r(r,z) = -\frac{1}{2} B_0 \frac{r}{D} e^{-\frac{r^2}{R^2}(1+\frac{z}{D})}. \qquad (9)$$

Here B_0 is the characteristic magnetic induction and R is the characteristic flux tube radius. D is the convergence length that corresponds to the depth where the flux tube cross-section decreases by a factor of two.

If the plasma velocity is zero, one can obtain pressure and density distributions using the equation of motion (see also Schlüter and Temesvary, 1958 for another similarity solution)

$$-\nabla p_m + \frac{1}{\mu}(\nabla \times \mathbf{B}) \times \mathbf{B} + \rho_m \mathbf{g} = 0. \qquad (10)$$

In order to understand how variations of the flux-tube parameters affect the hydrostatic properties of the scattering region, we show the distribution of pressure (Figure 1) and the ratio of the squares of Alfvén to sound speed (Figure 2) inside and outside the flux tube with $R = 2$ Mm, $B_0 = 1.4$ kG. The left and right panels are for "almost vertical" field (*i.e.* $D = 16$ Mm) and for "strongly convergent" field ($D = 0.5$ Mm), respectively. It is clear from this figure that the effect of the magnetic field is dominant in a rather shallow layer near the surface ($z = 0$) in both cases. However, in the case of a "vertical" field (Figure 1a) the pressure inside the flux tube is lower than the background pressure (p_a) while in the case of strong convergence (Figure 1b) when the field lines have substantial radial component, the pressure inside the tube near the surface ($z = 0$) is higher than the background pressure; this enhancement of pressure inside the tube may not be realistic when radiative transfer and steady flows are properly taken into account. The distribution of speeds also reveals a similar behaviour: the sound speed is depressed in the case of a vertical field (Figure 2a) and shows enhancement in the case of a strongly converging field (Figure 2b).

It is also interesting to note that the model with almost vertical field contains an equipartition layer (EPL) where the local sound speed (c) is equal to the local Alfvén speed (v_A) (denoted by EPL in Figure 2a). However, in the model with strongly converging field, sound speed c enhanced due to the presence of strong horizontal field is higher than v_A everywhere.

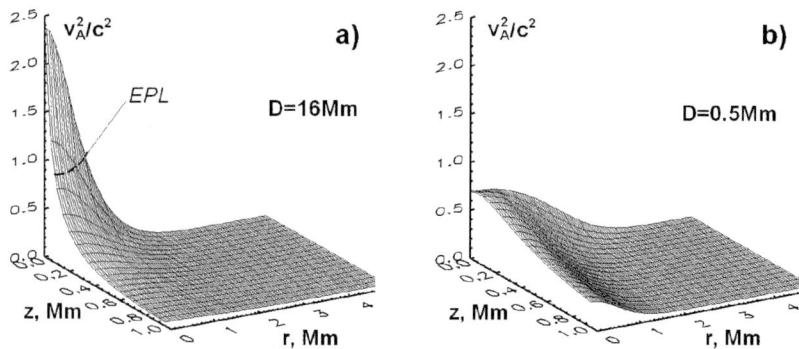

Figure 2 The ratio of squares of Alfvén to sound speed as a function of r and z for $B_0 = 1.4$ kG and (a) $D = 16$ Mm and (b) $D = 0.5$ Mm. Here $R = 2$ Mm. The equipartition layer (EPL) is indicated in (a).

It was shown by Gordovskyy and Jain (2007a, 2007b) that the inhomogeneous wave equation for pressure perturbation, derived using Equations (1) – (4), can be written as (see also Fan, Braun, and Chou, 1995)

$$\mathcal{L} p' + \frac{\omega^2}{(z+L)^{\alpha+1}} p' = \mathcal{S}, \qquad (11)$$

where $\mathcal{L}$ is a Sturm – Liouville differential operator

$$\mathcal{L} = \frac{g}{\alpha(z+L)^\alpha} \left[\nabla^2 - \left(\frac{\alpha}{z+L}\right)\frac{\partial}{\partial z} + \frac{\alpha}{(z+L)^2} \right], \qquad (12)$$

and $\mathcal{S}$ is the source term which can be written as

$$\mathcal{S} = \frac{g}{\alpha(z+L)^\alpha}\left[i\omega F + \frac{g}{i\omega}\frac{\partial F}{\partial z}\right] + \frac{1}{i\omega}\frac{g}{\alpha(z+L)^\alpha}\left\{\nabla \cdot \left[\left((\nabla \times (\nabla \times (\mathbf{v}' \times \mathbf{B}))) \times \frac{\mathbf{B}}{\mu}\right)\right.\right.$$
$$\left.\left. + \left(\left(\nabla \times \frac{\mathbf{B}}{\mu}\right) \times (\nabla \times (\mathbf{v}' \times \mathbf{B}))\right)\right]\right\}, \qquad (13)$$

where

$$F = \frac{1}{c_a^2(z)}(\mathbf{v}' \cdot \nabla) p_m(r, z) - (\mathbf{v}' \cdot \nabla) \rho_m(r, z) + \rho_a(z)\frac{c_m^2}{c_a^2}(\nabla \cdot \mathbf{v}'). \qquad (14)$$

Gordovskyy and Jain (2007a, 2007b) used the above governing equation to study p-mode scattering using the first-order Born Approximation. Figure 3 shows the distribution of zeroth and first-order source terms for the same flux tube as shown in Figure 1(a). The overall effect of inhomogeneity is dominant near the surface ($z = 0$) and decreases in size as higher-order source terms are considered.

We will compare the phase shift calculated using BA with the corresponding ones calculated numerically. Although the first-order BA can be used to study the variation of phase shifts, it cannot be used to study amplitude variation (see Fan, Braun, and Chou, 1995). Therefore, we study the absorption of p modes using numerical solution of Equations (1) – (4). It is assumed that pressure, density, velocity, and magnetic field wave perturbations vary with time as $e^{i\omega t}$. Thus, let us introduce p', $\mathbf{v}'$, ρ', and $\mathbf{B}'$ in the following form:

$$\rho' = \tilde{\rho} e^{i\omega t}, \qquad (15)$$

$$\mathbf{v}' = i\tilde{\mathbf{v}}e^{i\omega t}, \tag{16}$$

$$p' = \tilde{p}e^{i\omega t}, \tag{17}$$

$$\mathbf{B}' = \tilde{\mathbf{B}}e^{i\omega t}. \tag{18}$$

Substituting Equations (15)–(18) into Equations (1–4) and eliminating $\tilde{\rho}$ and $\tilde{\mathbf{B}}$ we obtain

$$\omega \rho \tilde{v}_r = \nabla \tilde{p} - \frac{1}{\omega} T_r, \tag{19}$$

$$\omega \rho \tilde{v}_z = \nabla \tilde{p} + \frac{1}{\omega}\left(\tilde{v}_r \frac{\partial \rho}{\partial r} + \tilde{v}_z \frac{\partial \rho}{\partial z}\right) g + \frac{1}{\omega} \rho \left(\frac{1}{r}\frac{\partial r\tilde{v}_r}{\partial r} + \frac{\partial \tilde{v}_z}{\partial z}\right) g - \frac{1}{\omega} T_z, \tag{20}$$

$$\omega \tilde{p} = -\left(\tilde{v}_r \frac{\partial p}{\partial r} + \tilde{v}_z \frac{\partial p}{\partial z}\right) - c_s^2 \rho \left(\frac{1}{r}\frac{\partial r\tilde{v}_r}{\partial r} + \frac{\partial \tilde{v}_z}{\partial z}\right). \tag{21}$$

Here T_r and T_z are the radial and vertical components of vector $\mathbf{T}$ accounting for the effect of the magnetic field

$$\mathbf{T} = \frac{1}{\mu}\left[(\nabla \times \mathbf{B}) \times (\nabla \times (\tilde{\mathbf{v}} \times \mathbf{B})) + (\nabla \times (\nabla \times (\tilde{\mathbf{v}} \times \mathbf{B})) \times \mathbf{B})\right]. \tag{22}$$

It is assumed that the solution of Equations (19)–(21) can be decomposed into two components – incoming and outgoing – everywhere in the model. In this numerical experiment, reflection of waves is not taken into account. Further we introduce two boundary conditions. At the right boundary ($r \gg R$) i.e. outside the flux tube, the pressure and velocity wave perturbations ($\tilde{p}$ and $\tilde{\mathbf{v}}$) for incoming wave components correspond to their eigenfunctions for undisturbed atmosphere. At the left boundary ($r = 0$) corresponding to the axis of the flux tube the relationship between incoming and outgoing wave components is

$$\tilde{p}_{\text{out}} = \tilde{p}_{\text{in}}; \quad \frac{\partial \tilde{p}_{\text{out}}}{\partial r} = -\frac{\partial \tilde{p}_{\text{in}}}{\partial r},$$

$$\tilde{v}_{r\text{out}} = -\tilde{v}_{r\text{in}}; \quad \frac{\partial \tilde{v}_{r\text{out}}}{\partial r} = -\frac{\partial \tilde{v}_{r\text{in}}}{\partial r},$$

$$\tilde{v}_{z\text{out}} = \tilde{v}_{z\text{in}}; \quad \frac{\partial \tilde{v}_{z\text{out}}}{\partial r} = -\frac{\partial \tilde{v}_{z\text{in}}}{\partial r}.$$

The set of Equations (19)–(21) together with boundary conditions is solved using a finite-difference scheme on uniform grid 512×256 meshes.

Now we study p-mode scattering in terms of phase shifts between outgoing and incoming waves and absorption in terms of the ratio of amplitudes of outgoing to incoming waves. For this, we assume that the inward and outward travelling wave field can be decomposed into components of the form (Braun, 1995)

$$\psi(r) = \mathcal{A}H^{(1)}(kr) + \mathcal{B}H^{(2)}(kr), \tag{23}$$

where $\mathcal{A}$ and $\mathcal{B}$ are complex coefficients representing phase and amplitude and k is the horizontal wavenumber. $H^{(1)}$ and $H^{(2)}$ are Hankel functions of the first and second kind, respectively.

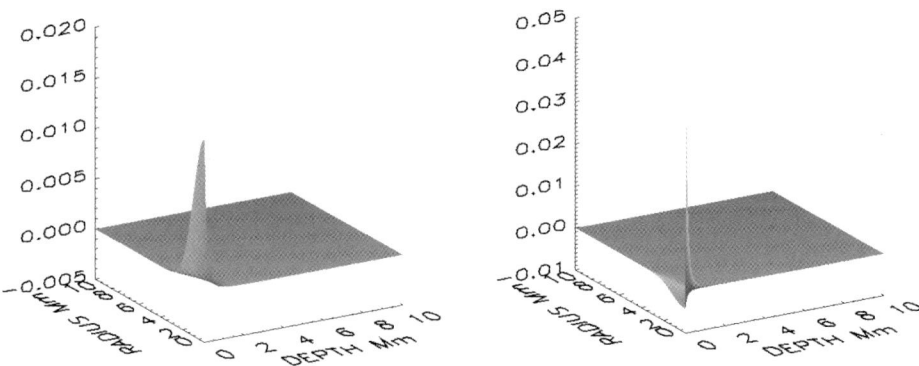

Figure 3 Zeroth (a) and first-order (b) source term (see Equation (11)) as a function of radius and depth for $D = 16$ Mm. All other parameters are the same as for Figure 1.

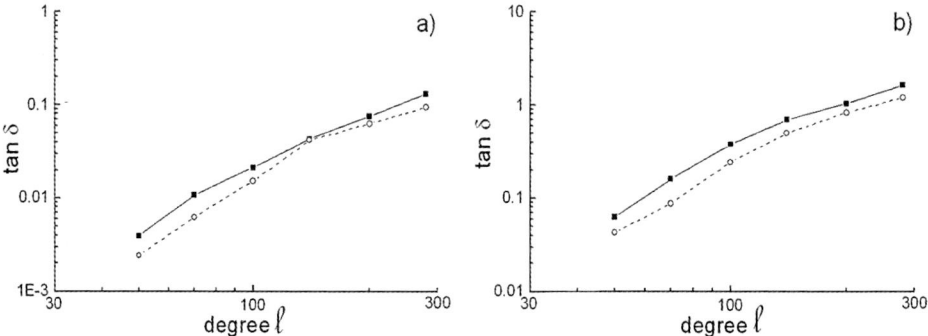

Figure 4 Comparison of phase shifts obtained using BA approach (dashed lines with circles) with those calculated numerically (solid lines with squares) for $B = 1.4$ kG, $n = 2$. (a) is for $R = 2$ Mm and (b) is for 8 Mm.

We describe the difference between the power of incoming and outgoing waves outside the magnetised region by defining absorption coefficient as:

$$a = 1 - \left(|\mathcal{A}|/|\mathcal{B}|\right)^2; \qquad (24)$$

and phase shift (δ) is defined as

$$\delta = \arg(\mathcal{A}/\mathcal{B}). \qquad (25)$$

3. Results and Discussion

Figure 4 shows phase shift (tangent of δ), in logarithmic scale, calculated numerically (solid line) and by using BA (dashed lines) as a function of degree (ℓ) for $R = 2$ Mm (Figure 4(a)) and 8 Mm (Figure 4(b)). Clearly, the agreement between the two phase shifts is good.

Figure 5 shows the phase shift (tangent of δ) in the left panel, in logarithmic scale, calculated numerically but now for a "convergent field" case as well. The phase shifts are proportional to ℓ^2 for low degrees with $\tan \delta$ lower by a factor of two for "strongly converging" field

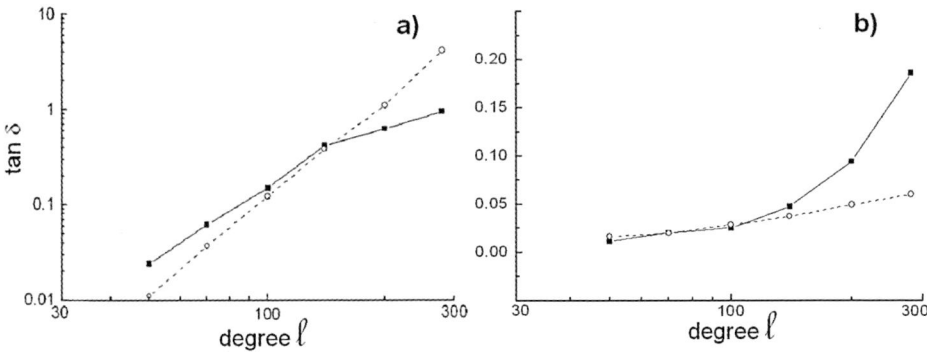

Figure 5 Phase shifts (a) and absorption coefficients (b) for $R = 2$ Mm, $B = 1.4$ kG, $n = 2$. Solid lines with squares are for the model with almost vertical field ($D = 16$ Mm) while dashed lines with circles are for the model with strongly converging field ($D = 0.5$ Mm).

than for "almost vertical" field. For higher degrees ($\ell > 150$), phase shifts for the "converging field" model reveal a rapid increase and become larger than the phase shifts for the model with vertical field. This could be due to an increase in the effective wave speed as a result of substantial radial component of magnetic fields (see also Gordovskyy and Jain, 2007b).

The absorption coefficients (Figure 5(b)) reveal two main features. First, they are significant only for $\ell > 150$ and substantial only for $D = 16$ Mm case. This suggests that the presence of equipartition layer (Figure 2) can play an important role in the absorption process; mode conversion at the equipartition layer has been studied in detail by Spruit and Bogdan (1992) and Cally and Bogdan (1993). Although, we do not explicitly study it here, it is possible that there is a loss of energy by p modes in the top, shallow layers of the inhomogeneity due to mode conversion. Second, the absorption coefficients are nonzero for larger ℓ (>150) in the case of strong convergence ($D = 0.5$ Mm) even when there is no equipartition layer in the model. However, the magnitudes (≈ 0.05 for $\ell > 200$) are only slightly higher than the precision error for a (about 10^{-2}) and, therefore, it is difficult to say if there is any *real* decrease in outgoing wave amplitude.

In order to investigate this further, we consider the case of "almost vertical" field ($D = 16$ Mm) with larger radius ($R = 4$ Mm) and smaller magnetic induction ($B_0 = 0.4$ kG). It can be seen, that although there is no equipartition layer in this model (Figure 6(a)), the resulting absorption coefficients are higher than when compared with $R = 2$ Mm and $D = 0.5$ Mm case (compare Figures 5(b) and 6(b)). It is possible that in this case also there is no *true* absorption and the energy is redistributed between p modes with the same frequency but different wavenumbers. The detailed investigation of these two effects is underway.

4. Conclusion

We investigated the role of an individual small-scale magnetic element on the scattering and absorption of p-mode power. The phase shifts are found to be significant for both "almost vertical" and "convergent" field cases. The increase of phase shifts for large ℓ for the "strongly convergent" case is believed to be due to an increase in the effective wave speed. Significant absorption for the "almost vertical" field case suggests that a "localised dissipation" mechanism such as mode conversion may play an important role in the apparent absorption of p-mode power seen in observational data (Braun, 1995). However, in

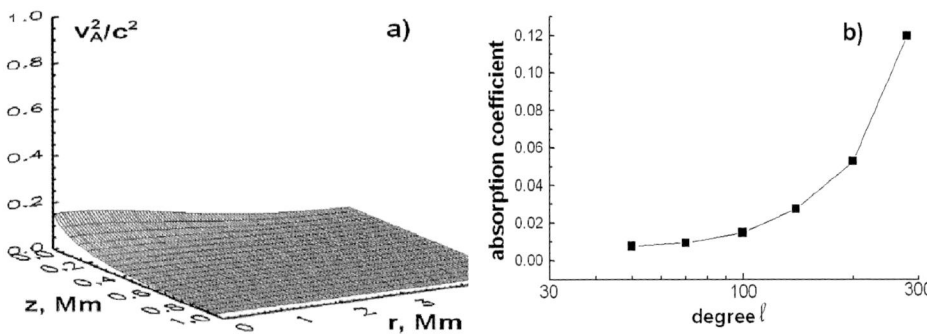

Figure 6 The distribution of the ratio of squares of Alfvén to sound speeds for $R = 4$ Mm, $B_0 = 0.4$ kG and $D = 16$ Mm (a) and absorption coefficients for this model (b).

our numerical experiment, we find significant absorption coefficient for the cases where the equipartition layer is absent. This may suggest that mechanisms other than mode conversion may also be present. Further work is underway to explore this possibility.

Acknowledgements The authors thank the organisers of GONG2007/SOHO19 (Monash University) and HELAS II (MPS, Lindau). R.J. appreciates many useful conversations with B.W. Hindman. This work is supported by the Engineering and Physical Sciences Research Council, grant EP/C548795/1.

References

Braun, D.C.: 1995, *Astrophys. J.* **451**, 859.
Cally, P.S., Bogdan, T.J.: 1993, *Astrophys. J.* **402**, 721.
Chou, D.Y., Chou, H.Y., Hsieh, Y.C., Chen, C.K.: 1996, *Astrophys. J.* **459**, 792.
Duvall, Jr., T.L., Birch, A.C., Gizon, L.: 2006, *Astrophys. J.* **646**, 553.
Fan, Y., Braun, D.C., Chou, D.Y.: 1995, *Astrophys. J.* **451**, 877.
Gordovskyy, M., Jain, R.: 2007a, *Astron. Nachr.* **328**, 309.
Gordovskyy, M., Jain, R.: 2007b, *Astrophys. J.* **661**, 586.
Rosenthal, C.S.: 1992, *Solar Phys.* **139**, 25.
Schlüter, A., Temesvary, S.: 1958, In: Lehnert, B. (ed.) *Electromagnetic Phenomena in Cosmical Physics*, IAU Symp. **6**, Cambridge University Press, Cambridge, 263.
Spruit, H.C., Bogdan, T.J.: 1992, *Astrophys. J.* **391**, L109.

Time – Distance Analysis of the Emerging Active Region NOAA 10790

S. Zharkov · M.J. Thompson

Originally published in the journal Solar Physics, Volume 251, Nos 1–2, 369–380.
DOI: 10.1007/s11207-008-9239-z © Springer Science+Business Media B.V. 2008

Abstract We investigate the emergence of Active Region NOAA 10790 by means of time – distance helioseismology. Shallow regions of increased sound speed at the location of increased magnetic activity are observed, with regions becoming deeper at the locations of sunspot pores. We also see a long-lasting region of decreased sound speed located underneath the region of the flux emergence, possibly relating to a temperature perturbation due to magnetic quenching of eddy diffusivity, or to a dense flux tube. We detect and track an object in the subsurface layers of the Sun characterised by increased sound speed which could be related to emerging magnetic-flux and thus obtain a provisional estimate of the speed of emergence of around 1 km s^{-1}.

Keywords Sun: Active Region emergence · Sun: time – distance helioseismology

1. Introduction

Local helioseismology provides a tool for studying the emergence and evolution of active regions and complexes of solar activity. Several theoretical scenarios for the formation of active regions on the solar surface have been suggested within the context of the dynamo theory explaining the formation of the "sunspot belt" at low latitudes. Recently developed "flux-transport" dynamo models include the magnetic flux transport by meridional circulation (Wang and Sheeley, 1991; Choudhuri, Schüssler, and Dikpati, 1995; Dikpati and Charbonneau, 1999; Nandy and Choudhuri, 2002; Krivodubskij, 2005), flux transport by an interface (Parker, 1993; Charbonneau and MacGregor, 1997) and overshoot dynamo with a positive radial shear beneath the surface (Ruediger and Brandenburg, 1995), or by a distributed dynamo with a negative shear (Roberts and Stix, 1972).

Helioseismology, Asteroseismology, and MHD Connections
Guest Editors: Laurent Gizon and Paul Cally.

S. Zharkov (✉) · M.J. Thompson
Department of Applied Mathematics, University of Sheffield, Hounsfield Road, Hicks Building, Sheffield S3 7RH, UK
e-mail: s.zharkov@sheffield.ac.uk

All of the models agree that the most favourable and promising place for the $\alpha\Omega$ dynamo is the region which includes the deepest layers of the convection zone, the convective overshoot layer and the tachocline. Nevertheless, it is difficult to keep flux tubes with strong magnetic field ($B > 100$ G) in the interior for a long time against magnetic buoyancy (Parker, 1979) and then deliver them to the surface within the sunspot belt.

In a flux-transport model it is possible to calculate the velocities of macroscopic diamagnetic transport as a function of depth in the convective zone by taking into account the magnetic-advection processes. Krivodubskij (2005) estimated that in the subsurface layers the transport velocity increases from 0.5 km s^{-1} at a depth of 30 Mm to around 0.9 km s^{-1} at 10 Mm depth. However, the way in which the flux travels to the surface is not fully understood. There are ongoing arguments in favour of each model, and they can only be tested by comparison with the observations of active regions and sunspots on the surface and their appearance in the solar interior by means of helioseismology.

Time–distance helioseismology (Duvall *et al.*, 1993) is a fast-growing area of helioseismology. The data used for time–distance inversions are the estimated wave traveltimes between different points on the surface of the Sun. The travel-time estimates are obtained by cross-correlating and averaging the Doppler-velocity signal at different locations at the Sun's surface and then obtaining the cross-correlation function. The travel times are then computed from cross-correlations either via Gabor wavelet fitting (Kosovichev and Duvall, 1997) or by comparing with a reference cross-correlation function (see Gizon and Birch, 2005 and references therein). Because of the stochastic nature of solar oscillations, substantial spatial and temporal averaging of the data is required to measure the frequencies and travel times accurately.

The properties of the solar interior are then determined by, first, establishing the relations between travel-time variations and internal properties (variations of the sound speed, flow velocity, magnetic field). This is followed by the inversion of these relations, which are typically cast as linear integral equations, to infer the internal properties.

Chang, Chou, and Sun (1999) applied the method of acoustic imaging to study the emergence of NOAA 7978, and reported the detection of the signature of upward-moving magnetic flux in the solar interior. The first time–distance investigation of an emerging active region was by Kosovichev, Duvall, and Scherrer (2000), wherein an active region, which emerged on the solar disk in January 1998, was studied with SOHO/MDI for eight days, starting before its emergence at the surface. The results have shown a complicated structure of the emerging region in the interior, and suggest that emerging flux ropes travel very quickly through the depth range of our observations, estimating the speed of emergence at about 1.3 km s^{-1}.

In Section 2 of this paper the method and the data used in this work are described. In Section 3 the results of sound-speed inversions are presented, together with the discussion. This is followed by conclusions in Section 4.

2. Data and Method

2.1. Time–Distance Helioseismology

The raw Doppler signal at any location is composed of velocity signals occurring due to convective motions on the scales of granulation and supergranulation, as well as the signal due to the superpositions of the wavepackets set up by resonant acoustic modes. The supergranular signal can largely be removed by filtering out frequencies less than 1.5 mHz. However even after removing the low frequencies and frequencies above the acoustic cut-off of about 5.3 mHz, the oscillation signals due to the nature of sources and sinks of acoustic waves set up by turbulent convection remain stochastic.

Due to the fact that acoustic waves with the same horizontal phase speed (ω/k) travel the same horizontal distance (Δ) the Gaussian phase-speed filter is tuned to select waves with horizontal phase speed (v_i) near the ray theory value corresponding to distance Δ_i (Duvall et al., 1997). We define a number of such filters (F_i) corresponding to different distances (Δ_i) as follows:

$$F_i(k, \omega; \Delta_i) = \exp\left[-\left(\frac{\omega}{k} - v_i\right)^2 \bigg/ 2\delta v_i^2\right],$$

where ω is the frequency and k is the horizontal wavenumber in the data power spectrum. Each filter is applied by pointwise multiplication of the Fourier transform $S(k_x, k_y, \omega)$ of the observed velocity data.

From the filtered data, we calculate point-to-annulus cross-correlation functions [$C(\mathbf{r}, \Delta_i, t)$] for each skip-distance and each position/pixel of our data, where $\mathbf{r}$ is the position vector, Δ_i is the distance between two surface bounces and t is time. The obtained set of cross-correlation functions is then averaged spatially over the quiet Sun to obtain an estimate for the reference cross-correlation function [$C^{\text{ref}}(\Delta_i, t)$]. This is used to measure one-way travel-time perturbations at each location following the Gizon and Birch (2002, 2004) and Jensen and Pijpers (2003) approach:

$$\tau_{\pm} = \frac{\sum_i \mp f(\pm t_i) \dot{C}^{\text{ref}}(t_i)[C(t_i) - C^{\text{ref}}(t_i)]}{\sum_i f(\pm t_i)[\dot{C}^{\text{ref}}(t_i)]^2}, \tag{1}$$

where $f(t)$ is a window function designed to pick out the signal from the first bounce (i.e., direct arrival times). Here τ_+ and τ_- correspond to waves travelling outwards and inwards from the measurement location, respectively. We define the mean travel times [$\tau(\mathbf{r}, \Delta)$] as $\tau(\mathbf{r}, \Delta) = (\tau_+(\mathbf{r}, \Delta) + \tau_-(\mathbf{r}, \Delta))/2$.

As a first approximation, perturbations [$\delta\tau(\mathbf{r}, \Delta)$] in these mean travel times are linearly related to the sound-speed perturbations [δc] in the wave propagation region,

$$\delta\tau(\mathbf{r}, \Delta) = \iint_S d\mathbf{r}' \int_{-d}^0 dz\, K(\mathbf{r} - \mathbf{r}', z; \Delta) \frac{\delta c^2}{c^2}(\mathbf{r}', z), \tag{2}$$

where S is the area of the region, d is its depth, and $\delta\tau(\mathbf{r}, \Delta)$ is defined as the difference between the measured travel time at a given location ($\mathbf{r}$) on the solar surface and the average of the travel times in the quiet Sun. The sensitivity kernel for the relative squared sound-speed perturbation is given by K.

For the forward problem we use wave-speed sensitivity kernels estimated using the Rytov approximation by Jensen and Pijpers (2003). The Rytov and ray approximations give very similar results, with the former being perhaps more reliable for inferring deeper structures (Couvidat et al., 2004).

The code for travel–time inversions is based on a multi-channel deconvolution algorithm (MCD) (Jensen, Jacobsen, and Christensen-Dalsgaard, 1998) enhanced by the addition of horizontal regularisation. Following Jensen (2001) we discretized Equation (2): $d_i = \delta\tau(\mathbf{k}, \Delta_i)$, $G_{ij} = K(\mathbf{k}, z_j; \Delta_i)$, $m_j = \delta s(\mathbf{k}, z_j)$. Then for each horizontal wavevector $\mathbf{k}$ we find the vector $\mathbf{m}$ that solves

$$\min\left\{\left\|(\mathbf{d} - G\mathbf{m})\right\|_2^2 + \epsilon(\mathbf{k})^2 \|L\mathbf{m}\|_2^2\right\}, \tag{3}$$

Table 1 Phase speed filter parameters.

i	Δ_i	Skip-Distance range, Mm	v_i (km s^{-1})	δv_i (km s^{-1})
1	5.83	3.553 – 8.107	14.21	2.63
2	7.29	5.013 – 9.567	15.40	2.63
3	11.664	9.387 – 13.941	17.49	3.33
4	16.038	13.761 – 18.315	25.82	3.86
5	20.412	18.135 – 22.689	30.46	3.86
6	24.786	22.509 – 27.063	35.46	5.25
7	29.16	26.883 – 31.437	39.71	3.05
8	33.66	31.383 – 35.937	43.29	3.15
9	37.91	35.633 – 40.1870	43.29	3.15
10	42.28	40.003 – 44.557	47.67	3.57
11	46.66	44.383 – 48.937	57.16	3.78
12	51.01	48.733 – 53.287	57.16	3.78

where L is a regularisation operator. In this work we chose to apply more regularisation at larger depths, where acoustic wave travel times provide less information, by setting $L = \mathrm{diag}(c_{0,1}, \ldots, c_{0,n})$, where $c_{0,i}$ is the sound speed in the ith layer of the reference model. We heavily regularise small horizontal scales (Couvidat et al., 2005) by taking $\epsilon(\mathbf{k})^2 = \epsilon_0^2 (1 + (|\mathbf{k}|/k_{\max})^2)^p$ with $p = 200$ and $\epsilon_0^2 = 5 \times 10^3$. We invert for 14 layers in depth located at 0.36, 1.17, 2.11, 3.28, 4.74, 6.41, 8.6, 11.22, 14.28, 17.78, 21.79, 26.31, 31.41, 36.95 Mm.

The errors due to realisation noise are estimated by processing the 24 quiet-Sun data cubes, calculating the noise covariance matrix in travel-time measurements (Jensen, 2001; Gizon and Birch, 2004; Couvidat et al., 2005), and then evaluating noise in inversions according to (Jensen, 2001).

2.2. Data

We investigate the active region NOAA 10790, which appeared on 13 July 2005. The full-disk MDI data with one-minute cadence are used. The time series start at 00:00 UT, 10 July 2005 and ends at 23:59 UT 13 July 2005. From this set we extract 44 512-minute series with the starting times every two hours. Each dataset is then re-mapped and de-rotated onto the heliographic grid using a Postel projection centred at 11° Carrington longitude and 11° latitude (South). The horizontal spread of the data is approximately 380 Mm. We use the Snodgrass rotation rate to remove the differential rotation (Snodgrass and Ulrich, 1990) for each processed data cube.

Each of the data cubes is then processed to obtain the travel-time measurements using the centre-to-annulus cross-correlations as described in Section 2.1. The phase-speed filter values and the range of skip-distances used in this paper are provided in Table 1. The averaging annuli thickness was set to three pixels wide, which corresponds to approximately 4.55 Mm.

3. Results and Discussion

3.1. The Active Region NOAA10790 Surface Emergence and Evolution

Signs of the increased magnetic activity at the site of the active-region emergence can be first observed on SOHO/MDI magnetogram images taken on 10 July 2005. At this time one can see the two points of flux at around 5° – 8° Carrington longitude. The flux slowly grows with the first pores appearing in the SOHO/MDI continuum images around 11 July 2005 06:30 UT (Figure 1). A number of different pores evolve and disappear at this location until 13 July 2005 06:30 UT, but no sunspots develop.

Also around 14:30 UT 13 July 2005, we see a new flux emergence to the East of the first group at around 14° longitude, 10 – 12° latitude (see Figure 1, panels (d) – (f)). At this new location, the first sunspot pores appear at about 17:30 UT 13 July 2005, which quickly evolve into the bipolar sunspot group (by 14 July 2005 14:30 UT) before disappearing behind the limb. Also, there is a bipolar region of relatively weak magnetic flux seen appearing at around 20 – 22° longitude on 10 July 2005 and 12 July 2005. No pores were observed at these locations.

3.2. The Subsurface Sound-Speed Inversions

Inversions of the sound-speed perturbation beneath the region of interest carried out in this work indicate that little change is detected at the first site of increased magnetic activity (6 – 10° longitude) prior to the emergence of magnetic flux at around 11:00 UT 10 July 2005. Then, once the magnetic flux appears on the surface, we see a consistent and substantial increase of the sound speed at shallow depths of up to 7 – 10 Mm directly beneath the locations of high magnetic-field strength, up until the end of the series used in this investigation. However, we have not been able to see any significant positive sound-speed perturbation in the deeper layers most of the time. This could be due to the fact that the spatial extent of the emerging flux at this location is quite small, 6 – 7 Mm in diameter for each footprint, and the signal cannot be resolved to better than half of the wavelength (Gizon and Birch, 2004). For the power spectrum produced by the phase-speed filters specified in Table 1, this gives us an estimated resolution of, at best, 7.5 Mm for waves travelling to 20 Mm depths. The averaging kernel at a target depth of 18 Mm, estimated at full width at half-maximum, extends from 8 to 22.5 Mm. This can also be a byproduct of the inversion regularisation, since due to our choice of operator, more regularisation is applied at deeper layers.

As the ratio of gas pressure to magnetic pressure (the plasma β) increases rapidly with depth, the sound speed increase due to magnetic field becomes very small, $\delta c/c$ of order 0.01 per cent at around 18 Mm below the surface. Using the relations $p_e = kT_e\rho_e/m$, $p_i = kT_i\rho_i/m$, $p_m = B^2/2\mu$ and $p_e = p_i + p_m$, and assuming that the measured wave-propagation speed is given by $c^2 = c_i^2 + c_A^2$, where the Alfvén speed is $c_A^2 = B^2/\mu\rho_i$, we can evaluate

$$\frac{\delta c^2}{c_e^2} = \frac{\delta T}{T_e} + \frac{2}{\gamma\beta}\left(1 + \frac{\delta T}{T_e}\right),$$

Here p is pressure, T temperature, ρ density, k Boltzman constant, m mean molecular weight, and indices e, i, m stand for *external*, *internal*, and *magnetic*; $\beta = p_i/p_m$, $\delta c^2 = c^2 - c_e^2$, and $\delta T = T_i - T_e$. Thus, at greater depths the measured perturbation will

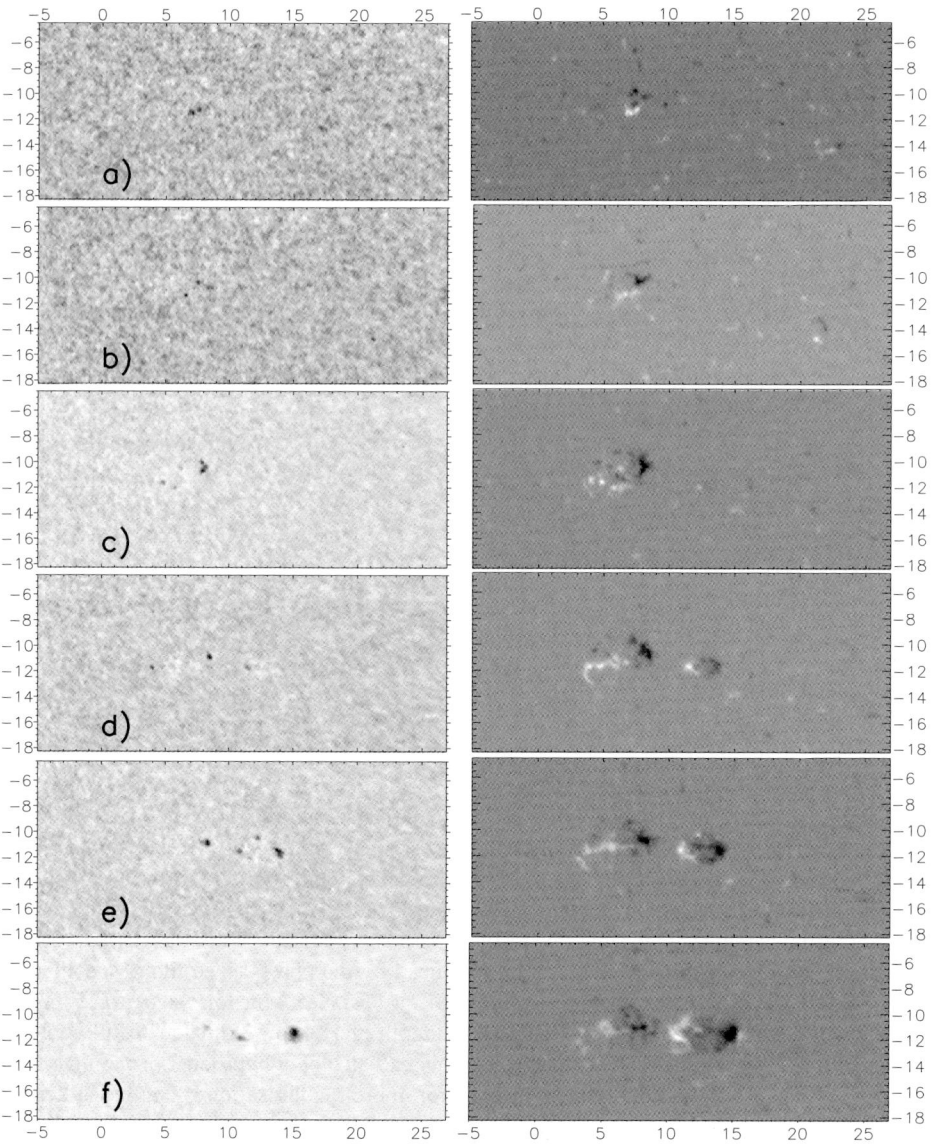

Figure 1 The evolution of active region NOAA 10790 as observed by MDI. Carrington longitude is plotted along the x-axis, latitude along the y-axis. MDI continuum images are in the left column, the greyscale for MDI magnetograms (right column) corresponds to -600 to 600 Gauss. The observation times for each row: (a) 11 July 2005 11:11:32 UT, (b) 12 July 2005 06:23:32UT, (c) 13 July 2005 06:23:32 UT, (d) 13 July 2005 17:35:32 UT, (e) 13 July 2005 23:59:32 UT, (f) 14 July 2005 11:11:32 UT.

be dominated by changes in temperature. This makes detecting and following localised perturbations due to emerging magnetic fields difficult.

Notwithstanding the argument above, in our measurements the positive perturbation below the surface regions of magnetic activity becomes deeper and better defined as sunspot pores begin to form at the surface, indicating perhaps the increase of magnetic-field strength.

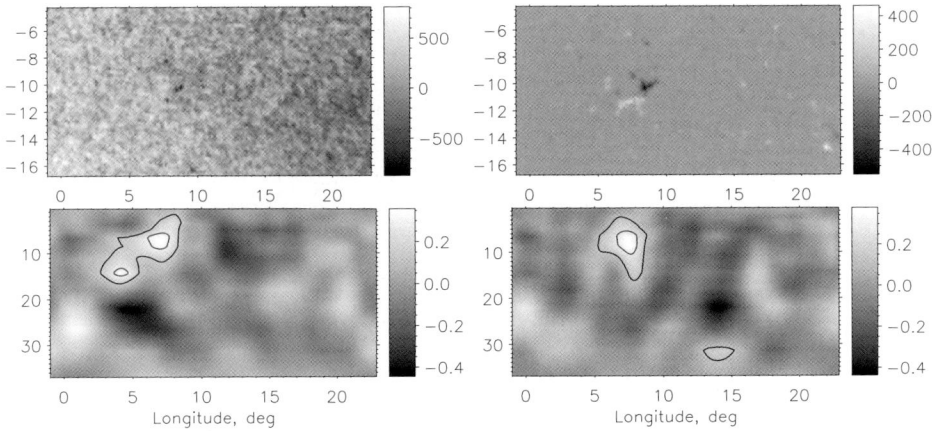

Figure 2 Sunspot pore emergence. Top panels: the SOHO/MDI intensity (left) and magnetogram (right) images, taken on 12 July 2005 06:23 UT and 09:35 UT. Bottom panels: the cuts at 11° latitude through the sound-speed inversion estimated from 512 minute dopplergram timeseries centred on 11 July 2005 08:00 UT (left) and 11 July 2005 10:00 UT (right). The greyscale is in km s^{-1}. Latitude is plotted on the left hand side along the y-axis and longitude along the x-axis. Depth in Mm is marked on the right hand side of y-axis. Contours correspond to sound-speed perturbation values of 0.25 and 0.3 km s^{-1}.

In the two bottom plots of Figure 2 we can see the region of the increased sound speed extending to depths of around 12–15 Mm at around 7° longitude that, we believe, is related to the appearance of a sunspot pore seen in SOHO/MDI intensity images. By the time of 13 July 2005 00:00 UT there are several pores observed at this location, and the flux configuration has a clear bi-polar structure (Figure 1(c)).

Three-dimensional visualisation of the sound-speed inversion for an eight-hour series centred around 04:00 UT on 13 July 2005 is presented in Figure 3. One can clearly identify the subsurface tooth-like structure of the increased sound speed directly underneath the first region of surface flux activity. In addition, an object characterised by increased sound-speed can be located deeper at the horizontal position corresponding to the point of the second flux emergence. Longitudinal cuts in depth taken at 11° latitude of sound-speed inversions for data cubes centred between 04:00 UT and 10:00 UT on 13 July 2005 are presented in Figure 4, with depth cuts at 21.8 Mm and 6.41 Mm of inversions for series centred on 04:00 UT and 08:00 UT plotted in Figure 5.

One can see the region of increased sound speed at depths of 20–25 Mm at about 14° longitude (see Figure 4, top left, and Figure 5, left). This region can be tracked rising on the next inversion as shown in Figure 4 (top right panel). The regions of increased sound speed can also be seen in the following two inversions (Figure 4, bottom panels) rising to the surface at the location of the second flux group formation. The time of the appearance on the surface (13 July 2005, 06:00–14:30 UT) corresponds to the first signs of the increased magnetic activity in the surface magnetograms. On the other hand, if the flux, by our assumption represented in the inversions as a region of the increased sound speed, travels from deeper layers to the surface, one should be able to see the traces of this flux presence at deeper layers in the inversions of eight-hour series centred at 08:00 and 10:00, while our inversions show only localised area of sound-speed increase. The averaged resolution kernels estimated for the depths layer around 21.79 Mm at half maximum extend from 8 to 26 Mm.

In order to check if the signals corresponding to sub-surface emerging flux add up while the noise structures cancel out, we have averaged all of our 44 3D sound-speed inversion

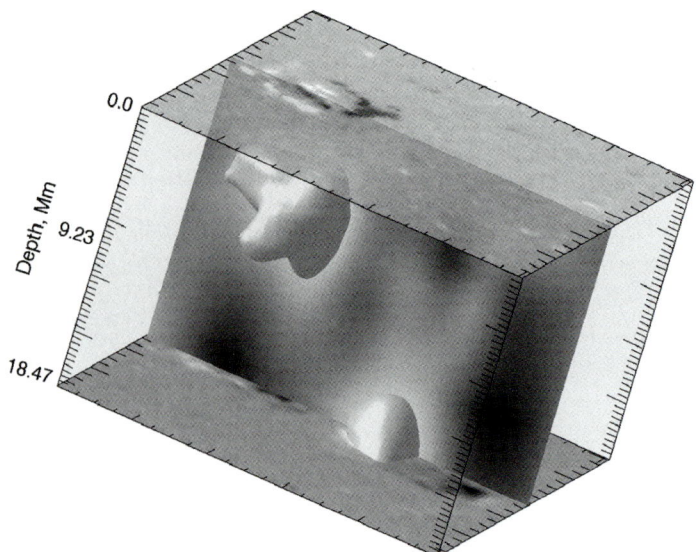

Figure 3 The sound-speed inversion of a 512 minute data cube centred around 13 July 2005 04:00 UT. The top is the surface magnetogram at the start of the time series. The bottom is the surface magnetogram taken around the time of the end of the whole series (14 July 2005 07:59 UT). The vertical panel is the absolute sound-speed perturbation at 11° latitude. The iso-surface corresponds to a value of 0.32 km s^{-1}.

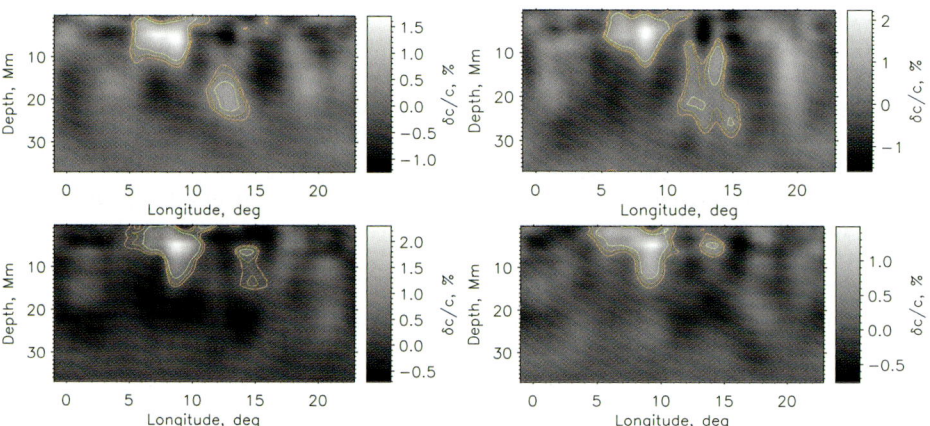

Figure 4 The magnetic-flux emergence. The sound-speed inversion as a function of depth and longitude. The cut is made at 11° latitude. The inversion is obtained for the 512-minute data cubes measured on 13 July 2005. The central times for the Dopplergram time series used to obtain the inversion are (top row): 04:00; 06:00 UT; and (bottom row) 08:00; 10:00 UT. Contours drawn for the features located between 4° and 16° longitude correspond to sound-speed perturbation values of 0.5, 0.7 and 1.0%.

maps. The averaging procedure is consistent within the linear limit used in the inversion scheme. The depth cut at 11° latitude of such an average is presented at the top of Figure 6, with daily averages plotted in other panels. One can clearly see the region of the increased sound speed corresponding to the first flux location. Interestingly, one can also observe a well-defined region of the decreased sound speed located beneath and to the West of the

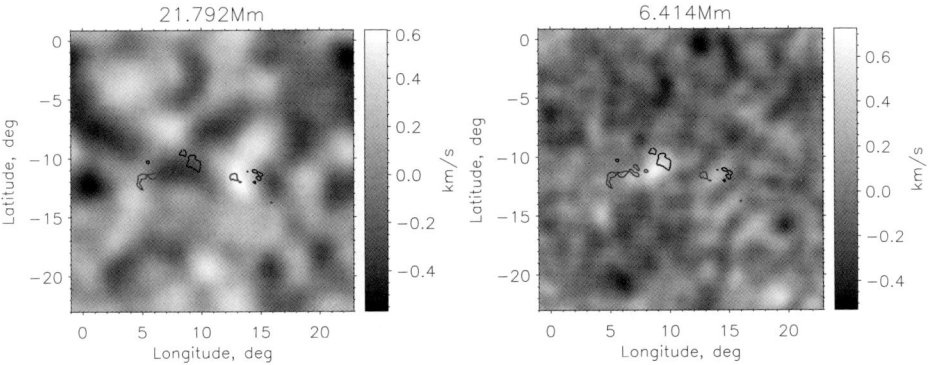

Figure 5 Left: the results of the inversion at 21.77 Mm for the data cube around 13 July 2005 04:00 UT. Right: results of the inversion at 6.41 Mm for the data cube centred around 13 July 2005 08:00 UT. The contours in both plots are lines of ±150 Gauss from the magnetogram taken around 13 July 2005 17:35 UT.

first flux region with a column of negative perturbation corresponding to the location of the second flux emergence. The structure is stable with respect to changes in the pre-processing (filtering) scheme and inversion parameters. It is also clearly present in the averages of the data cubes obtained on any of the days considered here, from 10 to 12 July 2005 presented in Figure 6. However, the shape of this decreased sound speed cannot be observed in individual eight-hour inversions, possibly owing to turbulent activity.

We are not certain how to interpret this region of negative sound-speed perturbation but presumably it is a region of cooler plasma. Under certain conditions such a signal could arise from a monolithic flux tube originating in the deeper interior, with internal density exceeding the matter density outside the tube. Alternatively, such an effect can be seen in the scenario described by Kitchatinov and Mazur (2000) in which a temperature instability leads to the formation of an active region due to eddy diffusivity and magnetic tension. If the former explanation is the correct one, then it is tempting to identify such emerging cool flux tubes with the flux tubes postulated by Strous and Zwaan (1999) to explain transient darkenings they observed in continuum intensity.

Thus, we can hypothesise that the sound-speed perturbation seen in Figure 4 is related to the emergence of a hot packet of plasma in the cold plasma column seen in Figure 6. In such circumstances, the absence of a positive-perturbation column in the bottom two images of Figure 4 may be due to the dynamic nature of the region considered, wherein the body of increased sound speed preceded and followed by regions of decreased sound speed may cancel each other in our inversion. We note that the inversions of the interior sound speed averaged over eight data cubes obtained on 13 July 2005 (Figure 6, bottom right panel) point to the presence of a decreased sound-speed structure column at the location of the second flux emergence with a positive perturbation directly below the surface. The latter feature can be expected given that these data are for a period after the flux emergence at the surface has become well established. So, if we assume that the region of the increased sound speed corresponds to the emerging magnetic flux and compare the sound-speed perturbation detected at the depth of 21.7 Mm on 13 July 2005 04:00 and appearing on the surface around 10:00 UT, this can provide an estimate of the speed of emergence at about 1.0 km s^{-1}. This estimate is slightly lower than the speed reported by Kosovichev, Duvall, and Scherrer (2000), but definitely higher than the velocities of a diamagnetic transport in near-surface layers estimated in Krivodubskij (2005). This can be an indication of some additional mechanisms affecting the flux emergence.

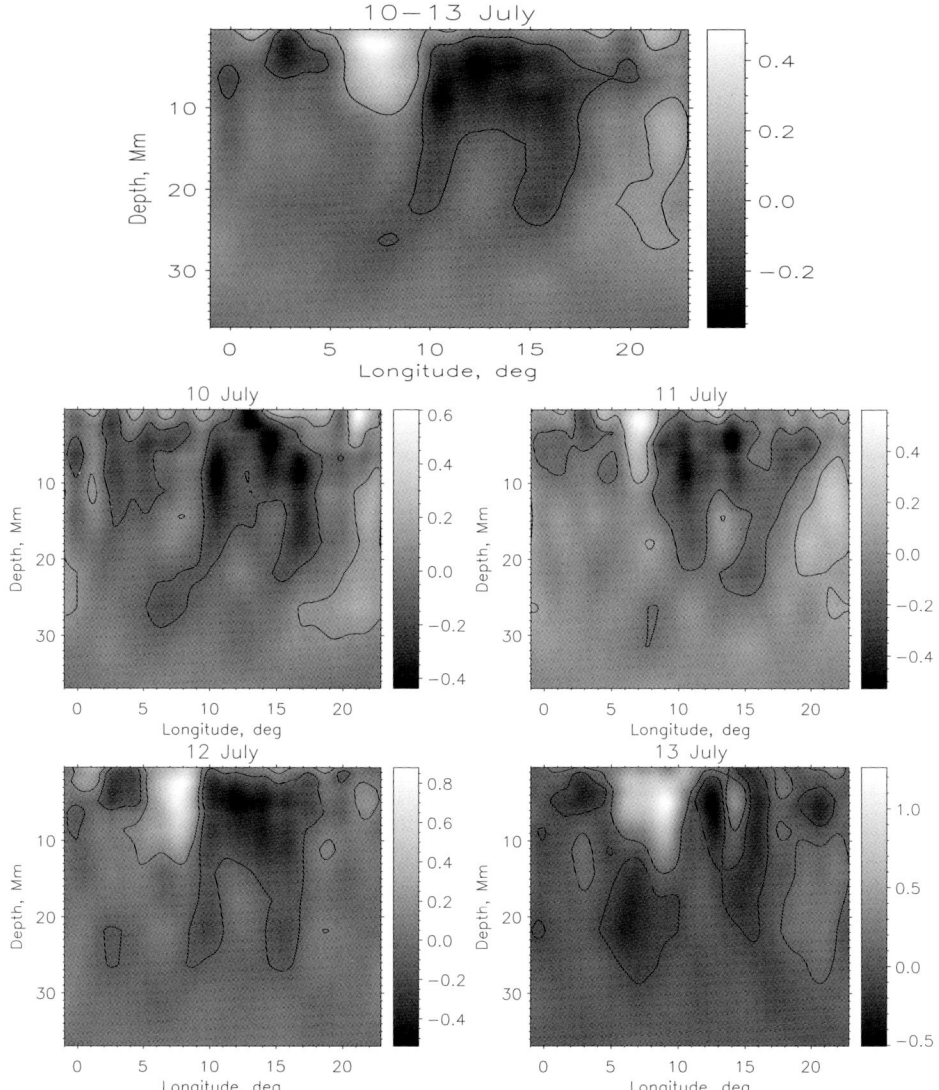

Figure 6 The mean sound-speed perturbation, averaged over all of the processed data, 44 data cubes, (top left) and the mean sound-speed perturbation for the selected dates (averages of 12 data cubes). The greyscale is $\delta c/c$ in percent. The contours correspond to sound-speed perturbation of 0.1%.

In our solutions, the errors due to realisation noise are less than 100 m s^{-1}: this is estimated by processing the 24 quiet-Sun datacubes and applying the approach described in Jensen (2001), Gizon and Birch (2004), and Couvidat *et al.* (2005).

4. Conclusions

We have investigated the emergence of the Active Region NOAA 10790. The region has a complex magnetic structure and can be viewed as an amalgamation of two magnetic region as described in Section 3.1. We see shallow regions of increased sound speed at the location of increased magnetic activity, with regions becoming deeper at the locations of sunspot pores. It appears that a subsurface sound-speed perturbation at depth is associated with the emergence on the surface of a sunspot and sunspot pores. We observe an object with increased sound speed which may be related to the subsurface flux prior to its emergence in the photosphere at the second location. We also observe a loop-like structure of the decreased sound-speed perturbation located below the investigated region, which is present on the averages of the inversions. It would be interesting to see if similar structures characterised by a drop in sound speed can be detected in other instances of emerging active regions.

Based on our results, one can speculate that magnetised regions without any sunspots or sunspot pores lead to increased subsurface sound speed at shallow depths but no perturbation detected at greater depths. On the other hand, the areas associated with sunspot appearance have increased finger-like sound-speed perturbations which extend to depths of 10–20 Mm. Then by tracking the times of the emergence of magnetic flux in the subsurface layers of the Sun one can estimate the speed at which flux emerges to be about 1 km s^{-1}.

We also note that the precise timing of the signature of emergence in the seismic imaging is very limited by the temporal resolution of the sound-speed inversions presented here, since each set is a representative of eight hours of solar oscillations.

In view of the dynamo models discussed in Section 1 (Nandy and Choudhuri, 2002; Krivodubskij, 2005), it could be interesting to compare such results for a variety of emerging active regions at different latitudes and stages of the solar cycle.

Acknowledgements We would like to thank an anonymous referee for valuable suggestions that have significantly improved this paper. We also acknowledge the support from UK Science and Technology Facilities Council grant number PP/C502914/1.

References

Chang, H.-K., Chou, D.-Y., Sun, M.-T.: 1999, *Astrophys. J.* **526**, 53.
Charbonneau, P., MacGregor, K.B.: 1997, *Astrophys. J.* **486**, 502.
Choudhuri, A.R., Schüssler, M., Dikpati, M.: 1995, *Astron. Astrophys.* **303**, 29.
Couvidat, S., Birch, A.C., Kosovichev, A.G., Zhao, J.: 2004, *Astrophys. J.* **607**, 554.
Couvidat, S., Gizon, L., Birch, A.C., Larsen, R.M., Kosovichev, A.G.: 2005, *Astrophys. J.* **158**, 217.
Dikpati, M., Charbonneau, P.: 1999, *Astrophys. J.* **518**, 508.
Duvall, T.L. Jr., Jefferies, S.M., Harvey, J.W., Pommerantz, M.A.: 1993, *Nature* **362**, 430.
Duvall, T.L. Jr., Kosovichev, A.G., Scherrer, P.H., Bogart, R.S., Bush, R.I., de Forest, C., Hoeksema, J.T., Schou, J., Saba, J.L.R., Tarbell, T.D., Title, A.M., Wolfson, C.J., Milford, P.N.: 1997, *Solar Phys.* **170**, 63.
Gizon, L., Birch, A.C.: 2002, *Astrophys. J.* **571**, 966.
Gizon, L., Birch, A.C.: 2004, *Astrophys. J.* **614**, 472.
Gizon, L., Birch, A.C.: 2005, *Living Rev. Solar Phys.* **2**, 6. http://solarphysics.livingreviews.org/Articles/lrsp-2005-6/
Jensen, J.: 2001, Ph.D. thesis. Univ. Aarhus.
Jensen, J., Pijpers, F.P.: 2003, *Astron. Astrophys.* **L5**, 461.
Jensen, J., Jacobsen, B.H., Christensen-Dalsgaard, J.: 1998, In: Korzennik, S., Wilson, A. (eds.) *Proc. SOHO 6/GONG98 Workshop* **SP-418**, ESA, Noordwijk, 635.
Kitchatinov, L.L., Mazur, M.V.: 2000, *Solar Phys.* **191**, 325.
Kosovichev, A.G., Duvall, T.L. Jr.: 1997, In: Pijpers, F.P., Christensen-Dalsgaard, J., Rosenthal, C.S. (eds.) *SCORe'96: Solar Convection and Oscillations and their Relationship*, ASSL **225**, Kluwer Academic, Dordrecht, 241.
Kosovichev, A.G., Duvall, T.L. Jr., Scherrer, P.H.: 2000, *Solar Phys.* **192**, 159.
Krivodubskij, V.N.: 2005, *Astron. Nachr.* **326**, 61.

Nandy, D., Choudhuri, A.R.: 2002, *Science* **296**, 1671.
Parker, E.N.: 1979, *Cosmical Magnetic Fields*, Clarendon Press, Oxford.
Parker, E.N.: 1993, *Astrophys. J.* **351**, 309.
Roberts, P.H., Stix, M.: 1972, *Astrophys. J.* **18**, 453.
Ruediger, G., Brandenburg, A.: 1995, *Astrophys. J.* **296**, 557.
Snodgrass, H.B., Ulrich, R.K.: 1990, *Astrophys. J.* **408**, 707.
Strous, L.H., Zwaan, Z.: 1999, *Astrophys. J.* **527**, 435.
Wang, Y.-M., Sheeley, N.R. Jr.: 1991, *Astrophys. J.* **375**, 761.

High-Resolution Mapping of Flows in the Solar Interior: Fully Consistent OLA Inversion of Helioseismic Travel Times

J. Jackiewicz · L. Gizon · A.C. Birch

Originally published in the journal Solar Physics, Volume 251, Nos 1–2, 381–415.
DOI: 10.1007/s11207-008-9158-z © The Author(s) 2008

Abstract To recover the flow information encoded in travel-time data of time–distance helioseismology, accurate forward modeling and a robust inversion of the travel times are required. We accomplish this using three-dimensional finite-frequency travel-time sensitivity kernels for flows along with a (2 + 1)-dimensional (2 + 1D) optimally localized averaging (OLA) inversion scheme. Travel times are measured by ridge filtering MDI full-disk Doppler data and the corresponding Born sensitivity kernels are computed for these particular travel times. We also utilize the full noise-covariance properties of the travel times, which allow us to accurately estimate the errors for all inversions. The whole procedure is thus fully consistent. Because of ridge filtering, the kernel functions separate in the horizontal and vertical directions, motivating our choice of a 2 + 1D inversion implementation. The inversion procedure also minimizes cross-talk effects among the three flow components, and the averaging kernels resulting from the inversion show very small amounts of cross-talk. We obtain three-dimensional maps of vector solar flows in the quiet Sun at horizontal spatial resolutions of 7 − 10 Mm using generally 24 hours of data. For all of the flow maps we provide averaging kernels and the noise estimates. We present examples to test the inferred flows, such as a comparison with Doppler data, in which we find a correlation of 0.9. We also present results for quiet-Sun supergranular flows at different depths in the upper convection zone. Our estimation of the vertical velocity shows good qualitative agreement with the horizontal vector flows. We also show vertical flows measured solely from f-mode travel times. In addition, we demonstrate how to directly invert for the horizontal divergence and

Helioseismology, Asteroseismology, and MHD Connections
Guest Editors: Laurent Gizon and Paul Cally.

J. Jackiewicz (✉) · L. Gizon
Max-Planck-Institut für Sonnensystemforschung, 37191 Katlenburg-Lindau, Germany
e-mail: jackiewicz@mps.mpg.de

L. Gizon
e-mail: gizon@mps.mpg.de

A.C. Birch
NWRA CoRA Division, Boulder, CO 80301, USA
e-mail: aaronb@cora.nwra.com

flow vorticity. Finally we study inferred flow-map correlations at different depths and find a rapid decrease in this correlation with depth, consistent with other recent local helioseismic analyses.

Keywords Helioseismology · Inverse modeling · Velocity fields · Photosphere · Supergranulation

1. Introduction

Time–distance helioseismology (Duvall *et al.*, 1993) comprises a set of tools that measures and interprets the travel times of seismic waves propagating from one point on the solar surface to any other point. It has been shown that these travel times contain information about solar flows (Kosovichev, 1996; Duvall *et al.*, 1997; Duvall and Gizon, 2000; Gizon, Duvall, and Larsen, 2000; Zhao, Kosovichev, and Duvall, 2001, among others). This paper focuses on the inversion of travel times to obtain high-spatial-resolution maps of near-surface vector flows in quiet-Sun regions. What is unique to this study is that it is the first fully consistent inversion in time–distance helioseismology. The consistency is described by several factors: *i*) We measure travel times with the same definition with which the travel-time sensitivity kernels are computed; *ii*) we use three-dimensional (3D) finite-frequency Born sensitivity kernels, which are necessary to detect flow structures that have spatial scales on the order of the mode wavelength – the regime where the commonly used ray approximation fails (Birch and Felder, 2004, for example); *iii*) we use the full noise-covariance properties of the travel times as an ingredient in the inversion; *iv*) the inversion procedure we choose to implement is "optimal," in that it simultaneously achieves the best possible spatial resolution while minimizing the magnification of the errors. Furthermore, the regularization is carried out in both the horizontal and vertical directions.

We have developed a novel two-plus-one-dimensional (2 + 1D) inversion scheme based on the well-known subtractive optimally localized averages (SOLA) technique (Pijpers and Thompson, 1992; Jackiewicz *et al.*, 2007b). For our purposes, 2 + 1D refers to first solving a 2D horizontal inverse problem followed by a 1D inversion in depth. The inversion procedure explicitly minimizes the cross-talk effects among the three flow components by imposing constraints on the averaging kernels. An important aspect that we introduce to this procedure is that of ridge-filtered travel-time measurements, whereby only wave packets of a particular radial order are used. Sets of point-to-annulus travel times are then computed for the f, p_1, p_2, p_3, and p_4 ridges. This is quite different than the usual phase-speed filtering implemented for travel-time measurements. Subsequently, the sensitivity kernels are also computed as corresponding ridge-filtered quantities to match the travel times. The 2 + 1D inversion is motivated by the observation that to a very good approximation the 3D Born ridge-filtered sensitivity kernels separate into the product of a 2D horizontal function and a 1D function in depth.

Together, these ingredients allow us to infer all three components of the vector flow in the near-surface layers of the quiet-Sun convection zone. In previous work, we determined for the first time the maximum amplitude of flows that can be reliably recovered from a linear model of the travel-time perturbations (Jackiewicz *et al.*, 2007a). The supergranular and other quiet-Sun flows (with velocities ≤ 400 m s^{-1}) that we detect in this study fall within this range, giving us confidence in the reliability of the method. The horizontal spatial resolution of the inversion presented here ($\approx 7-10$ Mm, depending on the observation time, depth, *etc.*) is on the order of, or even in some cases below, the wavelength of the waves used in the analysis (typically 5–20 Mm, depending on the dominant modes). In addition, we perform a direct measurement of the vertical component of the flow (without simply invoking mass conservation) and with confidence that the cross-talk between the horizontal and vertical components has been minimized. Interestingly, we find that we can determine the vertical velocity even from f-mode travel times. We also directly invert for the horizontal divergence of the flow as well as the vertical component of the flow vorticity.

As the main aim of this paper is to develop the inversion procedure and to perform tests of it on real solar data; we postpone the main interpretation of the results, as well as inversions carried out on artificial data, to a future publication. We typically show quiet-Sun flows that have been obtained from 24 hours of data, to maximize the signal-to-noise ratio from the supergranulation signal (Gizon and Birch, 2004). In all of the inferred flow maps we provide estimates of the noise and the spatial resolution.

The paper is organized as follows: In Section 2 we describe the data and the ridge-filtered travel-time measurements. This is followed in Section 3 by a brief discussion of the forward modeling, in other words, the computation of sensitivity kernels and the noise covariances that are consistent with the travel times. Since the multistep inversion procedure is somewhat complicated, we provide an overview in Section 4, followed by two sections that discuss in detail the 2D and 1D parts with several example calculations. Three-dimensional averaging kernels at different depths are presented in Section 7, and finally the various flow maps are presented, tested, and discussed in Section 8. We end with a summary of the results and a discussion of current and future work.

2. Data and Ridge-Filtered Travel-Time Measurements

For this study we use Dopplergram data from the Michelson Doppler Imager (MDI) (Scherrer *et al.*, 1995) onboard SOHO, which are full-disk images of 0.12° spatial sampling (≈ 1.46 Mm) and one-minute cadence. The region of interest is an area centered on NOAA AR 9787 observed between 20 and 28 January 2002 and tracked and remapped courtesy of T.L. Duvall Jr. (To download these data as well as the corresponding magnetograms and intensity images for analysis, visit http://www.mps.mpg.de/projects/seismo/NA4/DATA/data_access.html). The size of the full set of velocity data cubes is $512 \times 512 \times 1440 \times 9$ (two spatial dimensions, 1440 minutes, and nine days). We only use the middle seven days for the results shown here. These data are ideal for helioseismic analysis as there is a sunspot that is large, isolated, and quite stable as it traverses the disk. For our purposes, there are also large regions of relatively quiet Sun in these maps where we will focus our analysis.

We denote a Doppler velocity cube as $\phi(\mathbf{r}, t)$, where

$$\mathbf{r} = (x, y) \tag{1}$$

is the horizontal coordinate, x and y are the East–West and North–South directions, respectively, t is time, and we work in a Cartesian geometry. We filter the data by multiplying the Fourier transform of the data cube by a filter $[F_n(\mathbf{k}, \omega)]$ that selects all modes with the same radial order $[n]$ and removes all others. We call this ridge filtering. The ridge-filtered data $[\Phi_n]$ are then given by

$$\Phi_n(\mathbf{k}, \omega) = F_n(\mathbf{k}, \omega)\phi(\mathbf{k}, \omega), \tag{2}$$

where $\mathbf{k}$ is the horizontal wavevector and ω is the angular frequency. The ridges we filter and retain for this study are the surface-gravity wave (f-mode) ridge and the first four acoustic (p-mode) ridges. Throughout the text we carry the index n, which takes the possible values

$n = \{f, p_1, p_2, p_3, p_4\}$. We refer to these as mode ridges. It is important to note that this type of filtering is different from the phase-speed filtering that is typically done prior to any time–distance analysis.

The cross-covariance functions are computed from $\Phi_n(\mathbf{k}, \omega)$ for three different point-to-annulus geometries, denoted by "oi" (out minus in), "we" (West minus East), and "ns" (North minus South). The "oi" covariances are measured by using the wave signal at a given point and the wave signal averaged over a concentric annulus of radius Δ. The "we" ("ns") quantities use the central wave signal along with the wave signal averaged over the annulus but weighted by $\cos\theta$ ($\sin\theta$), where θ is the angle between the x-direction and each point on the annulus. This particular procedure was introduced in Jackiewicz et al. (2007b) and is similar to what is usually done in standard time–distance measurements (Duvall et al., 1997). The temporal cross-covariance functions are computed for 20 different annulus radii ($\Delta = 1.46$ to 29.2 Mm, incremented by $\delta\Delta = 1.46$ Mm) for each mode ridge and measurement type, which yields a set of functions $C^{\alpha,n}(\mathbf{r}, t; \Delta)$, where $\alpha = \{\text{oi, we, ns}\}$. Flows introduce asymmetries in time lag $[t]$ in the cross-covariance functions.

Travel times are then measured according to the procedure developed by Gizon and Birch (2002, 2004). This method establishes a linear relationship between the travel-time perturbations $[\tau]$ and the cross-covariances, given by

$$\tau_n^\alpha(\mathbf{r}, \Delta) = h_t \sum_t W^n(\Delta, t) \left[C^{\alpha,n}(\mathbf{r}, t; \Delta) - C^{\text{ref},n}(\mathbf{r}, t; \Delta) \right], \tag{3}$$

where the sum over t denotes a discrete sum over the entire observation time $[T]$ (for this work $T = $ one day in most cases), $h_t = $ one minute is the temporal sampling rate of the data, W are weight functions, and C^{ref} is a reference cross-covariance symmetric in time lag, which we choose to be the cross-covariance computed from our model power spectrum. This model is tuned to match the observed power spectrum of each individual ridge (see Section 3). C^{ref} is the inverse Fourier transform of the model power spectrum. Full details about the W function and Equation (3) can be found in Gizon and Birch (2004). The resulting three different types of travel-time measurements are constructed to have sensitivity to different flow geometries. The "oi" travel times are highly sensitive to horizontal flow divergence, whereas the "we" and "ns" travel times give information about directional flows.

3. Forward Modeling

We now briefly discuss the steps that we have taken to model the ridge-filtered travel-time measurements described in the previous section, as well as the measurement noise properties. Full details of forward modeling in time–distance helioseismology can be found in Gizon and Birch (2002, 2004), and Birch and Gizon (2007).

3.1. Travel-Time Sensitivity Kernels for Ridge Filtering

We consider travel-time measurements of type $\alpha = \{\text{oi, we, ns}\}$ for each distance $[\Delta]$ and for each mode ridge $[n]$ as described in the previous section. The travel-time perturbations are related to the small-amplitude flows through a linear relation:

$$\tau_n^\alpha(\mathbf{r}_i; \Delta) = h_r^2 h_z \sum_{j,z} \mathbf{K}^{\alpha,n}(\mathbf{r}_j - \mathbf{r}_i, z; \Delta) \cdot \mathbf{u}(\mathbf{r}_j, z) + \mathcal{N}_{\text{tt}}^{\alpha,n}(\mathbf{r}_i; \Delta), \tag{4}$$

where $\mathbf{K}$ denotes the three-dimensional travel-time sensitivity kernel, $\mathbf{u}$ is the real vector flow in the Sun, $\mathcal{N}_{\text{tt}}$ represents the noise in the travel times [tt], and the sum over the vertical

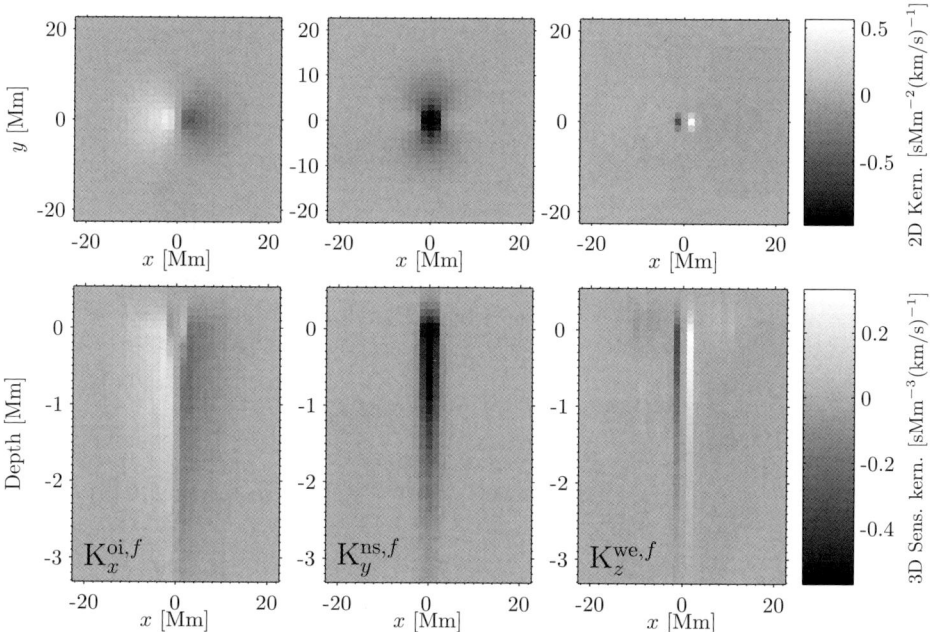

Figure 1 Examples of various sensitivity kernels $\mathbf{K}^{\alpha,f}(\mathbf{r},z;\Delta)$ used in this study and defined in Equation (4) obtained by computing weighted azimuthal averages of the point-to-point kernels for the α point-to-annulus geometries. The kernels in the left column give the sensitivity of "oi" f-mode travel times to u_x, and the kernels in the middle (right) column give the sensitivity of the "ns" ("we") f-mode travel times to u_y (u_z). The top row are 2D kernels obtained after integrating the 3D kernels over depth. The bottom row shows depth slices along $y = 0$ of the 3D kernels. The type [α] and mode ridge [n] of the kernels in each column is indicated in the bottom panels. For each case, $\Delta = 10.2$ Mm. The gray scale in the bottom row has been truncated to 75% of kernel maximum.

coordinate [z] denotes a discrete sum (where we note that $z = 0$ at the surface and is negative inside the Sun). The kernels are computed so that the horizontal grid spacing, $h_x = h_y = 1.46$ Mm and $h_r^2 = h_x h_y$, matches that of the travel-time measurements. The vertical grid (with spacing h_z) is taken from the model on which the kernels are computed. The kernel [$\mathbf{K}$] is computed in the first Born approximation (Gizon and Birch, 2002). We start with the point-to-point kernels for flows derived in Birch and Gizon (2007), computed for the f, p_1, p_2, p_3, and p_4 ridges, whose input power spectra have been tuned to match the ridge-filtered observed power spectra. For each ridge, 20 kernels are computed, one for each Δ. The 3D point-to-point kernels are then azimuthally averaged according to the three different point-to-annulus weighting geometries α used for the travel-time measurements. The total number of point-to-annulus kernels thus obtained is 300, with each kernel having three components. A few examples of f-mode kernels after weighted azimuthal averaging are shown in Figure 1 for $\mathbf{r}_i = 0$. From left to right, the columns show kernels that give the sensitivity to u_x, u_y, and u_z, respectively. The top row shows the resulting 2D kernel after integration over depth.

As mentioned in the introduction, the motivation behind the inversion we choose to perform is that the ridge-filtered kernels to a very good approximation separate into a horizontal 2D function of $\mathbf{r}$ times a 1D function of depth z. In other words, dropping the labels α, n, and Δ for the moment, we can write the ith component of the sensitivity kernel defined in

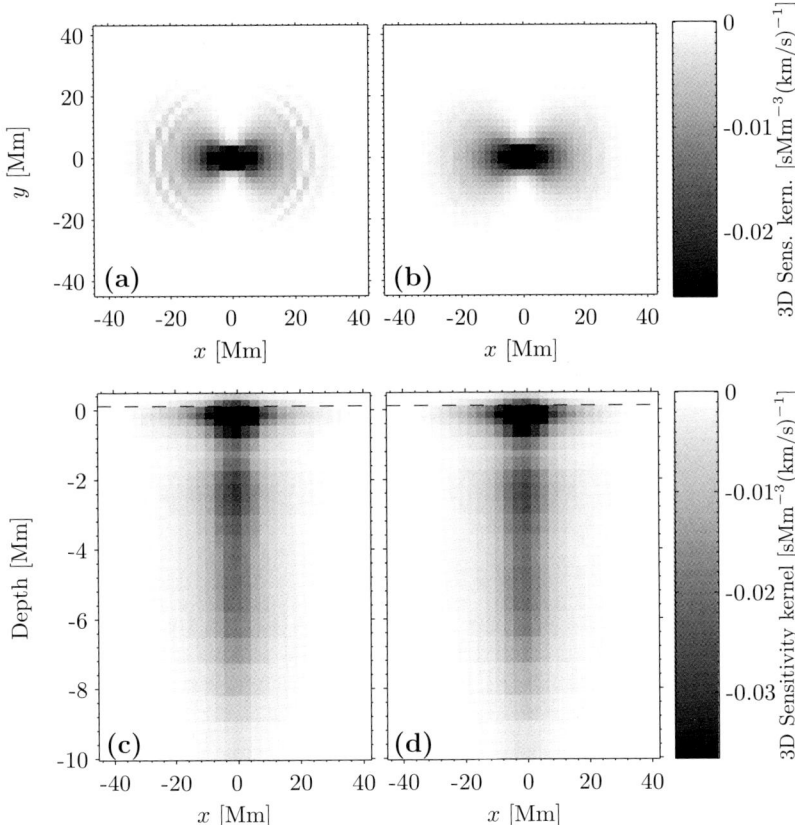

Figure 2 Separability of a sensitivity kernel for p_2 and for ridge filtering. This particular kernel, K_x^{we,p_2}, gives the sensitivity of "we" p_2-mode travel times to u_x for $\Delta = 23.4$ Mm. (a) Horizontal cut through the kernel at a height of about 100 km above the photosphere, denoted by the dashed line in the panel below. (b) Horizontal cut through the quantity obtained as the product of the two (normalized) functions, $f(\mathbf{r})$ and $g(z)$, computed by integrating the kernel in (a) horizontally and over depth, according to Equations (5) and (6). (c) A depth slice along $y = 0$ of the original kernel from (a). (d) A depth slice along $y = 0$ of the kernel in (b). The match between the left and right columns is quite good. The gray scale has been truncated to 40% of the maximum value for ease of comparison.

Equation (4) as the product

$$K_i(\mathbf{r}, z) \approx f_i(\mathbf{r}) g_i(z), \qquad (5)$$

where

$$f_i(\mathbf{r}) = \frac{h_z \sum_z K_i(\mathbf{r}, z)}{\sqrt{h_r^2 h_z \sum_{j,z} K_i(\mathbf{r}_j, z)}}, \qquad g_i(z) = \frac{h_r^2 \sum_j K_i(\mathbf{r}_j, z)}{\sqrt{h_r^2 h_z \sum_{j,z} K_i(\mathbf{r}_j, z)}}. \qquad (6)$$

In Figure 2 we demonstrate the separability for an example p_2 and "we"-averaged kernel. Figure 2(a) shows a slice of the K_x^{we,p_2} kernel at about 100 km above the photosphere, and in Figure 2(b) we show the same slice, but of the function obtained by Equation (5). In the bottom two panels, depth slices of these two kernels are shown for comparison. The match between the kernels is quite good. We have also studied the point-to-point kernels in this way. In general, the weighted azimuthally averaged kernels (the ones actually used in the inversion) separate "better" than the point-to-point ones, because much of the small-scale structure is averaged away. All of the other kernels for ridge filtering that we have studied separate this way to a very good approximation, giving us confidence in the type of inversion we choose to employ.

3.2. Noise-Covariance Matrix

Travel times contain a significant amount of realization noise and a good understanding of these noise properties allows us to assign accurate errors to the flow estimates. It has been shown in previous time–distance inversions that it is important to take into account the noise-covariance matrix (Jensen, Duvall, and Jacobsen, 2003; Couvidat *et al.*, 2005; Couvidat, Birch, and Kosovichev, 2006). Gizon and Birch (2004) showed in detail how to compute model noise covariances of travel times.

We assume that the solar oscillations are stationary and spatially homogeneous since we are restricting our study to the quiet Sun. We further assume that the noise between different ridge measurements n and n' is uncorrelated. This approximation is acceptable for this study because we carry out ridge filtering. The covariance matrix Λ of the noise components $\mathcal{N}_{tt}$ from Equation (4) is given by

$$\Lambda_n^{\alpha\beta}(\mathbf{r}_i - \mathbf{r}_j; \Delta, \Delta') = \text{Cov}[\mathcal{N}_{tt}^{\alpha,n}(\mathbf{r}_i; \Delta), \mathcal{N}_{tt}^{\beta,n}(\mathbf{r}_j; \Delta')]. \tag{7}$$

This quantity has units of s^2 and is computed according to Equation (28) in Gizon and Birch (2004). Example plots for the case when $n = f$ and for $\Delta = 5$ Mm were shown in Jackiewicz *et al.* (2007b) using the same model. Similar features are seen for all mode ridges and distances considered here. Note that the covariance matrix elements of Λ scale with the observation time T as T^{-1}.

4. Basic Strategy of the 2 + 1 Dimensional Subtractive Optimally Localized Averages Inversion

The problem we wish to solve is to estimate, for example, $u_x(\mathbf{r}, z)$ in Equation (4), given the travel-time measurements, the sensitivity kernels, and the noise-covariance matrix. We carry this out using a 2 + 1D SOLA inversion procedure. We formulate the problem in terms of the component u_x only for notational simplicity, noting that the procedure for finding all other flow components is completely equivalent. We first discuss the basic idea of this method, and then in the following two sections separately describe the 2D and 1D parts in more detail.

An OLA-type inversion for our purposes seeks a way to combine the sensitivity kernels to find a three-dimensional averaging kernel that is, roughly, spherical, centered horizontally about the origin $\mathbf{r} = 0$ and vertically about the target depth z_t located somewhere in the solar interior. The averaging kernel will be found from the 2 + 1D inversion and is defined by

$$v_x(\mathbf{r}; z_t) = h_r^2 h_z \sum_{j,z} {}^{2+1D}\mathcal{K}^{u_x}(\mathbf{r}_j - \mathbf{r}, z; z_t) \cdot \mathbf{u}(\mathbf{r}_j, z) + \text{noise}, \tag{8}$$

where v_x is an estimate of the real solar flow u_x (as is the case in the rest of the paper) and the noise term will be specifically quantified in the following. To clarify the notation, sensitivity kernels are written as K and averaging kernels are written as $\mathcal{K}$. Any superscript to the left of the averaging kernel denotes by which type of inversion it was computed, for example, $^{1D}\mathcal{K}$, $^{2D}\mathcal{K}$, and $^{2+1D}\mathcal{K}$. The superscripts to the right indicate for which component of $\mathbf{u}$ and ridge n the averaging kernel is computed.

It is important to observe from Equation (8) that the y- and z-components of the averaging kernel should be zero, so that the flow estimate $[v_x]$ is not contaminated by any cross-talk from u_y and u_z. As seen in the next section, the 2D inversion attempts to accomplish this by constraining the spatial integral of $\mathcal{K}_y$ and $\mathcal{K}_z$ to be zero. Ideally, one would want the x-component of the averaging kernel to be a δ function; however, noise and a finite set of travel times inhibit this. Nonetheless, if an acceptable averaging kernel is found, then the travel times can be properly averaged to give an estimate of the local flow $v_x \approx u_x$ in which we are interested.

Since the problem essentially separates into a 2D and a 1D problem because of ridge filtering, the three-dimensional averaging kernel $^{2+1D}\mathcal{K}^{u_x}$ from Equation (8) is derived in two main steps. The first step is to compute the 2D (horizontal) component of the averaging kernel, $^{2D}\mathcal{K}^{u_x}(\mathbf{r})$, such that its x-component is highly peaked about the point $\mathbf{r} = 0$, and the other components are zero. This typically involves trying to match the x-component to a target function that is a 2D Gaussian function in horizontal coordinates $\mathbf{r} = (x, y)$. The inversion coefficients, or weights w, that accomplish this averaging of the sensitivity kernels, are then used to combine the travel times in such a way that the estimated flow for any ridge measurements n is

$$v_x^n(\mathbf{r}; \Delta) = \sum_{j,\alpha} w^{\alpha,n}(\mathbf{r}_j - \mathbf{r}; \Delta) \tau_n^\alpha(\mathbf{r}_j; \Delta). \tag{9}$$

The resulting flow map $[v_x^n]$ is an average of the real flow $[u_x]$ over the depth that the dominant modes of ridge n probe. An intermediate step is to combine these maps over all distances $[\Delta]$ by an averaging procedure based on the correlated noise in the measurements. A set of weightings $[\gamma]$ is computed, described in detail in Section 5.1, such that the distance-averaged flows are given by

$$v_x^n(\mathbf{r}) = \sum_j \gamma_j^n v_x^n(\mathbf{r}; \Delta_j), \tag{10}$$

where γ_j is defined in Equation (21) and the sum j runs over all Δ used in the problem.

The second main step is to obtain localization of the 3D averaging kernel in the vertical direction about z_t by combining separate 2D ridge measurements. A 1D inversion in depth is thus performed that seeks a new set of inversion coefficients c^n. These new coefficients combine the 2D flow maps in Equation (10) in such a way that the final estimate of the flow around z_t is given by

$$v_x(\mathbf{r}; z_t) = \sum_n c^n(z_t) v_x^n(\mathbf{r}) \approx u_x(\mathbf{r}; z_t), \tag{11}$$

where $u_x(\mathbf{r}; z_t)$ denotes the real flow at a particular depth z_t. The whole procedure is in principle carried out to estimate each flow component (u_x, u_y, u_z) at many different target depths $[z_t]$ to infer the vector flow $[\mathbf{u}(\mathbf{r}, z)]$ throughout a desired interior region. We now describe in more detail how the 2D and 1D inversion weights are computed in the following two sections.

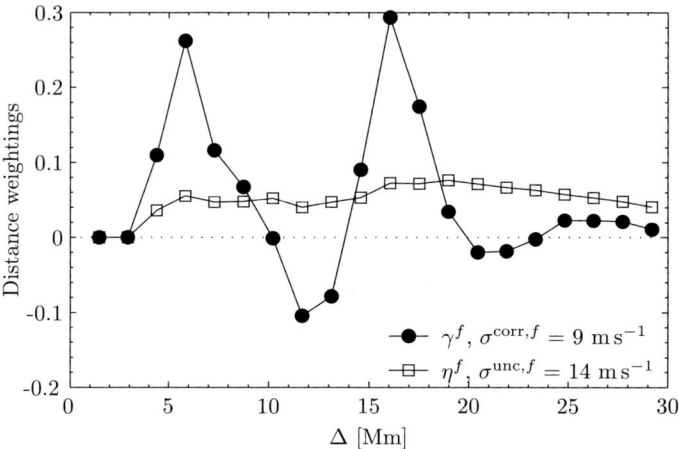

Figure 3 Example distance weightings from a 2D f-mode inversion for u_x using a target resolution of 11 Mm, as described in Section 5.1. The observation time is 24 hours. The open squares are the weights at each distance obtained by assuming that the estimated flows $v_x^f(\mathbf{r}; \Delta)$ are uncorrelated for different Δ (Equation (17)). The filled circles are the weights obtained by taking the correlations properly into account (Equation (21)). The noise σ for each case is indicated in the legend.

5. 2D Horizontal Inversion

Based on the separability of the sensitivity kernels discussed in Section 3, the 2D inversion is formulated to solve Equation (4) for $v_x(\mathbf{r})$ by using the 2D depth-integrated kernel

$$\mathbf{K}^{\alpha,n}(\mathbf{r}; \Delta) = h_z \sum_z \mathbf{K}^{\alpha,n}(\mathbf{r}, z; \Delta), \qquad (12)$$

which we compute for all α, n, and Δ available.

We can define the averaging kernel that we wish to find from the 2D inversion, $^{2D}\mathcal{K}^{u_x,n}$, by inserting Equation (4) into Equation (9) to obtain

$$v_x^n(\mathbf{r}; \Delta) = h_r^2 h_z \sum_{i,z} {}^{2D}\mathcal{K}^{u_x,n}(\mathbf{r}_i - \mathbf{r}, z; \Delta) \cdot \mathbf{u}(\mathbf{r}_i, z)$$
$$+ \sum_{j,\alpha} w^{\alpha,n}(\mathbf{r}_j - \mathbf{r}; \Delta) \mathcal{N}_{tt}^{\alpha,n}(\mathbf{r}_j; \Delta), \qquad (13)$$

where

$$^{2D}\mathcal{K}^{u_x,n}(\mathbf{r}, z; \Delta) = \sum_{j,\alpha} w^{\alpha,n}(\mathbf{r}_j; \Delta) \mathbf{K}^{\alpha,n}(\mathbf{r} - \mathbf{r}_j, z; \Delta). \qquad (14)$$

This shows explicitly that the averaging kernel gives only an estimate v_x^n of the flow that is an average of the real flows over some depth.

The details for obtaining the inversion weights w for a 2D SOLA inversion for flows were presented in Jackiewicz et al. (2007b); however, in that work only one mode ridge, one distance Δ, and two components of the sensitivity kernels (K_x and K_y) were utilized. The generalization to our present case is straightforward. To summarize this procedure, we first prescribe a target function $^{2D}\mathcal{T}^{u_x}$ that we wish the averaging kernel to resemble. It is a vector-valued function, chosen such that the x-component is typically a 2D Gaussian in $\mathbf{r}$ with dispersion σ, and the other components are zero:

$$^{2D}\mathcal{T}^{u_x}(\mathbf{r}) = \left(\frac{e^{-r^2/2\sigma^2}}{2\pi\sigma^2}, 0, 0\right), \tag{15}$$

where $r = \|\mathbf{r}\|$ is the 2D vector norm. The horizontal integral of the target function is normalized to one. The full width at half-maximum (FWHM $= 2\sigma\sqrt{2\ln 2}$) of the target function is a measure of the resolution of the inversion if the averaging kernel matches it well. For the sake of completeness, in an inversion for the jth component of $\mathbf{u}$, with the Gaussian function in Equation (15) denoted as $G(r)$, the ith component of the target function is $^{2D}\mathcal{T}_i^{u_j} = G(r)\hat{\mathbf{e}}_i\delta_{ij}$, where $\hat{\mathbf{e}}_i$ is the unit vector in the ith direction and δ is the Kronecker delta function.

Two quantities, one that measures the mismatch between the averaging kernel and target function and another that measures the noise propagation, are computed. Let the noise in the inversion for u_x for ridge n be denoted by $\mathcal{N}^{u_x,n}$; it is specifically defined in Section 5.1. A minimization (with respect to the inversion weights) is carried out according to

$$\min_w \sum_i \left\|^{2D}\mathcal{K}^{u_x,n}(\mathbf{r}_i) - {}^{2D}\mathcal{T}^{u_x}(\mathbf{r}_i)\right\|^2 + \beta\left(\mathcal{N}^{u_x,n}\right)^2, \tag{16}$$

where β is some regularization parameter that we choose typically to be quite small (Jackiewicz et al., 2007b). A large matrix is then regularized and inverted for each value of the trade-off parameter, which results in a unique set of weights at each point in this parameter space. We choose an "optimal" set of weights [w] from examining the trade-off curve (or "L" curve) as discussed in Jackiewicz et al. (2007b), such that the averaging kernel matches closely the target function. The 2D averaging kernel is constructed by convolution of the sensitivity kernels with the weights according to Equation (14).

Since the inversion in this example is carried out for u_x^n, it is important that $^{2D}\mathcal{K}_y^{u_x,n}$ and $^{2D}\mathcal{K}_z^{u_x,n}$ be as close to zero as possible, to minimize the "cross-talk" among all of the components. This is achieved in practice by constraining the total spatial integrals of the y- and z-components to be zero, although in practice there is usually some structure present even with this constraint. We are in the process of exploring other effective constraints. When a well-localized averaging kernel is found for each ridge [n], the weights are then suitable to be used to average the travel times to give an estimate of the flow $v_x^n(\mathbf{r})$ using Equation (9).

5.1. Combining All of the Distances

Throughout this inversion procedure, it is necessary to combine the quantities we obtain for different annulus radii [Δ], such as the estimated flow maps $v_x^n(\mathbf{r}; \Delta)$. This is accomplished by weighting each distance appropriately. One simple way of achieving this, typically used in helioseismology, is to assume that the noise in each measurement is independent and uncorrelated. Then the standard deviation in the estimated flow maps is used to determine the contribution of the errors at each distance. We denote the standard deviation of a set of flows [v_x^n] for each distance [Δ_j] and mode ridge [n] as $\sigma_j^{\text{unc},n}$. The "unc" superscript emphasizes the assumption of uncorrelated data. Finding the minimum variance of this set then gives weighting factors η_j^n according to

$$\eta_j^n = \frac{(1/\sigma_j^{\text{unc},n})^2}{\sum_{j=1}^{20}(1/\sigma_j^{\text{unc},n})^2}, \tag{17}$$

where the sum in the denominator runs over all 20 distances used in this problem. We carry out a 2D inversion as described earlier for u_x using f modes in a region of quiet Sun (the same region used in Section 8). The weights obtained from the estimated flows using Equation (17) are plotted in Figure 3 as open squares. For very small distances, where the noise is very high, the weights are nearly zero. From the total variance we obtain a noise estimation given by

$$\sigma^{\text{unc,f}} = \left[\sum_j (1/\sigma_j^{\text{unc,f}})^2\right]^{-1/2}, \tag{18}$$

which for this particular example is 14 m s^{-1}.

However, we know that the values of the flows at different Δ are correlated quite strongly due to noise (Gizon and Birch, 2004), and so we choose to average them in a way that takes these correlations into account. A covariance matrix C_n of the noise in the individual flow measurements at distances Δ and Δ' of ridge n is computed by using the 2D inversion weights as

$$C_n(\Delta, \Delta') = \sum_{i,j,\alpha,\beta} w^{\alpha,n}(\mathbf{r}_i; \Delta) \Lambda_n^{\alpha\beta}(\mathbf{r}_i - \mathbf{r}_j; \Delta, \Delta') w^{\beta,n}(\mathbf{r}_j; \Delta'), \tag{19}$$

where Λ is the covariance matrix of the noise in the travel times (Equation (7)) and w are the 2D inversion weights. Note that the matrix C has units of length^2 time^{-2}. The final measurement of any general quantity $[q]$ for each mode ridge is then obtained by averaging the 20 distances we use according to (for example, see Schmelling, 1995)

$$q^n(\mathbf{r}) = \sum_{j=1}^{20} \gamma_j^n q^n(\mathbf{r}; \Delta_j), \tag{20}$$

where the set of weightings $[\gamma_j^n]$ is given by

$$\gamma_j^n = \frac{(\sum_{i=1}^{20} C_n^{-1}(\Delta_i, \Delta_j))}{(1/\sigma^{\text{corr},n})^2}, \tag{21}$$

and the variance $(\sigma^{\text{corr},n})^2$ for the correlated case is

$$(\sigma^{\text{corr},n})^2 = \left(\sum_{ij} (C_n^{-1}(\Delta_i, \Delta_j))\right)^{-1}. \tag{22}$$

We can now identify $\mathcal{N}^{u_x,n} \equiv \sigma^{\text{corr},n}$, the noise in a measurement of v_x^n, introduced in Equation (16). A set of weights for the f-mode case $[\gamma_j^f]$ obtained this way is plotted as the filled circles in Figure 3 to compare with the uncorrelated case. What is interesting to note is that for some distances a negative contribution is needed to average the data properly, which is never the case for the uncorrelated data. The estimated noise is also lower than in the uncorrelated case. Equation (20) is quite general and has been used to obtain the distance-averaged flow maps presented in Section 8, as well as all of the averaging kernels shown in this paper. Quantities written without the distance argument Δ have been averaged according to Equation (20).

5.2. Averaging Kernels from the 2D Inversion

Example averaging kernels after combining all distances for a 2D inversion are shown in Figures 4–7. In all figures, the top row shows the 2D averaging kernel obtained by integrating over depth the adjacent 3D averaging kernel in the bottom row. The bottom panels show depth slices along $y = 0$. The inherent noise from the inversion corresponding to each figure is given in Table 1.

Figures 4 and 5 show averaging kernels for a 2D inversion for u_x for the f- and p_1-mode ridges, respectively, and Figures 6 and 7 are for an inversion for u_z. The absence of any dominant cross-talk is evident in the top panels of all figures, which are the 2D averaging kernels that come straight out of the 2D inversion. There is a completely negligible y-component in all cases for the inversions for u_x. The cross-talk is slightly more pronounced for the x- and y-components of the kernels in the inversions for u_z, but it is confined to the very near surface region, typically above the depths that are significant for our inversion results. In general, these averaging kernels are quite good; the only structure from the "off-diagonal" terms are of the order of about 5% of the diagonal terms. It is important to study averaging kernels such as these to have an idea of what the inversion is actually accomplishing in terms of sensitivity. Similar plots have been examined for all of the other mode ridges available and they exhibit similar features.

5.3. Minimum Variance

The averaging kernels from the 2D inversion in Figures 4–7 are computed for each separate mode ridge $[n]$. One possible way to combine them over all ridges is to use a simple minimum-variance treatment of the noise in the estimated flow component v_x^n, as described in Section 5.1 for the case of distance averaging. Because of ridge filtering, we consider the noise between mode ridges n and n' to be uncorrelated. Thus, any quantity $[q^n]$ that depends

Table 1 Values of the relevant quantities for some of the figures in this paper. Listed is the figure number, the type of inversion used to produce the figure, the 1D inversion weights c^n (or minimum variance weights d^n), and the estimated noise associated with each measurement (1σ values). Note that the 1D inversion weights do not apply for the 2D inversion.

Figure	Inversion type	c^f	c^{p_1}	c^{p_2}	c^{p_3}	c^{p_4}	Noise (m s^{-1})
4	2D	–	–	–	–	–	16
5	2D	–	–	–	–	–	26
6	2D	–	–	–	–	–	17
7	2D	–	–	–	–	–	24
8	2D (min. var.)	0.65	0.24	0.086	0.014	0.002	21
12	2 + 1D	1.02	− 0.04	− 0.01	10^{-4}	10^{-4}	27
13	2 + 1D	− 1.2	2.2	10^{-4}	− 0.04	10^{-4}	40
14	2 + 1D	− 0.17	− 1.7	2.3	0.5	0.05	57
18 (a)	2 + 1D	0	1	0	0	0	19
18 (b)	2 + 1D	0.33	–	0.47	0.17	0.05	17
20 (top)	2 + 1D	1.08	− 0.13	0.004	0.01	0.005	10
20 (middle)	2 + 1D	− 0.11	1.01	0.03	0.04	0.01	12
20 (bottom)	2 + 1D	− 0.92	1.26	0.63	0.02	0.01	22

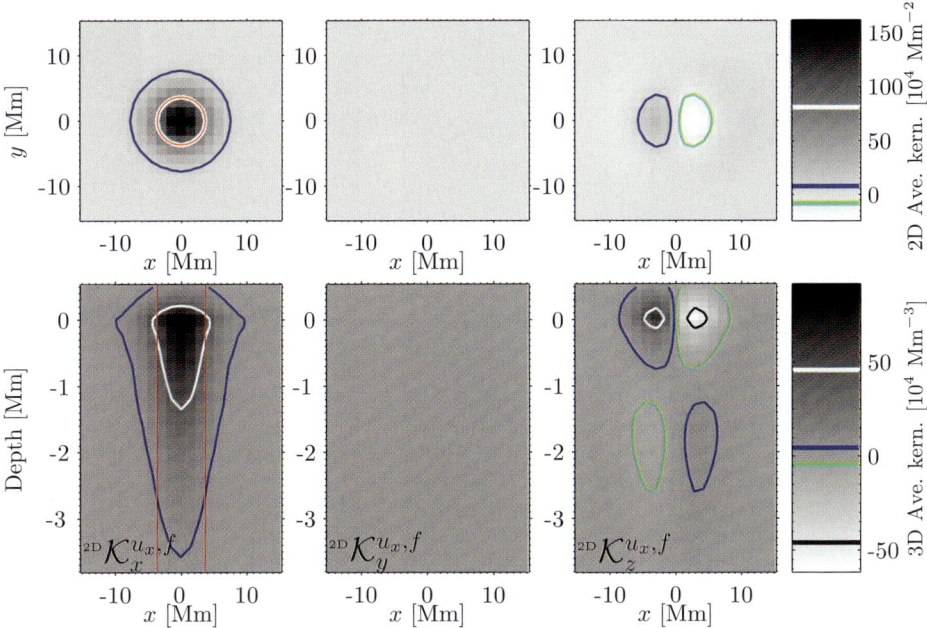

Figure 4 Two-dimensional (top panels) and three-dimensional (bottom panels) averaging kernels for a 2D inversion for u_x using only f modes. The panels in the top row show the integral over depth of the corresponding 3D kernel below it. The bottom row shows depth slices of the three components of the 3D kernel along the $y = 0$ line. The red line outlines the FWHM = 7.3 Mm of the 2D Gaussian target function used in the inversion (see Equation (15)). The overplotted color contours, which are also marked on the color bar for reference, denote the following: half-maximum of the kernel (white), negative of the half-maximum (black), and ±5% of the maximum value of the kernel (blue and green, respectively). Estimates of the noise from this and all other inversions are given in Table 1.

on the set of mode ridges can be averaged according to

$$q^{u_x} = \sum_n d^n q^{u_x,n}, \quad (23)$$

where the weights d^n are given by

$$d^n = \frac{(1/\sigma^{\text{corr},n})^2}{\sum_n (1/\sigma^{\text{corr},n})^2}, \quad (24)$$

and $\sigma^{\text{corr},n}$, the correlated noise estimated from the v_x^n measurements, is defined in Equation (22). In Figure 8 we show the three components of an averaging kernel $\mathcal{K}^{u_x}$ obtained by combining five kernels $\mathcal{K}^{u_x,n}$ from the minimum variance in v_x^n using the weights given in Equation (24). The minimum-variance weights for each ridge for this figure are given in Table 1. The noise level is low and the kernel is well localized horizontally; however, there is clearly have no control over the depth at which one wishes to have sensitivity.

In conclusion, the 2D inversion produces averaging functions (such as those shown in Figures 4–8) that are "optimally" localized in the horizontal direction, but not in depth. This depth localization is accomplished by a 1D inversion, to which we now turn.

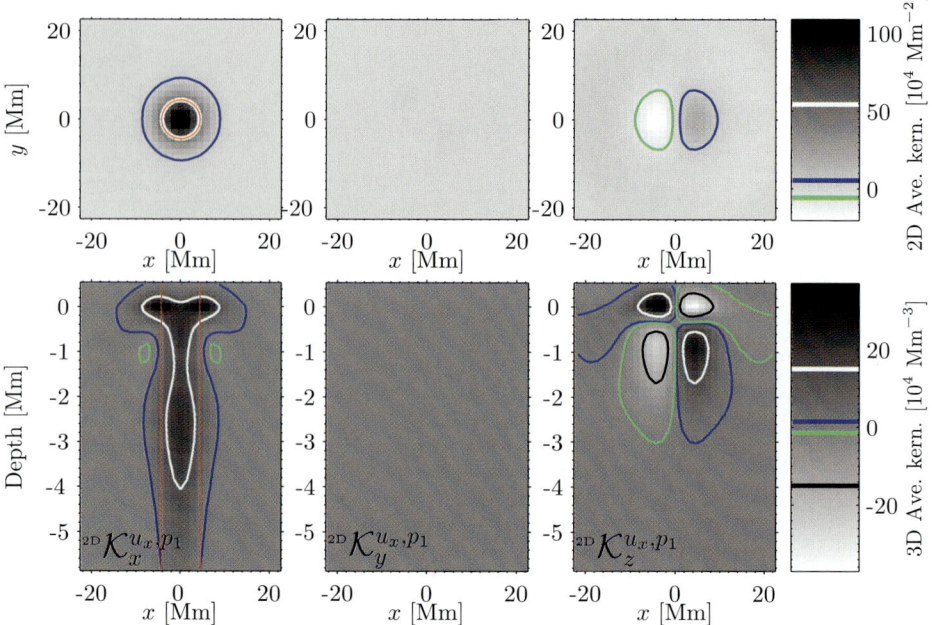

Figure 5 Two-dimensional (top panels) and three-dimensional (bottom panels) averaging kernels for a 2D inversion for u_x using only p_1 modes. The panels in the top row show the integral over depth of the corresponding 3D kernel below it. The bottom row shows depth slices of the three components of the 3D kernel along the $y = 0$ line. The red line outlines the FWHM = 7.3 Mm of the 2D Gaussian target function used in the inversion. The overplotted color contours, which are also marked on the color bar for reference, denote the following: half-maximum of the kernel (white), negative of the half-maximum (black), and ±5% of the maximum value of the kernel (blue and green, respectively).

6. 1D SOLA Depth Inversion

Up to this point, the 2D inversion has provided averaging kernels that average the real solar flows to give estimates for each particular ridge, according to Equation (13). If we assume that the real flows vary slowly in $\mathbf{r}$ over the horizontal extent of the averaging kernel, we can perform the summation in Equation (13) over $\mathbf{r}_i$ to obtain

$$v_x^n(\mathbf{r}) \approx h_z \sum_z {}^{1D}K_x^{u_x,n}(z) u_x(z) + \sum_{j,\alpha} w^{\alpha,n}(\mathbf{r}_j - \mathbf{r}) \mathcal{N}_{tt}^{\alpha,n}(\mathbf{r}_j), \qquad (25)$$

where

$$^{1D}K_x^{u_x,n}(z) \equiv h_r^2 \sum_i {}^{2D}\mathcal{K}_x^{u_x,n}(\mathbf{r}_i, z) \qquad (26)$$

is the 1D sensitivity kernel for an inversion for u_x and mode ridge n. Recall that the horizontal integrals of $^{2D}\mathcal{K}_y$ and $^{2D}\mathcal{K}_z$ are zero owing to the constraint imposed in the 2D inversion; therefore, only the x-component of the quantities remains in the right-hand sides of Equations (25) and (26).

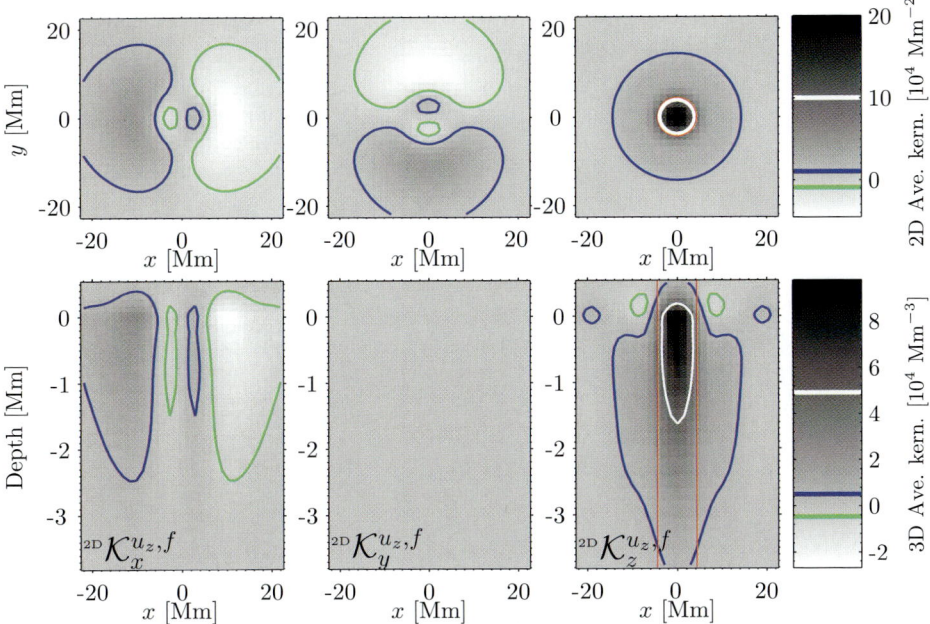

Figure 6 Two-dimensional (top panels) and depth slices of three-dimensional (bottom panels) averaging kernels for a 2D inversion for u_z using only f modes. The red line outlines the FWHM = 7.3 Mm of the 2D Gaussian target function used in the inversion. The overplotted color contours, which are also marked on the color bar for reference, denote the following: half-maximum of the kernel (white), negative of the half-maximum (black), and $\pm 5\%$ of the maximum value of the kernel (blue and green, respectively).

The five available 1D sensitivity kernels are shown in Figure 9. The 1D SOLA inversion seeks inversion coefficients $[c^n(z_t)]$ that combine each ridge measurement about target depth z_t, so that the final estimate of the flow using Equation (25) is

$$v_x(\mathbf{r}; z_t) = \sum_n c^n(z_t) v_x^n(\mathbf{r})$$

$$= h_z \sum_z {}^{1D}\mathcal{K}_x^{u_x}(z; z_t) u_x(\mathbf{r}, z) + \sum_{j,\alpha,n} c^n(z_t) w^{\alpha,n}(\mathbf{r}_j - \mathbf{r}) \mathcal{N}_{tt}^{\alpha,n}(\mathbf{r}_j),$$

where

$$^{1D}\mathcal{K}_x^{u_x}(z; z_t) = \sum_n c^n(z_t) {}^{1D}\mathrm{K}_x^{u_x,n}(z) \qquad (27)$$

is the 1D averaging kernel peaked about z_t. The 1D coefficients $[c^n]$ are obtained in an analogous way to the 2D case. A target function is chosen that is typically a 1D Gaussian in depth, centered about z_t:

$$^{1D}\mathcal{T}_x^{u_x}(z; z_t) = \frac{e^{-(z-z_t)^2/2\sigma^2}}{\sigma\sqrt{2\pi}}. \qquad (28)$$

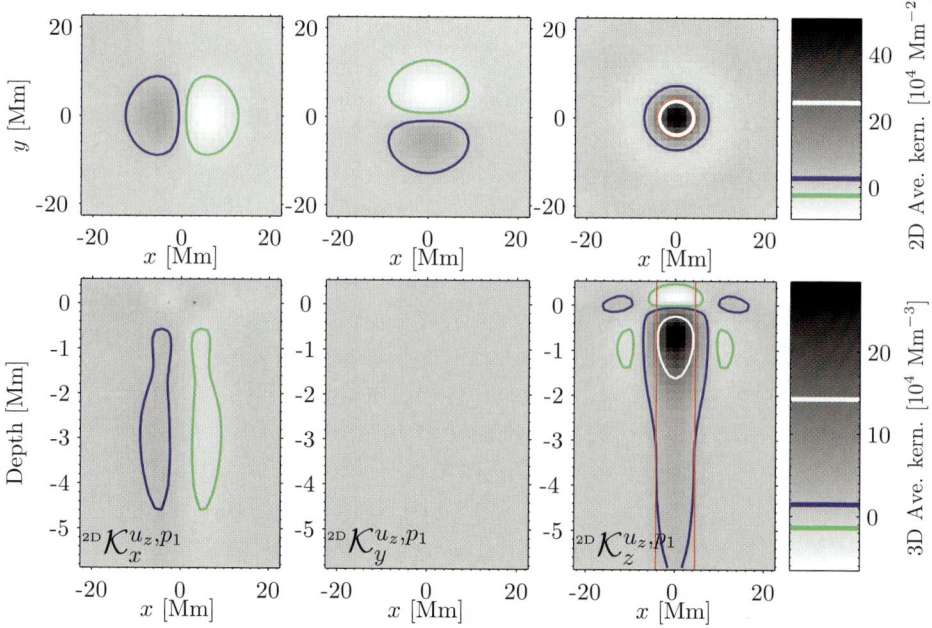

Figure 7 Two-dimensional (top panels) and depth slices of three-dimensional (bottom panels) averaging kernels for a 2D inversion for u_z using only p_1 modes. The red line outlines the FWHM = 7.3 Mm of the 2D Gaussian target function used in the inversion. The overplotted color contours, which are also marked on the color bar for reference, denote the following: half-maximum of the kernel (white), negative of the half-maximum (black), and ±5% of the maximum value of the kernel (blue and green, respectively).

A misfit quantity is constructed that measures the mismatch between the target and averaging functions:

$$\text{misfit}^2 = h_z \sum_z \left[{}^{1D}\mathcal{K}_x^{u_x}(z; z_t) - {}^{1D}\mathcal{T}_x^{u_x}(z; z_t) \right]^2. \tag{29}$$

In addition, a parameter that quantifies the error (noise) is included:

$$\text{error}^2 = \sum_n (c^n \mathcal{N}^{u_x, n})^2. \tag{30}$$

A regularization parameter [μ] is introduced, and a minimization procedure with respect to the weights [c^n] is carried out as

$$\min_c \left[\text{misfit}^2 + \mu \, \text{error}^2 \right]. \tag{31}$$

Computing the minimization results in a system of linear equations, which is solved by inverting a small matrix to obtain the coefficients for each value of the regularization parameter. Finally, we choose weights roughly in the "elbow" of the trade-off curve upon visual inspection. Several example trade-off curves from this procedure are shown in Figure 10 for two target depths [z_t] and different target widths.

In Figure 11 we provide examples of 1D averaging kernels for an inversion for u_x and for different target depths. Because of a limited mode set, there is a limited number of depths

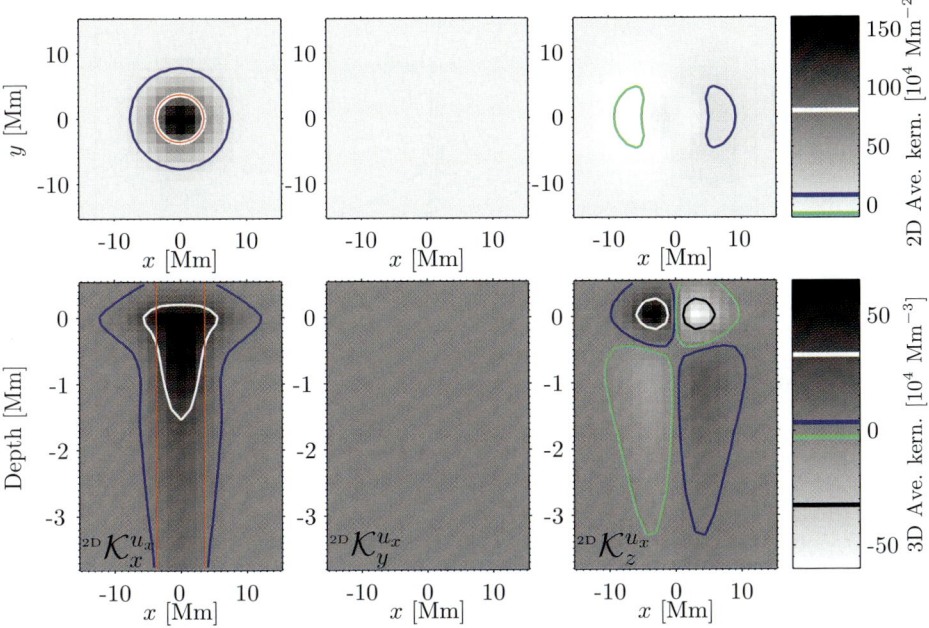

Figure 8 Three-dimensional minimum-variance averaging kernel for a 2D inversion for u_x. The top panels are the 2D averaging kernels obtained after integrating the 3D kernel over depth, and the bottom panels are slices at $y = 0$ through the 3D kernel. The method to obtain this kernel is discussed in Section 5.3. The resolution of the inversion is 7.3 Mm and is denoted by the red lines. The overplotted color contours, which are also marked on the color bar for reference, denote the following: half-maximum of the kernel (white), negative of the half-maximum (black), and $\pm 5\%$ of the maximum value of the kernel (blue and green, respectively).

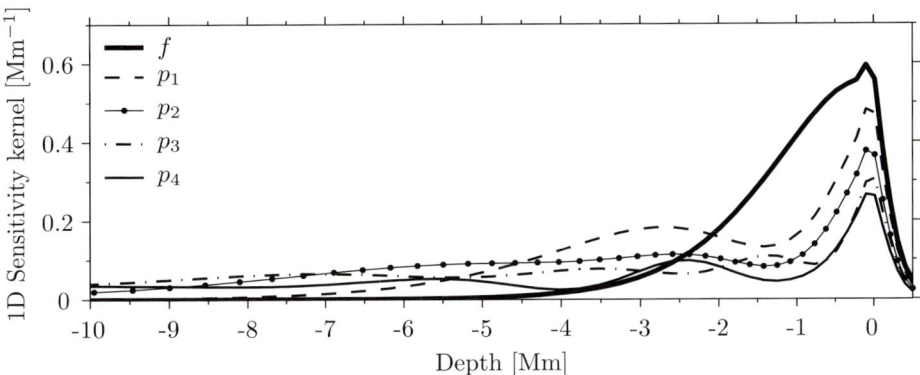

Figure 9 One-dimensional sensitivity kernels $[^{1D}K_x^{u_x,n}(z)]$ as a function of depth for each ridge $[n]$ (indicated in the legend) for an inversion for u_x. These are obtained according to Equation (26).

that can be targeted properly. There is also an obvious limit on the maximum depth with which we can probe with these modes. In addition, as with other helioseismology inversions, a surface component is present (see, *e.g.*, Basu, Antia, and Tripathy, 1999).

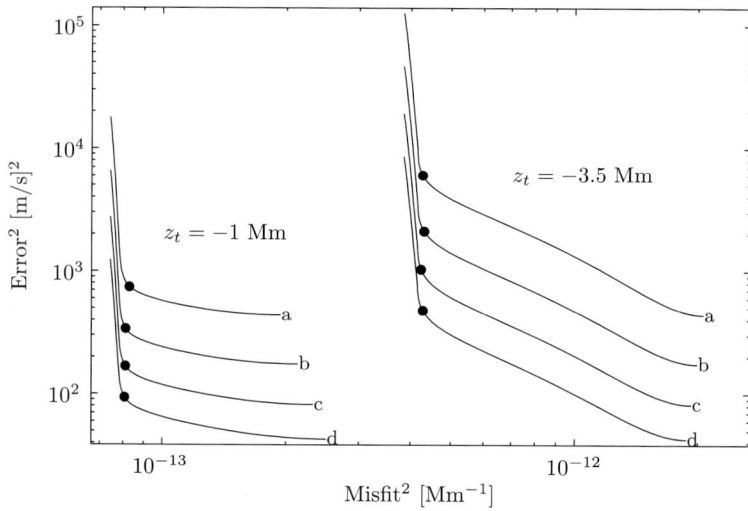

Figure 10 Example trade-off curves computed in the 1D inversion for two target depths. The individual curves in each set of depths are obtained by varying the horizontal target-function width. For the $z_t = -1$ Mm set, curves a, b, c, and d correspond to widths of 5.8, 7.3, 8.7, and 10.2 Mm, respectively. For the $z_t = -3.5$ Mm set, a, b, c, and d correspond to 7.3, 8.7, 10.2, and 11.6 Mm, respectively. In this scheme, sets of inversion weights would be chosen at the points given by the filled circles.

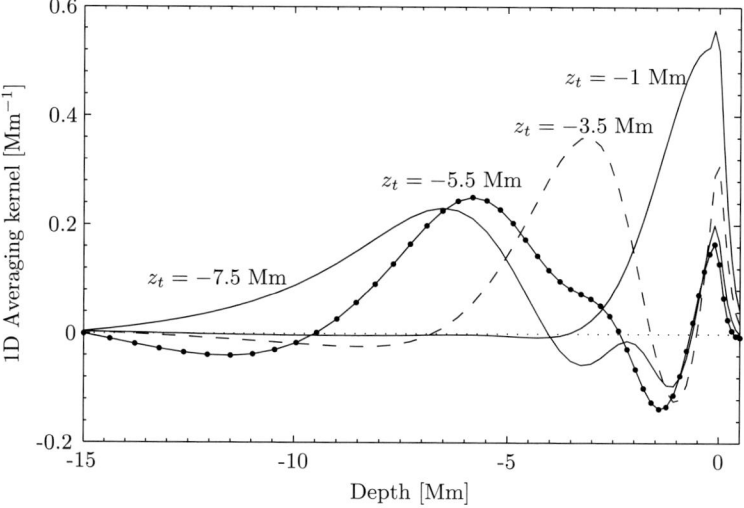

Figure 11 1D averaging kernels $^{1\mathrm{D}}\mathcal{K}^{u_x}(z; z_t)$ for an inversion for u_x for different target depths. These are obtained according to Equation (27).

7. 3D Averaging Kernels from the 2 + 1D Inversion

We denote the final 3D averaging kernel produced from the 2 + 1D inversion for u_x as $^{2+1\mathrm{D}}\mathcal{K}^{u_x}$. It has been defined in Equation (8) and can be constructed from both sets of inversion coefficients (w, c) in terms of the original sensitivity kernels by using Equations (11) and (13):

$$^{2+1D}\mathcal{K}^{u_x}(\mathbf{r}, z; z_t) := \sum_n c^n(z_t) {}^{2D}\mathcal{K}^{u_x,n}(\mathbf{r}, z) \tag{32}$$

$$:= \sum_{ij\alpha n} c^n(z_t) \gamma_i^n w^{\alpha,n}(\mathbf{r}_j; \Delta_i) \mathbf{K}^{\alpha,n}(\mathbf{r} - \mathbf{r}_j, z; \Delta_i), \tag{33}$$

where we emphasize that the weights w and c are obtained from a specific inversion for u_x. Recall that the weights γ are used to average the quantities over distance Δ.

We now check to see whether the final averaging kernels are as well localized as can be expected from the mode set used, which also justifies separating the problem into 2D and 1D parts. Performing the full $2 + 1$D inversion for u_x for different target depths produces example 3D averaging kernels such as those shown in Figures 12, 13, and 14. Plotted in the left and center panels in each case are depth slices along the $y = 0$ and $x = 0$ lines of the x-component of the kernel, $^{2+1D}\mathcal{K}_x^{u_x}$. The contour of the half-maximum value of the 3D target function is overplotted on the depth slices in red. The white contours show the half-maximum value of the 3D averaging kernel. Note that, in principle, in a noiseless inversion with a large set of available modes, the white contours would perfectly match the red contours. The blue and green contour lines denote $\pm 5\%$ of the maximum value. The 1D target and averaging functions are provided in the rightmost panel in each figure for comparison. For the shallowest target depth, instead of a simple 1D Gaussian we use a target function that goes to zero at $z = 0$. The 1D inversion coefficients c^n used to construct the kernels in Figures 12–14 are provided in Table 1. The shallowest depth we can reach with these modes is about -1 Mm, and the deepest is about -8 Mm. Of course, the noise begins to increase quickly with depth.

8. Results with MDI Data for Quiet-Sun Flows

In the rest of the paper we provide example flow maps in the quiet Sun from the 2D and $2 + 1$D inversion procedure discussed here. It is our main intention to demonstrate that the results obtained are sensible and consistent with what might be expected given the spatial resolution, observation time, and level of estimated noise. We note also that to study local flows, we typically remove a mean, large-scale, time-averaged flow from each retrieved map.

8.1. Tests of the 2D Inversion

The first simple test we perform is to compare the inferred flows from the 2D inversion to the direct MDI Doppler data. To accomplish this, the three components of the inferred 2D vector flows are projected onto the line-of-sight vector at each pixel. We use 24 hours of data from the seventh day (26 January 2002) of the nine-day data set available. There is a sunspot in the center of the map that is located vertically about the equator and at 30° toward the western solar limb for this day. Figure 15 shows the comparison with the Doppler map from a 2D inversion using only f modes. The correlation between the inferred flows and the Doppler flows is 0.9 for pixels with less than $|10|$ G of magnetic field. Also provided is a magnetogram to show the locations of the large sunspot, the surrounding plage, and the region of quiet Sun that is studied in all of the plots in the rest of the paper. The scatter plot of the two velocity maps shows that the magnetic field introduces an anomalous second component to the velocity field, reinforcing our reason to restrict all further analyses to the

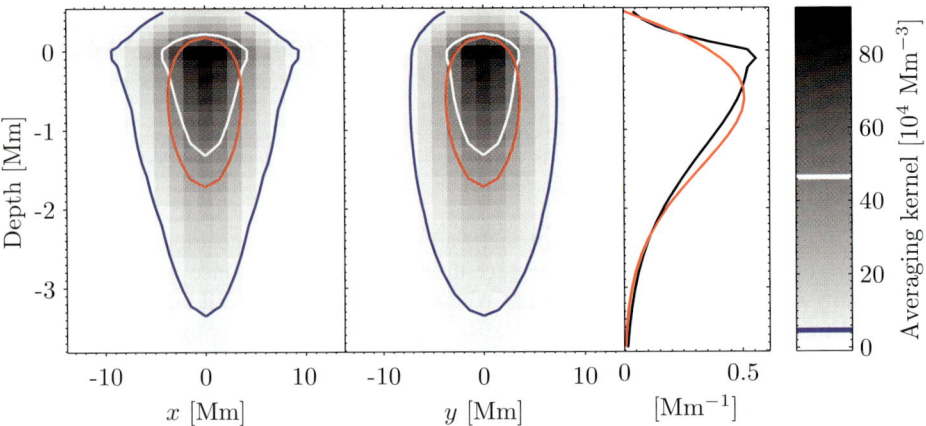

Figure 12 Vertical slice through a 3D averaging kernel [$^{2+1D}\mathcal{K}_x^{u_x}$] after performing a 2 + 1D inversion for u_x for a target depth of about -1 Mm below the solar surface. The left (middle) column is a slice along the $y = 0$ ($x = 0$) line. The red contours show the half-maximum value of the 3D target function, white contours show 50% of the maximum value of the averaging kernel, black contours show -50% of the maximum value of the averaging kernel, and the blue (green) contours denote $\pm 5\%$ of the maximum. In the rightmost panel, the red line is the 1D target function, and the black solid line is the 1D averaging kernel. The 1D weights c^n and the noise estimate are given in Table 1.

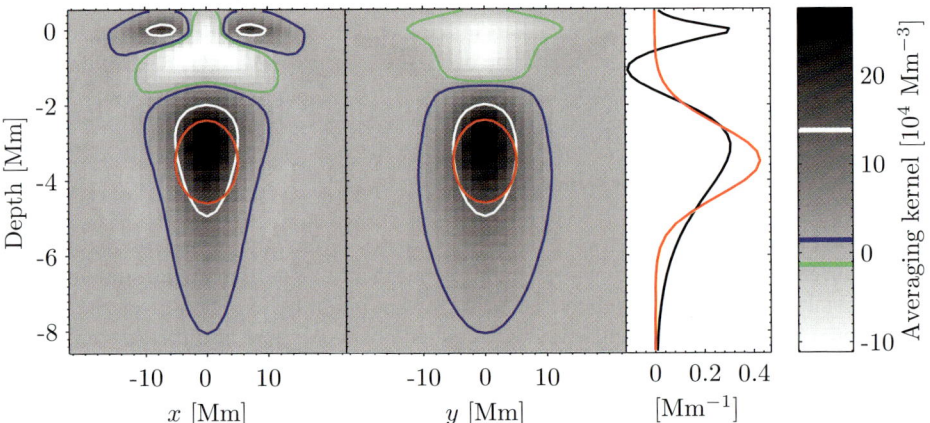

Figure 13 Vertical slices through an averaging kernel after performing a 2 + 1D inversion for u_x for target depth $z_t \approx -3.5$ Mm. The red contours show the half-maximum value of the 3D target function, white contours show 50% of the maximum value of the averaging kernel, black contours show -50% of the maximum value of the averaging kernel, and the blue (green) contours denote $\pm 5\%$ of the maximum. In the right panel, the red line is the 1D target function, and the black solid line is the 1D averaging kernel.

quiet Sun. That the inferred line-of-sight velocity is smaller than the Doppler velocity in not surprising. It is most likely attributable (Gizon, Duvall, and Larsen, 2000; Braun, Birch, and Lindsey, 2004) to the average depth over which the flows are measured. We cannot make a purely "surface" measurement with the available mode set.

A further example of the 2D inversion is shown in Figure 16. Here we show horizontal-flow maps inferred from inverting individual-ridge travel times for f, p_1, and p_2 and for 24 hours. The color scale of these images is the horizontal divergence, obtained from a

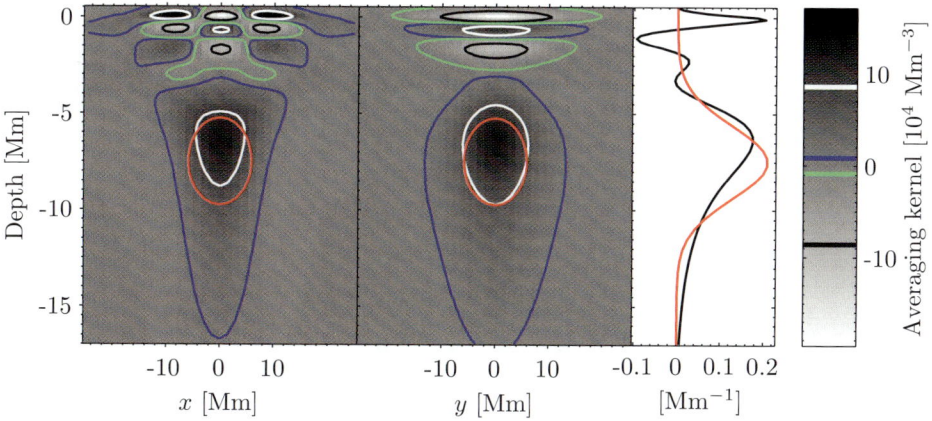

Figure 14 Vertical slices through an averaging kernel after performing a 2 + 1D inversion for u_x for target depth $z_t \approx -7.5$ Mm. The right panel shows the 1D averaging kernel and target function. The colors are the same as for Figure 13.

separate inversion as discussed in the following. The units of the horizontal divergence in this and all figures are inverse megaseconds [Ms^{-1}]. In the f-mode map we see signatures of supergranulation with strong horizontal divergence in the centers of the supergranules and flow convergence at the cell boundaries. The outflow is generally in the 200–350 m s^{-1} range. The horizontal flows obtained from p_1 and p_2 travel times are weaker, as is the supergranulation signature. Maps for the p_3 and p_4 ridges have also been studied, but the noise begins to increase quickly with these modes at this resolution.

Obtaining the horizontal-flow divergence from a direct inversion such as shown by the color scale in the plots in Figure 16 is conveniently done in an OLA inversion such as this one. All that is needed is to use a different 2D target function such that the quantity that is inverted for is simply the horizontal divergence. The function we use (which would replace the function defined in Equation (15) in the 2D inversion) is

$$\boldsymbol{T}^{\text{div}_h}(\mathbf{r}) = \left(\frac{x}{2\pi\sigma^4} \exp^{-r^2/2\sigma^2}, \frac{y}{2\pi\sigma^4} \exp^{-r^2/2\sigma^2}, 0 \right), \quad (34)$$

where div$_h$ is a superscript label that denotes the horizontal divergence of the flow [$\nabla_h \cdot \mathbf{v}_h$]. Another useful quantity that we have studied is the vertical component of the flow vorticity [$\hat{z} \cdot \nabla_h \times \mathbf{v}_h$]. The associated target function, labeled with the superscript vort$_z$, can be shown to be

$$\boldsymbol{T}^{\text{vort}_z}(\mathbf{r}) = \left(\frac{-y}{2\pi\sigma^4} \exp^{-r^2/2\sigma^2}, \frac{x}{2\pi\sigma^4} \exp^{-r^2/2\sigma^2}, 0 \right). \quad (35)$$

An example map of the vertical vorticity obtained directly from a 2D inversion is shown in Figure 17. We have studied plots of the divergence and vorticity at different resolutions and checked that the inverted quantities using these two types of targets and the direct numerical computation of these quantities using inferred horizontal vector flows agree reasonably well. The advantage of computing them directly from an inversion is that we are able to determine the noise and spatial resolution properly.

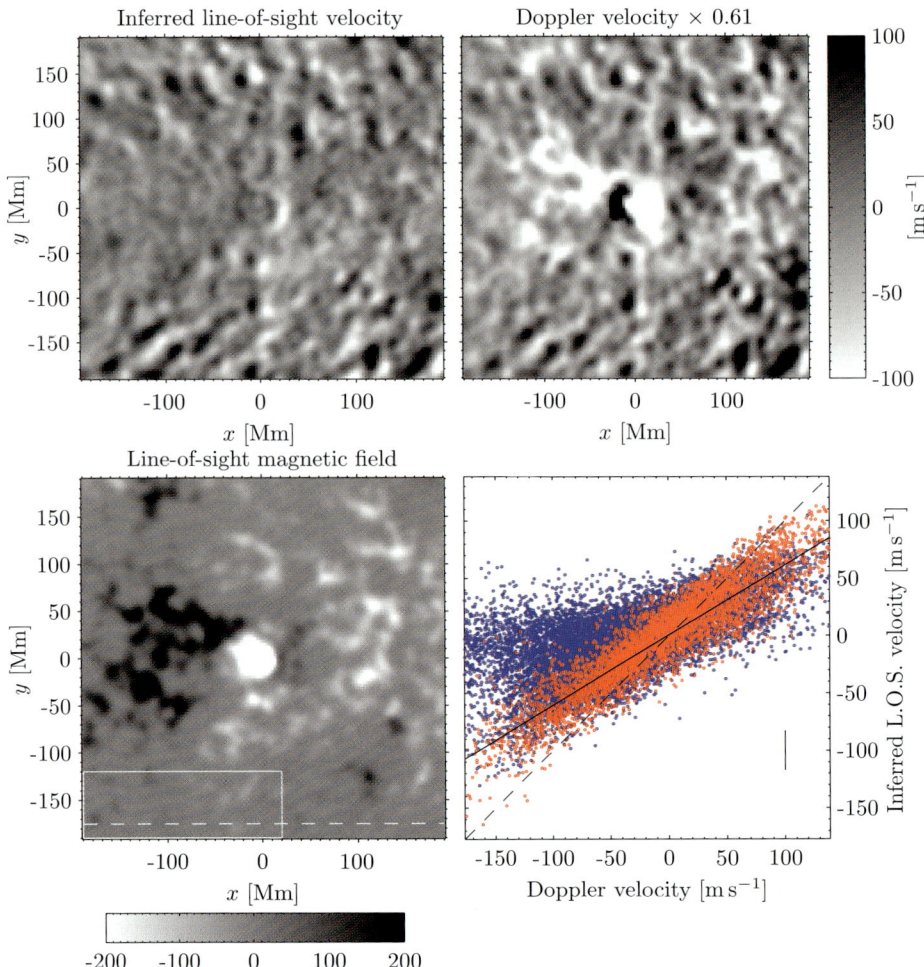

Figure 15 Test of the 2D inversion. The top panels show a comparison of the inverted flows with the direct Doppler data for a map centered at 30° west of disk center. The top-left panel is the inferred vector flow map after projection onto the line-of-sight vector at each pixel for an inversion using f modes and 24 hours of data. The resolution is about 7 Mm. The Doppler map is an average over one day of MDI full-disk data, smoothed with the 2D averaging kernel from the inversion and multiplied by the slope 0.61 of the best-fit line through the scatter plot (panel below) to allow for direct comparison with the inverted map. The lower left panel is the one-day-averaged (truncated) MDI magnetogram with a sunspot visible in the middle. The units are in gauss. The white box outlines the quiet-Sun region analyzed for all of the plots in the rest of the paper. The white dashed line shows the location of the slice for the flows in Figure 24. The lower right panel shows a scatter plot of the velocity maps, where red (blue) dots represent pixels of less (more) than |10| G. The correlation is 0.9 for the nonmagnetic data. The dashed line is $y = x$, and the solid line is a fit to the nonmagnetic data (with a slope of 0.61). The small vertical line represents the $\pm 1\sigma$ ($\sigma = 17$ m s^{-1}) noise in the flow estimation.

8.2. Tests of the 2 + 1D Inversion

A consistency test of the full 2 + 1D inversion is shown in Figure 18. We study flows in a quiet-Sun region obtained from two independent inversions using 24 hours of data. Shown are the horizontal flows computed from inversions for v_x and v_y along with the line-of-sight

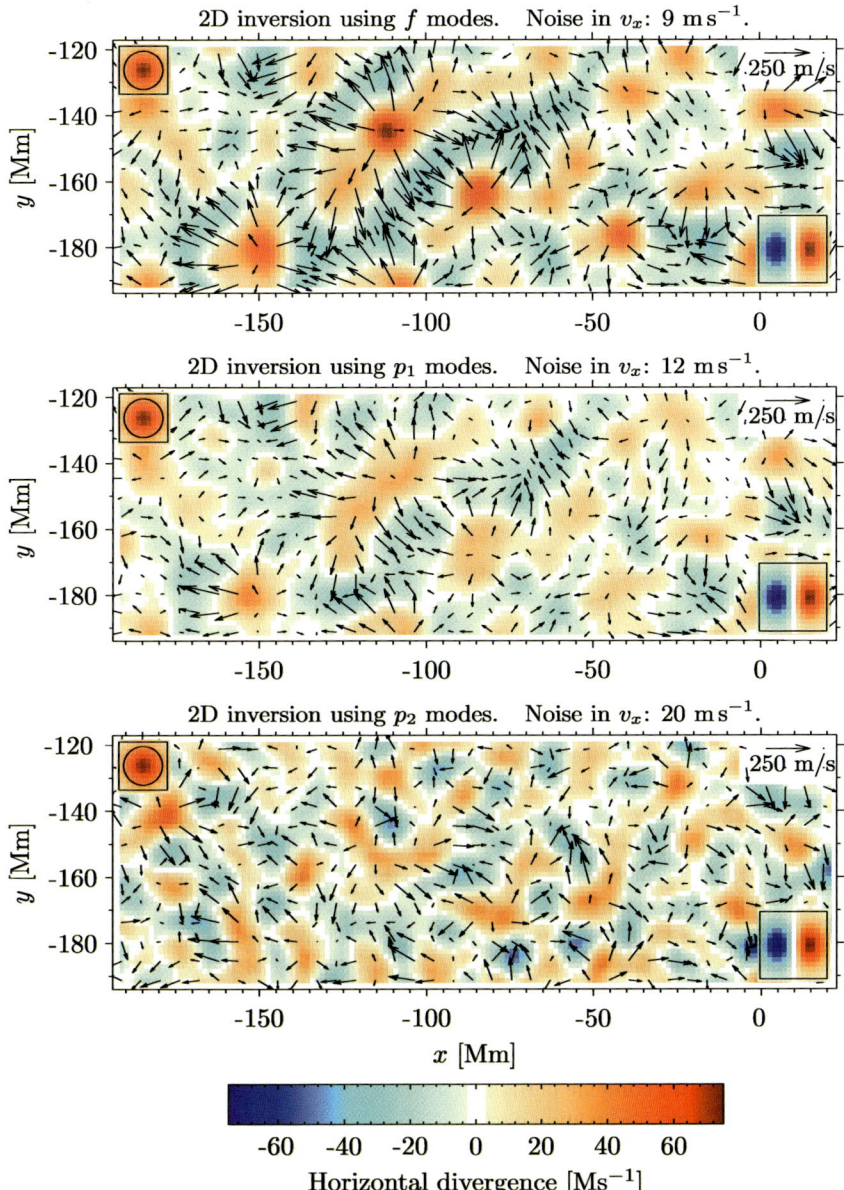

Figure 16 Horizontal flows and divergence from a 2D inversion using 24 hours of data for the f, p_1, and p_2 ridges. In each panel, the arrows denote the horizontal flows (obtained from inverting for v_x and v_y) for 24 hours of data, and the color scale is the horizontal flow divergence in units of inverse megaseconds, obtained from a separate inversion as described in the text at the end of Section 8.1. The x-component of the 2D averaging kernel from the flow inversion is plotted in the box in the upper left of each panel and the FWHM, outlined by the circle, is 11.6 Mm. The x-component of the 2D averaging kernel for the horizontal divergence is given by the quantity in the lower right box (see Equation (34)). Note the strong supergranular flows in the f-mode map, which gradually weaken as the modes probe deeper layers of the convection zone. The correlation of the f-mode map with the p_1-mode map is 0.88, and that between the f-mode and p_2-mode maps is 0.35. The noise in v_x is given for each panel. The noise in the horizontal divergence inversion is 10, 12, and 13 Ms^{-1} for the top, middle, and bottom panels, respectively.

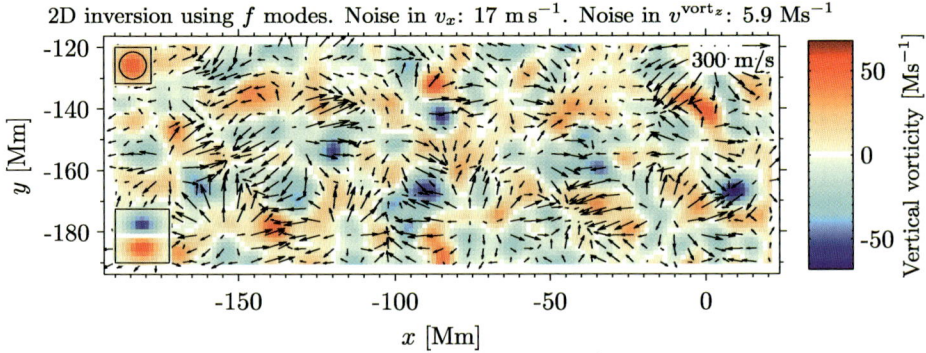

Figure 17 2D inversions for the vertical vorticity and horizontal flows in the quiet Sun. The arrows denote the horizontal flows and the color scale is the vertical vorticity obtained from a separate inversion using 24 hours of data. The x-component of the averaging kernel from the inversion for the horizontal flows is given in the upper left box, and the x-component of the averaging kernel for the vorticity inversion, which matches the target function from Equation (35), is shown in the lower left box.

magnetic field (color scale). In the inversion corresponding to panel (a), we only invert p_1 travel times. We then attempt to target the p_1 averaging kernel by using all of the other available mode-ridge kernels *except* p_1, in other words, f, p_2, p_3, and p_4. Once a similar averaging kernel is found, we invert the corresponding f, p_2, p_3, and p_4 travel times, and the resulting flows are shown in Figure 18(b). The 1D averaging kernels for each case are given in the panel on the bottom right. The maps are quite similar (correlation = 0.82), and the differences could be due simply to the differences in the averaging kernels. In these maps the supergranule-scale flows are evident, and the magnetic field is concentrated at the boundaries of the supergranules as expected. A best fit through a scatter plot of the data that takes into account the noise in both variables gives a slope of 0.83. The magnetic field does not introduce any anomalous component in the scatter as it did for the full-map study in Figure 15, confirming that this is a quiet-Sun region for our purposes. We have also used this test to see whether we can recover an f-mode map by inverting the four available acoustic-mode travel-time sets. For as closely as we are able to match the averaging kernels, it is successful. The same conclusion can be drawn for the other possible cases when noise is not a limiting factor.

In all of the plots studied so far, we have shown flows obtained from one day of travel times. In Figure 19, we compare horizontal flows at a depth of 1 Mm below the surface from inversions for different observation durations. The panels show inversions for 6 to 24 hours of data in six-hour intervals. Also shown is the horizontal divergence computed numerically (since we have not yet computed inversions directly for horizontal divergence at depth). What is evident is that even with six hours of data and a resolution of about 9 Mm, the noise level is reasonable and features are seen that have much in common with the one-day map. The correlation between the 6-hour and 24-hour maps is still about 0.8. It is encouraging that the supergranulation signal at this depth is not dominated by noise for six hours of data.

We now compare horizontal flows at three different depths from the full 2 + 1D inversion. Figure 20 shows the flow field at depths of 1 Mm (top), 2.6 Mm (middle), and 3.7 Mm (bottom) below the surface using 24 hours of data. The color scale is the horizontal divergence computed by numerical differentiation of v_x and v_y. The 1D inversion weights for each map are given in Table 1. The flows at the different depths in Figure 20 are not too unlike the individual ridge flows shown in Figure 16, and inspection of the 1D inversion

High-Resolution Mapping of Flows

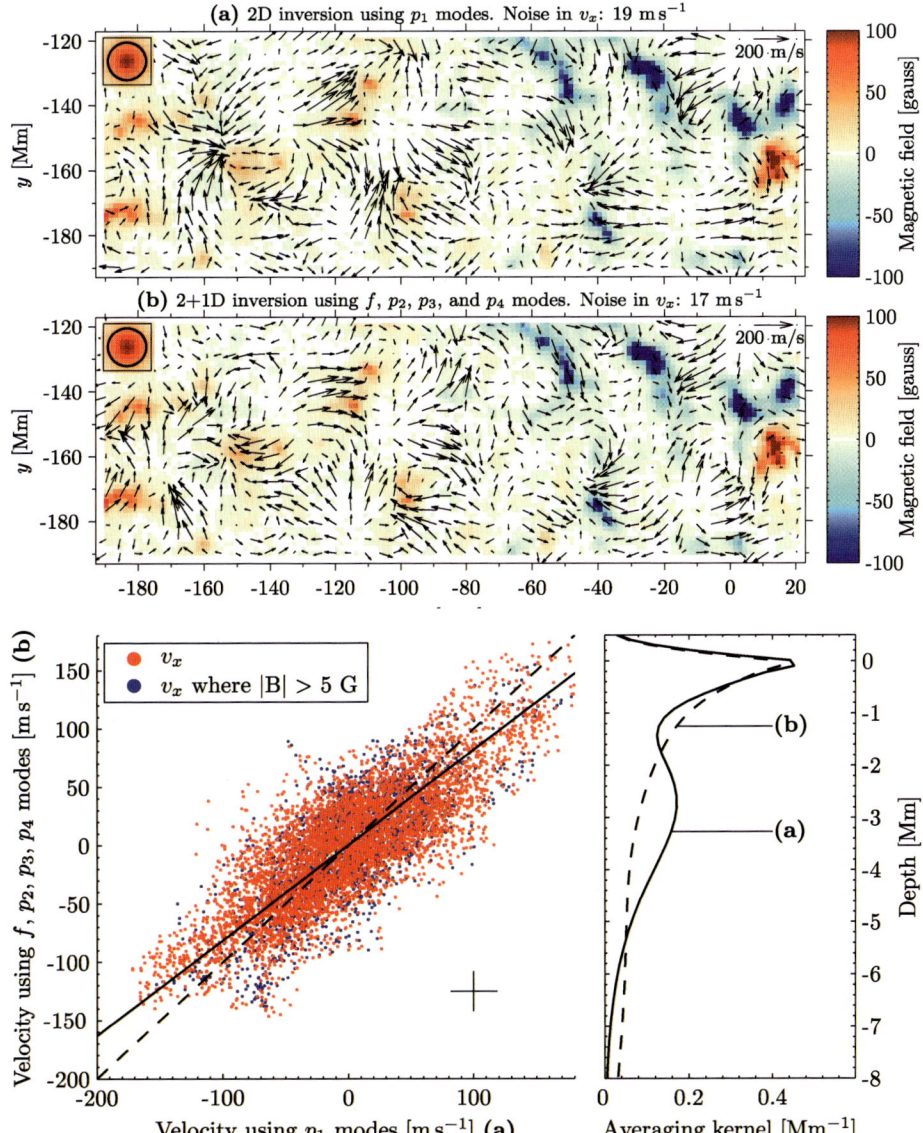

Figure 18 Test of the full $2+1\text{D}$ inversion for (a) the flows from a 2D inversion using only the p_1-mode ridge travel times and (b) the flows from a $2+1\text{D}$ inversion using travel times from all other mode ridges (no p_1 modes) by attempting to target the same averaging kernel. The arrows denote the horizontal flows (obtained from inverting for v_x and v_y) for 24 hours of data, and the color scale is the truncated magnetic field from MDI. The x-component of the 2D averaging kernel is plotted in the box in the upper left, and the FWHM = 11 Mm is outlined by the circle. The scatter plot shows the flows from panel (a) vs. (b), where the red (blue) dots are the values in the maps for which the magnetic field of the pixel is less (greater) than $|5|$ G. The dashed line is the $y = x$ line and the solid line is the best fit through the nonmagnetic data (slope $= 0.83$), taking account of the errors on each axis. The 1σ error bars for each measurement are denoted by the cross in the lower right of the plot. The correlation in the scatter is 0.82. The bottom right panel shows the 1D averaging kernel (solid line) for panel (a) and the 1D averaging kernel (dashed line) for panel (b). The 1D inversion weights for these two inversions are listed in Table 1.

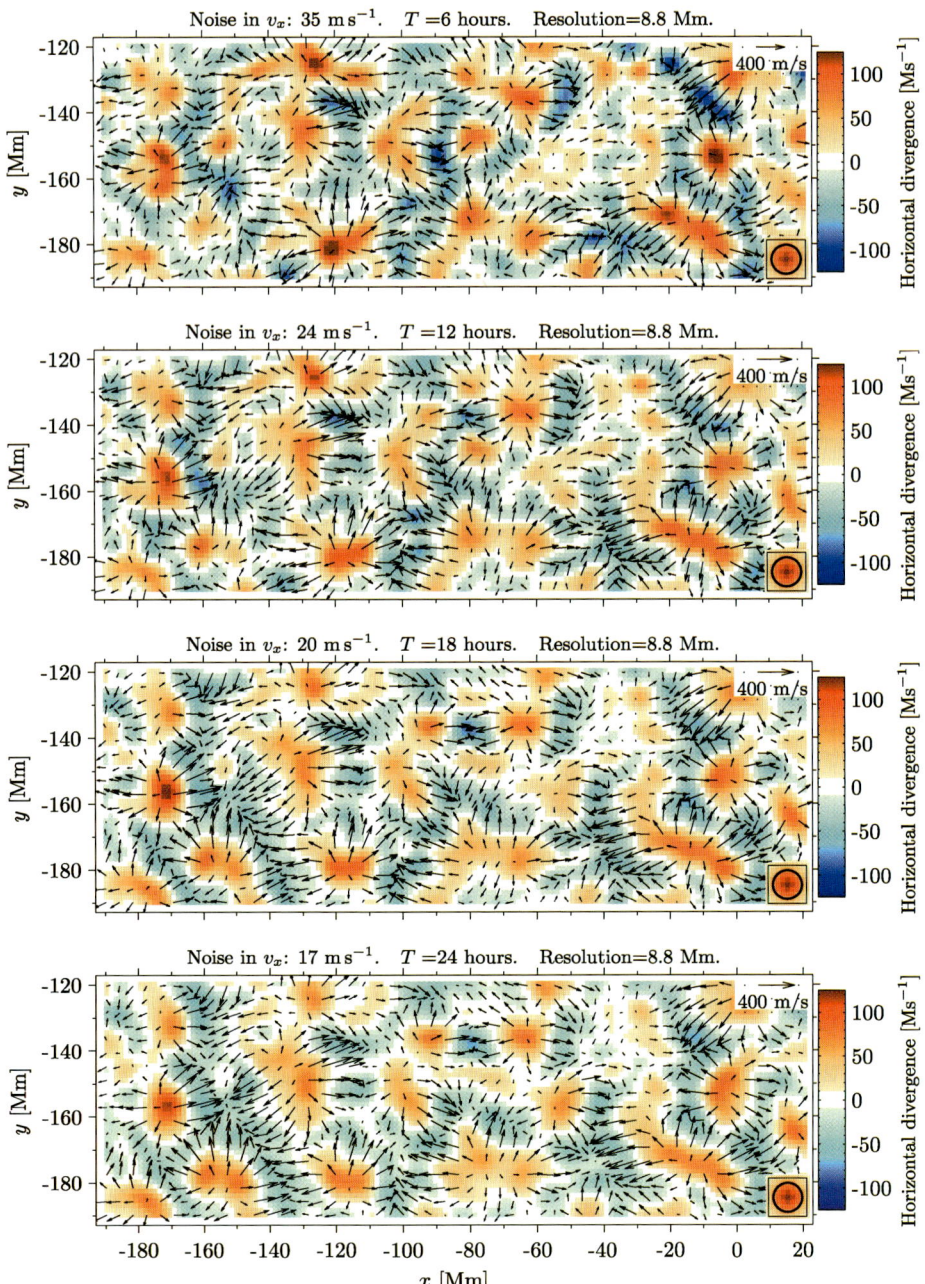

Figure 19 Comparison of flows for different observation durations [T] obtained from 2 + 1D inversions at a depth of 1 Mm beneath the surface. The arrows are the inferred horizontal flows and the color scale is the horizontal divergence computed numerically from v_x and v_y. The total observation time increases from top to bottom in six-hour intervals, but the horizontal resolution remains fixed at about 9 Mm. The correlation of the 6-hour map with the 24-hour map is 0.78, whereas the correlation of the 12-hour (18-hour) map with the one-day map is 0.9 (0.97). Note that the divergence scale is the same for each panel.

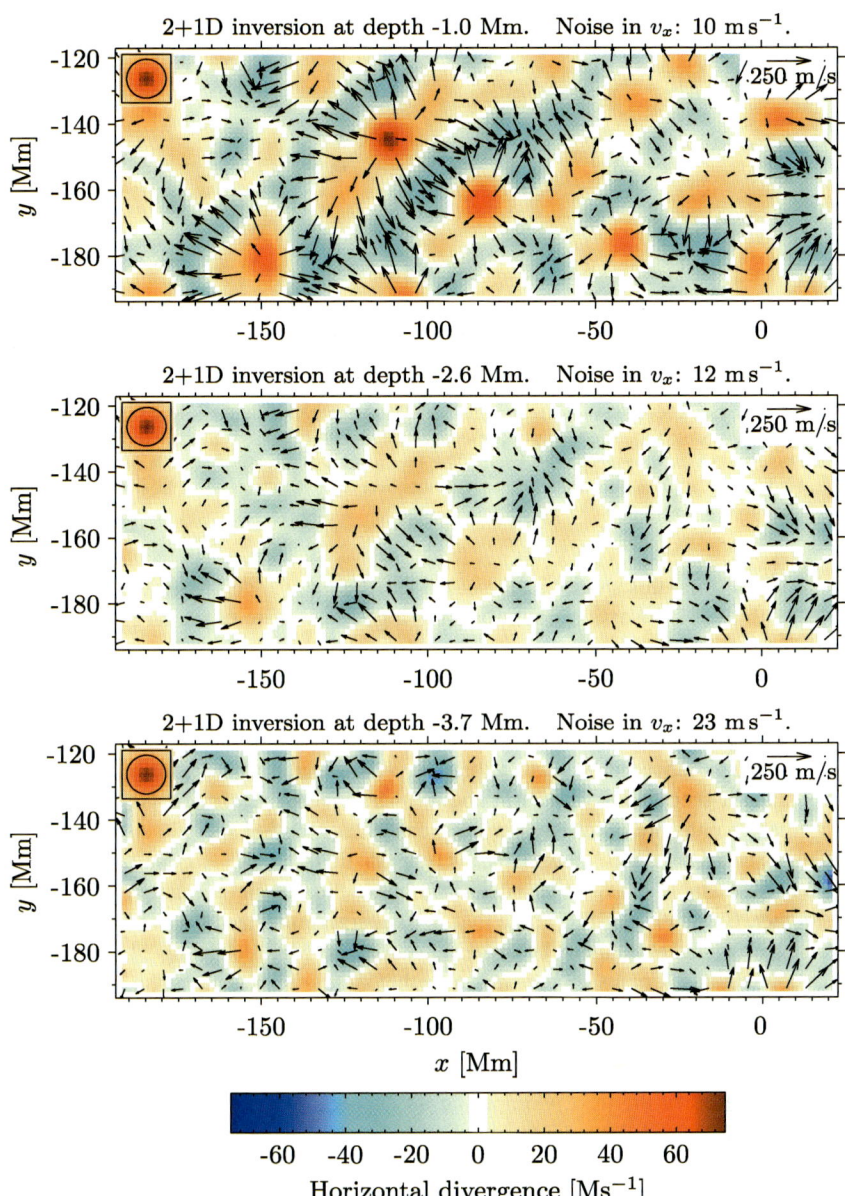

Figure 20 Comparison of horizontal flows (arrows) at different depths from a 2 + 1D inversion for one day of data. The background color scale is the horizontal divergence to emphasize the various flow structures and is obtained by numerical differentiation of the 2D flows. The x-component of the 2D averaging kernels are shown in the upper left and all have FWHM = 11.6 Mm. The correlation of the x-component of the flows at depths of -1 and -2.6 Mm is 0.77. The correlation of the x-component of the flows at depths of -1 and -3.7 Mm is 0.33. The 1D inversion coefficients for each depth inversion are provided in Table 1.

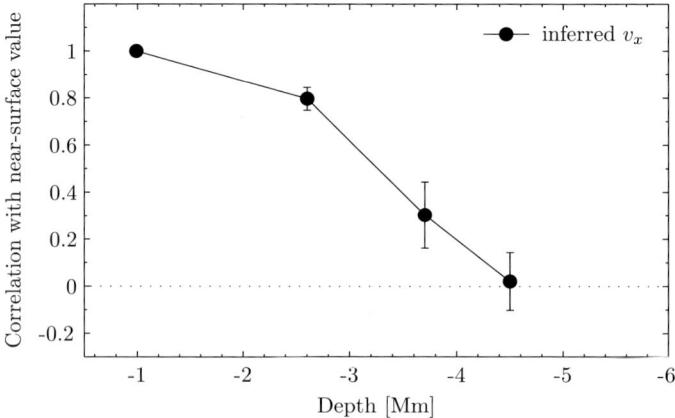

Figure 21 Correlation coefficient with respect to the near-surface value of the inferred v_x flows at different depths using 24 hours of data. The values represent the average over five days of the flows measured in the area of the quiet Sun used throughout this paper. The 1σ error bars are plotted at each depth point, obtained from the standard deviation in the correlation coefficients over the five days. The "near-surface value" for this case is at a depth of -1 Mm.

weights confirms that this should be the case. This figure also demonstrates that combining the maps with the 1D inversion not only gives a good estimate of the target depth but also generally lowers the noise levels.

We have studied the correlation of maps of v_x and v_y such as those in Figure 20 at different depths with the near-surface map and averaged over five days. The results are provided in Figure 21. Each measurement is for 24 hours of data, and the error bars are obtained by studying the variance in the correlation values. The correlation steadily decreases as we go deeper, and it seems to disappear at about 5 Mm below the surface. However, the noise levels at these depths are quite large and we can draw no other specific conclusions at this time. This is consistent with recent studies on realistic numerical simulations using time–distance helioseismology (Zhao *et al.*, 2007) and helioseismic holography (Braun *et al.*, 2007). In fact, Braun *et al.* (2007) note that "...supergranule-sized flows are essentially undetectable using current methods below depths around 5 Mm..." using 24 hours of data or less. We confirm this conclusion here and note that similar results have also been found with direct modeling techniques (Woodard, 2007).

We have also studied the day-to-day correlation of the v_x and v_y maps at various depths. If we were predominantly measuring noise, there would be no significant correlation from one 24-hour period to the next. Computing an average day-to-day correlation over seven days of data for the -1 Mm depth maps gives a value of about 0.4. For the -3.7 Mm depth, we find a 0.26 correlation, and at a depth of -6 Mm, about a 0.1 correlation. This again demonstrates that there is plenty of near-surface flow signal when 24-hour averages are studied, presumably owing to supergranulation (Gizon and Birch, 2004), which then quickly decreases with depth.

8.3. Vertical Flows

It has proven difficult in helioseismology to accurately measure the vertical component of the velocity near the surface owing in part to its small magnitude compared to the other components. In fact, in many helioseismic inversions for flows, v_z is obtained approximately by

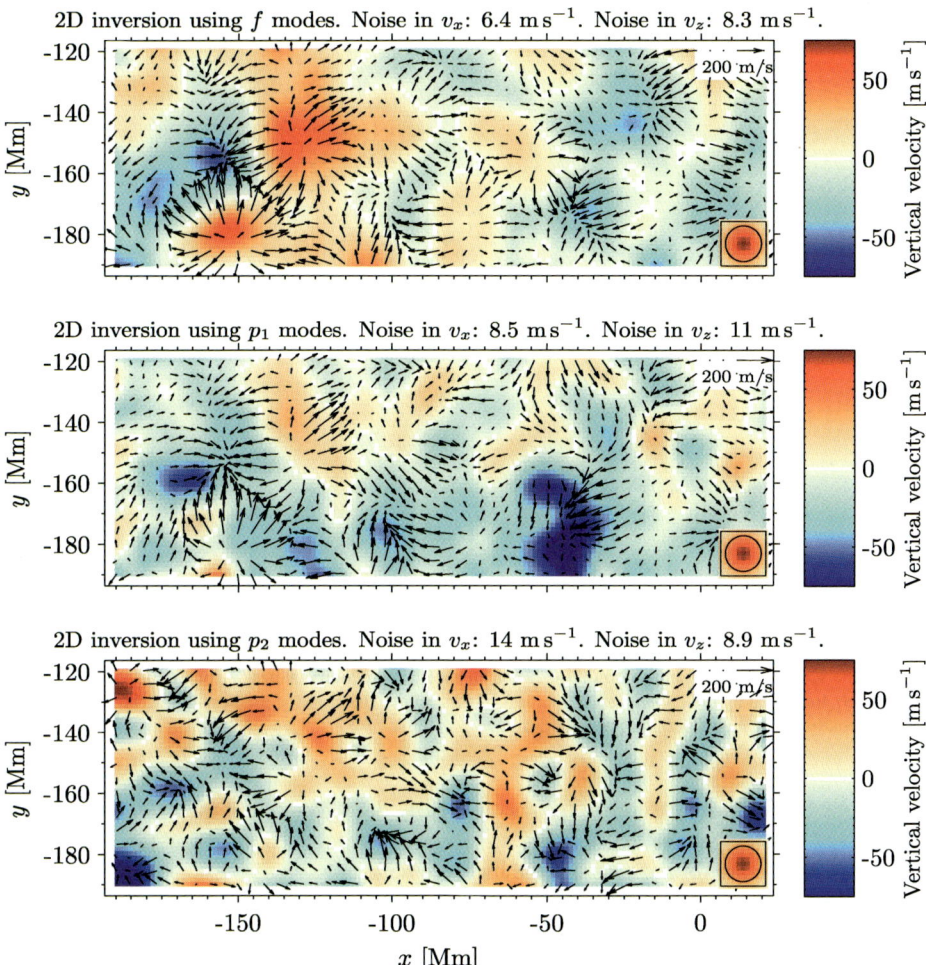

Figure 22 Vertical flows (color scale) and horizontal flows (arrows) in the quiet Sun for a 2D inversion using f modes (top), p_1 modes (middle), and p_2 modes (bottom). These flows were obtained by using 48 hours of data. A positive vertical velocity means an upflow. The z-component of the 2D averaging kernel from the inversion for u_z is given by the quantity in the box in the lower right. The noise for all of the measurements is indicated. The correlations of these particular p_1 and p_2 vertical flows with the f-mode map are about 0.3 and 0.2, respectively. The region of the Sun used here is outlined by the white box in Figure 15. The color scale is the same in each panel to ease comparison.

computing the horizontal-flow component and then invoking mass conservation from the continuity equation (see Komm *et al.*, 2004, for an example in ring-diagram analysis). Another source of difficulty in these measurements has been associated with cross-talk effects, whereby the inversion (or sensitivity kernel) becomes insensitive to differences between upflows and convergence and between downflows and divergence (Zhao *et al.*, 2007). These inversions, usually based on the ray approximation, have no obvious means of constraining the cross-talk. Since we have available Born sensitivity kernels for v_z and an inversion procedure that measures each flow component while minimizing the cross-talk with the others, we can obtain vertical flows directly and with the assurance that they are relatively indepen-

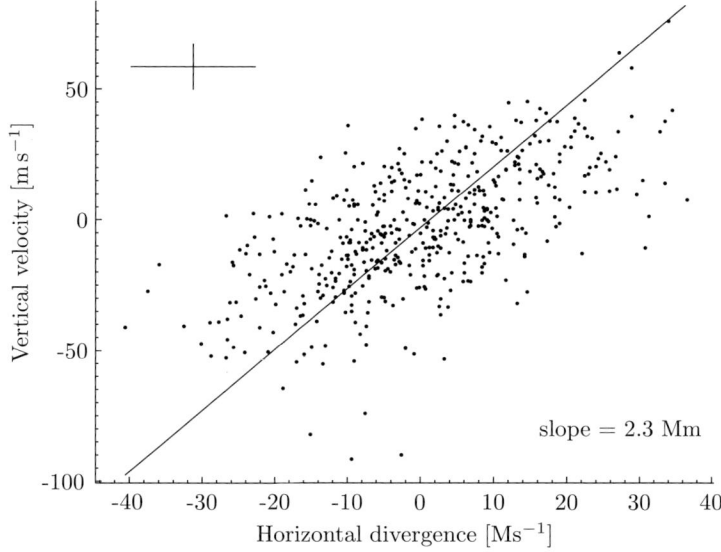

Figure 23 Vertical velocity *versus* horizontal divergence for flows using p_2 modes measured over two days for the same region of the Sun as shown in Figure 22. The vertical flows and the horizontal divergence were computed from two separate inversions. The correlation coefficient is 0.62. The line shows a best fit through the scatter by taking into account the noise on each axis, which is 9 m s^{-1} for v_z and 9 Ms^{-1} for the divergence, indicated by the cross in the upper left of the figure. The slope of the best-fit line is a rough proxy for the density scale height at the implied depth.

dent of the horizontal measurements. This is clearly demonstrated in the averaging kernels of Figures 6 and 7. We note that we have so far only tested the 2D inversion for v_z; thus, the maps shown here are for individual ridge measurements.

There tends to be much more relative noise in the measurements of vertical velocity, and therefore in Figure 22 we show the vertical component of the velocity as the color scale averaged over two days from a 2D inversion (with the noise going as $T^{-1/2}$, where T is the observation time). The top panel of Figure 22 is for the f-mode ridge, the middle panel for p_1, and the bottom panel for p_2. Also shown are the corresponding horizontal flows given by the arrows. One generally sees a good correspondence in all maps between the vertical upflows and horizontal outflows, as well as between downflows and horizontal inflows. Analysis of many similar maps shows that the speeds of the vertical flows in the center of supergranules near the surface are on average about 15 – 20% of the speeds of the horizontal outflow in the supergranules, slightly higher than recent observations might suggest (Hathaway *et al.*, 2002).

To understand whether the inferred vertical flows at these depths for p_2 are reasonable, we compare them with maps of the horizontal divergence, obtained from a separate and independent inversion of this quantity as explained in Section 8.1. The vertical component of the flow and horizontal divergence are proportional if one writes down an approximate continuity equation whereby one neglects the horizontal variations in the density and the vertical gradient of the vertical flow. The scaling factor is the density scale height. In Figure 23 we show a scatter plot of $v_z^{p_2}$ against the horizontal divergence inferred from inverting p_2 travel times for the same region of the Sun as in Figure 22. The correlation coefficient is 0.62. The noise in v_z is 9 m s^{-1} and the noise is 9 Ms^{-1} for the divergence measurement. The slope of the best-fit line, obtained by using the noise information in both variables, gives a value of about 2.3 Mm. This value is in the range of the density scale height for the implied depth range of these vertical flows. We have also studied the correlation of vertical flows maps with horizontal divergence maps for the f-mode and p_1-mode cases. The values are always in the range of 0.6 – 0.7.

Another interesting question is how well the near-surface vertical flows are correlated with deeper vertical flows. Since we have so far only implemented the 2D inversion scheme for vertical flows, we take different mode ridges as a proxy for depth. We correlate the 24-hour f-mode v_z map with the p_1 and p_2 maps, averaged over seven days, and find correlations of about 0.3 and 0.2, respectively. In addition, as was described previously for the horizontal component, we have also studied the day-to-day correlations of the vertical flows averaged over seven days of data. The average day-to-day f-mode map correlation is 0.15, and it is 0.2 for $v_z^{p_1}$ and 0.15 for $v_z^{p_2}$. This demonstrates again that the v_z^n inversions are measuring long-lived flow structures and not just noise.

Finally, in Figure 24 we show the culmination of our main results. It is a slice in depth through the horizontal divergence with overplotted v_z information. The inversion to obtain these flows used 72 hours of travel times. The slice is along a line through the quiet Sun (shown by the dashed white line in the magnetogram of Figure 15) chosen because of the presence of many near-surface large-scale flow structures. Inversions for v_x at different target depths were performed, the numerical divergence $\partial_x v_x$ was computed, and the results are given by the color scale. The color scale is such that a positive divergence means an outflow. The v_z flows were obtained by using f, p_1, and p_2 travel times, and the magnitudes and directions are shown by the arrows. Since we have not yet implemented a 1D depth inversion for v_z, we roughly determine the three depth locations by computing the average depth over which the dominant modes of these three ridges probe. We emphasize that these placements are only approximate. We also plot the 1σ noise levels of v_z at each depth. We see that over the whole depth range, the horizontal inflows (outflows) generally correspond to vertical downflows (upflows). What one would expect is to see relatively stronger vertical flows where the divergence is strongest in absolute value. This is for the most part the case. We emphasize that these good correlations are likely not due to cross-talk contamination, which tends to diminish as one moves further below the surface (see Section 5.2). Note also the presence of large-scale structures that live for at least three days.

9. Summary and Conclusions

We have presented in detail a fully consistent procedure for inverting helioseismic travel times to infer vector flows in the upper convection zone of the quiet Sun. Travel-time sets are measured for all modes that have the same radial order, in other words, along the ridges in the power spectrum. The travel times are constructed by using an analog to the common point-to-annulus geometry for 20 annulus radii (up to about 30 Mm). Three-dimensional Born sensitivity kernels for the same travel-time definition and ridge filtering are computed. In addition, the noise-covariance properties of the travel times are calculated. Based on the separability of the sensitivity kernels into horizontal and vertical components owing to the ridge filtering, the inversion is formulated in two steps: The first step solves the 2D horizontal problem and the second step solves the 1D depth inversion. Optimal sets of weights are chosen from both inversions, such that the final averaging kernel is regularized in the horizontal and vertical directions. We have provided examples of averaging kernels, which are useful for understanding where the sensitivity of the inversion is located, as well as for

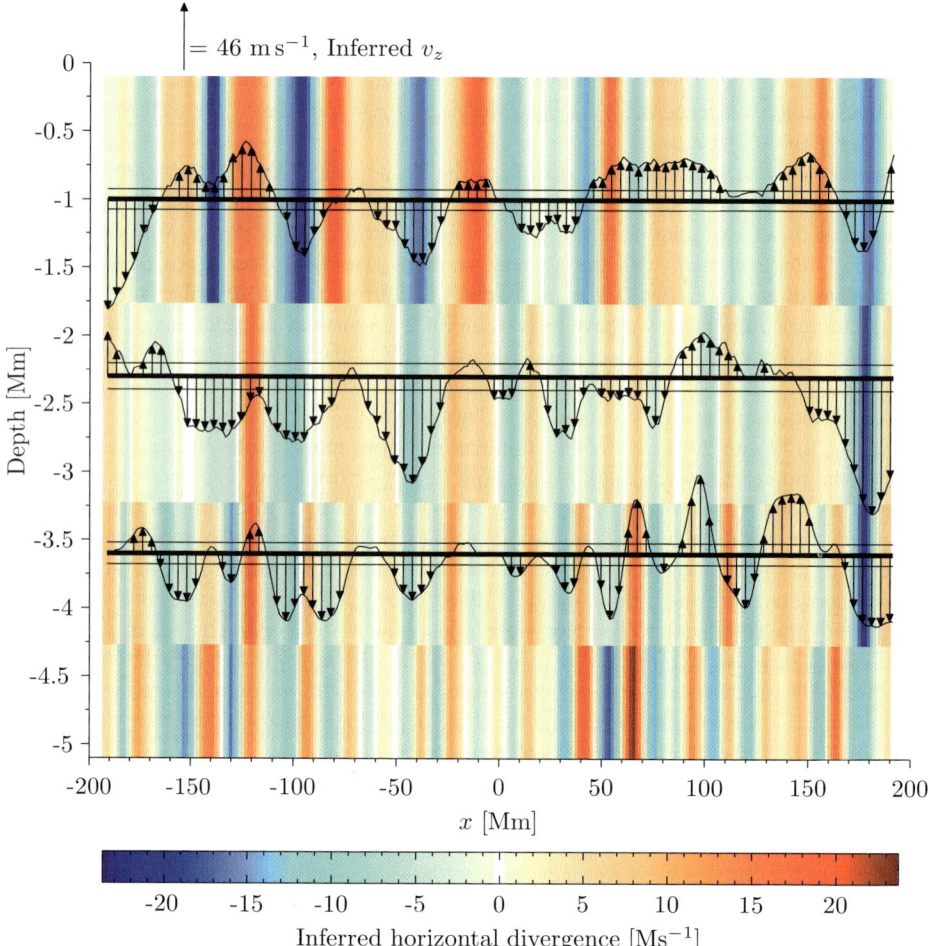

Figure 24 Depth slice through a flow map from inversions using three days of data. The horizontal divergence of v_x is shown by the color scale and the arrows represent the velocity of the vertical flows at three different depths. The three thick black horizontal lines denote the approximate depth locations of the v_z measurements, which are found by computing the center of mass of the 1D sensitivity functions for the first three ridges (see Figure 9). The top curve shows v_z^f, the middle v_z^{p1}, and the bottom v_z^{p2}. The divergence map was obtained by computing maps of v_x at several different depths where each depth block, centered about the target depth, has a horizontal spatial resolution of < 10 Mm. Then the numerical divergence was calculated. A positive divergence means an outflow. The $\pm 1\sigma$ noise for the three v_z measurements are given by the two thin lines above and below the thicker depth indication line, and all have values less than 10 m s^{-1}. Note at the top the reference scale arrow for v_z.

determining the amount of cross-talk among all of the flow components. It was found that the cross-talk is reasonably small because the inversion procedure attempts to minimize its effect by the use of certain constraints. We furthermore obtain consistent estimates of the noise and spatial resolution on the measured velocities. For practical reasons, the inversion technique is convenient because directly inverting for other quantities such as the horizontal divergence of the flows or the vertical vorticity only requires a change of the target function.

High-resolution example flow maps have been studied and tested. We have restricted ourselves to quiet-Sun cases. These maps all have horizontal spatial resolution of less than about 11 Mm, or about the p_1-mode wavelength at 3 mHz. The recovered flow speeds are below the limits for which a linearized theory of travel times is valid (Jackiewicz et al., 2007a). We have tested the inversion in several straightforward ways. We have shown that using independent measurements and similar averaging kernels gives consistent results. We have also been able to obtain high correlations (≈ 0.9) with the Doppler-velocity data after projecting the inferred horizontal flows onto the line-of-sight vector and ignoring pixels with strong magnetic fields.

We also found that the correlation coefficient of 24 hours of inferred horizontal flows from day to day on average is about 0.4 near the surface and about 0.1 down to about 6 Mm below the surface. This is consistent with the conclusion that we are not just measuring noise. However, we find that the correlation of flows at a particular depth with the surface flows falls off quite rapidly and disappears near ≈ 5 Mm beneath the surface, where we do not see any more evidence of supergranulation. Similar results have also recently been found with numerical simulations using time–distance helioseismology (Zhao et al., 2007) and holography (Braun et al., 2007). It could be that for 24 hours and at these depths the supergranulation signal is completely masked by noise (Braun et al., 2007; Woodard, 2007).

We have shown a direct inversion for the vertical component of the velocity using acoustic and surface-gravity waves. The results are in agreement with the overall behavior of the horizontal flows, and since the cross-talk among v_x, v_y, and v_z has been made small, we are fairly confident that the vertical flows are real. The vertical flows have also been compared to independent inversions for the horizontal divergence, and the values are in the expected ranges. We find that the upflow speeds in the center of supergranules are approximately 15–20% of the horizontal outflow speeds. Studying the day-to-day correlations of vertical flow maps also leads us to believe that the signal is above the noise.

Another way to validate many of the findings that we have reported would be to invert the available artificial velocity data from realistic numerical simulations of solar convection (Benson, Stein, and Nordlund, 2006). Even though for this inversion the averaging kernels give a complete picture of how the data are spatially averaged and therefore a good idea of what the final answer will look like – a nice feature of OLA-type inversions – we intend to carry this out in the near future to study the role that noise plays in the interpretation of the results.

Of course, we are undertaking many improvements to the inversion presented here. One obvious deficiency is the small set of modes that we have used. Such a limited number does not allow us to obtain many independent target depths, nor any substantially deep ones. More ridges, combined with utilizing the spatial-frequency content of the waves in each ridge in a more sophisticated way, would help us to obtain better, and deeper, averaging kernels.

There are several other improvements currently being studied, including ways to minimize the cross-talk among flow components as much as possible by constructing different types of constraints in the inversion procedure. Also, kernels that take into account the line-of-sight projection are almost certain to be necessary for inverting data well away from disk center. We already have some of these kernels available (Jackiewicz, Gizon, and Birch, 2006).

Acknowledgements We thank T. Duvall Jr. for helpful discussions and for providing the data set used in the analysis. We also gratefully acknowledge critical comments from a referee that significantly improved this paper. SOHO is a collaboration between NASA and ESA.

References

Basu, S., Antia, H.M., Tripathy, S.C.: 1999, Ring diagram analysis of near-surface flows in the sun. *Astrophys. J.* **512**, 458–470. doi:10.1086/306765.
Benson, D., Stein, R., Nordlund, Å.: 2006, Supergranulation scale convection simulations. In: Leibacher, J., Stein, R.F., Uitenbroek, H. (eds.) *Solar MHD Theory and Observations: A High Spatial Resolution Perspective* **CS-354**, Astron. Soc. Pac., San Francisco, 92.
Birch, A.C., Felder, G.: 2004, Accuracy of the Born and Ray approximations for time-distance helioseismology of flows. *Astrophys. J.* **616**, 1261–1264. doi:10.1086/424961.
Birch, A.C., Gizon, L.: 2007, Linear sensitivity of helioseismic travel times to local flows. *Astron. Nachr.* **328**, 228. doi:10.1002/asna.200610724.
Braun, D.C., Birch, A.C., Lindsey, C.: 2004, Local helioseismology of near-surface flows. In: Danesy, D. (ed.) *SOHO 14 Helio- and Asteroseismology: Towards a Golden Future* **SP-559**, ESA, Noordwijk, 337–340.
Braun, D.C., Birch, A.C., Benson, D., Stein, R.F., Nordlund, Å.: 2007, Helioseismic holography of simulated solar convection and prospects for the detection of small-scale subsurface flows. *Astrophys. J.* **669**, 1395–1405. doi:10.1086/521782.
Couvidat, S., Birch, A.C., Kosovichev, A.G.: 2006, Three-dimensional inversion of sound speed below a sunspot in the Born approximation. *Astrophys. J.* **640**, 516–524. doi:10.1086/500103.
Couvidat, S., Gizon, L., Birch, A.C., Larsen, R.M., Kosovichev, A.G.: 2005, Time–distance helioseismology: inversion of noisy correlated data. *Astrophys. J. Suppl.* **158**, 217–229. doi:10.1086/430423.
Duvall, T.L. Jr., Gizon, L.: 2000, Time–distance helioseismology with f modes as a method for measurement of near-surface flows. *Solar Phys.* **192**, 177–191.
Duvall, T.L. Jr., Jefferies, S.M., Harvey, J.W., Pomerantz, M.A.: 1993, Time–distance helioseismology. *Nature* **362**, 430–432. doi:10.1038/362430a0.
Duvall, T.L. Jr., Kosovichev, A.G., Scherrer, P.H., Bogart, R.S., Bush, R.I., de Forest, C., Hoeksema, J.T., Schou, J., Saba, J.L.R., Tarbell, T.D., Title, A.M., Wolfson, C.J., Milford, P.N.: 1997, Time–distance helioseismology with the MDI instrument: initial results. *Solar Phys.* **170**, 63–73.
Gizon, L., Birch, A.C.: 2002, Time–distance helioseismology: the forward problem for random distributed sources. *Astrophys. J.* **571**, 966–986. doi:10.1086/340015.
Gizon, L., Birch, A.C.: 2004, Time–distance helioseismology: noise estimation. *Astrophys. J.* **614**, 472–489. doi:10.1086/423367.
Gizon, L., Duvall, T.L. Jr., Larsen, R.M.: 2000, Seismic tomography of the near solar surface. *J. Astrophys. Astron.* **21**, 339–342.
Hathaway, D.H., Beck, J.G., Han, S., Raymond, J.: 2002, Radial flows in supergranules. *Solar Phys.* **205**, 25–38.
Jackiewicz, J., Gizon, L., Birch, A.C.: 2006, Sensitivity of solar f-mode travel times to internal flows. In: Lacoste, H. (ed.) *SOHO-17. 10 Years of SOHO and Beyond* **SP-617**, ESA, Noordwijk, 38–41.
Jackiewicz, J., Gizon, L., Birch, A.C., Duvall, T.L. Jr.: 2007a, Time–distance helioseismology: sensitivity of f-mode travel times to flows. *Astrophys. J.* **671**, 1051–1064. doi:10.1086/522914.
Jackiewicz, J., Gizon, L., Birch, A.C., Thompson, M.J.: 2007b, A procedure for the inversion of f-mode travel times for solar flows. *Astron. Nachr.* **328**, 234–239. doi:10.1002/asna.200610725.
Jensen, J.M., Duvall, T.L. Jr., Jacobsen, B.H.: 2003, Noise propagation in inversion of helioseismic time–distance data. In: Sawaya-Lacoste, H. (ed.) *GONG+ 2002. Local and Global Helioseismology: the Present and Future* **SP-517**, ESA, Noordwijk, 315–318.
Komm, R., Corbard, T., Durney, B.R., González Hernández, I., Hill, F., Howe, R., Toner, C.: 2004, Solar subsurface fluid dynamics descriptors derived from global oscillation network group and Michelson Doppler imager data. *Astrophys. J.* **605**, 554–567. doi:10.1086/382187.
Kosovichev, A.G.: 1996, Tomographic imaging of the Sun's interior. *Astrophys. J.* **461**, L55–L57. doi:10.1086/309989.
Pijpers, F.P., Thompson, M.J.: 1992, Faster formulations of the optimally localized averages method for helioseismic inversions. *Astron. Astrophys.* **262**, L33–L36.
Scherrer, P.H., Bogart, R.S., Bush, R.I., Hoeksema, J.T., Kosovichev, A.G., Schou, J., Rosenberg, W., Springer, L., Tarbell, T.D., Title, A., Wolfson, C.J., Zayer, I.; 1995, MDI Engineering Team: 1995, The solar oscillations investigation – Michelson Doppler imager. *Solar Phys.* **162**, 129–188.
Schmelling, M.: 1995, Averaging correlated data. *Phys. Scr.* **51**, 676–683.
Woodard, M.F.: 2007, Probing supergranular flow in the solar interior. *Astrophys. J.* **668**, 1189–1195. doi:10.1086/521391.
Zhao, J., Kosovichev, A.G., Duvall, T.L. Jr.: 2001, Investigation of mass flows beneath a sunspot by time–distance helioseismology. *Astrophys. J.* **557**, 384–388. doi:10.1086/321491.
Zhao, J., Georgobiani, D., Kosovichev, A.G., Benson, D., Stein, R.F., Nordlund, Å.: 2007, Validation of time–distance helioseismology by use of realistic simulations of solar convection. *Astrophys. J.* **659**, 848–857. doi:10.1086/512009.

Structure and Evolution of Supergranulation from Local Helioseismology

Johann Hirzberger · Laurent Gizon · Sami K. Solanki · Thomas L. Duvall, Jr.

Originally published in the journal Solar Physics, Volume 251, Nos 1–2, 417–437.
DOI: 10.1007/s11207-008-9206-8 © The Author(s) 2008

Abstract Supergranulation is visible at the solar surface as a cellular pattern of horizontal outflows. Although it does not show a distinct intensity pattern, it manifests itself indirectly in, for example, the chromospheric network. Previous studies have reported significant differences in the inferred basic parameters of the supergranulation phenomenon. Here we study the structure and temporal evolution of a large sample of supergranules, measured by using local helioseismology and SOHO/MDI data from the year 2000 at solar activity minimum. Local helioseismology with f modes provides maps of the horizontal divergence of the flow velocity at a depth of about 1 Mm. From these divergence maps supergranular cells were identified by using Fourier segmentation procedures in two dimensions and in three dimensions (two spatial dimensions plus time). The maps that we analyzed contain more than 10^5 supergranular cells and more than 10^3 lifetime histories, which makes possible a detailed analysis with high statistical significance. We find that the supergranular cells have a mean diameter of 27.1 Mm. The mean lifetime is estimated to be 1.6 days from the measured distribution of lifetimes (three-dimensional segmentation), with a clear tendency for larger cells to live longer than smaller ones. The pair and mark correlation functions do not show pronounced features on scales larger than the typical cell size, which suggests purely random cell positions. The temporal histories of supergranular cells indicate a smooth evolution from their emergence and growth in the first half of their lives to their decay in the second half of their lives (unlike exploding granules, which reach their maximum size just before they fragment).

Keywords Sun: supergranulation · Sun: photosphere · Helioseismology

---

Helioseismology, Asteroseismology, and MHD Connections
Guest Editors: Laurent Gizon and Paul Cally

J. Hirzberger (✉) · L. Gizon · S.K. Solanki
Max-Planck-Institut für Sonnensystemforschung, 37191 Katlenburg-Lindau, Germany
e-mail: hirzberger@mps.mpg.de

T.L. Duvall, Jr.
Laboratory for Solar and Space Physics, NASA Goddard Space Flight Center, Greenbelt, MD 20771, USA

1. Introduction

The velocity field in the solar photosphere is dominated by turbulent convective motions. These motions manifest themselves as two main horizontal scales – granulation and supergranulation. The existence of a distinct intermediate scale of mesogranulation is still under debate (see, *e.g.*, Rieutord *et al.*, 2000). The granular convective scales (≈ 1.5 Mm) are directly observable in brightness as a cellular pattern, which is closely related to the vertical motion of the gas determined from the spectral line shifts. Supergranular structures (≈ 30 Mm) are easily detectable away from the center of the solar disk in maps of the line-of-sight Doppler velocity (*e.g.*, Hathaway *et al.*, 2000): Supergranulation is associated with excess power near spherical harmonic degree $\ell = 120$. Supergranulation is outlined by the network-like distribution of quiet-Sun magnetic fields, visible as chromospheric emission in the Ca II H and K spectral lines. Supergranules, however, are barely detectable in intensity. Elaborate measurements by Meunier, Tkaczuk, and Roudier (2007) show a small intensity contrast that would correspond to a temperature difference of 0.8 – 2.8 K between cell centers and boundaries; these results are compatible with a convective origin of supergranulation. Supergranulation can also be detected indirectly by tracking the proper motions of surface features such as granules or magnetic bright points (*e.g.*, Leighton, 1964; Rimmele and Schröter, 1989; Shine, Simon, and Hurlburt, 2000; DeRosa and Toomre, 2004; Rieutord *et al.*, 2007).

Since the discovery of supergranulation (Hart, 1956; Leighton, Noyes, and Simon, 1962) several attempts have been made to explain its origin. Simon and Leighton (1964) proposed that a local minimum of the vertical adiabatic temperature gradient in the deep solar layers caused by He I or He II ionization would favor convective cells with horizontal sizes comparable to the depth of the ionization zones. Alternatively, supergranulation may be related to the spatial ordering of smaller convective cells. For example, Ploner, Solanki, and Gadun (2000) showed that mesogranular-scale structures are formed in two-dimensional (2D) simulations even if these are not sufficiently deep to reach the ionization zones. Rieutord *et al.* (2000) and Roudier *et al.* (2003) suggested that a pattern at supergranular scales could emerge from the nonlinear collective interaction of granules. Using a numerical n-body diffusion-limited aggregation model, Crouch, Charbonneau, and Thibault (2007) showed that the advection of magnetic elements by the granular flow may produce a supergranular network. The same simulation suggests that the supergranular cell size decreases with magnetic activity, which has recently been observed by Meunier, Roudier, and Tkaczuk (2007).

Various techniques of local helioseismology have been used to infer supergranular flows just below the photosphere (*e.g.*, Kosovichev and Duvall, 1997; Duvall and Gizon, 2000; Woodard, 2002; Zhao and Kosovichev, 2003; Braun, Birch, and Lindsey, 2004; Gizon and Birch, 2005, and references therein). Using f-mode time – distance helioseismology, Duvall and Gizon (2000) mapped supergranular flows and studied their evolution. This method of analysis provides information about the two horizontal components of velocity and the horizontal divergence of the flow. Duvall and Gizon (2000) described the distribution of the divergence values and noted that the spatial power spectrum of the divergence maps peaks at spherical harmonic degree 120. Using the same f-mode data, del Moro *et al.* (2004) found supergranular diameters in the range 10 – 40 Mm and determined a mean diameter of 27 Mm. Beck and Duvall (2001) measured the autocorrelation of the supergranulation pattern for time lags up to six days. This and a longer time series of observations led to the discovery of the wavelike properties of supergranulation (Gizon, Duvall, and Schou, 2003), which suggest that supergranulation is an example of traveling-wave convection.

Figure 5 Image segmentation based on Fourier band-pass filtering applied to the image displayed in Figure 2 (a) and to its negative (b). The white areas in panel (a) represent detected supergranular divergence centers; gray areas in panel (b) are inter-supergranular convergence centers.

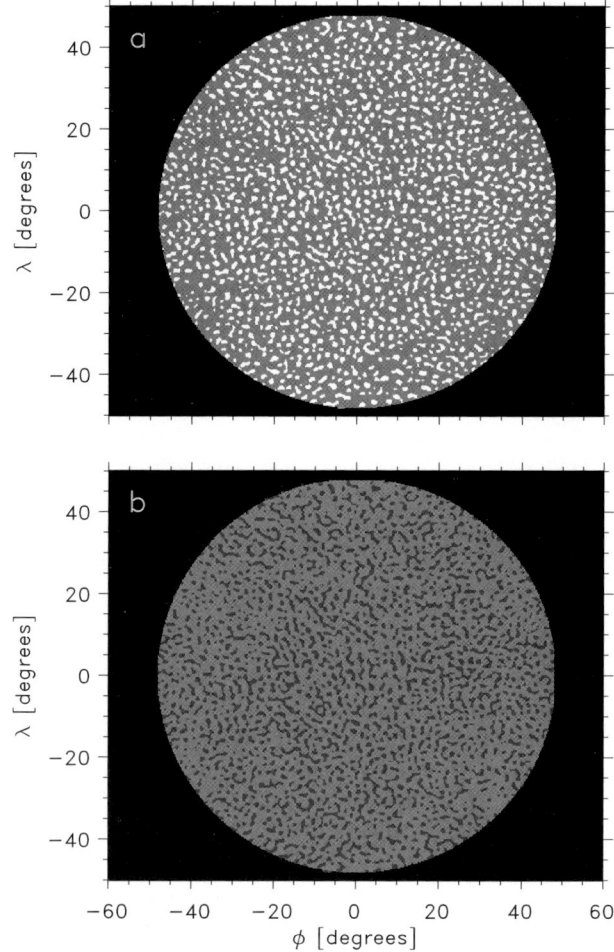

is consistent with the distribution of divergence values in nonsegmented data (*cf.* Figure 3) and granular intensities and velocities (see Hirzberger, 2002). This result indicates that there is a similarity between supergranulation and granulation.

4. Statistical Analysis of Cell Structure

After applying image segmentation, we can assign several parameters to each detected supergranular cell. Figure 10 shows distributions of cell areas [A] and maximum divergence values [$\xi \equiv (\text{div } \boldsymbol{v}_\text{h})_\text{max}$] in the cells. The cell area varies between 160 and about 2000 Mm^2, which corresponds to an effective cell diameter, defined as $d = \sqrt{4A/\pi}$, between 14 and 50 Mm. The median cell area is 650 Mm^2 and the median cell diameter is 28.8 Mm. These latter numbers were obtained by considering the entire sample of segmented supergranular cells.

We note that the shape of the histograms of cell areas depends on the parameters of the segmentation procedure and, in particular, on the filtering. The cutoff value of $0.8\,\sigma$

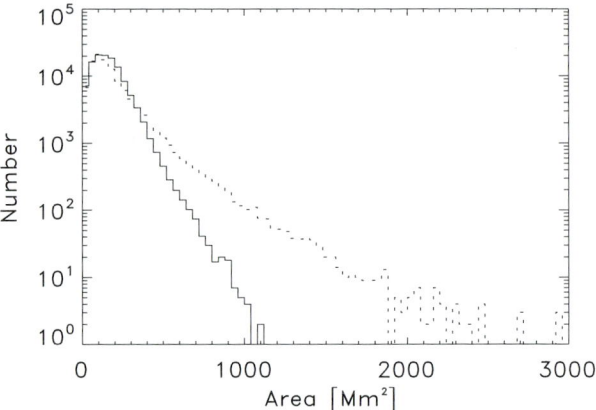

Figure 6 Histograms of areas of divergence centers (solid) and convergence centers (dotted) obtained by applying the image segmentation procedure based on Fourier pass-band filtering.

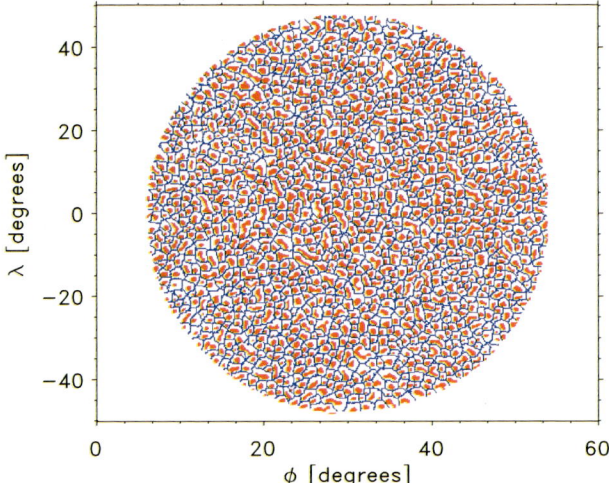

Figure 7 Dilation of detected supergranular areas of the map shown in Figures 2 and 5 toward their surrounding local divergence minima, leading to the delimitation of the supergranular cells (blue mesh). The red areas represent divergence centers detected by the segmentation algorithm.

(defined earlier) provides the maximum number of cells. Lowering this threshold results in merging structures, which reduces the number of detected cells and narrows the width of the histograms since the very small cells are less numerous. If the threshold is increased, then the very faint structures are ignored, which also reduces the total number of cells, whereas the histograms get broader. Smaller scales are present, but the regions where no structures are detected become filled by expanding the adjacent supergranules. This leads to several unrealistically large cells.

As can be seen in Figure 10 the distribution of cell sizes changes slightly with heliographic latitude [λ]. A monotonic decrease of the average cell size from the Equator toward higher latitudes can be detected. A similar trend, although weaker, can be seen for ξ shown in the lower left panel of Figure 10. Since this latter parameter only weakly depends on the definition procedure of a supergranular cell, this behavior is probably not caused by the segmentation algorithm. This conclusion is strengthened by the results shown in the right panels of Figure 10. Here the variation of the area and divergence histograms with heliocentric angle θ is plotted. The obtained trend is even more significant than the variation with λ.

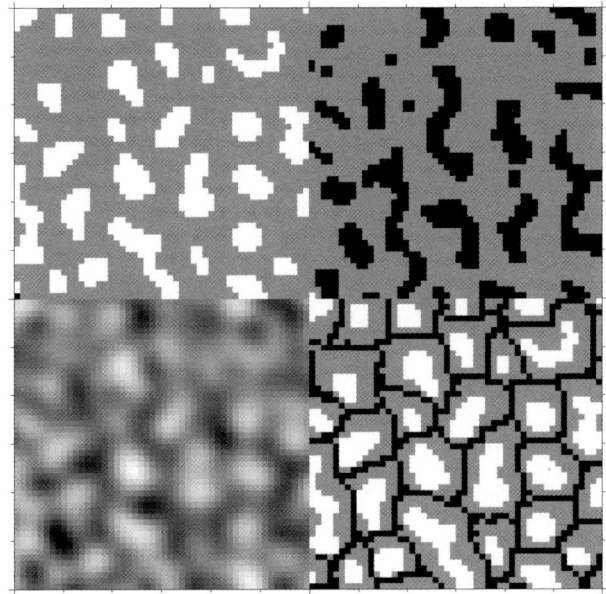

Figure 8 Illustration of the segmentation procedure: The lower left panel shows a detail (6° × 6°) of the filtered divergence map displayed in Figure 2; in the upper panels the corresponding segmented divergence (left) and convergence (right) centers (see Figure 5) are shown; in the lower right panel the divergence centers and their dilations to supergranular cells (see Figure 7) are depicted.

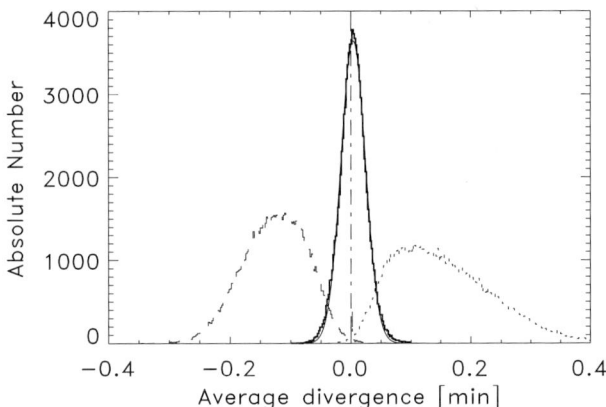

Figure 9 Histograms of average divergence for supergranular cells (thick solid line), for divergence centers (dotted), and for convergence centers (dashed). The (barely visible) thin solid line represents a Gaussian fitted to the cell histogram.

Consequently, it must be concluded that the observed latitudinal variation of supergranular size and divergence is not of solar origin only but is likely to be an intrinsic effect of the data processing using helioseismology techniques.

The shapes of the supergranular size distributions are characterized by a significant asymmetry, similar to the findings of, for example, Hagenaar, Schrijver, and Title (1997) and DeRosa and Toomre (2004). Compared to the resulting cell sizes reported in del Moro *et al.* (2004) the distribution shows a significantly steeper drop from the maximum toward small cells and also a slight shift of the location of the maximum value toward smaller cells. (We find $A = 575$ Mm^2 and $d = 27.1$ Mm, respectively, compared with $d = 33$ Mm reported in del Moro *et al.* (2004)). Another conspicuous feature of the histograms is the appearance of a minimum cell area of 158 Mm^2, which arises because of the applied minimum structure

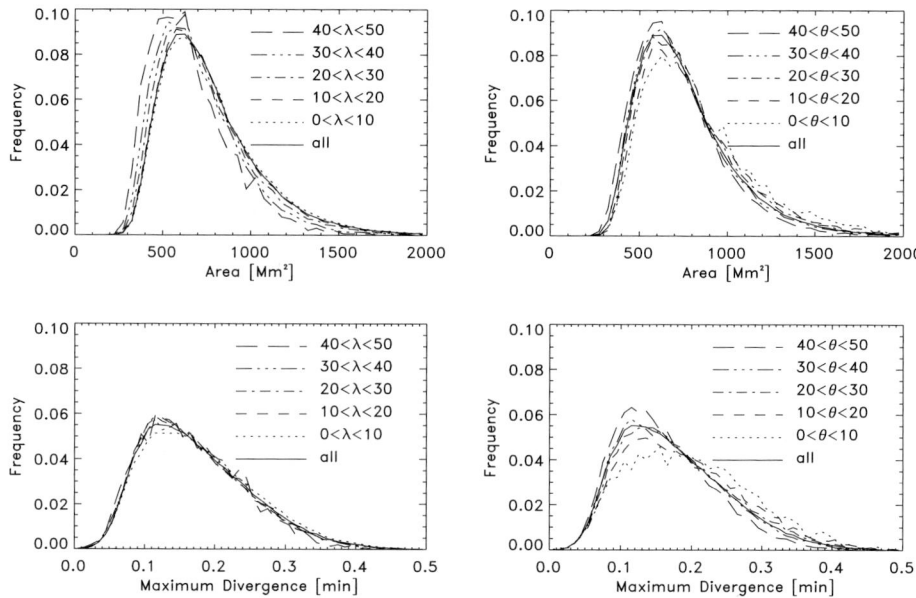

Figure 10 Histograms of areas A (upper panels) and maximum divergences ξ (lower panels) of supergranular cells obtained in different regions on the solar disk. The curves in the left panels have been obtained by considering stripes of different latitudes λ; the curves in the right panels are computed by changing heliocentric angle θ.

size in the Fourier segmentation. This fact, however, cannot explain the differences compared to the results of del Moro *et al.* (2004) since there a similar minimum size condition was applied. Another possibility may be the different (presumably larger) noise level in the data used by del Moro *et al.* (2004). Residual noise leads to a decrease of the obtained cell sizes. For instance, if the low-pass filter (allowing only structures larger than $\ell = 250$) is not applied to our data the median value of the cell areas decreases from 650.6 to 603.0 Mm^2. However, the obtained shapes of the histograms do not change from right-asymmetric to left-asymmetric as found in del Moro *et al.* (2004).

The close similarity of shapes of the area and divergence distributions visible in Figure 10 suggests a close relation of both parameters. Owing to the large number of structures in our sample, a 2D histogram is more appropriate for showing this relation than a simple scatter plot (see Figure 11). In the range $300 < A < 700$ Mm^2 a linear relation between the two parameters can be observed. Toward larger cells, the scatter of the distribution increases and the area–divergence relation becomes ambiguous. A close relation between size and flow divergence of supergranules is not surprising and was also reported by Meunier *et al.* (2007).

5. Temporal Evolution

Once supergranular cells have been identified in each image, it becomes possible to follow the temporal evolution of individual structures. We applied a feature-tracking code originally developed to track granules (Hirzberger *et al.*, 1999). A modified version of the code

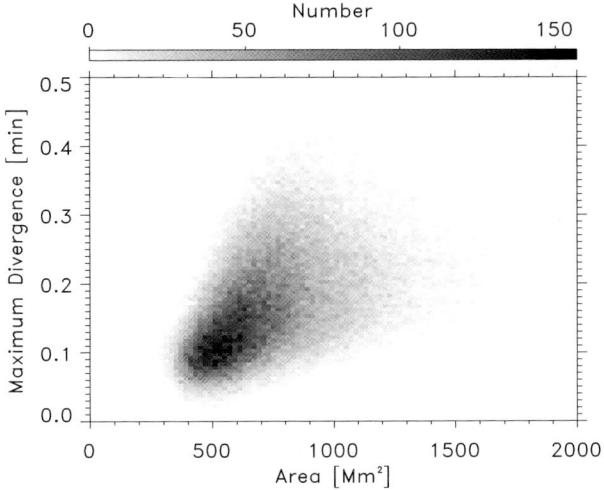

Figure 11 Two-dimensional histogram of maximum divergence versus supergranular cell area.

was also applied to mesogranules (Leitzinger et al., 2005). This code tracks a certain feature forward and backward in time until a stopping condition is reached, where the tracking is terminated. The length of time between the two termination points is the structure's lifetime [T]. In the present version of the code, the lifetime history of a structure was terminated if $A(t) > 2A(0)$ or if $A(t) < A(0)/2$, where $A(0)$ is the area of the cell when the tracking was started and $A(t)$ is its area at time t. The tracking was also terminated if the barycenter of a structure seems to jump more than three pixels between subsequent images. However, before applying this criterion, the structures where shifted by the proper motions caused by the mean rotation at the appropriate latitude. This second criterion avoids connecting the lifetime history of a feature with that of a neighboring one after the tracked feature vanishes. Finally, the resulting lifetime histories were smoothed by applying a parabolic fit to $A(t)$ and removing those points that are outside $\pm 1.5\sigma$ of the fit from the histories, where σ denotes the standard deviation of the fit from the data points. The obtained histories were cut when more than two subsequent data points were removed by the fit. If the stopping conditions were used in a restrictive way, residual noise would lead to an artificial shortening of the lifetime histories. The fitting procedure was implemented to mitigate the stopping conditions and to reconsider the performance of the tracking procedure. Redundantly tracked features and those histories that reach the boundaries of the data cube were deleted from the sample of evolutional histories. However, keeping only those histories that describe the entire lifetime of a supergranule cell from its birth to its death causes some bias to the obtained results since short-lived structures are favored. Altogether a sample of 3529 lifetime histories was obtained by applying this procedure. This number is small compared to the number of detected cells. This is because only a small fraction of histories is not truncated by the image borders or by rotating out of the field of view.

A second method to characterize the temporal evolution of supergranular features is to apply 3D version of the Fourier-based segmentation to the data. The pass-band filter (see Figure 4) was extended to three dimensions. We assumed ergodic conditions (i.e., the shapes of the supergranules are assumed to be statistically uniform in (ϕ, λ, t)-space). The time dimension was treated as a third spatial dimension and the filter described in Roudier and Muller (1987) was applied to the 3D data cube as was done before to the

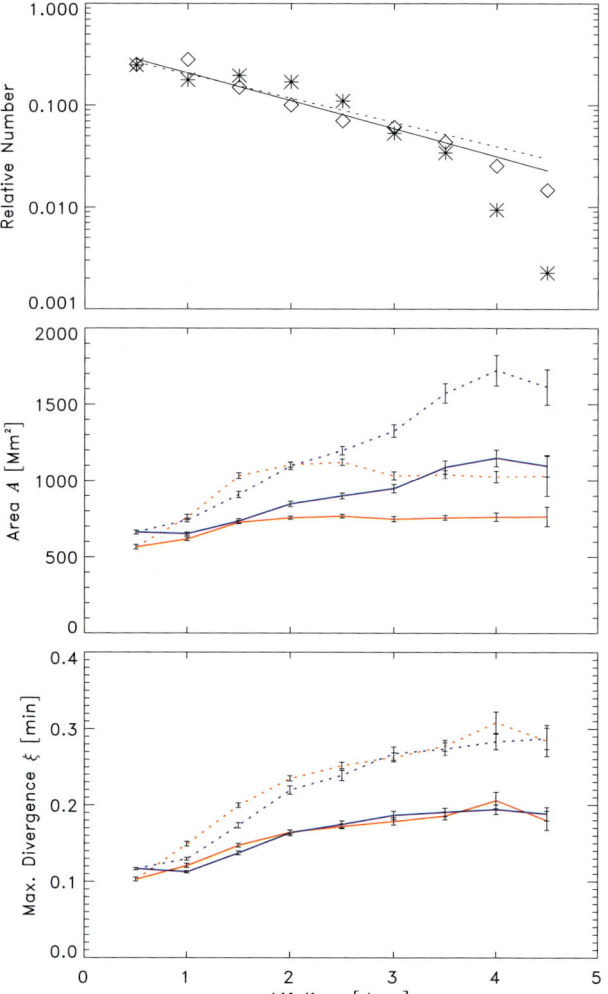

Figure 12 The upper panel shows histograms of supergranular lifetimes. Asterisks show the results from the feature-tracking method and diamonds from the 3D Fourier segmentation. The solid line represents an exponential fit to the feature-tracking data; the dotted curve shows a fit to the 3D segmented data. Also shown are the lifetime dependence of areas (middle panel) and divergences (lower panel) for both segmentation methods. Red solid lines give the mean A and ξ from feature tracking; red dashed lines give the maximum A and ξ from feature tracking; blue solid lines give the mean A and ξ from 3D Fourier segmentation; blue dashed lines give the maximum A and ξ from 3D segmentation. Error bars denote standard deviations of the mean ($\sigma/\sqrt{N}$).

2D divergence maps. After Fourier filtering and cutting at a level of $\pm 0.8\,\sigma$ above the mean (see Section 3), the lifetimes of resulting structures are simply given by their extent along the time axis. To retrieve supergranular cell areas, the data cube was then sliced into single images and each feature was extended toward the surrounding local divergence minima. Again all histories that reach the boundaries of the data cube have been deleted. In addition, those structures that are smaller than eight pixels in (ϕ, λ, t)-space were deleted. The supergranular areas computed by using 3D segmentation are almost identical to those found with the 2D version of the code. However, structures that have small extents, particularly in time, are deleted by the code. This leads to a significant reduction of the number of structures in the sliced images and, therefore, to an increase of the mean cell sizes when dilating the areas to the local divergence minima. Altogether, from the 3D image segmentation a sample of 4977 lifetime histories was obtained.

The inferred evolution parameters depend considerably on the segmentation and tracking methods (see, *e.g.*, Alissandrakis, Dialetis, and Tsiropoula, 1987, for a comparison of lifetime estimation procedures). The lifetime distributions (Figure 12) show a decrease of the relative frequency of cells toward longer lifetimes. The lifetime distribution obtained from the 3D segmentation may be modeled with an exponential distribution of the form $N(T) = N(0)e^{-T/\tau}$, where τ is the mean lifetime. We find $\tau = 1.59$ days, which is somewhat longer than the classical value of one day (Leighton, Noyes, and Simon, 1962) and longer than the value $\tau = 0.92$ days stated by del Moro *et al.* (2004). The standard deviation of τ is $\sigma_\tau = 0.74$ days. Note that for T longer than four days the effect of deleting histories that are truncated by the data boundaries leads to a deviation from the exponential form. If the first data point at $T = 0.5$ days is omitted from the fit, then the resulting distribution looks steeper and τ becomes 1.12 days (with a standard deviation $\sigma_\tau = 0.59$ days). Thus the classical lifetime of one day lies within the uncertainty of the present result.

The histogram obtained from the feature-tracking analysis shows stronger deviations from an exponential function than the Fourier-based histogram. A significant excess of structures at lifetimes around two days can be observed. Both short-lived ($T < 1.5$ days) and long-lived ($T > 3.5$ days) supergranules are comparatively less frequent than in the sample obtained from the 3D segmentation. An exponential fit gives $\tau = 1.83$ days with a standard deviation of $\sigma_\tau = 0.89$ days.

The middle and lower panels of Figure 12 show cell areas and maximum divergences (*i.e.*, the largest divergence within a supergranule at any given time) averaged over each lifetime history and, additionally, averaged over each bin of the histograms. The 3D segmentation shows a clear trend for larger and more divergent supergranular cells to live longer. The mean area even seems to be linearly related to the lifetime. The feature tracking reveals such a trend only for short-lived cells, while displaying a very similar trend to the 3D segmentation for the maximum divergence. Supergranules with lifetimes $T \geq 2$ days have constant mean areas of 750 Mm^2. This latter behavior is in almost perfect agreement with the lifetime–area relation presented in del Moro *et al.* (2004). The lifetime–area and the lifetime–divergence relations do not change significantly if the maximum area and divergence of each lifetime history is considered instead of using temporal means of each history (dashed and dash-dotted lines in Figure 12).

Typical time evolution histories of parameters A and ξ are shown in Figures 13 and 14. The curves represent averages of all histories included in a certain bin of the lifetime histograms derived from both 3D Fourier segmentation and feature tracking (see Figure 12). In the lifetime range $1 \leq T \leq 4.5$ days, the evolution shows typically convex shapes: A supergranular cell and its divergence are small when it is "born." After that, area and divergence increase until roughly half the lifetime is reached. After that, both parameters are decreasing again until the cell "dies." Very approximately, the time dependence of supergranule area and maximum divergence is quadratic. Within the achievable accuracy it is difficult to say whether there is a general asymmetry in the sense that the supergranules reach their maximum size in the first or the second half of their lifetime. This behavior is particularly well pronounced for the mean histories obtained from the 3D segmentation and more noisy in the feature-tracking results. Moreover, as already mentioned, the mean cell areas are significantly larger in the 3D than in the 2D segmentation. The mean evolution histories of the longest living supergranular cells do not seem to follow a clear trend, probably because of the low number of histories in the corresponding samples. The retrieved convex lifetime histories are considerably different from the histories of granulation cells (see, *e.g.*, Hirzberger *et al.*, 1999). They reach their maximum size at the end of their lifetimes (*i.e.*, the largest fraction of granular cells increase in size until they split into two or more fragments).

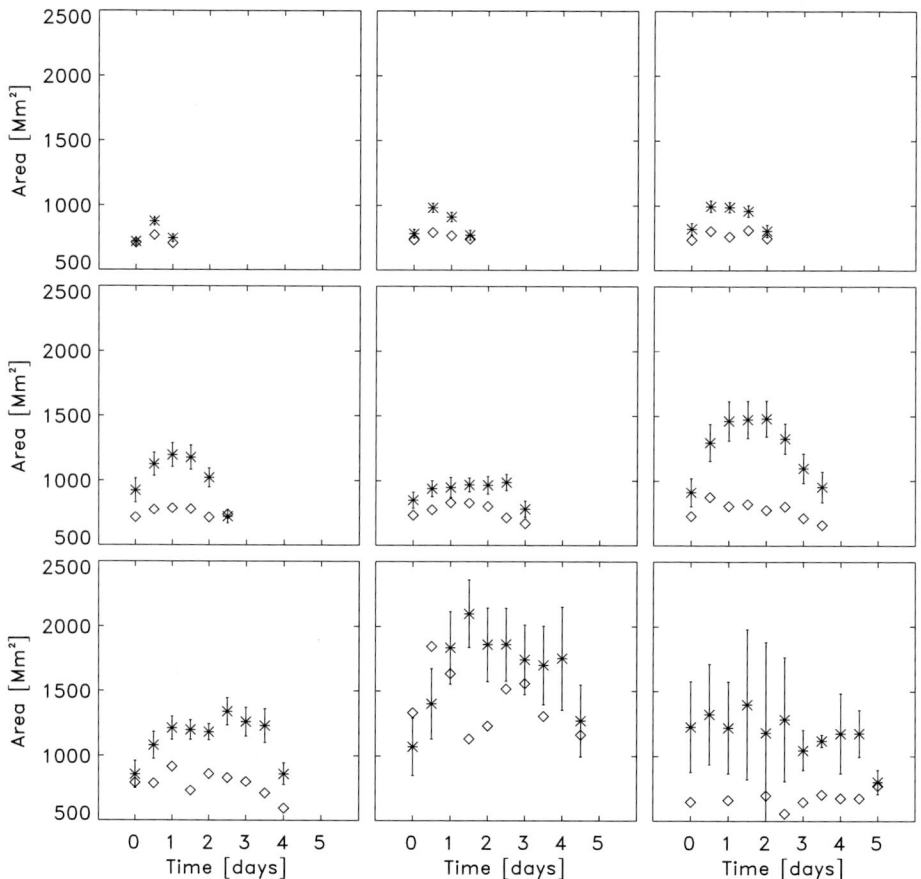

Figure 13 Mean area evolution of supergranules with different lifetimes. Each panel shows mean evolutions determined by averaging A over all histories in a single bin of the 3D segmentation (asterisks) and feature-tracking (diamonds) histograms shown in Figure 12. The error bars denote standard deviations of the mean.

6. Pair Correlation Functions

Important aspects of the supergranulation phenomenon are not only given by the properties of individual cells. For a detailed understanding of supergranulation the interaction between neighboring cells and, particularly, the arrangement of cells within a certain region might be highly instructive.

A useful tool to analyze the arrangement of structures in a given region is the so-called pair correlation function (PCF), which represents the probability of finding a supergranule located at a distance r from another supergranule. A comparison of the PCFs of granulation and supergranulation has already been performed by Berrilli *et al.* (2004). In the present work the position of a supergranular cell is represented by its barycenter, which is defined by its centroid weighted by its divergence. For a homogeneous and isotropic distribution of

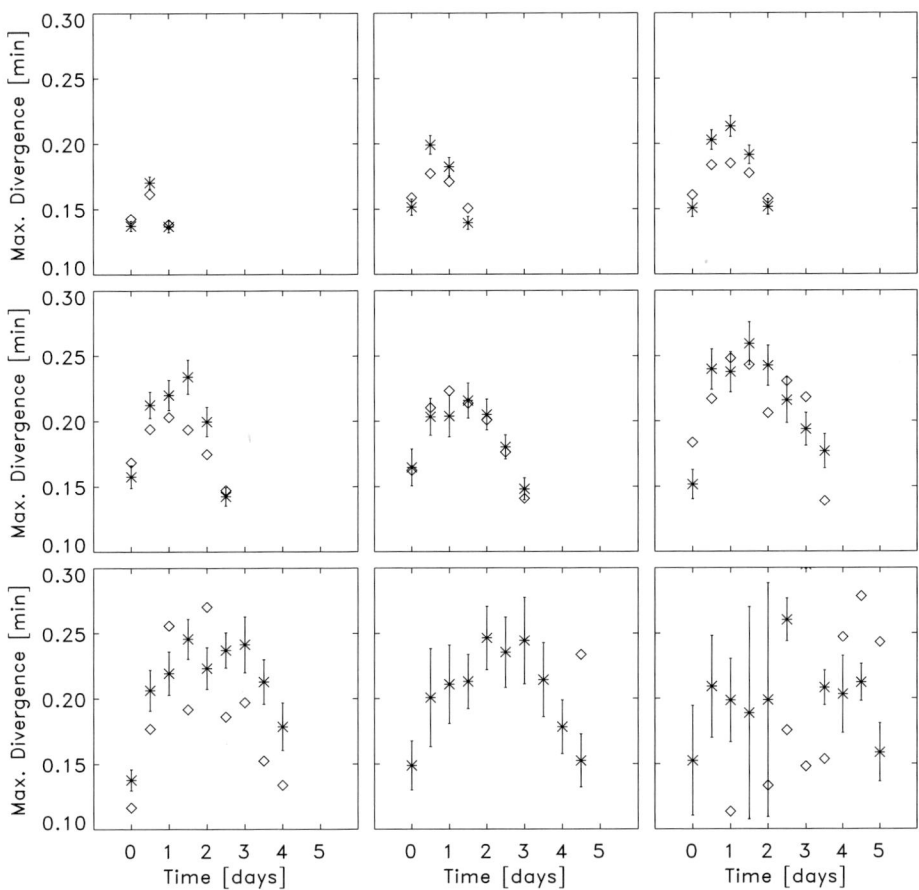

Figure 14 Mean evolution of supergranules with different lifetimes. Each panel shows mean evolutions determined by averaging ξ over all histories in a single bin of the 3D segmentation (asterisks) and feature-tracking (diamonds) histograms shown in Figure 12. The error bars denote standard deviations of the mean.

structures, the PCF $[g(r)]$ is defined as

$$g(r) = \frac{\rho(r)}{\eta^2}, \qquad (1)$$

where $\rho(r)$ is the product density, defined in the following. The function g describes the probability of finding exactly one structure in each of the two infinitesimal small circles at positions $\boldsymbol{x}'$ and $\boldsymbol{x}''$, separated by the distance r, and η is an intensity measure (*i.e.*, the mean number of structures in an area of size unity). The PCF depends on the product density (see Stoyan and Stoyan, 1994):

$$\rho(r) = \frac{1}{2\pi r} \sum_{i=1}^{n} \sum_{\substack{j=1 \\ (j \neq i)}}^{n} \frac{K_h(r - \|\boldsymbol{x}_j - \boldsymbol{x}_i\|)}{A(W_j \cap W_i)}, \qquad (2)$$

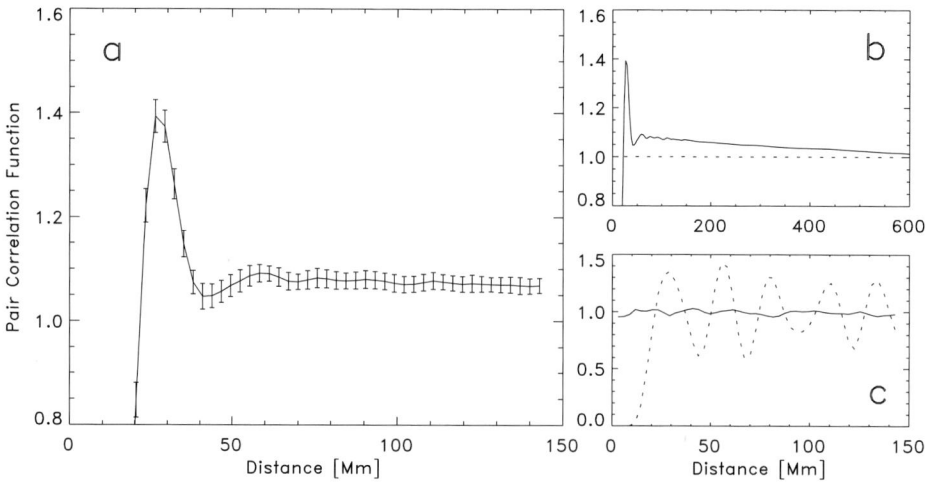

Figure 15 (a) Mean PCF obtained by averaging the PCFs of each 2D segmented divergence map. Error bars denote standard deviations. (b) The same PCF for distances up to $r = 600$ Mm. (c) For reference, the PCFs of a random configuration (solid) and a randomly disordered regular cell pattern (dotted) are shown.

where the Epanečnikov kernel is defined by

$$K_h(r) = \begin{cases} \frac{3}{4h}(1 - r^2/h^2) & \text{if } -h \leq r \leq h, \\ 0 & \text{else.} \end{cases} \quad (3)$$

In this kernel, the "bandwidth" h is a free parameter that has to be defined externally. In the present study a bandwidth of $h = 2$ pixels is used. The summation in Equation (2) runs over all n structures of the sample that are located at positions x_i and x_j, respectively. The area $A(W_j \cap W_i)$ is given by the intersection of maps shifted once by x_i and once by x_j. This form avoids bias from structures located close to the borders of the considered region where the distribution of structures within a circle with radius r is not homogeneous since the circle is divided by the border. The normalization of g is fixed by the factor $\eta^2 = n(n-1)/A(W)$, where $A(W)$ is the total area.

For each 2D segmented divergence map, a PCF was calculated. A mean PCF derived from averaging the functions calculated for each map is displayed in Figure 15. The plot shows a conspicuous peak at $r = 28$ Mm and a less distinct minimum at $r = 43$ Mm. This result gives strong evidence that a preferred distance between adjacent supergranular cells is 28 Mm and a slight tendency that distances of 43 Mm are rather unlikely. In contrast to the results of Berrilli *et al.* (2004), a clear second maximum is not visible and even the first minimum is only weakly pronounced (and above unity). The position of the maximum of the PCF corresponds well to the diameter of the preferred cell size (see Figure 10). The standard deviation of the peak position, obtained from polynomial fits to the individual PCFs, is $\sigma_{max} = 0.55$ Mm and that of the minimum position is $\sigma_{min} = 1.59$ Mm. The shape of the obtained PCF – only one conspicuous maximum and a weakly pronounced minimum – is typical for so-called soft-core fields (see Stoyan and Stoyan, 1994). Soft-core fields are good approximations for fields of nonoverlapping circles with variable diameters. This is unlike

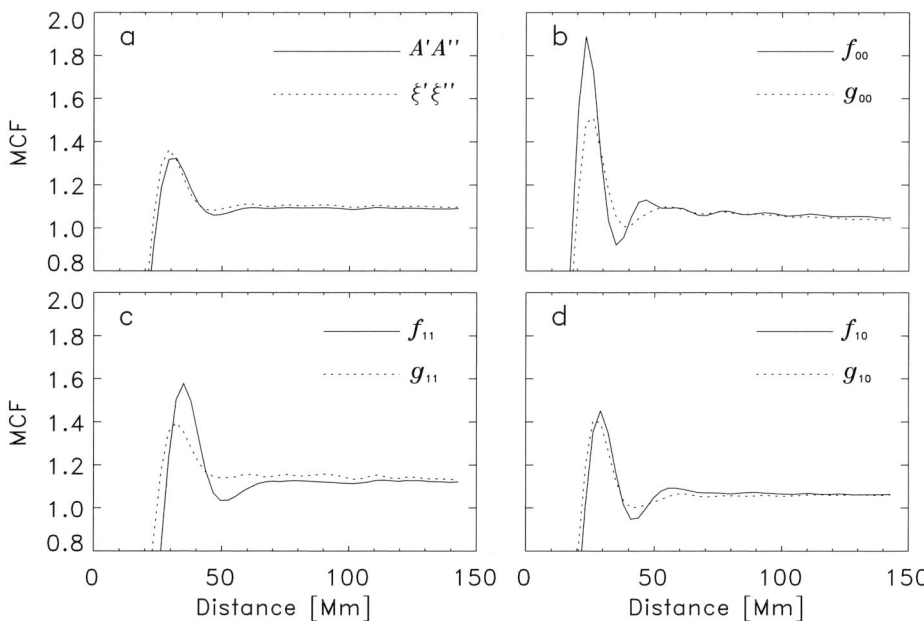

Figure 16 Mean MCFs obtained by averaging the MCFs of each 2D segmented divergence map. The four panels represent different test functions.

the results of Berrilli *et al.* (2004), who have assumed a hard-sphere close-packing model (*i.e.*, close-packed circles of uniform diameters).

The definition of $g(r)$ (Equation (1)) implies that it tends to one as r tends to infinity. Figure 15(b) shows a convergence to unity for large distances ($r \gtrsim 600$ Mm). For comparison, Figure 15(c) shows PCFs for a random distribution of cells (with a mean distance of 28 Mm) and for a regular 28-Mm grid for which the cell positions were randomly disordered (up to a maximum distance of 11.66 Mm, *i.e.*, four pixels in both the x- and y-directions). The distance needed to converge to unity is called the (correlation) range of the PCF. The long range for the supergranular cell positions indicates some kind of long-distance interaction of the cells.

The concept of PCFs can be extended by attributing a certain property (*i.e.*, a mark) to each structure in the map. If the marks m' and m'' are attributed to the positions x' and x'', separated by a distance r, then the so-called f-product density [$\rho_f(r)$] is equal to a nonnegative test function $f(m', m'')$ if exactly one structure is located in infinitesimally small circles around x' and x''. If this condition is not fulfilled then $\rho_f(r) = 0$. With the f-product density the so-called mark correlation function (MCF), $k_f(r)$, can be defined as

$$k_f(r) = \frac{\varkappa_f(r)}{\varkappa_f(\infty)}, \qquad (4)$$

where

$$\varkappa_f(r) = \frac{\rho_f(r)}{\rho(r)}. \qquad (5)$$

In practice, the f-product density and, hence, the MCF can be estimated (see Stoyan and Stoyan, 1994) with

$$\rho_f(r) = \frac{1}{2\pi r} \sum_{i=1}^{n} \sum_{\substack{j=1 \\ (j \neq i)}}^{n} \frac{f(m'_i, m''_j) K_h(r - \|x_j - x_i\|)}{A(W_j \cap W_i)}. \tag{6}$$

The shape of the test function is a free parameter. In the present work we use $f(m', m'') = m'm''$. The quantities m' and m'' can be either the area [A] or the maximum divergence [ξ] of a supergranule. The MCF is large if at a certain distance r the function f (e.g., the product $A'A''$ or the product $\xi'\xi''$) is large. In Figure 16 the mean MCFs are displayed for various choices of m and m'. Both MCFs show a distinct maximum for distances slightly larger than 30 Mm and a shallow minimum in the range between 40 and 50 Mm. At larger distances no conspicuous features are visible. The MCF for ξ seems to be slightly shifted toward smaller distances than that for A (cf. Figure 16(a)). This shows, in correspondence to the results shown in Figure 11, that for large supergranules a stringent linear relation between cell sizes and maximum divergences is not given. From the shapes of these MCFs it might be concluded that the positions of large and/or divergent supergranules are not different from that of the overall sample.

Instead of marking the structures with true areas and divergence values, both parameters can be discretized. If $\overline{A}$ and $\overline{\xi}$ are the mean values of the supergranular cell areas and maximum divergences, than the following discretized values can be defined:

$$A_0 = \begin{cases} 1 & \text{if } A \leq \overline{A}, \\ 0 & \text{if } A > \overline{A} \end{cases} \quad \text{and} \quad A_1 = \begin{cases} 1 & \text{if } A \geq \overline{A}, \\ 0 & \text{if } A < \overline{A} \end{cases} \tag{7}$$

and

$$\xi_0 = \begin{cases} 1 & \text{if } \xi \leq \overline{\xi}, \\ 0 & \text{if } \xi > \overline{\xi} \end{cases} \quad \text{and} \quad \xi_1 = \begin{cases} 1 & \text{if } \xi \geq \overline{\xi}, \\ 0 & \text{if } \xi < \overline{\xi}. \end{cases} \tag{8}$$

Using these definitions, we can consider the following test functions:

$$f_{00} = A'_0 A''_0, \quad f_{11} = A'_1 A''_1, \quad f_{10} = A'_1 A''_0 \tag{9}$$

and

$$g_{00} = \xi'_0 \xi''_0, \quad g_{11} = \xi'_1 \xi''_1, \quad g_{10} = \xi'_1 \xi''_0. \tag{10}$$

The corresponding MCFs are plotted in Figure 16. For large and highly divergent supergranular cells (Figure 16(c)) the MCFs are similar to the MCFs for the nondiscretized parameters. However, both the maximum and the first minimum are shifted to larger distances and are more pronounced for areas than for maximum divergences and the shift between the two curves is slightly larger. The minimum of the curve for g_{11} is even hardly detectable. This means that a regular ordering of strong divergence centers is clearly excluded. By applying f_{00} and g_{00} (Figure 16(b)) the heights of the first maxima as well as the depths of the first minima are more pronounced than for f_{11} and g_{11}. This is due to the finite resolution of the analyzed data, which results in a higher complexity of the shapes of large cells than that of small ones. For f_{00} even a shallow secondary maximum at $r = 45$ Mm might be guessed. This result indicates a dense packing of small cells. Surprisingly, the shift between the two curves is opposite to that for larger cells. Finally, by applying f_{10} and g_{10} (Figure 16(d)), the obtained MCFs are again similar to those of the nondiscretized ones. Even the position of the maxima and minima are almost identical. Thus, the probability that a large and strong divergence center is surrounded by small and weakly divergent structures (or *vice versa*) is not different from the probability of finding equally large and divergent structures at the same distances.

7. Conclusions

The application of helioseismic methods to study the supergranulation phenomenon makes possible a statistical analysis of the structure of supergranulation. We have analyzed a sample of more than 10^5 individual cells and several 10^3 lifetime histories. The inferred mean cell diameter of 27.1 Mm is in good agreement with former helioseismic studies (del Moro *et al.*, 2004) and the local correlation tracking of granulation (November, 1994), but it is considerably larger than that found in works based on the analysis of SOHO/MDI Dopplergrams (DeRosa and Toomre, 2004) and Ca II K filtergrams (Hagenaar, Schrijver, and Title, 1997). Hagenaar, Schrijver, and Title (1997) explained the systematically smaller cell sizes obtained from Ca II K filtergrams by the fact that autocorrelation (AC) methods preferentially select larger cells. Much earlier, Singh and Bappu (1981) already obtained a mean cell size of 23 Mm from Ca II K filtergrams and 32 Mm with an AC method. In the present study a helioseismic cross-correlation method was applied to detect divergence centers. The obtained sizes are comparable with results from AC studies.

Schrijver, Hagenaar, and Title (1997) and Hagenaar, Schrijver, and Title (1997) have analyzed the distribution of supergranular cell sizes in the context of a Voronoi tessellation. They found a good agreement with the right-asymmetric shape of a Voronoi distribution. In the present study the obtained distribution of cell sizes is also right-asymmetric. However, it differs from that presented by Hagenaar, Schrijver, and Title (1997) since we find a minimum supergranular cell size of approximately 250 Mm^2 whereas they did not find a gap at small scales. This difference can be explained by the fact that we applied a noise filter with a cutoff at $\ell = 250$. In contrast to these results, the size distribution found by del Moro *et al.* (2004) suggests a left-asymmetric shape.

From an analysis of the location of the cells on the solar disk using pair and mark correlation functions, we find some evidence that the supergranular cell locations represent a soft-core field, which has its closest analog of a field of nonoverlapping circles with variable diameters. The PCFs show only one conspicuous maximum and a weakly pronounced minimum. The results presented in Berrilli *et al.* (2004) are in disagreement with this result since in their analysis of PCFs they assume a hard-sphere random close packing model. The corresponding PCF shows several peaks, which can be roughly approximated by a damped cosine.

The inferred lifetime distributions are roughly exponential with mean decay times between 1.59 days (3D segmentation) and 1.83 days (feature tracking). These numbers are again larger than those obtained in most of the former studies, which are in a range around 20 hours (DeRosa and Toomre, 2004; del Moro *et al.*, 2004). However, the analysis of Beck and Duvall (2001) gives a lifetime distribution comparable to the one derived here.

The temporal evolution of individual supergranular cells shows rather smooth histories from birth until death by dissolution. This is a little different from the findings of DeRosa and Toomre (2004), who stated that only 35% of the supergranular cells do not interact with adjacent structures during their existence. This difference may be due to the less-than-ideal temporal resolution of our data (12 hours *versus* 31 minutes for DeRosa and Toomre, 2004).

The present study and the studies of Berrilli et al. (2004) and del Moro et al. (2004) have relied on similar data sets obtained from time–distance helioseismology. The two approaches, however, show significant differences in the distributions, lifetimes, and PCFs. This may be due to slightly different analyses or to differences in the computation of the divergence maps (here we applied a dispersion correction before measuring the travel times). The results presented in this paper are perfectly consistent with the findings of Gizon, Duvall, and Schou (2003), which also apply to the current data set.

Improvements are expected from the development of reliable inversions of the travel-time measurements (Jackiewicz et al., 2007; Jackiewicz, Gizon, and Birch, 2008). A better spatial sampling and, more importantly, a better temporal sampling will benefit studies of supergranulation evolution. We also plan to study a much longer data set to search for solar-cycle variations in the structure of supergranulation, as hinted at in the preliminary analysis of Gizon and Duvall (2004).

Acknowledgements This paper is based on data from the *Solar and Heliospheric Observatory*, a project of international cooperation between ESA and NASA. The authors are grateful to an anonymous referee for careful reading and valuable comments.

References

Alissandrakis, C.E., Dialetis, D., Tsiropoula, G.: 1987, *Astron. Astrophys.* **174**, 275.
Beck, J.G., Duvall, T.L.: 2001, In: Wilson, A. (ed.) *Helio- and Asteroseismology at the Dawn of the Millenium, Proc. SOHO 10/GONG 2000 Workshop* **SP-464**, ESA, Noordwijk, 577.
Berrilli, F., Ermolli, I., Florio, A., Pietropaolo, E.: 1999, *Astron. Astrophys.* **344**, 965.
Berrilli, F., del Moro, D., Conssolini, G., Pietropaolo, E., Duvall, T.L., Kosovichev, A.G.: 2004, *Solar Phys.* **221**, 33.
Braun, D.C., Birch, A.C., Lindsey, C.: 2004, In: Danesy, D. (ed.) *Helio- and Asteroseismology: Towards a Golden Future, Proc. SOHO 14/GONG 2004 Workshop* **SP-559**, ESA, Noordwijk, 337.
Crouch, A.D., Charbonneau, P., Thibault, K.: 2007, *Astrophys. J.* **662**, 715.
del Moro, D., Berrilli, F., Duvall, T.L., Kosovichev, A.G.: 2004, *Solar Phys.* **221**, 23.
DeRosa, M.L., Toomre, J.: 2004, *Astrophys. J.* **616**, 1242.
Duvall, T.L., Gizon, L.: 2000, *Solar Phys.* **192**, 177.
Gizon, L., Birch, A.C.: 2005, *Living Rev. Solar Phys.* **2**(6). http://www.livingreviews.org/lrsp-2005-6.
Gizon, L., Duvall, T.L.: 2004, In: Stepanov, A.V., Benevolenskaya, E.E., Kosovichev, A.G. (eds.) *Multi Wavelength Investigations of Solar Activity, IAU Symp.* **223**, Cambridge Univ. Press, Cambridge, 41.
Gizon, L., Duvall, T.L., Schou, J.: 2003, *Nature* **421**, 43.
Hagenaar, H.J., Schrijver, C.J., Title, A.M.: 1997, *Astrophys. J.* **481**, 988.
Hart, A.B.: 1956, *Mon. Not. Roy. Astron. Soc.* **116**, 38.
Hathaway, D.H., Beck, J.G., Bogart, R.S., Bachmann, K.T., Khatri, G., Petitto, J.M., Han, S., Raymond, J.: 2000, *Solar Phys.* **193**, 299.
Hirzberger, J.: 2002, *Astron. Astrophys.* **392**, 1105.
Hirzberger, J., Bonet, J.A., Vázquez, M., Hanslmeier, A.: 1999, *Astrophys. J.* **515**, 441.
Jackiewicz, J., Gizon, L., Birch, A.C.: 2008, *Solar Phys.* DOI:10.1007/s11207-008-9158-7.
Jackiewicz, J., Gizon, L., Birch, A.C., Thompson, M.J.: 2007, *Astron. Nachr.* **328**, 234.
Kosovichev, A.G., Duvall, T.L.: 1997, In: Pijpers, F.P., Christensen-Dalsgaard, J., Rosenthal, C.S. (eds.) *Proc. SCORe'96: Solar Convection and Oscillations and their Relationship, ASSL* **225**, Kluwer, Dordrecht, 241.
Leighton, R.: 1964, *Astrophys. J.* **140**, 1547.
Leighton, R., Noyes, R., Simon, G.W.: 1962, *Astrophys. J.* **135**, 474.
Leitzinger, M., Brandt, P.N., Hanslmeier, A., Pötzi, W., Hirzberger, J.: 2005, *Astron. Astrophys.* **444**, 245.
Mandelbrot, B.: 1977, *Fractals*, Freeman, San Francisco.
Meunier, N., Roudier, T., Tkaczuk, R.: 2007, *Astron. Astrophys.* **466**, 1123.
Meunier, N., Tkaczuk, R., Roudier, T.: 2007, *Astron. Astrophys.* **463**, 745.
Meunier, N., Tkaczuk, R., Roudier, T., Rieutord, M.: 2007, *Astron. Astrophys.* **461**, 1141.
November, L.J.: 1994, *Solar Phys.* **154**, 1.
Ploner, S.R.O., Solanki, S.K., Gadun, A.S.: 2000, *Astron. Astrophys.* **356**, 1050.

Rieutord, M., Roudier, T., Malherbe, J.M., Rincon, F.: 2000, *Astron. Astrophys.* **357**, 1063.
Rieutord, M., Roudier, T., Roques, S., Ducottet, C.: 2007, *Astron. Astrophys.* **471**, 687.
Rimmele, T., Schröter, E.H.: 1989, *Astron. Astrophys.* **221**, 137.
Roudier, T., Muller, R.: 1987, *Solar Phys.* **107**, 11.
Roudier, T., Lignières, F., Rieutord, M., Brandt, P.N., Malherbe, J.M.: 2003, *Astron. Astrophys.* **409**, 299.
Sánchez Cuberes, M., Bonet, J.A., Vázquez, M., Wittmann, A.D.: 2000, *Astrophys. J.* **538**, 940.
Scherrer, P.H., Bogart, R.S., Bush, R.I., Hoeksema, J.T., Kosovichev, A.G., Schou, J., Rosenberg, W., Springer, L., Tarbell, T.D., Title, A., Wolfson, C.J., Zayer, I., MDI Engineering Team: 1995, *Solar Phys.* **162**, 129.
Schrijver, C.J., Hagenaar, H.J., Title, A.M.: 1997, *Astrophys. J.* **475**, 328.
Shine, R.A., Simon, G.W., Hurlburt, N.E.: 2000, *Solar Phys.* **193**, 313.
Simon, G.W., Leighton, R.B.: 1964, *Astrophys. J.* **140**, 1120.
Singh, J., Bappu, M.K.V.: 1981, *Solar Phys.* **71**, 161.
Srikanth, R., Singh, J., Raju, K.P.: 2000, *Astrophys. J.* **534**, 1008.
Stoyan, D., Stoyan, H.: 1994, *Fractals, Random Shapes and Point Fields*, Wiley, Chichester.
Woodard, M.F.: 2002, *Astrophys. J.* **565**, 634.
Zhao, J., Kosovichev, A.G.: 2003, In: Sawaya-Lacoste, H. (ed.) *Local and Global Helioseismology: The Present and Future, Proc. SOHO 12/GONG 2002 Workshop* **SP-517**, ESA, Noordwijk, 417.

Probing the Subsurface Structures of Active Regions with Ring-Diagram Analysis

Richard S. Bogart · Sarbani Basu ·
Maria Cristina Rabello-Soares · H.M. Antia

Originally published in the journal Solar Physics, Volume 251, Nos 1–2, 439–451.
DOI: 10.1007/s11207-008-9213-9 © Springer Science+Business Media B.V. 2008

Abstract We analyze the variations in the near-surface profiles of sound speed and adiabatic constant between active regions and neighboring quiet-Sun areas using the technique of ring-diagram analysis and inversions of the frequency differences between the regions. This approach minimizes the systematic observational effects on the fitted spectral model parameters. The regions analyzed have been selected from a large sample of data available from both GONG and MDI and include a wide range of magnetic activity levels as measured in several respects. We find that the thermal-structure anomalies under active regions have a consistent depth profile, with only the magnitude of the effect varying with the intensity of the active regions. Both the sound speed and the first adiabatic index are depressed near the surface but enhanced at greater depths. The turnover for the sound speed occurs at a shallower depth than that for the adiabatic index. The amplitude of the thermal anomalies at all depths correlates more closely with the total magnetic flux of the active regions than with spot areas or flare activity levels. The depth of the turnover does not appear to depend on the strength of the region.

Keywords Active regions · Helioseismology · Ring diagrams

1. Introduction

To systematically explore the near-surface variations in thermal structure associated with regions of strong magnetic activity, we have analyzed a representative sample of solar active regions using the technique of differential ring-diagram analysis. Ring-diagram analysis,

R.S. Bogart (✉) · M.C. Rabello-Soares
Stanford University, Stanford, CA, USA
e-mail: rbogart@stanford.edu

S. Basu
Astronomy Department, Yale University, New Haven, CT, USA

H.M. Antia
Tata Institute of Fundamental Research, Mumbai, India

which relies on frequencies of high-degree modes, can be used to study properties of small regions on the Sun (Hill, 1988) by using 3D power spectra in temporal frequency and the two components of the horizontal wavenumber. The frequencies of these high-degree modes are determined by the structure of the underlying region. There are a number of systematic effects in ring-diagram analysis that make it difficult to determine the absolute structure of a given region (see, for example, Hindman *et al.*, 2000). Most of these systematic effects can be eliminated, however, by studying the differences in frequencies between different regions with the same observing geometry: This is the technique of differential ring-diagram analysis (Basu, Antia, and Bogart, 2004; hereafter BAB). To study the effect of magnetic fields, we can take the frequency differences between an active region and a quiet region, preferably at the same latitude and not too distant longitude.

It is well known that the frequencies of solar oscillations are affected by the presence of magnetic field (*e.g.*, Hindman *et al.*, 2000; Rajaguru, Basu, and Antia, 2001; Howe *et al.*, 2004). The frequencies are higher in the active regions and, furthermore, the frequency differences are correlated with the differences in the average surface magnetic field between the corresponding regions. These frequency differences can be related to differences in structure between the two regions by using an inversion technique. Common structural parameters obtained by these inversions are the sound-speed differences and the differences in the adiabatic index (Γ_1) between the two regions. However, these inversion techniques do not include the effects of the magnetic field, which play an important role in the frequency differences. Lin, Li, and Basu (2006) showed that the so-called sound-speed differences obtained by inverting frequency differences between magnetic and nonmagnetic regions are really differences in wave speed, though the differences in the adiabatic index inversions do correspond to the physical differences in that parameter. However, we continue to use the terminology "sound speed" differences or anomalies in this paper. Our results can also be compared with those obtained by using independent techniques, such as the time–distance analysis of Kosovichev, Duvall, and Scherrer (2000).

Our analysis technique has been successfully applied to a few active regions in the past using Doppler data from the MDI instrument on SOHO (BAB). Since the characteristics of active regions vary considerably, it is necessary to study a large number of active regions to find out how the inferred structure variations vary from one region to another. To examine a more comprehensive and representative set of active regions, however, we require continuous data availability for the disk passages of many active regions, not just those during the two or three months per year of MDI Dynamics Campaigns. We have therefore concentrated in recent years on the analysis of GONG Doppler data. The GONG data have somewhat poorer spatial resolution, and greater noise levels, associated in part with the merging of data and the need to calibrate out the more rapidly varying line-of-sight velocities from different sites. Nonetheless, at the high frequencies and modest spatial resolution appropriate for this analysis (with the spectra being averaged over areas of diameter 15° for both the active and comparison quiet regions), we find that we get sufficient mode sets for analysis. Two regions have been analyzed by using both MDI data and GONG data independently. Although the GONG inversions do not extend either as deep or as close to the surface, the agreement in the region of overlap is good. Since GONG data are available with almost continuous coverage since July 2001, it is possible to study many more active regions during the later half of the current solar cycle.

Table 1 Characteristics of the active regions selected for study. For each region, we give the date of its central meridian crossing (extrapolated backward for emerging region 9904), the heliographic coordinates of the region center at the time of central meridian crossing (including the relevant Carrington Rotation number), the longitude(s) of the comparison quiet region(s) selected (QL_0 and QL_1), and the various activity indices A_{max}, MAI, and FI(c) as discussed in the text. Units of MAI are in gauss and those of A_{max} in millionths of a hemisphere. The heliographic coordinates are in degrees.

AR	Lat.	CR:Long.	QL_0	QL_1	On CM	A_{max}	MAI	FI(c)
8040	+07	1922:016	341		1997.05.21	150	23	11.3
8193	−21	1934:082	067		1997.10.13	290	68[a]	1.0
8518	−14	1948:105	075		1999.04.23	170	53[a]	0.1
9026	+20	1963:071	041	126	2000.06.07	820	147[a]	422.1
9393	+19	1974:147	207		2001.03.29	2440	242[a]	568.0
9715	+05	1983:138	193	118	2001.12.01	990	113	66
9893	+19	1988:215	255		2002.04.10	490	87	57.0
9896	−11	1988:195	205		2002.04.11	110	27	0.0
9899	+18	1988:180	240		2002.04.12	230	58	30.0
9901	+20	1988:204	249		2002.04.10	350	112	24.0
9904	−16	1988:222	242	207	(2002.04.09)	60	27	3.0
9906	−15	1988:150	120	210	2002.04.14	850	133	129.0
9914	+04	1988:013	028		2002.04.25	260	98	4.1
10105	−08	1994:300	345	255	2002.09.14	1520	181	67
10197	+24	1996:129	139	069	2002.11.20	360	69	2
10226	−28	1997:127	142	072	2002.12.17	720	115	73
10296	+12	2000:170	200		2003.03.06	650	108	5.4
10349	−14	2002:154	174	109	2003.05.01	1030	180	36.9
10424	−18	2006:291	316	256	2003.08.07	760	162	39.0
10536	−11	2011:075	090	040	2004.01.07	980	161	37.6
10652	+08	2019:346	001	306	2004.07.23	2010	240	234.9
10696	+08	2022:026	046	011	2004.11.06	910	158	368.0
10783	−03	2031:100	115	070	2005.07.04	600	98	3.1
10822	−07	2036:080	095	050	2005.11.19	810	103	98.4
10838	+16	2038:321	331	301	2005.12.25	170	30	3.3
10865	−12	2041:110	145	055	2006.04.02	620	107	44.3
10904	−13	2046:120	170	090	2006.08.15	750	88	16.7
10930	−06	2050:010	035	335	2006.12.11	680	101	498.2

[a]Tracking of the magnetic flux was not confined to 4096 minutes.

2. Data Analysis

The active regions studied are described in Table 1. With the exception of AR 9715, all regions with numbers < 10 000 were those used in the previous study of BAB with MDI data. Those regions were necessarily confined to a few Carrington rotations per year by the MDI observing schedule. We have added AR 10652 and 10783 to the list of regions using MDI data. Generally, though, shorter periods of continuous data coverage combined with declining levels of solar activity in recent years have made it difficult to identify many suitable candidate regions for analysis in the MDI dataset.

The enhanced-resolution GONG+ network went into service in mid-2001, with the last "GONG Classic" data coming on 10 July, during CR 1978. Calibrated GONG+ data are currently available from 75 Carrington rotations, through CR 2052. To extend the range of regions studied, we selected the NOAA-SWPC active regions with the largest total spot area during each of these rotations, and from that sample we further selected a representative sample of 17 regions for analysis: AR 9715, 9906, and 10105–10930 in Table 1. The sampled regions are widely distributed in latitude and longitude, spot area, magnetic activity, and flare activity.

The GONG Dopplergrams are merged by selecting the data from a single site for each minute of the analysis interval. This avoids multiple spatial interpolations when data from sites with different image geometries are available. For this study we have consistently chosen the easternmost site when offered a choice. The effective duty cycle for the 17 regions analyzed with GONG data ranges from 69% to 99% with a mean of 85%. MDI data were selected for tracking only when the duty cycle exceeded 85%.

Each region analyzed is tracked for 8192 minutes centered on its central meridian crossing, at the photospheric rotation rate appropriate to its latitude. For the latitudes involved, ranging from 3° to 28°, the maximum displacement in longitude of the tracked region with reference to the Carrington system over the interval of tracking is 3.7°, considerably smaller than either the diameter of the regions tracked or the typical longitudinal extent of an active region. The tracked regions are mapped to a plane by using Postel's (azimuthal equidistant) projection. For the MDI data, we subtract a long-term mean (over a Carrington rotation) from the data; this removes a constant term in the slowly varying orbital velocity and first-order terms in solar rotation and spatial variations in the MDI calibration error. For the GONG data, to merge without discontinuities, we make explicit corrections for the Earth orbital and rotational velocity applicable to each site, as well as for a model solar rotational velocity. The power spectra of the resultant tracked cubes are fitted by using the 13-parameter model of Basu and Antia (1999), which takes into account the asymmetry of the peaks in the power spectrum.

The characteristics of all 28 regions studied are summarized in Table 1 and Figures 1 and 2. The Magnetic Activity Index (MAI) is determined by averaging area-weighted values of $|B_z| > 50$ G over all pixels in the MDI 96-minute magnetograms mapped and tracked over areas and intervals similar to those of the analysis regions. The mapping is done with an equivalent projection rather than Postel's azimuthal equidistant projection (used for the ring-diagram analysis), so that the sum represents total flux, but at the scale of the maps the differences in the area covered are negligible. Also, to minimize projection effects (see Bogart *et al.*, 2006), the tracking of the magnetic flux was generally confined to 4096 minutes centered on the meridian crossing of the region, rather than the 8192 minutes of tracking used for the ring-diagram analysis. The exceptions are noted in Table 1. As discussed in Bogart *et al.* (2006), because of instrumental noise and saturation, the MAI really represents a composite of the total magnetic flux through the regions and the fraction of the region covered by strong field (> 50 G).

The activity of the regions can also be characterized by the maximum total sunspot area attained during their disk transits ($A_{\max}$). This number, in units of millionths of the solar disk, is derived from the daily Joint USAF/NOAA Solar Region Summaries; it represents the maximum of the "snapshot" values as of 2400 UT each day, rather than any attempt at a fit to the growth/decay curve. We also determine a crude Flare Index [FI(c)] for each active region studied, defined as

$$\mathrm{FI}(c) \equiv 100 n_\mathrm{X} + 10 n_\mathrm{M} + n_\mathrm{C} + 0.1 n_\mathrm{B},$$

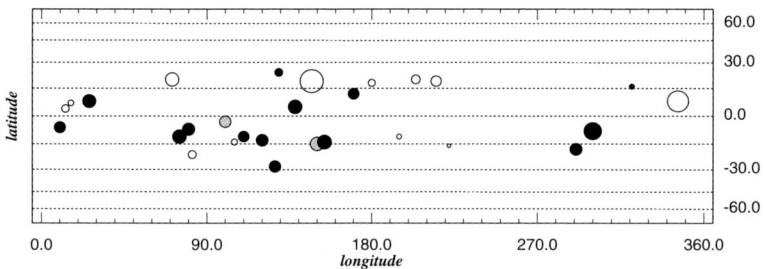

Figure 1 Heliographic locations of the active regions analyzed. The area of each symbol is equal to twice the maximum spot area for the corresponding active region at the scale of the map. Solid circles correspond to active regions selected from GONG data for analysis in this study; open circles correspond to regions selected from MDI data. The gray circles are for AR 9906 and 10783, analyzed from both sets.

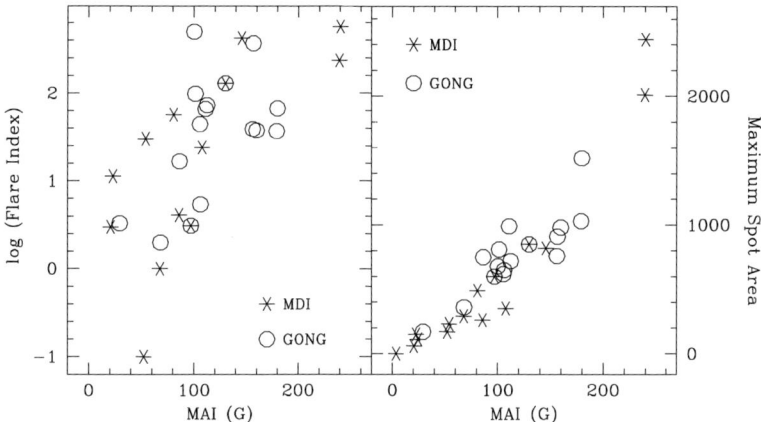

Figure 2 Distributions of the activity indices FI(c) (left) and A_{max} (right) *versus* MAI for the active regions studied. The various activity indices are described in the text.

where n_i is the number of flares of class i specifically identified with the active region in the GOES X-ray flare catalog. This is similar to the classical flare index introduced by Antalová (1990), except that it includes the contribution of M-class flares and applies to specific regions for the entirety of their disk transits rather than to the whole Sun for specific time intervals.

The maximum spot area (A_{max}) is quite well correlated with MAI, as can be clearly seen in Figure 2. To the extent that MAI is a measure of area covered by strong flux, this is not surprising. The flare index FI(c) appears to obey a power law in MAI, but it is not as closely correlated. This is expected, as the level of flare activity, although limited by the total magnetic energy, must also depend on the geometry of the active region. There is a hint of a bimodal distribution, with two groups following approximately the same power-law distribution, but with different constants. We have not attempted to separate our sample into different populations, however.

The MAI is used to select the quietest comparison regions for a given active region, from among those centered at the same latitude, and at longitudes within 60° of the active region. There are usually two comparison quiet regions used for each active region, one preceding

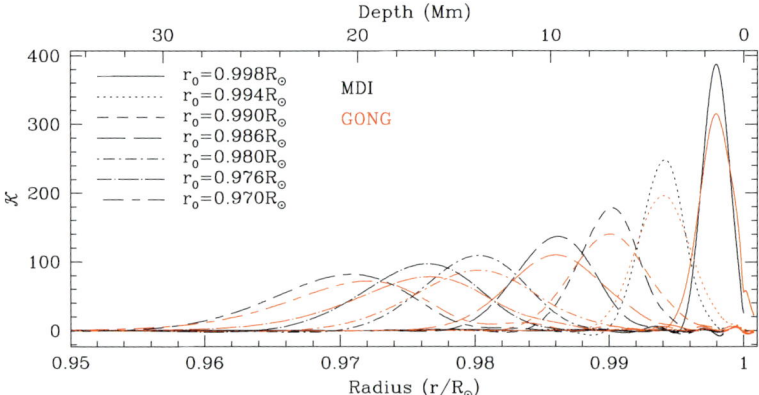

Figure 3 Samples of the mode-averaging kernels used for determining the sound-speed anomalies with the SOLA inversion technique. Each curve is labeled by the target radius for construction of the averaging kernel. Typical sets of averaging kernels for both GONG data and MDI data are shown. The differences in their detailed shapes are associated with the different noise characteristics of the data. As expected, the GONG kernels have perceptibly poorer depth resolution, thus limiting the extent of the depths over which inversions can be compared (see Figure 4).

and one following it in longitude; but occasionally it has only been possible to find one suitable quiet region for comparison.

3. Results

To determine the depth profiles of the anomalous thermal structure of the active regions, we invert the differences between their frequencies and those of the comparison quiet regions in common mode sets for the variations in sound speed (c) and of the adiabatic index (Γ_1). Our inversions make use of both the Regularized Least Squares (RLS) and the Subtractive Optimally Localized Averages (SOLA) techniques (*cf.* Antia and Basu, 2007).

Figure 3 shows a typical sample of the averaging kernels for the GONG data, and for comparison one for the MDI data as well. It is apparent that the somewhat higher errors in the data from the ground-based GONG project lead to slightly poorer resolution. In addition, the resolution kernels do not go as deep as the MDI ones. What could not be shown without crowding the figure is that MDI is somewhat more successful than GONG in inverting very close to the surface.

Figure 4 shows how sound-speed and Γ_1 inversion results obtained by using GONG and MDI data compare. We analyzed both GONG and MDI data for two active regions, AR 9906 and AR 10783. As the figure shows, the inversion results obtained by using data from both projects compare very well. As hinted in Figure 3, the GONG inversions do not go as deep as the MDI results, nor as shallow as the MDI results; however, there is a large degree of overlap. The GONG results have larger errors, but that is to be expected from ground-based data. The large overlap and the match between the two results give us confidence that the earlier results with only MDI data (BAB) were reliable and that we can continue the detailed study of solar active regions using GONG data, particularly since there has been very little MDI data suitable to study active regions since 2002.

Structure profiles for several active regions are shown in Figures 5 and 6. Similar profiles of some of the other regions were shown by BAB. In general we have tried to invert the

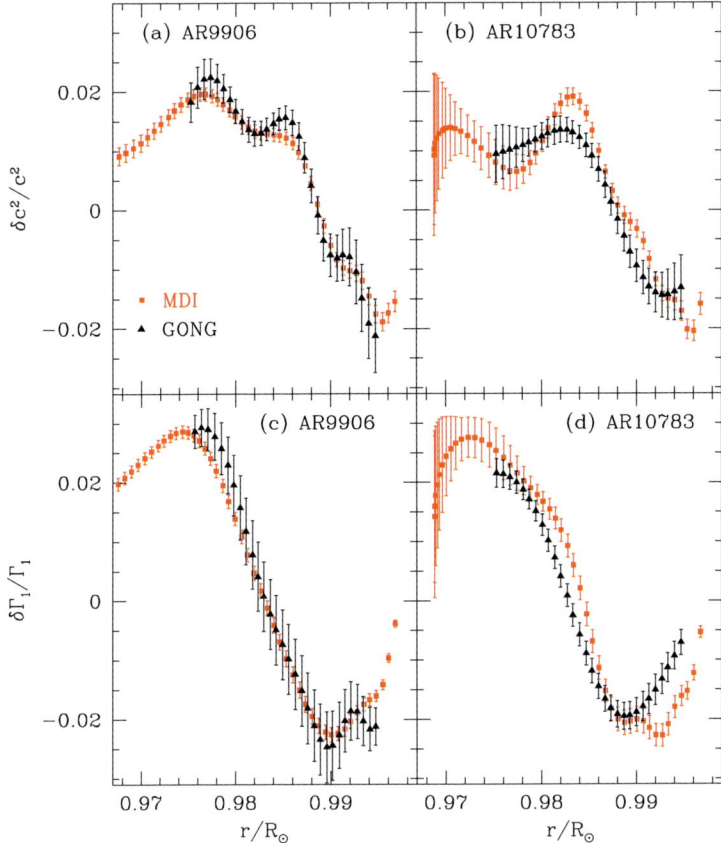

Figure 4 Comparisons of the inverted structure parameters c^2 and Γ_1 for AR 9906 and 10783 as determined separately from MDI and GONG data. For both active regions the inversions are the averages of those determined for two comparison quiet regions, one preceding and one following the active region at the same latitude.

differences for the two comparison quiet regions individually; this gives us an idea of the errors from noise and sampling. It is not always possible to do so, however, either because of the lack of suitably quiet nearby regions on both sides of the active region or because of poor fitting of the mode frequencies. When this happens the inversion results obtained from RLS and SOLA techniques are widely different; such poor fits have not been included in these results. In particular, we have not been able to successfully invert the structures of a few of the active regions originally included in the sample. One of these, AR 10720, was the most active in the sample as measured by MAI and total spot area; it is likely that the difficulties encountered are associated with problems observing the acoustic signal in the region itself, rather than with either of the comparison regions.

The general trend in the results is clear and confirms that of the earlier analysis of MDI data for the few comparatively strong active regions: a region just below the surface with negative sound-speed and adiabatic-index anomalies and a turnover to positive anomalies at greater depths. It is also clear from the figures that the turnover from negative to positive anomalies occurs deeper for the adiabatic index than the sound speed.

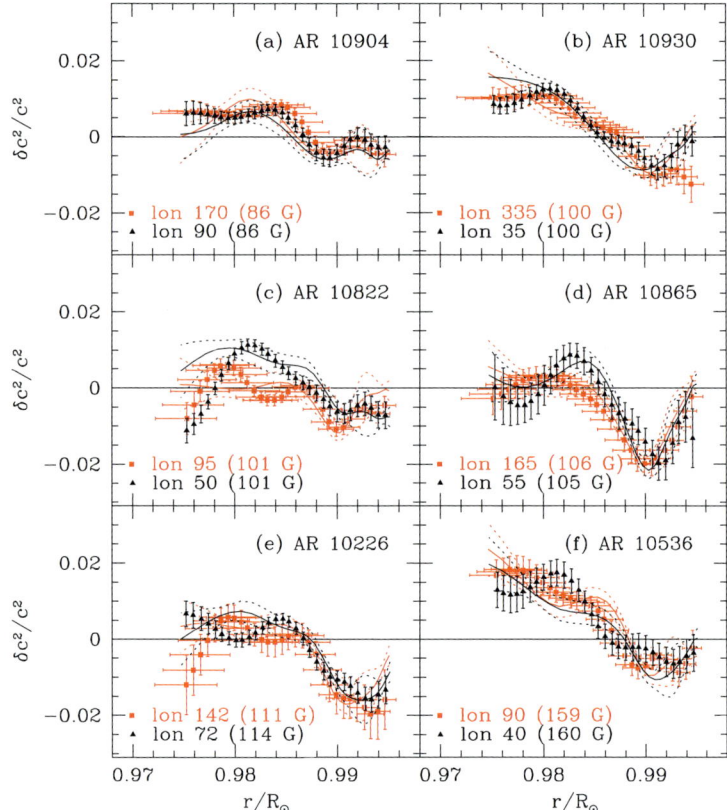

Figure 5 Inversions for the profiles of sound-speed anomalies for various active regions in the sample as determined solely from GONG data. Results of comparisons with each of the two quiet regions are shown separately as squares and triangles. The solid lines are the RLS inversions; the dashed lines represent the 1σ error levels. The points with individual error bars are from the SOLA inversions; the vertical error bars represent the 1σ error levels, and the horizontal error bars represent the distance between the quartile points of the averaging kernels. The MAI values given are the differences between that of the active region and those of each of the comparison quiet regions individually.

Correlations of the amplitude of the sound-speed anomaly and the adiabatic-index anomaly with various measures of the strength and size of the active region are shown in Figures 7 and 8, respectively. To show the correlations, we computed the average sound-speed and adiabatic-index anomalies $\langle \delta c^2/c^2 \rangle$ and $\langle \delta\Gamma_1/\Gamma_1 \rangle$ over different depth ranges and plotted them as a function of the MAI, the maximum spot area and the crude Flare Index defined in Section 2. The panels on the left-hand side of Figures 7 and 8 show the averages over radius ranges $(0.978 – 0.984)R_\odot$ and $(0.992 – 0.996)R_\odot$. These correspond to the largest and smallest depth ranges over which we can reliably invert and obtain averages. The panels on the right-hand side show results of two intermediate depth ranges, $(0.984 – 0.989)R_\odot$ and $(0.989 – 0.994)R_\odot$. The intermediate depth ranges were chosen such that the sound-speed inversion results changed sign between the two regions.

As is clear from Figure 7, the amplitude of the average anomaly in sound speed is roughly proportional to the MAI of the active region for all depth ranges that have been plotted. The averages are also correlated with the region's spot area, but the trend appears to be slower

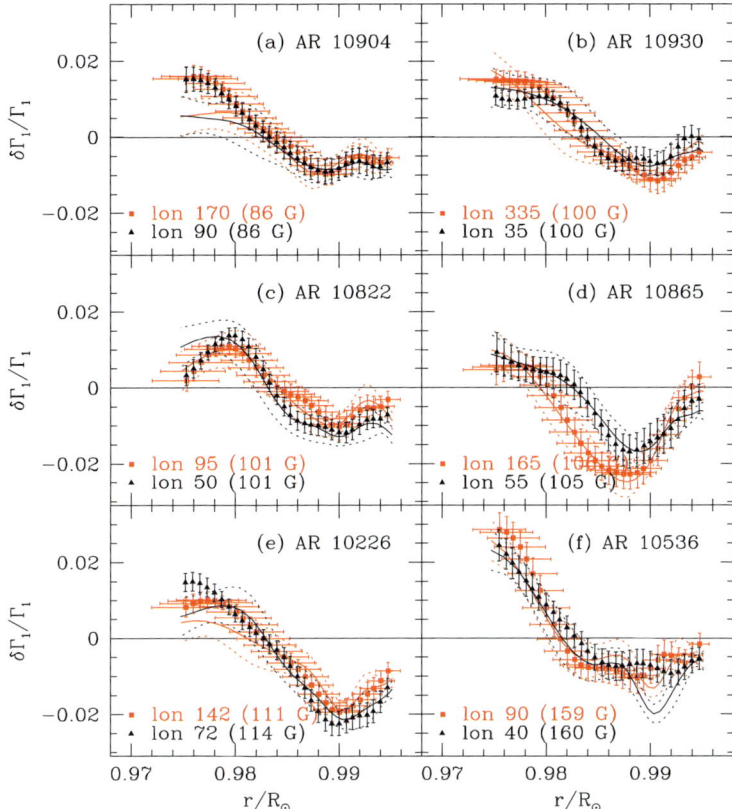

Figure 6 Inversions for the profiles of adiabatic-index anomalies for various active regions in the sample as determined solely from GONG data. Results of comparisons with each of the two quiet regions are shown separately as squares and triangles. The solid lines are the RLS inversions; the dashed lines represent the 1σ error levels. The points with individual error bars are from the SOLA inversions; the vertical error bars represent the 1σ error levels, and the horizontal error bars represent the distance between the quartile points of the averaging kernels. The MAI values given are the differences between that of the active region and those of each of the comparison quiet regions individually.

than linear. There does not appear to be any evidence for a correlation with the degree of flare activity. The behavior of the adiabatic index is quite similar, except that the turnover from negative to positive anomalies happens deeper than the $(0.984 - 0.989) R_{\odot}$ range, and hence the right-hand side panels of Figure 8 only show negative adiabatic-index anomalies.

Least-squares linear fits of the dependence of the variation in the thermal parameters on MAI at different depths are shown in Figure 9. The comparable fits for the dependence on the other activity measures show similar behavior, but with larger error bars.

4. Discussion

We have done a ring-diagram analysis of both MDI and GONG data to study the subsurface structure of solar active regions. We find that the GONG data can be used successfully to study the structure below active regions. However, we also find that they do not always yield

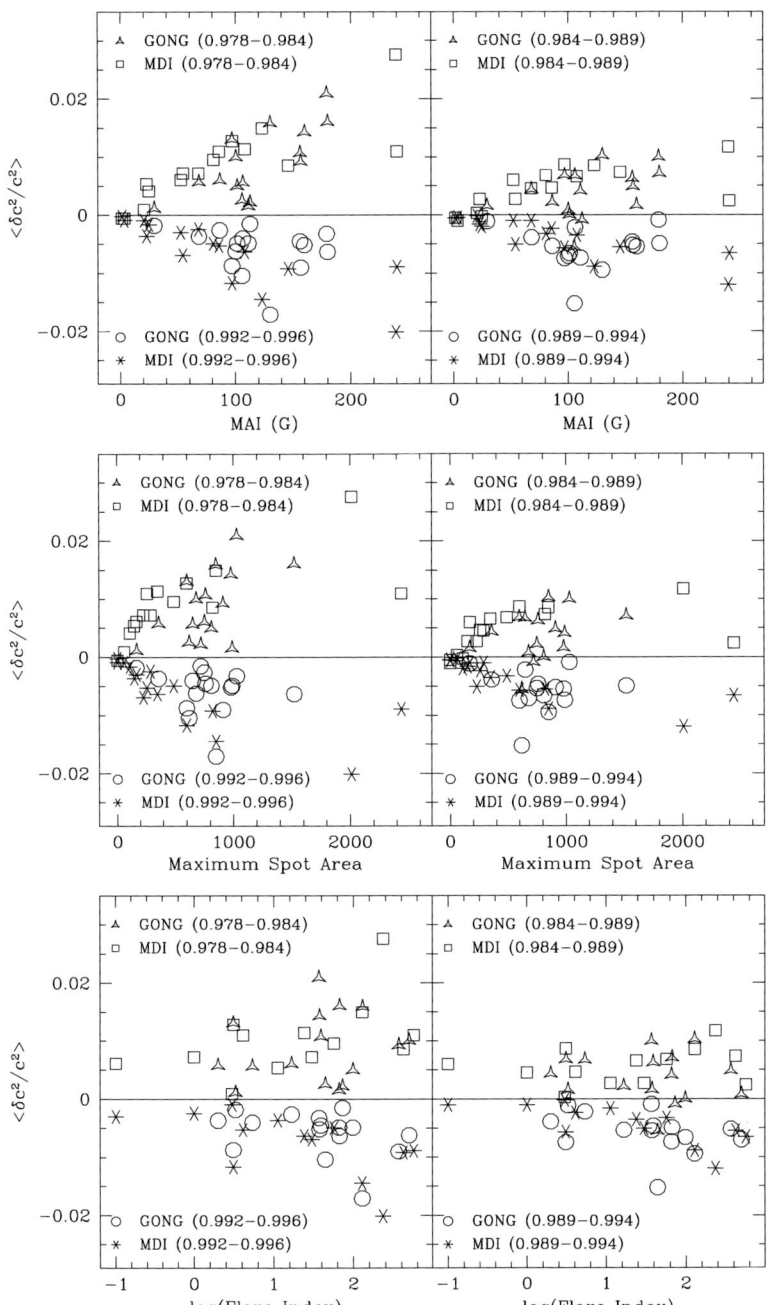

Figure 7 Variation of the active-region sound-speed anomalies at different depths, as functions of different measures of the strength of the active region. Anomalies at two depths are plotted in each panel. Note that the depth in each set of panels decreases clockwise from the upper left.

that are likely to be important but are ignored in our model. These include vertical subsurface motions associated with solar convection, atmospheric magnetic field, and a rapid but continuous decrease in density across the surface to a nonzero atmospheric value. In Part II of this paper (Kerekes, Erdélyi, and Mole, 2008) we include a vertical component of velocity and still find a significant decrease in frequency. A more realistic density profile would tend to reduce the frequency, whereas magnetic field tends to increase it (Campbell and Roberts, 1989; Evans and Roberts, 1990; Bogdan *et al.*, 1993; Rosenthal, 1995; Taroyan, 2003; Erdélyi, Kerekes, and Mole, 2005; Taroyan and Erdélyi, 2006). It seems likely that the magnetic field plays a role in limiting the decrease produced by the other effects. See, for example, Erdélyi (2006a) and Kerekes (2007) for more discussion of the possible impact of these various factors.

We also find that the turbulence has a damping effect on the surface gravity waves. Duvall, Kosovichev, and Murawski (1998) and Mędrek, Murawski, and Roberts (1999) obtained damping rates from numerical calculations, but they did not present any analytical results. Their results were similar to ours for the range of wavenumbers for which our results are valid. Murawski, Duvall, and Kosovichev (1998) gave an analytical result for the damping rate that differs from ours. At large wavenumber we find that the ratio of the damping rate to the frequency correction is of the order of the nondimensional wavenumber, which seems to be consistent with observations. Sazontov and Shagalov (1986) analysed the damping of surface gravity waves by turbulent scattering, ignoring the effect of the turbulence on the real part of the frequency. In the large-wavenumber approximation, their leading order result reduces exactly to that given in Equation (24).

Acknowledgements We wish to thank Misha Ruderman and Victor Shrira for helpful discussions, and we acknowledge the financial support by STFC (UK) and NFS Hungary (OTKA, K67746).

References

Abramowitz, M., Stegun, I.: 1965, *Handbook of Mathematical Functions*. Dover, New York.
Antia, H.M., Basu, S.: 1999, High-frequency and high-wavenumber solar oscillations. *Astrophys. J.* **519**, 400–406.
Batchelor, G.: 1953, *The Theory of Homogeneous Turbulence*, Cambridge University Press, Cambridge.
Bogdan, T.J., Brown, T.M., Lites, B.W., Thomas, J.H.: 1993, The absorption of p-modes by sunspots: variations with degree and order. *Astrophys. J.* **406**, 723–734.
Campbell, W.R., Roberts, B.: 1989, The influence of a chromospheric magnetic field on the solar p- and f-modes. *Astrophys. J.* **338**, 538–556.
Duvall, T.L., Kosovichev, A.G., Murawski, K.: 1998, Random damping and frequency reduction of the solar f-mode. *Astrophys. J.* **505**, L55–L58.
Erdélyi, R.: 2006a, Magnetic coupling of waves and oscillations in the lower solar atmosphere: can the tail wag the dog? *Philos. Trans. Roy. Soc. A* **364**, 351–381.
Erdélyi, R.: 2006b, Magnetic seismology of the lower solar atmosphere. In: Fletcher, K., Thompson, M. (eds.) *Proceedings of SOHO 18/GONG 2006/HELAS I, Beyond the spherical Sun* **SP-624**, ESA, Noordwijk, 15.1.
Erdélyi, R., Kerekes, A., Mole, N.: 2005, Influence of random magnetic field on solar global oscillations: the incompressible f-mode. *Astron. Astrophys.* **431**, 1083–1088.
Evans, D., Roberts, B.: 1990, The influence of a chromospheric magnetic field on the solar p- and f-modes. II. Uniform chromospheric field. *Astrophys. J.* **356**, 704–719.
Fabrikant, A.L., Raevsky, M.A.: 1994, The influence of drift flow turbulence on surface gravity wave propagation. *J. Fluid Mech.* **262**, 141–156.
Kerekes, A.: 2007, *Random effects on the solar f-mode*. PhD thesis, University of Sheffield.
Kerekes, A., Erdélyi, R., Mole, N.: 2008, Effect of random flows on the solar f mode: II. Horizontal and vertical flow. *Solar Phys.* in press.
Mędrek, M., Murawski, K., Roberts, B.: 1999, Damping and frequency reduction of the f-mode due to turbulent motion in the solar convection zone. *Astron. Astrophys.* **349**, 312–316.

Murawski, K., Roberts, B.: 1993a, Random velocity field corrections to the f-mode. I. Horizontal flows. *Astron. Astrophys.* **272**, 595–600.

Murawski, K., Roberts, B.: 1993b, Random velocity field corrections to the f-mode. II. Vertical and horizontal flow. *Astron. Astrophys.* **272**, 601–608.

Murawski, K., Duvall, T.L. Jr., Kosovichev, A.G.: 1998, Damping and Frequency Shift of the Solar f-mode Due to the Interaction with Turbulent Convection. In: Korzennik, S. (ed.) *Structure and Dynamics of the Interior of the Sun and Sun-like Stars* **SP-418**, ESA, Noordwijk, 825–828.

Nocera, L., Medrek, M., Murawski, K.: 2001, Solar acoustic oscillations in a random density field. *Astron. Astrophys.* **373**, 301–306.

Rosenthal, C.S.: 1995, The role of photospheric magnetic fields in the variation of solar oscillation eigenfrequencies. *Astrophys. J.* **438**, 434–444.

Rosenthal, C.S., Christensen-Dalsgaard, J.: 1995, The interfacial f-mode in a spherical solar model. *Mon. Not. Roy. Astron. Soc.* **276**, 1003–1008.

Rosenthal, C.S., Gough, D.O.: 1994, The solar f-mode as an interfacial mode at the chromosphere–corona transition. *Astrophys. J.* **423**, 488–495.

Sazontov, A.G., Shagalov, S.V.: 1986, Scattering of gravity waves by a turbulence of the upper ocean layer. *Izv. Atmos. Ocean. Phys.* **22**, 138–143.

Taroyan, Y.: 2003, *MHD waves and resonant interactions in steady states*. PhD thesis, University of Sheffield.

Taroyan, Y., Erdélyi, R.: 2006, Seismology of quiescent coronal loops. In: Bothmer, V., Hady, A.A. (eds.) *Solar Activity and its Magnetic Origin, IAU Symp.* **233**, Cambridge University Press, Cambridge, 191–192.

in $z > 0$, and

$$0 = -\frac{1}{\rho_2}\nabla p_{20} + g\hat{\mathbf{z}}, \quad (9)$$

in $z < 0$, where p_{i0} ($i = 1, 2$) denotes the equilibrium pressure. Contrary to the assumption of Paper I, the basic velocity – as Equation (4) clearly shows – is generally not irrotational. As a consequence, the pressure gradient and the gravitational force (both of which are conservative) cannot balance the inertial term on the left-hand side of Equation (8). In order to make our model work, we need to allow an additional, non-conservative force to be present in the equilibrium state. This random force $\mathbf{F}$ is assumed to be unaffected by the perturbations. The special case of $\mathbf{F} = 0$ corresponds to MR2. Their preliminary results were derived for a time-dependent flow ($u = u(x, t)$), but at a later stage the flow was assumed to be steady. Now we identify the type of flows that could satisfy the equilibrium state of MR2 and show that these represent a negligible subset of all flows given by Equation (4).

2.1.1. The Basic State of MR2

The equation of motion in MR2 [Equation (2.2)] reads as

$$\mathbf{V}_{0t} + \mathbf{V}_0 \cdot \nabla \mathbf{V}_0 = -\frac{1}{\rho_1}\nabla p_{10} + g\hat{\mathbf{z}}, \quad (10)$$

where $\mathbf{V}_0$ denotes the equilibrium velocity in MR2, which corresponds to Equation (2). A sufficient but not necessary condition for $\mathbf{V}_0$ to satisfy the equation of motion is that the basic flow has to be irrotational:

$$\nabla^2 \Psi_0(x, z, t) = \left[z\{u_{xx}(x, t) + \zeta^4 u(x, t)\} - 2\zeta^2 u(x, t)\right]e^{-\zeta^2 z} = 0. \quad (11)$$

It follows from Equation (11) that the equilibrium flow is irrotational only if $u(x, t) \equiv 0$.

The necessary condition is slightly more complicated. Certain rotational flows can satisfy Equation (10), provided that the inertial force remains conservative. This requirement may be written as

$$\nabla^2 \Psi_{0t}(x, z, t) + \left\{ze^{-\zeta^2 z}u_x(x, t)\nabla^2 \Psi_0(x, z, t)\right\}_z$$
$$- \left\{(1 - \zeta^2 z)e^{-\zeta^2 z}u(x, t)\nabla^2 \Psi_0(x, z, t)\right\}_x = 0. \quad (12)$$

The chosen random function, $u(x, t)$, has to satisfy Equation (12) in $z > 0$. After some manipulation Equation (12) can be reorganised into the form

$$F_2(x, z, t)z^2 + F_1(x, z, t)z + F_0(x, z, t) = 0, \quad (13)$$

where

$$F_0(x, z, t) = 2\zeta^2\left(e^{-\zeta^2 z}u_x u - u_t\right), \quad (14)$$

$$F_1(x, z, t) = u_{xxt} + \zeta^4 u_t + e^{-\zeta^2 z}(u_x u_{xx} - u u_{xxx}), \quad (15)$$

$$F_2(x, z, t) = \zeta^2 e^{-\zeta^2 z}(u_{xxx}u - u_x u_{xx}). \quad (16)$$

The above condition is satisfied only if $u_x = 0$ and $u_t = 0$, i.e., $u(x, t) = $ constant. Consequently, from Equations (5)–(6) we have that the horizontal velocity (V_1) is a function only of z and the vertical velocity (V_3) is zero.

In summary, we found that spatially varying flows with a vertical component require some external forcing in the basic state. We note that a similar conclusion was reached at the end of Section 2 in Paper I. Although in what follows we avoid specifying this random force, it is important to realise that this additional constraint may considerably narrow the scope of our model.

2.2. Wave Perturbation

Now we introduce small perturbations about the equilibrium state, Equations (7)–(8), and linearise the resulting equations. The perturbed quantities are denoted by $\tilde{\Psi}_i$, $\tilde{p}_i$, $i = 1, 2$ ($\Psi_i(x, z, t) = \Psi_{i0}(x, z) + \tilde{\Psi}_i(x, z, t)$). The z-components of the equations of motion can be written as

$$-\tilde{\Psi}_{1xt} - \tilde{\Psi}_{1z}\tilde{\Psi}_{1xx} + \tilde{\Psi}_{1x}\tilde{\Psi}_{1xz} = -\frac{1}{\rho_1}\tilde{p}_{1z} + g + F_z \tag{17}$$

in $z > 0$, and

$$-\tilde{\Psi}_{2xt} - \tilde{\Psi}_{2z}\tilde{\Psi}_{2xx} + \tilde{\Psi}_{2x}\tilde{\Psi}_{2xz} = -\frac{1}{\rho_2}\tilde{p}_{2z} + g \tag{18}$$

in $z < 0$. The boundary conditions at the perturbed surface, $z = \eta(x, t)$, are the kinematic boundary condition

$$\eta_t = (\tilde{\Psi}_{iz}\partial_x - \tilde{\Psi}_{ix}\partial_z)(z - \eta), \quad i = 1, 2 \tag{19}$$

and the pressure continuity

$$\tilde{p}_1 = \tilde{p}_2. \tag{20}$$

Since we are interested in surface waves only, we have to make sure that the energy of the waves is localised to the interface between the two fluids. Far away from the interface, the velocity perturbations must vanish

$$\Psi_1 \to 0 \quad \text{as } z \to +\infty, \tag{21}$$

$$\Psi_2 \to 0 \quad \text{as } z \to -\infty. \tag{22}$$

3. Derivation of the Dispersion Relation

The dispersion relation is obtained by solving the perturbed governing equations together with the boundary conditions. First we eliminate the pressure terms from Equations (17)–(18), using the continuity of pressure. The remaining equation, together with the kinematic boundary condition, is linearised about $z = 0$. Afterwards, we take the ensemble averages of the equations and separate the mean and fluctuating parts. Closure is achieved by the binary-collision approximation (Howe, 1971, more details below). Then we apply Fourier transforms and eliminate the variables one by one until we arrive at the dispersion relation.

Integration of Equations (17) and (18) gives the perturbed pressure at the boundary

$$\tilde{p}_1|_{z=\eta} = \rho_1 \int_u^\eta (\tilde{\Psi}_{1xt} + \tilde{\Psi}_{1z}\tilde{\Psi}_{1xx} - \tilde{\Psi}_{1x}\tilde{\Psi}_{1xz})\,dz + \rho_1 g\eta + f_1(x, t), \tag{23}$$

$$\tilde{p}_2|_{z=\eta} = \rho_2 \int_{z_l}^\eta (\tilde{\Psi}_{2xt} + \tilde{\Psi}_{2z}\tilde{\Psi}_{2xx} - \tilde{\Psi}_{2x}\tilde{\Psi}_{2xz})\,dz + \rho_2 g\eta + f_2(x, t), \tag{24}$$

where the functions $f_1(x,t)$ and $f_2(x,t)$ depend on the outer limits of the integration,

$$f_1(x,t) = \tilde{p}_1(z_u) - \rho_1 g z_u + \rho_1 \int_{z_u}^{\eta} F_z \, dz, \tag{25}$$

$$f_2(x,t) = \tilde{p}_2(z_l) - \rho_2 g z_l. \tag{26}$$

Application of the boundary condition (20) subsequently yields

$$\int_{z_u}^{\eta} (\tilde{\Psi}_{1xt} + \tilde{\Psi}_{1z}\tilde{\Psi}_{1xx} - \tilde{\Psi}_{1x}\tilde{\Psi}_{1xz}) \, dz$$

$$= \kappa \int_{z_l}^{\eta} (\tilde{\Psi}_{2xt} + \tilde{\Psi}_{2z}\tilde{\Psi}_{2xx} - \tilde{\Psi}_{2x}\tilde{\Psi}_{2xz}) \, dz - (1-\kappa) g \eta + \frac{f_2(x,t) - f_1(x,t)}{\rho_1}. \tag{27}$$

Now we let $z_u \to \infty$ and $z_l \to -\infty$, linearise about $z = 0$, and obtain

$$\int_{\infty}^{0} \{\Psi_{1xt} + \Psi_{0z}\Psi_{1xx} - \Psi_{0x}\Psi_{1xz} + \Psi_{1z}\Psi_{0xx} - \Psi_{1x}\Psi_{0xz}\} \, dz$$

$$= \kappa \int_{-\infty}^{0} \Psi_{2xt} \, dz - (1-\kappa) g \eta, \tag{28}$$

if we assume that the z-component of the unknown random force remains small close to the surface. In that case

$$\int_{0}^{\eta} F_z \, dz \approx 0. \tag{29}$$

In the horizontal flow case, the equilibrium state had to be supplemented by an additional random force so that we could consider a spatially varying flow. Assuming that this random force remains unchanged in the perturbed state was sufficient to eliminate it from the linearised equations. As Equation (29) clearly shows, this is not the case when a vertical flow is included in the basic state.

Substitution of the equilibrium profile (4) into Equation (28) gives

$$\int_{\infty}^{0} \{\Psi_{1xt} + e^{-\zeta^2 z}[(1-\zeta^2 z)(u\Psi_{1xx} - u_x \Psi_{1x}) + z(u_{xx}\Psi_{1z} - u_x \Psi_{1xz})]\} \, dz$$

$$= \kappa \int_{-\infty}^{0} \Psi_{2xt} \, dz - (1-\kappa) g \eta \tag{30}$$

at $z = 0$, where the remaining variables are Ψ_1, Ψ_2, and η. The two additional equations required to complete the set of equations are provided by the linearised kinematic boundary condition.

3.1. Linearisation of the Kinematic Boundary Condition

Now we present the linearisation of the kinematic boundary condition and show that the equivalent result of MR2 is incomplete. Linearisation is performed in two steps: first we obtain the linear approximation of Equation (19) at the perturbed boundary, $z = \eta(x,t)$, and afterwards determine the corresponding equation valid at $z = 0$.

We substitute $\tilde{\Psi}_1 = \Psi_0 + \Psi_1$ into Equation (19) and neglect the products of perturbed quantities to obtain

$$-\eta_t = \Psi_{0x} + \Psi_{1x} + \Psi_{0z}\eta_x = \Psi_{2x} \quad \text{at } z = \eta(x,t). \tag{31}$$

Since $\Psi_{0x} = ze^{-\zeta^2 z}u_x(x)$, for small values of z we can approximate Ψ_{0x} in Equation (31) with its first-order Taylor expansion:

$$\Psi_{0x}(x,z,t)|_{z=\eta} \approx \Psi_{0x}(x,z)|_{z=0} + \Psi_{0xz}(x,z)|_{z=0} \cdot \eta(x,t) = u_x(x)\eta(x,t). \tag{32}$$

Additionally, the last term in Equation (31) can be approximated as

$$\Psi_{0z}(x,z)\eta_x(x,t)|_{z=\eta} \approx u(x)\eta_x(x,t). \tag{33}$$

Since $\Psi_{1x}(z=\eta) \approx \Psi_{1x}(z=0)$ and $\Psi_{2x}(z=\eta) \approx \Psi_{2x}(z=0)$, the linearised kinematic boundary condition finally takes the form

$$-\eta_t = \Psi_{1x} + (u\eta)_x = \Psi_{2x} \quad \text{at } z = 0. \tag{34}$$

MR2 neglected the first-order term given by Equation (32). Nevertheless, as the differences between our results and those of MR2 arise from many factors, the direct impact of this missing term cannot be assessed.

3.2. Separation of Variables and Averaging

At this stage, the perturbed state can be described by Equations (30) and (34). Now we proceed by decomposing the perturbations Ψ_1, Ψ_2, and η into mean and fluctuating parts:

$$\Psi_1 = \langle \Psi_1 \rangle + \Psi_1', \quad \langle \Psi_1' \rangle = 0, \tag{35}$$

$$\Psi_2 = \langle \Psi_2 \rangle + \Psi_2', \quad \langle \Psi_2' \rangle = 0, \tag{36}$$

$$\eta = \langle \eta \rangle + \eta', \quad \langle \eta' \rangle = 0. \tag{37}$$

Afterwards, we take the ensemble averages of Equations (30) and (34) to obtain

$$\kappa \int_{-\infty}^{0} \langle \Psi_2 \rangle_{xt}\, dz - \int_{\infty}^{0} \langle \Psi_1 \rangle_{xt}\, dz - (1-\kappa)g\langle \eta \rangle$$
$$= \int_{\infty}^{0} e^{-\zeta^2 z}\left\{(1-\zeta^2 z)\left[\langle u\Psi_{1xx}' \rangle - \langle u_x \Psi_{1x}' \rangle\right] + z\left[\langle u_{xx}\Psi_{1z}' \rangle - \langle u_x \Psi_{1xz}' \rangle\right]\right\} dz \tag{38}$$

from the equation of motion (30), and

$$-\langle \eta \rangle_t = \langle \Psi_1 \rangle_x + \langle u\eta' \rangle_x = \langle \Psi_2 \rangle_x \tag{39}$$

from the kinematic boundary condition (34). The averaged equations are then subtracted from the full equations, and we arrive at

$$\kappa \int_{-\infty}^{0} \Psi_{2xt}'\, dz - \int_{\infty}^{0} \Psi_{1xt}'\, dz - (1-\kappa)g\eta'$$

$$= \int_\infty^0 e^{-\zeta^2 z}(1-\zeta^2 z)[u\langle\Psi_1\rangle_{xx} - u_x\langle\Psi_1\rangle_x$$
$$+\{u\Psi'_{1xx} - \langle u\Psi'_{1xx}\rangle\} - \{u_x\Psi'_{1x} - \langle u_x\Psi'_{1x}\rangle\}]\,\mathrm{d}z$$
$$+\int_\infty^0 e^{-\zeta^2 z}z[u_{xx}\langle\Psi_1\rangle_z - u_x\langle\Psi_1\rangle_{xz}$$
$$+\{u_{xx}\Psi'_{1z} - \langle u_{xx}\Psi'_{1z}\rangle\} - \{u_x\Psi'_{1xz} - \langle u_x\Psi'_{1xz}\rangle\}]\,\mathrm{d}z, \qquad (40)$$

determining the behaviour of the velocity fluctuations, and

$$-\eta'_t = \Psi'_{1x} + (u\langle\eta\rangle)_x + \{(u\eta')_x - \langle u\eta'\rangle_x\} = \Psi'_{2x}, \qquad (41)$$

describing the random fluctuations of the perturbed surface. The resulting system of four Equations (38)–(41) is not closed, therefore we cannot proceed any further without some closure assumptions that allow approximate solution of the equations.

3.2.1. Binary Collision Approximation

Closure is provided by the binary-collision approximation, which is equivalent to assuming that the terms in curly brackets in Equations (40) and (41) are negligible. This technique was described in detail by Howe (1971), who studied the propagation of a transverse wave along an infinite string with small random density inhomogeneities. Howe considered the interaction of an initially coherent wave with the density fluctuations, which produces a random wave component at the outset ("first or direct collision"). The result of further interaction between the generated random wave and the density fluctuations depends on the statistical properties of the interaction product; in general we expect contributions both to the coherent component ("second collision") and to the random component ("third collision") of the wave as well. Further interactions between the newly generated wave components and the random inhomogeneities lead to the continuous modification of the wave profile, although it is reasonable to assume that their overall impact on the wave is small.

In the binary collision approximation only the first-order modification of the mean wave is considered, which essentially means that the interaction between uncorrelated fluctuations is ignored. A similar approach is widely used in mean-field electrodynamics (*first-order smoothing* or *second-order correlation approximation*, see Krause and Roberts (1976) or Rädler (1976)), and also in scattering problems (Sazontov and Shagalov, 1986).

The propagation of transverse waves on an infinite, random string is a relatively simple problem which, at least in the binary-collision approximation, can be described by a system of two simple wave equations. The solution still requires a substantial amount of calculation, and the resulting dispersion relation is an integro-differential equation, which can only be solved after making quite a few additional assumptions. In comparison, the current problem regarding the effect of a random velocity field on a surface gravity wave involves more complicated physical processes. Therefore, even after applying the binary-collision approximation, the resulting system of equations remains much too complex to solve without some very strong assumptions on the perturbations.

3.2.2. Additional Assumptions

First we assume that the perturbations are separable in the x- and z-coordinates. In Paper I. the vertical variation of the perturbed velocity potential was determined by the Laplace equation, according to which the solution is $\sim e^{-kz}$ in $z > 0$. The equilibrium flow in the current model is not irrotational, hence the perturbed flow is normally not irrotational either. In the absence of the random flow, however, the velocity field associated with the theoretical f mode is irrotational. The random flow should not change the overall characteristics of the f mode as that would contradict the observations, which means that the deviation from irrotationality has to be relatively small.

MR2 simply prescribed the z-dependence of the perturbations without any comment. The chosen vertical profile was equivalent to assuming that the perturbations are irrotational. We do not suggest that the rotational part of the perturbations is indeed negligible but nevertheless, in what follows, we adapt the method of MR2. The consequences of this certainly crude assumption were analysed in some detail in Kerekes (2007), where the effect of a horizontal random magnetic profile on the f mode was studied using a different, more general approach.

3.3. Fourier Transformation

The system of Equations (38)–(41) is solved using Fourier transformation. The perturbations are written in an inverse Fourier form where the z-dependence is prescribed as exponential decay on both sides of the interface. The fluctuation parts of the perturbations are given as

$$\Psi'_1(x, z, t) = F^{-1}\{e^{-|k|z - i\omega t} \hat{\Psi}_1(k)\}, \tag{42}$$

$$\Psi'_2(x, z, t) = F^{-1}\{e^{|k|z - i\omega t} \hat{\Psi}_2(k)\}, \tag{43}$$

$$\eta'(x, t) = F^{-1}\{e^{-i\omega t} \hat{\eta}(k)\}. \tag{44}$$

Similarly, the mean parts of the perturbations are

$$\langle \Psi_1 \rangle(x, z, t) = F^{-1}\{e^{-|k|z - i\omega t} \bar{\Psi}_1(k)\}, \tag{45}$$

$$\langle \Psi_2 \rangle(x, z, t) = F^{-1}\{e^{|k|z - i\omega t} \bar{\Psi}_2(k)\}, \tag{46}$$

$$\langle \eta \rangle(x, t) = F^{-1}\{e^{-i\omega t} \bar{\eta}(k)\}, \tag{47}$$

where the Fourier transform pair is defined as

$$f(k) = F\{f(x)\} = \int_{-\infty}^{\infty} f(x) e^{-ikx} \, dx, \tag{48}$$

$$f(x) = F^{-1}\{f(k)\} = \frac{1}{2\pi} \int_{-\infty}^{\infty} f(k) e^{ikx} \, dk. \tag{49}$$

The factor $e^{-i\omega t}$ indicates that we seek normal modes of the system. This restriction is equivalent to carrying out a Laplace transform in time, solving the corresponding initial value problem, and later discarding the terms that depend on the initial conditions. The initial value problem needs to be considered only when the eigenvalues are complex, which is otherwise apparent from the singular form of the dispersion function. This method was discussed in sufficient detail in Section 4 of Paper I.

Fourier transformation of the mean and fluctuating parts of the kinematic boundary condition gives

$$\bar{\eta}(k) = \frac{k}{\omega} \bar{\Psi}_1(k) + \frac{k}{2\pi\omega} \iint_{-\infty}^{\infty} \langle \hat{\eta}(k_1) u(x) \rangle e^{-i(k-k_1)x} \, dx \, dk_1 = \frac{k}{\omega} \bar{\Psi}_2(k), \tag{50}$$

$$\hat{\eta}(k) = \frac{k}{\omega}\hat{\Psi}_1(k) + \frac{k}{2\pi\omega}\iint_{-\infty}^{\infty}\bar{\eta}(k_1)u(x)e^{-i(k-k_1)x}\,dx\,dk_1 = \frac{k}{\omega}\hat{\Psi}_2(k), \qquad (51)$$

and from the equation of motion we obtain

$$\omega\,\mathrm{sgn}(k)\big[\kappa\bar{\Psi}_2(k) + \bar{\Psi}_1(k)\big] - (1-\kappa)g\bar{\eta}(k)$$
$$= \frac{k}{2\pi}\iint_{-\infty}^{\infty}\frac{|k_1|(2k_1-k)}{(\zeta^2+|k_1|)^2}\langle\hat{\Psi}_1(k_1)u(x)\rangle e^{-i(k-k_1)x}\,dx\,dk_1, \qquad (52)$$

$$\omega\,\mathrm{sgn}(k)\big[\kappa\hat{\Psi}_2(k) + \hat{\Psi}_1(k)\big] - (1-\kappa)g\hat{\eta}(k)$$
$$= \frac{k}{2\pi}\iint_{-\infty}^{\infty}\frac{|k_1|(2k_1-k)}{(\zeta^2+|k_1|)^2}\bar{\Psi}_1(k_1)u(x)e^{-i(k-k_1)x}\,dx\,dk_1. \qquad (53)$$

We proceed by eliminating $\bar{\eta}(k)$ and $\hat{\eta}(k)$ from Equations (50)–(53). Later $\hat{\Psi}_2(k)$ is substituted from Equation (51) into Equation (53), so that we could express $\hat{\Psi}_1(k)$ as a function of $\bar{\Psi}_1(k)$ and $\bar{\Psi}_2(k)$. The resulting expression for $\hat{\Psi}_1(k)$ is then substituted into the mean equations (50) and (52). After some manipulation we obtain two simple algebraic equations for $\bar{\Psi}_1$ and $\bar{\Psi}_2$, from which the calculation of the dispersion relation is straightforward.

The derivation of the dispersion relation requires the evaluation of several integrals of the type

$$I(k) \qquad (54)$$
$$= \frac{1}{(2\pi)^2}\iiiint_{-\infty}^{\infty} g_1(k_1)g_2(K)\bar{\Psi}_j(K)\langle u(x)u(x_1)\rangle e^{-i\{(k-k_1)x+(k_1-K)x_1\}}\,dx\,dx_1\,dK\,dk_1, \qquad (55)$$

where $j = 1, 2$, and g_1 and g_2 are arbitrary functions. The velocity correlation between two points is assumed to be a function of the separation distance only, and defined as

$$r(x - x_1) \equiv \langle u(x)u(x_1)\rangle. \qquad (56)$$

After a suitable change of variables, the number of integrals in Equation (55) can be reduced to one,

$$I(k) = \frac{g_2(k)\bar{\Psi}_j(k)}{2\pi}\int_{-\infty}^{\infty} g_1(k_1)R(k-k_1)\,dk_1, \qquad (57)$$

where $R(k)$ denotes the Fourier transform of the correlation function $r(x)$.

Finally, we arrive at the dispersion relation

$$(1+\kappa)\omega^2 - (1-\kappa)gk = \frac{(1-\kappa)gk - \kappa\omega^2}{(\zeta^2+k)^2}kX_2(k)$$
$$+ k^2\big(\omega^2 X_3(k) - X_4(k)\big)\left\{1 + \frac{kX_2(k)}{(\zeta^2+k)^2}\right\}$$
$$+ \big(1 - kX_1(k)\big)\frac{\omega^2 k^2 X_5(k)}{(\zeta^2+k)^2} + k\omega^2 X_1(k), \qquad (58)$$

where we assumed that $k > 0$. The integrals $X_i(k)$ ($i = 1,\ldots,5$) can be written as

$$X_1(k) = \frac{1}{2\pi} \int_{-\infty}^{\infty} \frac{|k_1|}{D_0(k_1)} R(k-k_1)\,dk_1, \qquad (59)$$

$$X_2(k) = \frac{1}{2\pi} \int_{-\infty}^{\infty} \frac{k_1^2(2k-k_1)}{D_0(k_1)} R(k-k_1)\,dk_1, \qquad (60)$$

$$X_3(k) = \frac{1}{2\pi} \int_{-\infty}^{\infty} \frac{k_1(2k_1-k)}{D_0(k_1)(\zeta^2+|k_1|)^2} R(k-k_1)\,dk_1, \qquad (61)$$

$$X_4(k) = \frac{1}{2\pi} \int_{-\infty}^{\infty} \frac{|k_1|(2k_1-k)}{(\zeta^2+|k_1|)^2} R(k-k_1)\,dk_1, \qquad (62)$$

$$X_5(k) = \frac{1}{2\pi} \int_{-\infty}^{\infty} \frac{k_1|k_1|(2k-k_1)(2k_1-k)}{D_0(k_1)(\zeta^2+|k_1|)^2} R(k-k_1)\,dk_1, \qquad (63)$$

where $D_0(k)$ denotes the dispersion function in the absence of the random flow,

$$D_0(k) = (1+\kappa)\omega^2 \mathrm{sgn}(k) - (1-\kappa)gk. \qquad (64)$$

The RHS of the dispersion relation (58) provides a correction to the no-flow case described by $D_0(k) = 0$, or alternatively, by Equation (1).

4. Calculation of the Frequency Correction

In this section the implicit dispersion relation (58) is transformed into an explicit form, and eventually the frequency is determined as a function of the wavenumber. First, the dispersion relation is non-dimensionalised using the following variables:

$$K = kl_c, \qquad \Omega^2 = \frac{(1+\kappa)}{(1-\kappa)} \frac{l_c}{g} \omega^2, \quad \text{and} \quad p = \zeta^2 l_c, \qquad (65)$$

where K is the dimensionless wavenumber, Ω is the dimensionless frequency, l_c denotes the typical length-scale of the flow, and p is the dimensionless decay factor. Additionally, we define a non-dimensional parameter,

$$\epsilon^2 = \frac{\sigma_0^2}{\sqrt{\pi}(1-\kappa)gl_c}, \qquad (66)$$

where σ_0 corresponds to the characteristic flow speed of the turbulent motion. Here ϵ^2 is a Froude number similar to α^2 given by Equation (6) in Paper I,

$$\epsilon^2 = \frac{\alpha^2}{(1-\kappa)\sqrt{\pi}}. \qquad (67)$$

The correlation function was chosen to be a simple Gaussian,

$$r(x) = \sigma_0^2 e^{-x^2/4l_c^2}, \qquad R(k) = 2\sigma_0^2 l_c \sqrt{\pi} e^{-k^2 l_c^2}. \qquad (68)$$

This one-dimensional correlation function was incorrectly Fourier transformed in MR2, where the coefficient of $e^{-k^2 l^2}$ was given as $l^2\sigma_0^2/\pi$ (see Equation (4.13) of MR2). Their result suggests that a two-dimensional Fourier transform was carried out with the normalisation factor $1/2\pi$ (instead of 1).

We proceed by converting the correction integrals into non-dimensional form. Substitution of the correlation function (68) into Equations (59)–(63), and subsequent elimination of the absolute value function yields

$$X_i(K) = -C_i\, e^{-K^2} \int_0^\infty \frac{\beta_i(K, K_1)}{K_1 - \Omega^2}\, dK_1, \quad i = 1, 2, 3, 5, \tag{69}$$

$$X_4(K) = -C_4\, e^{-K^2} \int_0^\infty \beta_4(K, K_1)\, dK_1, \tag{70}$$

where the constants C_i in Equation (69) can be expressed as functions of ϵ^2

$$C_1 = \epsilon^2 l_c, \quad C_2 = \frac{\epsilon^2}{l_c}, \quad C_3 = \epsilon^2 l_c^2, \quad C_4 = \epsilon^2(1-\kappa)g l_c, \quad \text{and} \quad C_5 = \epsilon^2. \tag{71}$$

The auxiliary functions $\beta_i(K, K_1)$ can be written as

$$\beta_1(K, K_1) = 2K_1 e^{-K_1^2} \sinh(2K K_1), \tag{72}$$

$$\beta_2(K, K_1) = 2K_1^2 e^{-K_1^2}\left[2K \sinh(2K K_1) - K_1 \cosh(2K K_1)\right], \tag{73}$$

$$\beta_3(K, K_1) = \frac{2K_1 e^{-K_1^2}}{(p + K_1)^2}\left[2K_1 \sinh(2K K_1) - K \cosh(2K K_1)\right], \tag{74}$$

$$\beta_5(K, K_1) = \frac{2K_1^2 e^{-K_1^2}}{(p + K_1)^2}\left[5K K_1 \sinh(2K K_1) - 2(K^2 + K_1^2)\cosh(2K K_1)\right], \tag{75}$$

and $\beta_4(K, K_1) = -\beta_3(K, K_1)$. The singular integrals (X_1, X_2, X_3 and X_5) are now simplified using the contour integration method introduced in Paper I. For convenience, we define the following non-dimensional correction integrals

$$\tilde{X}_i(K) = \frac{X_i(K)}{C_i}, \quad i = 1, \ldots, 5. \tag{76}$$

We have shown in Section 4 of Paper I that if $\Omega^2 = k_0 = k_r + i\delta$ and $k_r \gg \delta$, then in the limit $\delta \to 0$ we have

$$\tilde{X}_i(K) \simeq -e^{-K^2}\left\{\fint_0^\infty \frac{\beta_i(K, k)}{k - K}\, dk + i\pi \beta_i(K, K)\right\}, \quad i = 1, 2, 3, 5. \tag{77}$$

The above expression has been derived by assuming that $k_r \sim K$, i.e., that the correction to the no flow case is small. The dispersion relation (58) can now be written as

$$\Omega^2 - K = \Omega^2 f_\Omega(K) + K f_K(K) \tag{78}$$

in dimensionless form, where the coefficient functions are

$$f_\Omega = \epsilon^2 \frac{K}{1+\kappa}\left\{K\tilde{X}_3 + \tilde{X}_1 + \frac{K\tilde{X}_5 - \kappa \tilde{X}_2}{(p+K)^2} + \epsilon^2 \frac{K^2}{(p+K)^2}(\tilde{X}_2\tilde{X}_3 - \tilde{X}_1\tilde{X}_5)\right\}, \tag{79}$$

$$f_K = \epsilon^2 K\left\{\frac{(1 - \epsilon^2 K \tilde{X}_4)}{(p+K)^2}\tilde{X}_2 - \tilde{X}_4\right\}. \tag{80}$$

4.1. Small ϵ^2 Approximation

In the absence of the random flow, the RHS of the dispersion relation (78) is zero and

$$\Omega_0 = K^{1/2}. \tag{81}$$

The presence of the random flow provides a small correction to Ω_0, which is quadratic in ϵ^2. A realistic estimate of ϵ^2 (for granular flows) is below 10^{-2}, therefore we can perform a simplified analysis of the dispersion relation. While the evaluation of Equation (78) is certainly possible for arbitrary ϵ^2, the algebra is tedious, and most importantly we would only find a significant difference between the general and the present case for larger ϵ^2. Such values of ϵ^2 either represent vigorous flows or extremely small-scale motions, none of which would be an appropriate representation of granular motions. Furthermore, this type of flow would probably cause strong damping, and that would invalidate the analysis of the previous section.

Neglecting all terms quadratic in ϵ^2 from Equation (78), we obtain that in the leading order

$$\Omega \simeq K^{1/2}\left\{1 + \frac{f_\Omega + f_K}{2}\right\} = \Omega_0 + \Omega_c, \tag{82}$$

where the correction to Ω_0 can be written as

$$\Omega_c = \epsilon^2 \frac{K^{3/2}}{2(1+\kappa)}\left\{K\tilde{X}_3 + \tilde{X}_1 - (1+\kappa)\tilde{X}_4 + \frac{K\tilde{X}_5 + \tilde{X}_2}{(p+K)^2}\right\}. \tag{83}$$

The real part of Ω_c gives a correction to the frequency, while the imaginary part of Ω_c is inversely proportional to the damping time of the mode (which is directly related to the observed linewidth).

In order to facilitate the separation of the real and imaginary parts of Ω_c, the non-dimensional integrals will be decomposed as follows:

$$\tilde{X}_i(K) = R_i(K) + iQ_i(K), \quad i = 1, \ldots, 5. \tag{84}$$

The calculation of Q_i is straightforward from Equation (77)

$$Q_1 = -\pi K\left(1 - e^{-4K^2}\right), \tag{85}$$

$$Q_2 = -\pi K^3\left(1 - 3e^{-4K^2}\right), \tag{86}$$

$$Q_3 = -\pi \chi^2\left(1 - 3e^{-4K^2}\right), \tag{87}$$

$$Q_5 = -\pi \chi^2 K^2\left(1 - 9e^{-4K^2}\right), \tag{88}$$

where the parameter $\chi = K/(p+K)$. $\tilde{X}_4$ is real, hence $Q_4 = 0$. The imaginary part of the frequency correction can be calculated by substituting Equations (85)–(88) into Equation (83). The damping of the wave is represented by the linewidth [$\Gamma = -2\,\text{Im}\{\Omega\}$], which can be expressed in the following form:

$$\Gamma = \epsilon^2 \frac{\pi K^{5/2}}{(1+\kappa)}\left[\left(1+\chi^2\right)^2 - \left(1+3\chi^2\right)^2 e^{-4K^2}\right]. \tag{89}$$

For large K the parameter $\chi \to 1$ and the exponential term becomes negligible, so asymptotically $\Gamma \to 4\pi\epsilon^2 K^{5/2}/(1+\kappa)$.

The calculation of $\text{Re}\{\Omega_c\}$ is slightly more complicated as the integrals (70) and (77) cannot be evaluated analytically. Formally, $\text{Re}\{\Omega_c\}$ can be written as

$$\text{Re}\{\Omega_c\} = \epsilon^2 \frac{K^{3/2}}{2(1+\kappa)} \left\{ K R_3 + R_1 - (1+\kappa) R_4 + \frac{K R_5 + R_2}{(p+K)^2} \right\}. \tag{90}$$

Using Equations (69)–(70) and (77) we obtain $R_i(K)$ in the form

$$R_i(K) = -e^{-K^2} \int_0^\infty \frac{\beta_i(K, K_1)}{(K_1 - K)} dK_1, \quad i = 1, 2, 3, 5, \tag{91}$$

$$R_4(K) = -e^{K^2} \int_0^\infty \beta_4(K, K_1) dK_1. \tag{92}$$

4.2. Numerical Integration

The integrals (91)–(92) were evaluated using the D01ARF NAG-routine in FORTRAN. The only non-singular integral, $R_4(K)$, can be integrated straight away. The rest of the integrals with a Cauchy principal part, however, need to be rewritten in a form more suitable for numerical evaluation. First, we change the variable of integration K_1 to $x = K_1/K$, and then use the following expansion:

$$\frac{x^n}{x-1} = \sum_{k=0}^{n-1} x^k + \frac{1}{x-1}. \tag{93}$$

As a result, the singular integrals split into several integrals, some of which can be evaluated analytically. After some manipulation we can rewrite the real part of the frequency correction as

$$\text{Re}\{\Omega_c\} = -\epsilon^2 \frac{K^{3/2}}{2(1+\kappa)} f_R(K). \tag{94}$$

Here the function $f_R(K)$ can be expressed as

$$f_R(K) = K^3 \left(D_1 - (1+\kappa) D_2 + \chi^2 D_3 + (2 + 5\chi^2) P_3 - (1 - 4\chi^2) P_4 \right)$$
$$+ (1+\chi^2) \sqrt{\pi} \, \text{erf}(K) + K \left((1 + 2\chi^2) P_1 - \chi^2 P_2 \right)$$
$$- \chi^2 K^{-2} \left(\sqrt{\pi}/2 + K e^{-K^2} \right), \tag{95}$$

with

$$D_1(K) = \int_0^\infty \frac{2(x+1) e_1(K, x) - e_2(K, x)}{(Kx + p)^2} dx, \tag{96}$$

$$D_2(K) = \int_0^\infty \frac{2x^2 e_1(K, x) - x e_2(K, x)}{(Kx + p)^2} dx, \tag{97}$$

$$D_3(K) = \int_0^\infty \frac{5(x^2 + x + 1) e_1(K, x) - 2(x^3 + x^2 + 2x + 2) e_2(K, x)}{(Kx + p)^2} dx, \tag{98}$$

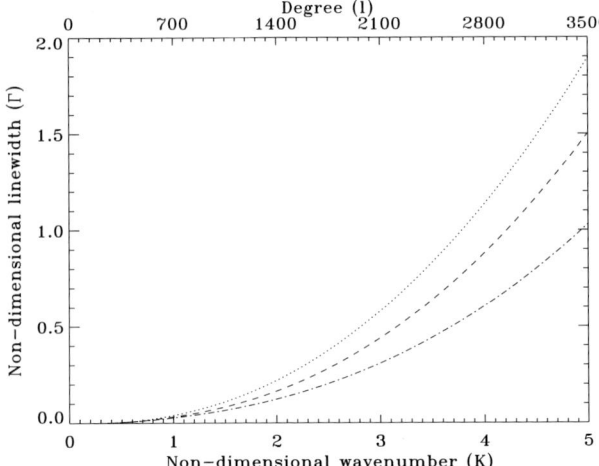

Figure 2 The non-dimensional damping rate $[\Gamma = -2\,\text{Im}\{\Omega\}]$ as a function of the non-dimensional wavenumber $[K]$ for $\epsilon^2 = 8.32 \times 10^{-3}$, $\kappa = 0.01$, $l_c = 10^3$ km, with $p = 1$ (dotted line), $p = 2$ (dashed line), and $p = 5$ (dash-dotted line).

and

$$P_1(K) = \int_0^\infty \frac{e_1(K,x)}{x-1}\,dx, \qquad P_2(K) = \int_0^\infty \frac{e_2(K,x)}{x-1}\,dx, \qquad (99)$$

$$P_3(K) = \int_0^\infty \frac{e_1(K,x)}{(Kx+p)^2(x-1)}\,dx, \qquad P_4(K) = \int_0^\infty \frac{e_2(K,x)}{(Kx+p)^2(x-1)}\,dx, \quad (100)$$

where

$$e_1(K,x) = e^{-K^2(x-1)^2} - e^{-K^2(x+1)^2}, \qquad (101)$$

$$e_2(K,x) = e^{-K^2(x-1)^2} + e^{-K^2(x+1)^2}. \qquad (102)$$

5. Results

The non-dimensional linewidth (Figure 2) was calculated analytically from Equation (89) for different values of the dimensionless decay factor p. This parameter represents the vertical confinement of the random flow. The characteristic thickness of the random layer was varied between 1000 km ($p = 1$) and 200 km ($p = 5$). We found that the damping increases with increasing thickness. The obtained linewidth for $p = 1$ (thick random layer) is very close to the corresponding result of the horizontal flow case (infinite random layer).

The linewidth was calculated under the assumption that the ratio of the imaginary and real parts of the complex frequency is very small. We note, that at very large wavenumbers ($K \gtrsim 4$) the magnitude of the imaginary part becomes comparable with the real frequency. The ratio $|\text{Im}\{\Omega\}|/\text{Re}\{\Omega\}$ is between 1/8 and 1/4 at $K = 4$, and the approximation applied in Section 4 breaks down at even larger wavenumbers. The numerical calculations were performed in the non-dimensional wavenumber range $0.1 < K < 4$, which corresponds to spherical degrees $70 < \ell < 2800$. Extending the calculations to higher degree would have required a different, more sophisticated numerical method. Since the behaviour of the f mode is highly uncertain for $\ell \gtrsim 2500$, and its identification is controversial for $\ell \gtrsim 3000$ (Antia and Basu, 1999), we believe that most degrees of interest were covered.

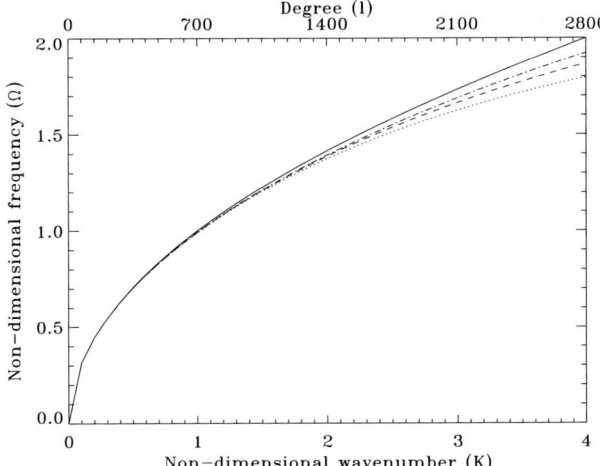

Figure 3 The non-dimensional frequency as a function of the non-dimensional wavenumber $[K]$ for $\epsilon^2 = 8.32 \times 10^{-3}$, $\kappa = 0.01$, $l_c = 10^3$ km, with $p = 1$ (dotted line), $p = 2$ (dashed line), and $p = 5$ (dash-dotted line). The solid line shows $K^{1/2}$.

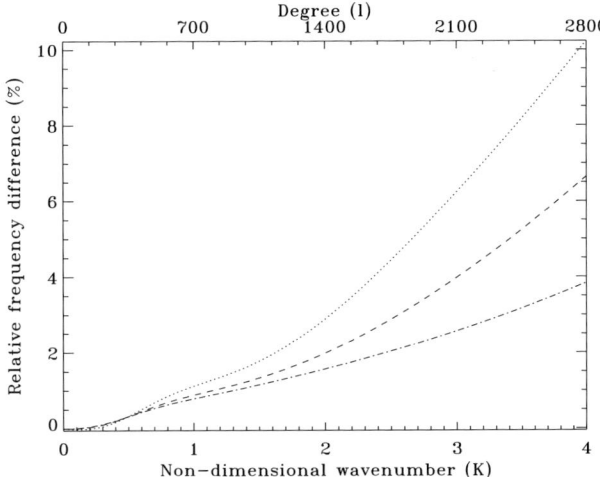

Figure 4 The relative frequency difference $[-\Delta\Omega/\Omega_0]$ in % as a function of the non-dimensional wavenumber $[K]$ for $\epsilon^2 = 8.32 \times 10^{-3}$, $\kappa = 0.01$, $l_c = 10^3$ km, with $p = 1$ (dotted line), $p = 2$ (dashed line), and $p = 5$ (dash-dotted line).

Figure 3 displays the calculated real frequencies for different values of the dimensionless decay factor, together with the frequencies given by the no-flow dispersion relation (1). Similarly to the linewidth, the magnitude of the frequency correction also increases monotonically with the wavenumber, and this tendency is more pronounced for thicker layers.

The obtained frequencies are almost indistinguishable from the no-flow curve at lower wavenumbers, therefore the relative frequency difference $-\Delta\Omega/\Omega_0$ is shown separately in Figure 4. The behaviour of the $p = 1$ curve is somewhat anomalous at $K \lesssim 0.3$ where we found a very small frequency increase ($> 0.1\%$). Since the present Cartesian model is aimed at higher-degree modes, only a spherical model may clarify whether this effect is real, or simply an artefact due to some numerical errors. Figure 4 suggests that, if indeed real, the positive shifts would increase for $p < 1$. Unfortunately, the applied numerical method breaks down just below $p = 1$, where the integrands with a Cauchy principal part become confined to a very narrow neighbourhood of the singularity.

6. Discussion

In summary, we found a good qualitative agreement with the data of Antia and Basu (1999) and Duvall, Kosovichev, and Murawski (1998), however the obtained frequency corrections and damping rates are somewhat larger than observed for realistic solar parameters. A better agreement could be achieved for smaller ϵ^2, which would correspond to slower and/or larger granules. The results of the present horizontal and vertical flow case are also very close to those of the horizontal flow case for smaller p, but the differences increase as the random flow becomes more confined to the surface.

6.1. Comparison with Random Magnetic Models

In an earlier study (Erdélyi, Kerekes, and Mole (2005) – hereafter EKM), we have examined the effect of a random atmospheric magnetic field on the solar f mode. The random flow in the present model (representing granular flows) and the random magnetic field of EKM (representing the solar magnetic carpet) was described with the same horizontal and vertical profiles, but on the opposite sides of the interface $z = 0$. (An extensive review of the solar magnetic field was given by Solanki, Inhester, and Schüssler (2006).) We found that, in the presence of the random magnetic atmosphere, the frequencies of the f mode were slightly increased above the values given by Equation (1). However, the effect of random magnetic field appears to be much weaker than that of random flows. This conclusion is also supported by Kerekes, Erdélyi, and Mole (2008), where the influence of a random magnetic atmosphere was considered using a different modelling technique.

In what follows, we compare the frequency shifts caused by random flows and magnetic field. Let Ω_M denote the frequency correction obtained in EKM. After some transformation (for more details see Kerekes (2007)), Ω_M can be expressed as

$$\Omega_M(K) = 2\pi \kappa \xi^2 K^{9/2} D_2(K), \tag{103}$$

where $D_2(K)$ is given by (97), and ξ^2 is a non-dimensional parameter that incorporates the characteristic properties of the random magnetic field,

$$\xi^2 = \frac{v_{Ar}^2}{\sqrt{\pi}(1-\kappa)gl_c}. \tag{104}$$

Here v_{Ar} denotes the characteristic Alfvén speed of the random magnetic field. We can see that the definition of ξ is almost identical to the definition of ϵ^2 given by Equation (66), only in Equation (104) we have v_{Ar} instead of the characteristic flow speed σ_0.

Figure 5 compares the frequency corrections caused by a subsurface random flow (Ω_F) with the corrections caused by a random atmospheric magnetic field (Ω_M). The ratio $|\Omega_F|/\Omega_M$ was calculated from Equations (94) and (103) for $\kappa = 0.01$ and different values of p, assuming that the characteristic flow speed in $z > 0$ equals the characteristic Alfvén speed in $z < 0$. In effect, we compare velocity and magnetic fields having identical profiles and equal energy. The curves in Figure 5 take their maxima at small wavenumbers whose exact locations also depend slightly on p. The ratio of the corrections was found to increase with increasing p (thinner layer), and in all cases asymptotically converges to the same constant value (≈ 50 in Figure 5) as K becomes large.

The apparent disparity between the two results is either a consequence of the different locations of the random fields (the flow is placed in the heavier interior while the magnetic field is situated in the lighter atmosphere), or a reflection of some inherent differences in

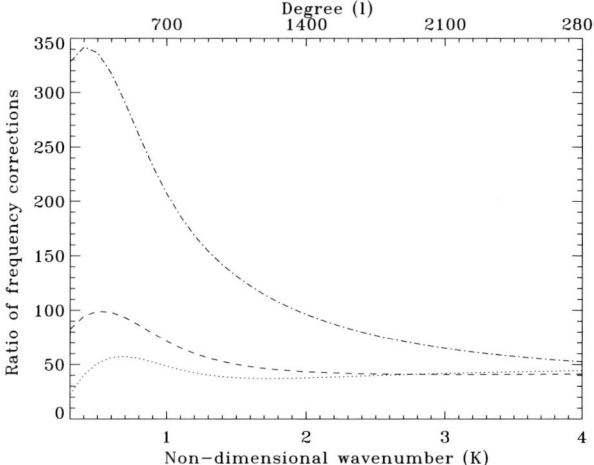

Figure 5 The ratio of frequency corrections $[|\Omega_F|/\Omega_M]$ as a function of the non-dimensional wavenumber $[K]$ for different values of the dimensionless decay factor: $p = 1$ (dotted), $p = 2$ (dashed), $p = 5$ (dash-dotted), with $\epsilon^2 = \xi^2$, $l_c = 10^3$ km, $\kappa = 0.01$.

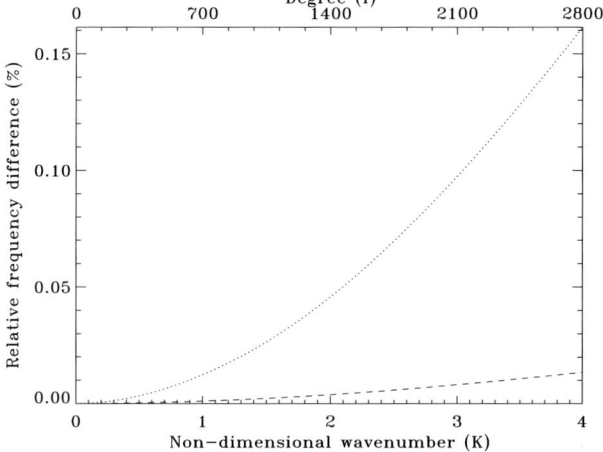

Figure 6 Relative frequency difference in % as a function of the non-dimensional wavenumber $[K]$, caused by either a random magnetic field $[\Omega_M^*/\Omega_0]$ (dotted) or a random flow $[|\Omega_F|/\Omega_0]$ (dashed), situated in the lower, heavier layer (interior) of the model. Parameters: $\epsilon^2 = \xi^2 = 8.32 \times 10^{-3}$, $l_c = 10^3$ km, $\kappa = 0.01$, and $p = 2$.

the dynamical effects of velocity and magnetic fields. Flows can directly accelerate or impede waves, while the influence of magnetic fields is indirect: the tension component of the Lorentz-force modifies the elasticity of the medium the waves travel through. The most conspicuous difference between the present results and those of EKM is that only random flows caused wave damping; the effect of the random magnetic field was limited to the real part of the frequency. We do not attempt to give an explanation for this discrepancy here, because only a thorough analysis of the energy equation could reveal why the nature of the energy transfer differs between the magnetic and flow cases.

To assess how the locations of the random fields affect the results, we now determine the frequency shifts caused by the *mirror images* of the velocity field (4) in $z < 0$ and the magnetic field (Equation (5) of EKM) in $z > 0$. Let Ω_F^* and Ω_M^* denote the frequency corrections caused by the atmospheric flow and interior magnetic field, respectively. Then Ω_F^* and Ω_M^* can be obtained by simply changing the sign of g and replacing κ with κ^{-1} in the expressions of Ω_F and Ω_M. Figure 6 compares Ω_F with Ω_M^*, while Figure 7 displays Ω_F^* and Ω_M together. The effect of the magnetic field appears to be more pronounced in both cases.

Figure 7 Relative frequency difference in % as a function of the non-dimensional wavenumber (K), caused by either a random magnetic field (Ω_M/Ω_0) (dotted) or a random flow $|\Omega_F^*|/\Omega_0$ (dashed), situated in the upper, lighter layer (atmosphere) of the model. Parameters: $\epsilon^2 = \xi^2 = 8.32 \times 10^{-3}$, $l_c = 10^3$ km, $\kappa = 0.01$, and $p = 2$.

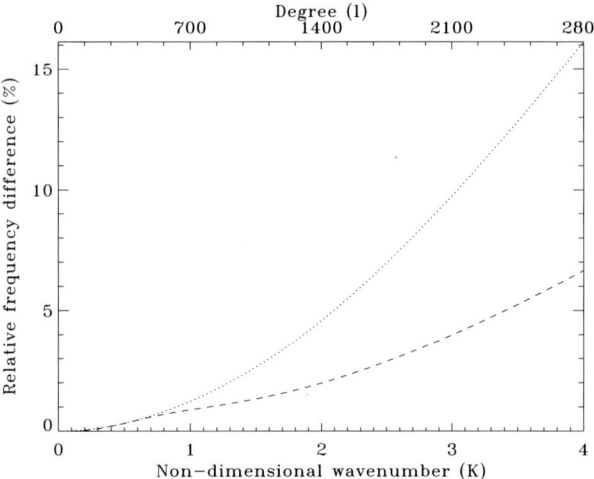

Figures 6 and 7 demonstrate that a random magnetic field has no less potential to influence the frequencies of the f mode than random flows. However, only models with more realistic density profiles could determine whether random effects are truly relevant to the solar case.

Acknowledgements We wish to thank Misha Ruderman, Victor Shrira, and Viktor Fedun for helpful discussions, and we acknowledge the financial support by (STFC) and NFS Hungary (OTKA, K67746).

References

Antia, H.M., Basu, S.: 1999, High-frequency and high-wavenumber solar oscillations. *Astrophys. J.* **519**, 400–406.
Christensen-Dalsgaard, J.: 2002, Helioseismology. *Rev. Mod. Phys.* **74**, 1073–1128.
Duvall, T.L., Kosovichev, A.G., Murawski, K.: 1998, Random damping and frequency reduction of the solar f-mode. *Astrophys. J.* **505**, L55–L58.
Erdélyi, R.: 2006, Magnetic coupling of waves and oscillations in the lower solar atmosphere: Can the tail wag the dog? *Philos. Trans. Roy. Soc. A* **364**, 351–381.
Erdélyi, R., Kerekes, A., Mole, N.: 2005, Influence of random magnetic field on solar global oscillations: The incompressible f-mode. *Astron. Astrophys.* **431**, 1083–1088.
Gruzinov, A.V.: 1998, Sound speed in a random flow and turbulent shifts of the solar eigenfrequencies. *Astrophys. J.* **498**, 458–464.
Howe, M.S.: 1971, Wave propagation in random media. *J. Fluid Mech.* **45**, 769–783.
Kerekes, A.: 2007, *Random effects on the solar f-mode*. PhD thesis, University of Sheffield.
Kerekes, A., Erdélyi, R., Mole, N.: 2008, A novel approach to the solar interior-atmosphere EVP. *Astrophys. J.*, accepted.
Krause, F., Roberts, P.H.: 1976, Some problems of mean field electrodynamics. *Astrophys. J.* **181**, 977–992.
Mędrek, M., Murawski, K., Roberts, B.: 1999, Damping and frequency reduction of the f-mode due to turbulent motion in the solar convection zone. *Astron. Astrophys.* **349**, 312–316.
Mole, N., Kerekes, A., Erdélyi, R.: 2008, Effect of random flows on the solar f-mode I.: Horizontal flow. *Solar Phys.*, accepted.
Murawski, K., Goossens, M.: 1993, Random velocity field corrections to the f-mode. III. A photospheric random flow and chromospheric magnetic field. *Astron. Astrophys.* **279**, 225–234.
Murawski, K., Roberts, B.: 1993a, Random velocity field corrections to the f-mode. I. Horizontal flows. *Astron. Astrophys.* **272**, 595–600.
Murawski, K., Roberts, B.: 1993b, Random velocity field corrections to the f-mode. II. Vertical and horizontal flow. *Astron. Astrophys.* **272**, 601–608.
Rädler, K.H.: 1976, Mean-field magnetohydrodynamics as a basis of solar dynamo theory. In: *Basic Mechanisms of Solar Activity, Proc. IAU Symposium* **71**, Reidel, Dordrecht, 323–344.

Sazontov, A.G., Shagalov, S.V.: 1986, Scattering of gravity waves by a turbulence of the upper ocean layer. *Izv. Atmos. Ocean. Phys.* **22**, 138–143.

Selwa, M., Skartlien, R., Murawski, K.: 2004, Numerical simulations of stochastically excited sound waves in a random medium. *Astron. Astrophys.* **420**, 1123–1127.

Skartlien, R.: 2002, Effects in the solar p-mode power spectrum from scattering on a turbulent background flow with stochastic wave sources. *Astrophys. J.* **578**, 621–647.

Solanki, S.K., Inhester, B., Schüssler, M.: 2006, The solar magnetic field. *Rep. Prog. Phys.* **69**, 563–668.

Spruit, H.C., Nordlund, A., Title, A.M.: 1990, Solar convection. *Ann. Rev. Astron. Astrophys.* **28**, 263–301.

Stein, R.F., Nordlund, A.: 1998, Simulations of solar granulation. I. General properties. *Astrophys. J.* **499**, 914–933.

Instrumental Response Function for Filtergraph Instruments

R. Wachter

Originally published in the journal Solar Physics, Volume 251, Nos 1–2, 491–500.
DOI: 10.1007/s11207-008-9197-5 © Springer Science+Business Media B.V. 2008

Abstract Dopplergrams and magnetograms arising from filtergraph instruments such as the Michelson Doppler Imager (MDI), the Helioseismic and Magnetic Imager (HMI), or the Hinode Narrow Band Filter Imager are generally associated with observation heights that are derived from the contribution function of the targeted absorption line, irrespective of the instrument characteristics. Observation heights are important for interpreting the phases of propagating waves, and for the diagnostics of the solar atmosphere. I show in this paper that the formalism presented by Ruiz Cobo and del Toro Iniesta (*Astron. Astrophys.* **283**, 129, 1994) provides a straightforward approach to associate an observation height for each observable given the instrumental algorithm, the transmission profiles and the local stratification at the point of observation. To demonstrate the method, I construct a simple radially symmetric sunspot model and calculate the mean observation height for various MDI observables as a function of horizontal location. It is shown that different ways of measuring the same quantity can result in different observation heights, that the offset velocity caused by the spacecraft motion has to be taken into account, and that observation heights in sunspots vary beyond the pure geometric effect of the Wilson depression.

Keywords Helioseismology, observations · Instrumental effects

1. Introduction

Recently, the observation of propagating waves in the solar atmosphere has drawn considerable attention. Despite the fact that solar five minute oscillations are eigenmodes trapped below the solar surface observed in their evanescent tail, wave propagation in the atmosphere is a widely observed phenomenon with large diagnostic capabilities and consequences for the atmospheric energy balance. First, solar p-modes can only be considered to be trapped modes below the peak of global acoustic cutoff frequency at about 5.2 mHz. The observed

---

R. Wachter (✉)
Hansen Experimental Physics Laboratory, Stanford University, 491 South Service Road, Stanford, CA 94305, USA
e-mail: richard@sun.stanford.edu

power at higher frequencies is referred to as "pseudomode spectrum" and shows relative phases of velocity and temperature that are characteristic for freely propagating waves. Further, the acoustic cutoff frequency depends on the local stratification which is substantially different in active regions. Observations show that in a magnetic atmosphere waves of lower frequencies are also able to propagate freely. McIntosh and Jefferies (2006) show that the acoustic cutoff frequency is substantially lowered for a strong inclined magnetic field, and Rajaguru *et al.* (2007) see a clear indication of propagating waves at *p*-mode frequencies by applying bisector analysis to a photospheric line. Even for the quiet Sun, Jefferies *et al.* (2006) have observed waves with frequencies much lower than 5.2 mHz propagating upwards into the atmosphere at the boundaries of supergranules. Stebbins and Goode (1987) found that the phase of the oscillation is not constant with height throughout the photosphere as expected for trapped eigenmodes observed in their evanescent tail. Interpreting these results, Goode, Gough, and Kosovichev (1992) have shown that the phase change is consistent with the impulsive nature of the excitation events and the finite propagation speed of the perturbations.

Whenever waves are able to propagate upwards, they are a source of energy flux into the higher atmosphere where they contribute to the heating of the layers they are finally dissipated in. It is still an open question to what degree upwards propagating waves contribute to the heating of the solar corona. Therefore, upwards propagating waves are of interest to understand the physical processes in the solar atmosphere. At the same time, they provide a diagnostic tool for the layers they actually travel through. Finsterle *et al.* (2004) have mapped the magnetic canopy with the help of the MOTH instrument (Finsterle *et al.*, 2004), which measures a velocity signal simultaneously in a chromospheric and a photospheric line.

Magnetograms and dopplergrams of the photosphere are conveniently obtained from filtergraph instruments. Generally, the line is sampled by a small number of equally spaced transmission profiles in different polarization states. These data represent the major source of data in the field of helioseismology, where uninterrupted long term and large scale dopplergrams of moderate spatial resolution are required to study the surface and the subsurface structure of the Sun. Important space instruments using filtergraph techniques are the Michelson Doppler Imager (MDI) described in Scherrer *et al.* (1995), the Narrow Band Filter Imager (NFI, see Shimizu, 2004), and the upcoming Helioseismic and Magnetic Imager onboard the Solar Dynamics Observatory (SDO).

Interpretation of these data requires a proper understanding of the formation height of the absorption lines which are used to detect the Doppler velocity. The formation height is usually provided by forward radiative transfer calculations. The so-called "contribution function" is a function of optical depth and wavelength. It results in the line profile when integrated over optical depth. In the presence of magnetic field, there is a contribution function for each Stokes profile. Generally, the core of the line is formed higher up than its wings, which means that the photons originate from a layer stretching from the formation height of the core down to the formation height of the nearby continuum. Each instrumental observable therefore provides a particularly weighted height average of the targeted atmospheric parameters. The center of gravity of this weight function will be called the "observation height".

In this paper I will outline a formalism to uniquely define the observation height of any filtergraph instrument and apply the method to the MDI instrument.

2. MDI Observables

The MDI instrument provides photospheric dopplergrams and line-of-sight magnetograms for the solar disk (or a central part of it in the so-called high resolution mode) by taking spatially resolved filtergrams in the Ni I 6768 Å magnetic absorption line. The different filtergrams (shown in Figure 1) are realized by tuning the Michelson interferometers[1] to equally spaced target wavelengths. The filtergrams are taken either in circular polarization (measuring both left-hand circularly polarized light (LCP) and right-hand circularly polarized light (RCP)), or in horizontally linearly polarized light (HP), measuring only one linear polarization state. MDI takes five filtergrams by tuning the Michelson Interferometers, using only four filtergrams to derive the dopplergrams and magnetograms. The fifth filtergram is used to derive a proxy for continuum intensity.

Given the filtergrams $F_1 \ldots F_4$, the velocity is derived from the filtergrams using the so-called MDI-algorithm. Calculating

$$\alpha_> = [(F_1 - F_3) + (F_2 - F_4)]/(F_1 - F_3),$$
$$\alpha_< = [(F_1 - F_3) + (F_2 - F_4)]/(F_4 - F_2), \quad (1)$$

with $\alpha_>$ for a positive numerator and $\alpha_<$ for a negative numerator, an on-board lookup table which presents the velocity as a nearly linear function of α is used to calculate the dopplergrams. Magnetograms are derived from the Zeeman splitting of the line and are obtained from the differences of the velocities in left-hand and right-hand circularly polarized light. Circular polarization velocity is the average of both velocities.

3. Instrument Response Function

In this section, I review the derivation of the formalism outlined by Ruiz Cobo and del Toro Iniesta (1994) and explain how it can be applied to filtergraph instruments. To give an example, the formalism will be applied to the MDI instrument.

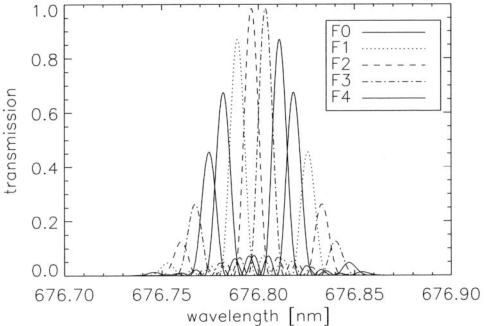

Figure 1 Spectral transmission profiles of MDI. The profiles are the theoretical profiles given by the tuning position of the Michelson interferometers.

---

[1] I am pointing out that there is a substantial uncertainty in the spectral transmission profiles of MDI, because they have not been measured and one therefore has to rely on theoretical values which probably show a substantial deviation from the actual profiles (see Wachter, Schou, and Sankarasubramanian, 2006). For HMI, measuring the transmission profile is part of the calibration process.

The solar spectrum in the different polarization states is given by the Stokes vector:

$$\mathbf{I} = (I, Q, U, V)^{\mathrm{T}}. \qquad (2)$$

$\mathbf{I}$ is a function of the wavelength λ and the optical depth τ (τ will always refer to the continuum optical depth at the reference wavelength 5000 Å throughout this paper). The instrument is located far away from the Sun, so the observed Stokes vector refers to $\tau = 0$. A small change in $\mathbf{I}$ can be written as

$$\delta \mathbf{I}(\tau = 0, \lambda) = \sum_i \int_0^\infty \mathbf{R}_i(\tau', \lambda) \delta x_i(\tau')\, \mathrm{d}\tau', \qquad (3)$$

where $\delta x_i(\tau)$ denotes a perturbation in a particular atmospheric quantity and $\mathbf{R}_i$ is the corresponding spectral response function. The sum runs over all perturbed parameters in the solar atmosphere.

Measuring the filtergrams $\mathbf{F}$ corresponds to applying a linear vector operator $\mathbf{\Psi}$ (operating in wavelength space) to the Stokes vector $\mathbf{I}$:

$$\mathbf{F} = \mathbf{\Psi}\bigl(\mathbf{I}(\tau = 0, \lambda)\bigr). \qquad (4)$$

Because $\mathbf{\Psi}$ is linear, a small perturbation in $\mathbf{F}$ is given by

$$\delta \mathbf{F} = \sum_i \int_0^\infty \mathrm{d}\tau\, \mathbf{\Psi}\bigl(\mathbf{R}_i(\tau, \lambda)\bigr) \delta x_i(\tau). \qquad (5)$$

In the case of MDI magnetograms or dopplergrams measured in circular polarization, $\mathbf{F}$ is the eight-component vector $(F_1^{\mathrm{LCP}}, \ldots, F_4^{\mathrm{LCP}}, F_1^{\mathrm{RCP}}, \ldots, F_4^{\mathrm{RCP}})$. For dopplergrams measured in linear polarization, $\mathbf{F}$ is the four-component vector $(F_1^{\mathrm{HP}}, \ldots, F_4^{\mathrm{HP}})$. $\mathbf{\Psi}$ is the wavelength integration of the product of transmission profiles and spectral line.

Any MDI observable o is a (nonlinear) function $o = \Omega(\mathbf{F})$ of the set of filtergrams. For MDI dopplergrams and magnetograms, Ω is given by Equation (1) and the on-board lookup table.

Linearizing Ω, a small perturbation in o is given by

$$\delta o = \sum_{i=1}^m \int_0^\infty \mathrm{d}\tau\, \nabla_{\mathbf{F}}(\Omega) \cdot \mathbf{\Psi}\bigl(\mathbf{R}_i(\tau, \lambda)\bigr) \delta x_i(\tau). \qquad (6)$$

Comparing Equations (6) and (3), the instrumental response function $R(\tau)$ can be defined as

$$R(\tau) = \nabla_{\mathbf{F}}(\Omega) \cdot \mathbf{\Psi}\bigl(\mathbf{R}_i(\tau, \lambda)\bigr). \qquad (7)$$

Whereas the spectral response function gives the response of the spectrum to a small perturbation in a physical quantity, the instrumental response function gives the response of a quantity *derived from the spectrum* (*i.e.*, the observable) to a small atmospheric perturbation. The instrumental observation height can then be defined as either the location of the peak or the first moment of the absolute value of the instrument response function.

Equation (7) states that the instrumental response function is given by the transmission profiles applied to the spectral response function and the partial derivatives of the observable algorithm with respect to the filtergrams. The same instrumental response function could be obtained by calculating the spectral line for a perturbed atmosphere, followed by evaluating the instrumental response to the perturbed atmosphere. Expression (7) however allows to obtain the instrumental response function from the spectral response function directly without the intermediate step of generating the spectrum. Ruiz Cobo and del Toro Iniesta (1994) point out that the direct calculation of the response function is much more effective.

It is important to note that Ω does not need to be differentiable for all values of **F**, but only for the values that can actually occur in a measurement. The different components of **F** are not independent and constrained by the transmission profile and the possible line shapes.

4. The Spectral Response Function

The instrumental response function is uniquely determined provided we know the local solar atmosphere, the way the instrument obtains the observable, and the spacecraft velocity.

The properties of the solar atmosphere are reflected in the spectral response functions $\mathbf{R}_i$. I used the radiative transfer code STOPRO (Solanki, 1987) which calculates spectral response functions and the Stokes profiles in local thermodynamic equilibrium in a plane parallel stratified magnetic atmosphere. Despite the fact that the sunspot model is naturally horizontally inhomogeneous, for simplicity we neglect the horizontal radiative transfer by performing one-dimensional radiative transfer calculations at each horizontal location.

The spacecraft velocity plays a role because the instrument is sensitive to the relative velocity of spacecraft and the observed solar atmosphere. Basically, different parts of the line are observed depending on the direction and magnitude of the spacecraft velocity. An identical effect has the long term drift of the absolute wavelength of the transmission profile. Whereas the relative positions of the transmission profiles are relatively stable, their absolute positions drift substantially with time and have to be regularly adjusted. Adding an offset velocity to the atmosphere or shifting the central wavelength of the filters is therefore able to capture both phenomena.

5. Sunspot Model

To obtain the atmosphere, I am constructing a sunspot model following the similarity assumption of Schlüter and Temesváry (1958). This assumption states that the sunspot configuration is cylindrically symmetric and the distribution of the vertical magnetic field geometrically similar in each horizontal plane. The distribution of the magnetic field is given by a function, which is normally chosen to be Gaussian. This geometrical constraint allows to determine the stratification of the entire sunspot given the stratification at the center of the spot and far away from the spot in the quiet Sun.

For the center and outer stratification, we use the semi-empirical models Maltby-SP (quiet Sun) and Maltby-M (umbra) presented in Maltby et al. (1986). The models are semi-empirical in the sense that they have been constructed to reproduce the observed intensity ratio of umbra and quiet Sun in a wide wavelength range, as well as its center-to-limb variation. Both models give the stratification with respect to the local $\tau = 1$ level. To determine their relative geometric displacement we take advantage of the observation by Maltby (1977) that the density is close to horizontally homogeneous at the height of quiet Sun optical depth unity. This is equivalent to setting the Wilson depression to roughly 550 km. A minor downward extrapolation has been accomplished by smoothly matching the models to a polytrope. The magnetic field is vertical at the bottom of the atmospheric model and spreads out by a rate that is controlled by the parameter representing the surface magnetic flux in the computational domain.

As in Schlüter and Temesváry (1958), the temperature follows the same shape function as the magnetic field, providing a smooth transition between the umbral and quiet Sun temperatures given by the sunspot models. Doing so, I avoid the complications of specifying a meaningful energy equation in the sunspot. The resulting model is therefore in mechanical, but not in energetic equilibrium. I also point out that the resulting stratification generally does not fulfill a solar equation of state between the core and the quiet Sun. To ensure consistency with the one dimensional radiative transfer calculations, I checked that the radiative flux as given in the diffusion approximation (*i.e.*, antiparallel to the temperature gradient) is predominantly upward below the local photospheric level. The model represents a large sunspot with a radius of roughly 30 Mm.

All these simplification can substantially influence the line formation and therefore the results in the next section. The stratifications in the spot center and the quiet Sun however are exactly given by a well established solar model, and atmospheric stratification in between is important to asses the effect of a horizontal magnetic field component.

Figure 2 gives the stratification as a function of radius and height. Height zero is defined as optical depth unity in the quiet Sun. This stratification is uniquely determined by the model of Maltby *et al.* (1986), the choice of a Gaussian shape function, the location of the lower boundary where the magnetic field is vertical, the spot radius, and the surface magnetic flux in the domain of computation.

6. Results

The instrumental response function has been calculated for each of the 150 equally spaced radial positions 67 Mm outwards from the sunspot center, which is well in the quiet Sun. We calculated the response of MDI linear polarization and MDI circular polarization dopplergrams to vertical velocity, and the response of MDI magnetograms to magnetic field. The response function for magnetic field and circular velocity are next to identical (difference below 1 km), which is not surprising as they are both measured from the same polarization states.

Figure 3 shows the response function for the sunspot center and the quiet Sun for linear polarization velocity as a function of height with respect to the $\tau = 1$ level. The observation height is determined as the first moment of the absolute value of $R(z) = R(\tau(z))$:

$$h = \frac{\int dz |R(z)| z}{\int dz |R(z)|}. \qquad (8)$$

Using the above definition of the observation height, we get a weighted average of the layers that contribute to the observable. The weight for each layer is given by the value of the amplitude of the observable's instrumental response function.

The left panel of Figure 4 shows the observation height h as a function of distance from the spot center. The variation across the spot is significant reflecting both the Wilson depression and the variation in line formation height. Circular polarization dopplergrams form slightly higher than linear polarization dopplergrams with a maximum deviation of 23 km.

The right panel of Figure 4 shows the same variation with respect to the local $\tau = 1$ level. Between the umbral value of 102 km and the quiet Sun value of 175 km, there are substantial variations leading to observation heights as much as 350 km above the local $\tau = 1$ level. I note that the line formation height values quoted in the literature (*e.g.*, Norton *et al.*, 2006 give a formation height of 291 km for the Maltby-M model) usually refer to the

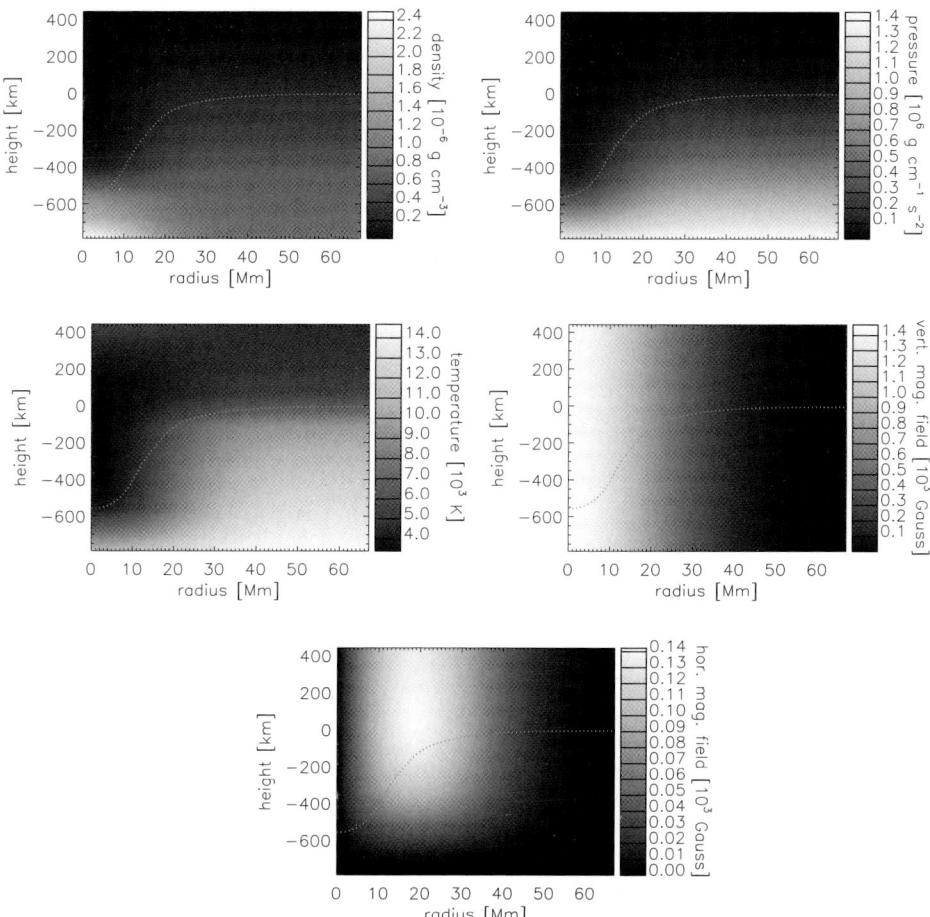

Figure 2 The panels show the stratification of the solar atmosphere as a function of height and sunspot radius. Height zero corresponds to optical depth unity at wavelength 5000 Å in the quiet Sun. Starting from the upper left corner, the panels show the density, the gas pressure, the temperature, the vertical and the horizontal magnetic field. The dotted line shows the location of $\tau = 1$.

formation height of the line center, which is an upper limit for the instrumental observation height this paper tries to determine. As the wings of the line are formed between the core observation height and the photospheric level, the spectral average implied in the formalism given above results in a value between the core formation height and the photospheric level.

Switching off the magnetic field in the radiative transfer calculations does not alter the picture outlined by Figure 4 substantially. This means that the major contribution to the change in observation height comes from the altered thermal stratification. The difference between linear and circular polarization measurements are by definition of magnetic origin, as the Stokes profiles measured in the two different ways are identical in the absence of a magnetic field.

Figure 5 shows the effect of an offset velocity on the observation height. I calculated the observation height for the spot center and the quiet Sun in circular and linear polarization. The observation height shows a somehow oscillatory behavior with a period that corresponds

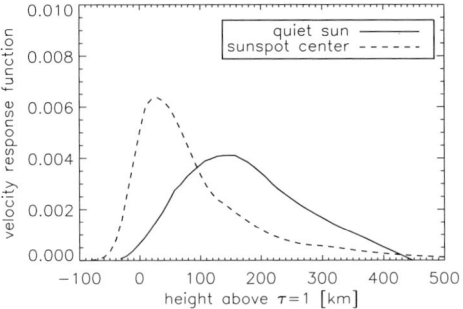

Figure 3 The instrumental response function for linear polarization dopplergrams given as a function of height, measured form the respective $\tau = 1$ level. The two curves represent the sunspot center and the quiet Sun.

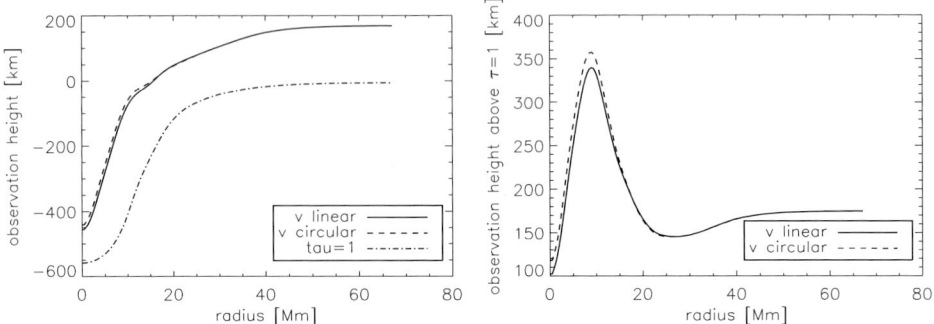

Figure 4 Left panel: Observation height h as a function of horizontal location in the model sunspot (height is measured positive outwards). The lower line represents the height of the local $\tau = 1$ level (Wilson depression). Right panel: Observation height h measured from the local $\tau = 1$ level.

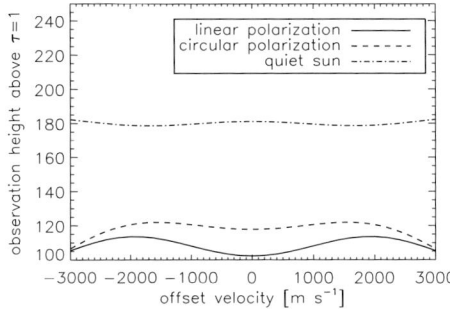

Figure 5 Observation height for the quiet Sun and the sunspot center for linear and circular polarization measurements. For the quiet Sun, linear and circular polarization are identical. The observation height is measured upwards from the local $\tau = 1$ level.

to the Doppler shift equivalent to the transmission profile spacing of 75 mÅ. As the filters shown in Figure 1 are off-centered from the line, the filter on the blue side of the spectrum is moved from the near continuum into the wing, whereas on the red side, the filter is moved from the wing to the near continuum. As these two effects basically compensate each other, there is only a remaining small effect (up to 20 km at the spot center) due to the discrete nature of the transmission profiles.

7. Discussion

Future refinement of the helioseismology tools used to investigate the subsurface structure of solar active regions will include the effect of waves that are not observed in their evanescent tail, but propagate upwards. With a dispersion relation of

$$\omega^2 = c^2 k^2 + \omega_{ac}^2 \qquad (9)$$

(where ω is the mode angular frequency, k is the wave number, c is the sound speed, and ω_{ac} is the acoustic cutoff frequency) for vertically traveling waves (see D'Silva and Duvall, 1995), we obtain a phase travel time difference Δt_p of

$$\Delta t_p = \frac{\Delta z \sqrt{\omega^2 - \omega_{ac}^2}}{\omega c} \qquad (10)$$

for a observation height difference Δz. A deviation of $\Delta z = 113$ km (equal to the difference between the value quoted by Norton *et al.* (2006) and the value found in this paper for the quiet Sun) would result in an error of the phase travel time of about 11 seconds at a frequency 2 mHz above the cutoff frequency (assuming a sound speed of 7 km s^{-1} and an acoustic cutoff frequency of 5.2 mHz as typical values for the quiet Sun).

Applying the formalism above to a state-of-the-art sunspot model will give a more meaningful measure of the variation in observation height than the oversimplified sunspot model I used to demonstrate the observation height method. Still, as real sunspots are complex and appear in a large variety, the model calculations cannot be directly applied to data analysis in helioseismology.

Information about the local atmospheric stratification of an active region – including the velocity stratification – only becomes available if the absorption line is measured with sufficient spectral resolution to properly resolve the line. Using bisector analysis on the line profiles, the phase of an upward traveling wave can be determined at different heights, and observables associated with a well defined observation height with respect to the local $\tau = 1$ level can be obtained. As the geometry used in local helioseismology generally assumes that waves are emitted, reflected, and received at the same geometrical height, finding a data set that actually fulfills this assumption would improve the consistency between the theoretical model and the observed data. Whenever data analysis techniques used in helioseismology actually include the phases of propagating waves, they would greatly profit from the availability of spectrally resolved data.

This paper presents a straightforward approach to give a meaningful number to the observation height associated with different filtergraph instruments for each particular observable. This approach can be particularly useful if signals from different instruments or different measurement modes are combined in the data analysis. Applying the approach to a sunspot model gives a rough estimate of the variation of formation height with location in the sunspot. Effects from the Wilson depression (which refers to the spectral continuum) must be carefully distinguished from line formation effects. The influence of different modes of measurement on the observation height has been investigated as well as the influence of offset velocities due to spacecraft motion and filter drifts. They have been found to be of minor importance in the case of MDI, but are supposedly significant for instruments orbiting the Earth like HMI or Hinode.

Acknowledgements I thank my colleagues S. Couvidat, S.P. Rajaguru, and S.M. Hanasonge for useful discussions, and J. Schou for valuable comments on the manuscript. I thank S. Solanki and A. Lagg for providing our group with the radiative transfer code STOPRO used to calculate the spectral response function of the absorption line. This research is supported by the NASA grants NAS5-02139 and NNX07AK36G, and the Lockheed Martin grant 810000791.

References

D'Silva, S., Duvall, T.L. Jr.: 1995, Time-distance helioseismology in the vicinity of sunspots. *Astrophys. J.* **438**, 454–462. doi:10.1086/175089.

Finsterle, W., Jefferies, S.M., Cacciani, A., Rapex, P., Giebink, C., Knox, A., Dimartino, V.: 2004, Seismology of the solar atmosphere. *Solar Phys.* **220**, 317–331. doi:10.1023/B:SOLA.0000031397.73790.7b.

Finsterle, W., Jefferies, S.M., Cacciani, A., Rapex, P., McIntosh, S.W.: 2004, Helioseismic mapping of the magnetic canopy in the solar chromosphere. *Astrophys. J. Lett.* **613**, L185–L188. doi:10.1086/424996.

Goode, P.R., Gough, D., Kosovichev, A.G.: 1992, Localized excitation of solar oscillations. *Astrophys. J.* **387**, 707–711. doi:10.1086/171118.

Jefferies, S.M., McIntosh, S.W., Armstrong, J.D., Bogdan, T.J., Cacciani, A., Fleck, B.: 2006, Magnetoacoustic portals and the basal heating of the solar chromosphere. *Astrophys. J. Lett.* **648**, L151–L155. doi:10.1086/508165.

Maltby, P.: 1977, On the difference in darkness between sunspots. *Solar Phys.* **55**, 335–346.

Maltby, P., Avrett, E.H., Carlsson, M., Kjeldseth-Moe, O., Kurucz, R.L., Loeser, R.: 1986, A new sunspot umbral model and its variation with the solar cycle. *Astrophys. J.* **306**, 284–303. doi:10.1086/164342.

McIntosh, S.W., Jefferies, S.M.: 2006, Observing the modification of the acoustic cutoff frequency by field inclination angle. *Astrophys. J.* **647**, L77–L81. doi:10.1086/507425.

Norton, A.A., Graham, J.P., Ulrich, R.K., Schou, J., Tomczyk, S., Liu, Y., Lites, B.W., Ariste, A.L., Bush, R.I., Socas-Navarro, H., Scherrer, P.H.: 2006, Spectral line selection for HMI: A comparison of Fe I 6173 Å and Ni I 6768 Å. *Solar Phys.* **239**, 69–91. doi:10.1007/s11207-006-0279-y.

Rajaguru, S.P., Sankarasubramanian, K., Wachter, R., Scherrer, P.H.: 2007, Radiative transfer effects on Doppler measurements as sources of surface effects in sunspot seismology. *Astrophys. J.* **654**, L175–L178. doi:10.1086/511266.

Ruiz Cobo, B., del Toro Iniesta, J.C.: 1994, On the sensitivity of Stokes profiles to physical quantities. *Astron. Astrophys.* **283**, 129–143.

Scherrer, P.H., Bogart, R.S., Bush, R.I., Hoeksema, J.T., Kosovichev, A.G., Schou, J., Rosenberg, W., Springer, L., Tarbell, T.D., Title, A., Wolfson, C.J., Zayer, I., MDI Engineering Team: 1995, The solar oscillations investigation – Michelson Doppler imager. *Solar Phys.* **162**, 129–188.

Schlüter A., Temesváry S.: 1958, The internal constitution of sunspots. In: Lehnert, B. (ed.) *Electromagnetic Phenomena in Cosmical Physics, IAU Symp.* **6**, 263.

Shimizu T.: 2004, SolarB solar optical telescope (SOT). In: Sakurai, T., Sekii, T. (eds.) *The Solar-B Mission and the Forefront of Solar Physics* **CS-325**, Astron. Soc. Pac., San Francisco, 3.

Solanki S.K.: 1987, The photospheric layers of solar magnetic flux tubes. PhD thesis No. 8309, ETH, Zürich.

Stebbins, R., Goode, P.R.: 1987, Waves in the solar photosphere. *Solar Phys.* **110**, 237–253.

Wachter, R., Schou, J., Sankarasubramanian, K.: 2006, Line shape changes and Doppler measurements in solar active regions. I. A method for correcting dopplergrams from SOHO MDI. *Astrophys. J.* **648**, 1256–1267. doi:10.1086/505930.

Seismology of a Sunspot Atmosphere

Y.D. Zhugzhda

Originally published in the journal Solar Physics, Volume 251, Nos 1–2, 501–514.
DOI: 10.1007/s11207-008-9251-3 © Government of Russian Federation 2008

Abstract Two competing theories of sunspot oscillations are discussed. It is pointed out that the normal mode (eigenoscillations) theory is in contradiction with a number of observations. The reasons for this are discussed. The revised filter theory of the three-minute sunspot oscillations is outlined. It is shown that the reason for the occurrence of the multi-passband filter for the slow waves is the interference that appears from the multilayer structure of the sunspot atmosphere. In contrast with Zhugzhda and Locans (*Sov. Astron. Lett.* **7**, 25–27, 1981) it is shown that along with the Fabry–Perot chromospheric passband the cutoff frequency passband and a number of high-frequency passbands occur. The effect of the nonlinearity of the sunspot oscillations in the upper chromosphere and the transition region is taken into account. The spectra of the distinct empirical models of the sunspot atmosphere are explored. An example of the interpretation of the sunspot oscillations based on the revised filter theory is presented. Only the filter theory can explain the complicated behavior of the oscillations across the sunspot. The observations provide evidence of the nonuniformity of the sunspot atmosphere.

Keywords Sun: sunspots · Sun: oscillations

A long time has elapsed since the discovery of the three-minute oscillations. Numerous studies of the oscillations have been performed. The story of the discovery and exploration of sunspot oscillations can be found in the review of Bogdan and Judge (2006) devoted to the observational aspects of the sunspot oscillations. We know a lot about the three-minute

Helioseismology, Asteroseismology, and MHD Connections
Guest Editors: Laurent Gizon and Paul Cally.

Y.D. Zhugzhda (✉)
Institute for Terrestrial Magnetism, Ionosphere and Radiowave Propagation of Russian Academy of Sciences, Troitsk, Moscow Region, 142190, Russia
e-mail: yzhugzhda@mail.ru

Y.D. Zhugzhda
Kiepenheuer-Institut für Sonnenphysik, Schöneckstr. 6, 79104 Freiburg, Germany

oscillations in the photosphere, chromosphere, and corona of sunspots. But so far no universally accepted theory of the three-minute oscillations exists. There are two competing theories: the filter theory and the theory of eigenoscillations of sunspots. The general theory of the eigenmodes of the atmosphere permeated by a uniform, vertical, magnetic field was developed by Cally and Bogdan (1993) and Bogdan and Cally (1997). This theory is believed to form a foundation for the theory of sunspot oscillations. But there are some contradictions between the concept of sunspot eigenoscillations and observations. It is found by many observers that the three-minute oscillations are not standing waves, as would be the case for normal modes. They are waves that are running from the sunspot photosphere through the chromosphere to corona. Besides, the local seismology of sunspots shows that the waves are going freely into and through the sunspot (Zhao and Kosovichev, 2006). This is again in contradiction with the concept of normal modes since the incident waves have to be captured by the sunspot cavity and re-emitted back after some time defined by the quality of the cavity. Thus, even in the case of a low-quality cavity, the delay of the reradiation of the captured waves has to be about a few periods of the oscillations. Nothing similar to this has been observed. Thus, there is no evidence that the sunspot is a cavity for the three-minute oscillations. This does not mean that the theory of normal modes of the atmosphere permeated by a magnetic field is mistaken. The problem is with the boundary conditions used for the treatment of normal modes. For example, Bogdan and Cally (1997) fixed the horizontal wavenumber. In fact it does mean that the condition of the complete reflection of the waves from the side walls of the sunspot magnetic tube was imposed. But the assumption of complete reflection does not work in subphotospheric layers since the jump of the plasma parameters across the boundary of the sunspot is not large enough to make a strong reflection. There is some concern about the boundary conditions at the bottom of the sunspot since the deep structure of the sunspot is still under discussion. Thus, if there is no trapping of the waves by the sunspot, the only way is to look for the propagation of the waves through the sunspot atmosphere. Zhugzhda and Locans (1981) were the first to explore wave propagation as an alternative to trapping of waves. They assumed that the three-minute oscillations are slow MHD waves, which was confirmed later by many observations. They revealed that the sunspot chromosphere can work as a Fabry–Perot filter for the slow waves. That is why we call this approach the "filter theory." Later on many calculations of slow-wave propagation through the sunspot atmosphere were carried out with modeling of the sunspot atmosphere (Settele, Staude, and Zhugzhda, 2001, and references there). In fact, it was revealed that there are numerous passbands for the slow waves (*i.e.*, the sunspot atmosphere works as a multiband filter for the slow waves). However, interpreting the wave filtering by the sunspot on the basis of the Fabry–Perot effect faces some problems. For example, the calculations and observations sometimes show close spectral peaks whereas the Fabry–Perot filters produce equidistant passbands. Furthermore, B.W. Lites and T.J. Bogdan (private communications) have raised the issue of the effect of the cutoff frequency, which can produce the spectral peak. Zhugzhda (2007) showed that the Fabry–Perot effect is not the only effect that affects slow-wave propagation in the sunspot atmosphere. The current paper is devoted to the complete revision of the filter theory. Another crucial point under discussion in this paper is whether the linear theory of wave propagation can be applied to the treatment of the three-minute oscillations since it was revealed that three-minute oscillations become nonlinear in the upper chromosphere and transition region (see, for example, Centeno, Collados, and Bueno, 2006). Finally, the interpretation of the observations of the three-minute oscillations in the framework of the filter theory is presented.

1. Basics of Slow-Wave Propagation

The propagation of MHD waves in a conducting atmosphere permeated by a vertical uniform magnetic field is governed by the following set of the equations (Cally and Bogdan, 1993):

$$\left[c_A^2 \frac{d^2}{dz^2} + \omega^2 - k_\perp^2 \left(c_S^2 + c_A^2\right)\right] v_\perp = -ik_\perp \left(c_S^2 \frac{d}{dz} - g_{\text{eff}}\right) v_\parallel, \tag{1}$$

$$\left[c_S^2 \frac{d^2}{dz^2} - \left(\gamma g_{\text{eff}} + c_S^2 \frac{d \ln \gamma}{dz}\right) + \omega^2 - c_S^2 \frac{dg_{\text{eff}}}{dz}\right] v_\parallel$$
$$= -ik_\perp \left(c_S^2 \frac{d}{dz} - g_{\text{eff}}(\gamma - 1) - c_S^2 \frac{d \ln \gamma}{dz}\right) v_\perp, \tag{2}$$

$$g_{\text{eff}} = g - \frac{1}{\rho} \frac{dp_{\text{turb}}}{dz}, \tag{3}$$

where c_A and c_S are the Alfvén and sound speeds, respectively, $v_\parallel$ and $v_\perp$ are the vertical and horizontal components, respectively, of the plasma velocity, p_{turb} is the turbulent pressure, and g_{eff} is the effective gravity taking into account the effect of turbulent pressure on the hydrostatic equilibrium of the atmosphere. The effect of turbulent pressure is essential in the chromosphere. The contribution from the turbulent pressure in maintaining hydrostatic equilibrium must be taken into account, because, otherwise, the energy-flux conservation law for waves is violated (Mihalas and Toomre, 1981). There is no surprise that this effect was not taken into account by the theory of normal modes since it is easy to miss the violation of energy conservation in this case. Another issue is with the treatment of the wave propagation. The effect of turbulent pressure was taken into account in all our explorations of the slow-wave propagation in the sunspot atmosphere.

Two coupled Equations (1) and (2) describe the propagation and coupling of the fast and slow waves. The right-hand parts of the equations define the wave coupling. In accordance with the general theory of MHD wave coupling (Zhugzhda, 1979; Zhugzhda and Dzhalilov, 1982) the coupling of the waves is negligible in the low-β plasma of the solar chromosphere. Moreover, the fast waves of three-minute period are evanescent in the chromosphere, temperature minimum, and upper photosphere of the sunspots whereas the slow waves are running waves. Thus, there is no question that the three-minute oscillations are slow waves since three-minute oscillations are running waves. The coupling between running and evanescent waves is very weak.

In the case of weak coupling of slow and fast waves the right-hand sides of Equations (1) and (2) can be dropped. The propagation of the slow waves in this case is governed by the approximate equation, which was obtained for the first time by Syrovatskii and Zhugzhda (1968):

$$\left[c_S^2 \frac{d^2}{dz^2} - \left(\gamma g_{\text{eff}} + c_S^2 \frac{d \ln \gamma}{dz}\right) + \omega^2 - c_S^2 \frac{dg_{\text{eff}}}{dz}\right] v_\parallel = 0, \tag{4}$$

$$g_{\text{eff}} = g - \frac{1}{\rho} \frac{dp_{\text{turb}}}{dz}. \tag{5}$$

Only the effective gravity was included later. The neglect of wave coupling is possible for the photosphere in the case when the horizontal wavelength of the slow waves is about the sunspot diameter. We specifically rewrite this equation in the final form because there is serious confusion in connection with it. Equation (4) looks identical to the equation for vertically

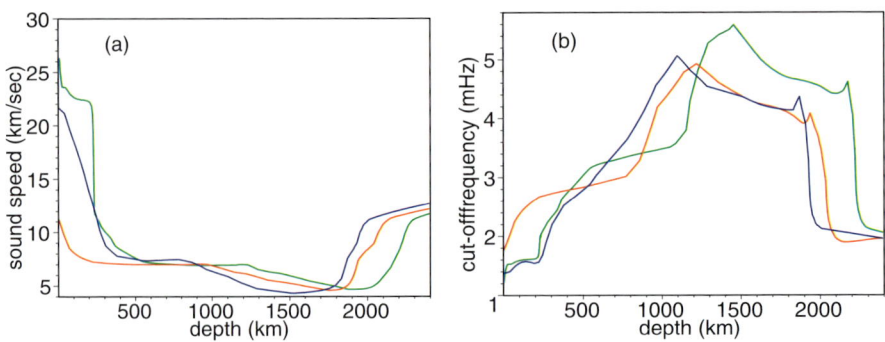

Figure 1 The sound speed (a) and cutoff frequency (b) as a function of depth for the empirical models Lites (green curves), Maltby (red curves), and Staude (blue curves).

propagating plane acoustic waves since there is no dependence on the horizontal wavenumber ($k_\perp$). This gives rise to an unfortunate mistake that is commonly encountered in the literature when no distinction is made between acoustic and slow magneto-acoustic-gravity waves. Sometimes the slow waves are called acoustic waves (see, for example, Brynildsen *et al.*, 1999; Brekke *et al.*, 2003). This is not correct. In fact, the slow waves propagate independently along each magnetic field line. Thus, Equation (4) can be applied separately to the treatment of the propagation of the slow waves along the separate field lines of the sunspot magnetic field if ones replaces the simplified model of the uniform vertical magnetic field by the diverging field or even by the real magnetic field of the individual sunspots. However, it is reasonable to mention that if the sunspot atmosphere is nonuniform (*i.e.*, the temperature and pressure of the plasma of the adjacent field lines are different), the slow waves propagate with a velocity less than the sound speed. If the sunspot atmosphere consist of thin flux tubes, the slow waves propagate with a tube speed $c_A c_S / \sqrt{c_A^2 + c_S^2}$, which is less than the sound speed (Zhugzhda and Goossens, 2001). The distinction between tube and sound speeds is negligible in the chromosphere but can be essential in the photospheric and subphotospheric layers of the sunspot.

One additional shortcoming of the filter theory, which has yet to be overcome, is the neglect of nonadiabatic effects. Taking these effects into account leads to a change of the phase velocity and the appearance of wave absorption.

2. Peculiarities of Sunspot Atmosphere

There are few empirical models of sunspot atmosphere. The different models show marked distinctions. Figures 1(a) and 1(b) show the sound speed and the cutoff frequency for the three well-known models of sunspot atmosphere, namely, the Staude (1981), Maltby *et al.* (1986), and Lites and Skumanich (1982) models. Differences among the empirical models based on observations can appear for different reasons, namely, differences among the individual sunspots, the inaccuracy of the observations, and the shortcomings of the modeling procedure. Seismology of the sunspot atmosphere based on observations of the three-minute oscillations can help elucidate the reason for the distinctions among the empirical models of the sunspot atmosphere. In connection with slow-wave propagation, two main features of the atmosphere are important: The first is the occurrence of the cutoff frequency, which is responsible for the low-frequency limit of the wave spectrum. The second is the nonmonotonic dependence of the temperature on altitude in the sunspot atmosphere, which causes the interference of waves. Only in the case of a nonmonotonic temperature profile does the transmission function of the slow waves display passbands. The waves running through the atmosphere with monotonic temperature gradient undergo the same reflection at all levels. In the case of a nonmonotonic temperature profile, the largest reflection of slow waves occurs at the levels of the maximum temperature gradient. This is a crucial point for the filter theory since the interference of the waves reflected from these levels can increase or decrease the transmission of the waves. The temperature profile for the different empirical models shown on Figure 1 is nonmonotonic. It has to be the enhancement of the reflection at the boundaries between the chromosphere, temperature minimum, photosphere, and subphotospheric layers. The wavelength of three-minute oscillations in the sunspot atmosphere is in the range $(1-2) \times 10^3$ km and the thickness of the transitions layers between the temperature minimum and temperature plateau does not exceed a few hundred kilometers. Thus, rather complicated interference of the slow waves occurs in the sunspot atmosphere. The Fabry–Perot effect predicted by Zhugzhda and Locans (1981) is only one among others. This is the reason why the interpretation of slow-wave propagation through the sunspot atmosphere (see Settele, Staude, and Zhugzhda, 2001, and references there) has to be revised. The use of empirical models of the sunspot atmosphere for the exploration of the nature of the distinct interference effect meets with difficulties since the parameters of the sunspot atmosphere, namely, the thickness and temperature of the different layers of the atmosphere, need to be varied. Thus, we arrive at the idea of exploring first a rather crude four-layer model of the sunspot atmosphere to understand the basics of the interference effect in a multilayer atmosphere with a temperature minimum.

3. Four-Layer Model of the Sunspot Atmosphere

To explore the basics of the slow-wave propagation the simple four-layer model of the sunspot atmosphere shown in Figure 1(a) has been used. The four isothermal layers correspond to the corona at temperature T_4, the chromosphere at temperature T_2, the temperature minimum at temperature T_1, and the photosphere at temperature T_3. The dimensionless thickness $(D = d/H_1)$ is measured in units of the pressure scale height H_1 of the temperature minimum. Two scenarios of wave propagation are explored, namely, for frequencies below and above the cutoff frequency of the temperature minimum The discontinuities between the isothermal layers produce a strong reflection in comparison with the real sunspot atmosphere. But this simple model helps us to explore the basic effects.

3.1. Fabry–Perot Chromospheric Filter

If the frequency of the slow waves is less than the cutoff frequency of the temperature minimum, the waves undergo strong reflection from the temperature minimum, and only a small fraction penetrates to the chromosphere by a tunnel effect (see the bottom scheme of arrows in Figure 1(a)). After partial reflection from the corona the waves penetrate back through the temperature minimum to the photosphere, where, as shown in Figure 2(a), they interfere with the waves reflected from the temperature minimum. In the case of destructive interference of the waves the reflection from the temperature minimum decreases and the transmission of waves to the solar corona increases at the frequency of the chromospheric resonance. Thus, the four-layer atmosphere works as a Fabry–Perot chromospheric filter for the slow waves.

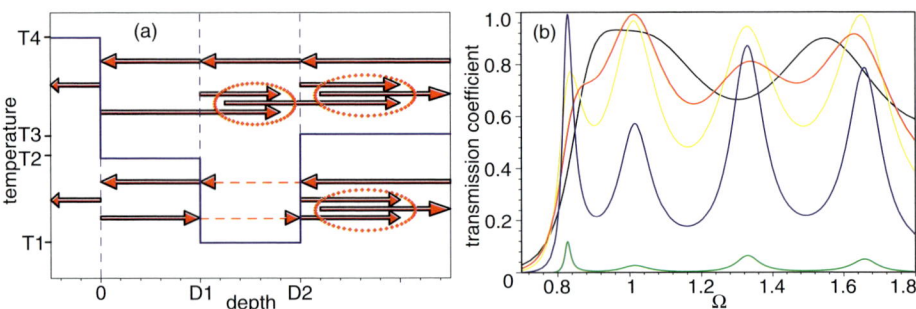

Figure 2 (a) The four-layer model of the sunspot atmosphere. The temperature profile as a function of the dimensionless depth is shown by the solid blue curve. Shown are the two scenarios of wave propagation, namely, for frequencies below (bottom) and above (top) the cutoff frequency of the temperature minimum. (b) The transmission coefficient for the distinct values of the amplitude reduction of the waves approaching the temperature minimum from the upper layers of the sunspot atmosphere $R = 1.0$ (green), 0.87 (blue), 0.55 (yellow), 0.29 (red), and 0.09 (black).

The frequency of the Fabry–Perot passband equals the frequency of the chromospheric resonance. But there is a crucial difference from classical Fabry–Perot filters. Unlike classical Fabry–Perot filters, four-layer atmospheres can work only as a single-band filter, whereas classical Fabry–Perot filters have a number of the equidistant passbands. The reason for the distinction is the strong frequency dependence of tunneling of the waves through the temperature minimum. Almost total transmission of slow waves through the sunspot chromosphere occurs when the reflection of the slow waves from the temperature minimum and the solar corona are equal each other, and the reflected waves cancel each other. But this condition is met only for a single frequency (*i.e.*, for the special choice of the thickness and temperature of the chromosphere). If the frequency of the chromospheric resonance decreases below this optimal frequency the transmission of waves decreases. However, the frequency of the Fabry–Perot passband cannot exceed the cutoff frequency of the temperature minimum owing to a decrease of the reflection from the temperature minimum. Thus, there is some range of frequencies where the chromospheric Fabry–Perot filter is effective. This range is so small that as a rule only one possible passband of the Fabry–Perot filter falls in the range. Moreover, if this solitary passband is near the edge of this range, the transmission of the passband can be so small that it does not produce a noticeable peak in the spectrum of the sunspot oscillations. Thus, we arrive at the conclusions that *i*) the chromospheric Fabry–Perot filter can be responsible for only one peak in the spectrum of three-minute oscillations, *ii*) the frequency of this peak has to be less than the cutoff frequency of the temperature minimum, and *iii*) the Fabry–Perot filter does not produce an observable peak in the spectrum of the oscillations in some cases, when the frequency of the chromospheric resonance does not fall within the prescribed range of frequencies. Therefore, the interpretation of the entire spectrum of three-minute oscillations by the Fabry–Perot filter effect is not possible.

The green curve in Figure 2(b) is the transmission coefficient for the slow waves running through the four-layer atmosphere. The transmission coefficient equals the ratio of the transmitted to incident energy fluxes of the slow waves. The calculations are performed for the following choice of sound-speed ratios: $c_{01}/c_{02} = 0.6$, $c_{03}/c_{02} = 2$, and $c_{04}/c_{02} = 14$, which roughly correspond to the ratios of the sound speeds of the corona, chromosphere, temperature minimum, and photosphere of the sunspots. The green curve is calculated for $D_1 = 2.5$, $D_2 = 12$. The transmission coefficient is a function of the dimensionless frequency $\Omega = \omega H_2/c_{02}$. The transmission coefficient is rather small because the reflection from the boundaries between layers is rather strong. The transmission band of the lowest frequency in the plot of Figure 2(b) is the Fabry–Perot passband.

3.2. Cutoff Passband and High-Frequency Passbands

The next passband appears at the cutoff frequency of the temperature minimum. A first naive glance would indicate that all of the waves of frequencies above the cutoff frequency of the temperature minimum propagate freely through the temperature minimum. But this is not the case since the waves undergo reflection from the boundaries of the temperature minimum. This effect is known in quantum mechanics as over-barrier reflection. The transmission function shown by the black curve in Figure 2(b) is calculated for the case of the three-layer atmosphere, when the corona is absent. The transmission of the slow waves for frequencies above the cutoff frequency is not constant. The interference bands appear from the interference of the waves reflected from the two boundaries of the temperature minimum. The frequency interval between them is defined by the travel time of waves through the temperature minimum. In the case of the sunspot atmosphere, the process is more complicated because of reflection from the corona. The scenario of wave propagation for frequencies above the cutoff frequency of the temperature minimum is shown in Figure 2(a) by the upper scheme of arrows. In this case, interference occurs in the temperature minimum and the photosphere. Numerous transmission bands for frequencies above the cutoff frequency appear because of this interference (Figure 2(b)). The mechanism of the origin of these passbands is similar to the antireflection effect for a multilayer coating, well-known in the optics and acoustics (Brekhovskikh and Godin, 1990). The sunspot atmosphere consists of a few layers, namely, the chromospheric temperature plateau, the temperature minimum, and, possibly, the photosphere. The sunspot atmosphere differs significantly from the optical multilayer coating in that it does not consist of exact quarter-wavelength layers for special values of the refraction index. Besides, the boundaries between the layers are smoothed out. As a consequence, the passbands are not equidistant and their transmission is not equal.

It should be emphasized that the frequency of the cutoff passband is not exactly equal to the cutoff frequency even in the simplest case of the ideal three-layer atmosphere when there is no reflection from the corona (Zhugzhda, 2007). But the frequency of the cutoff passband is subject to the adjacent interference passbands, namely, the Fabry–Perot passband and the first high-frequency passband. When they are located near the cutoff passband it is shifted to high or low frequencies. Thus, the frequency of the cutoff band coincides with the cutoff frequency with high accuracy only when it is far enough away from the adjacent passbands.

3.3. The Effect of Nonlinearity

In recent years it has become clear that the oscillations in the upper chromosphere and transition region above the sunspots are nonlinear. They have a sawtooth shape (*i.e.*, the fronts become steeper and shock waves are formed). This raises the question as to whether the filter theory based on the linear theory of wave propagation is applicable to the problem. In fact, the formation of the interference passbands for slow waves propagating through the sunspot atmosphere does not depend directly on nonlinear effects in the upper atmosphere, since the wave interference takes place in the lower atmosphere, more specifically, in the temperature minimum and the photosphere where the wave amplitude is low, and the applicability of the linear theory is beyond any doubt. However, the nonlinear effects indirectly affect the formation of the interference transmission bands. The nonlinearity leads to the nonlinear absorption of the waves, which causes a reduction of the waves coming from the upper layers to the regions where the interference takes place. The reduction of the wave amplitudes

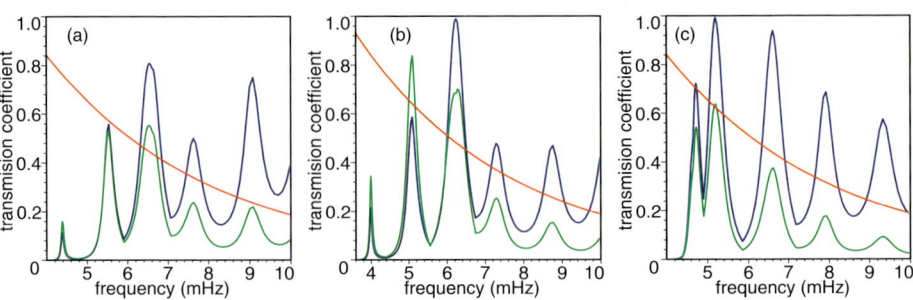

Figure 3 Model spectra (green), transmission functions (blue), and the decline of the spectrum of the incident noise in arbitrary units (red) for sunspot models of Lites (a), Maltby (b), and Staude (c).

affects the transmission of the waves through the atmosphere because the transmission of the passbands is defined by the ratio of the amplitudes of the interfering waves. The frequencies of the passbands can be shifted only by the change of the travel time of the waves from the upper chromosphere to the temperature minimum. However, the frequency shifts of passbands caused by the nonlinearity is hardly significant since the linear and nonlinear slow waves are moving approximately with the sound speed.

Unfortunately, it is difficult to evaluate the nonlinear absorption coefficient as a function of the amplitude of the incident waves. But it is simple to explore the effect of the nonlinearity by assuming different values for the reduction of the amplitude of the waves arriving at the temperature minimum from the upper layer of the sunspot atmosphere. The colored curves in Figures 2(b) are the transmission coefficients for the distinct values of amplitude reduction R that cover the entire range of possible absorptions from its absence (green curve) to complete absorption (black curve). A small nonlinear absorption makes the Fabry–Perot chromospheric filter almost perfect (see the blue curve on Figure 2(b)). Further increasing the nonlinear absorption makes the Fabry–Perot filter less and less transparent. It is interesting that the nonlinear absorption affects the cutoff passband as well. The proximity of the Fabry–Perot passband to the cutoff passband leads to involvement of interference effects in the formation of the cutoff passband. The nonlinear absorption leads, as a rule, to an increase in transmission for all high-frequency passbands. However, the complete absorption of the waves in the upper atmosphere changes drastically the transmission function (see the black curves on Figure 2(b)) since the interference in the temperature minimum disappears. This changes the frequency intervals between the high-frequency passbands. The frequency interval between the high-frequency interference passbands is defined by the travel time through the layers involved in their formation. In the case of a four-layer atmosphere, only two layers are involved in the formation of the high-frequency passbands. If the difference between travel times in the layers is significant, then the high-frequency passbands are not equidistant, and their transmission varies from band to band. The entire travel time through the layers defines in this case the mean interval between passbands. Absorption resulting from nonadiabaticity of the waves has to affect the transmission of waves, as well.

With knowledge of the interference effects, we arrive at an exploration of the filter properties of the empirical models of sunspot atmosphere, which without question provides a better description of the sunspot atmosphere than the four-layer atmosphere.

4. Spectra of Oscillations for Empirical Models of Sunspots

One should not directly compare the transmission functions with the spectrum of the oscillations in sunspots. To obtain the spectrum, we must multiply the transmission function by the spectrum of the waves that come from the subphotospheric layers. This spectrum is unknown, although it is clear that it must decrease rapidly with frequency. In addition, nonlinear wave absorption has to be taken into account. However, nonlinear effects are essential only for frequencies at which the wave amplitude is large enough. Consequently, nonlinear absorption must be taken into account only when the transmission function is calculated within the main transmission bands and must be disregarded at frequencies that correspond to the minima of the transmission function and at high frequencies. For simplicity, we assume in the calculations presented in Figure 2(b) that the nonlinear absorption does not depend on frequency.

To get an idea of how the oscillation spectra look for the empirical models, the calculations were performed under the assumption that the spectrum of the incident waves decreased exponentially with frequency and that the nonlinear absorption was proportional to the wave amplitude. Examples of such model spectra for the Lites, Maltby, and Staude empirical models are shown in Figures 3(a), 3(b), and 3(c). In general, the model spectra shown in these figures are similar to what is actually observed in the sunspots. The spectra for different sunspot models differ markedly. This is not surprising since, as has already been noted, the models themselves differ greatly.

The lowest frequency peaks in the spectra of the three sunspot models appear because of the Fabry–Perot chromospheric filter and their frequencies are determined by the thickness and temperature of the chromospheric plateau. The distinction between the frequencies of these peaks is most pronounced owing to the difference in the temperature plateau thickness for the three models. The oscillation spectrum for the Staude model exhibits another feature of the chromospheric interference filter that distinguishes it from an ordinary Fabry–Perot filter. As the cutoff frequency of the temperature minimum is approached by the Fabry–Perot passband, the chromospheric resonance frequency increasingly deviates toward lower frequencies from the frequency of the wave whose half-wavelength fits into the thickness of the temperature plateau. This is because the wave field in this case extends to an increasingly larger part of the temperature minimum. In this case, as previously mentioned, the cutoff passband shifts to higher frequencies because of the influence of interference effects. Thus, in the case of the appearance of the double peak, the nearness of the cutoff frequency to the chromospheric resonance frequency makes it difficult to use them for estimating the thickness of the temperature plateau and the cutoff frequency.

The frequency interval between the high-frequency interference passbands does not correspond to the travel time through the chromosphere and temperature minimum. They are located closer to each other than this travel time suggests. The frequency interval between these bands corresponds to the travel time through almost the entire sunspot atmosphere from the corona to the deep layers of the photosphere. Thus the three layers are involved in the formation of the high-frequency passbands. The lower layer is located between the temperature minimum and the deep layers of the photosphere where the temperature gradient increases. There is no surprise that it is not possible to reproduce perfectly the filter function of the empirical models by the choice of parameters of the four-layer atmosphere. The problem lies not only with the number of layers but with the strong discontinuities between layers, which leads to strong reflections of waves. This becomes clear from a comparison of the filtering functions for the four-layer atmosphere (the green curve of Figure 3(a)) with the empirical atmosphere (see Figure 2 in Settele, Staude, and Zhugzhda, 2001) in the case when

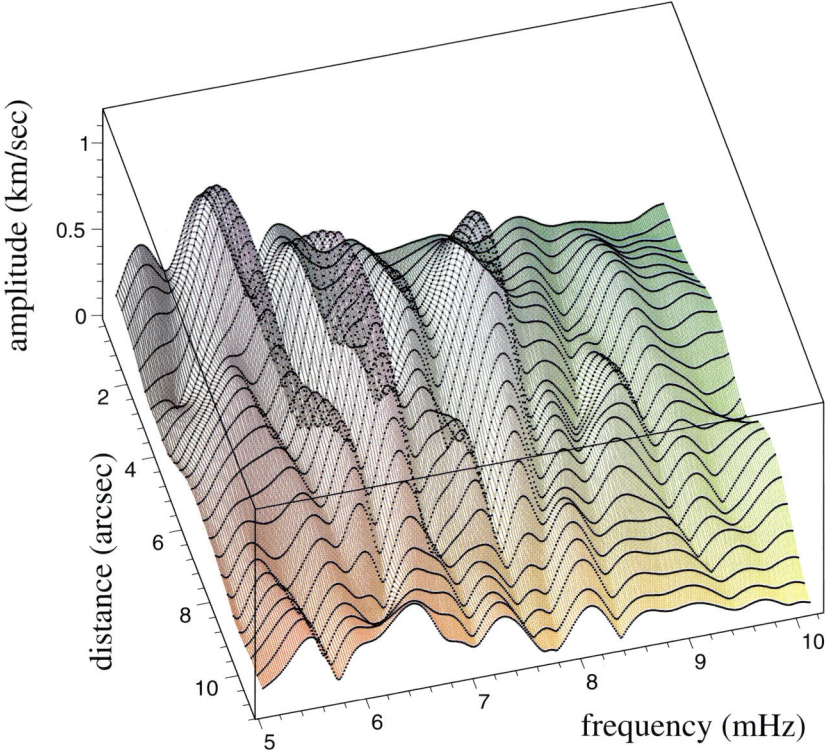

Figure 4 The amplitude spectrum of the chromospheric oscillations in the sunspot as a function of the distance along the slit.

nonlinear absorption of waves is absent. In general, the transmission of slow waves through the empirical atmosphere exceeds the transmission through the four-layer atmosphere. Thus, the four-layer atmosphere is just a tool to explore the interference effects in a sunspot atmosphere. Now we arrive at the next step of the exploration, that is, whether the observations can be explained in terms of the revised filter theory.

5. Interpretation of Observations by the Filter Theory

R. Centeno, M. Collados, and J. Trujillo Bueno provided me with the results of data processing of their unique observations of sunspot oscillations, making the current exploration of sunspot oscillations possible. (The details of the observations and the processing can be found in Centeno, Collados, and Bueno (2006).) The observations of the sunspot oscillations by Centeno, Collados, and Bueno (2006) were taken in the He I 10830 Å multiplet. In addition to the analysis of Centeno, Collados, and Bueno (2006) the variation of the spectrum of the oscillations along the slit is explored.

The goal of the current exploration is to check whether the sunspot oscillates as a whole. The theory of normal modes of the sunspots is based, in fact, on the assumption that the sunspot oscillates as a single entity. This is one of the reasons why it is usual to look for oscillations of the entire sunspot. Of course, the inhomogeneity of the sunspot is not included in the theoretical models for reasons of simplicity. However, the inhomogeneity of the sunspot atmosphere can lead only to the inhomogeneity of the amplitude of the normal modes across the sunspot but it cannot give rise to the frequency variations across the sunspot. It is quite another matter with the filter theory since, as already pointed out, the slow waves propagate independently along each magnetic field line. Thus, the distinctions of the temperature stratification along different field lines causes the changes of the passband frequencies across the sunspot. The distinctions of the temperature structure can appear because of the inhomogeneity of sunspot atmosphere as well as because of the divergence of the magnetic-field lines, which cause the distinctions of the temperature scale along the distinct field lines. Thus, it is difficult to assume that the transmission function for the slow waves is the same over the sunspot umbra.

To look for the variations of the spectrum of three-minute oscillations, the spectrum of sunspot oscillations for each of 26 points along the slit was obtained. The dependence of the spectrum on distance along the slit is presented in Figure 4. There is no doubt that the passband frequencies are not constant over the sunspot umbra. To trace the changes in the spectrum along the slit the dependence of the peak frequencies in the spectrum on the distance along the slit is presented in Figure 5(a). The frequency of the cutoff passband is not constant across the sunspot. It is about 5.9 mHz and almost constant along the slit in the range $0'' - 3''$. Then at the point $4''$ the high-frequency passband splits away from the cutoff passband. The frequency of the cutoff band shifts to lower frequencies because of the influence of the adjacent high-frequency passband. It varies in the range between 5.5 and 5.8 mHz. Near the far end of the slit, the high-frequency band disappears, and the frequency of the cutoff passband is about 5.9 mHz as it is on the opposite end of the slit. The cutoff frequency varies inversely with the sound speed. But it is difficult to believe that the temperature in the temperature minimum increases in the middle coolest parts of the sunspot umbra except in the case of the light bridges. Thus, the decrease of the frequency of the cutoff passband occurs mostly likely from the rearrangement of the high-frequency passbands. The first high-frequency passband comes close to the cutoff passband and pushes it toward lower frequencies. The decrease of the cutoff frequency affects in turn the Fabry–Perot passband. The frequency of the chromospheric Fabry–Perot passband is in between 5.3 and 5.4 mHz. But this band is absent in the middle of the sunspot, where it is likely united with the cutoff frequency. Thus, there is a variation of the sunspot atmosphere across the sunspot.

In addition to Figures 4 and 5(a), Figures 5(b)–5(d) display the amplitude spectrum of the sunspot oscillations at three positions along the slit. The spectrum in Figure 5(b) (where the location on the slit is at $1.3''$) looks perfect from the point of view of the filter theory. The Fabry–Perot passband is located on the low-frequency slope of the cutoff passband. On the high-frequency side of the cutoff a series of equidistant high-frequency interference passbands trace the exponential decline of the amplitude spectrum of the incident slow-wave noise, which is filtered by the multipassband atmospheric filter. The ideal equidistant location of the passbands can appear only in the case when the atmospheric layers responsible for the passbands formation have the same travel time as that of the slow waves across them. The quite different spectrum in Figure 5(c) is in the middle of the slit where there is the decrease of the frequency of the cutoff frequency. In this case the atmospheric layers do not have the same travel times across them. This makes the high-frequency passbands not equidistant and not equal in their transmission. Even the first high-frequency passband dominates over the cutoff passband. The number of high-frequency passbands increases from six to seven and the frequency difference between the adjacent passbands decreases. On the opposite side of the umbra (see Figure 5(d)) the number of high-frequency passbands decreases again to six, and they are far from being equidistant. It is clearly seen in Figure 5(d) that the

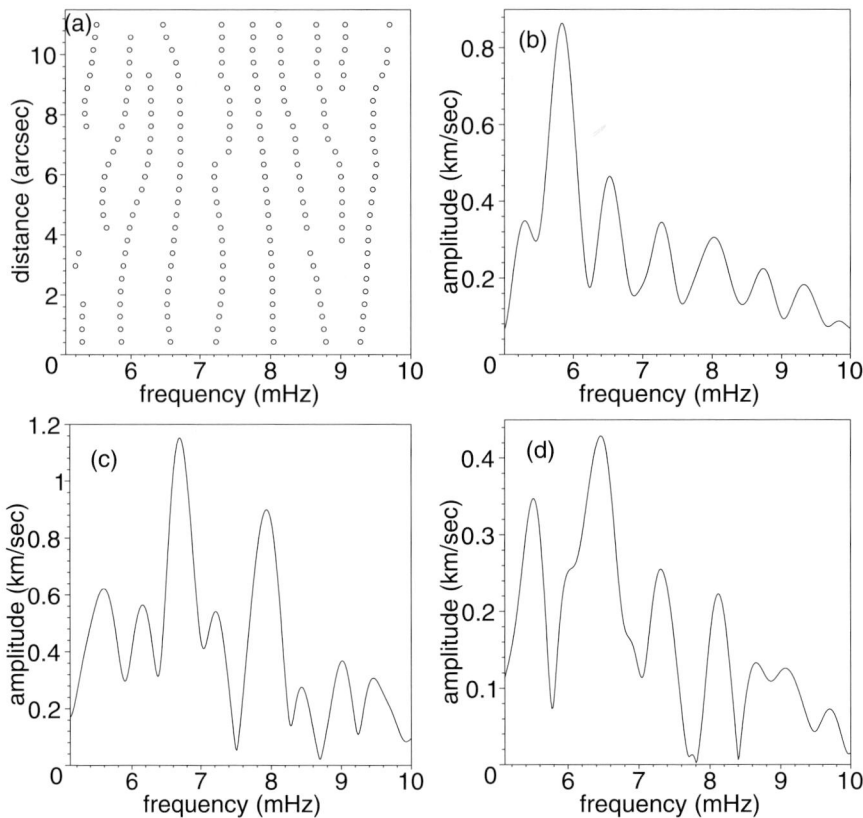

Figure 5 (a) The dependence of the peak frequencies in the spectrum as a function of distance along the slit. (b, c, d) The amplitude spectrum of the oscillations for the three positions on the slits: 1.3″ (b), 5.5″ (c), 11″ (d).

first high-frequency passband has merged with the cutoff passband and it is located on the low-frequency slope of the first high-frequency passband. This is why the frequency of the cutoff passband is not shown in Figure 5(a) for the far end of the slit. Thus, the variation of the spectrum of the sunspot oscillations across the umbra is revealed. This is a manifestation of the sunspot atmospheric variation across the umbra.

The exploration of empirical atmospheres reveals that the three layers have to be involved in the formation the high-frequency passbands. The number of high-frequency passbands and the mean frequency interval between them depends on the travel time of the slow waves through the entire sunspot atmosphere since all the atmospheric layers are involved in producing the interference effects. It is no surprise that the travel time through the atmosphere is less in the middle of the sunspot, since the central part of the sunspot is most likely cooler than peripheral parts. The distinction between the regular pattern of the equidistant high-frequency passbands on the one side of the sunspot (Figure 5(b)) and nonequidistant passbands on the other side (Figure 5(d)) is evidence that there are variations in some of the layers of the sunspot atmosphere. Therefore it comes as no surprise that the atmosphere varies across the sunspot, and it cannot be described satisfactory by the mean empirical model.

6. Discussion

The exploration of sunspot oscillations across the sunspot umbra reveals significant changes of the spectrum. This complicated behavior of the oscillations cannot be explained within the framework of the normal mode theory. Thus, this is one more argument for the filter theory, which can explain the multitude of effects appearing because of the interference of slow waves in the multilayer sunspot atmosphere. In connection with this the papers by Brynildsen *et al.* (2003) and Shibasaki (2001) should be mentioned. They explored sunspot oscillations, but because of technical problems time series of only 20-minute duration were available for their analysis. Such short time series do not permit the resolution of the structure of the spectrum – only one peak appears in the spectrum. But the authors of these papers ignored this point and made strong statements about the nature of sunspot oscillations and the occurrence of the chromospheric resonance. Unresolved spectra of sunspot oscillations cannot be used for the exploration of the nature of sunspot oscillations.

The revised version of the filter theory of the sunspot oscillations makes it possible to use the rich information about the structure of the sunspot atmosphere contained in the spectrum of three-minute oscillations. The observations provide solid proof that the sunspot atmosphere is nonuniform. The analysis of the observations presented here is just an example of how to use the filter theory for the exploration of sunspots. It is necessary to explore different sunspots and to use the observations of the oscillations of the entire sunspot.

Acknowledgements This work was supported by RFFI Grant No. 06-02-16359 and the European Helio- and Asteroseismology Network – HELAS funded by the European Union's Sixth Framework Programme. The author is grateful to Rebecca Centeno, Manuel Collados, and Javier Trujillo Bueno for providing the data and is indebted to the anonymous referee and Oscar van der Lühe for the helpful discussions and comments that helped to improve the paper.

References

Bogdan, T.J., Cally, P.S.: 1997, Waves in magnetized polytropes. *Proc. Roy. Soc. London, Ser. A* **453**, 943 – 961.
Bogdan, T.J., Judge, P.G.: 2006, Observational aspects of sunspot oscillations. *Philos. Trans. R. Soc. A* **364**, 313 – 331.
Brekke, P., Brynildsen, N., Maltby, P., Fredvik, T., Kjeldseth-Moe, O.: 2003, Sunspot oscillations and acoustic wave propagation. *Bull. Am. Astron. Soc.* **35**, 810.
Brekhovskikh, L.M., Godin, O.A.: 1990, *Acoustics of Layered Media*, Springer, New York.
Brynildsen, N., Leifsen, T., Kjeldseth-Moe, O., Maltby, P., Wilhelm, K.: 1999, Sunspot transition region oscillations in NOAA 8156. *Astrophys. J.* **511**, L121 – L124.
Brynildsen, N., Maltby, P., Brekke, P., Redvik, T., Kjeldseth-Moe, O.: 2003, Search for a chromospheric resonator above sunspots. *Adv. Space Res.* **32**(6), 1097 – 1102.
Cally, P.S., Bogdan, T.J.: 1993, Solar p-modes in a vertical magnetic field: trapped and damped π-modes. *Astrophys. J.* **402**, 721 – 732.
Centeno, R., Collados, M., Bueno, J.T.: 2006, Spectropolarimetric investigation of the propagation of magnetoacoustic waves and shock formation in sunspot atmosphere. *Astrophys. J.* **640**, 1153 – 1162.
Lites, B.W., Skumanich, A.: 1982, A model of a sunspot chromosphere based on OSO 8 observations. *Astrophys. J. Suppl. Ser.* **49**, 293 – 315.
Maltby, P., Avrett, E.H., Carlsson, Kjeldseth-Moe, O., Kurucz, R.L., Loeser, R.: 1986, A new sunspot umbral model and its variation with the solar cycle. *Astrophys. J.* **306**, 284 – 303.
Mihalas, B.W., Toomre, J.: 1981, Internal gravity waves in the solar atmosphere. I. Adiabatic waves in the chromosphere. *Astrophys. J.* **249**, 349 – 371.
Settele, A., Staude, J., Zhugzhda, Yu.D.: 2001, Waves in sunspots: resonant transmission and the adiabatic coefficient. *Solar Phys.* **202**, 281 – 292.
Shibasaki, K.: 2001, Microwave detection of umbral oscillations in NOAA ACTIVE REGION 8156: diagnostics of temperature minimum in sunspots. *Astrophys. J.* **550**, 1113 – 1118.

Staude, J.: 1981, A unified working model for the atmospheric structure of large sunspot umbrae. *Astron. Astrophys.* **100**, 284–290.
Syrovatskii, S.I., Zhugzhda, Y.D.: 1968, Oscillatory convection of a conducting gas in a strong magnetic field. *Sov. Astron.* **11**, 945–952.
Zhao, J., Kosovichev, A.G.: 2006, Surface magnetism effects in time–distance helioseismology. *Astrophys. J.* **643**, 1317–1324.
Zhugzhda, Y.D.: 1979, Magnetogravity waves in an isothermal conductive. *Sov. Astron.* **23**, 42–47.
Zhugzhda, Y.D.: 2007, Three-minute oscillations in sunspots: seismology of sunspot atmospheres. *Astron. Lett.* **33**(9), 622–643.
Zhugzhda, Y.D., Dzhalilov, N.: 1982, Transformation of magnetogravitationl waves in the solar atmosphere. *Astron. Astrophys.* **112**, 16–23.
Zhugzhda, Y.D., Goossens, M.: 2001, Hidden problems of thin-flux-tube approximation. *Astron. Astrophys.* **337**, 330–342.
Zhugzhda, Y.D., Locans, V.: 1981, Resonance oscillations in sunspots. *Sov. Astron. Lett.* **7**, 25–27.

Enhanced p-Mode Absorption Seen Near the Sunspot Umbral–Penumbral Boundary

Shibu K. Mathew

Originally published in the journal Solar Physics, Volume 251, Nos 1–2, 515–522.
DOI: 10.1007/s11207-008-9246-0 © Springer Science+Business Media B.V. 2008

Abstract We investigate p-mode absorption in a sunspot using SOHO/MDI high-resolution Doppler images. The Doppler power computed from a 3.5-hour data set is used for studying the absorption in a sunspot. The result shows an enhancement in absorption near the umbral–penumbral boundary of the sunspot. We attempt to relate the observed absorption with the magnetic-field structure of the sunspot. The transverse component of the potential field is computed by using the observed SOHO/MDI line-of-sight magnetograms. A comparison of the power map and the computed potential field shows enhanced absorption near the umbral–penumbral boundary where the computed transverse field strength is higher.

Keywords Solar oscillation · p-Mode absorption · Sunspot

1. Introduction

The interaction of solar oscillations with magnetic field has been reported by many authors. Two of the important findings are the reduced p-mode power in regions of strong magnetic field and the enhancement of power in higher frequencies surrounding the regions of strong fields (Woods and Cram, 1981; Tarbell *et al.*, 1988; Brown *et al.*, 1992). The magnetic field reduces p-mode power in active regions in the 3-mHz band while enhancing power of these modes in the 5-mHz band (Hindman and Brown, 1998; Venkatakrishnan, Kumar, and Tripathy, 2002). The mechanisms of p-mode absorption were reviewed by Spruit (1996) and he suggested a promising mechanism for reduced power as the conversion of p mode into a downward propagating slow mode along the magnetic flux tubes (Spruit and Bogdan, 1992; Cally and Bogdan, 1993; Cally, Bogdan, and Zweibel, 1994). The waves in the flux tube, once excited by the sound wave, can carry energy out of the p modes through the wave guide into the convection zone, thereby producing a reduction in observed power.

Helioseismology, Asteroseismology, and MHD Connections
Guest Editors: Laurent Gizon and Paul Cally.

S.K. Mathew (✉)
Udaipur Solar Observatory, Physical Research Laboratory, Badi Road, Udaipur 313001, India
e-mail: shibu@prl.res.in

Simulations by Cally (2000) show that the enhanced absorption takes place primarily in the more inclined magnetic-field regions toward the edge of the spot. A comparison of the spatial distribution of Doppler power and the magnetic-field configuration in a sunspot could reveal the variation of Doppler power with magnetic-field strength and the field inclination. In this paper, we study p-mode absorption in magnetic-field concentrations and compare that with the longitudinal and the computed transverse-field configuration in a sunspot.

2. Data Sets

For this analysis, we have used high-resolution (0.6″/pixel) Dopplergrams with one-minute cadence obtained with the *Solar and Heliospheric Observatory*/Michelson Doppler Imager (SOHO/MDI) instrument (Scherrer *et al.*, 1995). Complementing these data are a few line-of-sight magnetograms and intensity images from SOHO/MDI. The sunspot analyzed is a member of the active region NOAA 8395 and the observations were made on 1 December 1998 between 04:00 and 07:51 UT. The sunspot was near disk center ($\mu \approx 0.93$). Images were registered and remapped onto heliographic coordinates. Registration is carried out first in synodic rate and then by cross-correlating successive images with a reference image. The reference image is updated after every five minute to avoid errors in registration resulting from the evolution of the active region. The resulting images consist of 773 × 393 pixels, corresponding to $\approx 8 \times 4$ arcmin at the solar disk center. We selected a small subregion of 150 × 100 including one of the sunspots in the active region for our analysis.

The duration of 3 hours and 51 minutes and the cadence of one minute give a frequency resolution and Nyquist frequency of 72.1 µHz and 8.33 mHz, respectively. Remapped SOHO/MDI Dopplergrams and line-of-sight magnetograms were used for comparing the p-mode absorption with magnetic-field strength. An average of two magnetograms, one at the beginning and other toward the end of the observing period, are used for the comparison.

3. Data Analysis

We used a method similar to that described by Brown *et al.* (1992) for the Dopplergram analysis. From the accurately registered Doppler images, a time series is constructed. The power spectra is computed for every pixel in the image. Figure 1 shows the amplitude of

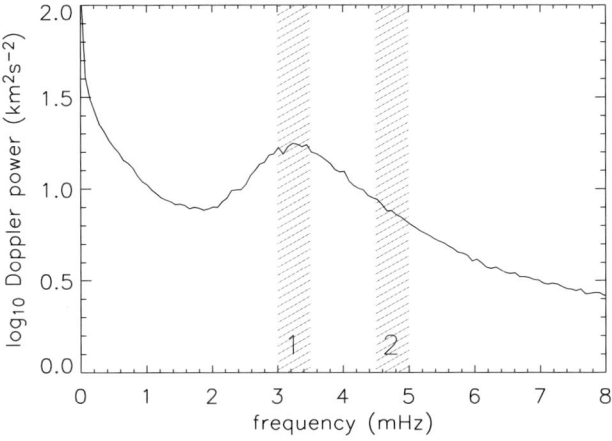

Figure 1 The spatially averaged Doppler power spectrum for the observed region. The hatched areas correspond to the frequency bands between 3–3.5 mHz (band 1) and 4.5–5 mHz (band 2).

the average power spectrum for the aforementioned subregion including the sunspot. The average power within the two frequency bands represented by the hatched areas is used for obtaining the results shown in the subsequent figures. The ranges of values included in the low-frequency (band 1) and high-frequency (band 2) bands are 3–3.5 mHz and 4.5–5 mHz, respectively.

From the line-of-sight magnetograms, the potential-field configuration of the entire active region is computed. All three components of the magnetic field (B_x, B_y, and B_z) are derived by assuming the potential-field approximation using the Fourier method (Sakurai, 1989). The mean field has been removed before computing the magnetic-field components and later added. The normal component (B_z) of the magnetic field is needed as the boundary condition for the potential-field extrapolation, whereas only the line-of-sight field is available from the MDI measurements. The measured line-of-sight field and the normal component differ when the sunspot is observed away from the disk center owing to the projection effect. In our analysis a first-order correction is carried out on the line-of-sight magnetic-field for the projection effect by taking the heliocentric angle into consideration (i.e., $B_z = B_{\mathrm{LOS}}/\cos\theta$). Since the observations of the analyzed sunspot were obtained when it was close to the disk center ($\mu = \cos\theta = 0.93$), this correction is approximately valid for strong vertical magnetic-field component. The transverse magnetic-field component is $\sqrt{B_x^2 + B_y^2}$ and the magnetic-field inclination is $\tan^{-1}(B_z/\sqrt{B_x^2 + B_y^2})$. Even though in the computed potential field the lowest magnetic-energy configuration of the sunspot is assumed, for obtaining a rather simplified idea about the magnetic structure the potential field calculation can be used. The computed magnetic-field structure (Figure 2(c)) of the observed sunspot is similar to the retrieved transverse-field configuration of sunspots from the vector magnetic field measurements (Westendorp Plaza et al., 2001; Keppens and Martinez Pillet, 1996; Mathew et al., 2003).

4. Results

Figure 2 shows the continuum image, the Dopplergram, the line-of-sight magnetogram, and the computed transverse field for the analyzed sunspot along with the average Doppler power for the frequency bands 1 and 2 marked in Figure 1. The Doppler power is displayed on a logarithmic scale to make the absorption clear in the lower levels. The dotted- and dashed-line contours are plotted for the umbral and penumbral boundaries, respectively. The Doppler power in both lower (band 1) and the upper (band 2) frequency bands show an enhanced absorption near the umbral–penumbral boundary; here the computed transverse field also becomes strong. In the upper panel of Figure 3, the continuum image of the analyzed sunspot and the equal-intensity contours are displayed. The inset in the upper panel shows the brightness-enhanced image of the sunspot. This gives a closer view of the split and irregular structure of the umbra. In the lower panel, variation of different parameters along two radial cuts (marked as a and b in the continuum image) through the sunspot are plotted. The dotted vertical lines in these figures mark the umbral boundary and the plotted Doppler power is for the lower frequency band. Around the umbral boundary, the line-of-sight magnetic field shows a smooth change, whereas the computed transverse field strength reaches a maximum value. The Doppler power in the lower frequency band shows a dip around this location in both the radial cuts.

In Figure 4 we plot the values averaged between equal intensity contours for the continuum, the line-of-sight magnetic field strength, the computed transverse magnetic field,

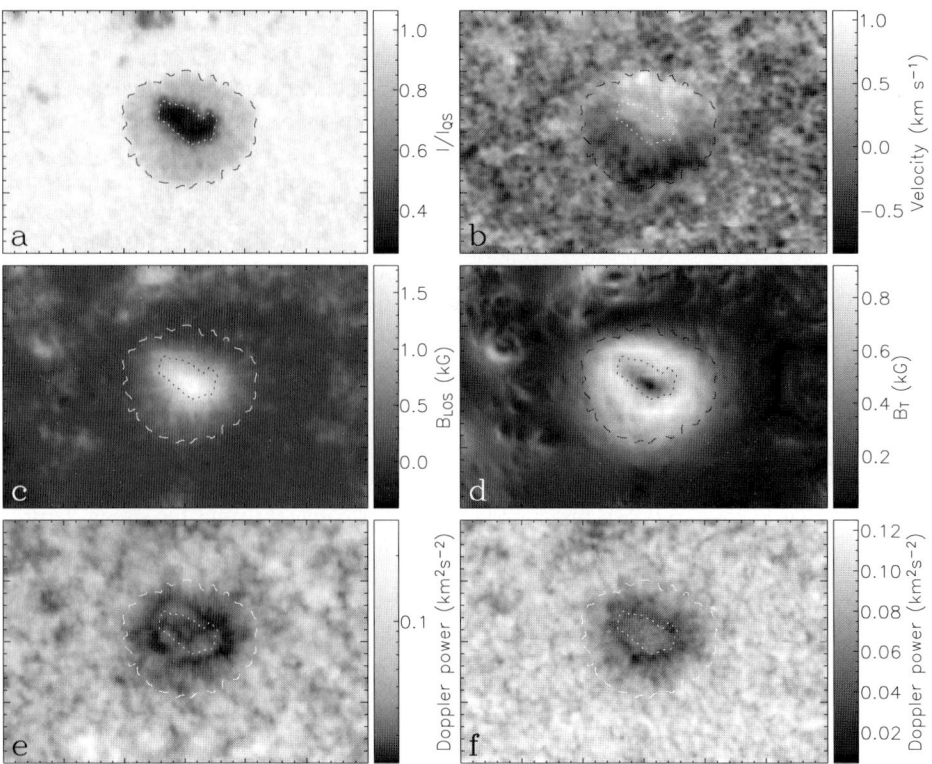

Figure 2 (a) Continuum intensity, (b) line-of-sight velocity, (c) line-of-sight magnetic field, (d) computed transverse magnetic field, and the total Doppler power in the (e) 1 and (f) 2 frequency bands for the analyzed sunspot. The dotted and dashed contours are the umbral and penumbral boundaries, respectively.

the computed inclination, the Doppler power in the lower frequency band, and the Doppler power in the higher frequency band. The contour numbers plotted are from the upper panel in Figure 3, starting from the inner umbra. Here also, a clear reduction in Doppler power near the umbral boundary is observed, where the strength of the computed transverse field is higher.

5. Discussion

It is well known that the Doppler power is considerably reduced in sunspots compared with the quiet Sun. Braun, Duvall, and LaBonte (1987) reported that sunspots absorb up to half of the incident p-mode power at favored frequencies and horizontal wavenumbers. From the analysis of the acoustic properties of two large sunspots using Fourier–Hankel decomposition of p-mode amplitudes, Braun (1995) showed that there is a peak in Doppler power absorption centered at around 3 mHz and an absence of absorption at 5-mHz frequencies. Lindsey and Braun (2005) reported a "penumbral acoustic anomaly" in which they found a ring of relatively depressed acoustic power in and around the sunspot penumbra. Study of active region oscillations by Muglach, Hofmann, and Staude (2005) also show (in their Figure 5) a reduction in Doppler power in the penumbra.

Enhanced p-Mode Absorption Seen Near the Sunspot

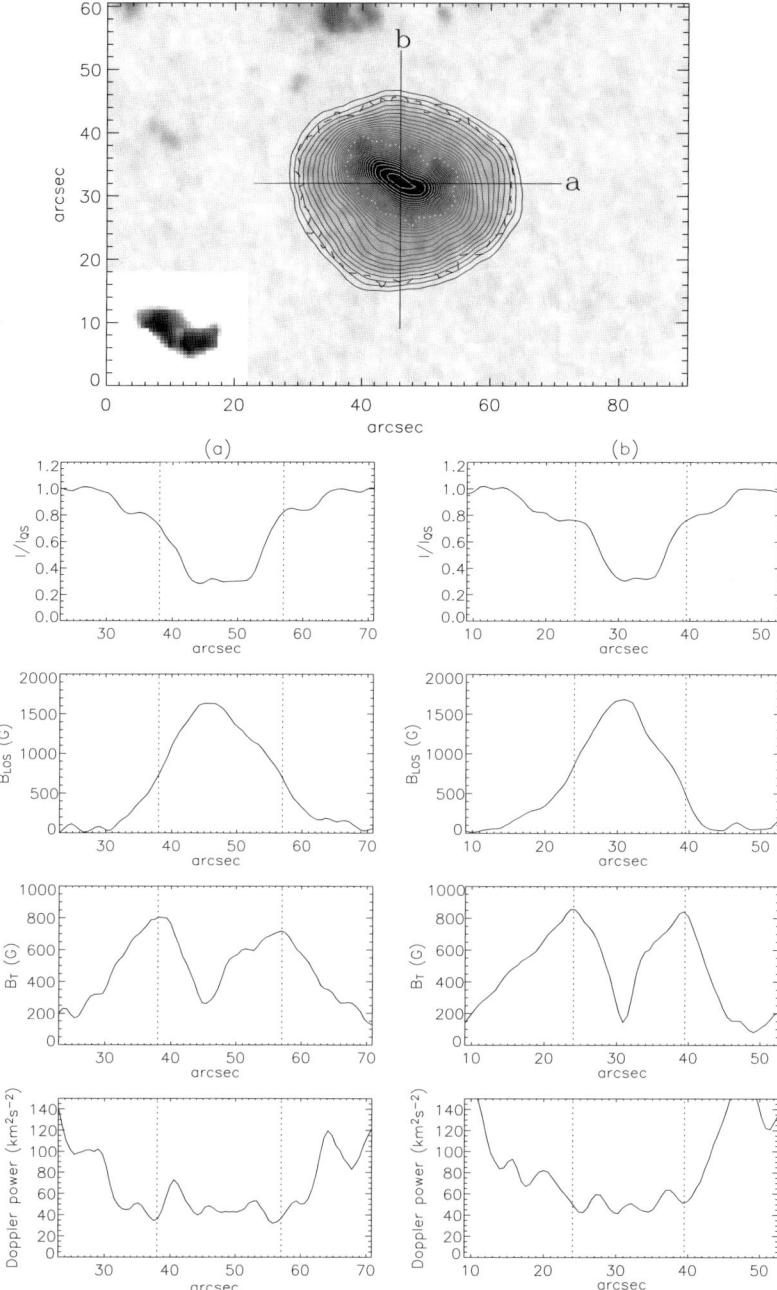

Figure 3 Various parameters (lower panel) for two cuts across the sunspot indicated by straight lines (a) and (b) in the upper panel. The two vertical dotted lines show the umbral–penumbral boundary. The average values between the adjacent solid contours are used to obtain the radial distribution of these parameters, shown in Figure 4.

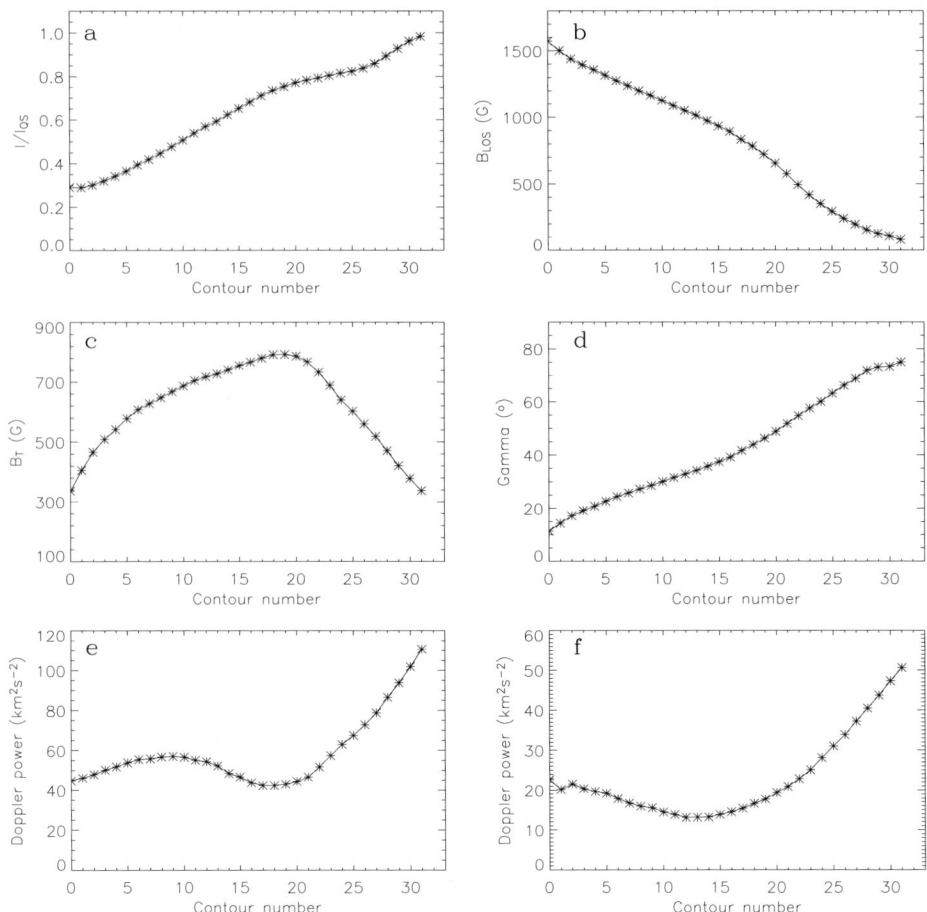

Figure 4 (a) Azimuthal averages of the intensity, (b) line-of-sight magnetic field, (c) computed transverse field, (d) computed field inclination, (e) Doppler power in band 1, and (f) Doppler power in band 2.

In our analysis, we present the spatial distribution of the oscillatory power in a sunspot. The total-power maps obtained for this sunspot in two frequency bands (3 – 3.5 and 4.5 – 5 mHz) show a structured ringlike depression in the penumbra. The reduction is visible in both of the frequency bands with a difference of a shift in the peak absorption toward the umbral – penumbral boundary in the higher frequency band. We also observe an absorption structure in the umbra in the lower frequency band that is not clearly visible in the higher frequency band. A comparison of the reduction in p-mode power with magnetic structure of the sunspot clearly indicates increased absorption near the locations where the transverse field is higher. The absorption structure in the umbra (also the decrease in Doppler power in umbra seen in the azimuthal average; Figure 4), with no correspondence to transverse-field structure could be due to the split structure of the umbra, which is clearly seen in the brightness-enhanced image of the sunspot displayed in the upper panel of Figure 3. We suggest that this absorption structure could be due to the increased field inclination resulting from the split structure of the umbra, which is not visible in the MDI line-of-sight magnetogram and thus also not in the computed transverse field.

Theoretical modeling of the influence of sunspot magnetic field on the incident oscillation has been described by several authors (Spruit, 1991; Spruit and Bogdan, 1992; Cally and Bogdan 1993; Cally, 2000; Rosenthal and Julien, 2000). In the presence of strong magnetic field, it is suggested that the fast wave is partially converted into a slow magneto-acoustic wave and is directionally influenced by the magnetic field. Simulations by Cally (2000) showed that the interaction of an acoustic wave and magnetic field has a strong dependence on the inclination of the magnetic field and found that the absorption peaks at 30° inclination angle (Cally, Crouch, and Braun, 2003). In our analysis we find that the computed inclination angles are around 45° and 35°, where the absorption peaks in the lower (band 1) and higher (band 2) frequency bands, respectively. These values should be considered cautiously since the exact inclination angle can only be derived from full vector magnetic field measurements corrected for the projection effect. In our case the potential field and thus the field inclination is computed from the measured line-of-sight magnetic field and a simple approach is used for correcting the observed field before computing the potential components, assuming the magnetic field is perpendicular to the photosphere. Here, without having full information on the vector magnetic field, the contribution of the transverse component to the line-of-sight field is unknown. However, it has been shown that the difference between the computed potential field from the observed and the corrected (for the projection effect using full vector magnetic field measurements) line-of-sight field is not very different from each other when the sunspot is located close to the disk center (Hagyard, 1987). From the analysis of an active region positioned within one-third solar radius of the disk center, Hagyard showed that the difference between the computed transverse field from the corrected and observed line-of-sight field is around 200 G and this difference occurs only in a localized area of the umbrae of large sunspot. In our case, since the sunspot is close to the disk center, we presume that the simple correction we employed is sufficient to represent the potential configuration of the sunspot.

More strikingly, we find that the computed transverse field reaches maximum values where the peak in absorption is observed. It is important to have full spectro-polarimetric vector magnetic-field observations to obtain a clear idea about the dependence of the field configuration on the spatial distribution of acoustic power. We have not carried out the study of effects of magnetic field in the Doppler measurement that is done on the Zeeman-sensitive spectral line profile. Rajaguru *et al.* (2007) presented a detailed study of the phase shifts on the acoustic waves when observed in a sunspot. They have shown that the phases of the acoustic waves within a sunspot, especially in the penumbra, could be altered because of the propagating nature of the p modes and also because of the Zeeman split of the spectral line, which could in effect produce reduction in Doppler power in a particular frequency band.

6. Conclusions

We analyzed the spatial distribution of Doppler power in a sunspot observed near disk center. We find a structured ringlike absorption pattern in Doppler power near the umbral–penumbral boundary. The computed transverse field is higher at those locations where the peak depression in Doppler power is observed. The computed inclination angle ranges between 35° and 45° at these locations. To understand the exact dependence of magnetic-field strength and inclination on the Doppler power absorption, full vector magnetic-field observations are required. Also, it is preferred to have Doppler observation in magnetically insensitive lines to avoid any cross-talk between the line-of-sight velocity measurement and Zeeman splitting of the spectral line.

Acknowledgements I would like to thank B. Ravindra for providing the code for potential field calculation and the anonymous referee for valuable comments that helped to improve the paper substantially. SOHO is a project of international cooperation between ESA and NASA.

References

Braun, D.C.: 1995, *Astrophys. J.* **451**, 859.
Braun, D.C., Duvall, T.L. Jr., LaBonte, B.J.: 1987, *Astrophys. J.* **319**, L27.
Brown, T.M., Bogdan, T.J., Lites, B.W., Thomas, J.H.: 1992, *Astrophys. J.* **394**, L65.
Cally, P.S., Bogdan, T.J.: 1993, *Astrophys. J.* **402**, 732.
Cally, P.S., Bogdan, T.J., Zweibel, E.G.: 1994, *Astrophys. J.* **437**, 505.
Cally, P.S.: 2000, *Solar Phys.* **192**, 395.
Cally, P.S., Crouch, A.D., Braun, D.C.: 2003, *Mon. Not. Roy. Astron. Soc.* **346**, 381.
Hagyard, M.J.: 1987, *Solar Phys.* **107**, 239.
Hindman, B.W., Brown, T.M.: 1998, *Astrophys. J.* **504**, 1029.
Keppens, R., Martinez Pillet, V.: 1996, *Astron. Astrophys.* **316**, 229.
Lindsey, C., Braun, D.C.: 2005, *Astrophys. J.* **620**, 1107.
Mathew, S.K., Lagg, A., Solanki, S.K., Collados, M., Borrero, J.M., Berdyugina, S., Krupp, N., Woch, J., Frutiger, C.: 2003, *Astron. Astrophys.* **410**, 695.
Muglach, K., Hofmann, A., Staude, J.: 2005, *Astron. Astrophys.* **437**, 1055.
Rajaguru, S.P., Sankarasubramanian, K., Wachter, R., Scherrer, P.H.: 2007, *Astrophys. J.* **654**, L175.
Rosenthal, C.S., Julien, K.A.: 2000, *Astrophys. J.* **532**, 1230.
Sakurai, T.: 1989, *Space Sci. Rev.* **51**, 11.
Scherrer, P.H., Bogart, R.S., Bush, R.I., Hoeksema, J.T., Kosovichev, A.G., Schou, J., Rosenberg, W., Springer, L., Tarbell, T.D., Title, A., Wolfson, C.J., Zayer, I., the MDI Engineering Team: 1995, *Solar Phys.* **162**, 129.
Spruit, H.C.: 1991, In: Toomre, J., Gough, D.O. (eds.) *Lecture Notes in Physics: Challenges to Theories of the Structure of Moderate Mass Stars* **388**, Springer, Berlin, 121.
Spruit, H.C.: 1996, *Bull. Astron. Soc. India.* **24**, 211.
Spruit, H.C., Bogdan, T.J.: 1992, *Astrophys. J.* **391**, L109.
Tarbell, T., Peri, M., Frank, Z., Shine, R., Title, A.: 1988, In: Rolfe, E. (ed.) *Seismology of the Sun and Sun-Like Stars* **SP-286**, ESA, Noordwijk, 315.
Venkatakrishnan, P., Kumar, B., Tripathy, S.C.: 2002, *Solar Phys.* **211**, 77.
Westendorp Plaza, C., Del Toro Iniesta, J.C., Ruiz Cobo, B., Martinez Pillet, V., Lites, B.W., Skumanich, A.: 2001, *Astrophys. J.* **547**, 1130.
Woods, D.T., Cram, L.E.: 1981, *Solar Phys.* **69**, 233.

Global Acoustic Resonance in a Stratified Solar Atmosphere

Y. Taroyan · R. Erdélyi

Originally published in the journal Solar Physics, Volume 251, Nos 1–2, 523–531.
DOI: 10.1007/s11207-008-9154-3 © Springer Science+Business Media B.V. 2008

Abstract The upward propagation of linear acoustic waves in a gravitationally stratified solar atmosphere is studied. The wave motion is governed by the Klein–Gordon equation, which contains a cutoff frequency introduced by stratification. The acoustic cutoff may act as a potential barrier when the temperature decreases with height. It is shown that waves trapped below the barrier could be subject to a resonance that extends into the entire unbounded atmosphere of the Sun. The parameter space characterizing the resonance is explored.

Keywords Waves: Acoustic · Oscillations: Solar · Chromosphere: Models · Heating: Chromospheric

1. Introduction

The vertical propagation of acoustic waves in a gravitationally stratified medium was studied by Poisson (1808), Rayleigh (1890), and Lamb (1909). The latter established that such waves could be described by a partial differential equation, which is now known as the Klein–Gordon (KG) equation. The KG equation is widely used in a range of fields including atmospheric physics of the Earth, cosmology, quantum field theory, solid state physics, and solar and stellar physics (Lamb, 1909; Detweiler, 1980; Poschl *et al.*, 1997; Carlson, Heiselberg, and Pandharipande, 1982; Rae and Roberts, 1982). An important feature introduced by vertical stratification is the cutoff frequency, which separates higher frequency propagating waves from lower frequency evanescent waves. Waves with cutoff frequencies are sometimes called "resonant." In an unbounded isothermal model atmosphere of

Helioseismology, Asteroseismology, and MHD Connections
Guest Editors: Laurent Gizon and Paul Cally.

Y. Taroyan (✉) · R. Erdélyi
SP²RC, Department of Applied Mathematics, University of Sheffield, Sheffield S3 7RH, UK
e-mail: y.taroyan@sheffield.ac.uk

R. Erdélyi
e-mail: robertus@sheffield.ac.uk

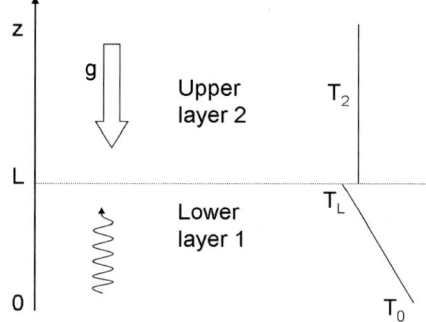

Figure 1 Two-layer model depicting a stratified solar atmosphere. The lower part of the atmosphere (index 1) is separated from the upper part (index 2) by a density and temperature discontinuity at $z = L$. Waves are launched at $z = 0$ and propagate in the vertical z-direction.

the Sun, the cutoff frequency characterizes the free atmospheric oscillations of the medium (Sutman, Musielak, and Ulmschneider, 1998). For the most recent reviews of waves in the solar atmosphere see Banerjee *et al.* (2007) and Erdélyi and Ballai (2007).

It had been thought that high-frequency waves could be responsible for the heating of the nonmagnetic chromosphere of the Sun as they develop into shocks. However, low-frequency waves were believed to play little role as far as the dynamics and energetics of the atmosphere are concerned owing to reflection from regions with steep temperature gradients. Recent work has changed these views (De Pontieu, Erdélyi, and James, 2004; De Pontieu, Erdélyi, and De Moortel, 2005; Fossum and Carlsson, 2005; Jefferies *et al.*, 2006). Millions of acoustic p modes are generated and trapped in the solar interior. De Pontieu, Erdélyi, and De Moortel (2005) have shown that these waves can penetrate and leak high into the solar atmosphere. Fossum and Carlsson (2005) established that the power of observed high-frequency propagating (> 5 mHz) waves is not enough to balance the radiative losses in the chromosphere. However, new observations by Jefferies *et al.* (2006) have shown that the energy flux carried by the low-frequency (< 5 mHz) waves into the chromosphere is about a factor of four greater than that carried by high-frequency waves. It was argued that these low-frequency waves could propagate and carry their energy into the higher layers of the atmosphere through portals formed by the inclined magnetic field lines. Further, De Pontieu, Erdélyi, and James (2004) demonstrated that dynamic features such as solar chromospheric spicules or fibrils could be directly associated with the leakage of global p modes into the atmosphere along inclined field lines. They also found a strong correlation between propagating intensity oscillations in the corona and p modes. These and other results have prompted renewed strong interest in the theory of low-frequency acoustic wave propagation in stratified media.

The present paper deals with a two-layer model (Figure 1) to study the vertical propagation of acoustic waves in a stratified atmosphere. The main result is the discovery of a resonance occurring at low frequencies that extends into the *entire* unbounded atmosphere. This previously unknown resonance may be responsible for the transfer of wave energy, which could have dynamic consequences and heat the higher atmospheric layers. The presented results may have wider applicability in distinct areas of physics and astrophysics as well.

2. Model and Governing Equations

The proposed one-dimensional model is shown in Figure 1. The atmosphere is stratified under gravity in the z-direction. The temperature $T = T(z)$ linearly decreases in the lower part and remains constant in the upper part of the atmosphere:

$$T = \begin{cases} T_0(1-az), & 0 < z < L, \\ T_2, & z > L, \end{cases} \quad (1)$$

where the constant a ($aL = 1 - T_L/T_0 > 0$) characterizes the steepness of temperature decrease from $T_0 = T(0)$ to $T_L = T(L)$, and L is the thickness of the nonuniform layer. The temperature discontinuity at $z = L$ is a general assumption of the model, which is removed (by setting $T_L = T_2$) when solar applications are discussed. The variations of equilibrium pressure (p_0) and density (ρ_0) are given by (see, for example, Roberts, 2004)

$$p_0(z) = p_0(0) \exp\left(-\int_0^z \frac{dz}{\Lambda(z)}\right), \quad z > 0,$$

$$\rho_0(z) = \rho_0(0) \frac{\Lambda(0)}{\Lambda(z)} \exp\left(-\int_0^z \frac{dz}{\Lambda(z)}\right), \quad z > 0, \quad (2)$$

Here Λ represents the pressure scale height in the atmosphere determined by

$$\Lambda(z) = \frac{p_0(z)}{g\rho_0(z)} = \frac{c^2(z)}{\gamma g}. \quad (3)$$

Further, $c(z) = [\gamma p_0(z)/\rho_0(z)]^{1/2} = (\gamma RT)^{1/2}$ denotes the sound speed and γ is the adiabatic index. In the upper layer ($z > L$), the pressure scale height is constant and, according to Equations (2), the pressure and density decay exponentially with increasing height. In the lower part of the atmosphere, Equations (2) reduce to

$$p_0(z) = p_0(0)(1-az)^{\nu+1}, \quad 0 < z < L,$$
$$\rho_0(z) = \rho_0(0)(1-az)^{\nu}, \quad 0 < z < L, \quad (4)$$

where $\nu = \gamma g/(ac_0^2) - 1$ is the polytropic index and $c_0 = c(0)$. The Schwarzschild condition for convective stability (see, *e.g.*, Lighthill, 1978),

$$\frac{\frac{1}{T}\frac{dT}{dz}}{(1-\frac{1}{\gamma})\frac{1}{p_0}\frac{dp_0}{dz}} = \frac{1}{(1-\frac{1}{\gamma})(\nu+1)} < 1, \quad (5)$$

implies that the polytropic index $\nu > 3/2$ when $\gamma = 5/3$. The proposed model may represent the photosphere–chromosphere environment of the solar atmosphere. Wave propagation in solar media with rapid variations on very short spatial and time scales has been examined by, for example, Murawski and Roberts (1993) and Erdélyi, Kerekes, and Mole (2005). The behavior of the temperature in the solar atmosphere is a subject of debate with the validity of time-averaged hydrostatic models as accurate representations of the atmosphere being challenged by dynamic models with height-dependent temperature profiles that are significantly different from time-averaged models such as VAL or FAL (Vernazza, Avrett, and Loeser, 1981; Fontenla, Avrett, and Loeser, 1990; Carlsson and Stein, 1995). In the present paper, it is assumed that the chromospheric temperature is constant and the changes in the equilibrium take place on time scales that are long compared to the typical wave periods.

The vertical propagation of small-amplitude waves is governed by the following set of linearized differential equations of continuity, momentum, and energy (Lighthill, 1978):

$$\frac{\partial \rho}{\partial t} + \frac{\partial}{\partial z}(\rho_0 u) = 0, \qquad (6)$$

$$\rho_0 \frac{\partial u}{\partial t} = -\frac{\partial p}{\partial z} - \rho g, \qquad (7)$$

$$\frac{\partial p}{\partial t} + \frac{\partial p_0}{\partial z} u = c^2(z) \left(\frac{\partial \rho}{\partial t} + \frac{\partial \rho_0}{\partial z} u \right). \qquad (8)$$

The density and pressure perturbations (ρ and p) can be eliminated from Equations (6)–(8). The result is a single second-order PDE for the vertical velocity perturbation $u = u(z, t)$:

$$\rho_0 \frac{\partial^2 u}{\partial t^2} = \frac{\partial}{\partial z}\left(\rho_0 c^2 \frac{\partial u}{\partial z} \right). \qquad (9)$$

By introducing a scaling variable

$$Q(z,t) = \sqrt{\frac{\rho_0(z) c^2(z)}{\rho_0(0) c^2(0)}} u(z,t), \qquad (10)$$

Equation (9) is reduced to the KG equation:

$$\frac{\partial^2 Q}{\partial t^2} - c^2(z) \frac{\partial^2 Q}{\partial z^2} + \Omega^2(z) Q = 0, \qquad (11)$$

where

$$\Omega^2 = \frac{c^2}{4\Lambda^2}\left(1 + 2\frac{d\Lambda(z)}{dz}\right). \qquad (12)$$

The quantity Ω represents the acoustic cutoff frequency, which imposes a time scale on the system. An extensive review on solar applications of the KG equation is presented by Roberts (2004). In the present work, the KG Equation (11) is applied to the study of waves driven at a boundary of a semi-infinite nonisothermal atmosphere:

$$\lim_{z \to 0} Q(z,t) = I(\omega) \cos(\omega t), \qquad (13)$$

where ω is the driver frequency and $I = I(\omega)$ is the frequency-dependent amplitude of the driver. For simplicity, we assume that Q is a complex variable. The boundary condition (13) is then replaced by

$$\lim_{z \to 0} Q(z,t) = I(\omega) \exp(-i\omega t). \qquad (14)$$

3. Results and Discussion

We seek stationary-state solutions of the form

$$Q(z,t) = \tilde{Q}(z) \exp(-i\omega t). \qquad (15)$$

In the lower part of the atmosphere, Equation (11) can be transformed to a Bessel equation, which possesses solutions of the form

$$Q(z,t) = \exp(-i\omega t)\sqrt{1-az}\left[A_1 J_\nu\left(\frac{2\omega}{c_0 a}\sqrt{1-az}\right)\right.$$
$$\left. + B_1 Y_\nu\left(\frac{2\omega}{c_0 a}\sqrt{1-az}\right)\right], \quad 0 < z < L, \tag{16}$$

where J_ν and Y_ν are the Bessel functions of the first and second kind, respectively. In the upper layer $z > L$, the solution of Equation (11) has the form

$$Q(z,t) = A_2 \exp(ikz - i\omega t), \quad z > L, \tag{17}$$

where

$$k = \begin{cases} i\frac{\sqrt{\Omega_2^2 - \omega^2}}{c_2}, & \omega < \Omega_2, \\ \frac{\sqrt{\omega^2 - \Omega_2^2}}{c_2}, & \omega > \Omega_2, \end{cases} \tag{18}$$

with c_2 and Ω_2 being, respectively, the sound speed and cutoff frequency in the upper layer. Equation (18) shows that solution (17) represents an outgoing wave when $\omega > \Omega_2$. This is consistent with the fact that the system contains no other sources of wave energy except at $z = 0$. However, for $\omega < \Omega_2$ the form of solution (17) is a manifestation of the requirement of finite wave energy density at $z = \infty$. The coefficients A_1, A_2, and B_1 can be uniquely determined by imposing certain physical requirements on the solution Q. The first condition follows from the requirement of continuity of the normal component of the Lagrangian displacement and total equilibrium pressure at $z = L$. These conditions together with Equation (10) imply

$$Q(z,t)|_{z=L_-} = Q(z,t)|_{z=L_+}; \tag{19}$$

that is, the solution $Q(z,t)$ must be continuous across $z = L$. By integrating Equation (9) with respect to z from $L - \varepsilon$ to $L + \varepsilon$ and taking the limit $\varepsilon \to 0$, we find that $\partial u/\partial z$ should be continuous across $z = L$. Therefore, the second matching condition for $Q(z,t)$ has the form

$$\left(\frac{\partial Q}{\partial z} + \frac{1}{2\Lambda_1}Q\right)\bigg|_{z=L_-} = \left(\frac{\partial Q}{\partial z} + \frac{1}{2\Lambda_2}Q\right)\bigg|_{z=L_+}. \tag{20}$$

The resulting set of algebraic Equations (14), (19), and (20) is solved for the coefficients A_1, A_2, and B_1 to give

$$A_2 = \frac{I(\omega)\omega \exp(ikL)}{c_0}\left[J_\nu(\beta)Y_{\nu+1}(\beta) - J_{\nu+1}(\beta)Y_\nu(\beta)\right]$$
$$\bigg/ \left[\frac{\omega}{c_L}(J_\nu(\alpha)Y_{\nu+1}(\beta) - J_{\nu+1}(\beta)Y_\nu(\alpha))\right.$$
$$\left. + \left(\frac{1}{2\Lambda_2} + ik\right)(J_\nu(\beta)Y_\nu(\alpha) - J_\nu(\alpha)Y_\nu(\beta))\right], \tag{21}$$

$$A_1 = \left[I(\omega)Y_\nu(\beta) - \frac{A_2 \exp(ikL)Y_\nu(\alpha)}{\sqrt{1-aL}}\right][J_\nu(\alpha)Y_\nu(\beta) - J_\nu(\beta)Y_\nu(\alpha)], \tag{22}$$

$$B_1 = [I(\omega) - A_1 J_\nu(\alpha)]/Y_\nu(\alpha), \tag{23}$$

where Λ_2 is the constant-pressure scale height in the upper layer 2 and

$$\alpha = \frac{2\omega}{c_0 a}, \qquad \beta = \frac{2\omega}{c_0 a}\sqrt{1-aL}. \tag{24}$$

The coefficient A_2 determines the wave amplitude in the region above $z = L$. According to Equations (18) and (21), it may become infinitely large when k is imaginary. Hence there may be resonant frequencies located below the cutoff frequency in the upper layer: $\omega < \Omega_2$. An analytical examination of Equation (21) could be instructive as it could reveal the functional dependence of the amplitude upon the driver frequency and other parameters of the model. The complicated form of Equation (21) suggests that only numerical analysis is possible. However, it turns out that an analytical examination is possible when the lower layer ($z < L$) is *thin*:

$$\frac{L}{\Lambda(z)} \ll 1, \quad 1 - \frac{T(z)}{T_0} \ll 1, \quad \kappa(z,\omega)L \ll 1, \quad \text{for } 0 < z < L, \tag{25}$$

where $\kappa(z,\omega) = \sqrt{\omega^2 - \Omega^2(z)}/c(z)$ represents the local wavenumber in the WKB limit. It is easy to check that the first two conditions in Equation (25) are equivalent to $|f_0' L/f_0| \ll 1$, where the prime denotes differentiation with respect to z and f_0 is any of the equilibrium quantities (p_0, ρ_0, T). The coefficient A_2 is reduced to

$$A_2 = \frac{I(\omega)\exp(ikL)}{1 - \frac{L}{2\Lambda_2} + L\frac{\sqrt{\Omega_2^2-\omega^2}}{c_2}} \tag{26}$$

when the lower layer is thin (*i.e.*, when conditions (25) are fulfilled). Expression (26) is derived from Equation (21) by using Taylor expansions and the well-known property of the Wronskian of the Bessel functions (Abramowitz and Stegun, 1980):

$$W(J_\nu(z), Y_\nu(z)) = J_{\nu+1}(z)Y_\nu(z) - J_\nu(z)Y_{\nu+1}(z) = \frac{2}{\pi z}. \tag{27}$$

Equation (26) shows that a necessary condition for the existence of a resonance is $L/2\Lambda_2 > 1$. The waves are resonantly amplified when

$$\omega = \frac{c_2}{L}\sqrt{\frac{L}{\Lambda_2} - 1}. \tag{28}$$

It can be shown that the resonant frequency determined by Equation (28) is located between the cutoff frequencies in the upper and lower layers: $\Omega_0 \ll \omega < \Omega_2$. This also means that in the lower thin layer is much higher than in the upper layer ($T_0/T_2 \gg 1$) whenever a resonance exists.

According to Equations (21)–(23), the resonance has a *global* nature as A_1, B_1 are proportional to A_2. In the following numerical examples based on Equation (21), we fix the temperature T_2 (and so the cutoff frequency (Ω_2) and the scale height (Λ_2)). In Figure 2, the scaled resonant frequency (ω) is plotted against the scaled thickness (L). Three different cases with different temperature ratios ($T_0/T_2 = 2, 6, 10$) are shown. In all three cases, $T_L = T_2$ is set. The resonant frequencies consecutively appear and decrease as the thickness

Figure 2 Scaled resonant frequency (ω) as a function of the thickness of the nonuniform layer L. Three different cases are shown: $T_0/T_2 = 2$ (solid line), $T_0/T_2 = 6$ (dashed line), and $T_0/T_2 = 10$ (dotted line). In all three cases, $T_L = T_2$ is set.

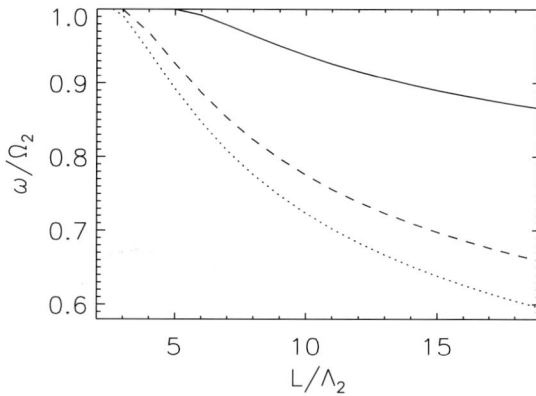

Figure 3 Dependence of scaled resonant frequency (ω) on scaled temperature (T_0) with $T_0/T_L = 2$ and $L/\Lambda_2 = 20$.

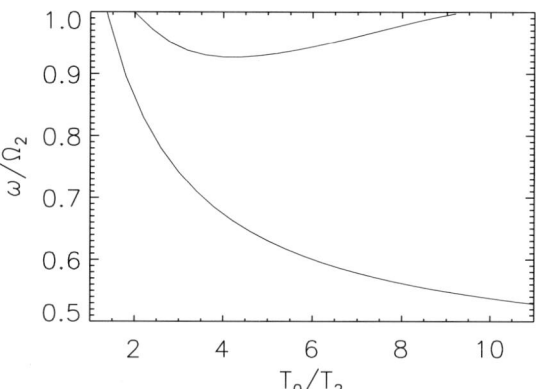

of the lower layer L increases. Figure 3 shows the variation of the resonant frequencies with increasing scaled temperature T_0. It is assumed that $T_0/T_L = 2$. The thickness L is fixed at $L/\Lambda_2 = 20$. There are two distinct curves, each appearing when the ratio T_0/T_2 is high enough. The lower curve in Figure 3 appears when $T_2 > T_L$. In general, the higher the length ratio L/Λ the lower the temperature ratio $T_0/T_2 > 1$ required for the existence of a resonance.

The physical mechanism responsible for wave amplification is the following: The decreasing temperature results in an increasing acoustic cutoff frequency $\Omega = \Omega(z)$, which forms a potential barrier similar to the one in quantum mechanics (Landau and Lifshitz, 1977). Low-frequency waves with $\Omega_0 < \omega < \Omega_2$ driven at $z = 0$ are reflected back from the barrier and are trapped in the lower layer 1. When the driver frequency matches the natural frequency of the cavity where the waves are trapped a standing wave is set up and is amplified resonantly. In the case of a thin layer, only the fundamental mode is present, with a frequency given by Equation (28). The frequency of the fundamental mode decreases and higher harmonics appear as the thickness L increases. For example, the upper curve in Figure 3 corresponds to the second harmonic. The resonance affects the evanescent tail of the waves in the upper atmosphere, leading to a global resonance. The temperature profile is discontinuous at $z = L$ in the numerical examples of Figure 3. Such a profile is perhaps unrealistic for the real Sun. Nevertheless, it is due to the discontinuity and the parameter values chosen in Figure 3 that we are able to clarify the physical mechanism associated with the resonance.

Solar acoustic p-mode wave packets are continually generated in the upper subphotospheric layers of the convection zone. The wave sources are thought to be located a few hundred kilometers below the visible surface. Kumar (1994) and Samadi et al. (2003) give depth estimates from 100 to 500 km. The temperature minimum occurs at a height of about 500 km, where the pressure scale height is about 120 km. The corresponding cutoff frequency is about 5 mHz. Assuming that the wave source is located at a depth of 200 km we obtain $T_0/T_L = 2$ and $L/\Lambda_2 = 5.9$. According to Figure 2, a resonance exists with a frequency close to $\Omega_2 = 5$ mHz and a corresponding period of three minutes. A numerical examination shows that the existence of the second harmonic requires a source located at a depth of more than 2 Mm (see Figure 3). Therefore, only the fundamental mode is expected to arise. The lower boundary of the present model should not be confused with the turning point of p modes resulting from temperature increase. It must be emphasized that the presented model is different from the chromospheric resonator introduced by Zhugzhda, Locans, and Staude (1983, 1984) in which efficient though partial reflection from the transition region occurs, leading to several finite peaks, equally spaced at ≈ 1 mHz in frequency. Observations usually indicate the presence of one dominant peak (see, *e.g.*, Brynildsen *et al.*, 2002). The present model predicts the existence of a single resonant peak. There is no wave transmission at the upper boundary and, as a consequence, an actual resonance occurs.

The present paper is aimed at unveiling the new resonant mechanism. The nonlinear development and dissipation of such waves generated in various physical systems and their energetic implications must be treated separately. Apart from the three-minute oscillations, these may include the generation of spicules or the heating of the lower atmosphere. An important extension of the present work is the treatment of the problem in two dimensions, the consideration of a nonharmonic driver, and the effects of temperature rise in the chromosphere. The present model neglects the transition region and regards the chromosphere as a semi-infinite layer. The resonance is a consequence of the wave reflection from a potential barrier resulting from the increasing cutoff frequency. It is well known that the transition region acts as a reflector on low-frequency waves because of the steep temperature gradient (see, *e.g.*, recent numerical works by Erdélyi *et al.*, 2007). Therefore, the inclusion of this region is most likely to add further reflectivity on the upward propagating waves. However, a more detailed examination of the role of the transition region is an important follow-up step.

The resonant waves presented here arise from the variation of the cutoff frequency introduced by stratification. However, such waves could operate in other systems with varying cutoff frequencies. The cutoff frequencies of waves, such as Alfvén, kink, and slow waves, in thin magnetic-flux tubes vary owing to, for example, cross-section expansion of the flux tube or the variations in the magnetic field strength and density (Spruit and Roberts, 1983).

Acknowledgements Y.T. thanks the Leverhulme Trust for financial support. R.E. acknowledges NSF, Hungary (OTKA, Ref. No. K67746).

References

Abramowitz, M., Stegun, I.A.: 1980, *Handbook of Mathematical Functions*, Dover, New York.
Banerjee, D., Erdélyi, R., Oliver, R., O'Shea, E.: 2007, *Solar Phys.* **246**, 3.
Brynildsen, N., Maltby, P., Fredvik, T., Kjeldseth-Moe, O.: 2002, *Solar Phys.* **207**, 259.

Carlson, J., Heiselberg, H., Pandharipande, V.R.: 1982, *Phys. Rev. C* **63**, 7603.
Carlsson, J., Stein, R.F.: 1995, *Astrophys. J.* **440**, L29.
De Pontieu, B., Erdélyi, R., James, S.P.: 2004, *Nature* **430**, 546.
De Pontieu, B., Erdélyi, R., De Moortel, I.: 2005, *Astrophys. J.* **624**, L61.
Detweiler, S.: 1980, *Phys. Rev. D* **22**, 2323.
Erdélyi, R., Ballai, I.: 2007, *Astron. Nachr.* **328**, 726.
Erdélyi, R., Kerekes, A., Mole, N.: 2005, *Astron. Astrophys.* **431**, 1083.
Erdélyi, R., Malins, C., Tóth, G., de Pontieu, B.: 2007, *Astron. Astrophys.* **467**, 1299.
Fossum, A., Carlsson, M.: 2005, *Nature* **435**, 919.
Fontenla, J.M., Avrett, E.H., Loeser, R.: 1990, *Astrophys. J.* **355**, 700.
Jefferies, S.M., McIntosh, S.W., Armstrong, J.D., Bogdan, T.J., Cacciani, A.B., Fleck, B.: 2006, *Astrophys. J.* **648**, L151.
Kumar, P.: 1994, *Astrophys. J.* **428**, 827.
Lamb, H.: 1909, *Proc. Lond. Math. Soc.* **7**, 122.
Landau, L.D., Lifshitz, E.M.: 1977, *Quantum Mechanics*, Pergamon Press, Oxford.
Lighthill, J.: 1978, *Waves in Fluids*, Cambridge University Press, Cambridge.
Murawski, K., Roberts, B.: 1993, *Astron. Astrophys.* **272**, 60.
Poisson, S.-D.: 1808, *J. l'École Polytech.* **7**, 319.
Poschl, W., Vretenar, D., Lalazissis, G.A., Ring, P.: 1997, *Phys Rev. Lett.* **79**, 3841.
Rae, I.C., Roberts, B.: 1982, *Astrophys. J.* **256**, 761.
Rayleigh, J.W.: 1890, *Phil. Mag.* **29**, 173.
Roberts, B.: 2004, In: Erdélyi, R., Ballester, J., Fleck, B. (eds.) *SOHO 13 Waves, Oscillations and Small-Scale Transients Events in the Solar Atmosphere: Joint View from SOHO and TRACE* **SP-547**, ESA, Noordwijk, 1.
Samadi, R., Nordlund, A., Stein, R.F., Goupil, M.J., Roxburgh, I.: 2003, *Astron. Astrophys.* **404**, 1129S.
Spruit, H.C., Roberts, B.: 1983, *Nature* **304**, 401.
Sutman, G., Musielak, Z.E., Ulmschneider, P.: 1998, *Astron. Astrophys.* **340**, 556.
Vernazza, J.E., Avrett, E.H., Loeser, R.: 1981, *Astrophys. J.* **45**, 635.
Zhugzhda, Iu.D., Locans, V., Staude, J.: 1983, *Solar Phys.* **82**, 369.
Zhugzhda, Iu.D., Staude, J., Locans, V.: 1984, *Solar Phys.* **91**, 219.

DOT Tomography of the Solar Atmosphere
VII. Chromospheric Response to Acoustic Events

R.J. Rutten · B. van Veelen · P. Sütterlin

Originally published in the journal Solar Physics, Volume 251, Nos 1–2, 533–547.
DOI: 10.1007/s11207-008-9116-9 © The Author(s) 2008

Abstract We use synchronous movies from the Dutch Open Telescope sampling the G band, Ca II H, and Hα with five-wavelength profile sampling to study the response of the chromosphere to acoustic events in the underlying photosphere. We first compare the visibility of the chromosphere in Ca II H and Hα, demonstrate that studying the chromosphere requires Hα data, and summarize recent developments in understanding why this is so. We construct divergence and vorticity maps of the photospheric flow field from the G-band images and locate specific events through the appearance of bright Ca II H grains. The reaction of the Hα chromosphere is diagnosed in terms of brightness and Doppler shift. We show and discuss three particular cases in detail: a regular acoustic grain marking shock excitation by granular dynamics, a persistent flasher, which probably marks magnetic-field concentration, and an exploding granule. All three appear to buffet overlying fibrils, most clearly in Dopplergrams. Although our diagnostic displays to dissect these phenomena are unprecedentedly comprehensive, adding even more information (photospheric Doppler tomography and magnetograms along with chromospheric imaging and Doppler mapping in the ultraviolet) is warranted.

Keywords Chromosphere · Granulation · Oscillations · Waves

Helioseismology, Asteroseismology and MHD Connections
Guest Editors: Laurent Gizon and Paul Cally.

Electronic supplementary material The online version of this article
(http://dx.doi.org/10.1007/978-0-387-89482-9_36) contains supplementary material, which is available to authorized users.

R.J. Rutten (✉) · B. van Veelen · P. Sütterlin
Sterrekundig Instituut, Utrecht University, Utrecht, The Netherlands
e-mail: r.j.rutten@astro.uu.nl

B. van Veelen
e-mail: b.vanveelen@astro.uu.nl

P. Sütterlin
e-mail: p.suetterlin@astro.uu.nl

R.J. Rutten
Institute of Theoretical Astrophysics, University of Oslo, Oslo, Norway

1. Introduction

"Acoustic events" (or "seismic events") is the term used by P. Goode and co-workers (Stebbins and Goode, 1987; Goode, Gough, and Kosovichev, 1992; Restaino, Stebbins, and Goode, 1993; Rimmele et al., 1995; Goode et al., 1998; Strous, Goode, and Rimmele, 2000) to describe localized, small-scale happenings in the granulation that produce excessive amounts of upward-propagating acoustic waves in the upper photosphere, with the claim that these indicate the kinetic sources of the global p modes, "the smoke from the fire exciting the solar oscillations" (Goode, 1995). Their technique was to sample the Doppler modulation of a suitable spectral line at different profile heights through Fabry–Perot imaging with rapid wavelength shifting, and then isolate the surface locations with the largest oscillatory amplitude at five-minute periodicity and upward-propagating phase. Their result was that this measure of "acoustic flux" tends to be maximum above intergranular lanes, in particular those in which a small granule has vanished in so-called granular collapse.

Acoustic excitation through granular dynamics was also addressed theoretically and through numerical hydrodynamics simulations by Rast (1995, 1999), Skartlien and Rast (2000), and Skartlien, Stein, and Nordlund (2000), confirming the picture of small vanishing granules, especially at mesogranular downdraft boundaries, acting as "collapsars" to excite upward-propagating waves, in particular at three-minute periodicity corresponding to the photospheric acoustic cutoff frequency.

In this paper we address the reaction of the overlying chromosphere to such events. Skartlien, Stein, and Nordlund (2000) suggested that they may also cause Ca II H_{2V} and K_{2V} cell grains, an enigma for many decades (see the review by Rutten and Uitenbroek, 1991) that was eventually solved by Carlsson and Stein (1994, 1997), who identified them as marking upward-propagating acoustic shocks. The announcement by Goode (2002) that indeed acoustic events cause such acoustic grains in the chromosphere was followed up by Hoekzema, Rimmele, and Rutten (2002), who found that, although extreme acoustic events do tend to correlate with the subsequent appearance of exceptionally bright Ca II K_{2V} internetwork grains, this correspondence is far from a one-to-one correlation.

We now turn to Hα rather than H and K as diagnostic of the chromospheric response to acoustic events. It is important to note and to explain that Hα is a much better chromospheric diagnostic than the Ca II H and K lines are. This is not the case for simple Saha–Boltzmann LTE partitioning in which H and K have larger opacity than Hα throughout the atmosphere (see Figure 6 of Leenaarts et al., 2006, and the second student exercise at http://www.astro.uu.nl/ rutten/education/rjr-material/ssa), and it is also not the case in the standard NLTE statistical and hydrostatic equilibrium VAL modeling of Vernazza, Avrett, and Loeser (1973, 1976, 1981) in which the line center of Ca II K is formed higher than the line center of Hα (see the celebrated Figure 1 of Vernazza, Avrett, and Loeser, 1981). We use displays from older data here to argue that the real Sun does not conform.

Figure 1 compares Dutch Open Telescope (DOT) images using the two diagnostics. The upper image is taken in Ca II H and shows a mixture of mid- and upper photospheric contributions (reversed granulation and acoustic grains) and chromospheric contributions (bright plage, network, and straws in hedge rows as described by Rutten, 2007). The lower panel shows the same area observed in Hα line center. Comparison of the two demonstrates unequivocally that in such filter imaging Hα presents a much more complete picture of the chromosphere. In our opinion, the chromosphere is best defined as the collective of ubiquitous fibrils seen in this line. Most of these are internetwork-spanning structures, which probably outline magnetic canopies. Only truly quiet internetwork areas (some of which are seen in the lower part of the image) are not masked by such canopy fibrils but show much

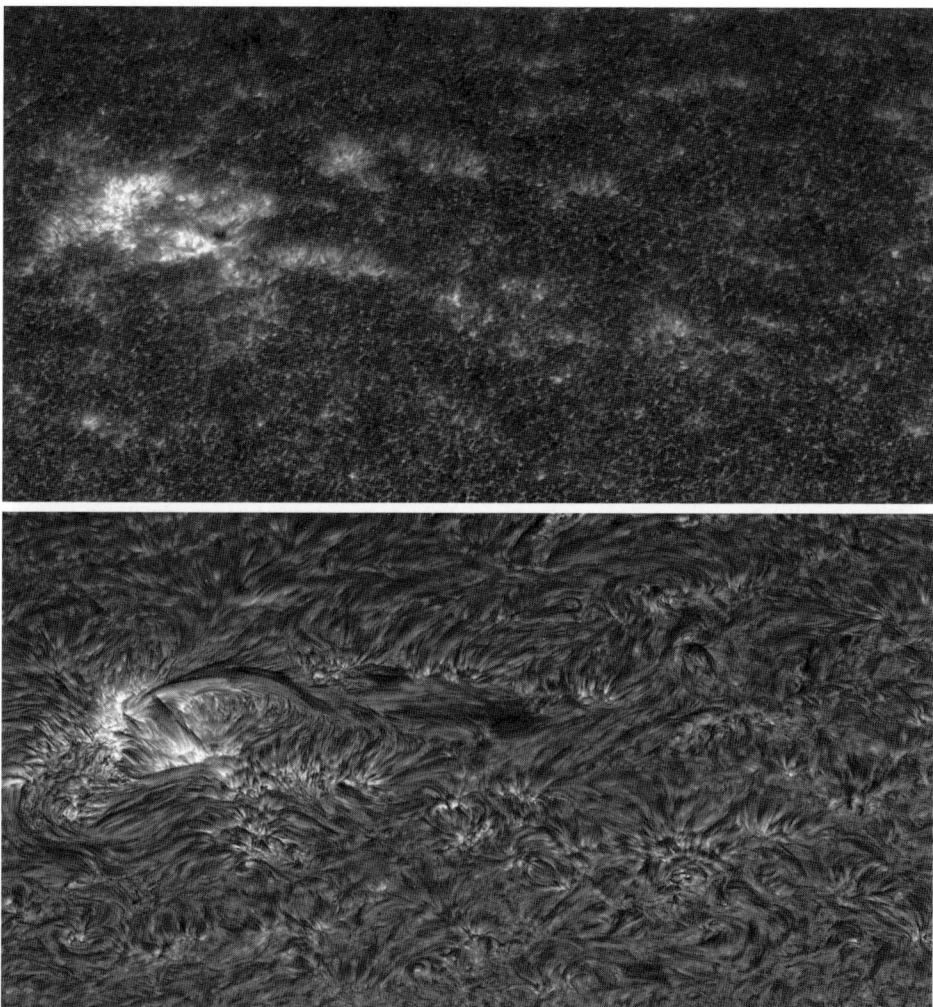

Figure 1 Simultaneous Ca II H and Hα image mosaics taken with the Dutch Open Telescope on 4 October 2005. The field of view is close to the limb (located off the top) and measures about 265×143 arcsec^2. The spectral passbands are FWHM 1.4 Å for Ca II H and 0.25 Å for Hα. The same area was observed with the SST and described by van Noort and Rouppe van der Voort (2006). High-resolution images of Figures 1–6 are available as Electronic Supplementary Material at http://dx.doi.org/10.1007/s11207-008-9116-9.

shorter, highly dynamic loops and grainy brightness patterns that are not connected to the network (see also Rouppe van der Voort *et al.*, 2007). None of these structures is observed in the upper panel, except for the straws that correspond to bright Hα fibril endings.

What about the spectral passbands in this comparison? The Doppler cores of H and K are much narrower than for Hα, but H and K filter bandpasses are usually wider with FWHMs of 3 Å for *Hinode*, 1.4 Å for the DOT, 0.6 Å for the Lyot filter at the German VTT (*e.g.*, Tritschler *et al.*, 2007), 0.3 Å for the Lockheed–Martin filter at the former SVST (Brandt *et al.*, 1992; Lites, Rutten, and Berger, 1999), and also 0.3 Å for the Halle filter at the NSO/SP/DST (*e.g.*, Bloomfield *et al.*, 2004, who assigned propagation speeds and

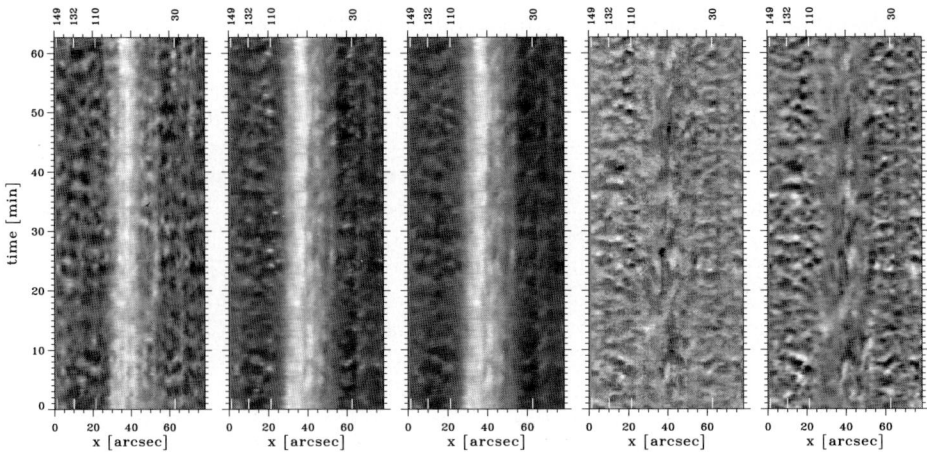

Figure 2 Various diagnostic measures of Ca II H plotted as space–time "time slice" evolution plots, assembled by B.W. Lites from a spectral sequence taken by him and W. Kalkofen in 1984 with the Dunn Solar Telescope at the National Solar Observatory/Sacramento Peak. The waves in the network part (bright strip at the center) were analyzed by Lites, Rutten, and Kalkofen (1993). The waves in the internetwork columns identified by numbers along the top were numerically simulated by Carlsson and Stein (1997). The spatial coordinate is measured along the slit of the spectrograph. The first three panels have logarithmic gray scales to reduce the large contrast between network and internetwork. Horizontal grayish "erasures" are due to poorer seeing. First panel: Intensity integrated over a spectral band of 0.9 Å width centered on line center. Second panel: The same for 0.16 Å bandpass. Third panel: Intensity of the core-profile minimum. Fourth panel: Wavelength variation of the core-profile minimum. Fifth panel: Dopplergram ratio $(R-V)/(R+V)$, with R and V being the intensities in 0.16 Å passbands at H_{2R} and H_{2V}.

mode conversions to chromospheric oscillations by simply adopting the VAL3C line-center height difference). Such H and K filter imaging adds considerable inner-wing contributions from the upper photosphere, including bright reversed granulation and yet brighter acoustic grains, which may wash out the line-center-only contribution from chromospheric fibrils in the internetwork except where these are very bright, as in straws. The narrowest band images with high spatial resolution published so far are R.A. Shine's 0.3 Å ones in Lites, Rutten, and Berger (1999), which only show reversed granulation and acoustic grains in the internetwork. Narrow-band K_3 spectroheliograms such as the 0.1 Å ones in the collection of Title (1966) have lower angular resolution but do not suggest the presence of similar masses of internetwork Ca II K fibrils as seen in Hα at similar resolution, except near active network and plage.

Figure 2 extends the Ca II H passband comparisons in Figure 1 of Rutten, De Pontieu, and Lites (1999) and Figure 9 of Krijger *et al.* (2001) with additional spectral measures from the ancient but high-quality Ca II H spectrogram sequence of Lites, Rutten, and Kalkofen (1993), which was also used by Carlsson and Stein (1997). The first three panels show that a narrower Ca II H passband produces larger network-to-internetwork brightness contrast but does not significantly change the morphology of the observed scene. The oscillation patterns in the internetwork, brightest at H_{2V}, remain visible even in the line-core intensity in the third panel. The line-core shift in the fourth panel confirms the acoustic nature of these patterns. Thus, even at full spectral resolution, the Ca II H core in this internetwork sample is dominated by radial three-minute waves, without obliteration by overlying fibrils or fibril-aligned motions. The final panel showing the $H_{2R} - H_{2V}$ Dopplergram ratio strengthens this conclusion. This measure comes closest to exhibiting chromospheric-fibril structure in the form of oscillatory branches jutting out from the network with time, but regular three-minute oscillation patterns dominate further away in the internetwork.

The lower left panel of Figure 21 of Krijger *et al.* (2001) demonstrated the close correspondence between Ca II H line-center dynamics and underlying photospheric dynamics in the internetwork parts of these spectrograms in Fourier terms. In fact, it would be strange if the line-center formation layer would not fully partake in the shock dynamics in these data because substantial in-phase upper atmosphere downdraft, above the next upcoming grain-causing acoustic shock, is required to give the H_{2V} and K_{2V} grains their characteristic spectral asymmetry, as shown in the formation breakdown diagrams of Carlsson and Stein (1997). They actually obtained their best sequence reproduction for near-network column 110 in Figure 2, suggesting that even there the field was too weak to upset H_3 acoustic shock signatures.

Thus, these well-studied Ca II H spectrograms contradict the presence of opaque chromospheric fibrils masking upper photosphere (or "clapotispheric") internetwork dynamics in H and K, whereas $H\alpha$ canopy fibrils ubiquitously do so. The notion that the Ca II core generally forms higher than the $H\alpha$ core, or that chromospheric structures should generally be more opaque in H and K than in $H\alpha$, seems questionable. The observations suggest instead that the fibrils that constitute the $H\alpha$ chromosphere are either transparent in H and K or, when opaque, are either pummeled by shocks from below without hindrance (such as magnetic tension) or represent a featureless, very dark blanket located well above the clapotisphere.

In summary, we suspect that much of the K_3 internetwork dynamics is dominated by shock modulation at heights around 1000 km, well below the VAL3C K_3 formation height of 1800–2000 km and far below most $H\alpha$ canopy fibrils.

This large discrepancy with equilibrium modeling may become understandable with the recent simulation of Leenaarts *et al.* (2007) implementing noninstantaneous hydrogen ionization and recombination. Three aspects combine in such an explanation. First, the simulated chromosphere is pervaded by shocks. Second, the large difference between hydrogen ionization/recombination balancing speed in hot shocks and in cool post-shock gas implies that instantaneous statistical equilibrium does not apply to the latter. Carlsson and Stein (2002) already explained why this is so. The large 10-eV excitation energy of its $n = 2$ level causes hydrogen to be ionized rapidly in shocks but to recombine slowly in the cool intershock phases so that the ionization degree remains high also in the latter. Third, the population of $H\alpha$'s lower $n = 2$ level is closely coupled to the ion population (through the Balmer continuum; see also Rutten and Carlsson, 1994) and therefore remains similarly high at low post-shock temperature. Thus, the large $n = 2$ excitation energy, being the very reason why $H\alpha$ has very small cool-gas opacity in Boltzmann balancing, slows down the hydrogen recombination in post-shock cool gas so much that its $H\alpha$ opacity exceeds the LTE value by many orders of magnitude. Therefore, fibrils can be opaque in $H\alpha$ whether they are shock-hot or post-shock cool, and they even may be more opaque in $H\alpha$ than in H and K at any temperature – in utter conflict with equilibrium modeling.

The chromospheric nonequilibrium balancing between Ca II (12-eV ionization energy) and Ca III is probably faster (with no top-heavy term diagram) but has not yet been analyzed in similar detail; it is of interest to evaluate the opacity ratio between H and K and $H\alpha$ in dynamic conditions. Another point of interest is that chromospheric fibrils seem to be more obvious in Ca II 8542 Å than in H_3 and K_3, possibly because of larger Doppler sensitivity (Cauzzi *et al.*, 2007).

Thus, $H\alpha$ is the key diagnostic of the internetwork chromosphere. Recent analyses of the chromosphere observed at high cadence in $H\alpha$ line center with the Swedish 1-m Solar

Telescope (SST) have shed new light on chromospheric fibrils, showing that near network and plage "dynamic fibrils" tend to have repetitive mass loading at three- to five-minute shock periodicity (De Pontieu *et al.*, 2007) whereas the little loops in quiet-Sun internetwork are even more dynamic, presumably because of stronger buffeting by shocks where the magnetic field is weak (Rouppe van der Voort *et al.*, 2007).

In this paper we use image sequences from the DOT with lower cadence than Hα really requires (van Noort and Rouppe van der Voort, 2006), but with Hα profile sampling and synchronous co-spatial imaging in other wavelengths including Ca II H. These images permit us to isolate specific events in the photosphere and inspect the chromospheric response in the apparent brightness of Ca II H and Hα, and particularly in Hα Doppler modulation. We do this here by presenting three exemplary cases: an acoustic grain, a persistent flasher, and an exploding granule.

2. Observations and Reduction

The image sequences used here were obtained with the DOT on La Palma on 14 October 2005 during 10:15:43 – 10:30:42 UT. The telescope, instrumentation, and data processing are described by Rutten *et al.* (2004).

Synchronous and co-spatial images were taken in blue and red continuum passbands, in the G band around 4305 Å, in Ca II H alternatingly at line center and at $\Delta\lambda = -1$ Å, and at five Hα wavelengths: line center, the H$\alpha \pm 350$ mÅ inner wing pair, and the H$\alpha \pm 700$ mÅ outer wing pair. The seeing had mean Fried parameter $r_0 = 7$ cm for the G band, just about good enough that the standard DOT speckle reconstruction produced angular resolution close to the diffraction limit of the 45-cm aperture (0.2 arcsec in the blue). The field of view measured 79×60 arcsec^2 (56×43 Mm^2), covering a very quiet area at disk center containing only few and sparse weak-network clusters of magnetic elements (bright points in the G band, bright grain clusters in Ca II H), as demonstrated by the first two sample images in Figure 3. This area was sufficiently quiet that Hα shows only short, highly dynamic fibrils concentrated around the magnetic clusters.

The image sequences had slightly irregular cadence and were interpolated to 30-image sequences at strict 30-second cadence. The G-band images were subsonically Fourier-filtered in (k, ω) space, passing only components with apparent motion below the sound speed of 7 km s^{-1} to remove the brightness modulation owing to the p-mode oscillations since in this study granular dynamics is emphasized as a pistoning agent. Horizontal flows were measured by using the cross-correlation tracking algorithm of November (1986) and November and Simon (1988), with Gaussian FWHM widths of 1.5 arcsec and 330 seconds as boxcar averaging parameters. Differences between neighboring pixels were used to evaluate derivatives to determine the local horizontal divergence and vorticity in the flows.

The Hα sequences were also cone-filtered in (k, ω) space but in this case reversely, only passing the supersonic components because here the oscillatory response of the chromosphere is of interest. Our intention was to remove stable fibrils to enhance the visibility of oscillatory modulations such as spreading rings, but in this very quiet area the fibrils are so dynamic that the differences between the filtered and unfiltered sequences are small. Finally, Hα Dopplergrams were constructed from the filtered H$\alpha \pm 700$ mÅ and H$\alpha \pm 350$ mÅ image pairs.

We used a live, multipanel, on-screen display similar to Figures 4 – 6 to inspect the various diagnostics simultaneously and in mutual correspondence at large local magnification, varying both the location and time coordinates of the point of scrutiny within these image granule had already a large dark center and was about to become three small granules with some shards in between. A minute later, the Ca II H line became very bright at that location, and after two more minutes very dark (top row). The same dark blob is seen along the top row in the Hα wing columns, implying large updraft. The Hα ± 350 mÅ Doppler time slice in the last panel again shows much more spatially extended oscillatory behavior than the other diagnostics, suggesting fibril buffeting. The Hα line-center intensity time slice again mimics the Doppler behavior partially. The exploding granule seems to cause an oscillation amplitude increase above it that modulates the Hα core intensity.

6. Discussion

Each example shows a large morphological difference among the low-photosphere scene in the G band, the high-photosphere scene in Ca II H, and the chromospheric scene in Hα. The Hα cutouts do not contain internetwork-spanning fibrils in this quiet region; each of the three events therefore shows chromospheric response that would otherwise have been blocked by overlying canopy fibrils. The responses are not very striking; in no case may one locate the photospheric happening uniquely from its subsequent Hα signature.

The Hα chromosphere sampled by the image cutouts comes close to our tentative description of the Ca II H_3 and K_3 internetwork chromosphere as a featureless opaque blanket pummeled from below. They may indeed be about the same, as suggested by the Hα downdrafts in Figure 4 above grains requiring H_3 downdraft for H_{2V} profile asymmetry. The much smaller temperature sensitivity of the H and K opacity then indeed diminishes the H and K feature contrast. Because Ca II 8542 is an excited and narrower line, it combines larger temperature and Doppler sensitivity to such pummeling but at smaller blanket opacity.

Fibrilar structuring appears most clearly in the Hα Dopplergram time slices. Our impression is that indeed this quiet Hα chromosphere is continuously pummeled by internetwork waves, gaining much dynamics from these in the quietest areas where the magnetic fields have low tension but yet enough to impose slight fibrilarity. We suspect that slanted internal-gravity waves contribute significantly in this pummeling (see Lighthill, 1967, pp. 440–443) and, together with the primarily vertically growing acoustic shocks, cause intricate and fast-changing interference patterns at granular to mesogranular scales. Only the fiercest events punch through to be individually recognizable directly above their sources.

Our three examples whet the appetite for even more comprehensive multidiagnostic solar-atmosphere tomography. The complexity of the panel layout in Figures 4–6 already demonstrates that the dynamical coupling between photosphere and chromosphere can only be addressed holistically with diverse diagnostics. The complexity of the solar scenes shown in these panels and their large differences among diagnostics strengthen this conclusion. In fact, our data are yet incomplete: we lack the principal acoustic-event measure of "acoustic flux" in this analysis, requiring Fabry–Perot Doppler mapping of the photosphere at multiple heights. Sensitive photospheric magnetic field mapping and ultraviolet chromospheric image and Doppler diagnostics would also be welcome – preferably all at similar or better angular resolution, cadence, duration, and field size than used here.

In this paper we have limited the discussion to three, hand-picked cases from an early set of tomographic DOT image sequences. The large DOT database collected in the meantime (openly available at http://dotdb.phys.uu.nl/DOT) now permits wider statistical study of such various types of events. Rapid-cadence Fabry–Perot imaging at the SST promises unprecedented Hα diagnostics. Space-mission co-pointing adds EUV diagnostics of the transition region. Pertinent numerical simulations including realistic formation of Hα are also coming into reach.

Acknowledgements The DOT is owned by Utrecht University and located at the Spanish Observatorio del Roque de los Muchachos of the Instituto de Astrofísica de Canarias. We are deeply indebted to V. Gaizauskas for the DOT Hα filter. We thank M. Carlsson, Ø. Langangen, S.E.M. Keek, J.M.D. Kruijssen, and A.G. de Wijn for inspiring debates. This research made much use of SST hospitality and of NASA's Astrophysics Data System. R.J. Rutten thanks the Leids Kerkhoven-Bosscha Fonds and the organizers of the SOHO 19/GONG 2007 meeting for travel support.

References

Babcock, H.W., Babcock, H.D.: 1955, *Astrophys. J.* **121**, 349.
Bloomfield, D.S., McAteer, R.T.J., Mathioudakis, M., Williams, D.R., Keenan, F.P.: 2004, *Astrophys. J.* **604**, 936.
Brandt, P.N., Rutten, R.J., Shine, R.A., Trujillo Bueno, J.: 1992, In: Giampapa, M.S., Bookbinder, J.A. (eds.) *Cool Stars, Stellar Systems, and the Sun, Proc. Seventh Cambridge Workshop* **26**, ASP, San Francisco, 161.
Brandt, P.N., Rutten, R.J., Shine, R.A., Trujillo Bueno, J.: 1994, In: Rutten, R.J., Schrijver, C.J. (eds.) *Solar Surface Magnetism, NATO ASI Series C* **433**, Kluwer, Dordrecht, 251.
Carlsson, M., Stein, R.F.: 1994, In: Carlsson, M. (ed.) *Chromospheric Dynamics. Proc. Miniworkshop*, Inst. Theor. Astrophys., Oslo, 47.
Carlsson, M., Stein, R.F.: 1997, *Astrophys. J.* **481**, 500.
Carlsson, M., Stein, R.F.: 2002, *Astrophys. J.* **572**, 626.
Cauzzi G., Reardon K.P., Uitenbroek H., Cavallini F., Falchi A., Falciani R., Janssen K., Rimmele T., Vecchio A., Woeger F.: 2007, *Astron. Astrophys.* submitted. http://arxiv.org/abs/0709.2417.
Cheung, M.C.M., Schüssler, M., Moreno-Insertis, F.: 2007, *Astron. Astrophys.* **461**, 1163.
De Pontieu, B., Hansteen, V.H., Rouppe van der Voort, L., van Noort, M., Carlsson, M.: 2007, *Astrophys. J.* **655**, 624.
De Wijn, A.G., Rutten, R.J., Haverkamp, E.M.W.P., Sütterlin, P.: 2005, *Astron. Astrophys.* **441**, 1183.
Goode, P.: 1995, STI/Recon Technical Report N **96**, NASA, Washington, 10005.
Goode P.R.: 2002, *Bull. Am. Astron. Soc.* **34**, 730.
Goode, P.R., Gough, D., Kosovichev, A.G.: 1992, *Astrophys. J.* **387**, 707.
Goode, P.R., Strous, L.H., Rimmele, T.R., Stebbins, R.T.: 1998, *Astrophys. J.* **495**, L27.
Hoekzema, N.M., Rimmele, T.R., Rutten, R.J.: 2002, *Astron. Astrophys.* **390**, 681.
Krijger, J.M., Rutten, R.J., Lites, B.W., Straus, T., Shine, R.A., Tarbell, T.D.: 2001, *Astron. Astrophys.* **379**, 1052.
Leenaarts, J., Wedemeyer-Böhm, S.: 2005, *Astron. Astrophys.* **431**, 687.
Leenaarts, J., Rutten, R.J., Sütterlin, P., Carlsson, M., Uitenbroek, H.: 2006, *Astron. Astrophys.* **449**, 1209.
Leenaarts, J., Carlsson, M., Hansteen, V., Rutten, R.J.: 2007, *Astron. Astrophys.* **473**, 625.
Lighthill, M.J.: 1967, In: Thomas, R.N. (ed.) *Aerodynamical Phenomena in Stellar Atmospheres, IAU Symp.* **28**, Academic Press, New York, 429.
Lites, B.W., Rutten, R.J., Kalkofen, W.: 1993, *Astrophys. J.* **414**, 345.
Lites, B.W., Rutten, R.J., Berger, T.E.: 1999, *Astrophys. J.* **517**, 1013.
November, L.J.: 1986, *Appl. Opt.* **25**, 392.
November, L.J., Simon, G.W.: 1988, *Astrophys. J.* **333**, 427.
Rast, M.P.: 1995, *Astrophys. J.* **443**, 863.
Rast, M.P.: 1999, *Astrophys. J.* **524**, 462.
Restaino, S.R., Stebbins, R.T., Goode, P.R.: 1993, *Astrophys. J.* **408**, L57.
Rimmele, T.R., Goode, P.R., Harold, E., Stebbins, R.T.: 1995, *Astrophys. J.* **444**, L119.
Rouppe van der Voort, L.H.M., De Pontieu, B., Hansteen, V.H., Carlsson, M., van Noort, M.: 2007, *Astrophys. J.* **660**, L169.
Rutten R.J.: 2007, In: Heinzel, P., Dorotovič, I., Rutten, R.J. (eds.) *The Physics of Chromospheric Plasmas, Conf. Series* **368**, ASP, San Francisco, 27.
Rutten R.J., Carlsson M.: 1994, In: Rabin, D.M., Jefferies, J.T., Lindsey, C (eds.) *Infrared Solar Physics, IAU Symposium* **154**, Kluwer, Dordrecht, 309.
Rutten, R.J., Uitenbroek, H.: 1991, *Solar Phys.* **134**, 15.
Rutten, R.J., De Pontieu B., Lites B.: 1999, In: Rimmele, T.R., Balasubramaniam, K.S., Radick, R.R. (eds.) *High Resolution Solar Physics: Theory, Observations, and Techniques, Conf. Series* **183**, ASP, San Francisco, 383.
Rutten, R.J., De Wijn, A.G., Sütterlin, P.: 2004, *Astron. Astrophys.* **416**, 333.

Rutten, R.J., Hammerschlag, R.H., Bettonvil, F.C.M., Sütterlin, P., De Wijn, A.G.: 2004, *Astron. Astrophys.* **413**, 1183.
Skartlien, R., Rast, M.P.: 2000, *Astrophys. J.* **535**, 464.
Skartlien, R., Stein, R.F., Nordlund, Å.: 2000, *Astrophys. J.* **541**, 468.
Stebbins, R., Goode, P.R.: 1987, *Solar Phys.* **110**, 237.
Strous, L.H., Goode, P.R., Rimmele, T.R.: 2000, *Astrophys. J.* **535**, 1000.
Title A.M.: 1966, *Selected Spectroheliograms*, Mount Wilson and Palomar Observatories, Pasadena.
Tritschler, A., Schmidt, W., Uitenbroek, H., Wedemeyer-Böhm, S.: 2007, *Astron. Astrophys.* **462**, 303.
van Noort, M.J., Rouppe van der Voort, L.H.M.: 2006, *Astrophys. J.* **648**, L67.
Vernazza, J.E., Avrett, E.H., Loeser, R.: 1973, *Astrophys. J.* **184**, 605.
Vernazza, J.E., Avrett, E.H., Loeser, R.: 1976, *Astrophys. J. Suppl. Ser.* **30**, 1.
Vernazza, J.E., Avrett, E.H., Loeser, R.: 1981, *Astrophys. J. Suppl. Ser.* **45**, 635.

Velocity and Intensity Power and Cross Spectra in Numerical Simulations of Solar Convection

G. Severino · T. Straus · M. Steffen

Originally published in the journal Solar Physics, Volume 251, Nos 1–2, 549–562.
DOI: 10.1007/s11207-008-9156-1 © Springer Science+Business Media B.V. 2008

Abstract Fitting observed power and cross spectra of medium-degree p modes in velocity (V) and intensity (I) has been widely used for getting information about the p-mode excitation process and, in particular, for trying to determine the type and location of the acoustic sources. Numerical simulations of solar convection allow one to "observe" velocity and temperature (T, used as proxy for I) fluctuations in different reference frames. Sampling the oscillations on planes of constant optical depth (τ-frame) closely corresponds to the observer's point of view, whereas sampling the oscillations at constant geometrical height (z-frame) is more appropriate for comparison with predictions from theoretical models based on Eulerian hydrodynamics. The results of the analysis in the two frames show significant differences. Considering the effects introduced on oscillations by the steep temperature gradient of the photosphere and by the temperature- and pressure-dependent continuum opacity, we develop a new model for fitting the simulated V and T power and cross spectra both in the τ- and z-frames and discuss its merits and limitations.

Keywords Helioseismology: Direct modeling · Waves: modes · Velocity fields: Photosphere

---

Helioseismology, Asteroseismology, and MHD Connections
Guest Editors: Laurent Gizon and Paul Cally.

G. Severino (✉) · T. Straus
INAF Osservatorio Astronomico di Capodimonte, Via Moiariello 16, 80131 Napoli, Italy
e-mail: severino@oacn.inaf.it

T. Straus
e-mail: straus@oacn.inaf.it

M. Steffen
Astrophysikalisches Institut Potsdam, An der Sternwarte 16, Potsdam, Germany
e-mail: msteffen@aip.de

1. Introduction

In addition to mode frequencies used by helioseismology to probe the solar interior, solar p-mode line profiles carry further information that can help to improve our knowledge of the physical processes giving rise to the five-minute oscillations, generally believed to be stochastically excited by solar convection. Mode line profile diagnostics may rapidly become an interesting field also in stellar studies, since we expect in the near future a strong development of the asteroseismology of solar-like stars, where a similar excitation mechanism should be at work.

Duvall *et al.* (1993) showed that p-mode line profiles are asymmetric, with the velocity (V) peak having a stronger low-frequency side, and the asymmetry is opposite in intensity (I). Line asymmetry was explained in terms of interference between the p mode and a fraction of its background correlated with the mode. Thereafter, a large amount of work has be done to infer from line asymmetries the type of the acoustic sources (monopole, dipole, *etc.*) and their location, sometimes with controversial results (*e.g.*, Gabriel, 1995; Abrams and Kumar, 1996; Roxburg and Vorontsov, 1997; Nigam *et al.*, 1998; Nigam and Kosovichev, 1999a, 1999b; Kumar and Basu, 2000; Skartlien and Rast, 2000; Wachter and Kosovichev, 2005). In the meantime, using the technique of correlation analysis between V and I, which in addition to power spectra also produces the ϕ_{I-V} phase difference and C_{I-V} coherence spectra, Straus *et al.* (1999) and Oliviero *et al.* (1999) showed that *i*) the convective background of solar helioseismic spectra, previously modeled by Harvey (1985) and Harvey *et al.* (1998), is largely coherent, with a negative $I-V$ phase extending to low-ℓ degrees and corresponding to the interridge-plateau phase regime discovered by Deubner *et al.* (1990), and *ii*) line asymmetries in power have their distinctive counterparts in phase and coherence spectra.

Goode *et al.* (1998) and Strous, Goode, and Rimmele (2000) described the solar seismic events that they identified as the convective phenomena producing acoustic waves at the top of the convection zone. The strong darkening in the intergranular lane and the subsequent fast downdraft that precede the generation of acoustic flux may represent the fraction of the convective background that is correlated with the mode, whereas the resulting outward acoustic wave should contribute to the intrinsic or natural correlated background for the resonant mode. A scenario for acoustic energy production compatible with the observation of seismic events appeared from the analysis of numerical simulations of solar surface convection (Stein and Nordlund, 2001).

Severino *et al.* (2001) developed the first model capable of fitting simultaneously the four observed helioseismic spectra (V and I power and $I-V$ phase difference and coherence). This model is based on coherent components, which have a fixed phase relation between the associated V and I fluctuations, and on correlated components, which have a fixed phase relation to the mode. The coherent (c) component includes the p mode, a fraction of the solar background that is correlated with the p mode, and a fraction of the background that is still coherent but uncorrelated with the mode; finally, there is an incoherent component or noise.

This model was used successfully for fitting individual p-mode line profiles in a limited frequency range around the mode (*e.g.*, Jefferies *et al.*, 2003; Barban, Hill, and Kras, 2004). However, without additional constraints on the excitation processes, it is not possible to disentangle intrinsic from convective correlated background (Jefferies *et al.*, 2003; Wachter and Kosovichev, 2005), since both interfere with the mode in a similar way. Moreover, fitting cannot be extended straightforwardly to many modes at the same time because of the large number of model parameters necessary for each mode.

Recently, an important contribution to this field came from numerical simulations of solar surface convection. Numerical simulations of solar convection allow us to perform the cross-spectrum analysis of the solar surface V and I fluctuations, both in the τ frame (sampling the oscillations at constant optical depth), corresponding to the observer's point of view, and in the z frame (sampling the oscillations at constant geometrical height), where the hydrodynamical equations are naturally written. The results of the analysis in the two frames show definite differences, as pointed out by Georgobiani, Stein, and Nordlund (2003) and Straus, Severino, and Steffen (2006). The former authors suggest that the opacity variations are playing a major role in the τ-frame, and the latter propose that the temperature signal (with T used as a proxy of I) in the z-frame is mainly controlled by the presence of the steep temperature gradient at the base of the photosphere.

In this paper, Sections 2 and 3 give a description of the temperature-gradient and opacity effects, respectively, in the case of linear adiabatic acoustic-gravity waves. The results of these sections are also used to build a test version of the Severino *et al.* (2001) model of the four helioseismic spectra. In Section 4 this model is tried on the V and T power and cross spectra simulated by Straus, Severino, and Steffen (2006) in both the τ- and z-frames. Finally, Section 5 contains our conclusions.

2. The T-Gradient Effect

Following Straus, Severino, and Steffen (2006), in the case of evanescent waves the temperature fluctuations at fixed geometrical height are dominated by the contribution from the combination of the wave vertical displacement with the steep temperature gradient present in the low photosphere. In this section we describe this effect starting from the equations for linear stellar oscillations. The goal is to present a quantitative basis for reviewing the interpretation of V and T helioseismic spectra in both the τ- and z-frames.

In the case of an isothermal atmosphere with constant gravity and in adiabatic conditions, the evanescent wave commonly used to represent a p mode has a $+90°$ phase difference between T and V_z fluctuations (Marmolino and Severino, 1991). This isothermal model ignores the existence of a steep temperature gradient at the base of the photosphere. In a nonuniform medium the equation of continuity, the equation of motion, and the adiabatic relations can be written in the following form (Christensen-Dalsgaard, 2003):

$$\rho' + \nabla \cdot (\rho_0 \delta \mathbf{r}) = 0, \tag{1}$$

$$\rho_0 \frac{\partial^2 \delta \mathbf{r}}{\partial t^2} = -\nabla P' + \rho' \mathbf{g}_0, \tag{2}$$

$$\delta P = c_0^2 \delta \rho, \tag{3}$$

$$\delta T = \nabla_a \frac{T_0}{P_0} \delta P, \tag{4}$$

where the prime denotes an Eulerian perturbation, δ denotes Lagrangian perturbation, and the subscript $_0$ indicates an unperturbed quantity. Moreover, $c_0^2 = \Gamma_{1,0} \frac{P_0}{\rho_0}$, and $\nabla_a = \frac{(\Gamma_{2,0}-1)}{\Gamma_{2,0}}$, where Γ_1 and Γ_2 are the first and second adiabatic exponents, respectively.

The relations between Lagrangian and Eulerian perturbations are

$$\delta P = P' + \delta \mathbf{r} \cdot \nabla P_0,$$

$$\delta \rho = \rho' + \delta \mathbf{r} \cdot \nabla \rho_0,$$

$$\delta T = T' + \delta \mathbf{r} \cdot \nabla T_0.$$

By eliminating the Lagrangian perturbations the adiabatic relations, Equations (3) and (4), become

$$P' = c_0^2 \rho' - \delta \mathbf{r} \cdot \left(\nabla P_0 - c_0^2 \nabla \rho_0\right), \tag{5}$$

$$T' = \nabla_a \frac{T_0}{P_0} P' + \nabla_a \frac{T_0}{P_0} \delta \mathbf{r} \cdot \nabla P_0 - \delta \mathbf{r} \cdot \nabla T_0. \tag{6}$$

Equations (1), (2), (5), and (6) comprise six equations in the six unknowns ρ', $\delta \mathbf{r}$, P', and T'. After the system (1), (2), and (8) is solved as a function of height, giving $\delta \mathbf{r}(z)$, $\rho'(z)$, and $P'(z)$, one can compute $T'(z)$ from Equation (6). The solution in a realistic solar model, which may be carried out along the lines of Mihalas and Toomre (1981), will be part of a forthcoming work. Here, we define as the isothermal T fluctuation the first two terms in the right-hand side of Equation (6), which dominate when the atmospheric temperature gradient is small,

$$T'_{\text{iso}} = \nabla_a \frac{T_0}{P_0} P' + \nabla_a \frac{T_0}{P_0} \delta \mathbf{r} \cdot \nabla P_0, \tag{7}$$

and hence we write the equation for the temperature perturbation as

$$T' = T'_{\text{iso}} - \delta \mathbf{r} \cdot \nabla T_0. \tag{8}$$

Following Equation (8), we assume that the wave's (complex) vertical displacement (δz) combined with the atmospheric temperature gradient produces an additional temperature fluctuation T_g, with respect to the isothermal case, of the form

$$T_g = \left|\frac{dT_0}{dz}\right| \delta z. \tag{9}$$

Note that the absolute value of the gradient is present since the maximum of T_g corresponds to maximum upward displacement.

Although Equation (8) is valid for adiabatic waves (and the rest of the paper is limited to the purely adiabatic case), it may be worthwhile to discuss the expression of the T-gradient effect by assuming a damped or growing harmonic time variation with complex frequency $\omega_c = \omega - i\eta$. For the time growth rate $\eta \neq 0$ this corresponds to a nonadiabatic situation, with $\eta > 0$ in the case of amplification. By the definition of vertical velocity V_z we have

$$T_g = \left|\frac{dT_0}{dz}\right| \frac{V_z}{i\omega_c}, \tag{10}$$

which can be written

$$T_g = \left|\frac{dT_0}{dz}\right| V_z \frac{\eta - i\omega}{\eta^2 + \omega^2}. \tag{11}$$

This relation implies a gain $\left|\frac{T_g}{V_z}\right|$ depending on frequency as

$$\left|\frac{T_g}{V_z}\right| = \left|\frac{dT_0}{dz}\right| \frac{1}{\sqrt{\eta^2 + \omega^2}} \tag{12}$$

and a $T_g - V_z$ phase difference of

$$\phi_{T_g-V_z} = \arctan\left(\frac{-\omega}{\eta}\right), \qquad (13)$$

which belongs to $]-90°, 0°]$ if $\eta > 0$, equals $-90°$ when $\eta = 0$, and falls within $[-180°, -90°[$ if $\eta < 0$.

3. The Opacity Effect

Georgobiani, Stein, and Nordlund (2003) attributed the cause of the asymmetry reversal between the T and V p-mode line profiles in simulated power spectra at optical depth $\tau_{\text{cont}} = 1$ to radiative-transfer effects and, more specifically, to "the nonlinear amplitude of the displacement in the simulation and the nonlinear dependence of the H$^-$ opacity on temperature." Although the particular case of nonlinear effects must be considered, we think that the *linear* radiative transfer effects from waves deserve a quantitative assessment. Therefore, we consider here in detail the consequences introduced by the continuum opacity fluctuations associated with the wave temperature and pressure fluctuations.

The observer's reference frame is the one where the instantaneous optical depth (τ) is kept constant during the wave motion. This condition enforces a fluctuation of the geometrical height one observes. Since the unperturbed atmospheric model and the wave perturbation are functions of the atmospheric level, the height fluctuation produces additional wave perturbations, which we refer to as "opacity effects."

Consider two instants (t_1 and t_2) and the corresponding observed heights (h_1 and h_2) such that $\tau_1(h_1) = \tau_2(h_2) = 1$. By using Equation (8), the vertical velocity and the temperature at these levels are

$$\begin{aligned}
V_{z,1} &= V_z(h_1, t_1), \\
V_{z,2} &= V_z(h_2, t_2), \\
T_1 &= T_0(h_1) + T'_{\text{iso}}(h_1, t_1) - \delta z(h_1, t_1)\frac{dT_0}{dz}(h_1), \\
T_2 &= T_0(h_2) + T'_{\text{iso}}(h_2, t_2) - \delta z(h_2, t_2)\frac{dT_0}{dz}(h_2).
\end{aligned} \qquad (14)$$

Therefore, the fluctuations in the τ-frame are

$$\begin{aligned}
V'_z(\tau = \text{constant}) &= V_z(h_2, t_2) - V_z(h_1, t_1), \\
T'(\tau = \text{constand}) &= T_0(h_2) - T_0(h_1) + T'_{\text{iso}}(h_2, t_2) - T'_{\text{iso}}(h_1, t_1) \\
&\quad - \left[\delta z(h_2, t_2)\frac{dT_0}{dz}(h_2) - \delta z(h_1, t_1)\frac{dT_0}{dz}(h_1)\right].
\end{aligned} \qquad (15)$$

Then, the height fluctuation in the τ-frame introduces two additional contributions to wave fluctuations with respect to the z-frame: *i*) a spatial phase change in all wave perturbations and *ii*) a height variation of the corresponding parameter in the unperturbed model, which obviously does not affect the velocity. In the case of an adiabatic evanescent wave, the spatial change in V_z is

$$V'_z(\tau = \text{constant}) = V_0(x)e^{i\omega t_1 + k_z h_1} \cdot \left[e^{i\omega(t_2-t_1)+k_z(h_2-h_1)} - 1\right], \qquad (16)$$

where we assumed for simplicity $1/k_z$ as the same wave scale height at both h_1 and h_2 and $V_0(x)$ includes amplitude and horizontal phase. The amplitude modulation $e^{k_z(h_2-h_1)}$ may contribute to line asymmetry; however, since $k_z(h_2 - h_1)$ is small for evanescent waves and, moreover, the frequency resolution of the simulated spectra used for testing the model in the next section does not allow us to see clearly line asymmetry, we neglect this modulation in the following.

Let us discuss the height fluctuation induced by the opacity. If k is the opacity (per centimeter), at the two instants t_1 and t_2 in the wave, the corresponding observed heights h_1 and h_2 are given by

$$\tau_1(h_1) = \int_{h_1}^{\infty} k_1(h')\,dh' = 1,$$
$$\tau_2(h_2) = \int_{h_2}^{\infty} k_2(h')\,dh' = 1. \tag{17}$$

Taking the difference and considering only the first-order variations in k and h, which allows us to neglect the integral of $\Delta k = k_2 - k_1$ between h_1 and h_2, one gets

$$\int_{h_1}^{h_2} k_1(h')\,dh' = \int_{h_1}^{\infty} \Delta k(h')\,dh', \tag{18}$$

which gives approximately

$$\Delta h = h_2 - h_1 \approx \frac{1}{k_1(h_1)} \int_{h_1}^{\infty} \Delta k(h')\,dh'. \tag{19}$$

As expected, this implies that at time t_2 one observes higher in the atmosphere than at t_1 when the opacity is increased owing to the wave.

In terms of T and P fluctuations, the opacity variation is

$$\Delta k = \frac{\partial k}{\partial T}\Delta T + \frac{\partial k}{\partial P}\Delta P. \tag{20}$$

According to Cox and Giuli (1968), for a Population I star and $5040 \leq T \leq 10\,080$ K, the total continuum opacity per gram k_g from both H and H^- absorption varies as

$$k_g = k_{g,1} P^{0.74} T^{-0.74-s}, \tag{21}$$

where $k_{g,1}$ = constant. The value of s depends on gas pressure and is -7.5 at $P = 10^5$ dyne cm^{-2}, which is close to the value of the gas pressure at the base of the photosphere. This scaling holds because i) H^- is the dominant opacity source, ii) $P_e \ll P_g$, and iii) partial ionization of H is the main source of electrons. To get the opacity per centimeter we need the density, which is, by neglecting ionization,

$$\rho = \mu \frac{P}{k_B T} \approx 1.27 m_H \frac{P}{k_B T}, \tag{22}$$

where μ is the mean molecular weight and k_B is Boltzmann's constant. Therefore the opacity per centimeter, k, scales as

$$k = k_{cm,1} P^{1.74} T^{5.76}, \tag{23}$$

with $k_{\mathrm{cm},1}$ constant. This result, combined with Equation (20), gives

$$\Delta k(h) = k(h)\left[5.76\frac{\Delta T}{T}(h) + 1.74\frac{\Delta P}{P}(h)\right]. \qquad (24)$$

The continuum opacity fluctuation Δk is, then, in phase with the T and P fluctuations. ΔT is, in turn, in phase with ΔP when it is dominated by the T-gradient effect, which scales with the wave vertical displacement δz (Equation (9)). Therefore, according to Equation (19), the height fluctuation in the τ-frame, Δh, follows the wave vertical displacement δz, and this opacity effect tends to reduce the T-gradient effect.

In the simple case that $\frac{\Delta T}{T}$ and $\frac{\Delta P}{P}$ are constant with h, we can derive an explicit, approximated expression for the observed height fluctuation:

$$\Delta h \approx \left(5.76\frac{\Delta T}{T} + 1.74\frac{\Delta P}{P}\right)\frac{1}{k_1(h_1)}. \qquad (25)$$

Finally, we are in a position to write the expressions linking the fluctuations in the z-frame with those in the τ-frame. Since we neglected the spatial phase change introduced in all perturbations by the height fluctuation occurring in the τ-frame (see Equation (16) and discussion), we conclude that the vertical velocity fluctuation is the same in both frames. Furthermore, the temperature fluctuation in the z-frame is due to the sum of an isothermal wave contribution and the T-gradient effect (Equations (8) and (9)), whereas in the τ-frame the opacity effect is also at work, producing an additional contribution to the temperature fluctuation, which is given to first order by the atmospheric temperature gradient times a height variation expressed, for example, by Equation (25). Therefore we can write the following relations:

$$\begin{aligned} V'_z(\tau = \text{constant}) &= V'_z(z = \text{constant}), \\ T'(z = \text{constant}) &= T'_{\text{iso}}(z = \text{constant}) - \delta z \frac{dT_0}{dz} \\ &= T'_{\text{iso}}(z = \text{constant}) - \frac{V'_z(z = \text{constant})}{\iota\omega}\frac{dT_0}{dz}, \\ T'(\tau = \text{constant}) &= T'_{\text{iso}}(z = \text{constant}) - (\delta z - \Delta h)\frac{dT_0}{dz}. \end{aligned} \qquad (26)$$

4. Results

In this section we use Equation (26) as a basis to model the temperature signal in both the z- and τ-frames. The results of this model are then compared with the helioseismic spectra computed by Straus, Severino, and Steffen (2006) from their numerical simulation of solar surface convection.

4.1. The Numerical Simulation

The helioseismic spectra in the z- and τ-frames computed by Straus, Severino, and Steffen (2006) are based on a 3D, time-dependent radiation-hydrodynamics simulation of the solar granulation, including a detailed treatment of radiative transfer and a realistic equation of state accounting for partial ionization. Designed to represent the solar surface layers

($T_{\text{eff}} = 5770$ K, $\log g = 4.44$), the model was computed for the particular purpose of analyzing photospheric oscillations. The computational domain is a Cartesian box measuring $11.2 \times 11.2 \times 3.13$ Mm ($200 \times 200 \times 150$ cells), ranging from $\log \tau_{\text{Ross}} \approx -6.5$ to $+6.5$ in optical depth. A system of 360 000 rays is used for the calculation of the 3D radiation field (grey LTE). The simulation covers a time sequence of more than 30 000 seconds, with a sampling interval of 30 seconds. Details of the employed CO^5BOLD code can be found in Freytag, Steffen, and Dorch (2002) and Wedemeyer *et al.* (2004) and at the code's Web site (http://www.astro.uu.se/~bf/co5bold_main.html). In particular, for comparison with our model we selected the harmonic degree $\ell = 400$, which, for the simulated box width of 11.2 Mm, corresponds to the first horizontal mode.

4.2. Comparison with the Model

Before comparing our model for the temperature fluctuation with the numerical simulation of Straus, Severino, and Steffen (2006), it is worthwhile to show the behavior of the model for realistic values of its two parameters (*i.e.*, the atmospheric temperature gradient and the continuum opacity at fixed geometrical depth).

Figure 1 displays the different contributions to the temperature fluctuation and the total temperature signals in both the z- and τ-frames.

In the evanescent region (ν in the range between 2 and 4 mHz), where the isothermal T fluctuation tends to vanish on the f mode, the total T fluctuation in the z-frame is dominated by the T-gradient effect. The total phase with respect to vertical velocity is $\phi_{T-V} = -90°$. However, the variation of the total amplitude with frequency is steeper than the one of the T-gradient effect alone, because the isothermal T fluctuation is added below the f mode and is subtracted above, owing to the 180° phase jump of the isothermal temperature at the f mode (Marmolino and Severino, 1991). In contrast, in the region of propagating waves, the T fluctuation in the isothermal atmosphere prevails and the amplitude and phase of the sum follow the isothermal trends with frequency. The amplitude of the opacity effect has a variation with frequency similar to the total T fluctuation in the z-frame. This is because the temperature fluctuation dominates the opacity fluctuation. (In fact, even if it is $|\Delta T/T| < |\Delta P/P|$, Equation (25) shows that the contribution of the T fluctuation to the displacement owing to opacity is a factor 5.76/1.74 greater than that of the P fluctuation.) Also, the T fluctuation owing to the opacity effect is opposite to the total T fluctuation in the z-frame. The T fluctuation in the τ-frame has again a variation with frequency similar to the total T fluctuation in the z-frame but with an amplitude and a phase depending on the difference between the opacity effect and the total T signal in the z-frame.

Figure 1 was computed for a value of -19 K km^{-1} of the atmospheric temperature gradient, representative of the base of the solar photosphere. Changing the value of atmospheric temperature gradient from $\frac{dT_0}{dz} = -10$ to -30 K km^{-1} increases the total T fluctuation in the z-frame proportionally in a gradually larger frequency range of the evanescent region, because of the increasing contribution of the T-gradient effect to the sum.

The same variation of $\frac{dT_0}{dz}$ produces a more complex behavior of the T fluctuation in the τ-frame. In fact, the total T fluctuation in the τ-frame is minimum for $\frac{dT_0}{dz} \approx -18$ K km^{-1}. For $\frac{dT_0}{dz} > -18$ K km^{-1}, the T-gradient effect prevails over the opacity effect, yielding $\phi_{T-V} = -90°$ and a total amplitude decreasing with decreasing $\frac{dT_0}{dz}$. In contrast, for $\frac{dT_0}{dz} < -18$ K km^{-1}, the opacity effect prevails, with $\phi_{T-V} = +90°$ and the amplitude of the total T fluctuation increases. This behavior occurs because, unlike the T-gradient effect,

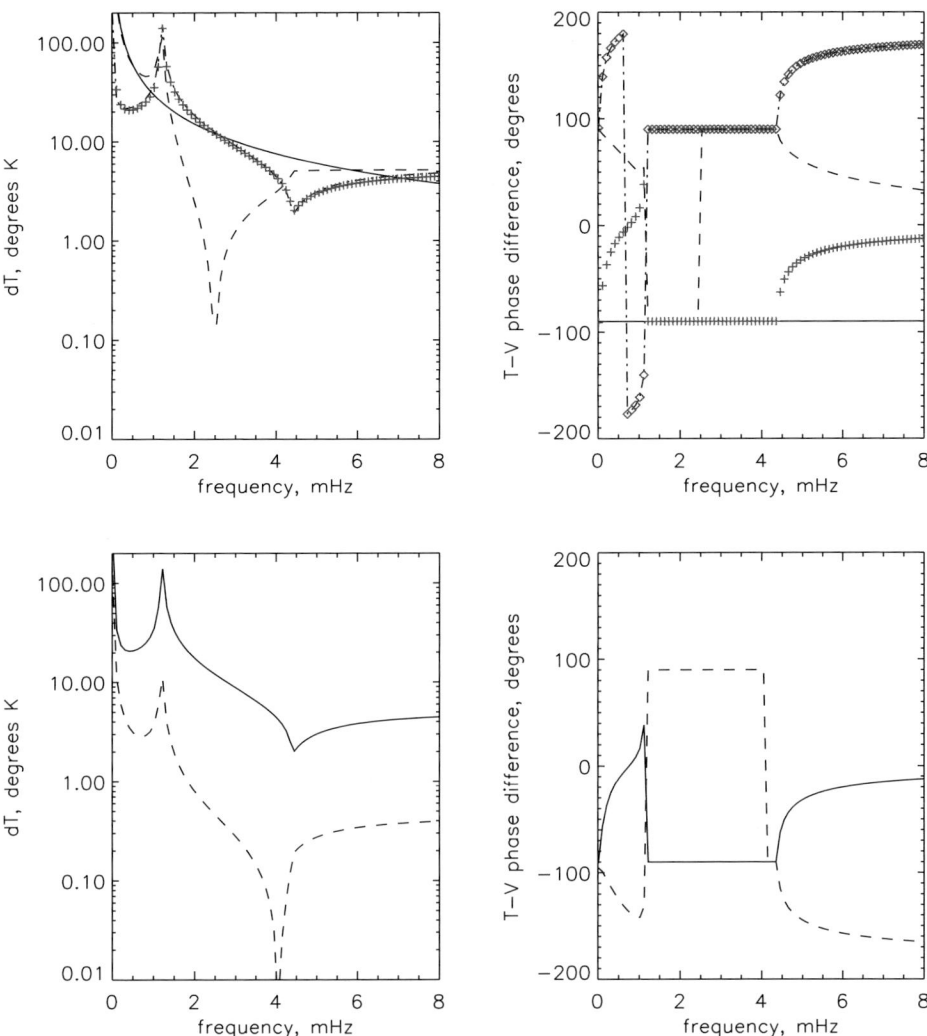

Figure 1 Temperature fluctuation as a function of frequency computed from Equation (26) with the atmospheric temperature gradient $dT_0/dz = -19$ K km^{-1}, the continuum opacity is $k = 2.75 \; 10^{-7}$ cm^{-1}, and the atmospheric temperature is 6420 K, which all represent photospheric values close to $z = 0$. We also assumed a fixed vertical velocity $V_z = 10^{-2}$ km s^{-1}, independent of frequency, and $\ell = 620$, to account for the order of magnitude of the V amplitude and the frequency of the simulated f mode, respectively, in the subsequent comparison with the simulated spectra. The upper panels display the different contributions to the temperature fluctuation (amplitude in the left panel and phase in the right panel). The solid line is the T-gradient effect, the dashed line is T in the isothermal atmosphere (for energy propagation upward), and crosses are the sum of the two; the dotted-dashed line is the opacity effect, whose phase is overplotted with diamonds in the upper right panel. The lower panels represent the temperature in the z-frame (solid line) and in the τ-frame (dashed line), both amplitude (left panel) and phase (right panel).

which scales linearly with the atmospheric temperature gradient, the opacity effect is proportional to the square of the atmospheric temperature gradient because it scales as $\Delta h \frac{dT_0}{dz}$, where Δh is in turn proportional to $\frac{dT_0}{dz}$ through the wave temperature fluctuation.

As a further quantitative experiment, we discuss a new version of the Severino et al. (2001) model for the helioseismic spectra that includes the basic ideas on the T-gradient and opacity effects we developed in the previous sections. In Figures 2 and 3 we compare this model with the helioseismic spectra calculated by Straus, Severino, and Steffen (2006) for both the atmospheric levels $z = 0$ and $\tau_{cont} = 1$ and in the frequency range from 0 to 8 mHz. In particular, the model assumed that there are mode components from both the f and p_1 modes and that these modes are not correlated. The fitting parameters in the velocity power spectrum for the f and p_1 modes are their peak amplitudes and line widths. The T signal for the modes is obtained from the corresponding V signal by means of Equations (26) with no further free parameter (i.e., in the z-frame it is the sum of the T-gradient effect and the isothermal T fluctuation, and in the τ-frame it contains also the additional contribution of the opacity effect).

The present model has an exploratory nature. It would be interesting to discuss in detail the question of p-mode line asymmetries. Unfortunately, the scatter of the simulated spectra, owing to both the limited frequency resolution and the small number of horizontal wavenumbers contributing to the average, does not permit us to fully validate the modeled line asymmetries. Therefore, we assumed that all of the convective background is represented by two coherent components in V and T, which are uncorrelated with the modes, that have a phase difference $\phi_{T-V} = 0°$ and Lorentzian frequency profiles with amplitudes to be fitted and widths corresponding to a common lifetime of ten minutes. The T-gradient contribution is ineffective for this convective background. Moreover, in this model there is no correlated background owing to convection, which in the original Severino et al. model was associated with acoustic sources. Finally, there is also no noise, which is supported by the simulated coherence close to unity in the z-frame everywhere below $\nu \approx 4$ mHz.

Let us discuss first the scenario of the z-frame, which is reported in Figure 2. Just as one expects, the T fluctuation associated with the wave displacement of the atmospheric temperature gradient dominates the mode signals. Its characteristic asymmetry is due to the amplitude dependence on the inverse of the frequency. The isothermal temperature fluctuation vanishes for the f mode but it is roughly one order of magnitude lower than the contribution of the T-gradient effect on the p_1 mode. Therefore, in the spirit of the Severino et al. model, we may describe the p_1 resonance with a mode component resulting from the T-gradient effect and a particular correlated background represented by the isothermal T fluctuation, which has a large-amplitude variation around the p_1 resonance, where in this example it is stronger than the convective background. Figure 2 shows that the model can match very closely the simulation in the z-frame, reproducing in particular the similarity of modes in temperature with those in velocity, as well as $T - V$ phase differences close to $-90°$ at resonances, both of which were the surprising characteristics of this frame. However, there are still some minor differences, such as a smaller coherence around $\nu = 1.5$ mHz and slightly lower phases at the resonances. These residual discrepancies indicate that the model may be further improved as far as the description of the waves and of the background are concerned.

The model results in the τ-frame strongly depend upon the precise balance between the opacity effect and the T-gradient effect in the modal components, according to Equation (26). For the selected value of the atmospheric temperature gradient, the opacity effect, slightly prevailing over the T-gradient effect, can reproduce both the strong reduction of the T power and the flipping of the $T - V$ phase difference displayed by the simulated data passing from the z- to the τ-frame. Moreover, we remark that the asymmetries of both the modeled modes in temperature are reversed in this frame, in agreement with the observed I line asymmetries, because of the 180° phase jump in the T fluctuation caused by the opacity

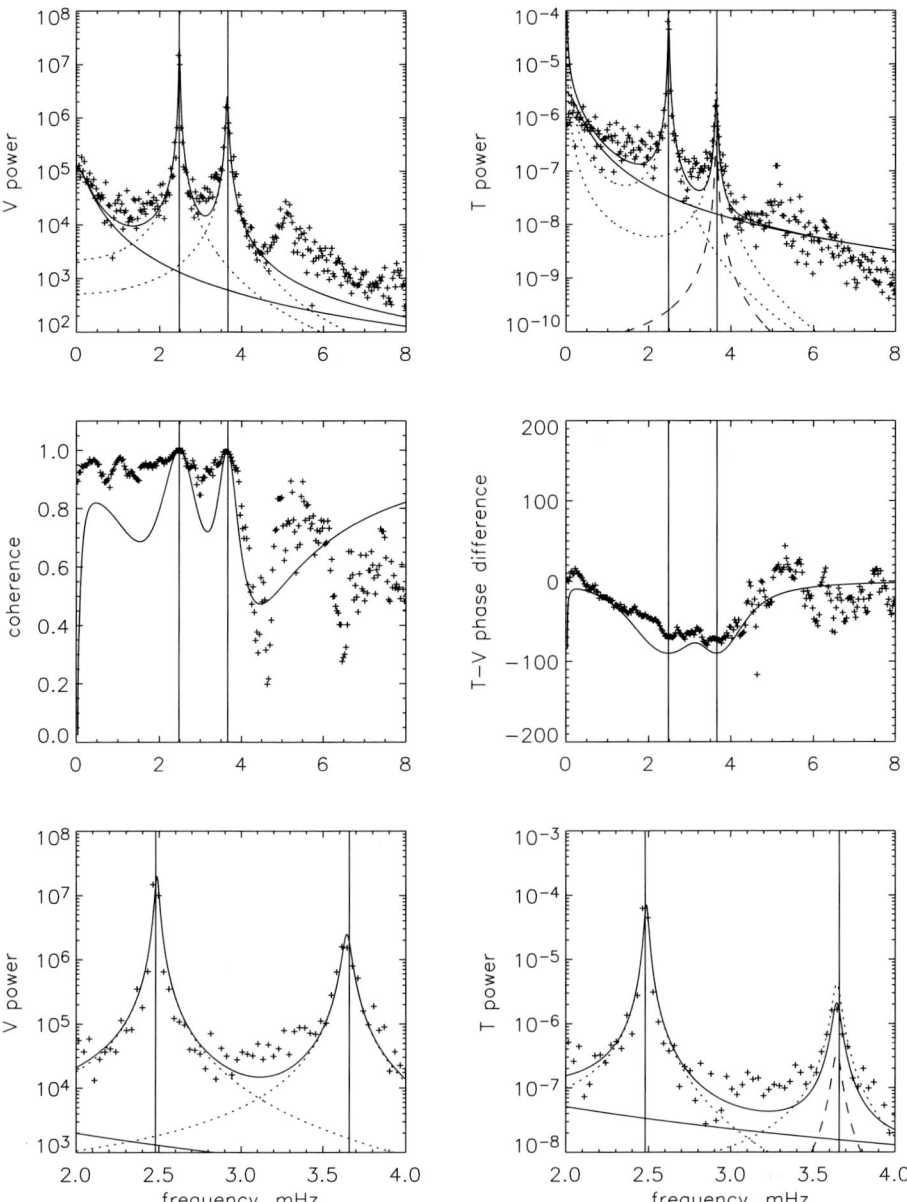

Figure 2 Comparison of simulated and modeled spectra in the z-frame. Panels represent, from top left to lower right, V power in (m s^{-1})^2, T power (dimensionless power of relative fluctuations), $T - V$ coherence and phase difference in degrees, and blowups on the modes of V and T power. The x-axes display frequency in mHz. The simulated spectra (crosses) refer to degree $\ell = 400$ and were computed by Straus, Severino, and Steffen (2006). Thin vertical lines mark the locations of f and p modes. Solid lines represent the total signals that the new version of the Severino *et al.* model (described in Section 4.2) provides in the z frame for a value of the atmospheric temperature gradient of -19 K km^{-1}. Dashed lines refer to the mode component in the V power panels and to the T-gradient contribution to modes in the T power panels. Long-dashed lines correspond to the isothermal temperature fluctuation only. Thin solid curves represent the convective background. Note that there is no isothermal T fluctuation for the f mode.

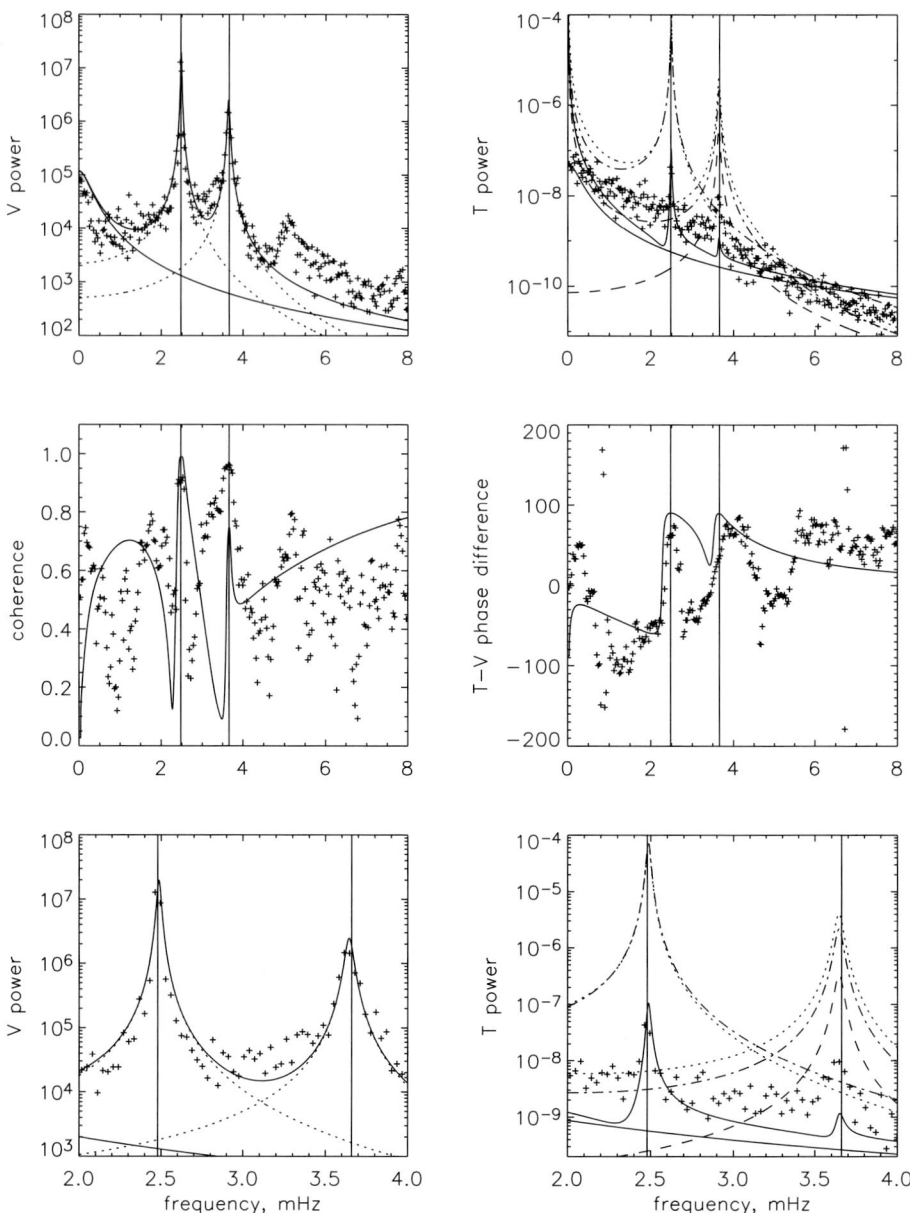

Figure 3 Comparison of simulated and modeled spectra in the τ-frame. Panels represent, from top left to lower right, V power in $(\text{m s}^{-1})^2$, T power (dimensionless power of relative fluctuations), $T - V$ coherence and phase difference in degrees, and blowups on the modes of V and T power. The x-axes display frequency in mHz. The simulated spectra (crosses) refer to degree $\ell = 400$ and were computed by Straus, Severino, and Steffen (2006). Thin vertical lines mark the locations of f and p modes. Solid lines represent the total signals that the new version of the Severino *et al.* model (described in Section 4.2) provides in the τ frame for a value of the atmospheric temperature gradient of -19 K km^{-1}. Dash-dot lines in the panels for T power refer to the opacity effect. Dashed lines refer to the mode component in the V power panels and to the T-gradient contribution to modes in the T power panels. Long-dashed lines correspond to the isothermal temperature fluctuation only. Thin solid curves represent the convective background.

effect. On the whole, the fit of the V and T power and cross spectra is rather less accurate in this frame than it was in the z-frame. This behavior suggests that the simulated spectra in the τ-frame may also be sensitive to details of the continuum opacity, which were neglected in this preliminary model.

5. Conclusions

The ultimate goal of our analysis of the V and T power and cross spectra computed from numerical simulations of solar surface convection is to put on solid ground the large amount of work done up to now for inferring some characteristics of the solar acoustic sources from the corresponding observed spectra. Once established for the Sun, these diagnostic techniques may become applicable to solar-like stars owing to the rapid observational improvement of this field.

In the present paper we have discussed the important role that the presence of a strong temperature gradient in the low photosphere and the opacity fluctuations associated with oscillations can play in a correct interpretation of helioseismic spectra.

In Sections 2 and 3 we have given an extensive description of both of these effects for linear acoustic-gravity waves in the pure adiabatic case. In a forthcoming paper, this approach will be extended to nonadiabatic waves, which are a more adequate representation in the low photosphere, by means of the Newtonian cooling approximation.

The comparison of modeled and simulated spectra, which was the subject of Section 4, is the first quantitative testing of the T-gradient and opacity effects with simulated power and cross spectra. Our results clearly demonstrate that the T fluctuation associated with the atmospheric temperature gradient is fundamental for interpreting the simulated spectra involving temperature fluctuations in the z-frame. Furthermore, our experiment suggests that we observe (in the τ-frame) the residual effect of the opacity and the temperature gradient fluctuations, which happen to cancel each other to a large degree but dominate what we expect to see in the z-frame of an isothermal atmosphere. Therefore, we believe that the linear radiative transfer effects from waves are equally important for understanding the τ-frame, even though nonlinear opacity effects associated with the combined wave and convective motions also deserve to be considered in detail, as suggested by Georgobiani, Stein, and Nordlund (2003). In addition, our analysis based on linear and adiabatic waves, strongly suggests that in both the z- and τ-frames the T signals of the modes are largely insensitive to the T fluctuations owing to an isothermal atmosphere that we would have naively called the intrinsic T fluctuations.

To conclude, we consider the approach developed in the present paper to be very promising, since it is a step in the direction of fitting simultaneously the simulated spectra in both the z- and the τ-frames and this will certainly help in building models for the observed helioseismic spectra with a reduced number of free parameters. In addition it allows us to trace a clear road for improvements. On the one hand, we are working to allow for a realistic wave model in the photosphere, where the gas temperature and pressure can vary dramatically on scales smaller than the vertical wavelengths of evanescent waves, and, as mentioned, strong wave damping is at work (Mihalas and Toomre, 1982). On the other hand, soon we will have a numerical simulation of solar convection available that is much longer than the about ten solar hours of Straus, Severino, and Steffen (2006). This will allow us to compute spectra with a high frequency resolution and, hence, will permit us to also tackle the question of line asymmetries, especially as preliminary results show that this simulation matches the real Sun with high precision when we limit the analysis to high ℓ values.

Acknowledgements One of the authors (GS) is grateful to the HELAS Network for support. We thank the anonymous referee for making a number of very useful comments.

References

Abrams, D., Kumar, P.: 1996, *Astrophys. J.* **472**, 882.
Barban, C., Hill, F., Kras, S.: 2004, *Astrophys. J.* **602**, 516.
Christensen-Dalsgaard, J.: 2003, *Lecture Notes on Stellar Oscillations*, 5th edn. http://astro.phys.au.dk/~jcd/oscilnotes/.
Cox, J.P., Giuli, R.T.: 1968, *Principles of Stellar Structure*, Gordon and Breach, New York.
Deubner, F.-L., Fleck, B., Marmolino, C., Severino, G.: 1990, *Astron. Astrophys.* **236**, 509.
Duvall, T.L., Jr., Jefferies, S.M., Harvey, J.W., Osaki, Y., Pomerantz, M.A.: 1993, *Astrophys. J.* **410**, 829.
Freytag, B., Steffen, M., Dorch, B.: 2002, *Astron. Nachr.* **323**, 213.
Gabriel, M.: 1995, *Astron. Astrophys.* **299**, 245.
Georgobiani, D., Stein, R.F., Nordlund, Å.: 2003, *Astrophys. J.* **596**, 698.
Goode, P.R., Strous, L.H., Rimmele, T.R., Stebbins, R.: 1998, *Astrophys. J.* **495**, L27.
Harvey, J.W.: 1985, In: Rolfe, E.J., Battrick, B. (eds.) *Future Missions in Solar, Heliospheric and Space Plasma Physics* **SP-235**, ESA, Noordwijk, 199.
Harvey, J.W., Jefferies, S.M., Duvall, T., Jr., Osaki, Y., Shibahashi, H.: 1998, In: Provost, J., Schmider, F.-X. (eds.) *Sounding Solar and Stellar Interiors, IAU Symp.* **181**, OCA and Université de Nice, Nice, 177.
Jefferies, S.M., Severino, G., Moretti, P.F., Oliviero, M., Giebink, C.: 2003, *Astrophys. J.* **596**, L117.
Kumar, P., Basu, S.: 2000, *Astrophys. J.* **519**, 389.
Marmolino, C., Severino, G.: 1991, *Astron. Astrophys.* **242**, 271.
Mihalas, B.W., Toomre, J.: 1981, *Astrophys. J.* **249**, 349.
Mihalas, B.W., Toomre, J.: 1982, *Astrophys. J.* **263**, 386.
Nigam, R., Kosovichev, A.G.: 1999a, *Astrophys. J.* **510**, L149.
Nigam, R., Kosovichev, A.G.: 1999b, *Astrophys. J.* **514**, L53.
Nigam, R., Kosovichev, A.G., Scherrer, P.H., Schou, J.: 1998, *Astrophys. J.* **496**, L115.
Oliviero, M., Severino, G., Straus, Th., Jefferies, S.M., Appourchaux, T.: 1999, *Astrophys. J.* **516**, L45.
Roxburg, I.W., Vorontsov, S.V.: 1997, *Mon. Not. Roy. Astron. Soc.* **292**, L33.
Severino, G., Magrì, M., Oliviero, M., Straus, T., Jefferies, S.M.: 2001, *Astrophys. J.* **561**, 444.
Skartlien, R., Rast, M.: 2000, *Astrophys. J.* **535**, 464.
Stein, R.F., Nordlund, Å.: 2001, *Astrophys. J.* **546**, 585.
Straus, T., Severino, G., Steffen, M.: 2006, In: Lacoste, H., Ouwehand, L. (eds.) *SOHO 17 Workshop: 10 Years of SOHO and Beyond* **SP-617**, ESA, Noordwijk, 4.1, Published on CDROM.
Straus, T., Severino, G., Deubner, F.-L., Fleck, B., Jefferies, S.M., Tarbell, T.: 1999, *Astrophys. J.* **516**, 939.
Strous, L.H., Goode, P.R., Rimmele, T.R.: 2000, *Astrophys. J.* **535**, 1000.
Wachter, R., Kosovichev, A.G.: 2005, *Astrophys. J.* **627**, 550.
Wedemeyer, S., Freytag, B., Steffen, M., Ludwig, H.-G., Holweger, H.: 2004, *Astron. Astrophys.* **414**, 1121.

3D MHD Coronal Oscillations about a Magnetic Null Point: Application of WKB Theory

J.A. McLaughlin · J.S.L. Ferguson · A.W. Hood

Originally published in the journal Solar Physics, Volume 251, Nos 1–2, 563–587.
DOI: 10.1007/s11207-007-9107-2 © Springer Science+Business Media B.V. 2008

Abstract This paper is a demonstration of how the WKB approximation can be used to help solve the linearised 3D MHD equations. Using Charpit's method and a Runge–Kutta numerical scheme, we have demonstrated this technique for a potential 3D magnetic null point, $\mathbf{B} = [x, \epsilon y, -(\epsilon + 1)z]$. Under our cold-plasma assumption, we have considered two types of wave propagation: fast magnetoacoustic and Alfvén waves. We find that the fast magnetoacoustic wave experiences refraction towards the magnetic null point and that the effect of this refraction depends upon the Alfvén speed profile. The wave and thus the wave energy accumulate at the null point. We have found that current buildup is exponential and the exponent is dependent upon ϵ. Thus, for the fast wave there is preferential heating at the null point. For the Alfvén wave, we find that the wave propagates along the field lines. For an Alfvén wave generated along the fan plane, the wave accumulates along the spine. For an Alfvén wave generated across the spine, the value of ϵ determines where the wave accumulation will occur: fan plane ($\epsilon = 1$), along the x-axis ($0 < \epsilon < 1$) or along the y-axis ($\epsilon > 1$). We have shown analytically that currents build up exponentially, leading to preferential heating in these areas. The work described here highlights the importance of understanding the magnetic topology of the coronal magnetic field for the location of wave heating.

Keywords Magnetohydrodynamics: waves, propagation · Magnetic fields: models · Heating: coronal

1. Introduction

The WKB approximation is an asymptotic approximation technique that can be used when a system contains a large parameter (see, *e.g.*, Bender and Orszag, 1978). Hence, the WKB

Guest Editors: Laurent Gizon and Paul Cally

Electronic supplementary material The online version of this article (http://dx.doi.org/10.1007/978-0-387-89482-9_38) contains supplementary material, which is available to authorized users.

J.A. McLaughlin (✉) · J.S.L. Ferguson · A.W. Hood
School of Mathematics and Statistics, University of St Andrews, St Andrews, Fife, KY16 9SS, UK
e-mail: james@mcs.st-and.ac.uk

method can be used in a system where a wave propagates through a background medium that varies on some spatial scale that is much longer than the wavelength of the wave. The SOHO and TRACE satellites have recently observed MHD wave motions in the corona (*i.e.*, fast and slow magnetoacoustic waves and Alfvén waves; see the reviews by Nakariakov and Verwichte, 2005; De Moortel, 2005, 2006). The coronal magnetic field plays a fundamental role in their propagation and to begin to understand this inhomogeneous magnetised environment, it is useful to look at the structure (topology) of the magnetic field itself. Potential-field extrapolations of the coronal magnetic field can be made from photospheric magnetograms. Such extrapolations show the existence of an important feature of the topology: *null points*. Null points are points in the field where the magnetic field, and hence the Alfvén speed, is zero. Detailed investigations of the coronal magnetic field, using such potential-field calculations, can be found in Beveridge, Priest, and Brown (2002) and Brown and Priest (2001).

McLaughlin and Hood (2004) found that, for a single 2D null point, the fast magnetoacoustic wave was attracted to the null and the wave energy accumulated there. In addition, they found that the Alfvén-wave energy accumulated along the separatrices of the topology. They solved the 2D linearised MHD equations numerically and compared the results with a WKB approximation: The agreement was excellent. From their work and other examples (*e.g.*, Galsgaard, Priest, and Titov, 2003; McLaughlin and Hood, 2005, 2006a; Khomenko and Collados, 2006) it has been clearly demonstrated that the WKB approximation can provide a vital link between analytical and numerical work and often provides the critical insight to understanding the physical results. This paper demonstrates the methodology of how to apply the WKB approximation in linear 3D MHD. We believe that with the vast amount of 3D modelling currently being undertaken, applying this WKB technique to three dimensions will be very useful and beneficial to modellers in the near future.

The work undertaken by Galsgaard, Priest, and Titov (2003) deserves special mention here. They performed numerical experiments on the effect of twisting the spine of a 3D null point and described the resultant wave propagation towards the null. They found that when the field lines around the spine are perturbed in a rotationally symmetric manner, a twist wave (essentially an Alfvén wave) propagates towards the null along the field lines. Whilst this Alfvén wave spreads out as the null is approached, a fast-mode wave focuses on the null and wraps around it. They concluded that the driving of the fast wave was likely to come from a nonlinear coupling to the Alfvén wave (Nakariakov, Roberts, and Murawski, 1997). They also compare their results with a WKB approximation and find that, for the $\beta = 0$ fast wave, the wavefront wraps around the null point as it contracts towards it. They perform their WKB approximation in cylindrical polar coordinates and thus their resultant equations are two dimensional (since a simple 3D null point is essentially two dimensional in cylindrical coordinates). In contrast, we solve the WKB equations for three Cartesian components, and thus we can solve for more general disturbances and more general boundary conditions. This also allows us to concentrate on the transient features that are not always apparent when only cylindrically symmetric solutions are permitted.

More recently, Pontin and Galsgaard (2007) and Pontin, Bhattacharjee, and Galsgaard (2007) have performed numerical simulations in which the spine and fan of a 3D null point are subject to rotational and shear perturbations. They found that rotations of the fan plane lead to current sheets in the location of the spine and rotations about the spine lead to current sheets in the fan. In addition, shearing perturbations lead to 3D localised current sheets focused at the null point itself. This general behaviour is in good agreement with the work presented in this paper (*i.e.*, current accumulation at certain parts of the topology). However, the primary motivation in Pontin and Galsgaard (2007) and Pontin, Bhattacharjee, and Galsgaard (2007) was to investigate current-sheet formation and reconnection rates, whereas the techniques described in this paper focus on MHD wave-mode propagation and interpretation.

The propagation of fast magnetoacoustic waves in an inhomogeneous coronal plasma has been investigated by Nakariakov and Roberts (1995), who showed how the waves are refracted into regions of low Alfvén speed. In the case of null points, the Alfvén speed actually drops to zero.

The paper has the following outline: In Section 2, the basic equations are described. Section 3 details the 3D WKB approximation utilised in this paper. The results for the fast wave and Alfvén waves are shown in Sections 4 and 5, respectively. The conclusions and discussion are presented in Section 7. There are four appendices that complement the results in the main text.

2. Basic Equations

The usual resistive, adiabatic MHD equations for a plasma in the solar corona are used:

$$\rho \frac{\partial \mathbf{v}}{\partial t} + \rho(\mathbf{v} \cdot \nabla)\mathbf{v} = -\nabla p + \mathbf{j} \times \mathbf{B} + \rho \mathbf{g}, \tag{1}$$

$$\frac{\partial \mathbf{B}}{\partial t} = \nabla \times (\mathbf{v} \times \mathbf{B}) + \eta \nabla^2 \mathbf{B}, \tag{2}$$

$$\frac{\partial \rho}{\partial t} + \nabla \cdot (\rho \mathbf{v}) = 0, \tag{3}$$

$$\frac{\partial p}{\partial t} + \mathbf{v} \cdot \nabla p = -\gamma p \nabla \cdot \mathbf{v}, \tag{4}$$

$$\mu \mathbf{j} = \nabla \times \mathbf{B}, \tag{5}$$

where $\mathbf{v}$ is the plasma velocity, ρ is the mass density, p is the gas pressure, $\mathbf{B}$ is the magnetic induction (usually called the magnetic field), $\mathbf{j}$ is the electric current, $\mathbf{g}$ is gravitational acceleration, γ is the ratio of specified heats, η is the magnetic diffusivity, and μ is the magnetic permeability.

2.1. Basic Equilibrium

We choose a 3D magnetic null point for our equilibrium field of the form

$$\mathbf{B}_0 = \frac{B}{L}\left[x, \epsilon y, -(\epsilon + 1)z\right], \tag{6}$$

where B is a characteristic field strength, L is the length scale for magnetic-field variations, and the parameter ϵ is related to the predominate direction of alignment of the field lines in the fan plane. Parnell *et al.* (1996) investigated and classified the different types of linear magnetic null points that can exist (and our ϵ parameter is called p in their work). Topologically, this 3D null consists of two key parts: The z-axis represents a special, isolated field line called the *spine*, which approaches the null from above and below (Priest and Titov, 1996), and the xy-plane through $z = 0$ is known as the *fan* and consists of a surface of field lines spreading out radially from the null. Figure 1 shows two examples of 3D null points: $\epsilon = 1$ (left) and $\epsilon = 1/2$ (right). Titov and Hornig (2000) have investigated the steady-state structures of magnetic null points.

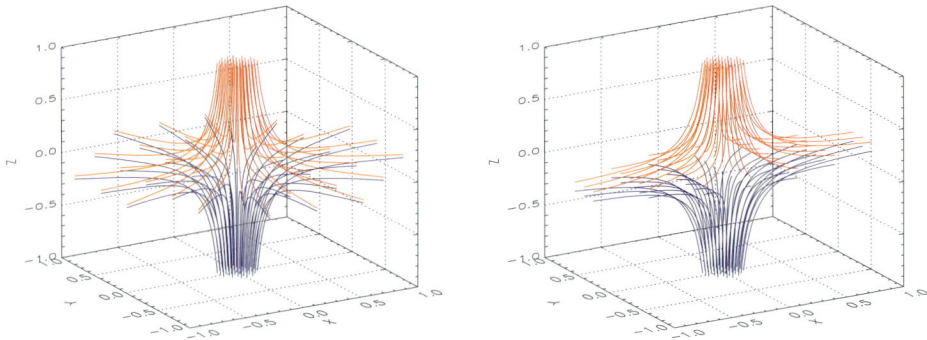

Figure 1 Left: Proper radial null point, described by $\mathbf{B} = (x, y, -2z)$ (i.e., $\epsilon = 1$). Right: Improper radial null point, described by $\mathbf{B} = (x, \frac{1}{2}y, -\frac{3}{2}z)$ (i.e., $\epsilon = \frac{1}{2}$). Note, for $\epsilon = \frac{1}{2}$, the field lines rapidly curve such that they run parallel to the x-axis along $y = 0$. In both figures, the z-axis indicates the spine and the xy-plane at $z = 0$ denotes the fan. The red field lines have been tracked from the $z = 1$ plane; the blue lines from $z = -1$.

Equation (6) is the general expression for the linear field about a potential magnetic null point (Parnell *et al.*, 1996: Section IV). In this paper, we only consider $\epsilon \geq 0$ and so all nulls we describe are *positive* nulls; that is, the spine points into the null and the field lines in the fan are directed away. In addition, all potential nulls are designated *radial*; that is, there are no spiral motions in the fan plane. In general, there are three cases to consider:

- $\epsilon = 1$: This case describes a *proper null* (Figure 1, left). This magnetic null has cylindrical symmetry about the spine axis.
- $\epsilon > 0$, $\epsilon \neq 1$: This case describes an *improper null* (Figure 1, right). Field lines rapidly curve such that they run parallel to the x-axis if $0 < \epsilon < 1$ and parallel to the y-axis if $\epsilon > 1$.
- $\epsilon = 0$: Equation (6) reduces to the X-point potential field in the xz-plane and forms a null line along the y-axis through $x = z = 0$. MHD wave propagation in this 2D configuration has been studied extensively by McLaughlin and Hood (2004, 2005, 2006a).

2.2. Assumptions and Simplifications

In this paper, the linearised MHD equations are used to study the nature of wave propagation near the null point. By using subscripts of 0 for equilibrium quantities and 1 for perturbed quantities, Equations (1)–(5) become

$$\rho_0 \frac{\partial \mathbf{v}_1}{\partial t} = -\nabla p_1 + \mathbf{j}_0 \times \mathbf{B}_1 + \mathbf{j}_1 \times \mathbf{B}_0 + \rho_1 \mathbf{g}, \tag{7}$$

$$\frac{\partial \mathbf{B}_1}{\partial t} = \nabla \times (\mathbf{v}_1 \times \mathbf{B}_0) + \eta \nabla^2 \mathbf{B}_1, \tag{8}$$

$$\frac{\partial \rho_1}{\partial t} + \nabla \cdot (\rho_0 \mathbf{v}_1) = 0, \tag{9}$$

$$\frac{\partial p_1}{\partial t} + \mathbf{v}_1 . \nabla p_0 = -\gamma p_0 \nabla \cdot \mathbf{v}_1, \tag{10}$$

$$\mu \mathbf{j}_1 = \nabla \times \mathbf{B}_1. \tag{11}$$

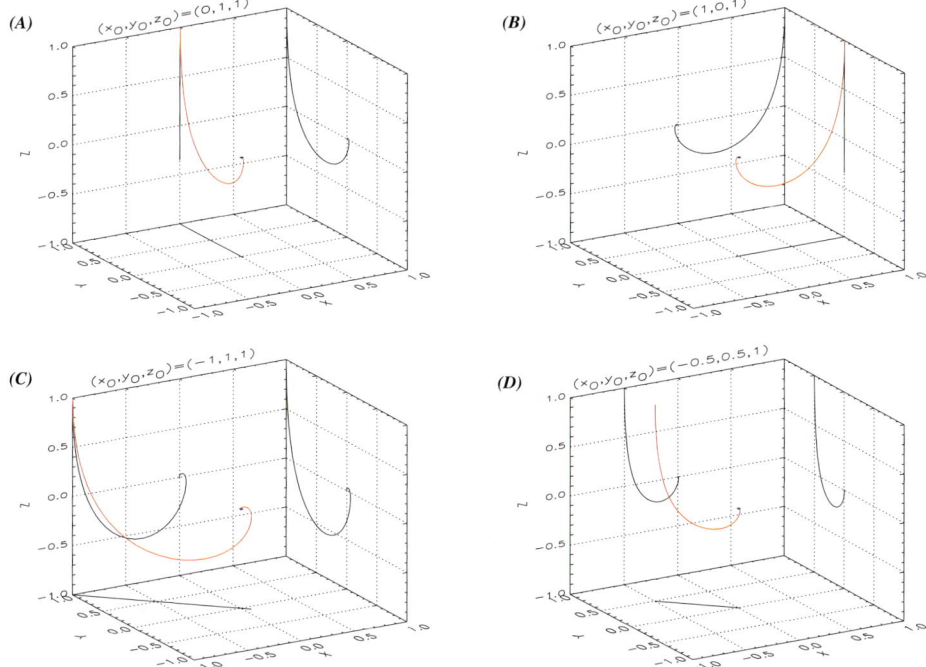

Figure 3 ($\epsilon = 1$) Ray paths for fluid elements that begin at (x_0, y_0, z_0) = (A) $(0, 1, 1)$, (B) $(1, 0, 1)$, (C) $(-1, 1, 1)$, and (D) $(-0.5, 0.5, 1)$. This figure is also available as an mpg animation in the electronic edition of *Solar Physics* (http://dx.doi.org/10.1007/978-0-387-89482-9_38), showing all $-1 \leq x_0 \leq 1$, $z_0 = 1$ along $y = -1$ and $y = 1$, and all $-1 \leq y_0 \leq 1$, $z_0 = 1$ along $x = -1$ and $x = 1$. Here, red indicates the 3D ray path and black indicates the xy-, yz-, and xz-projections of this ray path onto the respective planes. The blue dot indicates the position of the magnetic null point.

4.2. Planar Fast Wave Starting at $y_0 = 1$

We again solve Equation (16) but now subject to the initial conditions

$$\phi_0 = 0, \qquad \omega_0 = 2\pi, \qquad -1 \leq x_0 \leq 1, \qquad y_0 = 1, \qquad -1 \leq z_0 \leq 1,$$
$$p_0 = 0, \qquad q_0 = \omega_0/\sqrt{x_0^2 + \epsilon^2 y_0^2 + (\epsilon + 1)^2 z_0^2}, \qquad r_0 = 0,$$
(19)

These initial conditions correspond to a fast wave being sent in from the side boundary (along $y = y_0$). This choice of planar fast wave is incident perpendicular to the spine (z-axis).

Let us first consider $\epsilon = 1$. Figure 5 shows surfaces of constant ϕ at four values of t, showing the behaviour of the (initially planar) wavefront that starts at $-1 \leq x \leq 1$, $y = 1$, and $-1 \leq z \leq 1$. Again, we see the deformation of the (initially planar) wave caused by the refraction effect and, again, the wave accumulates at the null point. However, the nature of this refraction is different from that seen in Figure 2, since the refraction varies in magnitude in different planes. We see that the wavefront is initially "pinched" preferentially in the yz-plane [since $v_A(y, z) > v_A(x, y)$ for $|2z| > |x|$].

In Figure 6, we can see the ray paths for fluid elements that begins at four different starting points in the $y_0 = 1$ plane. Again, we can clearly see the refraction wrapping the fast wave elements around the null point, and the ray paths accumulate at the null point.

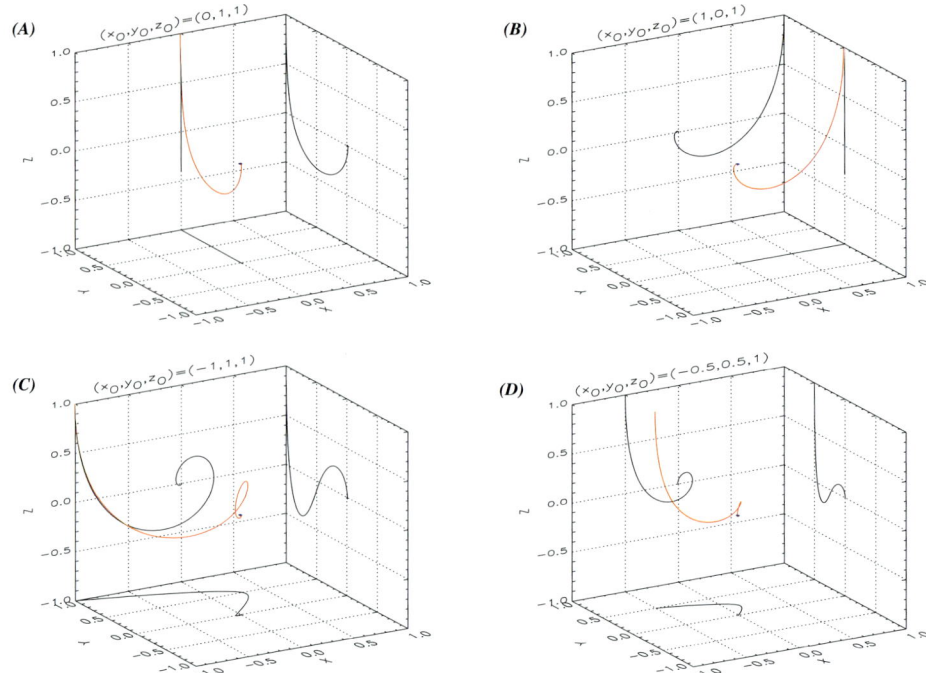

Figure 4 ($\epsilon = 1/2$) Ray paths for fluid elements that begin at $(x_0, y_0, z_0) =$ (A) $(0, 1, 1)$, (B) $(1, 0, 1)$, (C) $(-1, 1, 1)$, and (D) $(-0.5, 0.5, 1)$. This figure is also available as an mpg animation in the electronic edition of *Solar Physics* (http://dx.doi.org/10.1007/978-0-387-89482-9_38), showing many more starting points. Red indicates the 3D ray path and black indicates the xy, yz, and xz-projections of this ray path onto the respective planes. The blue dot indicates the position of the magnetic null point.

We do not show the ray paths corresponding to a planar fast wave starting at $x_0 = 1$ and approaching the null point, because these ray paths behave identically to those in Figure 6 under the transformation $(x, y) \to (-y, x)$ (for the $\epsilon = 1$ configuration). The ray paths corresponding to a planar fast wave starting at $y_0 = 1$ and starting at $x_0 = 1$ in the $\epsilon = 1/2$ magnetic configuration can be found in Figures 11 and 12 in Appendix B.

Thus, Sections 4.1 and 4.2 have shown that the fast wave experiences a refraction effect in the neighbourhood of a 3D magnetic null point and that, in all of these cases, the main result is the same: The ray paths accumulate at the null point. Of course, the actual paths taken vary depending upon initial conditions and choice of ϵ. Hence, we conclude that the fast wave and thus the fast wave energy eventually accumulate at the 3D null point for all ϵ and all initial conditions that generate a wave approaching the null.

Finally, it should be noted that the behaviour of the fast wave is entirely dominated by the Alfvén-speed profile, and since the magnetic field drops to zero at the null point, the wave will never actually reach there. However, there is still current accumulation and hence nonideal effects may be able to extract the wave energy in a finite time. This is investigated in the next section.

4.3. Current Buildup

From Section 3, we know that for the fast wave $\mathbf{v} \cdot \mathbf{B}_0 = 0$ and $\mathbf{v} \cdot (\mathbf{B}_0 \times \mathbf{k}) = 0$. This gives us two equations for the three velocity variables:

3D MHD Coronal Oscillations about a Magnetic Null Point

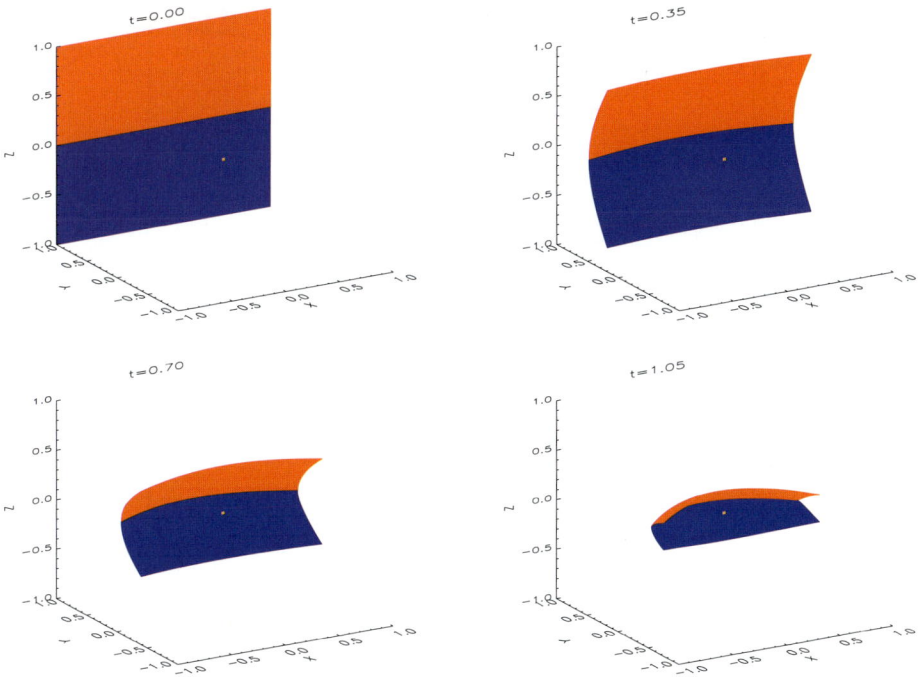

Figure 5 ($\epsilon = 1$) Surfaces of constant ϕ at four values of t, showing the behaviour of the (initially planar) wavefront that starts at $-1 \leq x \leq 1$, $y = 1$, and $-1 \leq z < 0$ (blue) and $-1 \leq x \leq 1$, $y = 1$, and $0 \leq z \leq 1$ (red). The (arbitrary) colouring has been added to aid the reader in tracking the wave behaviour. The yellow dot indicates the position of the magnetic null point.

$$xv_x + \epsilon y v_y - (\epsilon + 1)z v_z = 0,$$
$$\left[\epsilon yr + (\epsilon + 1)zq\right]v_x - \left[xr + (\epsilon + 1)zp\right]v_y + [xq - \epsilon yp]v_z = 0.$$

Thus, we can express two of the velocity components in terms of the third.

Recall from Section 2 that the perturbed electric current is given by $\mathbf{j}_1 = \nabla \times \mathbf{B}_1$. Thus,

$$\frac{\partial}{\partial t}\mathbf{B}_1 = \nabla \times (\mathbf{v} \times \mathbf{B}_0) \Rightarrow -\omega \mathbf{B}_1 = \mathbf{k} \times (\mathbf{v} \times \mathbf{B}_0)$$
$$\Rightarrow \mathbf{j}_1 = i\mathbf{k} \times \mathbf{B}_1 = -i\mathbf{k} \times \left[\mathbf{k} \times (\mathbf{v} \times \mathbf{B}_0)\right]/\omega = i|\mathbf{k}|^2 (\mathbf{v} \times \mathbf{B}_0)/\omega,$$

where we have made use of $\mathbf{v} \cdot (\mathbf{B}_0 \times \mathbf{k}) = 0$. From Equation (15), we can substitute for $|\mathbf{k}|^2$ to obtain

$$\mathbf{j}_1 = i\omega(\mathbf{v} \times \mathbf{B}_0)/|\mathbf{B}_0|^2$$
$$= i\omega \frac{[-(\epsilon + 1)z v_y - \epsilon y v_z, (\epsilon + 1)z v_x + x v_z, \epsilon y v_x - x v_y]}{x^2 + \epsilon^2 y^2 + (\epsilon + 1)^2 z^2}$$
$$\Rightarrow |\mathbf{j}_1| = \omega |\mathbf{v}|/|\mathbf{B}_0|, \tag{20}$$

where we have used $\mathbf{v} \cdot \mathbf{B}_0 = 0$ to simplify $|\mathbf{v} \times \mathbf{B}_0|$. Thus, since $|\mathbf{v}|$ is bounded [from our assumed form of $\mathbf{v}$ seen in Equation (13)] we can see that the current associated with the

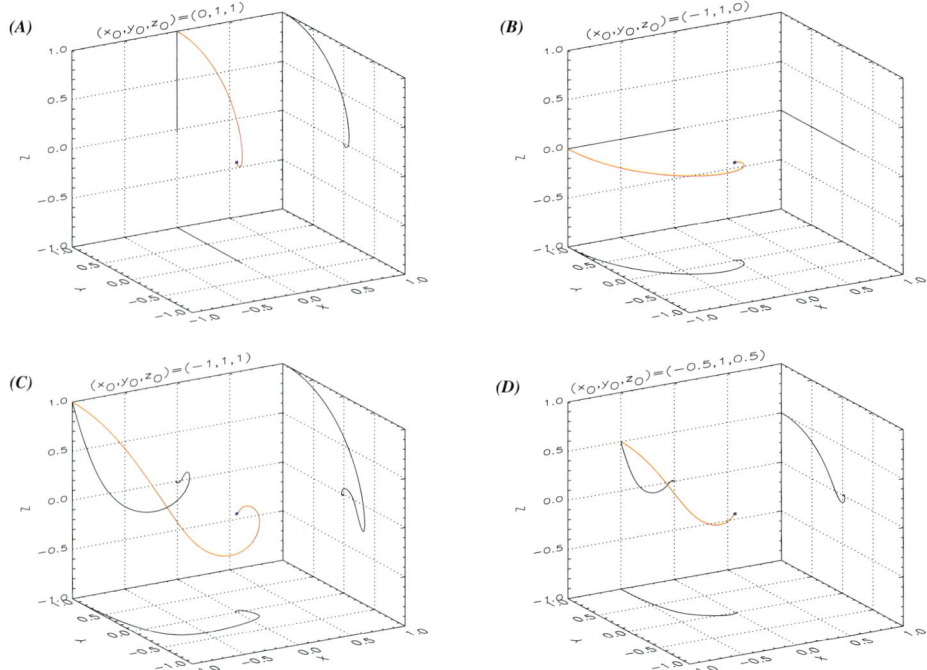

Figure 6 ($\epsilon = 1$) Ray paths for fluid elements that begin at $(x_0, y_0, z_0) =$ (A) $(0, 1, 1)$, (B) $(-1, 1, 0)$, (C) $(-1, 1, 1)$, and (D) $(-0.5, 1, 0.5)$. This figure is also available as an mpg animation in the electronic edition of *Solar Physics* (http://dx.doi.org/10.1007/978-0-387-89482-9_38), showing many more starting points. Red indicates the 3D ray path and black indicates the xy, yz, and xz-projections of this ray path onto the respective planes. The blue dot indicates the position of the magnetic null point.

fast wave will grow $\sim 1/|\mathbf{B}_0|$. Equivalent behaviour was found for the fast wave in the 2D case, that is, $\epsilon = 0$ (McLaughlin and Hood, 2004). Moreover, we can place limits on the magnitude of the current buildup. From Equation (37) in Appendix C, we can place limits on $|\mathbf{j}_1|$ such that

$$\frac{\omega |\mathbf{v}|}{(\epsilon + 1) R_0} e^{\alpha \epsilon^2 t/\omega_0} \leq |\mathbf{j}_1| \leq \frac{\omega |\mathbf{v}|}{\epsilon R_0} e^{\alpha (\epsilon+1)^2 t/\omega_0}, \tag{21}$$

where $R_0^2 = x_0^2 + y_0^2 + z_0^2$ and $\alpha = x_0 p_0 + y_0 q_0 + z_0 r_0$ (see Appendix C) and we have assumed $0 \leq \epsilon \leq 1$. Thus, we can see that the current buildup is bounded by two exponentially growing functions.

We now demonstrate this current buildup for two particular cases. Firstly, consider a planar fast wave starting at $z_0 = 1$ (Section 4.1). Here, we can solve Charpit's equations for the fast wave [Equation (16)] analytically for the initial conditions $p_0 = q_0 = x_0 = y_0 = 0$, that is, along $x = y = 0$, which is the path along which we expect the maximum current buildup to occur. Under these conditions, Equation (16) reduces to

$$x = 0, \quad y = 0, \quad \frac{dz}{ds} = -(\epsilon + 1)^2 r z^2, \quad p = 0, \quad q = 0, \quad \frac{dr}{ds} = (\epsilon + 1)^2 z r^2,$$

where we have used initial conditions (18). We also note that our conserved quantity [Equation (17)] states $zr = z_0 r_0 = \omega_0/(\epsilon + 1)$. Thus

$$z = z_0 e^{-(\epsilon+1)\omega_0 s} = z_0 e^{-(\epsilon+1)t}, \qquad r = r_0 e^{(\epsilon+1)\omega_0 s} = \frac{\omega_0}{(\epsilon+1)z_0} e^{(\epsilon+1)t}. \qquad (22)$$

As mentioned previously, the Alfvén speed drops to zero at the null point, indicating that the wave will never actually reach there, but the length scales (which can be thought of as the distance between the leading and trailing edges of the wave pulse) rapidly decrease, indicating that the current (and all other gradients) will increase. As an illustration, consider the wavefront as it propagates down the z-axis along $x = y = 0$. From Equation (22), the leading edge of the wave pulse is located at a position $z = z_0 e^{-(\epsilon+1)t}$, when the wave is initially at $z = z_0$. If the trailing edge of the wave pulse leaves $z = z_0$ at $t = t_1$ then the location of the trailing edge of the wave pulse at a later time is $z_2 = z_0 e^{-(\epsilon+1)(t-t_1)}$. Thus, the distance between the leading and trailing edges of the wave is $\delta z = z_0 e^{-(\epsilon+1)t}[e^{(\epsilon+1)t_1} - 1]$ and this decreases with time, suggesting that all gradients will increase exponentially.

We can also find analytical solutions for the velocity and polarisation of the fast wave. $\mathbf{v} \cdot \mathbf{B}_0 = 0$ along $x = y = 0$ implies $v_z = 0$, and hence using Equation (22) we obtain

$$\mathbf{k} = \left(0, 0, r_0 e^{(\epsilon+1)t}\right), \qquad \mathbf{v} = (v_x, v_y, 0) e^{i\phi_0}.$$

Using these forms in Equation (20) gives

$$\mathbf{j}_1 = -\frac{i\omega_0}{(\epsilon+1)z}(-v_y, v_x, 0) e^{i\phi_0} = -\frac{i\omega_0}{(\epsilon+1)z_0} e^{(\epsilon+1)t}(v_y, -v_x, 0) e^{i\phi_0}, \qquad (23)$$

where we have substituted for z from Equation (22). Thus, along the z-axis current builds up exponentially: $|\mathbf{j}_1| \sim z^{-1} \sim e^{(\epsilon+1)t}$. Comparing to Equation (21) we see that this exponent is the same as that of our theoretical maximum current buildup [under these initial conditions $\alpha = \omega_0/(\epsilon+1)$]. The coefficient is slightly smaller than our theoretical maximum, but this is most likely because the limits we assumed for Equation (35) (see Appendix C) were not very strong.

Secondly, for a planar fast wave starting at $y = y_0$ (Section 4.2), we can perform the same analysis along $x = z = 0$. Using the appropriate initial conditions [Equation (19)] and following the same analysis as before, we obtain

$$y = y_0 e^{-\epsilon t}, \qquad \mathbf{v} = (v_x, 0, v_z) e^{i\phi_0}, \qquad \mathbf{k} = \left(0, q_0 e^{\epsilon t}, 0\right)$$

$$\Rightarrow \mathbf{j}_1 = -\frac{i\omega_0}{\epsilon y_0} e^{\epsilon t}(v_z, 0, -v_x) e^{i\phi_0}.$$

Hence, we have exponential current buildup: $|\mathbf{j}_1| \sim y^{-1} \sim e^{\epsilon t}$. Again, this exponential buildup is within our theoretical limits [Equation (21)].

5. Alfvén Wave

We now consider the second root to Equation (14), which corresponds to the Alfvén wave. Hence, we assume $\omega^2 \neq |\mathbf{B}_0|^2 |\mathbf{k}|^2$ and simplify Equation (14) to

$$\mathcal{F}(\phi, x, y, z, p, q, r) = \omega^2 - (\mathbf{B}_0 \cdot \mathbf{k})^2$$

$$= \omega^2 - \left[xp + \epsilon yq - (\epsilon+1)zr\right]^2 = 0. \qquad (24)$$

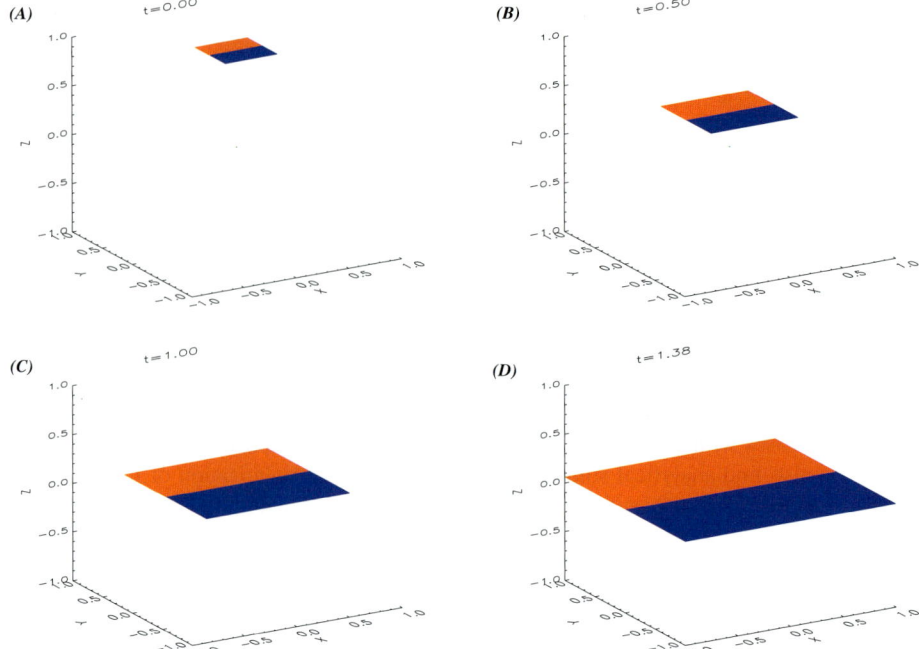

Figure 7 ($\epsilon = 1$) Surfaces of constant ϕ at four values of t, showing the behaviour of the (initially planar) wavefront that starts at $-0.25 \leq x \leq 0.25$, $-0.25 \leq y < 0$, and $z_0 = 1$ (blue) and $-0.25 \leq x \leq 0.25$, $0 \leq y \leq 0.25$, and $z_0 = 1$ (red). The (arbitrary) colouring has been added to aid the reader in tracking the wave behaviour.

Charpit's equations relevant to Equation (24) are

$$\frac{d\phi}{ds} = 0, \quad \frac{dt}{ds} = \omega, \quad \frac{dx}{ds} = -x\xi, \quad \frac{dy}{ds} = -\epsilon y\xi, \quad \frac{dz}{ds} = (\epsilon+1)z\xi,$$
$$\frac{d\omega}{ds} = 0, \quad \frac{dp}{ds} = p\xi, \quad \frac{dq}{ds} = \epsilon q\xi, \quad \frac{dr}{ds} = -(\epsilon+1)r\xi, \quad (25)$$

where $\xi = xp + \epsilon yq - (\epsilon+1)zr$. Thus, we can see that $\phi = \text{constant} = \phi_0$ and $\omega = \text{constant} = \omega_0$. In addition, $t = \omega s$, where we have set $t = 0$ at $s = 0$.

5.1. Planar Alfvén Wave Starting at $z_0 = 1$

We now solve Equation (25) as before, subject to the initial conditions

$$\phi_0 = 0, \quad \omega_0 = 2\pi, \quad -1 \leq x_0 \leq 1, \quad -1 \leq y_0 \leq 1, \quad z_0 = 1,$$
$$p_0 = 0, \quad q_0 = 0, \quad r_0 = \omega_0/[(\epsilon+1)z_0], \quad (26)$$

where we have (arbitrarily) chosen $\omega_0 = 2\pi$ and $\phi_0 = 0$. This corresponds to a planar Alfvén wave initially at $z = z_0$.

We can see the behaviour of the Alfvén wavefront in Figure 7 (where we have plotted surfaces of constant ϕ as in Section 4.1). We have also only plotted the wavefronts originating from $-0.25 \leq x_0, y_0 \leq 0.25$ so as to better illustrate the wavefront evolution. We

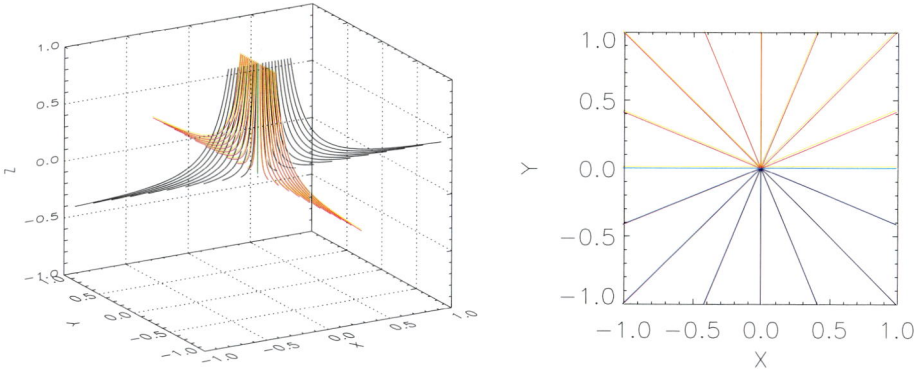

Figure 8 ($\epsilon = 1$) Left: Ray paths for fluid elements that begin at points $-0.25 \leq x_0 \leq 0.25$ along $y_0 = 0$, $z_0 = 1$ (indicated in black) and at $-0.25 \leq y \leq 0.25$ along $x_0 = 0$, $z_0 = 1$ (indicated in red) after a time $t = \pi/2$. The ray path from $x_0 = y_0 = 0$ is indicated in green and corresponds to the spine field line. Right: Projection of ray paths onto the xy-plane; red indicates $y > 0$, blue $y < 0$, and green $y = 0$.

can clearly see that the initially planar wavefront expands (in the xy-plane) as it approaches the null point and keeps its original shape (*i.e.*, planar with no rotation). The Alfvén wave eventually accumulates along the fan plane and never enters the $z < 0$ domain.

In Figure 8 (left), we can see the ray paths for fluid elements that begin at points $-0.25 \leq x_0 \leq 0.25$, $y_0 = 0$, $z_0 = 1$ and $-0.25 \leq y_0 \leq 0.25$, $x_0 = 0$, $z_0 = 2$, after a time $t = \pi/2$. Here, we see that the fluid elements travel along and are confined to the field lines that they start on (*i.e.*, the Alfvén wave spreads out following the field lines). This explains the expansion of the wavefront seen in Figure 7. A similar effect was seen in the 2D case (McLaughlin and Hood, 2004). As noted for the wavefront, all of the elements have travelled a different distance along their respective field lines but still form a planar wave. This is explained in Section 6.

5.2. Planar Alfvén Wave Starting at $y_0 = 1$

We again solve Equation (24) but now subject to the initial conditions

$$\phi_0 = 0, \quad \omega_0 = 2\pi, \quad -1 \leq x_0 \leq 1, \quad y_0 = 1, \quad -1 \leq z_0 \leq 1, \\ p_0 = 0, \quad q_0 = \omega_0/(\epsilon y_0), \quad r_0 = 0. \tag{27}$$

This corresponds to an Alfvén wave being sent in from the side boundary (along $y = y_0$).

We can see the behaviour of the Alfvén wavefront in Figure 9 (where surfaces of constant ϕ are plotted). We have plotted the wavefronts starting at $-1 \leq x_0 \leq 1$, $y_0 = 1$, and $-0.5 \leq z_0 \leq 0.5$ to more clearly show the Alfvén wave propagation. We can see that the (initially rectangular) wavefront expands in the z direction but is also squeezed in the x direction as it approaches the null (*i.e.*, as y decreases). We have imposed maximum and minimum values of unity in the z direction, purely for illustrative purposes. The Alfvén wave and hence the wave energy eventually accumulate along the spine. Again, the wave remains planar as it propagates.

In Figure 10 (left), we can see the ray paths for fluid elements that begin at points $-1 \leq x_0 \leq 1$, $y_0 = 1$, and at $z = -0.25, 0, 0.25$ after time $t = 2\pi$, where we have imposed maximum and minimum values of unity in the z direction (again purely for illustrative purposes). The ray paths in the fan plane all focus towards the null, which is expected as they

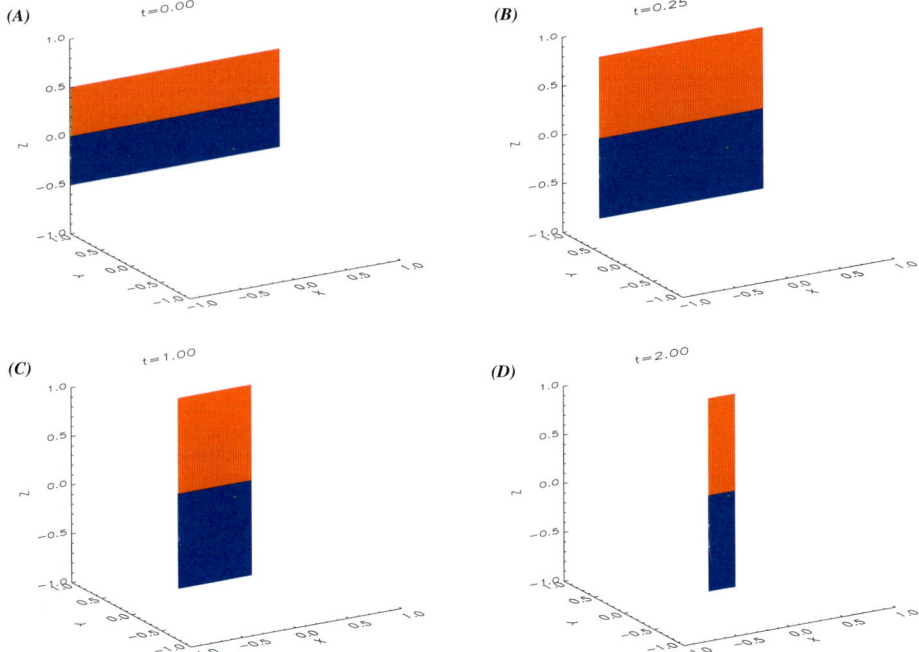

Figure 9 ($\epsilon = 1$) Surfaces of constant ϕ at four values of t, showing the behaviour of the (initially planar) wavefront that starts at $-1 \leq x_0 \leq 1$, $y_0 = 1$, and $-0.5 \leq z_0 < 0$ (blue) and $-1 \leq x_0 \leq 1$, $y_0 = 1$, and $0 \leq z_0 \leq 0.5$ (red). The (arbitrary) colouring has been added to aid the reader in tracking the wave behaviour. The green dot indicates the position of the magnetic null point. We have imposed maximum and minimum values of unity in the z-direction, purely for illustrative purposes.

follow the fan field lines. In contrast, the fluid elements on field lines above and below the fan plane propagate away from the null point, but they are simply following their respectively field lines. This is also clearly seen in Figure 10 (right), which shows various ray paths in the yz-plane along $x = 0$. This behaviour explains the narrowing and stretching effect seen in Figure 9: The Alfvén wave crosses the fan plane in this scenario and thus travels along the radially converging fan plane field lines. Meanwhile, the stretching effect comes from the diverging field lines the wave initially crosses. This work highlights the importance of understanding the magnetic topology of a system.

6. Analytical Solution for the Alfvén Wave

We can also solve Charpit's equations for the Alfvén wave (25) analytically. Firstly, let us consider a planar wave starting at $z = z_0$. Using the appropriate initial conditions [Equation (26)], we find

$$\frac{d\xi}{ds} = \frac{d}{ds}\left(xp + \epsilon yq - (\epsilon + 1)zr\right) = 0$$
$$\Rightarrow \xi = x_0 p_0 + \epsilon y_0 q_0 - (\epsilon + 1)z_0 r_0 = -\omega_0, \quad (28)$$

where $\xi = xp + \epsilon yq - (\epsilon + 1)zr$ as before, and where the values of x_0, p_0, y_0, q_0, z_0, and r_0 and the sign of ω_0 are taken from Equation (26). Thus, Equation (25) can be solved

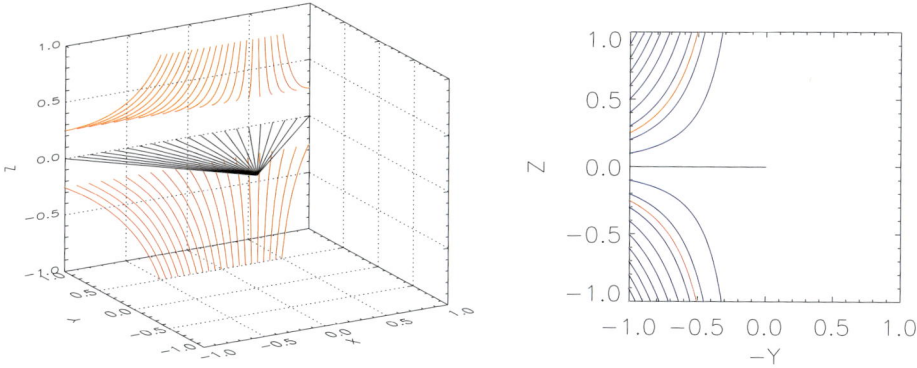

Figure 10 ($\epsilon = 1$) Left: Ray paths for fluid elements that begin at points $-1 \leq x_0 \leq 1$, $y_0 = 1$, and $z_0 = \pm 0.25$ (indicated in red) and $z_0 = 0$ (indicated in black) after time $t = 2\pi$. We have imposed maximum and minimum values of unity in the z direction, purely for illustrative purposes. Right: Ray paths for fluid elements that begin at points $-1 \leq z_0 \leq 1$ in the yz-plane along $x = 0$ (with blue indicating starting points of $-1 \leq z_0 \leq 1$ in divisions of 0.1, red indicating $z_0 = \pm 0.25$, and black indicating $z_0 = 0$). We have also plotted $y \to -y$ to aid the comparison between the left and right figures.

analytically:

$$p = p_0 e^{-t}, \qquad q = q_0 e^{-\epsilon t}, \qquad r = r_0 e^{(\epsilon+1)t}, \qquad (29)$$
$$x = x_0 e^{t}, \qquad y = y_0 e^{\epsilon t}, \qquad z = z_0 e^{-(\epsilon+1)t},$$

where $t = \omega_0 s$. This solution is valid for all ϵ. For $\epsilon = 0$, we recover the 2D solution of McLaughlin and Hood (2004).

From these equations we can see why an initially planar wave remains planar: If p and q are initially zero ($p_0 = q_0 = 0$) they remain zero for all time. In addition, z is independent of starting position x_0 and y_0. Thus, after a given time, different elements have travelled different distances along their respective field lines, but all have the same z (*i.e.*, all remain planar if originally planar). In addition, it can be shown that the volume occupied by the Alfvén-wave pulse is conserved (Appendix D).

Consider a circular wavefront at $z = z_0$, such that $x_0^2 + y_0^2 = r^2$, where r is some chosen radius. Let x_1, y_1, z_1 represent the position of the wavefront after some time t. Thus, the change in length scales ($\delta \mathbf{x}$) can be represented as

$$\delta x = (x_0 - x_1) e^{t}, \qquad \delta y = (y_0 - y_1) e^{\epsilon t}, \qquad \delta z = (z_0 - z_1) e^{-(\epsilon+1)t}.$$

Thus, the wave eventually accumulates along the fan plane (*i.e.*, $\delta x \to \infty$, $\delta y \to \infty$, $\delta z \to 0$). Furthermore, the circular wavefront evolves as

$$x_0^2 + y_0^2 = r^2 \quad \Rightarrow \quad \left(\frac{x}{e^t} \right)^2 + \left(\frac{y}{e^{\epsilon t}} \right)^2 = r^2;$$

that is, the circle becomes an ellipse (with semimajor axis in the direction of x if $0 < \epsilon < 1$ and in the direction of y if $\epsilon > 1$ but remaining circular for $\epsilon = 1$). Thus, the wave only accumulates over the whole fan plane for $\epsilon = 1$ and instead accumulates along a preferential axis for $\epsilon \neq 1$.

Charpit's equations (25) can also be solved by using the initial conditions for a planar wave starting at $y = y_0$, that is, Equation (27). Following the same techniques as before, we see that the length scales evolve as

$$\delta x = (x_0 - x_1)e^{-t}, \qquad \delta y = (y_0 - y_1)e^{-\epsilon t}, \qquad \delta z = (z_0 - z_1)e^{(\epsilon+1)t}.$$

In this case, the wave eventually accumulates along the spine (*i.e.*, $\delta x \to 0$, $\delta y \to 0$, $\delta z \to \infty$), for all values of ϵ. As before, an initially circular wavefront becomes elliptical for $0 < \epsilon \neq 1$ and evolves according to

$$x_0^2 + z_0^2 = r^2 \quad \Rightarrow \quad \left(\frac{x}{e^{-t}}\right)^2 + \left(\frac{z}{e^{(\epsilon+1)t}}\right)^2 = r^2.$$

6.1. Wavevector and Velocity

From Section 3, we know that for the Alfvén wave $\mathbf{v} \cdot \mathbf{B}_0 = 0$, $\mathbf{v} \cdot \mathbf{k} = 0$, and $\mathbf{v} \cdot (\mathbf{B}_0 \times \mathbf{k}) \neq 0$. Consider a planar wave starting at $z = z_0$; using Equation (29) we obtain

$$v_x x_0 e^t + v_y \epsilon y_0 e^{\epsilon t} - v_z(\epsilon + 1)z_0 e^{-(\epsilon+1)t} = 0,$$

$$v_x p_0 e^{-t} + v_y q_0 e^{-\epsilon t} + v_z r_0 e^{(\epsilon+1)t} = 0.$$

Using the initial conditions from Equation (26) gives $p_0 = q_0 = 0$ and so $v_z = 0$. Thus

$$\mathbf{k} = \left(0, 0, r_0 e^{(\epsilon+1)t}\right), \qquad \mathbf{v} = v_y\left(-\frac{y_0}{x_0}e^{(\epsilon-1)t}, 1, 0\right)e^{i\phi_0}. \tag{30}$$

Thus, the angle between v_x and v_y changes with time. There is one special case: For $\epsilon = 1$, we have $v_x/v_y = -y/x = -\tan\theta$. Recall that in cylindrical coordinates $v_x = v_r \cos\theta - v_\theta \sin\theta$, $v_y = v_r \sin\theta + v_\theta \cos\theta$, and so we must have $v_r = 0$ and $v_\theta \neq 0$. Hence, for $\epsilon = 1$ we have circular rotation of the field lines.

Similarly, for a planar Alfvén wave starting at $y = y_0$ [Equation (27)] and using the same derivation, we obtain

$$\mathbf{k} = \left(0, q_0 e^{\epsilon t}, 0\right), \qquad \mathbf{v} = v_z\left((\epsilon + 1)\frac{z_0}{x_0}e^{(\epsilon+2)t}, 0, 1\right)e^{i\phi_0}.$$

Finally, for a planar Alfvén wave starting at $x = x_0$, we obtain

$$\mathbf{k} = \left(p_0 e^t, 0, 0\right), \qquad \mathbf{v} = v_z\left(0, \frac{\epsilon + 1}{\epsilon}\frac{z_0}{y_0}e^{(2\epsilon+1)t}, 1\right)e^{i\phi_0}.$$

These velocity and polarisation solutions will be used in the next section.

6.2. Current Buildup

Recall from Section 2 that the perturbed electric current is given by $\mathbf{j}_1 = \nabla \times \mathbf{B}_1$. Now that we have an analytic solution for $\mathbf{v}$ we can solve Equation (8) for $\mathbf{B}_1$. Hence, $\mathbf{j}_1$ can be found:

$$\frac{\partial}{\partial t}\mathbf{B}_1 = \nabla \times (\mathbf{v} \times \mathbf{B}_0) \Rightarrow -\omega_0 \mathbf{B}_1 = \mathbf{k} \times (\mathbf{v} \times \mathbf{B}_0) = (\mathbf{B}_0 \cdot \mathbf{k})\mathbf{v} - (\mathbf{k} \cdot \mathbf{v})\mathbf{B}_0$$

$$\Rightarrow \mathbf{j}_1 = i\mathbf{k} \times \mathbf{B}_1 = -i(\mathbf{k} \times \mathbf{v})(\mathbf{B}_0 \cdot \mathbf{k})/\omega_0,$$

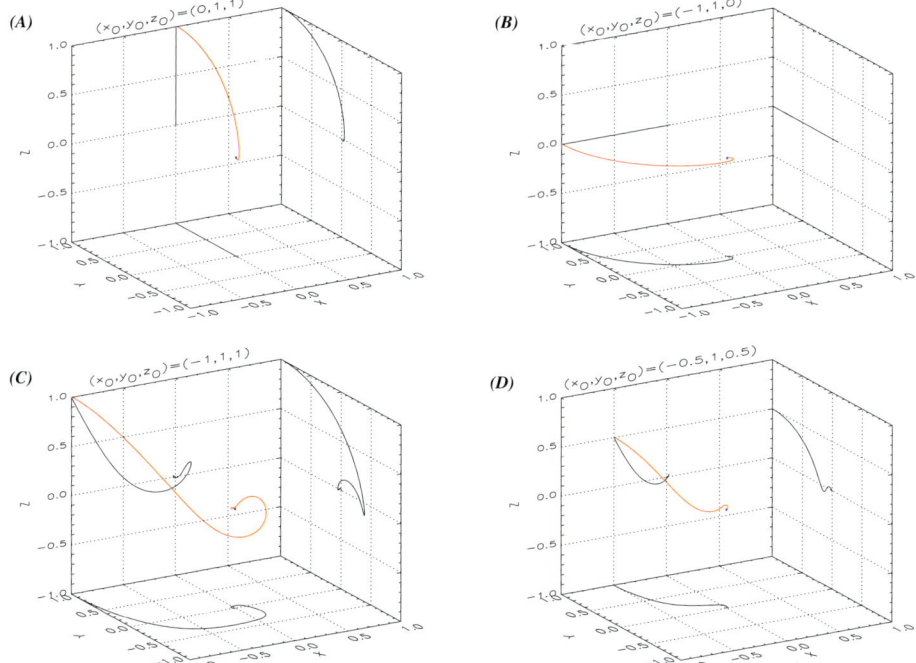

Figure 11 ($\epsilon = 1/2$) Ray paths for fluid elements that begin at $(x_0, y_0, z_0) =$ (A) $(0, 1, 1)$, (B) $(-1, 1, 0)$, (C) $(-1, 1, 1)$, and (D) $(-0.5, 1, 0.5)$. This figure is also available as an mpg animation in the electronic edition of *Solar Physics* (http://dx.doi.org/10.1007/978-0-387-89482-9_38), showing many more starting points. Red indicates the 3D ray path and black indicates the xy-, yz-, and xz-projections of this ray path onto the respective planes. The blue dot indicates the position of the magnetic null point.

We can integrate and invert this inequality to obtain

$$R_0^2 e^{-2\alpha(\epsilon+1)^2 s} \leq R^2 \leq R_0^2 e^{-2\alpha\epsilon^2 s} \quad \Rightarrow \quad \frac{1}{R_0^2} e^{2\alpha\epsilon^2 s} \leq \frac{1}{R^2} \leq \frac{1}{R_0^2} e^{2\alpha(\epsilon+1)^2 s}, \quad (36)$$

where R_0 is a constant that depends upon starting position: $R_0^2 = x_0^2 + y_0^2 + z_0^2$. Hence, inverting Equation (35) and combining it with Equation (36) gives

$$\frac{1}{(\epsilon+1)^2 R_0^2} e^{2\alpha\epsilon^2 s} \leq \frac{1}{(\epsilon+1)^2 R^2} \leq \frac{1}{|\mathbf{B}_0|^2} \leq \frac{1}{\epsilon^2 R^2} \leq \frac{1}{\epsilon^2 R_0^2} e^{2\alpha(\epsilon+1)^2 s}.$$

Finally, we recall $t = \omega_0 s$ and thus

$$\frac{1}{(\epsilon+1)^2 R_0^2} e^{2\alpha\epsilon^2 t/\omega_0} \leq \frac{1}{|\mathbf{B}_0|^2} \leq \frac{1}{\epsilon^2 R_0^2} e^{2\alpha(\epsilon+1)^2 t/\omega_0}. \quad (37)$$

Appendix D: Volume

Assume we generate an initially rectangular wave pulse of volume $V_0 = (x_1 - x_2) \times (y_1 - y_2) \times (z_1 - z_2)$, where x_1, x_2, y_1, y_2, z_1, and z_2 define the starting points at the edges of our

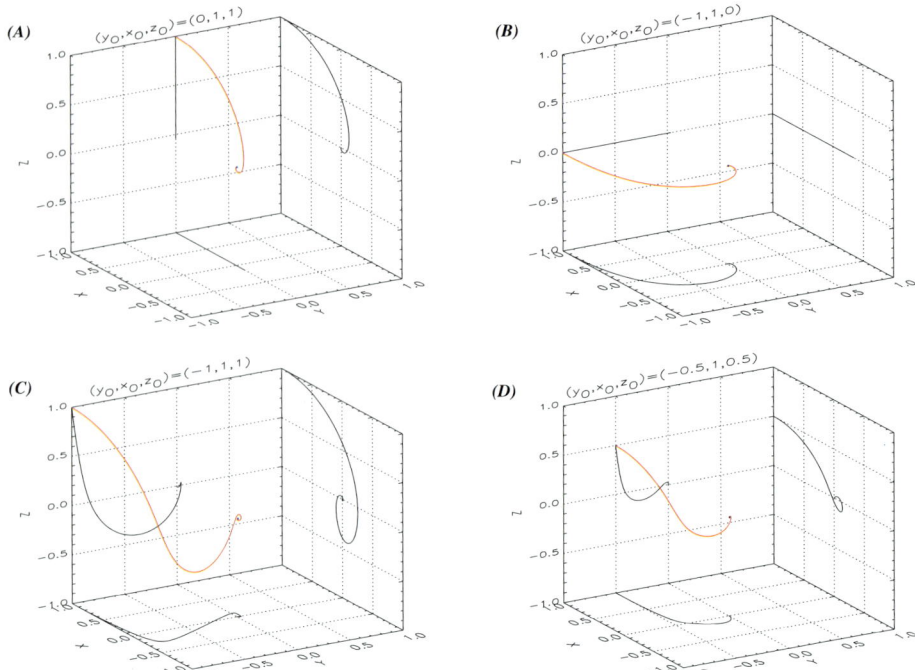

Figure 12 ($\epsilon = 1/2$) Ray paths for fluid elements corresponding to a fast wave starting at $x_0 = 1$. We have made the transformation $(x, y) \rightarrow (-y, x)$ to more easily compare with Fig. 11. Thus, this figure shows ray paths that begin at $(y_0, x_0, z_0) =$ (A) $(0, 1, 1)$, (B) $(-1, 1, 0)$, (C) $(-1, 1, 1)$, and (D) $(-0.5, 1, 0.5)$. This figure is also available as an mpg animation in the electronic edition of *Solar Physics* (http://dx.doi.org/10.1007/978-0-387-89482-9_38), showing many more starting points. Red indicates the 3D ray path and black indicates the xy, yz, and xz-projections of this ray path onto the respective planes. The blue dot indicates the position of the magnetic null point.

domain. The wave will evolve according to Equation (29) and thus, after travelling distance s along the characteristic curve, will occupy a volume

$$V_{\text{end}} = \left(x_1 e^{\omega s} - x_2 e^{\omega s}\right) \times \left(y_1 e^{\epsilon \omega s} - y_2 e^{\epsilon \omega s}\right) \times \left(z_1 e^{-(\epsilon+1)\omega s} - z_2 e^{-(\epsilon+1)\omega s}\right)$$
$$= (x_1 - x_2) \times (y_1 - y_2) \times (z_1 - z_2) = V_0.$$

Thus, volume is conserved for an Alfvén wave in this system.

Acknowledgements J.S.L.F. acknowledges financial assistance from a Cormack Vacation Research Scholarship awarded by the Royal Society of Edinburgh. J.A.M. wishes to thank the Royal Astronomical Society for awarding him an RAS grant to travel to the SOHO19 conference (where this work was first presented). J.A.M. also acknowledges financial assistance from the St Andrews STFC Rolling Grant and from the Leverhulme Trust. J.A.M. wishes to thank Jesse Andries, Ineke De Moortel, and Jaume Terradas for insightful discussions. A.W.H. and J.A.M. also wish to thank Clare Parnell for helpful suggestions regarding this paper.

References

Bender, C.M., Orszag, S.A.: 1978, *Advanced Mathematical Methods for Scientists and Engineers*, McGraw-Hill, Singapore.

Beveridge, C., Priest, E.R., Brown, D.S.: 2002, *Solar Phys.* **209**, 333–347.
Brown, D.S., Priest, E.R.: 2001, *Astron. Astrophys.* **367**, 339–346.
Cairns, R.A., Lashmore-Davies, C.N.: 1983, *Phys. Fluids* **26**, 1268–1274.
Craig, I.J., McClymont, A.N.: 1993, *Astrophys. J.* **405**, 207–215.
Craig, I.J., Watson, P.G.: 1992, *Astrophys. J.* **393**, 385–395.
De Moortel, I.: 2005, *Philos. Trans. Roy. Soc. A* **363**, 2743–2760.
De Moortel, I.: 2006, *Philos. Trans. Roy. Soc. A* **364**, 461–472.
De Moortel, I., Hood, A.W., Ireland, J., Arber, T.D.: 1999, *Astron. Astrophys.* **346**, 641–651.
Evans, G., Blackledge, J., Yardley, P.: 1999, *Analytical Methods for Partial Differential Equations*, Springer, London.
Galsgaard, K., Priest, E.R., Titov, V.S.: 2003, *J. Geophys. Res.* **108**, 1–12.
Heyvaerts, J., Priest, E.R.: 1983, *Astron. Astrophys.* **117**, 220–234.
Hood, A.W., Brooks, S.J., Wright, A.N.: 2002, *Proc. Roy. Soc. A* **458**, 2307–2325.
Khomenko, E.V., Collados, M.: 2006, *Astrophys. J.* **653**, 739–755.
McDougall, A.M.D., Hood, A.W.: 2007, *Solar Phys.* **246**, 259–271.
McLaughlin, J.A., Hood, A.W.: 2004, *Astron. Astrophys.* **420**, 1129–1140.
McLaughlin, J.A., Hood, A.W.: 2005, *Astron. Astrophys.* **435**, 313–325.
McLaughlin, J.A., Hood, A.W.: 2006a, *Astron. Astrophys.* **452**, 603–613.
McLaughlin, J.A., Hood, A.W.: 2006b, *Astron. Astrophys.* **459**, 641–649.
Nakariakov, V.M., Roberts, B.: 1995, *Solar Phys.* **159**, 399–402.
Nakariakov, V.M., Roberts, B., Murawski, K.: 1997, *Solar Phys.* **75**, 93–105.
Nakariakov, V.M., Verwichte, E.: 2005, *Living Reviews in Solar Physics* **2**, http://www.livingreviews.org/lrsp-2005-3 (cited August 2007).
Parnell, C.E., Smith, J.M., Neukirch, T., Priest, E.R.: 1996, *Phys. Plasmas* **3**, 759–770.
Pontin, D.I., Galsgaard, K.: 2007, *J. Geophys. Res.* **112**, 3103–3116.
Pontin, D.I., Bhattacharjee, A., Galsgaard, K.: 2007, *Phys. Plasmas* **14**, 2106–2119.
Priest, E.R., Titov, V.S.: 1996, *Philos. Trans. Roy. Soc.* **354**, 2951–2992.
Titov, V.S., Hornig, G.: 2000, *Phys. Plasmas* **7**, 3350–3542.
Weinberg, S.: 1962, *Phys. Rev.* **6**, 1899–1909.

Nonlinear Numerical Simulations of Magneto-Acoustic Wave Propagation in Small-Scale Flux Tubes

E. Khomenko · M. Collados · T. Felipe

Originally published in the journal Solar Physics, Volume 251, Nos 1–2, 589–611.
DOI: 10.1007/s11207-008-9133-8 © Springer Science+Business Media B.V. 2008

Abstract We present results of nonlinear, two-dimensional, numerical simulations of magneto-acoustic wave propagation in the photosphere and chromosphere of small-scale flux tubes with internal structure. Waves with realistic periods of three to five minutes are studied, after horizontal and vertical oscillatory perturbations are applied to the equilibrium model. Spurious reflections of shock waves from the upper boundary are minimized by a special boundary condition. This has allowed us to increase the duration of the simulations and to make it long enough to perform a statistical analysis of oscillations. The simulations show that deep horizontal motions of the flux tube generate a slow (magnetic) mode and a surface mode. These modes are efficiently transformed into a slow (acoustic) mode in the $v_A < c_S$ atmosphere. The slow (acoustic) mode propagates vertically along the field lines, forms shocks, and remains always within the flux tube. It might effectively deposit the energy of the driver into the chromosphere. When the driver oscillates with a high frequency, above the cutoff, nonlinear wave propagation occurs with the same dominant driver period at all heights. At low frequencies, below the cutoff, the dominant period of oscillations changes with height from that of the driver in the photosphere to its first harmonic (half period) in the chromosphere. Depending on the period and on the type of the driver, different shock patterns are observed.

Helioseismology, Asteroseismology, and MHD Connection
Guest Editors: Laurent Gizon and Paul Cally.

E. Khomenko (✉) · M. Collados · T. Felipe
Instituto de Astrofísica de Canarias, C/ Vía Láctea, s/n, 38205 Tenerife, Spain
e-mail: khomenko@iac.es

M. Collados
e-mail: mcv@iac.es

T. Felipe
e-mail: tobias@iac.es

E. Khomenko
Main Astronomical Observatory, NAS, Zabolotnogo str. 27, 03680 Kyiv, Ukraine

Keywords Photosphere · Chromosphere · Magnetohydrodynamics · Oscillations · Magnetic fields

1. Introduction

Solar magnetic structures show a continuous distribution of fluxes and sizes. In active regions, the field strength concentrated in large-scale structures (sunspots and pores) can be as large as $2-4$ kG. In plage and network areas, the average flux decreases to several hundred gauss. However, individual flux tubes in these areas can have intrinsic magnetic field strength similar to that of solar pores (*i.e.*, $1-2$ kG). The decrease of the average flux is caused mainly by the decrease of the size of the magnetic features (*i.e.*, their filling factor).

The observed photospheric brightness of the magnetic structures is in close relationship with their flux and size (see, *e.g.*, Berger *et al.*, 2007). Large-scale features such as sunspots and pores appear dark in photospheric observations. Plage and facular flux tubes appear as bright features (see, *e.g.*, Figure 7 in Berger *et al.*, 2007).

Because of their larger size, sunspots show horizontal gradients of the magnetic field strength and inclination. Instead, small-scale flux tubes possess a more homogeneous field with a rather sharp transition between the magnetized and nonmagnetized surroundings. From the point of view of wave propagation, it is important to realize that the temperature at a given geometrical height is different in the different structures. Temperature and magnetic field define the characteristic wave propagation speeds, that is, the sound speed (c_S) and the Alfvén speed (v_A), as well as the acoustic cutoff frequency (ω_C) and the height of the transformation layer, where c_S is equal to v_A and different wave modes can interact. These parameters can be rather different in large-scale dark structures and small-scale bright structures. Thus, it is not surprising that the observed properties of waves in sunspots, network, and plage areas are rather different.

1.1. Sunspots and Pores

Several decades of studies of sunspot oscillations can be summarized as follows. Waves observed at different layers of the atmosphere of sunspots umbrae and penumbrae seem to be the manifestation of the same phenomenon (see, *e.g.*, Maltby *et al.*, 1999, 2001; Brynildsen *et al.*, 2000, 2002; Christopoulou *et al.*, 2000, 2001; Rouppe van der Voort *et al.*, 2003; Tziotziou *et al.*, 2006). In the photosphere, five-minute oscillations are observed in sunspot umbrae and penumbrae and in solar pores (Bogdan and Judge, 2006). Wave propagation is linear with amplitudes of several hundred meters per second and is parallel to the magnetic field lines (see, *e.g.*, Gurman and Leibacher, 1984; Lites, 1984; Collados *et al.*, 2001; Centeno *et al.*, 2006a; Bloomfield *et al.*, 2007). Magnetic field oscillations of a few gauss are detected in the deeper layers and attempts have been made to interpret these oscillations in terms of fast and slow MHD waves (*e.g.*, Rüedi *et al.*, 1998; Lites *et al.*, 1998; Bellot Rubio *et al.*, 2000; Khomenko *et al.*, 2003).

In the chromosphere, the dominant period of oscillations decreases and shock waves are observed with a three-minute periodicity in sunspot umbrae and in pores (Lites, 1984, 1986, 1988; Socas-Navarro *et al.*, 2000; Centeno *et al.*, 2006a; Tziotziou *et al.*, 2007). The amplitudes of the observed shocks are in the range of $5-15$ km s^{-1}, depending on the size of the structure. In pores, shocks are weaker (Centeno *et al.*, 2006b). Umbral flashes also show a three-minute periodicity. In contrast, in the penumbra, the shock behavior of waves with five-minute periodicity is detected for running penumbral waves even at chromospheric heights (Tziotziou *et al.*, 2006). Waves observed in the transition region and the corona conserve these properties (De Moortel *et al.*, 2002; Marsh and Walsh, 2005, 2006; Maltby *et al.*, 1999, 2001; Brynildsen *et al.*, 2000, 2002; Christopoulou *et al.*, 2000, 2001). Wave propagation from the photospheric pulse into the chromosphere and higher layers is along the magnetic field lines for both three- and five-minute perturbations.

1.2. Plages, Facular Regions, and Network

Bright magnetic field structures such as plages, facular regions, and network manifest five-minute linear oscillations in the photosphere, and the oscillations maintain the five-minute periodicity in the chromosphere and at greater heights (Lites *et al.*, 1993; Krijger *et al.*, 2001; De Pontieu *et al.*, 2003; Centeno *et al.*, 2006b; Bloomfield *et al.*, 2006; Vecchio *et al.*, 2007). Shock occurrence is much lower in these quiet-Sun regions with enhanced flux (Krijger *et al.*, 2001; Centeno *et al.*, 2006b). The velocity oscillation amplitudes in the chromosphere do not reach values as large as in sunspots and stay within $2-4$ km s^{-1}. Waves at photospheric and chromospheric layers seem to be correlated and vertical propagation along the vertical magnetic fields seems to dominate (Krijger *et al.*, 2001; Centeno *et al.*, 2006b; Vecchio *et al.*, 2007). Curiously, there are areas with enhanced three-minute power surrounding the strong magnetic field concentrations, possibly related to flux tube canopies (aureoles; see Braun and Lindsey, 1999; Thomas and Stanchfield, 2000; Krijger *et al.*, 2001; Judge *et al.*, 2001). In the higher coronal layers, loops connecting sunspots oscillate with three-minute periods and the loops connecting network and plage regions oscillate with five-minute periods (De Moortel *et al.*, 2002).

Thus, the wave properties are distinctly different in large-scale and small-scale magnetic features. The difference in oscillation properties is present through the whole solar atmosphere and should be a consequence of the thermal and magnetic properties of these structures.

1.3. Theoretical Models

Numerous theoretical mechanisms have been proposed to explain the oscillation spectra at different heights in magnetic structures and to interpret the observed oscillations in terms of MHD waves. Zhugzhda and Locans (1981), Gurman and Leibacher (1984), and Zhugzhda (2007) argue that the observed spectrum of sunspot umbral oscillations in the chromosphere is due to the temperature gradients of the atmosphere acting as an interference filter for linear three-minute period acoustic waves. However, it has been demonstrated by Fleck and Schmitz (1991) that the change of the period with height from five to the three minutes is a basic phenomena occurring even in an isothermal atmosphere for linear waves owing to the resonant excitation at the atmospheric cutoff frequency. In the case of the solar atmosphere, the temperature minimum gives rise to the three-minute cutoff period. Later, the same authors found that nonlinear shock interaction and overtaking cause the frequency of acoustic oscillations to also change with height (Fleck and Schmitz, 1993). The response of the solar atmosphere to an adiabatic shock wave leads to a natural appearance of the three-minute peak in the oscillation power spectra, in the case when the underlying photosphere has a five-minute periodicity. A third mechanism was shown to explain the three-minute periodicity of calcium bright grains in the quiet Sun: With a spectrum of periods in the photosphere with a peak at five minutes, this distribution changes to one with a maximum at the acoustic cutoff because at longer periods the energy falls off exponentially with height (Carlsson and Stein, 1997). These mechanisms may explain the shift with height of the period of oscillations from five to three minutes observed in sunspots.

The question remains as to why is there no such shift in small-scale structures of plages and network regions. De Pontieu *et al.* (2004) argue that the inclination of the magnetic field may play an important role. If (acoustic or slow MHD) waves have a preferred direction of propagation defined by the magnetic field, the effective cutoff frequency is lowered by the cosine of the inclination angle with respect to the vertical. This allows evanescent waves to propagate. It should be recalled, however, that mainly vertical propagation is observed in the photosphere and chromosphere in plage regions (see, *e.g.*, Krijger *et al.*, 2001; Centeno *et al.*, 2006b); thus it is unclear whether the mechanism suggested by De Pontieu *et al.* (2004) is at work. Alternatively, the decrease of the effective acoustic cutoff frequency can be produced by taking into account the radiative losses of acoustic oscillations (Roberts, 1983; Centeno *et al.*, 2006b). If the radiative relaxation time is effectively small (as expected in small-scale magnetic structures, which are transparent to radiation), the cutoff frequency can be reduced and otherwise evanescent waves can propagate.

All the works cited so far apply the theory of acoustic waves in idealized atmospheres to explain the observed oscillation properties in magnetic structures. In magnetized atmospheres, different wave types can exist and different modes can be observed depending on the magnetic field configuration and the height of the transformation layer ($c_S = v_A$) relative to the height of the formation of the spectral lines used in observations. Bogdan *et al.* (2003), Rosenthal *et al.* (2002), Hasan and Ulmschneider (2004), and Hasan *et al.* (2003, 2005) addressed this question in their simulations of magneto-acoustic waves in nontrivial magnetic configurations. They pointed out the importance of defining the mode type and the mode-conversion height when interpreting the observations. Because of mode mixing at the transformation layer, no correlation may be observed between perturbations below and above the $c_S = v_A$ height.

All of these arguments point toward the need to have realistic, self-consistent, and systematic models of the wave spectrum in magnetic structures. It is necessary to include nontrivial magnetic field configurations to allow the existence of different mode types and mode conversion. Nonlinearities should be taken into account for waves propagating into the chromosphere. The thermal structure should be realistic to make possible the reflection of waves with frequencies below the cutoff. Radiative damping also plays an important role, in particular in estimating the amount of energy that may be deposited in the higher atmosphere from the waves. In the present work we address several of these questions. We perform nonlinear numerical modeling of magneto-acoustic waves in small-scale magnetic flux tubes with internal structure. We consider waves with realistic periods of three and five minutes, which has never been done before for nontrivial magnetic configurations. We study the mode transformation and oscillation spectra at different heights from 0 to 2000 km as a function of the type of driver and its period. At present, radiative damping of oscillations is not taken into account. This question will be addressed in a separate paper.

2. Numerical Model

We solve the basic equations of the ideal MHD, written in conservative form as

$$\frac{\partial \rho}{\partial t} + \nabla \cdot (\rho \mathbf{V}) = 0, \qquad (1)$$

$$\frac{\partial (\rho \mathbf{V})}{\partial t} + \nabla \cdot \left[\rho \mathbf{V}\mathbf{V} + \left(P + \frac{B^2}{8\pi} \right) \mathbf{I} - \frac{\mathbf{B}\mathbf{B}}{4\pi} \right] = \rho \mathbf{g}, \qquad (2)$$

Figure 3 Time series of snapshots of the transverse (right) and longitudinal (left) velocities in the simulation with a horizontal driver at 50 seconds, showing the early stages of the evolution. Numbers give the elapsed time in seconds since the start of the simulation. The transverse component is normalized to $\sqrt{v_A \rho_0}$ and the longitudinal component is normalized to $\sqrt{c_S \rho_0}$. Each snapshot covers 900 km horizontally and 2000 km vertically.

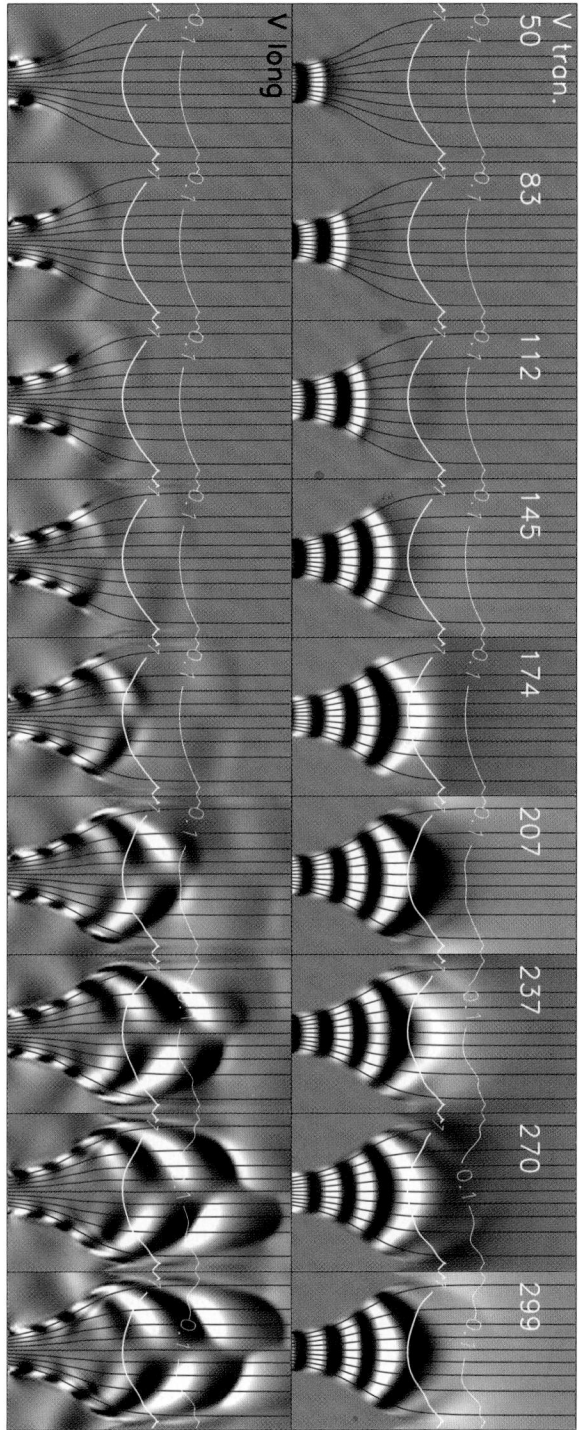

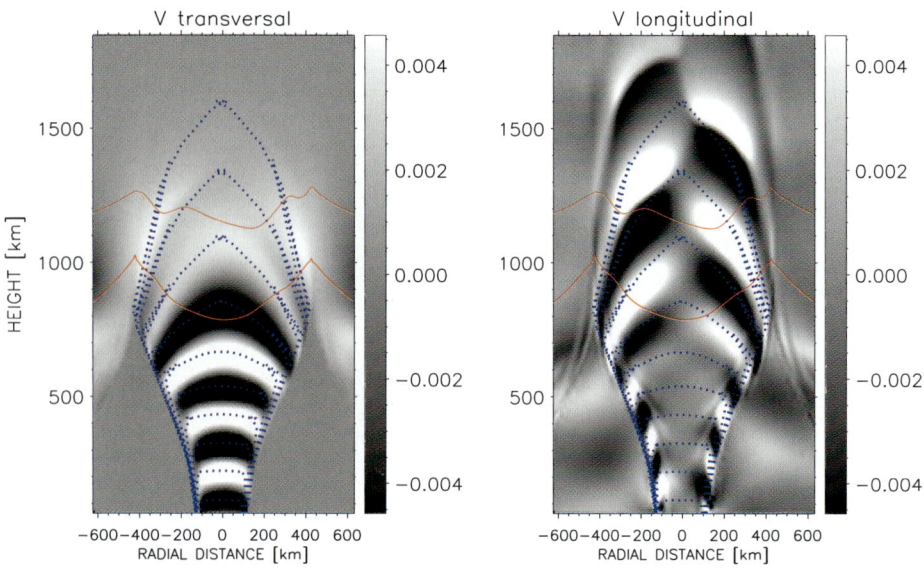

Figure 4 WKB solution for the simulations with a horizontal driver at 50 seconds. The format of the figure is the same as Figure 2. Blue contours are the positions of constant phase of the WKB slow-mode solution at the different equidistant time moments.

from their simulation that a significant part of the slow-mode energy goes into the fast mode, thus preserving the magnetic nature of the wave at all heights. To better understand our simulations, we present in Figure 4 the WKB solution for the slow mode. The WKB approach is linear and assumes that the wavelength of the perturbation is much smaller than the characteristic scale of the variations of the background atmospheric parameters, and thus can be applied in the case of the short-wavelength simulations with the 50-second driver. The equations describing the WKB approach in this particular case can be found in Khomenko and Collados (2006). The blue contours in Figure 4 describe the position of the wavefront at different time moments separated by equal intervals. The initial wavefront at the base of the atmosphere is assumed planar (*i.e.*, the horizontal wave vector k_x is equal to zero). The WKB solution describes rather well the position of the wavefront at every time moment; that is, the waves propagate at the slow-mode speed. As can be appreciated from the figure, as the wave propagates upward, the horizontal gradients of its phase speed produce the deformation of the wavefront, so that its central part propagates faster. Thus at greater heights the wave vector (k) is no longer directed along the vertical but forms a significant angle with respect to the magnetic field vector B_0. According to Cally (2006), if the angle between k and B_0 is close to zero, the most effective transformation is from the slow to the fast mode. However, as this angle increases, the slow to slow mode transformation becomes more important. Since in our simulations the slow mode reaches the transformation layer with k significantly inclined with respect to B_0, it explains the effectiveness of the slow to slow mode transformation. This does not contradict the fact that the wave energy may propagate longitudinally, which is a result of the strong anisotropy of the medium.

The slow acoustic mode develops shocks with amplitudes of about 5 km s^{-1} that propagate straight up into the chromosphere. Owing to the clear nonlinear behavior of the acoustic mode at the greater heights, harmonics are generated in addition to the frequency of the

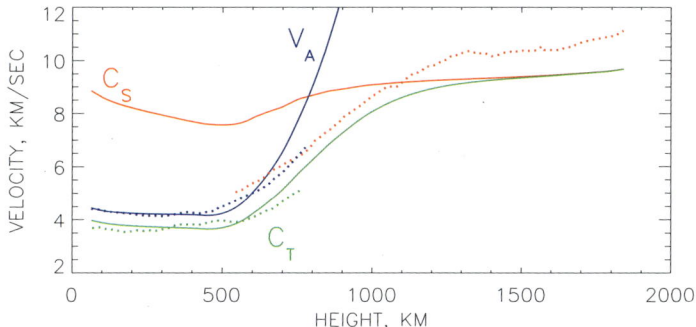

Figure 5 Height dependence of phase speeds of the slow mode (blue dotted line), surface mode (green dotted line), and acoustic mode (red dotted line) measured from the simulations compared to the characteristic propagation speeds calculated in the magnetostatic model (solid lines). The values of the Alfvén and sound speed (v_A and c_S) are taken at the flux tube axis. The value of the tube speed (c_T) is computed along the trajectory of the surface mode.

original perturbation of the driver. However, the maximum power at all heights is kept at the driving frequency.

Figure 5 gives the characteristic speeds of the waves in the simulation. Using the time series of the simulations, we have calculated the phase speed of the slow magnetic mode (at heights from 0 to about 800 km, where this mode exists), the slow acoustic mode (at heights from 500 km upward), and the surface mode. In the latter case, the phase speed along the curved trajectory of the surface mode was calculated. The retrieved phase speeds are displayed in Figure 5 as a function of height. The stratifications of v_A, c_S, and tube speed $[c_T^2 = c_S^2 v_A^2/(c_S^2 + v_A^2)]$ are also presented.

The comparison of the dotted and solid line curves in Figure 5 leads to coherent results. The slow magnetic mode propagates with the Alfvén speed (v_A). The surface mode propagates with the tube phase speed (c_T) up to 400 km. Despite the fact that the tube and the Alfvén velocities are rather similar in the lower photosphere, the phase speed of the surface wave is distinct from that of the slow magnetic wave. The former propagates with its own characteristic speed, as suggested from the analytical theory (see, *e.g.*, Roberts, 1981). At heights about 700 km (where $v_A \approx c_S$) the speeds of all modes become close to one another and the energy can be easily transferred among the different wave types. It can be seen that the slow magnetic mode and the surface mode give their energy to the slow acoustic mode. As the slow acoustic mode propagates upward, shock formation occurs and the propagation becomes supersonic. The phase speed of the shocks increases with height in agreement with the increase of their amplitudes.

The following items summarize the most important features of the simulations with the horizontal driver at 50 seconds:

- Slow to slow mode transformation occurs at the $v_A = c_S$ height.
- There is an antiphase left–right acoustic shock wave pattern.
- Surface modes are excited.
- The period of the waves is maintained at all heights.

The simulation with this 50-second driver follows the spirit of the calculations by Rosenthal *et al.* (2002), Bogdan *et al.* (2003), and Hasan *et al.* (2005), where small-scale flux tubes were perturbed by horizontal high-frequency motions at the lower boundary. Similarly to these works, we also find that such motions generate acoustic shocks at chromospheric heights. The differences in the wave behavior between these works and ours are mainly due to the difference in the initial magnetostatic situation. Once more, this suggests that the background magnetostatic state should be taken into account while interpreting observations, since different wave types can be generated.

6. Horizontal Driving at 180 Seconds

Here, we show the results of the simulations using a 180-second driver, still above the cutoff frequency. For this reason, it is not surprising that we get a behavior similar to the previous case. The wave pattern that appears in the simulations is basically the same as in the case of the 50-second driver, but the wavelength of the perturbations is larger. The driver mainly excites in the lower layers a transverse slow magnetic mode inside the flux tube, together with a surface mode at the interface with the nonmagnetic surroundings. As before, the surface mode is mainly longitudinal. Figure 6 shows some snapshots of the transverse and longitudinal velocities during one period of the wave after the simulations reached the stationary state. In contrast to the case of the short-wavelength 50-second-driver simulation, we find the normalization of the velocity components to $\sqrt{c_S \rho_0}$ or $\sqrt{v_A \rho_0}$ to be inappropriate for the 180- and 300-second-driver simulations since the group velocity of waves follows a more complicated behavior than either c_S or v_A owing to the larger wavelength and nonlinear effects. Instead we normalize both velocity components with an exponential function ($\rho_0^{1/4}$) that allows a better visualization of the results.

Around the $c_S = v_A$ layer, mode conversion takes places and most of the energy goes again to the slow acoustic mode. The antisymmetric pattern with respect to the flux tube axis is still present in longitudinal velocity above this layer, although the antisymmetry is not as perfect as in the 50-second-driver case. Shocks clearly develop during the evolution. Very clear examples are apparent at time 379 in the left part of the flux tube and at time 452 in the right part. These nonlinear effects are more pronounced in the higher layers. The shock formation occurs above a height of about 1000 km. The temporal evolution of the transverse velocity shows the refraction of the fast magnetic mode at the higher layers where $v_A > c_S$. The horizontal motions of the flux tube field lines in the photosphere and chromosphere resulting from this mode can also be seen during the time evolution.

In Figure 7, the temporal evolution of the longitudinal velocity at different points in the flux tube, corresponding to different heights and horizontal distances from the axis, is shown. The first two upper panels on the left correspond to two points located near the boundary of the flux tube, one on the right and the other on the left of the axis, 200 km above the driver. The almost sinusoidal oscillation pattern is clear, since nonlinearities have not had time to develop. When one compares the position in the horizontal axis of maxima and minima for the velocity time series from both points, it is apparent that both points oscillate in antiphase. The two middle panels on the left correspond to two points located at the same distance from the axis as the previous ones, but at a larger height (704 km). This is the height where $v_A \approx c_S$ and the wave transformation starts. The amplitude of the oscillation has increased by almost a factor of four. It can be seen how the wave takes some time to reach this layer. The perturbation reaches this layer about four minutes after the driver starts. Up to there, nonlinearities are still negligible, and the sinusoidal aspect of the wave is maintained, as well as this antisymmetry of the velocity pattern. At higher layers (1664 km in the plots) velocity discontinuities appear, indicating the development of shocks, with amplitudes of several kilometers per second (two bottom panels of Figure 7). The corresponding power spectra of this velocity temporal series are plotted in the right-hand part, and they all demonstrate

Figure 6 Time series of snapshots of the transverse (right) and longitudinal (left) velocities in the simulation with a horizontal driver at 180 seconds. Numbers give the elapsed time in seconds since the start of the simulation. After removal of the average velocity increase with height produced by the density falloff, all images have the same scale. Each snapshot covers 900 km horizontally and 2000 km vertically.

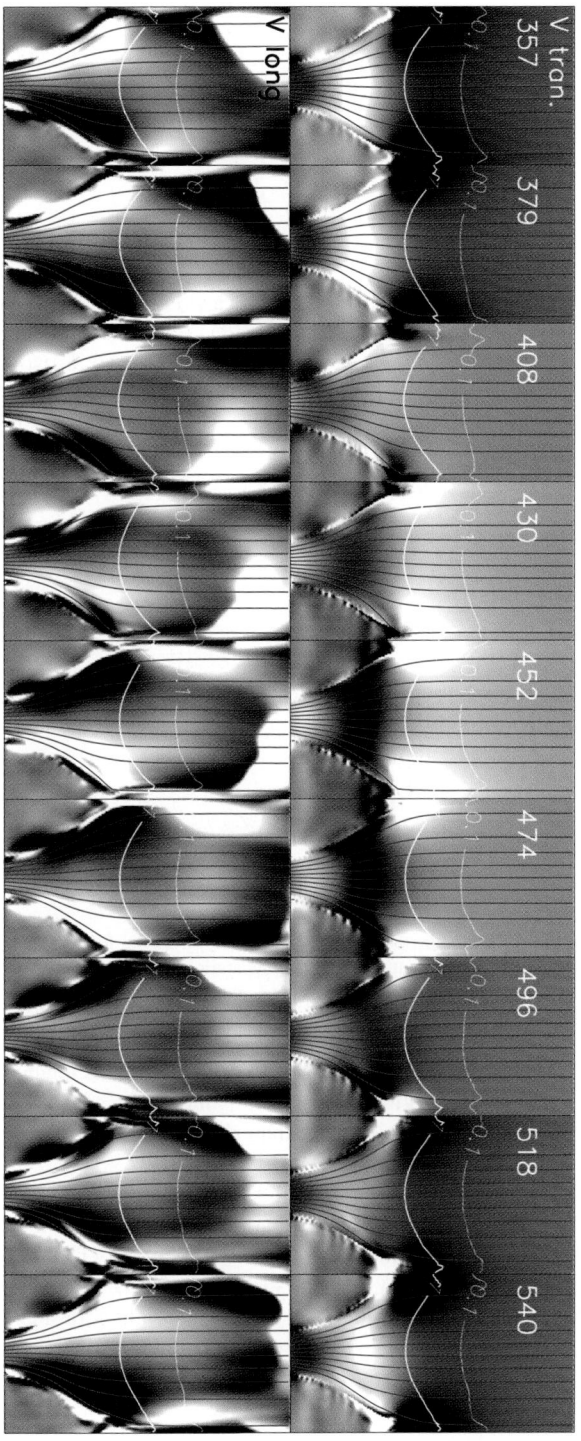

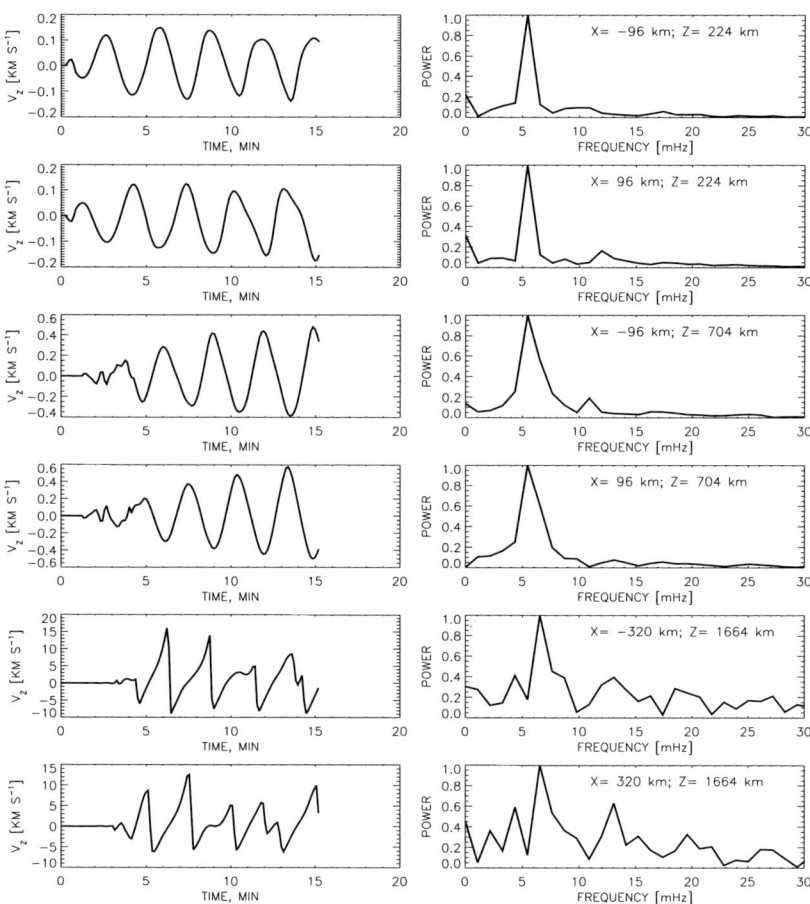

Figure 7 Examples of the temporal evolution of the longitudinal velocity (left) and their corresponding power spectra (right) at different heights and positions inside the flux tube for the simulation with horizontal driving at 180 seconds. The coordinates are indicated on the right panels.

that the frequency of the driver (5.56 Hz) remains unaltered at all heights. The presence of several harmonics in higher layers is indicative of the nonlinearities giving rise to the sawtooth velocity profiles, rather than sinusoidal ones. Note that the shocks are in antiphase on both sides of the tube.

Probably, the most important difference with respect to the previous case with a driver of 50 seconds lies in the fact that the slow to slow mode conversion is not as efficient now (see Cally, 2006), and a larger part of the energy of the slow mode excited by the driver is transferred to the fast (transverse, and hence mainly magnetic) mode. The atmosphere above the $v_A = c_S$ oscillates uniformly to left and right with small phase differences. The horizontal motions are in phase in the whole atmosphere owing to the large oscillation wavelength.

In summary, the most important features of the simulations with the horizontal driver at 180 seconds are the following:

- Partial slow to slow (acoustic) transformation occurs at $v_A = c_S$ height.
- There is a horizontal flux tube motion at 2000 km because of the fast (magnetic) wave.

- There is an antiphase left–right shock wave pattern.
- Surface modes are excited.
- The period of the wave is maintained with height.

7. Horizontal Driving at 300 Seconds

Figure 8 shows the temporal evolution of the transverse and longitudinal velocities over one wave period, after the simulations reach the stationary state, and Figure 9 gives the temporal variation of the longitudinal velocity at selected points, together with their corresponding power spectra, for the simulations with a horizontal driver with a period of 300 seconds. The frequency of the driver is below the cutoff frequency. As in two previous cases, the driver excites the slow and surface modes. Note that, unlike the shorter period simulations, there are significant motions excited outside the flux tube in the nonmagnetic atmosphere as well. The nature of these motions is partly acoustic and partly convective since the lower layers of the photosphere are still unstable to convection.

The slow to slow mode transformation is only partial, and part of the energy of the slow mode is transferred to the fast mode above the $v_A = c_S$ layer, producing a horizontal shaking of the tube that propagates upward with the Alfvén speed. The behavior of the fast mode seen in the transverse velocity above $v_A = c_S$ height is similar to the three-minute period simulations, but the left–right symmetry pattern is lost.

Most of the energy in high layers is nonetheless still carried by the slow (acoustic) mode, giving rise to shock waves with amplitudes of 10–15 km s^{-1} above 1000 km height. The antisymmetric nature of these shocks has disappeared. Left and right parts of the tube are almost in phase now. However, the most conspicuous difference lies in the period of the wave. As it corresponds to an evanescent wave, the five-minute oscillation is rapidly damped, and the residuals coming from nonlinear effects at 6 Hz, twice the driver frequency, already dominate at a height of 700 km. This double frequency is clear in the temporal series of the velocity, and in the corresponding power spectra, at this height (two middle panels in Figure 9). From 700 km upward, the velocity increases again, owing to the density decrease, giving rise to shocks in the upper part of the simulation domain. The main frequency is maintained around 6 mHz, once the five-minute oscillation has been damped. Kalkofen et al. (1994) and Fleck and Schmitz (1991) showed that the change of the dominant period of acoustic oscillations with height, from five to three minutes, is due to the resonant excitation of waves at the atmospheric cutoff frequency (i.e., the cutoff frequency corresponding to the temperature minimum). However, this effect may not be dominant in our simulations since the acoustic mode only appears above 1000 km (i.e., well above the temperature minimum). At heights above 1000 km, the cutoff frequency is again around 3 mHz and the resonant excitation cannot produce oscillations at the higher frequency. We conclude that the generation of oscillations at the first harmonic of the driver in our simulations is mostly a nonlinear phenomenon.

The most important properties of the wave propagation from the simulations with the horizontal driver at 300 seconds are the following:

- Partial slow to slow (acoustic) transformation occurs at $v_A = c_S$ height.
- There is a horizontal flux tube motion at 2000 km because of the fast (magnetic) wave.
- There is an in-phase left–right shock wave pattern.
- Surface modes are excited.
- The period of the wave is not constant with height.

Figure 8 Time series of snapshots of the transverse (right) and longitudinal (left) velocities in the simulation with a horizontal driver at 300 seconds. Numbers give the time elapsed in seconds since the start of the simulation. After removal of the average velocity increase with height produced by the density falloff, all images have the same scale. Each snapshot covers 900 km horizontally and 2000 km vertically.

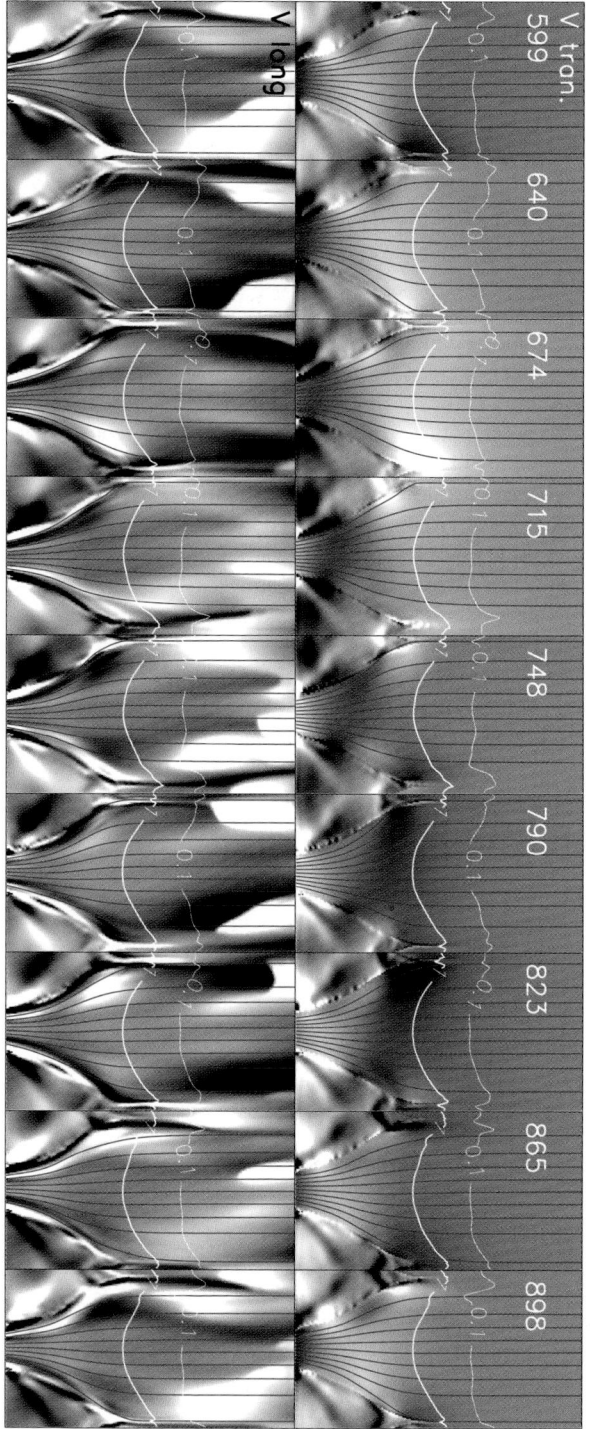

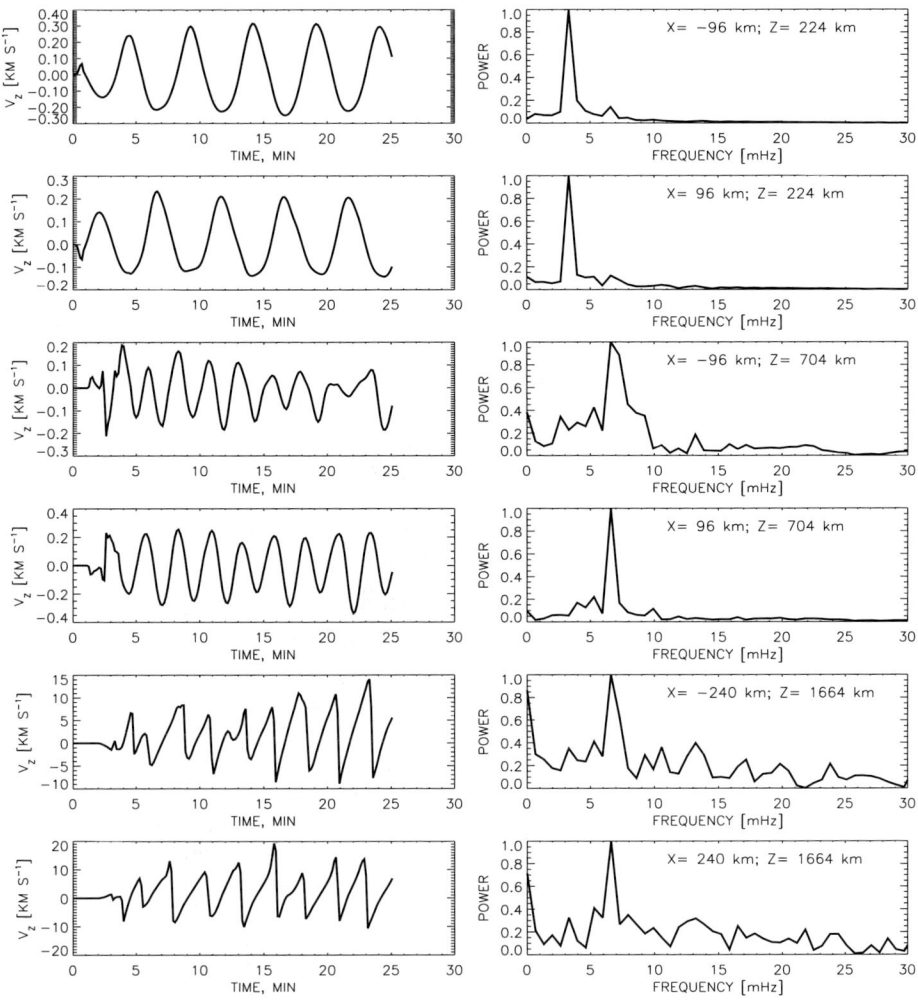

Figure 9 Examples of the temporal evolution of the longitudinal velocity (left) and their corresponding power spectra (right) at different heights and positions inside the flux tube for the simulations with horizontal driving at 300 seconds. The coordinates are indicated on the right panels.

8. Vertical Driving at 300 Seconds

Finally, the results after exciting the flux tube with a 300-second vertical driver are presented. Figure 10 gives some snapshots of the velocity evolution during one period with the same format as in the previous figures. The time series shown in the figure is taken after the simulations reach the stationary state. Now, the fast (acoustic) wave is excited directly by the driver inside the flux tube in the deeper layers (see lower panels of Figure 10). The surface wave exists as well at the magnetic/field-free interface. There are some acoustic disturbances that can be observed in the field-free atmosphere surrounding the tube at the lower layers. After some time has elapsed since the start of the simulation, convective vortices appear in the nonmagnetic atmosphere.

Figure 10 Time series of snapshots of the transverse (right) and longitudinal (left) velocities in the simulation with a vertical driver at 300 seconds. Numbers give time in seconds since the start of the simulation. After removal of the average velocity increase with height produced by the density falloff, all images have the same scale. Each snapshot covers 900 km horizontally and 2000 km vertically.

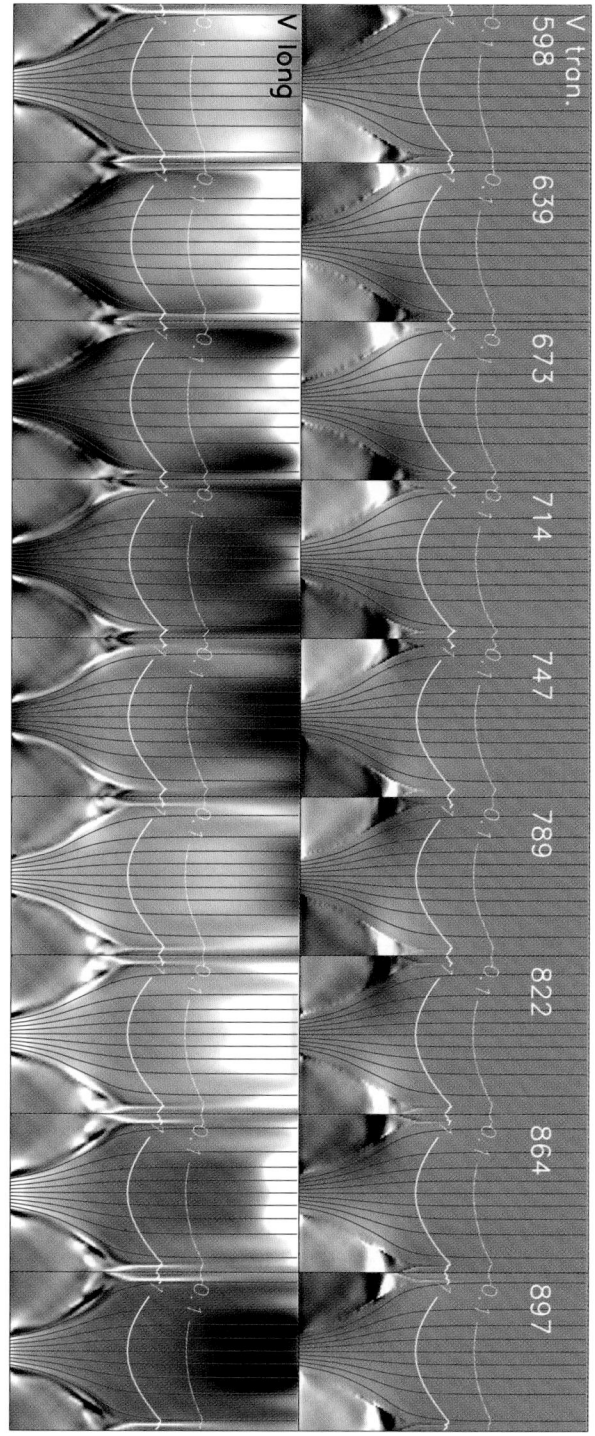

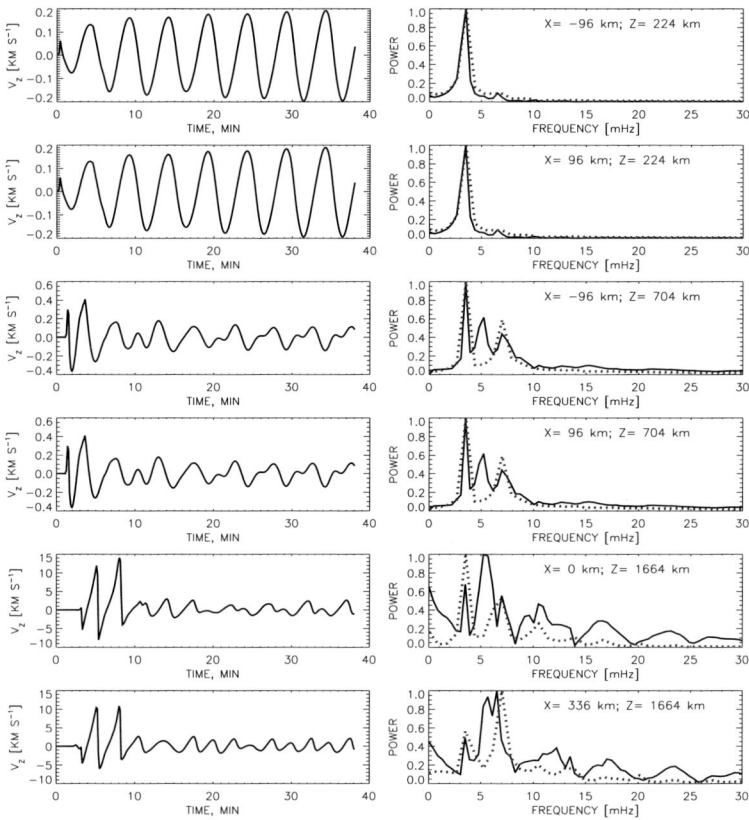

Figure 11 Examples of the time evolution of the longitudinal velocity (left) and their corresponding power spectra (right) at different heights and positions inside the flux tube in simulations with vertical driving at 300 seconds. The coordinates are indicated on the right panels. The blue dotted lines on the right panels give the power spectra only for the last half of the simulations after reaching the stationary state.

At the $v_A = c_S$ layer, the fast to slow transformation occurs, keeping the acoustic nature of the wave. Almost no energy is transferred to the fast magnetic mode in high layers. It can be seen in the upper panels of Figure 10 that the transverse velocity inside the flux tube remains close to zero all of the time. The magnetic field lines are compressed and expanded as the wave passes through the atmosphere, but no horizontal motions are observed.

The perturbation is almost perfectly symmetric at all heights with respect to the tube axis, as demonstrated in the four upper panels of Figure 11, where the time series in symmetric positions inside the flux tubes are given. The most striking fact comes from the different behavior of the velocity at points near the tube axis or at the tube boundary in the upper heights. At deep layers, the propagation is mainly linear, showing the five-minute sinusoidal pattern of the driver (see the upper panels in Figure 11). From there upward, two shock waves develop at the beginning of the series, which are followed by a highly damped oscillation. The period of these shocks is three minutes. After the simulation reaches the stationary state, the dominant periods are different at the tube axis and off axis. For points near the axis, the five-minute oscillation does not suffer such a large amplitude damping and becomes the dominate period once the oscillation reaches a stationary state (see the power spectra corresponding to the second half of the simulation as a blue line in Figure 11). For off-axis points, there is a clear change with height of the dominating frequency. The contribution of the five-minute fundamental frequency is reduced as the wave propagates, increasing the relative importance of the first harmonic at 6 mHz (lower panel of Figure 11). Thus, in this simulation, 3- and 6-mHz oscillations coexist in different parts of the tube. In any case, the amplitudes are rather small when compared to those derived from horizontal drivers and only weak eventual shocks are observed with amplitudes of $2-3$ km s^{-1}.

In summary, the most important features of the simulations with a horizontal driver of 300 seconds are the following:

- Fast to slow transformation occurs at $v_A = c_S$ height.
- There are weak eventual shocks.
- The on-axis period is five minutes.
- The off-axis period is three minutes.

If the simulations with horizontal drivers can be understood in terms of mode transformation, and wave amplification and damping, this is not the case for this simulation with a 300-second vertical driver. Why does the main period of the wave above the $v_A = c_S$ layer depend on the distance to the axis? Why do shocks develop at the beginning of the series and not later? More simulations are needed in different flux tubes, with varying initial conditions, to understand this complex behavior.

9. Discussion and Conclusions

We have performed simulations of waves in flux tubes with nontrivial magnetic field configuration. The model flux tube has horizontal and vertical variations of magnetic field strength and gas pressure and of the ratio between the characteristic wave speeds v_A and c_S. These properties have allowed us to study wave propagation and mode transformation inside the flux tube. Waves are excited by a photospheric driver with a period of three and five minutes, for the first time with a magnetic field configuration of this kind. Different wave patterns are observed depending on the period and on the type of the driver. The following items summarize our findings:

- Horizontal motions of the flux tube at the bottom photospheric boundary generate a slow magneto-acoustic mode inside the tube and a surface mode at the magnetic/field free interface.
- After the slow magneto-acoustic mode and the surface mode reach the height where $v_A \approx c_S$, their energy is effectively transformed into a slow acoustic mode in the high atmosphere where $v_A > c_S$. Only a small part of the driver energy is returned to the photosphere by the fast magneto-acoustic mode.
- The slow acoustic mode propagates vertically along the magnetic field lines in the atmosphere where $v_A > c_S$, forms shock waves above 1000 km with amplitudes of $5-15$ km s^{-1}, and remains always within the same flux tube. Thus, it can deposit effectively most part of the energy of the driver into the chromosphere.
- If the frequency of the horizontal driver is above the acoustic cutoff, nonlinear wave propagation in the tube occurs with the dominant period of the driver at all heights. Nonlinear effects produce higher harmonics, but their amplitude is small.
- If the frequency of the horizontal driver is below the acoustic cutoff, the dominant period of the longitudinal velocity changes with height from five to three minutes owing to nonlinear generation of the higher harmonics.

- The five-minute vertical perturbation of the flux tube at the bottom photospheric boundary generates a fast acoustic mode that propagates upward through the transformation $v_A = c_S$ layer, without changing its nature and only later forming shocks in the chromosphere.
- After reaching the stationary state in the simulations with a vertical driver at 300 seconds, both three- and five-minute oscillations coexist inside the flux tube at chromospheric heights. The dominant period at the axis is five minutes and the off-axis dominant period is three minutes.

Our simulations suggest once again that the properties of waves observed in magnetic structures are the direct consequence of their magnetic configuration, temperature, magnetic field strength, and the height where the mode transformation occurs. When analyzing observations, it is important to know whether they correspond to the level below or above the layer $v_A = c_S$. If the motions that excite oscillations inside flux tubes are purely horizontal, no correlation may be observed between velocity variations measured in the photosphere and those in the chromosphere. This is because only an acoustic mode can be detected in observations (at least at disk center) since it produces significant vertical velocities, unlike the slow magnetic mode existing in the photosphere. The acoustic mode is only generated above 700 – 1000 km in our simulations with a horizontal driver. In contrast, if the driver that excites oscillations has a vertical component, the acoustic mode is generated already in the photosphere. In this case, the oscillations in vertical velocity are coherent at photospheric and chromospheric heights.

In general, both simulations with horizontal and vertical drivers with 300-second periodicity show the change of the dominant wave period with height. This property is different from observations of plage and network regions where the five-minute periodicity is preserved also in the chromosphere (Lites *et al.*, 1993; Krijger *et al.*, 2001; De Pontieu *et al.*, 2003; Centeno *et al.*, 2006b; Bloomfield *et al.*, 2006; Vecchio *et al.*, 2007). Although our study may help to identify the wave modes observed in small-scale magnetic structures, further work is needed to explain the behavior of the oscillatory spectra with height.

Acknowledgements The authors are grateful to the anonymous referee for suggestions that helped to improve the manuscript. Financial support by the European Commission through the SOLAIRE Network (MTRN-CT-2006-035484) and by the Spanish Ministry of Education through projects AYA2007-66502 and AYA2007-63881 is gratefully acknowledged.

References

Bellot Rubio, L.R., Collados, M., Ruiz Cobo, B., Rodríguez Hidalgo, I.: 2000, Oscillations in the photosphere of a sunspot umbra from the inversion of infrared stokes profiles. *Astrophys. J.* **534**, 989 – 996.
Berenger, J.P.: 1994, A perfectly mached layer for the absorption of electromagnetic waves. *J. Comput. Phys.* **114**, 185 – 200.
Berger, T.E., Rouppe van der Voort, L., Löfdahl, M.: 2007, Contrast analysis of solar faculae and magnetic bright points. *Astrophys. J.* **661**, 1272 – 1288.
Bloomfield, D.S., Lagg, A., Solanki, S.K.: 2007, Observations of running waves in a sunspot chromosphere. In: Heinzel, P., Dorotovic, I., Rutten, R.J. (eds.) *The Physics of Chromospheric Plasmas*, ASP Conf. Ser. **368**, ASP, San Francisco, 239.
Bloomfield, D.S., McAteer, R.T.J., Mathioudakis, M., Keenan, F.P.: 2006, The influence of magnetic field on oscillations in the solar chromosphere. *Astrophys. J.* **652**, 812 – 819.
Bogdan, T.J., Judge, P.G.: 2006, Observational aspects of sunspot oscillations. *Phil. Trans. Roy. Soc.* **364**(1839), 313 – 331.
Bogdan, T.J., Carlsson, M., Hansteen, V., McMurry, A., Rosenthal, C.S., Johnson, M., Petty-Powell, S., Zita, E.J., Stein, R.F., McIntosh, S.W., Nordlund, A.: 2003, Waves in the magnetized solar atmosphere. ii. waves from localized sources in magnetic flux concentrations. *Astrophys. J.* **599**, 626 – 660.

Braun, D.C., Lindsey, C.: 1999, Helioseismic images of an active region complex. *Astrophys. J.* **513**, L79 – L82.
Brynildsen, N., Maltby, P., Leifsen, T., Kjeldseth-Moe, O., Wilhelm, K.: 2000, Observations of sunspot transition region oscillations. *Solar Phys.* **191**, 129 – 159.
Brynildsen, N., Maltby, P., Fredvik, T., Kjeldseth-Moe, O.: 2002, Oscillations above sunspots. *Solar Phys.* **207**, 259 – 290.
Cally, P.: 2006, Dispersion relations, rays and ray splitting in magnetohelioseismology. *Roy. Soc. Lond. Trans. Ser. A* **364**, 333 – 349.
Cally, P.: 2007, What to look for in the seismology of solar active regions. *Astron. Nachr.* **328**, 286.
Carlsson, M., Stein, R.F.: 1997, Formation of Solar Calcium H and K Bright Grains. *Astrophys. J.* **481**, 500.
Centeno, R., Collados, M., Trujillo Bueno, J.: 2006a, Spectropolarimetric investigation of the propagation of magnetoacoustic waves and shock formation in sunspot atmospheres. *Astrophys. J.* **640**, 1153 – 1162.
Centeno, R., Collados, M., Trujillo Bueno, J.: 2006b, Oscillations and wave propagation in different solar magnetic features. In: Casini, R., Lites, B.W. (eds.) *Solar Polarization 4, ASP Conf. Ser.* **358**, ASP, San Francisco, 465 – 470.
Christopoulou, E.B., Georgakilas, A.A., Koutchmy, S.: 2000, Oscillations and running waves observed in sunspots. *Astron. Astrophys.* **354**, 305 – 314.
Christopoulou, E.B., Georgakilas, A.A., Koutchmy, S.: 2001, Oscillations and running waves observed in sunspots. iii. multilayer study. *Astron. Astrophys.* **375**, 617 – 628.
Collados, M., Trujillo Bueno, J., Bellot Rubio, L.R., Socas-Navarro, H.: 2001, In: Ballester, J.L., Roberts, B. (eds.) *INTAS Workshop on MHD Waves in Astrophysical Plasmas*, Universitat de les Illes Balears, 151 – 154.
De Moortel, I., Ireland, J., Hood, A.W., Walsh, R.W.: 2002, The detection of 3 & 5 min period oscillations in coronal loops. *Astron. Astrophys.* **387**, L13 – L16.
De Pontieu, B., Erdelyi, R., de Wijn, A.G.: 2003, Intensity oscillations in the upper transition region above active region plage. *Astrophys. J.* **595**, L63 – L66.
De Pontieu, B., Erdelyi, R.J., Stewart, P.: 2004, Solar chromospheric spicules from the leakage of photospheric oscillations and flows. *Nature* **430**, 536 – 539.
Fleck, B., Schmitz, F.: 1991, The 3-min oscillations of the solar chromosphere – a basc physical effect? *Astron. Astrophys.* **250**, 235 – 244.
Fleck, B., Schmitz, F.: 1993, On the interactions of hydrodynamic shock waves in stellar atmospheres. *Astron. Astrophys.* **273**, 671.
Gurman, J.B., Leibacher, J.W.: 1984, Linear models of acoustic waves in sunspot umbrae. *Astrophys. J.* **283**, 859 – 869.
Hasan, S.S., Ulmschneider, P.: 2004, Dynamisc and heating of the magnetic network on the sun. efficiency of mode transformation. *Astron. Astrophys.* **422**, 1085 – 1091.
Hasan, S.S., Kalkofen, W., van Ballegooijen, A.A., Ulmschneider, P.: 2003, Kink and longitudinal oscillations in the magnetic network of the sun: Nonlinear effects and mode transformation. *Astrophys. J.* **585**, 1138 – 1146.
Hasan, S.S., van Ballegooijen, A.A., Kalkofen, W., Steiner, O.: 2005, Dynamics of solar magnetic network: two-dimensional mhd simulations. *Astrophys. J.* **631**, 1270 – 1280.
Judge, P.G., Tarbell, T.D., Wilhelm, K.: 2001, A study of chromospheric oscillations using the soho and trace spacecraft. *Astrophys. J.* **554**, 424 – 444.
Kalkofen, W., Rossi, P., Bodo, G., Massaglia, S.: 1994, Propagation of acoustic waves in a stratified atmosphere. *Astron. Astrophys.* **284**, 976 – 984.
Khomenko, E., Collados, M.: 2006, Numerical modeling of magnetohydrodynamic wave propagation and refraction in sunspots. *Astrophys. J.* **653**, 739 – 755.
Khomenko, E.V., Collados, M., Bellot Rubio, L.R.: 2003, Magnetoacoustic waves in sunspots. *Astrophys. J.* **588**, 606 – 619.
Korn, G.A., Korn, T.M.: 2000, *Mathematical Handbook for Scientists and Engineers*, Dover, New York.
Krijger, J.M., Rutten, R.J., Lites, B.W., Straus, T., Shine, R.A., Tarbell, T.D.: 2001, Dynamics of the solar chromosphere. III. Ultraviolet brightness oscillations from trace. *Astron. Astrophys.* **379**, 1052 – 1082.
Lites, B.W.: 1984, Photoelectric observations of chromospheric sunspot oscillations. II. Propagation characteristics. *Astrophys. J.* **277**, 874 – 888.
Lites, B.W.: 1986, Photoelectric observations of chromospheric umbral oscillations. IV. The Ca II h line and He I 1830. *Astrophys. J.* **301**, 1005 – 1017.
Lites, B.W.: 1988, Photoelectric observations of chromospheric umbral oscillations. V. Penumbral oscillations. *Astrophys. J.* **334**, 1054 – 1065.
Lites, B.W., Rutten, R.J., Kalkofen, W.: 1993, Dynamics of the solar chromosphere. I. Long-period network oscillations. *Astrophys. J.* **414**, 345 – 356.

Lites, B.W., Thomas, J.H., Bogdan, T.J., Cally, P.S.: 1998, Velocity and magnetic field fluctuations in the photosphere of a sunspot. *Astrophys. J.* **497**, 464–482.
Maltby, P., Brynildsen, N., Fredvik, T., Kjeldseth-Moe, O., Wilhelm, K.: 1999, On the sunspot transition region. *Solar Phys.* **190**, 437–458.
Maltby, P., Brynildsen, N., Kjeldseth-Moe, O., Wilhelm, K.: 2001, Plumes and oscillations in the sunspot transition region. *Astron. Astrophys.* **373**, L1–L4.
Marsh, M.S., Walsh, R.W.: 2005, Observed wave propagtion along the sunspot magnetic field through the chromosphere, transition region and corona. In: *Chromospheric and Coronal Magnetic fields* **596**, ESA SP, Noordwijk, 75.1.
Marsh, M.S., Walsh, R.W.: 2006, p-mode propagation through the transition region into the solar corona. I. Observations. *Astrophys. J.* **643**, 540–548.
Pneuman, G.W., Solanki, S.K., Stenflo, J.O.: 1986, Structure and merging of solar magnetic fluxtubes. *Astron. Astrophys.* **154**, 231–242.
Roberts, B.: 1981, Wave propagation in a magnetically structured atmosphere. I. Surface waves at a magnetic interface. *Solar Phys.* **69**, 27–38.
Roberts, B.: 1983, Wave propagation in intense flux tubes. *Solar Phys.* **87**, 77–93.
Rosenthal, C.S., Bogdan, T.J., Carlsson, M., Dorch, S.B.F., Hansteen, V., McIntosh, S.W., McMurry, A., Nordlund, A., Stein, R.F.: 2002, Waves in the magnetized solar atmosphere. I. Basic processes and internetwork oscillations. *Astrophys. J.* **564**, 508–524.
Rüedi, I., Solanki, S.K., Stenflo, J., Tarbell, T., Scherrer, P.H.: 1998, Oscillations of sunspot magnetic fields. *Astron. Astrophys.* **335**, L97–L100.
Socas-Navarro, H., Trujillo Bueno, J., Ruiz Cobo, B.: 2000, Anomalous circular polarization profiles in sunspot chromospheres. *Astrophys. J.* **544**, 1141–1154.
Thomas, J.H., Stanchfield, D.C.H.: 2000, Fine-scale magnetic effects on p-modes and higher frequency acoustic waves in a solar active region. *Astrophys. J.* **537**, 1086–1093.
Tziotziou, K., Tsiropoula, G., Mein, N., Mein, P.: 2006, Observational characteristics and association of umbral oscillations and running penumbral waves. *Astron. Astrophys.* **456**, 689–695.
Tziotziou, K., Tsiropoula, G., Mein, N., Mein, P.: 2007, Dual-line spectral and phase analysis of sunspot oscillations. *Astron. Astrophys.* **463**, 1153–1163.
Rouppe van der Voort, L.H.M.R., Rutten, R.J., Sütterlin, P., Sloover, P.J., Krijger, J.M.: 2003, La Palma observations of umbral flashes. *Astron. Astrophys.* **403**, 277–285.
Vecchio, A., Cauzzi, G., Reardon, K.P., Janssen, K., Rimmele, T.: 2007, Solar atmospheric oscillations and the chromospheric magnetic topology. *Astron. Astrophys.* **461**, L1–L4.
Vernazza, J.E., Avrett, E.H., Loeser, R.: 1981, Structure of the solar chromosphere. III. Models of the EUV brightness components of the quiet sun. *Astrophys. J.* **45**, 635–725.
Zhugzhda, Y.D.: 2007, *Astron. Lett.* **44**(9), 622–643.
Zhugzhda, Y.D., Locans, V.: 1981, Resonance oscillations in sunspots. *Sov. Astron. Lett.* **7**, 25–27.

Seismic Emissions from a Highly Impulsive M6.7 Solar Flare

J.C. Martínez-Oliveros · H. Moradi · A.-C. Donea

Originally published in the journal Solar Physics, Volume 251, Nos 1–2, 613–626.
DOI: 10.1007/s11207-008-9122-y © Springer Science+Business Media B.V. 2008

Abstract On 10 March 2001 the active region NOAA 9368 produced an unusually impulsive solar flare in close proximity to the solar limb. This flare has previously been studied in great detail, with observations classifying it as a type 1 white-light flare with a very hard spectrum in hard X-rays. The flare was also associated with a type II radio burst and coronal mass ejection. The flare emission characteristics appeared to closely correspond to previous instances of seismic emission from acoustically active flares. Using standard local helioseismic methods, we identified the seismic signatures produced by the flare that, to date, is the least energetic (in soft X-rays) of the flares known to have generated a detectable acoustic transient. Holographic analysis of the flare shows a compact acoustic source strongly correlated with the impulsive hard X-rays, visible continuum, and radio emission. Time–distance diagrams of the seismic waves emanating from the flare region also show faint signatures, mainly in the eastern sector of the active region. The strong spatial coincidence between the seismic source and the impulsive visible continuum emission reinforces the theory that a substantial component of the seismic emission seen is a result of sudden heating of the low photosphere associated with the observed visible continuum emission. Furthermore, the low-altitude magnetic loop structure inferred from potential-field extrapolations in the flaring region suggests that there is a significant anti-correlation between the seismicity of a flare and the height of the magnetic loops that conduct the particle beams from the corona.

Keywords Flares · Sun quakes · Particle acceleration · Helioseismology

1. Introduction

Recent developments in the study of flare acoustic emissions (Donea and Lindsey, 2005; Donea *et al.*, 2006; Moradi *et al.*, 2007; Martínez-Oliveros *et al.*, 2007) have bolstered the

---

Helioseismology, Asteroseismology, and MHD Connections
Guest Editors: Laurent Gizon and Paul Cally

J.C. Martínez-Oliveros (✉) · H. Moradi · A.-C. Donea
Centre of Stellar and Planetary Astrophysics, Monash University, Clayton, Victoria 3800, Australia
e-mail: juan.oliveros@sci.monash.edu.au

view that seismic emission from flares offers major new insights into both flare physics and helioseismology, ranging from a greatly improved understanding of flare dynamics and kinematics to an understanding of how seismic emission is generated differently by turbulence in magnetic subphotospheres from inside the quiet Sun.

Sunquakes emanate from compact sources that represent only a small fraction of the energy released by flares. The surface manifestation of these sources appears as circular (or near-circular) waves propagating outward from the solar surface, approximately 20–60 minutes after the impulsive phase of the flare. Donea and Lindsey (2005) considered the possibility that relatively weak flares might be able to produce sunquakes and that acoustically active flares may indeed be more common than previously thought. This was confirmed soon after by Besliu-Ionescu et al. (2006) following comprehensive helioseismic observations of flares using helioseismic holography and the data from the Michelson Doppler Imager (MDI) onboard the *Solar and Heliospheric Observatory* (SOHO).

Although the majority of the flare acoustic transients discovered to date have been released by the more energetic X-class flares, recently, however, a number of strong acoustic emissions from M-class flares have been discovered. Donea et al. (2006) analyzed the helioseismic properties of the strong seismic transient produced by the M9.5-class flare of 9 September 2001 and, very recently, Martínez-Oliveros et al. (2007) performed a comprehensive electromagnetic and acoustic analysis of the M7.4-class flare of 14 August 2004 – the smallest flare known to have produced a detectable acoustic transient prior to the discovery of the sunquake reported in this paper. They, along with Donea and Lindsey (2005) and Moradi et al. (2007), have identified a number of distinct observational characteristics that distinguish acoustically active flares from others:

1. The sites of seismic emission generally coincide spatially with impulsive hard X-ray (HXR) and microwave (MW) emissions, suggesting a relation to thick-target heating of the chromosphere by energetic particles.
2. The sites of seismic emission similarly coincide spatially with impulsive continuum emission, suggesting acoustic emission associated with extra heating and ionization of the low photosphere.
3. The seismicity of the active region appears to be closely related to the heights of the coronal magnetic loops that conduct high-energy particles.

In this paper, we will examine the last of these characteristics – introduced by Martínez-Oliveros et al. (2007) – in greater detail, with further evidence from the flare of 10 March 2001. The evidence we provide reinforces the theory that shorter coronal loops are more likely to be conducive to a more rapid injection of trapped, high-energy electrons into the chromosphere at their footpoints. This enhances the magnitude and suddenness of the chromospheric heating that gives rise to the intense visible continuum emission seen in all acoustically active flares. This mechanism appears to be a prospective source of the energy required to drive a powerful acoustic transient into the solar interior.

2. The Helioseismic Signatures

The helioseismic analysis relies on data from SOHO/MDI. The data consist of full-disk Doppler, magnetogram, and continuum images for the photospheric line Ni I 6768 Å, obtained at a cadence of one minute. The Dopplergrams were corrected for small effects attributed to reduced oscillatory amplitudes in magnetic regions, following the method outlined by Rajaguru et al. (2006). We utilize two different, but in our opinion complementary,

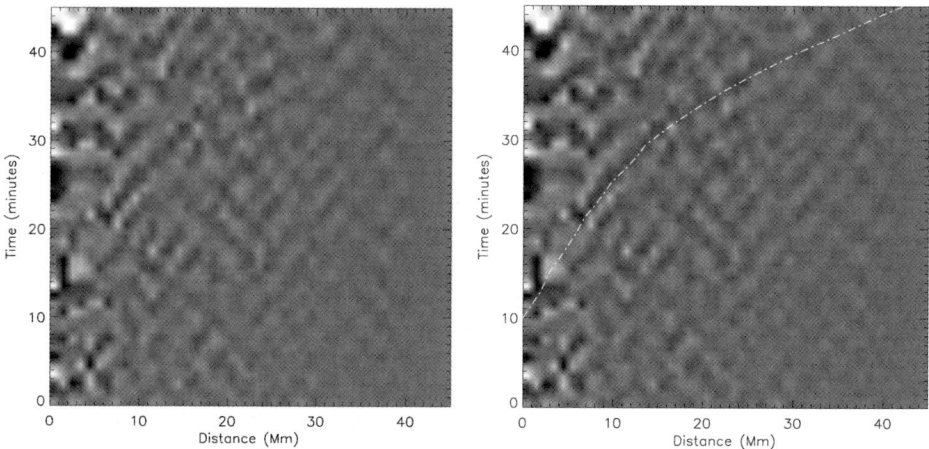

Figure 1 Time–distance plot of the amplitude of the surface ridge averaged over curves of constant radius in the azimuth range +135° to +225° (rendered in gray in both frames). The white curve superimposed on the right frame represents the wave time travel for a standard model of the solar interior. The time represented as 0 along the vertical axis of the plot is 04:07 UT.

helioseismic techniques to analyze the seismicity of the acoustic emission produced by the flare.

The first method employed was the time – distance technique described by Kosovichev and Zharkova (1998). We generate the time – distance plot over a selected range of azimuths from the primary HXR and magnetic-transient sources, in this case +135° to +225°, to gauge the expanding signal from this region and compare this signal with a curve that represents the theoretical group travel time. The resulting signature, manifested as a "ridge" in the time – distance diagram, was significant but, as was expected, appeared to be quite weak (see Figure 1). This is more than likely a consequence of the relatively small energy released by the flare (class M6.7 in X-rays) that produced the sunquake.

The theoretical curve appears to match the observed ridge with a delay of approximately five minutes from the time of the flare maximum. A temporal delay of such nature was contemplated by Zharkova and Zharkov (2007), in our case being of slightly longer duration. According to Zharkova and Zharkov (2007), this delay is due to the time required for the electrons to move along the magnetic field lines and hit the upper photosphere or chromosphere. The velocity and acceleration of the expanding wave packet were also computed. The velocity of the wave front between 5 and 9 Mm was calculated to be ≈ 13 km s^{-1} and between 29 and 33 Mm, ≈ 66.67 km s^{-1}. The mean acceleration of the wave front was estimated to be ≈ 3.35 km s^{-2}.

The second method employed in our analysis, computational seismic holography, was used to image the acoustic source of the sunquake. This method has been used extensively in the analysis of acoustically active flares, with great success in identifying numerous seismic sources from solar flares (Donea, Braun, and Lindsey, 1999; Donea and Lindsey, 2005; Donea et al., 2006; Moradi et al., 2007; Martínez-Oliveros et al., 2007). Helioseismic holography can be described as essentially the phase-coherent reconstruction of acoustic waves observed at the solar surface into the solar interior to render stigmatic images of subsurface sources that have given rise to the surface disturbance. In general, the acoustic reconstruction can be done either forward or backward in time. When it is backward in time, we call the extrapolated field the "acoustic egression." In the case of subjacent-vantage holography this represents waves emanating from a surface focus downward into the solar interior that have subsequently refracted back to the surface in an annular pupil surrounding the source. For the sake of brevity, we direct the reader to Lindsey and Braun (2000) for a more in-depth discussion of holographic techniques.

To assess the seismic emission from the flare, we computed both the acoustic and egression power over the neighborhood of the active region at one-minute intervals, mapping them for each minute of observation. It is important to distinguish between the "egression" power – wherein each pixel is a coherent representation of acoustic waves that have emanated downward from the focus, deep beneath the solar surface, and reemerged into a surrounding annular pupil – and the local "acoustic" power – wherein each pixel represents local surface motion as viewed from directly above the photosphere.

The resulting acoustic and egression power movies and "snapshots" (acoustic/egression power sampled over the solar surface at any definite time) are computed over 2-mHz bands, centered at 3 and 6 mHz. The higher-frequency band has a number of advantages because it avoids the much greater ambient noise of the quiet Sun that predominates the 2 – 4 mHz frequency band, and because of its shorter wavelength, it also provides us with the images that have a finer diffraction limit.

Acoustic and egression power snapshots at the maximum of the flare are shown in Figure 2. In these computations, the pupil was an annulus of radial range 15 – 45 Mm centered on the focus. To improve the statistics, the original egression power snapshots are smeared by convolution with a Gaussian with a 1/e half-width of 3 Mm. The egression power images and the continuum image are also normalized to unity at respective mean quiet-Sun values. The acoustic signature of the flare – consisting of a bright compact source – is clearly visible at 6 mHz in both the acoustic and egression power snapshots at 04:05 UT (indicated by the arrows in Figure 2). At 3 mHz the egression and local acoustic power snapshots show a less conspicuous signature than at 6 mHz owing to the much greater background acoustic power at 3 mHz.

The temporal profiles of the seismic source, seen in the acoustic/egression time series in Figure 3, correspond closely with other compact manifestations of the flare including significant white-light (WL) emission with a sudden, impulsive onset as discussed by Li, Ding, and Liu (2005) and Uddin et al. (2004). The spatial and temporal features of the seismic source observed also coincides closely with the HXR signature reported by Li, Ding, and Liu (2005), indicating that high-energy particles accelerated above the chromosphere contribute to the generation of the seismic source. We will discuss their observations in more detail in the next section.

3. Multiwavelength Analysis

The multiwavelength properties of the extremely impulsive white-light flare (WLF) of 10 March 2001 have previously been studied in detail by a number of authors (Liu, Ding, and Fang, 2001; Ding et al., 2003; Uddin et al., 2004; Li, Ding, and Liu, 2005), all emphasizing the impulsiveness of the flare and the very good spatial and temporal coincidence of the HXR emission with the enhanced continuum emission.

The observations of Uddin et al. (2004) showed that the flare possessed a very hard spectrum in HXR, a type II radio burst, and a coronal mass ejection. GOES SXR observations classified it as M6.7 class, beginning at 04:00 UT, reaching its maximum at 04:05 UT, and ending at 04:07 UT. A very important characteristic of the flare is its duration, which was approximately seven minutes, indicating that the physical processes associated with the flare

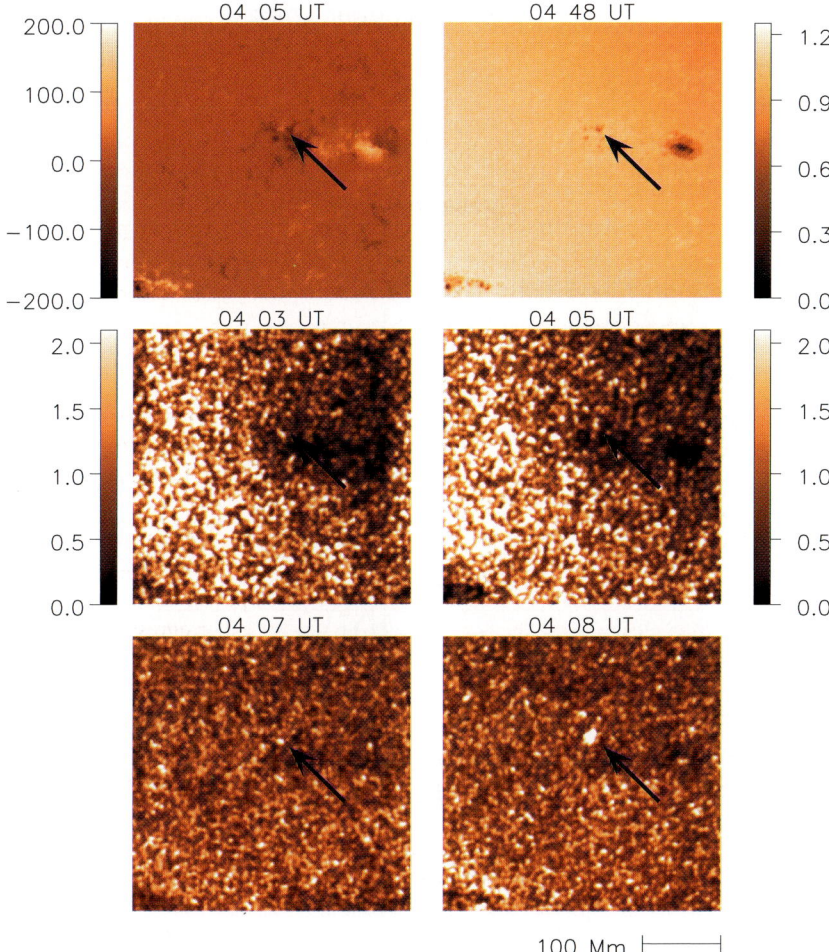

Figure 2 Egression and acoustic power snapshots of AR 9348 on 10 March 2001 integrated over 2.0–4.0 mHz and 5.0–7.0 mHz frequency bands and taken at the maximum of the correspondence frequency. Top frames show MDI magnetogram of the active region (right) at 04:05 UT and a visible continuum image at 04:08 UT (left). Second row shows egression power at 3 mHz (left) and 6 mHz (right) at the respective maxima. The bottom row shows acoustic power. Times are indicated above the respective panels, with arrows inserted to indicate the location of the seismic source.

also had a very short duration. Uddin *et al.* (2004) made a detailed study of this flare at different wavelengths and determined that all three main phases of the flare could be observed clearly in different temporal profiles in HXR at different energy bands (Figure 4). The precursor phase was observed to occur at 04:03 UT with a duration of 15 seconds, the impulsive phase between 05:03:15 and 04:03:40 UT, and the gradual phase after 04:03:40 UT. Also, they calculated the column emission measure, the spectral index of the flare signal, and the temporal variation of the temperature. They found that the emission has a nonthermal component before 04:04 UT and a thermal component after 04:05 UT. From the observed profiles, they concluded that a very rapid acceleration of the electrons occurs during the impulsive phase.

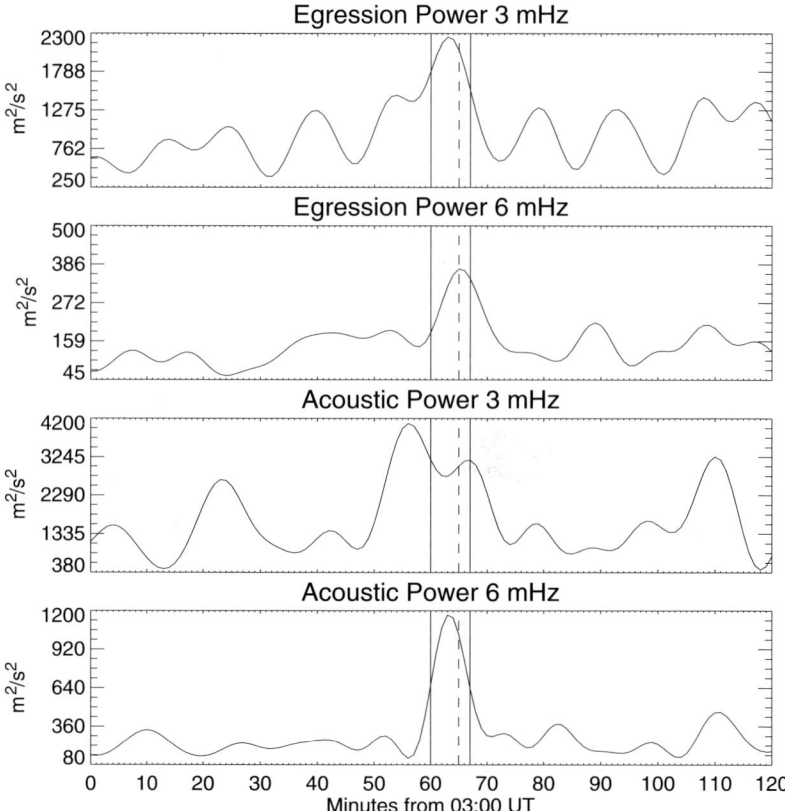

Figure 3 The 3- and 6-mHz egression and acoustic power time series, integrated over the neighborhood of the egression power signatures. The vertical lines represent the beginning (04:00 UT), maximum (04:05 UT), and end (04:07 UT) times of the GOES X-ray flare.

Uddin *et al.* (2004) also emphasized the spatial and temporal correlation of the HXR source and the continuum emission. They also commented on the change of magnetic flux that they detected, concluding that it indicates that the WLF was triggered by a new emerging flux that induces a flux cancellation (see Sudol and Harvey, 2005). As a result, they conclude that magnetic reconnection occurred in the upper atmosphere of the sunspot region, thereby precipitating high-energy electrons along magnetic-field lines and depositing energy at the sunspot region, which produce the HXR and continuum enhancement.

The importance of this particular type of spatial and temporal correlation among the different types of multiwavelength signatures described here, in the presence of a seismic source, was first identified and discussed in depth by Martínez-Oliveros *et al.* (2007). They identified a significant temporal correlation between the fluxes at different frequencies and energy bands (for the M7.4-class flare of 14 August 2004), which were seen to be directly related to two electron populations: one trapped in the magnetic field and another precipitating into the chromosphere. The highly impulsive character of this flare indicates that the trapped population of electrons in the magnetic field was injected into the chromosphere very quickly. The electrons had no time to thermalize in the coronal loop but were evacuated by rapid precipitation; therefore they did not produce a significant emission in MW. Indeed, this type of emission is absent in the MW profile reported by Uddin *et al.* (2004). The radio

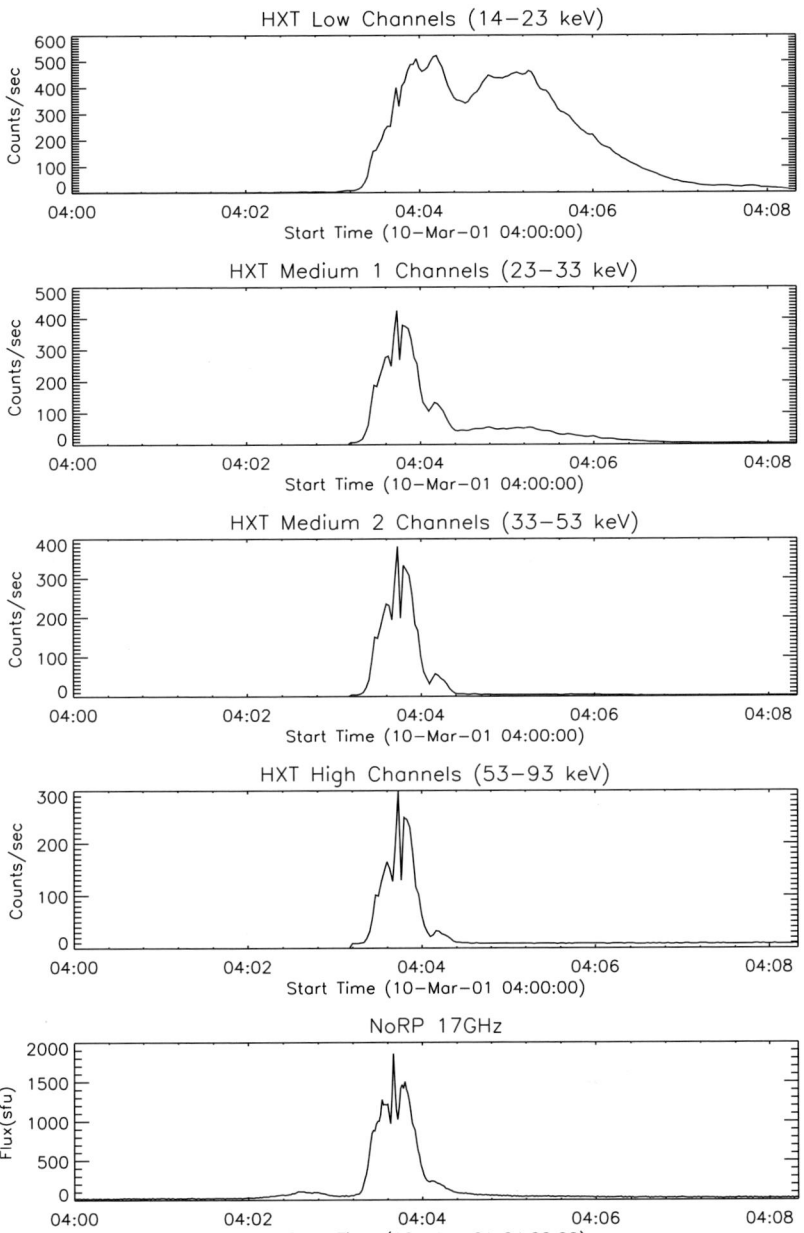

Figure 4 HXR and MW time profiles. The HXR fluxes were taken by *Yohkoh* in the L (14–23 keV), M1 (23–33 keV), M2 (33–53 keV), and H (53–93 keV) channels. The NoRP flux plotted corresponds to the 17-GHz channel.

emission does not show a long exponential decay, implying that high-energy electrons that are generally trapped for a significant amount of time in long coronal loops that extend to great heights are evacuated by rapid precipitation in short, low-lying loops.

Li, Ding, and Liu (2005) also observed the WL properties of the 10 March 2001 flare, detecting an infrared continuum enhancement of 4–6% compared to pre-flare values. The study of the continuum images shows that the WL source is located over the magnetic neutral line and that the source is most likely composed of the two footpoints of the magnetic loop, which are too close together to be resolved by the *Yohkoh*/HXR observations. They also detected a HXR source near the sunspot. The authors also concluded from their observations that the temporal and spatial coincidence of the HXR emission with the continuum emission indicates that electron precipitation may have been the main energy source of the chromospheric heating, producing the excess continuum emission. Furthermore, they suggest that the electron-beam bombardment, coupled with radiative back-warming effects, plays the main role in the heating of the sunspot atmosphere. This is significant because all instances of seismic emissions to date have exhibited very similar WLF characteristics, characterized in particular by the sudden appearance of the WL signature during the impulsive phase of the acoustically active flare.

The images in Figure 5 show a number of the multiwavelength signatures emitted by the 10 March 2001 solar flare. Figures 5(a) and (b) show the position of the magnetic transients, represented by the yellow and green circles, over the MDI intensity continuum and magnetogram, respectively. The magnetic neutral line is overplotted (red line) in all frames for reference. Figure 5(c) shows the magnetic difference maps at the time of the maximum of the flare (04:04:01.61 UT). We can clearly see that one transient coincides well with the region of HXR emission (denoted by the contours), lying across the magnetic neutral lines. In Figure 5(d) we have plotted the Doppler differences for the same time. Here we can see two photospheric signatures (spatially coinciding with the magnetic transients) that can be associated with surface perturbations of the solar photosphere. We also note that observations by Li, Ding, and Liu (2005) show that the WL signature is composed of two sources, both of them being well correlated (spatially) with the magnetic transients. One strong and extended source lies in the region of the HXR and seismic source; the second one appears to correlate well with the second magnetic transient.

Uddin *et al.* (2004) extensively analyzed the temporal and spatial behavior of the solar flare. The maximum time in both HXR and MW emission reported by them and by Li, Ding, and Liu (2005) (who undertook a very similar analysis) coincides very well with the maximum of the seismic emission [following the already well-known delay of approximately three–four minutes (Moradi *et al.*, 2007)]. Uddin *et al.* (2004) also discuss the spatial correlations among the different sources, showing that all three forms of emissions (WL, HXR, and MW) are located in the region of maximum magnetic shearing. Chandra *et al.* (2006), in a similar work, reported the locations of two Hα kernels in the flaring region. One of these kernels is (spatially) well correlated with the HXR source (observed by *Yohkoh*) and the observed seismic source, suggesting the precipitation of electrons in the chromosphere. The second Hα kernel is however not correlated with any HXR source, possibly indicating proton precipitation in this region. (See Zharkova and Gordovskyy (2004) for a discussion about the partial separation of electrons and protons into the loop legs.)

4. The Magnetic Field

The magnetic-field topology of the active region has also been studied by other authors (Uddin *et al.*, 2004; Li, Ding, and Liu, 2005) and was correlated with other emissions produced by the flare. Using vector magnetograms from the Mitaka Solar Observatory (Figure 6), we can see that the shearing of the magnetic field lines is close to 80° at the location of the

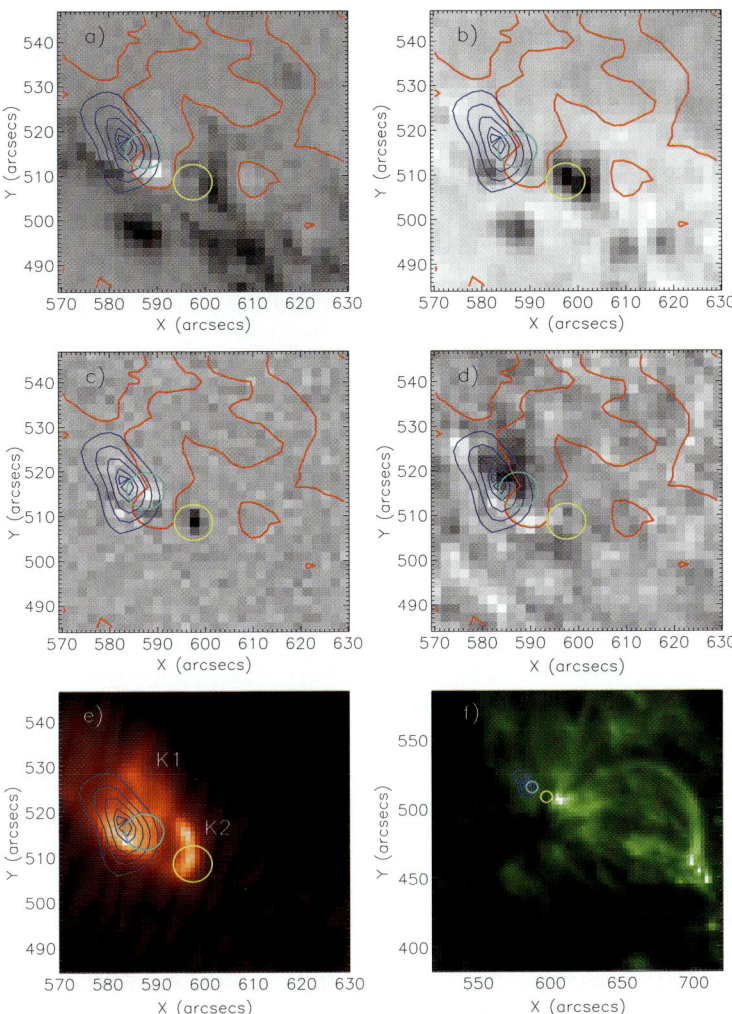

Figure 5 HXR contours of the flare at 04:03:38 UT overlaid over (a) MDI intensity continuum, (b) MDI magnetogram, (c) MDI magnetogram difference at the flare maximum, (d) Doppler difference at the maximum, (e) Hα, and (f) SOHO/EIT at 171 Å. The background images all correspond to the same time (04:04:01.61 UT). The HXR contour levels are 20%, 40%, 60%, 80%, and 90% of the maximum emission in the M2 (22–53 keV) channel. The MDI magnetogram neutral line (red line) is overlaid in frames a, b, c, and d. The blue and yellow circles in all frames represent the relative position of the main magnetic transients. The seismic source coincides spatially with the blue circle, where there is also HXR emission.

seismic source (see the white arrow), which would imply that a vast amount of energy was stored in the magnetic field prior to the flare. The area where the shearing is significant is very small. The seismic source itself has been proven to be of small size (19×25 Mm). The magnetic energy released by the flare is used to accelerate particles, heat the chromosphere and also drive the coronal mass ejection (see Uddin *et al.*, 2004; Li, Ding, and Liu, 2005), and produce the compact seismic source.

To verify the magnetic field configuration of the active region (particularly in the corona), we computed the nonlinear force-free field (NLFFF) coronal-magnetic-field extrapolations

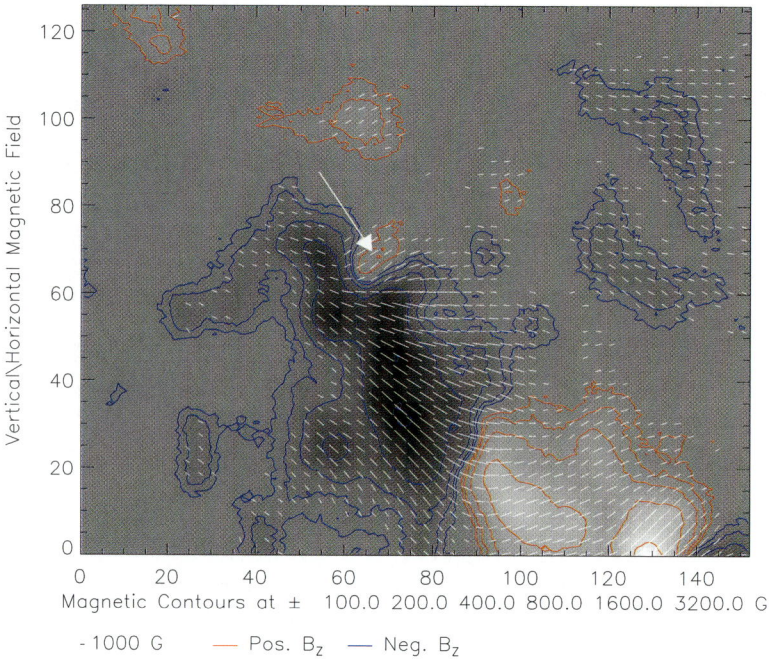

Figure 6 Vector magnetogram of the active region taken by the Mitaka Observatory (NAOJ) at 00:10:16 UT. The sunquake region is indicated with an arrow.

of the active region using vector magnetograms from the Mitaka Solar Observatory. The resulting extrapolations (seen in Figure 7) clearly show high-altitude magnetic-field lines connecting the two leading sunspots of the group, whereas between the leading and the following sunspots, only low-lying loops are visible (see arrow in Figure 7). A comparison between the extrapolations with SOHO/EIT images at 171 Å (Figures 5(f) and 7) shows that our derived coronal-magnetic-field extrapolations are in agreement with the observed magnetic field. Because of the close proximity of the sunspot to the solar limb and other observational constraints, it is not entirely possible to fully reconstruct the complete configuration of the magnetic field (in the flaring region). Nonetheless, we can qualitatively infer the overall structure of the coronal magnetic field from our estimates.

In a closely related work, Chandra *et al.* (2006) conducted a detailed study of the dynamics of the 10 March 2001 flare. As mentioned previously, they identified two Hα kernels, with only one kernel (K1) found to be spatially correlated with the HXR emission (see Figure 5(e)) and therefore with the seismic source. The second Hα kernel, labeled K2, has an elongated structure. No HXR emission has been correlated with this source nor have we detected any seismic source from this region despite the WL signature present at $\approx$ 04:04 UT. These findings, along with observations of the flaring region made by the SXR telescope onboard *Yohkoh*, led the authors to propose a possible configuration of the magnetic field composed of two magnetic loops sharing one footpoint ("three-legged" configuration) and associated with the single HXR source observed by *Yohkoh*. One of the loops appears to be connecting the shared footpoint with an opposite-polarity region associated by Chandra *et al.* (2006) with a secondary, stronger, yet distant MW source. The second loop is a low-lying loop connecting the shared footpoint with another located inside the region with a high degree of magnetic shearing. Furthermore, it is important to state that the two kernels

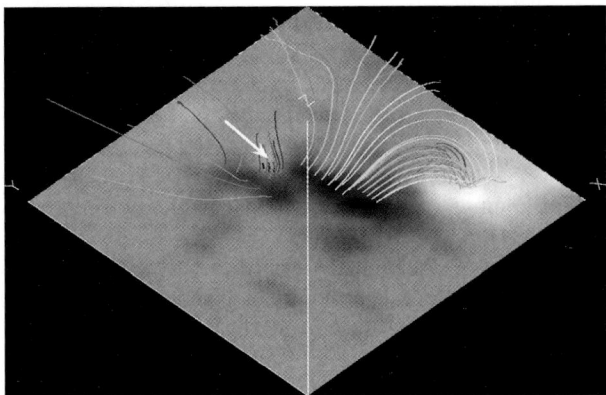

Figure 7 NLFFF magnetic field extrapolation. The arrow shows the low-lying magnetic field region associated with the seismic emission.

observed by Chandra *et al.* (2006) spatially coincide remarkably well with the magnetic transients observed in Figure 5(e), with only one of them also being well correlated with the HXR emission and the Doppler signature.

The existence of a relationship between the height of the coronal magnetic loops and the seismicity of active regions has previously been proposed in Martínez-Oliveros *et al.* (2007). The idea behind this assertion was that electrons in short, low-altitude magnetic loops precipitate more effectively than long, high-altitude loops because of enhanced scattering by thermal electrons ablated from the chromosphere. Electrons whose pitch angles are greater than the loss-cone threshold are trapped in the corona until they are scattered into the loss cone. Eventually, these electrons precipitate into the chromosphere and, depending on their energy, into the photosphere, efficiently transferring energy and momentum to the system. This scattering rate is greatly increased when the population of thermal electrons in the loop is large. This generally depends on the ablation of chromospheric gas into the corona by the fraction of electrons that were initially injected into the loss cone. The volume of a short, low-lying loop is much smaller than that of a long high-altitude loop. The electron density that results from a given mass of the chromosphere having been ablated is thus inversely proportional to this volume. Hence, given these understandings, we propose that short, low-lying loops become efficient scattering environments promptly, greatly expediting precipitation on time scales conducive to seismic emission.

The collapse or relaxation of a high-altitude loop into a low-altitude one owing to reconnection can greatly expand the loss cone, which would then enhance the precipitation distribution if pitch angles were left unchanged (Aschwanden, 2004). As we understand it, such a collapse facilitates electrons, initially trapped in the coronal magnetic field, precipitating into the chromosphere and photosphere. Observation in Hα (Uddin *et al.*, 2004) of this flare show the evolution of the filaments in the flaring region, changing from a potential configuration to a sigmoidal structure because of the high shearing of the magnetic field, with a post-flare relaxation of the magnetic field lines also observed in Hα. This suggests that the scenario of electron injection described here could very well take place, making the electron precipitation process much more efficient.

5. Discussion

The standard flare scenario divides the flare process into a number of phases. In this scenario, the flare particles are accelerated to relativistic or super-relativistic velocities in the corona and injected into magnetic field loops whose footpoints are in active-region chromospheres. Inevitably, some particles are going to be trapped in the coronal magnetic field, whereas others – those in the magnetic loss cone – will precipitate directly into the chromosphere. Eventually the majority of the trapped particles are either scattered into the loss cone and precipitated or get thermalized (or both) by thermal plasma in the magnetic loop. In the case of the very sudden and impulsive flare of 10 March 2001, the hypothesis is that acceleration and injection of particles into the magnetic loop occurred in a short period of time (Uddin et al., 2004).

This kind of phenomenon can be described by using the trapping and injection model proposed by Aschwanden (2004). In this model the rate of precipitation of charged particles into the chromosphere is controlled by the relaxation time of the system. The aperture angle of the loss cone changes with time, significantly opening as the magnetic field collapses to a more potential configuration. It is important to note that, in this model, the times of acceleration and injection of the particles into the magnetic field are almost the same and relatively short compared with the precipitation and trapping times. It is fair to assume that if the relaxation time is short, the aperture of the loss cone also will change rapidly, allowing more particles to reach the chromosphere in a short period of time. This depends on efficient scattering of high-energy electrons into the expanded loss cone, which is greatly enhanced by chromospheric ablation of thermal plasma into short, low-lying loops. Rapid evacuation of trapped electrons is suggested by observations of a rapid decay in nonthermal MW emission. As a general rule, thermalization of particles in a magnetic trap is small compared to losses from precipitation. Hence, high-energy electrons evacuated from the coronal loop in this way contribute to HXR bremsstrahlung emission substantially as well as their counterparts that were initially injected into the loss cone.

A much more complex model of particle precipitation, which includes processes such as nonthermal excitation and ionization of hydrogen atoms and nonthermal plasma heating (coulomb and ohmic), is explored by Aboudarham and Henoux (1986), Zharkova and Kobylinskii (1993), and Zharkova and Zharkov (2007). Interestingly, the latter show that ohmic heating of the corona by the electron beams is so effective that the corresponding particle-induced downward propagating shocks are almost depleted of energy, leaving very little energy to reach the photosphere and induce any kind of seismic activity. Perhaps this is the explanation for why we did not see any seismic sources at the location of the Hα K2 kernel in Figure 3. However, we also want to emphasize here the possibility that photospheric heating also contributes to flare acoustic emission.

As the whole flaring process occurred relatively rapidly (and given the highly impulsive properties of the 10 March 2001 flare, it is not unreasonable to assume so), the solar chromosphere was heated quite suddenly. As we have already seen, the multiwavelength emissions of the flare indicate this; furthermore the strong spatial and temporal correspondence among the different types of emissions point to radiative "back-warming" (as described by Machado, Emslie, and Avrett, 1989) playing a significant role in the heating mechanism. This conclusion was in fact drawn by both Li, Ding, and Liu (2005) and Ding et al. (2003) to explain the origin of the continuum feature of the 10 March 2001 flare in terms of an "electron-beam-heated flare model," with chromospheric radiative back-warming suspected of being the chief heating agent, originating in the temperature-minimum region.

These conclusions, when viewed in conjunction with those of Donea and Lindsey (2005), Donea et al. (2006), Moradi et al. (2007), and Martínez-Oliveros et al. (2007), provide direct evidence of flare acoustic emission being driven, in part, by heating of the low photosphere. The basic principle here is that the chromospheric radiation further heats up the photosphere, resulting in optically thick H$^-$ bound-free absorption, which then introduces a pressure transient directly to the underlying medium. The photospheric-heating hypothesis is well supported by our observations and previous ones – which all indicate that instances of flare seismic emission have been characterized by a close spatial correspondence between the seismic emission and sudden WL emission during the impulsive phase of the flare. Radiative fluxes characteristic of WL emission seen in all acoustically active flares, if emitted downward from the chromosphere (as well as upward), are probably sufficient to heat the photosphere a few percent within a few seconds of the onset of the incoming radiative flux, a process described by Machado, Emslie, and Avrett (1989) as "back-warming."

According to rough models described by Donea *et al.* (2006) and Moradi *et al.* (2007), such heating (if applied suddenly) should cause a pressure transient in the heated layer that drives a seismic transient whose energy flux is of the order of those estimated for acoustically active flares. The energy invested into the seismic transient is in proportion to

$$\varepsilon \sim \left(\frac{\Delta I_c}{I_c}\right)^2, \tag{1}$$

a fraction of order $\Delta I_c / I_c$ times the radiative energy suddenly emitted by the flare. We therefore expect acoustic emission from photospheric heating to be inefficient in flares whose WL signatures are weak, diffuse, or not very sudden, and this is consistent with examples we have encountered to date.

In conclusion, to date, no one, single mechanism can fully explain the mechanics of flare acoustics and their observational signatures – because to do so would be a gross oversimplification of the problem. What these results have shown is that the study of flare mechanics, as well as helioseismology, would greatly benefit from the development of detailed models of solar-flare-induced seismic emission – from the corona, down to the photosphere, including modeling of realistic active region subphotospheres. Credible models would need to include realistic subphotospheric thermal anomalies to represent penumbral and perhaps umbral subphotospheres, along with realistic photospheric magnetic fields extrapolated to depths of a few hundred kilometers beneath the photosphere. The latter should also include an account for the highly inclined magnetic fields that characterize sunspot penumbra – where a significant majority of the seismic emissions observed to date have been detected.

Acknowledgements The authors would like to sincerely thank Profs. Paul Cally, Markus Aschwanden, and Valentina Zharkova, Drs. Charlie Lindsey and Wahab Uddin, and Diana Besliu-Ionescu for their helpful and interesting comments that contributed directly to the development and/or improvement of this article.

References

Aboudarham, J., Henoux, J.C.: 1986, *Astron. Astrophys.* **186**, 73.
Aschwanden, M.: 2004, *Astrophys. J.* **608**, 554.
Besliu-Ionescu, D., Donea, A.-C., Cally, P.S., Lindsey, C.: 2006, In: Fleck, B. (ed.) *A New Era in Helio- and Asteroseismology. Proceedings of the 2006 SOHO-18/GONG-2006/HELAS I Meeting*, ESA Publications, Darmstadt, CDROM, 67.1.
Chandra, R., Jain, R., Uddin, W., Yoshimura, K., Kosugi, T., Sakao, T., Joshi, A., Deshpande, M.R.: 2006, *Solar Phys.* **239**, 239.
Ding, M.D., Liu, Y., Yeh, C.-T., Li, J.P.: 2003, *Astron. Astrophys.* **403**, 1151.
Donea, A.-C., Lindsey, C.: 2005, *Astrophys. J.* **630**, 1168.
Donea, A.-C., Braun, D.C., Lindsey, C.: 1999, *Astrophys. J.* **513**, L143.
Donea, A.-C., Beşliu-Ionescu, D., Lindsey, C., Zharkova, V.V.: 2006, *Solar Phys.* **239**, 113.
Kosovichev, A., Zharkova, V.V.: 1998, *Nature* **393**, 317.
Li, J.P., Ding, D., Liu, Y.: 2005, *Solar Phys.* **229**, 115.
Lindsey, C., Braun, D.C.: 2000, *Solar Phys.* **192**, 261.

Liu, Y., Ding, M.D., Fang, C.: 2001, *Astrophys. J.* **563**, L169.
Machado, M.E., Emslie, A.G., Avrett, E.H.: 1989, *Solar Phys.* **124**, 303.
Martínez-Oliveros, J.C., Moradi, H., Besliu-Ionescu, D., Donea, A.-C., Cally, P.S.: 2007, *Solar Phys.* **245**, 121.
Moradi, H., Donea, A.-C., Lindsey, C., Beşliu-Ionescu, D., Cally, P.S.: 2007, *Mon. Not. Roy. Astron. Soc.* **374**, 1155.
Rajaguru, S.P., Birch, A.C., Duvall, T.L. Jr., Thompson, M.J., Zhao, J.: 2006, *Astrophys. J.* **646**, 543.
Sudol, J.J., Harvey, J.W.: 2005, *Astrophys. J.* **635**, 647.
Uddin, W., Jain, R., Yoshimura, K., Chandra, R., Sakao, T., Kosugi, T., Joshi, A., Despande, M.R.: 2004, *Solar Phys.* **225**, 325.
Zharkova, V.V., Gordovskyy, M.: 2004, *Astrophys. J.* **604**, 884.
Zharkova, V.V., Kobylinskii, V.A.: 1993, *Solar Phys.* **143**, 259.
Zharkova, V.V., Zharkov, S.: 2007, *Astrophys. J.* **664**, 573.

Mechanics of Seismic Emission from Solar Flares

C. Lindsey · A.-C. Donea

Originally published in the journal Solar Physics, Volume 251, Nos 1–2, 627–639.
DOI: 10.1007/s11207-008-9140-9 © Springer Science+Business Media B.V. 2008

Abstract Instances of seismic transients emitted into the solar interior in the impulsive phases of some solar flares offer a promising diagnostic tool, both for understanding the physics of solar flares and for the general development of local helioseismology. Among the prospective contributors to flare acoustic emission that have been considered are: *i*) chromospheric shocks propelled by pressure transients caused by impulsive thick-target heating of the upper and middle chromosphere by high-energy particles, *ii*) heating of the photosphere by continuum radiation from the chromosphere or possibly by high-energy protons, and *iii*) magnetic-force transients caused by magnetic reconnection. Hydrodynamic modeling of chromospheric shocks suggests that radiative losses deplete all but a small fraction of the energy initially deposited into them before they penetrate the photosphere. Comparisons between the spatial distribution of acoustic sources, derived from seismic holography of the surface signatures of flare acoustic emission, and the spatial distributions of sudden changes both in visible-light emission and in magnetic signatures offer a possible means of discriminating between contributions to flare acoustic emission from photospheric heating and magnetic-force transients. In this study we develop and test a means for estimating the seismic intensity and spatial distribution of flare acoustic emission from photospheric heating associated with visible-light emission and compare this with the helioseismic signatures of seismic emission. Similar techniques are applicable to transient magnetic signatures.

Keywords Flares, dynamics · Helioseismology

Helioseismology, Asteroseismology, and MHD Connections
Guest Editors: Laurent Gizon and Paul Cally

C. Lindsey (✉)
NorthWest Research Associates, Boulder, CO 80301, USA
e-mail: clindsey@cora.nwra.com

A.-C. Donea
Centre for Stellar and Planetary Astrophysics, School of Mathematical Sciences, Monash University, Melbourne, Victoria 3800, Australia
e-mail: alina.donea@sci.monash.edu.au

1. Introduction

Some solar flares are known to drive strong seismic transients into the subphotospheres of the magnetic regions that produce them (Kosovichev and Zharkova, 1998). These waves appear to be generated during the impulsive phase of the flare and usually emanate from a relatively compact region (Donea, Braun, and Lindsey, 1999; Donea and Lindsey, 2005). Most of this energy is refracted back to the solar surface within an hour of the onset of the flare, where it makes a surface ripple that can be discriminated by various techniques in local helioseismology. These "sunquakes" offer a powerful diagnostic of both the magnetohydrodynamics of seismic wave generation and propagation in the active-region chromosphere and photosphere and of the structure and dynamics of the active-region subphotosphere.

From a diagnostic point of view it can very well be argued that the discovery of seismic emission from flares is one of the most important developments in local helioseismology:

1. Flare acoustic transients represent the most localized coherent sources that we are aware of, temporally as well as spatially.
2. They are the "hardest" acoustic radiation known so far (*i.e.* the most intense at high frequencies).
3. They are the only acoustic waves that are known to be generated in plain view above the solar surface.

At the same time, there are significant unanswered questions about the physics of seismic emission from flares. For several years it appeared that flare acoustic emission was a relatively rare occurrence, the only known instance being that of the X2.6-class flare of 09 July 1996 discovered by Kosovichev and Zharkova (1998). However, a comprehensive survey by Donea *et al.* (2006a) uncovered more than a dozen instances of significant seismic transient emission in the declining phase of solar activity cycle 23, emanating from flares as small as class M6.7 (Martinez-Oliveros, Moradi, and Donea, 2000). This established that flare acoustic emission was a relatively common phenomenon. Some of these flares were observed by an impressive array of other space-borne and ground-based facilities. This has greatly increased diagnostic prospects of flare acoustic emission.

A number of mechanisms have been considered as possible contributors to flare acoustic emission:

1. Chromospheric shocks driven by sudden, thick-target heating of the upper and middle chromosphere (Kosovichev and Zharkova, 1995, 1998, 2006, 2007; Donea and Lindsey, 2005). Evidence of chromospheric shocks derived from chromospheric line profiles suggests that there is more than sufficient energy flux in the chromosphere to account for the energy seen in flare acoustic transients. However, hydrodynamic modeling of waves driven by thick-target heating of the chromosphere indicates that these waves are heavily damped by radiative losses, such that an insufficient amount of energy penetrates through the photosphere to explain the helioseismic observations (Fisher, Canfield, and McClymont, 1985; Ding and Fang, 1994; Allred *et al.*, 2005).
2. Wave-mechanical transients driven by heating of the photosphere. This contribution was initially suggested by Donea and Lindsey (2005), motivated by the strong spatial correspondence between sudden excess visible continuum emission emanating from active regions during the impulsive phases of flares and the source distributions of flare acoustic emission shown by seismic holography (Lindsey and Braun, 2000; Donea, Braun, and Lindsey, 1999; Donea and Lindsey, 2005) applied to helioseismic observations of the flares. Figure 1 shows an example. Seismic emission from flares has invariably emanated from within or near sunspot penumbrae in instances encountered

to date. The sources tend to be relatively compact (Donea, Braun, and Lindsey, 1999; Donea and Lindsey, 2005; Kosovichev, 2006) and are the site of sudden continuum emission as well as overlying chromospheric line emission. Considerable chromospheric line emission and significant continuum emission are seen from regions well outside of the seismic sources. However, as this study will confirm, the sudden, compact component of continuum emission is heavily concentrated in the region from which the high-frequency seismic emission emanates. Donea and Lindsey (2005), Donea *et al.* (2006b), and Moradi *et al.* (2007) suggested that the continuum emission is causally associated one way or another with photospheric heating, which would give rise to a pressure increase that would drive the acoustic transient. Once the transient penetrates substantially beneath the photosphere, significant radiative losses are supposed to be blocked by highly opaque ionized hydrogen, and the transient is supposed to proceed undamped until its next encounter with the solar surface.

3. Lorentz-force transients resulting from magnetic reconnection in the corona. Transient shifts in magnetic signatures have been detected in a number of flares, some of which were acoustically active and others of which were not (detectably). Zharkova and Kosovichev (2002) considered magnetic transients as a source of waves, both coronal, chromospheric and helioseismic, paying particular attention to magnetic observations of the flare of 14 July 2000. This flare was acoustically inactive as far as helioseismic analyses to date have been able to discern. Sudol and Harvey (2005) measured localized transients in the line-of-sight magnetic field in a variety of flares, including the acoustically active flare of 29 October 2003.[1] Donea *et al.* (2006b) found a strong local transient in the line-of-sight magnetic signature of the M9.5-class flare of 09 September 2001 coincident with the source region of strong transient acoustic emission. Hudson, Fisher, and Welsch (2008) have formally introduced the hypothesis that transient shifts in magnetic signatures during the impulsive phases of acoustically active flares are the result of flare-related magnetic reconnection and a source of flare acoustic emission, estimating the mechanical work that would be done on the photosphere by a sudden shift in magnetic inclination consistent with the magnetic signatures. They found values roughly consistent with energy estimates based on helioseismic observations.

It must be admitted that any mechanism that proposes to express flare acoustics in terms of any single one of these mechanisms would have to be a great oversimplification of the reality. Among detailed models of chromospheric waves driven by thick-target heating, only Allred *et al.* (2005) take back-warming into account. This may reduce radiative losses in waves generated by thick-target heating. Modeling efforts to date have yet to include an account of the inclined magnetic fields that dominate sunspot penumbrae, from which strong seismic emission has invariably emanated in instances we know of so far.

Donea *et al.* (2006b) and Moradi *et al.* (2007) published rough estimates of the energy in flare acoustic transients generated by sudden photospheric heating based on intensity images. These were roughly consistent with energy estimates based on the helioseismic observations they analyzed. However, these estimates are relatively crude and contain no account of penumbral magnetic fields.

---

[1] As in the rest of this study, "transient" refers to shifts on a time scale of order $\tau \sim 2H/c$ or shorter, where H is the density scale height of the solar atmosphere and c is the sound speed. This is approximately 40 seconds in the photosphere. It should be understood that the magnetic signatures actually observed are both transient and long-lasting. It is particularly the transient component that is understood to be relevant for the excitation of flare acoustic emission.

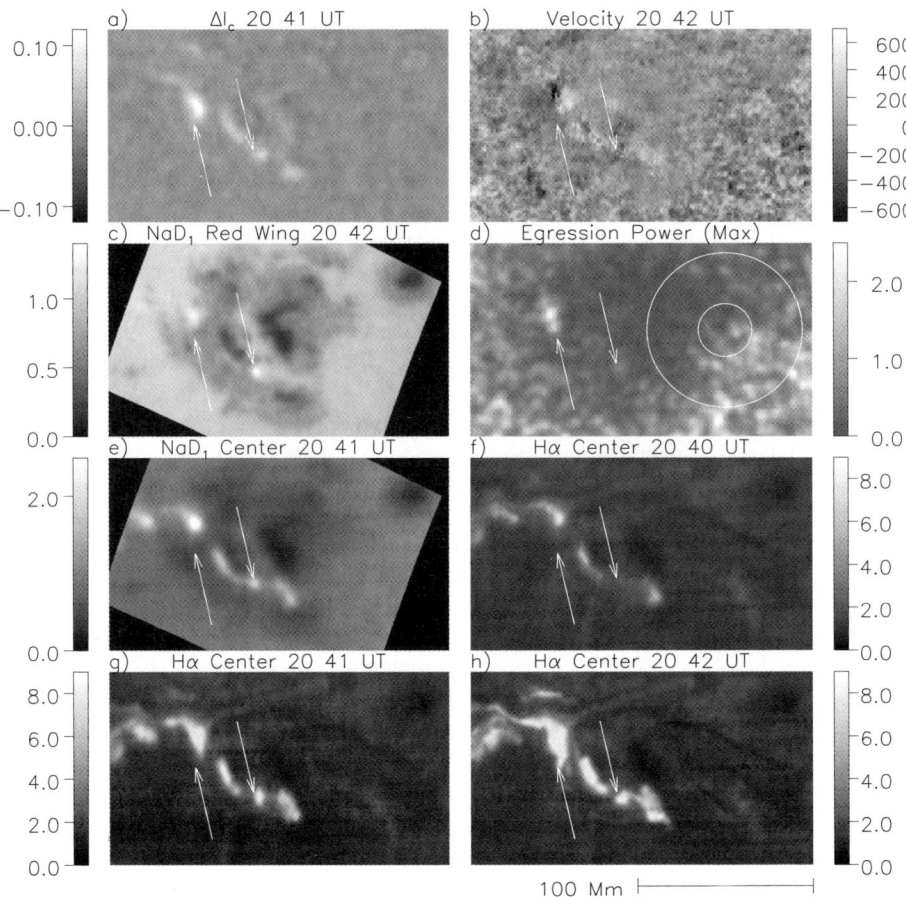

Figure 1 Co-spatial comparison of acoustic emission, Hα, Na D_1-line and continuum emission, and Doppler disturbance in AR 10486 in the impulsive phase of the flare of 29 October 2003, taken from Beşliu-Ionescu *et al.* (2007). Intensities are normalized to unity for the quiet Sun at disk center. (a) Difference between two visible-intensity images taken a minute before and a minute after the time indicated above the panel. (b) Transient downward Doppler disturbance (in meters per second) appearing in AR 10486 during the impulsive phase of the flare. (c) Emission in the red wing of the Na D_1 line, 0.14 Å from line center, showing compact downdrafts. Left and right arrows reproduced in all frames locate these for reference. (d) 5 – 7 mHz egression power map of AR 10486. The annular pupil of the egression computation is drawn at the right of this panel. (e) Impulsive-phase emission in the center of the Na D_1 line. (f) – (h) Line-center ISOON Hα images of AR 10486 at one-minute intervals beginning at 20:40 UT. Times are indicated above respective panels.

2. Mechanics of Seismic Emission Driven by Photospheric Heating

We review estimates by Donea *et al.* (2006b) and Moradi *et al.* (2007) of seismic emission driven by photospheric heating, referring to the diagram in Figure 2. In the exercise to follow we consider the transient energy flux that would result in a horizontally invariant, gravitationally stratified atmosphere if we could engage a massless piston at some depth (z_0) to apply a sudden, step-function increment (δp) in the pressure, initially p_0 at this depth, represented by step 1 in Figure 2. We suppose this excess to be maintained until the surface settles to a new static equilibrium, indicating that the transient driven by the resulting sud-

Mechanics of Seismic Emission from Solar Flares 629

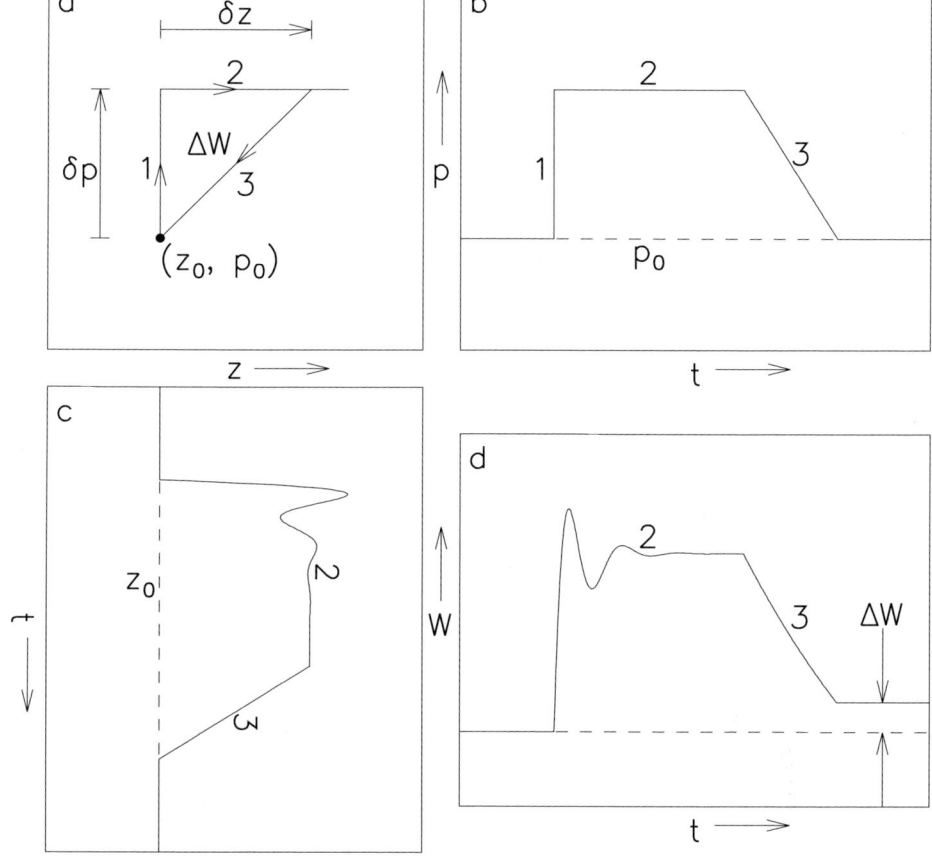

Figure 2 The pressure (frames a and b), vertical displacement (frames a and c) and work done (frame d) as a function of time (frames b–d) by a piston that introduces a sudden increment (δp) in pressure, initially (p_0), at depth (z_0) in a horizontally invariant, gravitationally stratified atmosphere (step 1 in all frames). The pressure increment is maintained while the transient driven by the resulting sudden impulse escapes downward into the medium (step 2). Once the piston has come to a displaced equilibrium, a quasi-static relaxation of the pressure back to that of the undisturbed medium is carried out (step 3). The work (ΔW) represented by the area of the triangular figure in frame a is equal to the energy carried away by the transient during step 2 of the cycle.

den impulse has escaped downward into the underlying medium, a development that occurs during step 2 in Figure 2. We now suppose that the depth ($z_0 + \delta z$) at which the piston eventually settles is that at which the pressure in the undisturbed atmosphere had previously been $p_0 + \delta p$. For a pressure increment that is a relatively small fraction of p_0, this would be accurately approximated by

$$\delta p = \rho g \delta z, \qquad (1)$$

where ρ is the density of the solar medium and g is the gravitational acceleration in the neighborhood of z_0.

Finally, by quasi-statically relaxing the pressure to its initial value, a process represented by step 3 in the diagram, and equating the area (ΔW) of the resulting triangular hysteresis loop that appears in Figure 2a to that which escapes in the transient, we find that

$$\Delta W = \frac{1}{2}\delta p \delta z = \frac{(\delta p)^2}{2\rho g}. \qquad (2)$$

For a rough estimate of the pressure perturbation to be expected from sudden heating of the low photosphere, we proceed along lines somewhat parallel to Wolff (1972). Based on Boyle's law, we suppose that

$$\frac{\delta p}{p_0} = \frac{\delta T}{T_0}, \qquad (3)$$

wherein the heating is expressed in terms of a characteristic temperature increase (δT) divided by a reference temperature (T_0) characterizing the preflare photosphere. It must be realized that even for an initially isothermal atmosphere Equation (3) can be an accurate representation of the mechanics only to within about a factor of two, since for it to accomplish a depression of the underlying medium requires the photosphere to expand, which results in a reduction in density. Moreover, such a pressure excess can only be maintained for a period comparable to the recoil time,

$$\tau_r \sim \frac{2H}{c}, \qquad (4)$$

of the heated layer, where H is the density e-folding height of the medium and c is the sound speed.[2] This is approximately 40 seconds in the low photosphere.

After Equation (3) of Wolff (1972), we approximate the heating associated with an excess ($\delta F_{\rm rad}$) in continuum flux by the Stefan–Boltzmann law:[3]

$$\frac{\delta T}{T_0} \sim \frac{1}{4}\frac{\delta F_{\rm rad}}{F_{\rm rad0}}. \qquad (5)$$

It follows that the energy deposited into transient emission is

$$\Delta W \sim \frac{p_0^2}{32\rho_0 g}\frac{(\delta F_{\rm rad})^2}{F_{\rm rad0}^2}. \qquad (6)$$

Donea *et al.* (2006b) and Moradi *et al.* (2007) found reasonable quantitative agreement between energy fluxes derived from egression power maps and Equation (6) applied to intensity observations from the Global Oscillation Network Group (GONG). However, the outstanding uncertainties are much too large to assure that anything like the entirety of the seismic emission measured is due to photospheric heating. This is not only because of the crudeness of the hydrodynamics, including the neglect of magnetic forces, but also because of significant uncertainties in how concentrated the intensity distribution actually is. The

---

[2]This point is reinforced by bearing in mind that the pressure averaged over an extended time is the column mass density multiplied by g, which in a horizontally invariant geometry is unchanged by local heating of the medium. We note that τ_r in Equation (4) is the reciprocal of $\omega_{\rm ac}$, the familiar acoustic cutoff frequency.

[3]The general accuracy of the approximation stated by Equation (5) rests on the close thermal connection between the photosphere and the ambient radiation field generally expressed by the term local thermodynamic equilibrium (LTE), a condition from which the overlying chromosphere deviates radically. Its accuracy in the photosphere depends on whether the continuum excess is a direct result of photospheric heating alone or the result of back-warming, in which case we understand $\delta T/T_0$ to be less by a factor of approximately two.

GONG images show this with limited resolution. For any given spatially integrated excess (δF_{rad}) the integrated square-excess can be regarded as inversely proportional to the characteristic area from which the emission emanates. If this area is significantly less than the resolution of the observations, the transient would be proportionally more powerful than estimates based on the intensity maps.

If seismic emission from solar flares is substantially a result of photospheric heating, then Equation (6) explains why the energy delivered to the seismic transient is a small fraction of that which emanates from the region in the visible continuum, as indicated by the seismic energy estimates of Donea and Lindsey (2005). For purposes of general comparison we note that the energy flux carried by a propagating wave in a uniform medium is similarly proportional to the square of the velocity amplitude (v) of the motion of the medium,

$$F_{\text{ac}} = \rho c v^2, \tag{7}$$

as the wave passes through it. This offers a basis for spatially detailed comparisons between holographic maps of acoustic emission and projections of the energy flux in acoustic transients driven by photospheric heating associated with excess continuum emission. This will be the subject of the next section.

3. The GONG Intensity Observations

3.1. General Description

We already have a considerable database of intensity observations during acoustically active flares by benefit of GONG (Donea and Lindsey, 2005; Donea *et al.*, 2006b; Moradi *et al.*, 2007). In previous analyses, the use of the GONG intensity observations has been limited to some degree by the effects of the terrestrial atmosphere, through which the GONG instruments must observe the Sun. The primary technical question this study confronts is how the major effects of the terrestrial atmosphere on helioseismic observations of active regions can be characterized and whether these can be substantially discriminated from those of actual brightness variations in active regions. We are convinced that such a discrimination can be accomplished to a highly useful degree.

The GONG observations are made in the photospheric line Ni I λ6768 Å. The GONG instruments make a single full-disk Doppler image, a full-disk line-of-sight magnetogram, and a full-disk intensity image with a cadence of one minute during clear weather.

The GONG intensity images represent radiation integrated over an approximately Gaussian passband whose FWHM is 0.75 Å centered on the line. The line itself has a FWHM of 0.011 Å and an equivalent width of 0.07 Å (Debouille, Roland, and Neven, 1973; Donea *et al.*, 2006b). In this study, we treat the GONG intensity observations as representative of the continuum, understanding that this is subject to errors to the extent that the equivalent width of the line varies with continuum-intensity variations. Comparisons between GONG intensity images of white-light flares and concurrent irradiance variations by SORCE/TIM show agreement to within approximately 20% if the GONG intensity excess is attributed to a variation in the local temperature of a blackbody spectrum at 6768 Å. Figure 10 of Donea and Lindsey (2005) shows this comparison for the X10-class flare of 29 October 2003.

The pixel size in the GONG intensity images is 2.5 arcsec. However, the spatial resolution of the observations depends significantly on atmospheric conditions, which can change perceptibly from one image to the next.

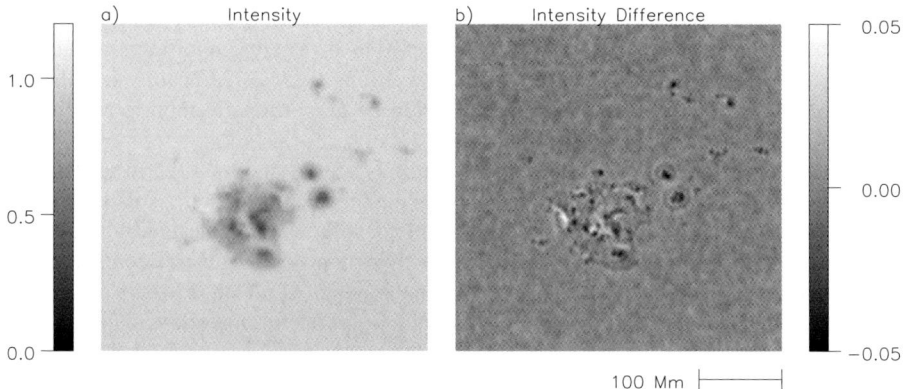

Figure 3 GONG intensity maps of AR 10486 on 29 October 2003. (a) The intensity at 20:42 UT, the impulsive phase of the flare. (b) The intensity difference between the map shown in the left frame and the intensity one minute earlier. The maps are normalized to unity for the quiet Sun.

3.2. Effects of Atmospheric Smearing

If the atmospheric smearing were independent of time, its effect on helioseismic measurements could be characterized simply in terms of limiting spatial resolution. Even with significant variations from one minute to the next, smearing does little in the quiet Sun in the way of introducing spurious temporal intensity variations. However, variable smearing of an active region with large *spatial* intensity variations introduces unacceptable spurious *temporal* variations. The character of these is illustrated in Figure 3. The left frame shows a intensity map of AR 10486 during the impulsive phase of the flare of 29 October 2003. The right frame shows the same intensity map but with the previous intensity map, one minute earlier, subtracted from it. For any particular pixel in the quiet Sun surrounding the active region, the intensity difference can be regarded as significantly representative of that which would have been observed if atmospheric smearing had undergone no change. However, within the active region, the difference manifested by a variation in the degree of smearing can easily predominate over true intensity variations. The result is a pattern in the intensity difference that reflects the morphology of the active region but flickers stochastically from one intensity difference to the next. This pattern can sometimes be minimized by selecting times at which the difference in the degree of atmospheric smearing was accidentally relatively small if not entirely null. The two-minute intensity difference represented in Figure 1a is a happy example of such an occurrence. However, routine helioseismic analysis cannot rely on such accidents, and this provides the motivation for the remedial exercise that follows.

3.3. Correcting Variations in Atmospheric Smearing

Among the many momentary effects terrestrial atmospheric turbulence is known to exert on astronomical images, two are particularly familiar: *i*) a local stochastic translation of the region of interest, which we will represent here by a vector displacement (α) in the image plane, and *ii*) a smearing of the image, such as that already discussed, which we will represent by a scalar parameter (β) whose character we will specify shortly. When the smearing is isotropic over the region of interest, moreover, the point spread function of the smearing has a finite upper bound with a characteristic radius that is significantly less than the predominant scales of the source to be examined, and, finally, (α) is similarly less than the source scales; the effects of the smearing can be approximated by a field of the following form:

$$I'(\mathbf{r}) = I(\mathbf{r}) - \alpha \cdot \nabla I(\mathbf{r}) + \beta \nabla^2 I(\mathbf{r}). \tag{8}$$

Here $I'(\mathbf{r})$ represents the intensity map of the translated and smeared source at location $\mathbf{r}$ in the image plane such that $I(\mathbf{r})$ would represent that of the un-smeared, untranslated source.

The characteristic lifetime of atmospheric scintillation is only a fraction of a second. For intensity maps integrated over many times this lifetime, during which the solar disk image is continually stabilized by limb tracking, α averages to only a small fraction of what it could be for a single, instantaneous snapshot. This is the case for the GONG observations, in which each pixel represents radiation integrated for a full minute. The smearing that characterizes the integrated image might be significantly greater than for an instantaneous snapshot, but it is more likely isotropic. In any case, Figure 3 makes it clear that atmospheric smearing integrated over a full minute varies significantly from one minute to the next, and it is straightforward to confirm that the pattern that appears in Figure 3b conforms closely to some constant times the Laplacian of either of the two intensity maps from which the difference was computed.

In practice, what is more important than whether Equation (8) accurately represents the overall smearing introduced by the terrestrial atmosphere is that it can apply to just the variation in smearing. This can be represented by applying a relatively small-β differential smearing to an intensity map [$I(\mathbf{r})$] that represents a source that has already been smeared by the atmosphere to a nominal degree that is constant. It then must simply be recognized that the differential smearing can be negative, meaning a differential resharpening of the image, which is accomplished by β itself being negative rather than positive.

The procedure we prescribe, then, is to adjust α and β so as to optimize the fit of each image in the time series to a single reference image in a region that excludes that in which significant white-light emission actually occurs during the flare. The results of this clean-up operation is illustrated in Figure 4. In this case, the intensity in the right frame is the difference between the intensity shown in the left frame and preflare intensity averaged over

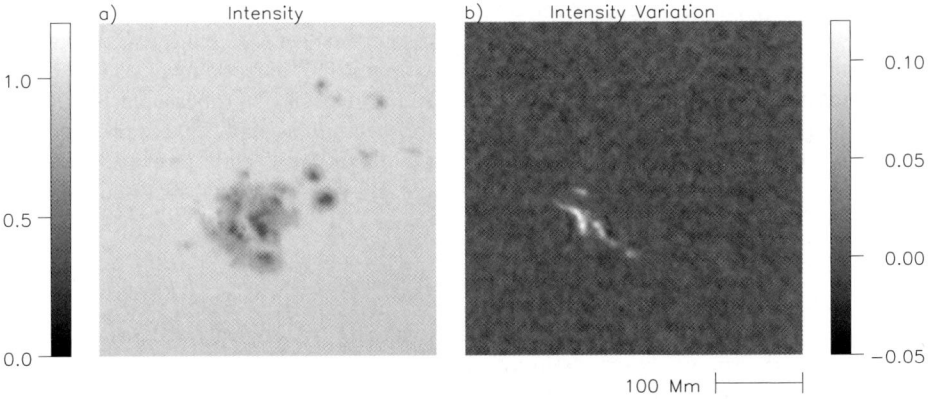

Figure 4 GONG intensity maps of AR 10486 normalized to unity for the quiet-Sun intensity and corrected for variations in smearing by the terrestrial atmosphere. (a) The intensity map shown in Figure 3a corrected for variable smearing. (b) The intensity difference between the map shown in Figure 4a and the preflare intensity averaged over a 300-second period similarly corrected.

a 300-second period. This image cleanly shows impulsive-phase emission from the active region that was relatively subtle in Figure 3.

With the intensity maps stabilized to this degree, it is now possible to apply essentially the same spectral-analysis techniques to GONG intensity observations of active regions as those applied to the MDI Doppler images for helioseismic applications. This is the object of the section that follows.

3.4. The High-frequency Spectrum of the Intensity

A careful comparison of the acoustic source distribution shown in Figure 1d with the distribution of intensity map shown in Figure 1a shows considerable excess radiation emanating from regions from which no significant seismic emission is detected. There are two major reasons for anticipating this:

1. To drive seismic emission, the heating must be sudden. The significant onset must be accomplished in a time not far in excess of the natural recoil time, $\tau_r \sim 40$ seconds [see Equation (4)], of the heated layer. The overwhelming preponderance of intensity excess from the flare of 29 October 2003 emanated from far outside of the acoustic source, but also on a time scale much longer than τ_r.
2. The efficiency of seismic emission driven by photospheric heating is greatly enhanced if the available heating flux is concentrated into a relatively compact region. This is simply because $(\delta F_{rad})^2$ integrated over the solar surface is greater for a given δF_{rad} integrated over the same surface if the flux is concentrated into a smaller region. High-frequency seismic emission is proportional to the square of the sudden component of δF_{rad}, which significantly emphasizes excesses that are more compact, spatially as well as temporally.

To address the issue of suddenness, we note that the range of characteristic times $\tau_c \sim 1/(2\pi \nu)$, for variations in the 5–7 mHz spectrum are 32–23 seconds, comfortably less than τ_r, qualifying heating in this part of the temporal spectrum as a prospective source of seismic emission. This may account in large part for the relative "hardness," of flare acoustic emission mentioned in the introduction, as compared with seismic waves produced by subphotospheric convection.

Motivated by this observation, we proceed by computing the power in the 5–7 mHz intensity-excess from the smearing-corrected GONG intensity observations and comparing maps of this with concurrent maps of the acoustic-source power (*i.e.*, the egression power), of which Figure 1d shows the latter at seismic maximum. This exercise relies on the simple assumption that the relationship between the specific intensity (I_c) measured by GONG and the energy flux (F_{rad}) integrated over the entire spectrum as applicable to photospheric heating is such that relatively small variations in the latter are linearly proportional to the former, and vice versa. The result, shown in Figure 5, is the subject of the discussion that now follows.

4. Discussion and Conclusions

Notwithstanding differences on a microscopic level, the morphologies of the 5–7 mHz acoustic-emission and intensity-excess powers shown in Figure 5 are remarkably closely aligned. Figure 5 also graphically illustrates how high-frequency seismic emission driven by photospheric heating connected to excess visible-light emission can be relatively unresponsive to local excesses that are clearly seen in individual GONG images if these are insufficiently concentrated spatially or insufficiently sudden. This is why the brightest kernels in

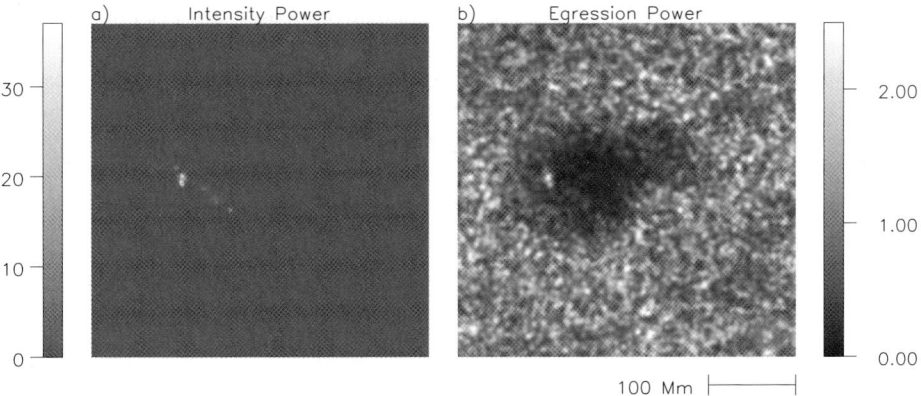

Figure 5 Comparison between 5–7 mHz intensity-excess power (a) and holographic egression power (b) maps for the flare of 29 October 2003. The intensity-excess power is derived from GONG observations corrected for variable smearing by the terrestrial atmosphere, as described in the text. The intensity-excess power is expressed in millionths of the square of the quiet-Sun intensity. The egression power is normalized to a mean of unity for the quiet Sun.

Figure 1a are so heavily represented in Figure 5a and hence, we suppose, in high-frequency seismic emission.

In fact, somewhat similar behavior can be expected to apply to magnetic-force transients (Hudson, Fisher, and Welsch, 2008), which we have not properly begun to consider in this study. Donea *et al.* (2006b) express concern about the accuracy of magnetic measurements made in the radiative environment of a white-light flare. We understand that shifts in the radiative environment during a white-light flare can spuriously introduce a significant shift, even possibly a reversal, in line-of-sight magnetic signature (Patterson, 1984). Other analysts acknowledge this concern but note that the magnetic signature often remains shifted for some time after the impulsive phase, when the radiative environment has relaxed to something like the preflare values. Donea *et al.* (2006b) express a concern that molecular contamination in the sunspot spectrum may affect magnetic signatures. If the molecular abundances of the photosphere are perturbed by excess continuum and UV radiation during the impulsive phase of the flare, it is not entirely clear how long it should take for molecular balance to recover after the impulsive phase of the flare. This is a particular concern for molecules whose atomic components are many times the mass of hydrogen and orders of magnitude less abundant.[4] Our understanding of this would benefit from spectral observations of an active region in appropriate molecular lines in the hour or so succeeding a white-light flare. This would be just one of a possibly significant number of useful applications for a white-light flare alarm that could be provided by the Helioseismic-Magnetic Imager (HMI) onboard the *Solar Dynamics Observatory* in solar activity cycle 24.

A comparison between Donea and Lindsey (2005) and Sudol and Harvey (2005) suggests that magnetic-transient signatures do not always exactly coincide with excess visible-light emission. This suggests that a careful spatial comparison benefiting from the local discrimination that high-frequency holographic observations can deliver may discriminate magnetically driven transient emission from that resulting from photospheric or chromospheric heat-

---

[4]Some authorities regard this to be improbable, as most molecular lines in sunspot penumbrae, from which most white-light flares and transient acoustic emission emanate, are much weaker than in sunspot umbrae.

ing. What is needed for such a discrimination is simultaneous high-quality Doppler-seismic, Stokes-magnetic, and visible-continuum observations of acoustically active flares.

A major obstacle to helioseismology from its inception has been the relative sparcity of controls, in that the object of study is mostly hidden from view in the electromagnetic spectrum, and the means of viewing it acoustically, for the most part, has yet to be developed. Four prospective sources of acoustic waves in the solar interior have been considered since the advent of helioseismology: *i*) convection (Goldreich and Keeley, 1977b), *ii*) overstable oscillations (Ando and Osaki, 1975; Antia, Chitre, and Kale, 1977; Goldreich and Keeley, 1977a), *iii*) comets occasionally plunging into the Sun (Kosovichev and Zharkova, 1995), and *iv*) flares (Wolff, 1972; Haber, Toomre, and Hill, 1988; Kosovichev and Zharkova, 1995). Of these, only flares can substantially be seen in the electromagnetic spectrum above the photosphere. The discovery of seismic emission from flares thus presents local helioseismology with an exceptionally useful prospective control resource. What is needed to develop this resource to its full potential is detailed modeling of how acoustic transients are generated, how they propagate through the outer solar atmosphere, and how they are injected into the solar interior, with a careful account of non-LTE radiative transfer and Lorentz forces in inclined magnetic fields and a realistic assessment of their observational manifestations.

Acknowledgements This research has benefited greatly from time and resources contributed by D. Beşliu-Ionescu, H. Moradi, J. Martinez-Oliveros, and P.S. Cally. We greatly appreciate the insight of the anonymous referee. This research was supported by grants from the Astronomy and Stellar Astrophysics Branch of the US National Science Foundation.

References

Allred, J.C., Hawley, S.L., Abbett, W.P., Carlsson, M.: 2005, *Astrophys. J.* **630**, 573.
Ando, H., Osaki, Y.: 1975, *Pub. Astron. Soc. Japan* **27**, 518.
Antia, H.M., Chitre, S.M., Kale, M.: 1977, *Solar Phys.* **56**, 275.
Beşliu-Ionescu, D., Donea, A.-C., Mariş, G., Cally, P., Lindsey, C.: 2007, *Adv. Space Res.* **40**, 1921.
Debouille, L., Roland, G., Neven, L.: 1973, *Spectrophotometric Atlas of the Solar Spectrum from* λ3000 *to* λ10000, Inst. d'Astrophysique, Liège.
Ding, M.D., Fang, C.: 1994, *Astrophys. Space Sci.* **213**, 233.
Donea, A.-C., Lindsey, C.: 2005, *Astrophys. J.* **630**, 1168.
Donea, A.-C., Braun, D.C., Lindsey, C.: 1999, *Astrophys. J.* **513**, L143.
Donea, A.-C., Besliu-Ionescu, D., Cally, P.S., Lindsey, C.: 2006a, In: Uitenbroek, H., Leibacher, J., Stein, R. (eds.) *Solar MHD: Theory and Observations – a High Spatial Resolution Perspective*, *CS*-**354**, Astron. Soc. Pac., San Francisco, 204.
Donea, A.-C., Besliu-Ionescu, D., Cally, P.S., Lindsey, C., Zharkova, V.V.: 2006b, *Solar Phys.* **239**, 113.
Fisher, G.H., Canfield, R.C., McClymont, A.N.: 1985, *Astrophys. J.* **289**, 434.
Goldreich, P., Keeley, D.A.: 1977a, *Astrophys. J.* **211**, 934.
Goldreich, P., Keeley, D.A.: 1977b, *Astrophys. J.* **212**, 243.
Haber, D.A., Toomre, J., Hill, F.: 1988, In: Christensen-Dalsgaard, J., Frandsen, S. (eds.) *Adv. Helio- and Asteroseismology*, *IAU Symp.* **123**, Kluwer, Dordrecht, 59.
Hudson, H.S., Fisher, G.W., Welsch, B.J.: 2008, In: Howe, R., Komm, R., Balasubramaniam, K.S., Petrie, G.J.D. (eds.) *Subsurface and Atmospheric Influences on Solar Activity*, *CS*-**383**, Astron. Soc. Pac., San Francisco, 221.
Kosovichev, A.G.: 2006, *Solar Phys.* **238**, 1.
Kosovichev, A.G.: 2007, *Astrophys. J.* **670**, L65.
Kosovichev, A.G., Zharkova, V.V.: 1995, In: Hoeksema, J.T., Domingo, V., Fleck, B., Battrick, B. (eds.) *Helioseismology*, *Proc. 4th SOHO Workshop* **SP-376**, ESA, Noordwijk, 341.
Kosovichev, A.G., Zharkova, V.V.: 1998, *Nature* **393**, 317.
Lindsey, C., Braun, D.C.: 2000, *Solar Phys.* **192**, 261.
Martinez-Oliveros, J., Moradi, H., Donea, A.-C.: 2000, *Solar Phys.* **192**, 261.

Moradi, H., Donea, A.-C., Lindsey, C., Besliu-Ionescu, D., Cally, P.S.: 2007, *Mon. Not. Roy. Astron. Soc.* **374**, 1155.
Patterson, A.: 1984, *Astrophys. J.* **280**, 884.
Sudol, J.J., Harvey, J.W.: 2005, *Astrophys. J.* **635**, 647.
Wolff, C.L.: 1972, *Astrophys. J.* **176**, 833.
Zharkova, V.V., Kosovichev, A.G.: 2002, In: Wilson, A. (ed.) *From Solar Min to Max: Half a Solar Cycle with SOHO, Proc. SOHO 11 Symposium* **SP-508**, ESA, Noordwijk, 159.